Oxford Handbook of

Infectious Diseases and Microbiology

Oxford Handbook of Infectious Diseases and Microbiology

Dr M. Estée Török

Consultant in Infectious Diseases,
Addenbrooke's Hospital
Cambridge, UK

Dr Fiona J. Cooke

HPA Specialist Registrar in Medical Microbiology
Addenbrooke's Hospital
Cambridge, UK

Dr Ed Moran

Specialist Registrar in Infectious Diseases
and General Medicine
Oxford Radcliffe Hospitals NHS Trust, UK

OXFORD
UNIVERSITY PRESS

OXFORD
UNIVERSITY PRESS

Great Clarendon Street, Oxford OX2 6DP

Oxford University Press is a department of the University of Oxford.
It furthers the University's objective of excellence in research, scholarship,
and education by publishing worldwide in

Oxford New York

Auckland Cape Town Dar es Salaam Hong Kong Karachi
Kuala Lumpur Madrid Melbourne Mexico City Nairobi
New Delhi Shanghai Taipei Toronto

With offices in

Argentina Austria Brazil Chile Czech Republic France Greece
Guatemala Hungary Italy Japan Poland Portugal Singapore
South Korea Switzerland Thailand Turkey Ukraine Vietnam

Oxford is a registered trade mark of Oxford University Press
in the UK and in certain other countries

Published in the United States
by Oxford University Press Inc., New York

First published 2009
Reprinted 2010

British Library Cataloguing in Publication Data
Data available

Library of Congress Cataloging in Publication Data
Data available

Typeset by Cepha Imaging Private Ltd., Bangalore, India
Printed in China
on acid-free paper through
Asia Pacific Offset.

ISBN 978–0–19–856925–1

10 9 8 7 6 5 4 3

Foreword

Every so often a new concept, idea or gadget is developed that proves itself so indispensible that you wonder how we managed before it was available. The same applies to textbooks. The Oxford Handbook series fits into this category and the latest member of the series, *The Oxford Handbook of Infectious Diseases and Microbiology*, is an immensely welcome companion, and true to the burgeoning reputation of this series. The range of topics covered is staggering and the quality and common sense holistic advice distilled in this pocket sized book makes it a resource of value to anyone, from senior infectious diseases specialists to paramedical staff, from medical students to clinical scientists wherever they work in the world. Combining infectious diseases and microbiology together in one book is visionary. For too long textbooks have divided these two essential components of the same topic. I have found the information inside incredibly helpful in my day to day clinical work based in a large tropical infectious diseases hospital in Ho Chi Minh City, Vietnam. I am sure it would be equally valuable to someone challenged by infectious diseases issues either as a specialist or non-specialist anywhere in the world. The authors are all practicing clinical scientists with extensive experience of a diversity of illnesses and with truly global reach. They are all already overly committed with huge personal workloads. It is a tribute to their enthusiasm and commitment that they have found the time to put this wonderful book together.

Jeremy Farrar
January 2009

Preface

Infectious diseases have troubled humankind for several thousand years. Some diseases such as tuberculosis and malaria have been described for centuries and continue to affect and kill millions of people today. Others such as HIV, SARS and H5N1 avian influenza have emerged relatively recently but have nonetheless been responsible for alarming epidemics. Some relatively rare pathogens such as anthrax have recently been used as agents of bioterrorism.

Microbiology, or the study of pathogenic micro-organisms, really began in the 19th century with the pioneering work of Robert Koch and Louis Pasteur, who were among the first to stain and culture bacteria. Although many microbiology laboratories continue to rely on these traditional methods the advent of molecular techniques in the 20th century has revolutionised the diagnosis of infectious diseases.

The introduction of antimicrobial agents in the 1950s had a major impact on the treatment and survival of patients with infectious diseases. Resistance to antimicrobial agents was observed shortly after their introduction and continues to increase globally, threatening the successful treatment and control of many infections. Multi-drug-resistant organisms are a major concern as some such as extensively drug-resistant tuberculosis are virtually untreatable with currently available therapies. This has highlighted the importance of appropriate use of antimicrobials, particularly in view of the paucity of new antibacterial agents in development.

The importance of infection control methods has long been recognised. In the 19th century, Ignaz Semmelweiss demonstrated the efficacy of hand washing in reducing puerperal sepsis and Robert Lister was the first to use antiseptics to reduce surgical infections. More recently hospital-associated infections caused by meticillin-resistant *Staphylococcus aureus*, *Clostridium difficile*, and multi-drug-resistant Gram-negative bacteria have dominated the medical literature.

A basic knowledge of infectious diseases and microbiology is therefore crucial for all medical practitioners. The purpose of this book is to provide a concise overview of the infectious diseases and their causative organisms. It is not intended to be an exhaustive reference and cannot possibly cover all syndromes or potential pathogens. However, we hope that it will provide a useful introduction and cater to the needs of interested readers.

ET, FC and EM
February 2009

Contents

Detailed contents

Senior reviewers

Dr Christopher Conlon
Consultant in Infectious Diseases
Oxford Radcliffe Hospitals NHS Trust
Oxford, UK

Dr Nicola Jones
Consultant in Infectious Diseases
Oxford Radcliffe Hospitals NHS Trust
Oxford, UK

Dr Bridget Atkins
Consultant in Microbiology and Infectious Diseases
Oxford Radcliffe Hospitals NHS Trust
Oxford, UK

Symbols and abbreviations

Ab	antibody
ABG	arterial blood gas
ACDP	Advisory Committee on Dangerous Pathogens
AFB	acid-fast bacilli
AFLP	amplified fragment length polymorphism
AFP	alphafetoprotein
Ag	antigen
ALP	alkaline phosphatase
ALT	alanine aminotransferase
AMP	adenosine monophosphate
ANA	antinuclear antibody
ANCA	antineutrophil cytoplasmic antibody
API	Analytical Profile Index
ARDS	acute respiratory distress syndrome
ARF	acute renal failure
ASOT	anti-streptolysin O titre
AST	aspartate transaminase
ATP	adenosine triphosphate
ATS	American Thoracic Society
AUC	area under the curve
AV	atrioventricular
BAL	bronchoalveolar lavage
BCG	bacillus Calmette–Guérin
bd	twice daily
BMT	bone marrow transplant
bp	base pair
BSE	bovine spongiform encephalopathy
BSI	bloodstream infection
BTS	British Thoracic Society
cAMP	cyclic AMP
CA-MRSA	community-acquired MRSA
CAP	community-acquired pneumonia
CCDC	consultant in communicable disease control
CDAD	*C. difficile*-associated disease/diarrhoea
CDC	Centers for Disease Control
CDSC	HPA Communicable Disease Surveillance Centre

CF	cystic fibrosis
Cfi	HPA centre for infections
CFT	complement fixation test
CFU	colony-forming unit
CJD	Creutzfeldt–Jakob disease
CK	creatine kinase
CL	containment level
CLSI	Clinical Laboratory Standards Institute
CMV	cytomegalovirus
CNS	central nervous system
CO_2	carbon dioxide
CoNS	coagulase-negative staphylococci
COSHH	control of substances hazardous to health
CrCl	creatinine clearance
CRP	C-reactive protein
CSF	cerebrospinal fluid
CSSD	Centre for Surgical Sterilization and Disinfection
CT	computerized tomography
CTL	cytotoxic T lymphocyte
CVC	central venous catheter
CVS	cardiovascular system
CXR	chest x-ray
DH	Department of Health
DIC	disseminated intravascular coagulation
DIPC	Director of infection prevention and control
DNA	deoxyribonucleic acid
DNase	deoxyribonuclease
EAGA	Expert Advisory Groups on AIDS
EBNA	Epstein–Barr virus nuclear antigen
EBV	Epstein–Barr virus
ECDC	European Centre for Disease Prevention and Control
ECG	electrocardiogram
EEG	electroencephalogram
eGFR	estimated GFR
EIA	enzyme immunosorbent assay
ELISA	enzyme-linked immunosorbent assay
EM	electron microscopy
ENT	ear, nose, and throat
EPP	exposure-prone procedure
ERCP	endoscopic retrograde cholangiopancreatography
ESβL/ESBL	extended-spectrum β-lactamase

ESR	erythrocyte sedimentation rate
ESRD	end-stage renal disease
FBC	full blood count
FDA	Food and Drug Administration
FNA	fine needle aspiration
FTA-Abs	fluorescent treponemal antibody antibody-absorption test
G6PD	glucose-6-phosphate dehydrogenase
GAS	group A *Streptococcus*
GBS	group B *Streptococcus*
G-CSF	granulocyte-colony-stimulating factor
GFR	glomerular filtration rate
GGT	gamma-glutamyl transferase
GI	gastrointestinal
GISA	glycopeptide-intermediate *Staphylococcus aureus*
GMC	General Medical Council
GM-CSF	granulocyte-macrophage-colony-stimulating factor
GNR	Gram-negative rod
GP	general practitioner
GRE	glycopeptide-resistant enterococcus
GU	genitourinary
GUM	genitourinary medicine
GVHD	graft-versus-host disease
HA	haemagglutinin
HAART	highly active antiretroviral therapy
HAI	hospital-acquired infection
HAP	hospital-acquired pneumonia
HAV	hepatitis A virus
Hb	haemoglobin
HBcAg	hepatitis B core antigen
HBeAg	hepatitis B e antigen
HBIG	hepatitis B immune globulin
HBsAg	hepatitis B surface antigen
HBV	hepatitis B virus
HCAI	healthcare-associated infection
HCV	hepatitis C virus
HCW	healthcare worker
HDV	hepatitis D virus
HEPA	high-efficiency particulate air
HEV	hepatitis E virus
HHV	human herpesvirus

Hib	*H. influenzae* type b (vaccination)
HICPAC	Healthcare Infection Control Practices Advisory Committee
HII	High Impact Intervention: from 'Saving Lives' programme
HIV	human immunodeficiency virus
HLA	human leucocyte antigen
HNIG	human normal immunoglobulin
HPA	Health Protection Agency
HPLC	high-pressure liquid chromatography
HPV	human papillomavirus
HSE	Health and Safety Executive
HSV	herpes simplex virus
HTLV	human T-cell lymphotropic virus
HTM	health technical memorandum
HUS	haemolytic uraemic syndrome
IM	intramuscular
IV	intravenous
ICC	infection control committee
ICD	infection control doctor
ICN	infection control nurse
ICP	intracranial pressure
ICT	infection control team
ICU	intensive care unit
ID	infectious diseases
IDSA	Infectious Diseases Society of America
IDU	injecting drug user
IE	infective endocarditis
IFA	immunofluorescence assay/indirect fluorescent antobody
IFAT	immunofluorescence antibody test
IFN	interferon
Ig	immunoglobulin (IgA, IgE, IgG, IgM)
IGRA	interferon-gamma release assay
IL	interleukin
IPV	inactivated polio vaccine
ITP	idiopathic thrombocytopaenic purpura
ITU	intensive therapy unit
IUCD	intrauterine contraceptive device
IUGR	intrauterine growth restriction
IVC	intravenous catheter
IVDU	intravenous drug user
IVIG	intravenous immunoglobulin

KS	Kaposi's sarcoma
LDH	lactate dehydrogenase
LFT	liver function test
LGV	lymphogranuloma venereum
LOS	lipo-oligosaccharide
LPS	lipopolysaccharide
LRTI	lower respiratory tract infection
MAC	*M. avium* complex
MAC	membrane attack complex
MAI	*Mycobacterium avium intracellulare*
MALT	mucosa-associated lymphoid tissue
MBC	minimum bactericidal concentration
MC&S	microscopy, culture, and sensitivity
MDA	Medical Devices Agency
MDR	multi-drug-resistant
MDT	multi-drug therapy
MGE	mobile genetic element
MHC	major histocompatibility complex
MIC	minimum inhibition concentration
MLEE	multi-locus enzyme electrophoresis
MLS_B	macrolide-lincosamide-streptogramin B (resistance)
MLST	multiple locus sequence typing
MLVA	multiple loci variable number repeat tandem analysis
MMR	measles, mumps, rubella (vaccination)
MOTT	mycobacteria other than tuberculosis
MRAB	multi-resistant *Acinetobacter baumanii*
MRI	magnetic resonance imaging
mRNA	messenger RNA
MRSA	meticillin-resistant *Staphylococcus aureus*
MSM	men who have sex with men
MSSA	meticillin-sensitive *Staphylococcs aureus*
MSU	midstream urine
MTB	*Mycobacterium tuberculosis*
NA	neuraminidase
NAD	nicotinamide adenine dinucleotide
NADP	NAD phosphate
NADPH	nicotinamide adenine dinucleotide phosphate-oxidase
NASBA	nucleic acid sequence-based amplification
NCCLS	National Committee on Clinical Laboratory Standards
neb	nebulized
NG	nasogastric

NGU	non-gonococcal urethritis
NICE	National Institute for Health and Clinical Excellence
NINSS	Nosocomial Infection National Surveillance Scheme (UK)
NNIS	National Nosocomial Infection Surveillance (US)
NNRTI	non-nucleoside reverse transciptase inhibitor
NPA	nasopharyngeal aspirate
NRTI	nucleoside/nucleotide analogue reverse transcriptase inhibitor
NTM	non-tuberculous mycobacteria
NTS	non-typhoidal *Salmonella* spp.
nvCJD	new variant CJD
od	once a day
OPAT	outpatient antimicrobial therapy
OPV	oral polio vaccine
PABA	para-aminobenzoic acid
PAIR	puncture, aspiration, injection, and re-aspiration
PaO_2	partial pressure of arterial oxygen
PAS	para-aminosalicylic acid
PAS	periodic acid–Schiff
PBMC	peripheral blood mononuclear cell
PBP	penicillin-binding protein
PCNSL	primary central nervous system lymphoma
PCP	pneumocystis pneumonia
PCR	polymerase chain reaction
PCV	pneumococcal conjugate vaccine
PE	pulmonary embolism
PEG-IFN	pegylated interferon
PEP	post-exposure prophylaxis
PFGE	pulsed-field gel electrophoresis
PGL	persistent generalized lymphadenopathy
PI	protease inhibitor
PICU	paediatric intensive care unit
PID	pelvic inflammatory disease
PML	progressive multifocal leucoencephalpathy
PO	orally
PPD	purified protein derivative
PPE	personal protective equipment
PR	per rectum
PRA	PCR restriction enzyme assay
PROM	premature rupture of the membranes
PrP	prion protein

PSA	prostate-specific antigen
PT	prothrombin time
PTLD	post-transplant lymphoproliferative disorder
PUO	pyrexia of unknown origin
PV	per vagina
PVC	peripheral venous catheter
PVE	prosthetic valve endocarditis
PVL	Panton–Valentine leucocidin
qds	four times a day
RBC	red blood cell
RCT	randomized controlled trial
REA	restriction endonuclease analysis
RFLP	restriction fragment length polymorphism
RIBA	recombinant immunoblot assay
RIDDOR	reporting of injuries, diseases and dangerous occurrences regulations
RM	regional microbiologist
RMAR	rapid micro-agglutination test
RMN	regional microbiology network
RMSF	Rocky Mountain spotted fever
RNA	ribonucleic acid
ROAST	rigidity, opisthotonus, autonomic dysfunction, spasms, and trismus
rRNA	ribosomal RNA
RSSE	Russian spring–summer encephalitis
RSV	respiratory syncytial virus
RT-PCR	reverse transcription polymerase chain reaction
RVF	Rift Valley fever
s/c	subcutaneous
SaO_2	arterial oxygen saturation
SARS	severe acute respiratory syndrome
SCID	severe combined immunodeficiency disease
sCJD	sporadic CJD
SDD	selective decontamination of the digestive tract
SENIC	Study on Efficacy of Nosocomial Infection Control
SHEA	Society for Healthcare Epidemiology of America
SHOT	Serious Hazards of Transfusion
SIADH	syndrome of inappropriate antidiuretic hormone secretion
SIRS	systemic inflammatory response syndrome
SLE	systemic lupus erythematosus

SNP	single nucleotide polymorphism
SRSV	small round structured virus
SSI	surgical site infection
SSPE	subacute sclerosing panencephalitis
STD	sexually transmitted disease
STI	sexually transmitted infection
SVR	sustained virological response rate
TB	tuberculosis
tds	three time a day
Th (cell)	helper cell
TIG	tetanus immunoglobulin
TK	thymidine kinase
TNF	tumour necrosis factor
TOE	transoesophageal echocardiography
TOP	topically
TPHA	treponema pallidum haemagglutination assay
TPN	total parenteral nutrition
tRNA	transfer RNA
TSE	transmissible spongiform encephalopathy
TSS	toxic shock syndrome
TSST	toxic shock syndrome toxin
TST	tuberculin skin test
TSU	theatre supplies unit
U&E	urea and electrolytes
UICP	universal infection control precautions
URTI	upper respiratory tract infection
USS	ultrasound scan
UTI	urinary tract infection
UV	ultraviolet
VA	ventriculo-atrial
VAP	ventilator-associated pneumonia
vCJD	variant CJD
VDRL	Venereal Disease Research Laboratory (test)
VHF	viral haemorrhagic fever
VIP	visual infusion phlebitis (score)
VISA	vancomycin-intermediate *Staphylococcus aureus*
VNTR	multiple-locus variable-number tandem repeats
VP	ventriculoperitoneal
VRE	vancomycin-resistant enterococci
VRSA	vancomycin-resistant *Staphylococcus aureus*

VUR	vesico-ureteric reflux
VZIG	varicella zoster immune globulin
VZV	varicella zoster virus
WCC	white cell count
WHO	World Health Organization
XDR	extensively drug resistant
ZN	Ziehl–Neelsen (stain)

Chapter 1

Basic principles

Host defence

Immunity is concerned with recognizing and disposing of foreign material entering the body. Our defence against infection starts with physical barriers such as skin, mucous membranes etc. If these are breached, the consequent immune response falls into two distinct yet interacting branches: the innate and the adaptive immune systems. Both have cellular and humoral (antibody-mediated) components.

Innate defence mechanisms

It takes time to raise an adaptive response. The innate system provides a high-speed, on-the-scene reaction to foreign material. Its main components are complement proteins, phagocytes, and natural killer cells. The cytokines (proteins released by cells that influence other cells) and chemokines (chemicals that attract cells) produced by these players initiate inflammation, and both activate and control the manner of the adaptive response.

Complement pathways

One of the oldest parts of the immune system (sea urchins have had it for 700 million years), it comprises around 20 different protein that operate in cascade to destroy invaders and signal other immune system components. It can be activated in three ways:

- classical pathway – activation by antibody bound to its antigen. Immunoglobulin M (IgM) is particularly good at this, which is logical since it is one of the early antibodies made by the adaptive system
- alternative pathway – unlike the classical this is non-specific. One of the complement proteins, C3b binds amino or hydroxyl groups on, for example, bacterial cells. Once stabilized in this way it is able to activate the cascade
- lectin activation pathway – mannose-binding lectin (a lectin is a protein that binds carbohydrates), present in the tissues and circulation binds mannose on the surface of, for example, a bacterium. This triggers the production of C3b with the consequences described above.

The activated complement cascade produces membrane attack complexes (MAC), which punch holes in the surface of the foreign organism. In addition, complement proteins opsonize invaders facilitating phagocytosis, and act as chemoattractants for other immune system players.

Phagocytes

Macrophages and neutrophils are the key innate phagocytes. Macrophages can live for months and are based in tissues. Their role is that of sentinel – continually phagocytosing, watching for invaders, and signalling attacks. When activated they can present antigen to T cells. Neutrophils comprise around 70% of circulating white blood cells. In the blood they are inactive. In response to cytokines indicating a local immune response, they leave the circulation and become active, highly phagocytic killing machines.

Natural killer cells

Natural killer (NK) cells are in the same family as lymphocytes. They leave the circulation at sites of infection and can kill tumour cells, virus-infected cells, parasites, and fungi. They act by injecting enzymes into the target cell, or by interacting with a protein called 'Fas' on the surface of a target, prompting 'suicide'. They may be activated by substances such as lipopolysaccharide (LPS) from bacterial cell walls, and by interferon-α and -β from virus-infected cells. They are able distinguish self from foreign to a limited degree.

Adaptive defence mechanisms

B cells

B lymphocytes are produced and mature in the bone marrow. Each is able to produce only one antibody. It does this by rearranging the deoxyribonucleic acid (DNA) encoding the antibody into one of an estimated 100 million possibilities. This antibody is displayed on the cell surface as the B cell receptor (BCR). Activation of a naive B cell requires binding of the BCR to its cognate antigen and (generally) a co-stimulatory signal (usually provided by a helper T cell). It then proliferates to produce a clone of identical B cells, most of which mature into antibody factories: plasma B cells. Others may become memory cells.

Antibodies

These large proteins (immunoglobulins) opsonize foreign material, marking it for phagocytosis. Some are able to bind and block the proteins on a virus' surface required for uptake into cells, 'neutralizing' antibodies. Some are specific for toxins produced by an infecting organism. There are four main classes, each with an antigen-binding region (Fab) and a constant 'tail' region (Fc) (Fig. 1.1). Variations in the Fc region determine the class, its function, and to which immune system cells it will bind. The class a B cell produces is determined by the cytokine environment, particularly those from T helper cells.

- IgM – the first antibody (Ab) made when naive B cells activated. Like a pentameric IgG it is very effective at activating complement, and at viral neutralizing.
- IgG – 75% of Ab in the blood. Good at opsonizing, neutralizing viruses and can increase NK cells' cytotoxic activity (antibody-dependent cell cytotoxicity). Longest-lived Ab (half life about 3 weeks) and able to cross the placenta.
- IgA – the most abundant Ab in the body (as opposed to circulation). Protects all mucosal surfaces. Much like a dimeric IgG, it has four binding sites and can bind pathogens into clumps aiding their expulsion (e.g. in faeces). Resistant to acid and enzymes in gastrointestinal tract. Secreted in breast milk. No complement activity.
- IgE – role in defence against parasites. Pathological role in anaphylactic shock and allergy.

T cells

T lymphocytes are produced in the bone marrow but mature in the thymus. They bear T-cell receptors of similar variety to BCR and produced by a comparable DNA-juggling mechanism. Unlike BCR (which can recognize nearly any organic molecule), they recognize solely protein antigens and only if presented by another cell in association with a major histocompatability complex (MHC) molecule. There are two main types of T cell:

- cytotoxic T lymphocyte (CTL) – recognizes Ag presented on MHC class I molecules which are found on nearly all cells in the body. MHC class I acts much like an advertising hording, displaying to the CTL what is 'on' inside the cell, be it benign native proteins, or foreign material (e.g. a cell producing viral protein). CTLs can then make contact with potentially infected cells and trigger their demise. CTLs bear the MHC class I co-receptor, CD8

- helper T cell (Th cell) – recognize Ag presented on MHC class II which is made only by certain 'antigen-presenting cells' (APCs). The proteins presented are derived from material the APC has taken up from its environment – e.g. bacterial proteins, opsonized and phagocytosed viral particles. Th cells bear the MHC class II co-receptor, CD4.

Th cells are activated when their TCR recognizes its Ag–MHC and the cell receives a co-stimulatory signal – this is the role of APCs: macrophages, dendritic cells, and acitivated B cells. Once properly activated, they begin dividing to produce a clone of cells. These produce cytokine mixes appropriate to the insult (e.g. in a viral infection the cocktail might stimulate increased B-cell production of IgG, and activate CTLs and NK cells).

Virgin CTL activation requires T-cell receptor (TCR)/Ag–MHC recognition and help from a helper T cell. Once activated, the CTL proliferates and the offspring move to the area of infection where they identify and kill infected cells.

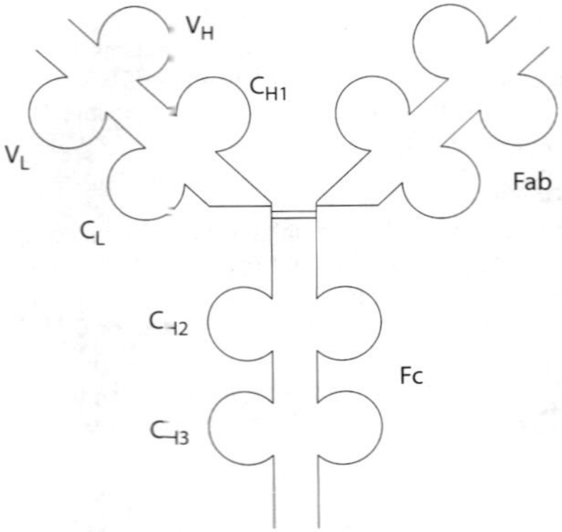

Fig. 1.1 Antibody structure. Immunoglobulin consists of a light chain containing a variable domain V_L and a constant domain C_L. The heavy chain consists of one variable V_H and three constant domains C_{H1-3}. The segment containing the C_{H2-3} of the heavy chains is the F_C portion and the two segments containing V_L, V_H, C_L, and C_H are each termed Fab fragments. Reproduced from Wilkins et al. *Oxford Handbook of Medical Sciences* (2005), with permission from Oxford University Press.

Immunity against viruses

The key process in viral infection is intracellular replication, hence the adaptive system and CTLs play a key role. The immune system can subvert the virus in a number of ways as detailed in the following text and in Fig. 1.2.

Defence against viral cell entry

- Antibodies may bind viral surface antigens required for cell entry, thereby neutralizing the virus (1).

Reducing the chance of cell infection

- Virus infection stimulates the production of interferons, a group of proteins which stimulate cells to block transcription of virus, and thus protect them from infection (7).

Recognizing and destroying virus-infected cells

- In a new infection, CTL recognize foreign viral peptides presented by MHC class I on the surface of infected cells (2). They are activated if they experience a co-stimulatory signal from a Th cell and proliferate. These activated CTLs can then trigger cell death on future encounters with their specific Ag–MHC complex (6)
- Certain viruses cause a reduction in surface expression of MHC I in cells they have infected as a means of avoiding CTLs. NK cells become suspicious of cells with absent MHC I and are capable of killing these cells early in infection (8).

Extracellular virus

- Virus enters the circulation from infected cells. Viral components presented on APCs activate Th cells (3), enabling them to produce appropriate cytokines to coordinate the immune response, recruit further cells, and provide co-stimulatory signals for CTLs' and B cells' activation.
- B cells are activated on recognizing their cogent Ag (4) together with T-cell help. They proliferate into plasma cells (5) producing antibody. Some form memory.

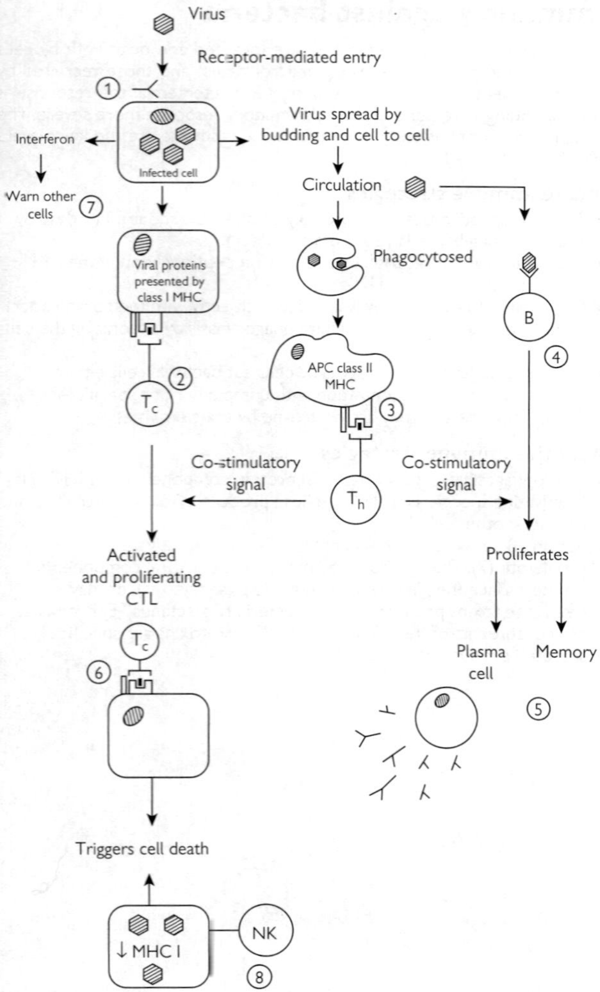

Fig. 1.2 Overview of viral defence.

Immunity against bacteria

Most bacterial infections, e.g. a boil, are localized and dealt with by cells residing in the affected tissue (e.g. macrophages) and those recruited by the consequent inflammatory response. In bacteraemia, the response is correspondingly greater and the inflammatory response more severe. The complement proteins and humoral immune response are the key players (Fig. 1.3).

Innate immune strategies

- The high-speed innate response may be all that is required to destroy a few infecting bacteria (e.g. from a splinter).
- Lysozyme is a 'natural antibiotic'. It acts on peptidoglycan in the bacterial cell wall causing lysis (2).
- Complement proteins may lyse bacteria directly with membrane attack complexes (see above) or facilitate phagocytosis by opsonizing the cell (3, 4).
- Macrophages and other phagocytic cells eat bacterial cells either directly or aided by opsonization with complement or specific Ab (4). Once inside the cell they are destroyed by toxic enzymes.

Adaptive immune strategies

- Macrophages can present bacterial protein components on MHC class II activating specific Th cells (6). These produce cytokines recruiting and activating other cells.
- Specific B cells recognize Ag and with T-cell help activate and proliferate (7). The Ab produced may be to structural components on the surface (helping complement and phagocytosis), or may neutralize toxins produced by the bacteria (e.g. tetanus). Endotoxins are constituents of the cell wall (e.g. LPS); exotoxins are specifically secreted products (1).

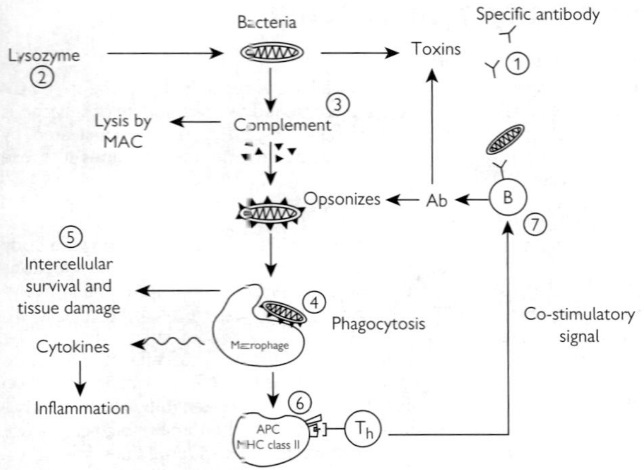

Fig. 1.3 Overview of bacterial defence.

Immunity against fungi

Superficial infections

The commonest fungal infections are usually superficial (e.g. the dermatophytes). In a healthy host these are controlled by constituents of sebaceous secretions, and cell-mediated immune responses (as demonstrated by the increased rate of severe skin/mucous membrane infections in those patients with AIDS).

Invasive infections

Other species can cause serious systemic infection, often entering the body as spores breathed into the lung. The manifestations of such infections depend on the nature of the immune response, and the immune avoidance mechanisms exhibited by the organism, e.g. *Histoplasma capsulatum* pulmonary infection may heal spontaneously, disseminate, or lead to the generation of chronic granulomas and fibrosis. This reflects the nature of the immune response and the organism's ability to resist digestion within macrophages. Host resistance may be weakened by specific immunodeficiency such as HIV infection or an inherited immunological defect. However, infections with organisms such as *Candida albicans* are significantly increased in those people with more general weakening in immune resistance caused by malnourishment, alcoholism, diabetes, iron deficiency, and age. The key players in the immune response to fungal infection include:

- antibody and complement responses – these opsonize fungal cells and facilitate phagocytosis (e.g.*Cryptococcus neoformans* possesses an anti-phagocytic capsule and can resist phagocytosis unless opsonized by antibody or complement). Antibody responses may also contribute to pathology (e.g. the hypersensitivity seen as part of infection with *Aspergillus* spp.)
- neutrophils – important phagocytic cells. Recurrent or severe fungal infections may be seen in those with defective neutrophil responses. They are a key part of first-line defence in organs such as the lung
- T-cells – severe infections of the skin and mucous membranes, as well as the lung and elsewhere are common in those with T-cell deficiencies (e.g. patients with AIDS). The response to chronic infection may lead to the development of granulomas.

Immunity against parasites

Protozoa

Protozoa owe their success as pathogens to a combination of strategies aimed at avoiding the full force of the immune response. These include adopting an intracellular habitat (e.g. toxoplasmosis), rapid antigenic variation (e.g. African trypanosomes use a gene-switching mechanism to repeatedly replace their outer coat as fast as antibodies are raised against it), immunosuppression (e.g. *Toxoplasma* suppresses T-cell function to facilitate its intracellular survival), and the generation of non-specific polyclonal B-cell responses which serve to inhibit the generation of effective specific immunity and may precipitate autoimmunity.

Worms

Eosinophils and immunoglobulin E (IgE) production are important players in the immune response to infection with worms, but as a result hypersensitivity reactions in the skin and lung may occur. It is unusual for worms to be eliminated by the immune response, and at most it serves to control their number. Eosinophilia results in part from mast cell and T-cell factors. Eosinophils phagocytose the antigen–antibody complexes that circulate in enormous quantities in worm infection, and modulate hypersensitivity. They may also kill certain worms. IgE production is stimulated by the presence of worms, and the resulting inflammatory response is thought to reduce worm attachment and gut entry. Chronic infection with certain flukes (e.g. schistosomes) can lead to fibrosis of the liver or bladder as a result of a T-cell-mediated reaction to their trapped eggs.

Basic principles of bacteriology

Taxonomy

Taxonomy is the art of dividing into ordered groups or categories. With respect to bacteria it refers to two main concepts:

- classification – the division of organisms into related groups based on similar characteristics. The species is the most definitive level of classification. Organisms may be reclassified from time to time as new information (e.g. genetic relatedness) becomes available
- nomenclature – the naming of groups and members of a group. This is governed by the International Committee on Systematics of Prokaryotes (www.the-icsp.org). The most recent revision of the International Code of Nomenclature of Bacteria was published in 1992. Amendments are published in journal form (the International Journal of Systematic and Evolutionary Microbiology). The basic rules for naming are outlined in Boxes 1.1 and 1.2.

Identification

Taxonomy is dynamic and throughout its history has been dependent on the techniques of identification available – originally phenotypic characteristics, and more-recently methods of determining the genetic 'relatedness' (phylogenetics) of a group of organisms, which should hopefully lead to a more-stable classification with fewer revisions in the future. Changes are overseen by the Judicial Commission of the International Union of Microbiological Societies.

- Phenotypic characteristics – cellular morphology, staining (e.g. Gram, or acid-fast, Box 1.3), motility, growth characteristics (speed, requirements, colonial appearance), biochemical characteristics (e.g. acid from specific carbohydrates), serology, analysis of metabolic end-products).
- Phylogenetic identification – nucleic acid hybridization (denaturation of double-stranded DNA into single strands and assessing their ability to anneal to the single strands of another related organism), 16S ribosomal ribonucleic acid (RNA) sequence analysis.

For identification purposes, simple phenotypic characteristics continue to be used which often (but not always) correlate with genotypes. These methods of phenotypic characterization have been developed and codified over years to facilitate laboratory identification of organisms (collected in texts such as *Bergey's Manual of Systematic Bacteriology*).

Box 1.1 Microorganism nomenclature rules

- Each organism should have only one correct name. Where more than one exists, the oldest legitimate name takes precedence.
- Confusing names should be abandoned.
- Regardless of origin, all names are in Latin or are Latinized.
- The first word (genus) always starts with a capital letter.
- The second word (species) is in small letters.
- The genus and species name are underlined or italicized when printed.

Box 1.2 The naming hierarchy

- Order – names ending -ales
- Families – names ending -aceae
- Tribes – names ending -eae
- Genus
- Species – a collection of strains sharing common characteristics
- Strain – a bacterial culture derived from a pure isolate

Box 1.3 The Gram stain

Named after the Danish bacteriologist who devised it in 1884, the Gram stain remains a useful test – the outcome of which is determined largely by the structure of the cell wall. The procedure is simple:

- cells are stained with crystal violet
- then they are treated with iodine, forming a crystal violet/iodine complex in the cell
- next they are washed with an organic solvent (acetone-alcohol)
- then they are stained with a red counterstain, e.g. safranin
- Gram-positive organisms retain the crystal violet/iodine complex within the cell because of the thick peptidoglycan cell wall, and appear dark purple. In Gram-negative organisms the stain is leached from the cell due to disruption of the lipid-rich outer membrane by the organic solvent and they appear pink.

Structure and function of bacteria

Bacteria are prokaryotic – they have a single chromosome that is not enclosed in a nuclear membrane. They are around 0.2–2 micrometres by 1–6 micrometres long and exist in four basic shapes: cocci (spheres), bacilli (rods), spirillia (spirals), and vibrios (comma shaped).

Cytoplasm

Cytoplasm is a gel containing the enzymes, ions, subcellular organelles, and energy reserves of the organism. Energy and food are stored in membrane-bound granules. Glycogen is the major storage material of enteric bacteria. Ribosomes are the sites of protein synthesis. Bacterial ribosomes are 70S (the 'S' referring to a unit of sedimentation on ultracentrifugation) and are formed from 2 subunits – 30S (which contains 16S RNA) and 50S. Ribosomes are formed from specific ribosomal proteins and ribosomal RNA (they account for 80% of total cell RNA). They complex with a messenger RNA transcript from DNA to form polyribosomes (polysomes). Extrachromosomal DNA is often found within the cytoplasm in the form of plasmids. These are covalently closed double-stranded DNA (dsDNA) circles, and are capable of replication and are inherited by progeny cells. They may contain genetic information encoding structure or functions relating to bacterial virulence (antibiotic resistance, adhesions, toxins etc).

Cytoplasmic membrane

The cytoplasm is surrounded by the cytoplasmic membrane, a phospholipid bilayer into which various proteins are inserted. The membrane is involved in the synthesis and secretion of enzymes and toxins and active transportation of materials into the cytoplasm.

Bacterial cell wall

This provides rigidity and a physical barrier to the outside world. Peptidoglycan provides the strength and is found in all bacterial species except *Mycoplasma* and *Ureaplasma* spp. It comprises a carbohydrate backbone cross-linked by short peptides. Variations in the peptide linkages are responsible for different cell wall characters (see Box 1.3).

Gram-positive cell walls

These are composed of several layers of peptidoglycan, within which are trapped a variety of proteins, polysaccharides, and teichoic acids (polymers of glycerol or ribitol), which stabilize the cell wall and maintain its association with the cell membrane as well as having roles in cellular interaction and growth. They are antigenic in some organisms. Certain organisms will possess cell wall structures that confer virulence characteristics e.g. M protein of group A streptococci.

Gram-negative cell walls

These are thinner but more complex than those of Gram-positive organisms. Outside the cytoplasmic membrane is a periplasmic space. The outer part of this is bounded by a single peptidoglycan layer, beyond which lies the outer membrane – a phospholipid bilayer within which lie other large molecules. Lipoproteins link this membrane to the peptidoglycan below. Unique to gram-negative bacteria are the lipopolysaccharides (LPS) in the outer membrane. These are the key surface antigens and endotoxins of Gram-negative organisms. They are composed of a lipid A (principally responsible for the endotoxin activity) attached to a core polysaccharide with side chains which vary within a species and confer the serological identity (O antigen) of individual strains. Other components of the outer membrane include: porin proteins (allow entry to the periplasmic space from the outside), and non-porin proteins (such as penicillin-binding proteins)

'Acid-fast' cell walls

Mycobacterium spp., *Nocardia* spp., and *Corynebacterium* spp. have a modified Gram-positive cell wall. They have a higher cell-wall lipid content which is due to mycolic acids, which can also confer virulence characteristics. Acid-fast organisms are so-called because once stained with red carbofuschin dye they are resistant to decolourization with acid-alcohol – a property conferred by the cell wall lipids.

Bacterial surface structures

Capsules

Certain bacteria possess a capsule around their cell wall, usually composed of polysaccharide, but polypeptide in some organisms. They are manufactured at the cell membrane. They serve to protect cells from toxins, desiccation, complement proteins, and antibodies, and play a role in adherence (e.g. the glucan capsule of *Streptococcus mutans* forms the matrix of dental plaque). Capsules are antigenic and can be used to identify certain organisms (e.g. *Haemophilus influenzae* type B), and may be detectable in body fluids.

Flagellae

These are long thin appendages that are anchored in the cytoplasmic membrane and extend through the cell wall into the surrounding medium; they are responsible for cellular motility. They are usually found on Gram-negative rods, but motile Gram-positive organisms also exist. Flagellar number and arrangement vary from single (monotrichous) to multiple over the whole surface (peritrichous). The filament is composed of multiple flagellin proteins which have the capacity to self-assemble. The cell membrane-anchored base rotates as part of an energy-dependent reaction causing the rigid flagella to rotate. They are antigenic and several genuses, e.g. *Salmonella* spp. are able to alter the antigenic type of flagella they produce (phase variation) by the differential expression of the genes coding various flagellin proteins.

Fimbriae

These are smaller appendages (~15–20 micrometres in length) composed of fibrilin and found on many Gram-negative bacteria. They form hollow tubes and are involved in attachment to cells or mucosal surfaces (also called adhesins). Different adhesins display different binding properties (e.g. mannose), which are partly responsible for the tissue tropism seen with certain species of bacteria. They are also involved in bacterial conjugation and the exchange of DNA from one cell to another. The term 'pili' is given to the fimbriae used by Gram-negative bacteria for DNA transfer in conjugation.

Bacterial genetics

Bacterial DNA

Bacterial genetic information is encoded in the cell's DNA, of which there are 2 types:

- chromosomal DNA – prokaryotic organisms have a single, covalently closed, circular chromosome of dsDNA. It lies in a supercoiled state within the cytoplasm, not enclosed but attached to the bacterial cell membrane at certain points. Individual genes are arranged linearly. In *E. coli* the chromosome contains around 5 million base pairs. DNA replication and transcription to mRNA occur continually (unlike eukaryotes)
- extra-chromosomal DNA – plasmids are small DNA molecules consisting of circular dsDNA. Replication is autonomous and occurs independently of the host cell. Multiple copies of the same plasmid and many different plasmids can coexist in the same cell. Plasmids pass to daughter cells and some are capable of transferring to other bacteria of the same (or other) species. They code for many different functions and structures, for example antibiotic resistance.

Genetic material can move between plasmids and from plasmid to chromosome (and vice versa) via transposons. These are DNA sequences that can copy themselves to new site, carrying associated genes with them.

In manufacturing proteins, single-stranded 'messenger' RNA (mRNA) is synthesized from dsDNA during transcription by a DNA-dependent RNA polymerase using the 'sense' strand of the DNA as a template. The mRNA forms a complex with several ribosomes (a polysome–mRNA complex). The mRNA is translated as transfer-RNA (tRNA) molecules bearing the appropriate amino acid, and its 'sense' bases associate with the mRNA 'anti-sense' bases.

Genetic variation

Genetic variation can occur by mutation or direct gene transfer.

Mutation

This occurs when one or more bases in the DNA sequence changes. It is permanent (barring remutation to the original sequence) and will be inherited by any progeny. Such changes may alter the amino acid sequence of the encoded protein or may change the circumstances in which a normal protein is produced (transcription changes). Mutations can be:

- deletion – losing a base will cause a frame-shift mutation, changing the amino-acids represented by the sequence from the point of mutation onwards. Deletions can involve several bases
- insertion – additional base or bases will also cause a frame-shift
- substitution – change of a single base to one of the other three changes the amino acid represented by the code.

Gene transfer

This is the main means by which bacteria achieve their rapid genetic variability. There are three mechanisms:

- transformation – the uptake of free bacterial DNA from the surrounding environment into recipient cells. Cells able to take up and incorporate free DNA are termed 'competent'. This state is usually transient, occurring towards the late exponential phase of growth with the expression of surface receptors for DNA. DNA that enters can only be incorporated into the genome if there are homologous regions with which it can integrate (only DNA from related species is likely to achieve it) and requires the presence of the *recA* gene

- transduction – the exchange of genes by bacteriophages (or simply 'phages'). Phages are viruses that infect only bacteria. Certain phages integrate their genetic material into the bacterial host DNA. During phage replication, excision of viral sequence from the host DNA may result in fragments of bacterial DNA becoming enclosed within the viral particle. When this particle infects a new bacterial cell the DNA fragment recombines into the chromosome of the second bacterium. Transduction may be generalized (random accidental host DNA is transferred) or specialized (specific host genes are transferred as the phage DNA integrates at specific sites). Phage conversion refers to the phenomenon of phage DNA becoming integrated into the bacterial chromosome and bringing about a change in the bacterial phenotype – for example, toxin production in *Corynebacterium diphtheriae*. This is due to phage DNA precipitating the expression of otherwise unexpressed bacterial genes
- conjugation – the only mechanism that requires cell-to-cell interaction, and the major means by which bacteria acquire additional genes. Gram-negative cells achieve this by means of the sex pilus which is encoded on a specific plasmid (the F plasmid). The pilus establishes contact with another cell and is the tube through which DNA is passed. Some organisms integrate the F plasmid into chromosomal DNA – such cells are termed Hfr cells (high-frequency recombination). Gram-positive cells achieve conjugation by aggregating in response to the production of pheremones by the donor bacterium.

Bacterial growth and metabolism

Bacterial growth requires, logically enough, materials for the manufacture of cell components and a source of energy.

Materials

Some bacteria can synthesize all they require from simple raw materials. However, most pathogenic bacteria require a ready-made supply of the organic compounds they need for growth. Most of these nutrients diffuse freely across the cell membrane to enter the cell. Some are required at high concentration and uptake is energy dependent. Enzymes involved in these processes may be inducible (produced in the presence of the substrate) or constitutive (produced constantly and independent of the substrate).

Carbon

Lithotrophic bacteria are able to use carbon dioxide as the sole source of carbon, and use it as the basis of their organic metabolites. Thus the only other materials needed are water, inorganic salts, and energy. Organotrophic bacteria require organic carbon such as glucose – thus their energy source is also used in the synthesis of materials. Different bacterial species can utilize different organic carbon sources, *Pseudomonas* spp. being among the most versatile.

Nitrogen

Ammonium ions (NH_4^+) provide the nitrogen required by bacterial cells. This is turned into glutamate and glutamine, which in turn are processed into certain amino acids, purines etc. Certain bacterial species and blue-green algae can make ammonium direct from atmospheric nitrogen – predominantly soil-dwelling organisms, but certain human pathogens such as *Klebsiella* and *Clostridium* species can 'fix' nitrogen in this manner. Other organisms produce their ammonium ions by nitrate reduction or from the deamination of amino acids released from proteins.

Growth factors

Substances such as B vitamins, minerals, certain amino acids, purine, and pyrimidines are required by many bacteria, although not all are capable of synthesizing their own. An organism is described as prototrophic for a growth factor if it is capable of synthesizing it and does not require an exogenous source. All bacteria need certain inorganic ions such as magnesium and calcium, and some need zinc and copper among others.

Environmental conditions

As well as (of course) water and carbon dioxide, bacteria have specific optimal environmental requirements for growth, including temperature and pH. Oxygen requirements are discussed in the next section.

Energy

Bacterial metabolism is a balance between biosynthesis (anabolic) and degradation (catabolic reactions). Catabolic reactions power the biosynthetic processes as hydrolysis of substances being broken down liberates energy which is captured in the formation of the phosphate bonds of adenosine triphosphate (ATP).

An organism's ability to utilize certain carbohydrates (e.g. sucrose, mannose) and convert them to glucose (the starting point for both aerobic and anaerobic catabolism) for metabolism is a useful feature for characterizing bacteria. Many tests in clinical microbiology detect the acidic end-products of bacterial metabolism in controlled conditions.

The oxygen requirement of a specific organism reflects the means it uses to meet its energy needs.

- Obligate anaerobes – grow only in conditions of high reducing intensity, and oxygen is toxic.
- Aerotolerant anaerobes – anaerobic metabolism but not killed by the presence of oxygen.
- Facultative anaerobes – can grow in anaerobic and aerobic conditions.
- Obligate aerobes – need oxygen to grow.
- Microaerophilic organisms – best growth is seen in low oxygen levels; high levels may be inhibitory.
- Aerobes produce a free radical superoxide ($O_2^{-}\bullet$) which is reduced to oxygen and hydrogen peroxide. Catalase enzymes convert the latter to water and oxygen.

Anaerobic metabolism

Glucose use in anaerobic conditions is fermentation. This occurs via glycolysis, producing pyruvate and two molecules (net) of ATP per glucose molecule. Pyruvate can then enter several different pathways producing different end-products (e.g. lactic acid, acetaldehyde, ethanol, etc.)

Aerobic metabolism

Glucose use in aerobic conditions is respiration. Pyruvate forms in glycolysis as in anaerobic conditions but then enters the Krebs' cycle. Complete oxidation of glucose by these paths results in 38 molecules (net) of ATP per glucose molecule. The Krebs' cycle also produces precursors for several other important cellular components such as purines, pyrimidines, amino acids, and lipids.

Bacterial growth

Bacterial growth occurs as the mass of cellular constituents increases. Cell division starts once a critical mass is reached and occurs by binary fission. In a liquid medium, bacteria display a uniform growth curve.

- Lag phase – the cell synthesizes new enzymes and cofactors, and imports nutrients from the media.
- Increasing growth phase – enzymatic reaction rates approach steady state and cell growth begins.
- Logarithmic growth phase – cell growth and division is at maximum. This is influenced by temperature, the carbon source, oxygen, nutrient availability, and so on.
- Declining growth phase – nutrients are exhausted and growth slows.
- Stationary phase – new organisms produced equal those dying.
- Death phase – cells die off.

Bacterial virulence and pathogenicity

Definitions

Pathogenicity is defined as the ability of an organism to cause disease, whereas virulence is the the degree of pathogenicity within a group of organisms. Virulence is determined by several factors related to the organism and the host, most particularly the infectivity of the bacteria and the severity of the condition it produces. To be considered to be pathogenic, organisms will have strains of varying degrees of virulence.

Pathogenicity

Infection of the host is the necessary first step – infection does not, however, equate with disease. We are all colonized with many bacteria. These however only become disease causing in certain abnormal situations.

The organism must enter the host and attach to the mucous membrane surfaces. Some go no further than this and disease is caused by exotoxins (e.g. *Vibrio cholerae*), others penetrate deeper and multiply causing tissue damage and eventually gaining access to the blood and potentially disseminating. Some species such as mycobacteria are able to reside within cells, taking up a long-term residence within the host. Still others are highly specific in the organs they will infect (e.g. *Neisseria gonorrheae*). This may be related to the presence of specific receptors for bacterial attachment, or the presence of nutrients (e.g. *Brucella abortus* has a requirement for erythritol, which is found in the bovine placental tissue and results in localization of infection to this site).

Virulence factors

Adhesins

Bacterial cell surface adhesins adhere to complementary structures on the surface of susceptible cells. These adhesins may be fimbriae (see 🕮 Structure and function of bacteria, p.14), components of the bacterial capsule (see below), and other cell surface antigens. The adherence process is a prerequisite if a microorganism is to infect a cell.

Aggressins

These substances allow the cell to evade host defence mechanisms. These may act to prevent initial attack and phagocytosis or to enable to cell to survive once phagocytosed (with the added benefit that once settled within a phagocytic cell they are safe from continued exposure to antibody and complement). They include:

- *capsules* – enable organisms to avoid phagocytosis by preventing interaction with the bacterial cell surface and the phagocytic cell, or concealing surface antigens. Specific antibodies against the capsular material will opsonize the organism and allow its ingestion. Examples include *Streptococcus pneumoniae* and *Haemophilus influenzae* type B. Mycobacteria have components in their cell wall that prevent lysosome/phagosome fusion once ingested
- extracellular slime substances – these are surface proteins or carbohydrates (polysaccharides). Examples include the M protein of group A streptococci (*Strepococcus pyogenes*,) which impairs complement function, protein A of *Staphylococcus aureus*, which binds

IgG by the Fc region, interfering with the phagocytosis of opsonized organisms, and the LPS of Gram-negative bacteria, which may delay or blunt the acute inflammatory response
- enzymes – some bacteria produce proteases which can hydrolyse and inactivate IgA, aiding mucosal colonization. *Listeria monocytogenes* secretes enzymes that inhibit destruction by the myeloperoxidase system of phagocytic cells. *Staphylococcus aureus* produces hyaluronidase, an enzyme that depolymerizes hyaluronic acid (responsible for cell-to-cell adhesion), thereby easing the spread of the organism. Group A streptococci produce streptokinases that lyse fibrin clots, *Clostridum perfringens* produces collagenases contributing to its ability to produce necrotizing skin infection
- siderophores – are molecules produced by most pathogenic bacteria and they scavenge iron from the host. Iron is required for virulence by several bacteria, and siderophores seem to protect bacteria from the killing effects of human serum
- plasmids – although these are not convetional virulence factors, they may code for a wide range of additional features promoting virulence such as antibiotic resistance, sex pili, and chromosomal mobilization (allowing the transfer of genetic material such as antibiotic resistance, toxin, etc to other cells)

Toxins

- Exotoxins are are among the most potent biological toxins and are mainly produced by Gram-positive organisms. They are usually heat-labile proteins, and many can be inactivated by proteolytic enzymes or neutralized by specific antibodies. The effects are highly varied, e.g. the clinical manifestations of tetanus, botulism, and diphtheria are all due to toxins. Some toxins are secreted in the active form, whereas others require cleavage to become active, e.g. *Corynebacterium diphtheriae*[1].
- Endotoxins are only produced by Gram-negative bacteria and consist primarily of LPS. They are heat stable and are only partly neutralized by specific antibodies. They are relatively low toxicity compared to exotoxins. They cause fever, hypotension, haemorrhage, and disseminated intravascular coagulation, and stimulate cytokine release from macrophages.

1 Interestingly only those strains that contain a lysogenic bacterophage (β-corynephage) are able to produce the toxin. The toxin genes exist on the phage genome.

Basic principles of virology

Classification

Viruses are classified on the basis of a number of criteria. These include:

- type of nucleic acid (DNA, RNA) and strand number (single, double)
- conformation of the nuclear material (linear, circular etc)
- whether the genetic information is positive or negative sense
- nucleocapsid symmetry
- presence or absence of an envelope
- their antigenic or genetic similarity.

They are grouped into orders, families, subfamilies, and genera. There are no consistent rules governing the naming of individual viruses. Some are named according to the disease they produce (poxvirus), others by acronyms (papovavirus – papilloma polyma vacuolating virus), some by appearance (coronavirus), and still others after the location in which they were first identified (Marburg). They may rarely be called after their discoverers (Epstein–Barr). Official names should be Latinized and printed in italic.

Properties and structure

Viruses are small (20–150 nm in diameter) protein packages, containing genetic material (DNA or RNA); some also contain enzymes. Viruses depend on living cells for their existence, genome expression, and replication. They have colonized most life forms including bacteria, plants, insects, and animals. The viral particle is composed of structural proteins. A capsid (protein coat) protects the nucleic acid contents and facilitates viral entry to a host cell. It is composed of many capsomeres (protein subunits). The term nucleocapsid refers to the capsid and viral nucleic acid. Certain viruses contain enzymes (e.g. reverse transcriptase in HIV). Some viral capsids have an outer envelope (derived from the plasma membrane of the infected cell from which it was released), into which are embedded protein spikes. Beneath the envelope, some viruses may have a stabilizing membrane protein. The entire particle is referred to as a virion.

Nucelocapsids may take several geometric forms:

- helical (like a spiral staircase) – the nucleic acid forms the central core with the nucleocapsid proteins forming the steps e.g. ssRNA viruses such as influenza and rabies. These viruses are enveloped, the envelope itself resting on the underlying membrane protein shell
- icosahedral (20 triangular faces with 12 corners) – e.g. all human DNA viruses. Each capsomere may itself be made of several peptides. DNA-herpesviruses have icosahedral structure and are additionally surrounded by a lipid envelope
- complex – these viruses do not fall into neat structural categories and often have large genomes e.g. poxvirus.

Viral genomes

The genetic material may be DNA or RNA (RNA viruses tend to have smaller genomes). Nucleic acid conformation varies widely between viral families (double-stranded, single-stranded, linear, circular, etc). Genomes vary widely in size but are limited by the space available in the virion. Bacteria may have several thousand genes but even the largest viruses have less than 200 and the smallest perhaps only four (such viruses may produce more than one protein from the same gene by means of RNA splicing or frame shifting). Viruses evolve rapidly due the high number of genome duplications undergone in short spaces of time. RNA viruses have high error rates, with genomes diverging by as much as 2% in the course of a year – 1 million times the rate of eukaryotic cell DNA genomes. Many mutations are non-functional but some will allow the virus to evade host immune responses and medical therapies.

The genome encodes both structural and non-structural (NS) proteins (enzymes required for viral expression and replication). The manner in which expression occurs depends on the nature of the nucleic acid:

- DNA viruses make RNA copies of the relevant segments of their DNA to direct protein synthesis – they may use host enzymes to achieve this or rarely carry them within the virion
- positive sense RNA viruses produce messenger RNA directly
- negative sense RNA viruses possess enzymes that produce positive strand copies that are used as mRNA
- retroviruses produce DNA from their RNA. This is integrated into the host's chromosomal DNA, transcription then taking place in broadly the same way as host mRNA is made from host DNA.

Viral replication

The manner in which a virus infects a cell varies but generally involves the interaction of a viral protein and host cell receptor (proteins, glycoproteins, or glycolipids intended for other functions and simply exploited by the virus), precipitating internalization. Once in the cell the virus uncoats (sheds its protein shell) and frees the nucleic acid, at least partially. It needs to achieve two things: the production of its enzymes and structural proteins; and replication of the viral genome.

Transcription

The manner in which mRNA is produced depends on the nature of the genome (see above). Ribosomes translate the viral mRNAs. Proteins may be produced in phases: 'early' proteins may be involved in DNA synthesis, or act as transcriptional activators to speed viral expression over host proteins. 'Late' proteins are produced from mRNA transcribed from newly synthesized viral nucleic acid, and tend to be structural. Some viruses produce a single long polypeptide which is then cleaved by proteases into individual proteins. Proteins often undergo post-translational processes such as glycosylation.

Viral genome replication

RNA viruses produce an RNA polymerase, either packaged with the virion (negative-sense viruses) or manufactured upon infection (positive-sense viruses). This rapidly produces RNA copies for incorporation into the viral particles. Genetic variation may arise by two mechanisms:

- mutations can occur due to errors in replication and the absence of proof-reading activity in enzymes such as RNA replicase and reverse transcriptase. DNA viruses replicate their genomes in the host nucleus where the necessary host enzymes can be exploited. The exceptions are the poxviruses which carry DNA polymerases with them and are thus capable of working entirely in the cytoplasm
- recombination of genetic material can occur either within a genome or between two viruses of the same kind if the host cell is co-infected with both viruses. RNA viruses are also capable of gene reassortment which, although not true mutation, may result in progeny with a quite different phenotype from parental strains, e.g. influenza pandemics may be a consequence of reassortment between human, avian, and pig flu.

Viral assembly

This may occur predominantly in the nucleus (e.g. adenovirus), or in the cytoplasm (e.g. poliovirus). Viral release then occurs by budding from the cell surface (e.g. measles), lysis of the cell (e.g. polio), or cell-to-cell spread. Some, such as HIV, may require a phase of post-release maturation. Overall, a complete viral lifecycle typically takes 6–8 hours, with the potential to produce thousand of viruses from each infected cell.

Viral pathogenesis

The effect of viral infection on the host ranges from asymptomatic infection to devastating disease with a wide variety of clinical manifestations. Viral species vary in pathogenicity and different strains of the same species may vary in virulence.

Entry

Viruses enter the body through the skin (usually via some degree of trauma) or via mucous membranes (where they adsorb directly to epithelial cells, in which they undergo primary replication). Resulting infections may be localized (e.g. papillomavirus and warts, conjunctivitis) or generalized, in which case pathology is not necessarily focused at the organ initially infected (e.g. enteroviruses are spread faeco-orally yet cause encephalitis). Transmission may also occur vertically (mother to child) and iatrogenically (organ transplants, blood transfusions etc).

Cytopathic effect

Viruses may disrupt the function of the cells they infect. This may result in inhibition of host-cell protein manufacture and lead to the death of the cell, lysis, and the release of virion. Alternatively they may precipitate cell fusion, forming multinucleated giant cells, or syncitia (e.g. respiratory syncytial virus (RSV)). Others may form inclusion bodies within the cell (eosinophilic or basophilic staining areas within the cell, representing aggregations of virions, sites of viral synthesis, or degenerative change). These inclusions can occur within the cytoplasm or nucleus.

Extent of infection

Some viral infections remain confined to tissues at, or continuous with, the site of entry. They may form focal lesions (e.g. skin and papillomavirus) or affect large areas of specific mucous membranes (e.g. viral gastroenteritis), and tend to have short incubation periods. Other viruses produce generalized infections, an initial phase of local replication near to the site of entry being followed by haematogenous spread (primary viraemia) to regional lymph nodes. This allows infection of large reticuloendothelial organs resulting in a secondary viraemia. The virus may travel free in the blood or within infected blood cells. It travels to organs distant from the site of entry and may infect specific organs preferentially. Such infections tend to have longer incubation periods.

Target organs

Symptoms depend on the target organ:

- skin – rashes are a common feature of viral infections and may be due to virus replication in the skin (vesicular rashes of Herpes simplex and Varicella zoster), the killing of infected cells (measles) or more general features of infection (disseminated intravascular coagulation (DIC), thrombocytopenia)
- respiratory tract – the lung may be involved in local infections (e.g. influenza) but can also be involved in generalized viral infections (e.g. chickenpox or measles pneumonitis)

- liver – the hepatitis viruses (hepatitis A to E) are tropic for the liver, which may also be affected as part of a more generalized infection (e.g. Epstein–Barr virus (EBV), cytomegalovirus)
- central nervous system (CNS) – the CNS can be invaded either as a result of viral passage along nerves (e.g. rabies) or by haematogenous spread (e.g. polio).

Illness duration

Viral illness may be acute or chronic.

- Acute viral illnesses present within a relatively short period of time. Most such infections are mild with a quick spontaneous recovery, e.g. chickenpox, measles, mumps, rubella. Some may have delayed serious features after an apparent recovery (e.g. encephalitis), or cause a rapid decline and possible death (e.g. rabies, viral haemorrhagic fevers).
- Chronic or persistent infections require the survival of viral DNA within the host cell, either integrated within the host DNA or in episomal form separate from it. They may be latent with no apparent illness or virus but occasional periods of reactivation (e.g. herpes simplex, varicella zoster) or chronic with continuous production of infectious virus (hepatitis B, hepatitis C, HIV).

Transformation

Some viruses have the potential to induce malignant change. For example EBV is associated with Burkitt's lymphoma, nasopharyngeal carcinoma, primary cerebral lymphoma and post-transplant lymphoproliferative disorder (PTLD).

Chapter 2

Antimicrobials

Definitions

Antimicrobials/anti-infectives

These are umbrella terms for drugs with activity against microorganisms. They include antibacterials, antivirals, antifungals, and antiparasitic agents.

Antibiotic

This is strictly defined as a chemical compound made by a microorganism that inhibits or kills other microorganisms at low concentrations (Waksman 1945). This does not include synthetic agents, although in practice it is often used for any antibacterial.

Antiparasitic

Antiparasitic agents are used to treat parasitic diseases, and include anti-protozoals and antihelminthics.

Development of antibiotics

- Substances with some form of anti-infective action have been used since ancient times. The Chinese used 'mouldy' soybean curd to treat boils and carbuncles. The South American Indians chewed cinchona tree bark (contains quinine) for malaria. They also wore 'mouldy' sandals for foot infections.
- In Europe one of the earliest recorded examples was the use of mercury to treat syphilis in the 1400s.
- In 1877 Louis Pasteur showed that injections of extracts of soil bacteria cured anthrax in animals.
- In 1908–1910, Paul Erlich, known as the father of chemotherapy and a Nobel Prize winner, synthesized arsenic compounds that were effective against syphilis.
- In 1924 the compound actinomycetin, so named because it is produced by actinomycetes, was discovered.
- In 1932 Domagk discovered the dye prontosil which cured streptococcal infections in animals. The active group turned out to be the sulfonamide attached to the dye, and by 1945 over 5000 sulfonamide derivatives had been developed. However, clinical use of these compounds has been limited by adverse effects and drug resistance.
- In 1939 Dubos isolated two agents that were active against Gram-positive organisms (gramicidin and tyrocidin) from *Bacillus brevis*.
- In 1944–1945, Waksman isolated streptomycin from the soil microbe, *Streptomyces griseus*. It was active against *Mycobacterium tuberculosis* and some Gram-negative organisms, and Waksman was awarded the Nobel Prize.

History of penicillin

Alexander Fleming returned to St Mary's Hospital after a weekend away in 1928 to discover that the mould *Penicillium notatum* had contaminated his culture plates. He observed that the colonies of *S. aureus* nearest the mould had lysed, while those further away had not and hypothesized that the *Penicillium* had released a product that caused bacterial cell lysis. He called this product penicillin. Although Fleming discovered penicillin, he was unable to purify sufficient quantities for clinical trials. In 1939 Howard Florey, Ernst Chain, and Norman Heatley, working in Oxford, obtained the *Penicillium* fungus from Fleming. They overcame the technical difficulties and conducted clinical trials to demonstrate the efficacy of penicillin. Mass production soon began in the UK and USA. Initially penicillin was used almost exclusively for soldiers injured during the Second World War. It became widely available by 1946. As soon as he discovered penicillin, Fleming warned of the development of penicillin resistance. He was right, and resistance was seen almost immediately. Scientists have chemically modifed the drug to create derivatives (e.g. ampicillin) that are less susceptible to enzymatic degradation.

Table 2.1 shows a timeline of the discovery or introduction of many antibiotics.

Table 2.1 Timeline of the discovery/introduction of some antibiotics

Year	Antibiotic	Class of antibiotic
1929	Penicillin discovered	β-lactam
1932	Prontosil discovered	Sulfonamide
1942	Penicillin introduced	β-lactam
1943	Streptomycin discovered	Aminoglycoside
1945	Cephalosporins discovered	β-lactam
1947	Chloramphenicol discovered	Protein synthesis inhibitor
1947	Chlortetracycline discovered	Tetracycline
1949	Neomycin discovered	Aminoglycoside
1952	Erythromycin discovered	Macrolide
1956	Vancomycin discovered	Glycopeptide
1960	Flucloxacillin introduced	β-lactam
1961	Ampicillin introduced	β-lactam
1963	Gentamicin discovered	Aminoglycoside
1964	Cephalosporins introduced	β-lactam
1964	Vancomycin introduced	Glycopeptide
1966	Doxycycline introduced	Tetracycline
1971	Rifampicin introduced	Rifamycin
1974	Co-trimoxazole introduced	Sulfonamide and trimethoprim
1976	Amikacin introduced	Aminoglycoside
1984	Ampicillin/clavulanate introduced	β-lactam/β-lactamase inhibitor
1987	Imipenem/cilastin introduced	Carbapenem
1987	Ciprofloxacin introduced	Quinolone
1993	Azithromycin and clarithromycin	Macrolide
1999	Quinupristin/dalfopristin introduced	Streptogramin
2000	Linezolid introduced	Oxazolidinone
2003	Daptomycin introduced	Lipopeptide
2004	Telithromycin introduced	Ketolide
2005	Tigecycline introduced	Glycylcycline

Global antibiotic use

National data on the quantity and trends of antibiotic usage are usually not available. However, it is generally thought that about 50% of all antimicrobials are used in human medicine, and about 50% for animals and crops. The total amount of antimicrobials used varies between countries; developed countries use proportionately more than developing countries. Different countries have different antibiotics available, e.g. in Japan there are a large number of carbapenems, whereas the UK only has three (imipenem, meropenem, and ertapenem). Just because an antibiotic is widely prescribed in one country, it does not mean it will be licensed in another, e.g. teicoplanin is not available in the USA.

Human use of antibiotics

Most infections are treated in the community, by general practitioners (GPs) or through outpatient clinics; this accounts for the greatest number of antibiotic prescriptions. Some agents with antimicrobial activity are available over the counter (OTC) (e.g. Canesten® for vaginal thrush and topical aciclovir for cold sores). The quantity of antibiotics used varies between hospitals, depending on the specialist units there (e.g. hospitals with large intensive care units or a renal unit are likely to use more antibiotics). Sometimes hospitals with similar patient populations have different approaches to prescribing antibiotics, which may be influenced by local antimicrobial susceptibility data. In the first instance you should follow your local hospital policy and consult microbiology/infectious diseases specialists for advice.

Problems with human use of antibiotics

- Overuse of antibiotics has led to rising rates of resistance to antimicrobials, with the result that a few organisms have become virtually untreatable (e.g. vancomycin-resistant *S. aureus* and extensively drug-resistant *M. tuberculosis*).
- The impact of resistance in terms of increased morbidity and mortality, and the economic and social consequences are considerable. In the USA, the Centers for Disease Control and Prevention (CDC) estimates that one-third of all outpatient antibiotic prescriptions are not required. Hence many organizations and bodies have been working to try to reduce inappropriate prescribing, e.g. public education (e.g. most upper respiratory tract infections (URTIs) are caused by viruses so antibiotics will not work), changing the practice of healthcare prescribers (e.g. limiting treatment duration for an uncomplicated urinary tract infection to 3 days).
- Other problems associated with the overuse of antibiotics include superinfection (e.g. with *Candida albicans*, *Clostridium difficile*) and the unnecessary risk of adverse effects and drug interactions and expense.
- In the developing world a number of additional problems exist:
  - antibiotics are often available without prescription from a corner shop
  - if the patient does take an appropriate antibiotic they may take the wrong dose for the incorrect duration
  - the purchased drug may be inappropriate, so the infection remains untreated and patients may therefore develop a more-severe or complicated infection before consulting a doctor
  - some drugs are 'fake', substandard, or past their expiry date
  - patients may take a combination of traditional therapy along with antibiotics that may lead to interactions and toxicity.

Non-human use of antibiotics

Antimicrobials have been used increasingly for the prevention and treatment of infections in animals (e.g. on farms, in fish factories, and for domestic pets) and in the environment (e.g. crop production). Since the discovery of their growth-promoting abilities, they have also been added to animal feed (particularly for pigs and poultry). In addition, some antibiotics increase feed efficiency (the amount of feed absorbed by an animal), increasing the chance that the animal reaches its target weight on time. Other examples of non-human use include:

- tetracycline is sprayed on apple plantations to treat fire-blight;
- oxytetracycline is added to water in commercial fish farms to treat infections;
- antibiotics are used to eliminate bacterial growth inside oil pipelines

When farm animals consume antibiotics in their feed, they may excrete them into the environment. This may select for antibiotic-resistant organsisms that may then infect other animal species. Antibiotics in the environment undoubtedly contribute to increasing antibiotic resistance, particularly amongst food-borne pathogens such as *Salmonella* spp. and *Campylobacter* spp. Many organizations have developed strategies to try to control the non-human use of antibiotics (particularly in agriculture and animal husbandry) and thus reduce the development of resistance. These include the Alliance for the Prudent Use of Antibiotics (APUA) and the World Health Organization (WHO).

Table 2.2 gives sources of information on antimicrobial resistance.

Table 2.2 Sources of information on antimicrobial resistance

Information	Web address
Alliance for the Prudent Use of Antibiotics	http://www.tufts.edu/med/apau
World Health Organization	http://www.who.int/foodborne_disease/resistance/en/
Health Protection Agency, UK	http://www.hpa.org.uk
Department for Environment, Food, and Rural Affairs, UK	http://www.defra.gov.uk
European data (Eurosurveillance)	http://www.eurosurveillance.org/
National Antimicrobial Resistance Monitoring System (USA)	http://www.cdc.gov/narms/
Drug resistance updates	http://health.surfwax.com/files/Antibiotic Resistance.html
History of antibiotic resistance	http://www.fda.gov/fdac/features/795_antibio.html
The Alexander project	http://www.ncbi.nlm.nih.gov/pubmed/12865398
The MYSTIC program	http://www.blackwellpublishing.com/eccmid17/abstract.asp?id=57603

Mechanisms of action

Antimicrobial agents are classified by their specific modes of action against bacterial cells. The modes of action of antimicrobial agents against Gram-positive and Gram-negative bacteria are very similar and can be divided into five categories:

- inhibition of cell wall synthesis
- inhibition of protein synthesis
- inhibition of nucleic acid synthesis
- inhibition of folate synthesis
- disruption of the cytoplasmic membrane.

Inhibition of cell wall synthesis

Antimicrobial agents that interfere with cell wall synthesis block peptidoglycan synthesis. They are active against growing bacteria and are bactericidal.

- Gram-negative bacteria – β-lactam antimicrobials enter the cell through porin channels in the outer membrane and bind to penicillin-binding proteins (PBPs) on the surface of the cytoplasmic membrane. This blocks their function, causing weakened or defective cell walls and leads to cell lysis and death.
- Gram-positive bacteria lack an outer membrane, so β-lactam antimicrobials diffuse directly through the cell wall and bind to PBPs, which results in weakened cell walls and cell lysis.

Inhibition of protein synthesis

- Tetracyclines bind to the 30S ribosomal subunit and block attachment of transfer RNA (tRNA) and addition of amino acids to the protein chain. Tetracyclines are bacteriostatic.
- Aminoglycosides also bind to the 30S ribosomal subunit and prevent its attachment to messenger RNA (mRNA). They can also cause misreading of the mRNA resulting in insertion of the wrong amino acid or interference in the ability of amino acids to connect with each other. The combined effect of these two mechanisms is bactericidal.
- Macrolides and lincosamides attach to the 50S ribosomal subunit causing termination of the growing protein chain. They are bacteriostatic.
- Chloramphenicol also binds to the 50S ribosomal subunit and inteferes with binding of amino acids to the growing chain. It is also bacteriostatic.
- Linezolid (an oxazolidinone) binds to the 23S ribosomal RNA of the 50S subunit and prevents formation of a functional 70S initiation complex which is necessary for protein synthesis. It is bacteriostatic.

Inhibition of nucleic acid synthesis

- Fluoroquinolones interfere with DNA synthesis by blocking the enzyme DNA gyrase. This enzyme binds to DNA and introduces double-stranded breaks that allow the DNA complex to unwind. Fluoroquinolones bind to the DNA gyrase–DNA complex and allow broken DNA strands to be released into the cell, resulting in cell death.
- Rifampicin binds to DNA-dependent RNA polymerase, which blocks synthesis of RNA and results in cell death.

Inhibition of folate synthesis (fig 2.4)

- For many organisms para-aminobenzoic acid (PABA) is an essential metabolite which is involved in the synthesis of folic acid, an important precursor to the synthesis of nucleic acids.
- Sulfonamides are structural analogues of PABA and compete with PABA for the enzyme dihydropteroate synthetase.
- Trimethoprim acts on the folic acid synthesis pathway at a point after the sulfonamides, inhibiting the enzyme dihyrofolate reductase.
- Both trimethoprim and the sulfonamides are bacteriostatic. When they are used together (e.g. co-trimoxazole) they produce a sequential blockade of the folic acid synthesis pathway and have a synergistic effect.

Disruption of the cytoplasmic membrane

- Polymyxin molecules diffuse through the outer membrane and cell wall of susceptible cells to the cytoplasmic membrane. They bind to the cytoplasmic membrane and disrupt and destabilize it. This causes the cytoplasm to leak out of the cell, resulting in cell death.

Mechanisms of resistance

There are a number of ways by which microorganisms become resistant to antimicrobial agents. These include:

- production of enzymes
- alteration in outer membrane permeability
- alteration of target sites
- efflux pumps
- alteration of metabolic pathways.

Production of enzymes

- β-lactamases are enzymes that hydrolyse β-lactam drugs. In Gram-negative bacteria, the β-lactam drug enters the cell through the porin channels and encounters β-lactamases in the periplasmic space. This results in hydrolysis of the β-lactam molecules before they reach their penicillin binding protein (PBP) targets. In Gram-positive bacteria, the β-lactamases are secreted extracellularly into the surrounding medium and destroy the β-lactam molecules before they enter the cell.
- Aminoglycoside-modifying enzymes – Gram-negative bacteria may produce adenylating, phosphorylating or acetylating enzymes that modify an aminoglycoside so that it is no longer active.
- Chloramphenicol acetyl transferase – Gram-negative bacteria may produce an acetyl transferase that modifies chloramphenicol so that it is no longer active.

Alteration in outer-membrane permeability

Gram-negative bacteria may become resistant to β-lactam antibiotics by developing permeability barriers.

- Mutations resulting in the loss of porin channels in the outer membrane no longer allow the entrance and passage of antibiotic molecules into the cell.
- Alterations in proton motive force may result in reduced inner-membrane permeability

Alteration of target sites

- PBPs in Gram-positive and Gram-negative bacteria may be altered through mutation so that β-lactams can no longer bind to them.
- Methylation of ribosomal RNA confers resistance to macrolides, linconsamides, and streptogramins
- Mutations in the chromosomal genes for DNA gyrase and topoisomerase IV confer quinolone resistance.

Efflux pumps

A wide variety of efflux pumps produce antimicrobial resistance in both Gram-positive and Gram-negative bacteria. Active efflux of antibiotics is mediated by transmembrane proteins which form channels that actively export an antimicrobial agent out of the cell as fast as it enters. This is the main mechanism of resistance to tetracyclines.

Alteration of metabolic pathways

Some microorganisms develop an altered metabolic pathway that bypasses the reaction inhibited by the antimicrobial. Mutations that inactivate thymidylate synthetase block the conversion of deoxyuridylate to thymidylate. These mutants use exogenous thymine or thymidine for DNA synthesis and therefore are resistant to folate synthesis antagonists.

Molecular genetics of resistance

Genetic variability is essential for microbial evolution and may occur by a variety of mechanisms:

- point mutations
- rearrangements of large segments of DNA from one location of a bacterial chromosome or plasmid to another
- acquisition of foreign DNA from other bacteria, via a mobile genetic element (MGE).

Acquisition of resistance, in the presence of the antibiotic, confers a survival advantage on the host, thus leading to the emergence of resistant clones. The overuse, incorrect use, or injudicious use of antibiotics contributes to this problem, and explains why new resistance patterns tend to emerge in areas of the hospital with the greatest antibiotic consumption (e.g. intensive care units). This antibiotic resistance may be passed vertically to future generations through clonal expansion, resulting in a bacterial population that is resistant to an antibiotic.

Point mutations

These are often referred to as single nucleotide polymorphisms (SNPs). The bacterial genome is dynamic, in that bacteria are constantly undergoing mutations. Some of these mutations will (by chance) result in a survival advantage to the organism, e.g. due to increased virulence or antibiotic resistance. In certain environments, these will be preferentially selected for, thus the mutation that arose by chance will be preferentially retained and passed to future generations. Examples of point mutations include the generation of β-lactamases and fluoroquinolone resistance.

Mobile genetic elements (MGE)

These are pieces of DNA that can move around between genomes. They may thus be involved in the horizontal transfer of resistance genes between bacteria, as opposed to the vertical transfer of resistance by clonal expansion (see list below). Bacterial genomes consist of core genes and accessory genes: it is the latter which are defined by acquisition and loss. There are several different MGEs described in the following list.

- Plasmids – these are extrachromosomal pieces of circular DNA, which vary in size from 10 kb to over 400 kb. In addition to carrying resistance genes, they may determine other functions, e.g. virulence factors and metabolic capabilities. They are autonomous self-replicating genetic elements that possess an origin for replication and genes that facilitate their maintenance in the host bacteria. Conjugative plasmids require additional genes to initiate self-transfer.
- Insertion sequences (IS) – these are short DNA sequences that are usually only 700 to 2500 bp (base pairs) long. They encode for an enzyme needed for transposition (i.e. to excise a segment of DNA from one position in the chromosome and insert it elsewhere) and a regulatory protein, which either stimulates or inhibits the transposition activity. They are thus different from transposons, which also carry accessory genes such as antibiotic resistance genes. The coding region in an insertion sequence is usually flanked by inverted repeats.

- Transposons – these are often called 'jumping genes' and may contain insertion sequences. They cannot replicate independently, but can move between one replicating piece of DNA to another, e.g. from a chromosome to a plasmid. Conjugative transposons mediate their own transfer between bacteria whereas non-conjugative transposons need prior integration into a plasmid to be transferred.
- Integrons may be defined as a genetic element that possesses a site (*attI*) at which additional DNA in the form of gene cassettes can be integrated by site-specific mutation. They also encode a gene, integrase, that mediates these site-specific recombination events. Gene cassettes normally consist of an antibiotic resistance gene and a 59 base element that functions as a site-specific recombination site. The largest integrons (e.g. in *V. cholerae*) can contain hundreds of gene cassettes.
- Bacteriophages – a bacteriophage is a virus which infects bacteria and may become integrated into the bacterial chromosome (and is then called a prophage). They typically consist of an outer protein enclosing genetic material (which may be single- or double-stranded DNA or RNA). Bacteriophages may be considered MGEs, but are rarely involved in the transfer of resistance genes. They have been used as an alternative to antibiotics (phage therapy) in eastern Europe and the former USSR for ~60 years.

Clonal expansion

Clonal Expansion refers to the multiplication of a single 'ancestor' cell. This may result in the propagation of antibiotic resistance into daughter cells. The antibiotic resistance genes will be passed from one generation of bacteria to the next, which is also called vertical transfer of resistance. If an organism becomes resistant to an antibiotic, either by mutation or acquisition of a mobile genetic element, it will have a survival advantage in an environment where that antibiotic is present. Thus the daughter cells that are generated will be positively selected for, over daughter cells from another antibiotic sensitive strain of the bacteria, and future generations will be resistant to that agent.

A bacterial clone (see below) refers to all organisms that are likely to have arisen from a common ancestor. This may not be immediately obvious, e.g. in the case of *S. Typhi*, the common origin has been estimated by molecular techniques (multi locus sequence typing MLST) to have existed 50,000 years ago. Examples of more recent clonal expansion relating to the spread of antibiotic resistance genes are MRSA and penicillin resistant pneumococci.

Important definitions

Isolate: this refers to a pure culture. It says nothing about typing.

Clone: this refers to bacterial cultures which have been isolated independently, from different sources, in different places, and maybe at different times, but are so similar phenotypically and genotypically, that the most likely explanation is that they arose from a common ancestor.

Strain: this refers to a phenotypically and/or genotypically distinctive group of isolates. It is dependent on the typing scheme used, and some experts suggest avoiding the use of this term.

Type: this refers to organisms with the same pattern or set of markers displayed by a strain, when the bacteria are subject to a particular typing system.

Breakpoints

Definition

Breakpoints are antibiotic concentrations that have been established for most antimicrobial/organism combinations to interpret the results of susceptibility testing. Thus strains are categorized as susceptible, resistant, or intermediate to each drug tested. When new breakpoints are set, clinical, microbiological, and pharmacological factors must be considered.

Practical aspects

Each isolate cultured in the microbiology laboratory that is deemed significant (rather than a likely contaminant) is tested against a panel of appropriate antibiotics. This panel may vary between hospitals, depending on local antibiotic policies and microbiologist's preferences.

Many different methods exist for sensitivity testing e.g. Stoke's method, Kirby–Bauer method, CLS (Clinical and Laboratory Standards Institute) formerly National Committee on Clinical Laboratory Standards (NCCLS)) but many UK labs use the British Society of Antimicrobial Chemotherapy (BSAC) method. (http://www.bsac.org.uk/susceptibility_testing/guide_to_antimicrobial_susceptibility_testing.cfm). The principle of this method is that the diameter of any zone of clearing around an antibiotic-containing disc is compared to the 'zone breakpoint diameter' for that specific species/drug combination. These published zone breakpoints are related to the minimum inhibition concentration (MIC) of the organism. If the diameter of clearance is greater than or equal to the susceptible breakpoint, the isolate is classified as sensitive to that antibiotic. If there is no zone of clearing, or the diameter of the zone is less than or equal to the resistant breakpoint, the isolate is classified as resistant to that antibiotic.

Each isolate is reported as sensitive or resistant to each antibiotic tested (the term 'intermediate' is generally avoided as interpretation may be difficult). Some microbiologists only report certain sensitivity results (usually consistent with the hospital antibiotic policies) to encourage use of these drugs. It is worth remembering that most isolates are tested against at least six drugs, so if none of the sensitivity results available are suitable, contact a microbiologist. Also remember that when an organism is 'fully sensitive' this only means 'fully sensitive to the antibiotics tested', not all agents!

In some circumstances the exact MIC of an isolate is determined in the laboratory, usually by Etest®. This applies to organisms with borderline resistance, unusual resistances, and some invasive infections e.g. *S. pneumoniae* and infective endocarditis. The MIC value is then compared to the published MIC breakpoints for that species and antibiotic.

Heteroresistance

This is defined as the growth of one bacterial subpopulation at a higher antibiotic concentration than predicted by the MIC for most cells. Heteroresistance may be difficult to diagnose and result in poor response to treatment. Examples include:

- *S. aureus* and vancomycin
- *C. neoformans* and azoles
- *M. tuberculosis* and rifampicin
- *E. faecium* and vancomycin
- *A. baumanii* and carbapenems and colistin
- *H. pylori* and metronidazole and amoxicillin
- *S. pneumoniae* and penicillin.

Pharmacokinetics

This is what the body does to the drug; it comprises absorption, distribution, metabolism, and excretion (mnemonic ADME)

Absorption

To be effective, a drug must reach the site of the infection. In some cases this is possible by topical application (e.g. nystatin pastilles for oral candidiasis), but in most cases drugs are transported around the body by the circulation.

- Some drugs are poorly absorbed when given by mouth (e.g. aminoglycosides and glycopeptides) and are therefore given parenterally. Occasionally oral drugs may be used to treat luminal infections, e.g. vancomycin for *C. difficile*
- If a drug is absorbed when given by mouth, the proportion that is absorbed into the systemic circulation is called the bioavailability. Drugs given intravenously have 100% bioavailability. The time profile of absorption versus elimination is usually more important than the total amount of drug absorbed (see 📖 Pharmacodynamics, p.48).
- Absorption may be affected by interactions with other drugs or food that may bind the drug (e.g. tetracyclines should not be given with milk). Altered physiology (e.g. diarrhoea) may reduce absorption. None of the commonly prescribed antibiotics are subject to significant first-pass metabolism in the liver.

Distribution

The volume of distribution relates the drug concentration in the blood to the amount of drug given. A drug with a small volume of distribution is largely confined to the plasma. A drug with a large volume of distribution is widely distributed, e.g. fat-soluble drugs.

Metabolism

Some antibiotics are metabolized in the liver by isoforms of cytochrome P-450, of which the CYP3A4 isoform is the most abundant (Box 2.1). Rifampicin induces the activity of CYP3A4 leading to increased metabolism (and reduced efficacy) of drugs that share this pathway, e.g. HIV protease inhibitors. In contrast, the azole antifungals and the macrolides inhibit the activity of CYP3A4, which will reduce the metabolism (and may increase toxicity) of drugs also metabolized by this isoform. Always check for interactions before starting or stopping antibiotics and seek expert advice if unsure.

Box 2.1 Antimicrobials metabolized by CYP3A4

- Inducers – rifamipicin
- Inhibitors – ketoconazole, itraconazole, erythromycin, clarithromycin
- Substrates – ritonavir, saquinavir, indinavir, nelfinavir

Excretion

This can be divided into renal (e.g. aminoglycosides, glycopeptides) and non-renal (biliary tree, e.g ceftriaxone; gastrointestinal tract (GI) tract, e.g. azithromycin). Clearance determines the half-life of the drug (the time for the blood concentration to decrease by half). Steady-state generally occurs when a patient has taken the drug for a period of time equal to 5–7 half-lives. Calculating creatinine clearance (CrCl, see Antimicrobials in renal impairment) as a measure of renal function can be essential for safe dosing of some renally excreted drugs, such as once-daily aminoglycosides (see 🕮 Aminoglycosides, p 78).

Pharmacodynamics

This is what the drug does to the body – it includes the drug's mechanism of action and biochemical and physiological effects.

Bacteriostatic

Antibiotics that inhibit growth and replication of bacteria, but are non-lethal (e.g. drugs that inhibit folic acid synthesis, see 📖 pp. 96–99).

Bactericidal

Antibiotics that cause bacterial cell death by inhibition of (a) cell wall synthesis, (b) nucleic acid synthesis, or (c) protein synthesis. Some antibiotics may be bacteriostatic at low concentrations but bactericidal at higher concentrations.

Minimum inhibitory concentration (MIC)

This is the concentration of antimicrobial required to inhibit the overnight growth of 90% (MIC_{90}) or 50% (MIC_{50}) of a particular bacterial isolate *in vitro*. Dose regimens in common use generally produce plasma concentrations 2 to 4 times the MIC. Clinically the MIC is used to assign an organism to a susceptibility category (sensitive, intermediate, resistant).

Minimum bactericidal concentration (MBC)

This is the concentration of antimicrobial required to kill 90% (MBC_{90}) or 50% (MBC_{50}) of a bacterial isolate *in vitro*. The MBC is 2 to 4 times the MIC for the same isolate. This used to be determined for isolates that caused endocarditis, but is now rarely performed.

Synergism

This occurs when the activity of two drugs together is greater than the sum of their actions if each were given separately. An example is the use of ampicillin and gentamicin for enterococcal infections (see 📖 Enterococci, p. 285). where ampicillin acts on the cell wall to enable gentamicin to gain entry to the cell and act on the ribosome.

Antagonism

One drug diminishes the activity of another drug, so giving both antibiotics together may result in a worse clinical outcome than just giving one antibiotic. For example co-administration of a bacteriostatic agent (e.g. tetracycline) with a β-lactam may inhibit cell growth, and prevent the bactericidal activity of the β-lactam.

Concentration-dependent killing

The antibiotic kills the organism when its concentration is well above the MIC of the organism. The greater the peak, the greater the killing, e.g. once-daily dosing of gentamicin (see 📖 Aminoglycosides, p. 78).

Time-dependent killing

The antibiotic *only* kills the bacteria when its concentration is above the MIC of the organism, but increasing the concentration does not lead to increased killing. If the concentration rises above four times the MIC, any additional effect is negligible. Most recommended dosing schedules account for this (see 📖 Penicillins, p.64; 📖 Cephalosporins, p.66; 📖 Macrolides, p. 82). On a practical level it is important when adjusting the dose of glycopeptides (see 📖 Glycopeptides, p.76).

Post-antibiotic effect (PAE)

The post-antibiotic effect (PAE) is defined as the time during which bacterial growth is inhibited after antibiotic concentrations have fallen below the MIC. The mechanism is unclear, but it may be due to a delay in the bacteria re-entering a log-growth period. Several factors influence the presence or duration of the PAE, including the type of organism, type of antimicrobial, concentration of antimicrobial, duration of antimicrobial exposure, and antimicrobial combinations. *In vitro*, β-lactam antimicrobials demonstrate a PAE against Gram-positive cocci but not against Gram-negative bacilli. Antimicrobials that inhibit RNA or protein synthesis produce a PAE against Gram-positive cocci and Gram-negative bacilli. The clinical relevance of the PAE is probably most important when designing dosage regimens. The presence of a long PAE allows aminoglycosides to be dosed infrequently; the lack of an *in vivo* PAE suggests that β-lactam antimicrobials require frequent or continuous dosing.

Eagle effect

This is a paradoxical effect, first described by Eagle in 1948, whereby higher concentrations of penicillin resulted in decreased killing of staphylococci and streptococci. Eagle also showed that this paradoxical effect seen *in vitro* correlated with an adverse outcome *in vivo*. This effect has since been described with a number of other antimicrobials and organisms e.g. ampicillin and *E. faecalis*, carbenacillin and *P. mirabilis*, mecillinam and *P. stuartii*, cefotaxime and *S. aureus* and *P. aeruginosa*, and aminoglycosides and Gram-negative bacteria.

Preventing the development of resistance

Studies are focusing on defining the breakpoints that predict the emergence of resistance. An ideal antibiotic should have a low rate of resistance mutation, high fitness cost of resistance, and low rate of fitness restoring complementary mutation. Novel parameters that are being investigated include:

- mutant prevention concentration – the ability to restrict selection of resistant mutants
- mutant selection window – the concentration range between the minimal concentration required to block growth of wild-type bacteria, up to the concentration needed to inhibit the growth of the least susceptible, single-step mutant. There are different concentation ranges for each organism/drug combination.

Choosing an antimicrobial

Before prescribing an antimicrobial the following factors should be considered:

- host factors – medical history (e.g. immunocompromise), severity of infection, allergies, renal, and hepatic function, pregnancy, breast feeding, age, weight, ability to tolerate drugs by mouth, compliance
- organism factors – known/likely organism(s) and known/likely antimicrobial susceptibility. Always consult your local antimicrobial policy or seek advice from an infection specialist
- drug factors – pharmacokinetics (see 🕮 Pharmacokinetics, p.46), pharmacodynamics (see 🕮 Pharmacodynamics, p.48), side-effects, interactions with other medications, previous therapy
- dose – may depend not just on the severity of the infection but also on host factors (age, weight, renal function). An inadequate dose may result in a suboptimal clinical response and increase the likelihood of antibiotic resistance. Conversely, too large a dose may result in unnecessary toxicity and expense
- route of administration – antimicrobials may be given by a variety of routes e.g. orally (PO), intravenously (IV), intramuscularly (IM), per rectum (PR), per vagina (PV), aerolized/nebulized (NEB), or topically (TOP)
- duration of treatment – this is usually 5 to 7 days for community-acquired infections. Prolonged courses are discouraged as they may lead to increased antibiotic resistance, side-effects, and expense. There is evidence to support a single-dose therapy for uncomplicated urinary tract infections. At the other extreme, several months of treatment are needed for tuberculosis (TB) and chronic osteomyelitis (see 🕮 Duration of antibiotic therapy, p.60).

Routes of administration

Oral administration (PO)

Most antibiotics used in human medicine are given orally in the community. If a drug is absorbed when given by mouth, the proportion that is absorbed into the systemic circulation is called the bioavailability (see Pharmacokinetics, p.46). This depends on the formulation of the drug, and how it is taken, e.g. some tetracyclines should not be taken with milk or antacids as these decrease their absorption. Some drugs are not absorbed when given orally. This can be advanatageous when treating luminal infections, e.g. oral vancomycin for *C. difficile* and neomycin in hepatic failure.

Intravenous administration (IV)

The intravenous route enables higher doses to be given, and results in higher, more reliable drug concentrations (see Box 2.2).

Indications

Antimicrobials are given intravenously in the following situations:

- life-threatening infections – e.g. meningitis, septicaemia, endocarditis require intravenous therapy. Antibiotics may be given as infusions or bolus doses, depending on the drug
- inability to take/absorb oral medications – e.g. nil by mouth, severe vomiting or diarrhoea, oesophageal or intestinal obstruction, postoperative ileus
- poor oral bioavailability – some drugs are not absorbed if given orally, e.g. aminoglycosides, glycopeptides, colistin.

Disadvantages

Intravenous therapy may be associated with a number of problems:

- side-effects, which may be local (e.g. phlebitis) or systemic (e.g. rapid infusion may result in anaphylactoid reations such as the 'red man syndrome' with vancomycin)
- line infections, which may be local (e.g. exit site, tunnel or pocket infections) or systemic (e.g. bacteraemia, endocarditis)
- inconvenience to patient
- need to stay in-hospital. This may be overcome by the use of outpatient antimicrobial therapy (OPAT) which is now available in some regions of the UK
- intravenous antibiotics are usually considerably more expensive than the oral formulation.

IV to oral switch

As a result of the problems associated with IV therapy, many hospitals employ 'IV to oral switch' protocols for certain conditions which encourage clinicians to change to oral antibiotics as soon as is safe. Criteria include:

- suitable oral agent available
- patient can tolerate, swallow, and absorb oral antibiotics

- no symptoms or signs of ongoing sepsis
- some conditions are specifically excluded (e.g. meningitis and endocarditis). If in doubt consult an infection specialist.

Box 2.2 Practical points

- Many people believe that IV antibiotics are somehow 'stronger' than oral antibiotics. This is not necessarily the case (e.g. ciprofloxacin is as effective when given orally as when given IV and much cheaper).
- The oral and IV doses of the same antibiotic may be different (e.g. metronidazole).

Intramuscular administration (IM)

This is an infrequent method of administration, largely because absorption is unpredictable and the injection may be painful. Local side-effects include irritation and development of a sterile abscess. The advantages are that there is no question of compliance and the agent can easily be administered in the community. Intramuscular administration is commonly used for vaccinations, in genitourinary clinics, and for TB treatment in the developing world.

Never give IM injections to patients with bleeding/clotting disorders, e.g. thrombocytopaenia, haemophilia.

Examples of drugs given intramuscularly

- Benzylpenicillin should be given immediately if a GP suspects bacterial meningitis (especially meningococcal disease) in the community, before transferring the patient urgently to hospital. If it cannot be administered IV then deep IM injection is recommended.
- Procaine penicillin (procaine benzylpenicillin) is given as daily IM injections in early syphilis or late latent syphilis. It is only available on a named patient basis.
- Cefotaxime IM may be given as secondary prophylaxis for contacts of meningococcal disease, if the individual is unable to take rifampicin or ciprofloxacin (although it is not licensed for this indication).
- Ceftriaxone IM is used for gonococcal infection (particularly pharyngeal or conjunctival infection).
- Spectinomycin IM is occasionally given to patients who cannot take cephalosporins or quinolones (e.g. pregnant women with β-lactam allergy).
- Streptomycin IM is commonly given for the treatment of tuberculosis in the developing world.
- Gentamicin IM is sometimes given before changing a urinary catheter in a patient in the community.

Topical administration (TOP)

Many antimicrobials are available as topical preparations. They are most commonly used in general practice and dermatology. However, they are not without risk and should be used with caution. Before prescribing a topical drug consider the following:

- does the condition require treatment? Not all skin conditions that are oozing, crusted or pustular are infected. Would improving hygiene resolve the situation? Even if an organism is cultured from a swab, it may represent colonization and not require treatment
- would systemic antibiotics be more appropriate? Some skin infections (e.g. erysipelas, cellulitis) require systemic antibiotics as the infection is too deep for topical antibiotics to penetrate adequately
- development of resistance – topical antibacterials should be limited to those not used systemically, in order to prevent development of resistance
- duration of treatment – topical agents should only be used for short periods in defined infections.

Examples of drugs given topically

- Fusidic acid may be used to treat impetigo, although oral therapy is often required. It should not be used for more than 7–10 days.
- Mupirocin may also be used to treat impetigo (if MRSA-positive) or given as part of MRSA decolonization regimens. It should not be used for more than 7–10 days.
- Neomycin is also used to treat skin infection but may cause ototoxicity if large areas of skin are treated, and sensitization.
- Chloramphenicol may be given as eye drops or ear drops for conjunctivitis or otitis externa respectively.
- Aciclovir cream may be used for the treatment of oral and genital herpes simplex infections.
- Nystatin pastilles can be used for oral candidiasis.
- Clotrimazole cream is used for vulvovaginal candidiasis or athlete's foot.
- Permethrin and malathion are used for scabies.
- Malathion, pyrethroids, or dimeticone may be used for headlice.

Aerosolized administration (NEB)

Aerosolized antibiotics are usually given for treatment or prophylaxis of respiratory infections. They are administered directly to the site of action, and may have fewer systemic adverse effects. However, they are usually more difficult to give and there may still be some systemic absorption. One of the main groups to benefit from aerosolized antibiotics are cystic fibrosis (CF) patients, who may acquire multiresistant organisms (see 🕮 Cystic fibrosis, p.668).

Examples of antimicrobials given by inhalation

- Tobramycin (see 🕮 Aminoglycosides, p.78) is an aminoglycoside often given by nebulizer for chronic pulmonary infection with *P. aeruginosa* in CF patients. It is usually given cyclically (twice daily for 28 days, followed by a 28-day tobramycin-free period). Not all patients respond to treatment, and some become less responsive as drug resistance develops.

- Colistin (see 📖 Polymyxins, p.106) is a polymyxin antibiotic active against many Gram-negative organisms including *P. aeruginosa* and *Acinetobacter* spp. It is not absorbed orally, and is toxic when given systemically, so inhalation of a nebulized solution is the preferred route for treating respiratory infections. It is mainly used as an adjunct to standard antibiotics in CF patients. It has also been used for the prevention and treatment of ventilator-associated pneumonia due to *Acinetobacter* spp, although this practice is controversial.
- Pentamidine isetionate (see 📖 Antiprotozoal drugs (2), p.158) used as a 2nd-line agent for the treatment of *Pneumocystis jiroveci* pneumonia (see 📖 Pneumocystis jiroveci, p.530). In mild disease, inhaled pentamidine may be used in patients who are unable to tolerate co-trimoxazole, but systemic absorption may occur. In severe disease, IV pentamidine is used in patients who are unable to tolerate or have not responded to co-trimoxazole. Side-effects include hypotension following administration, and severe, sometimes fatal, reactions due to hypotension, hypoglycaemia, pancreatitis, and arrhythmias. Intermittent inhaled or IV pentamidine may be used as prophylaxis in patients unable to tolerate co-trimoxazole (although inhaled pentamidine does not protect against extrapulmonary disease).
- Ribavirin (see 📖 Antivirals for respiratory syncytial virus, p.134) is licensed for the treatment of severe respiratory synctial virus (RSV) bronchiolitis in infants and children, especially if they have other serious diseases. There is no evidence of mortality benefit. Side-effects include worsening respiration, bacterial pneumonia, and pneumothorax. CAUTION: ribavirin is teratogenic, and exposure should be avoided in pregnant and breast-feeding women.

Antimicrobials in renal impairment

General principles

- Avoid nephrotoxic drugs in patients with renal impairment.
- Keep antibiotic prescriptions to a minimum for patients with severe renal disease.
- Some intravenous antibiotic preparations contain sodium (e.g. Tazocin®), which may cause difficulties in patients with renal impairment.
- The use of drugs in patients with renal impairment can cause several problems:
  - reduced excretion of a drug or its metabolites may cause toxicity
  - increased sensitivity to some drugs
  - many side-effects are poorly tolerated in patients with renal impairment
  - some drugs are not effective when renal function is impaired.
- Some of these problems may be avoided by reducing the dose (or using alternative drugs).

Assessment of renal function

Renal function can be assessed in a number of ways:

- serum creatinine is the most commonly used parameter. It is affected by muscle mass which may be reduced in elderly patients (resulting in underestimation of renal impairment), or increased in certain races (e.g. blacks). Serum creatinine does not rise until 60% of total kidney function is lost.
- creatinine clearance (CrCl) can be measured using a 24-h urine collection. Estimated CrCl is calculated using the Cockcroft and Gault formula (Box 2.4), which is based on age, weight, sex, and serum creatinine. Thus, estimated CrCl may be inaccurate in patients who are obese or have acute renal failure.

Box 2.4 Cockcroft and Gault formula for estimated creatinine clearance

$$\text{CrCl (mL/min)} = \frac{(140 - \text{age}) \times \text{lean body weight (kg)} \times \text{N}}{\text{serum creatinine (micromol/L)}}$$

N=1.23 for males and 1.03 for females

- Glomerular filtration rate (GFR) is the volume of fluid filtered from the renal glomerular capillaries into Bowman's capsule per unit time, and can be measured by injecting inulin into the plasma. Estimated GFR (eGFR) can be calculated using the modification of diet in renal disease (MDRD) formula (Box 2.5), which is based on serum creatinine, age, sex, and race. Renal impairment was previously classified into three grades (mild, moderate, and severe); this has now been superseded by the use of CrCl or GFR.

Box 2.5 MDRD formula for estimated glomerular filtration rate

eGFR (mL/min/1.73m^2) = 186 x (serum creatine/88.4)$^{-1.154}$ x (Age)$^{-0.203}$ x (0.742 if female) or x (1.21 if black)

Dose modification in renal impairment

- The level of renal function below which the dose of a drug must be reduced depends on the proportion of the drug eliminated by renal excretion and its toxicity.
- For drugs with only minor or no dose-related side-effects, very precise modification of the dose regimen is unnecessary and a simple scheme for dose reduction is sufficient.
- For more toxic drugs with a small safety margin, dose regimens based on GFR should be used.
- When both efficacy and toxicity are closely related to plasma-drug concentration, recommended regimens should be regarded only as a guide to initial treatment; subsequent doses must be adjusted according to clinical response and plasma-drug concentration (e.g. vancomycin, gentamicin).
- The total daily maintenance dose of a drug can be reduced either by reducing the size of the individual doses or by increasing the interval between doses. For some drugs, although the size of the maintenance dose is reduced, it is important to give a loading dose if an immediate effect is required. The loading dose should usually be the same size as the initial dose for a patient with normal renal function.
- Seek specialist advice from your hospital pharmacist for patients on haemodialysis, haemofiltration, or chronic ambulatory peritoneal dialysis.

Drugs to be used with caution

- For up-to-date guidance always consult your hospital pharmacist, the British National Formulary http://www.bnf.org/, or the electronic Medicines Compendium http://emc.medicines.org.uk/.
- A wide range of antimicrobials should be used with caution in patients with renal impairment including:
  - Antibacterials, e.g. aminoglycosides, aztreonam, cephalosporins, carbapenems, chloramphenicol, colistin, ethambutol, isoniazid, linezolid, macrolides, ketolides, penicillins, quinolones, sulfonamides, tetracyclines, trimethoprim, vancomycin
  - Antifungals, e.g. amphotericin B, flucytosine, fluconazole, itraconazole, voriconazole
  - Antimalarials, e.g. atovaquone, chloroquine, Malarone®, proguanil, pyrimethamine, quinine, Riamet®, sulfadiazine
  - Antivirals, e.g. aciclovir, adefovir, amantadine, antiretrovirals, famciclovir, foscarnet, ganciclovir, oseltamivir, pentamidine, ribavirin, valaciclovir, valganciclovir.

Antimicrobials in liver disease

Metabolism by the liver is the main route of elimination for many drugs, but hepatic reserve is large and liver disease has to be severe before important changes in drug metabolism occur. Routine liver functions tests are a poor guide to metabolic capacity, and it is not possible to predict the extent to which the metabolism of a particular drug may be impaired in an individual patient. Drug prescribing should be kept to a minimum in all patients with severe liver disease.

Effect of liver disease on response to drugs

Liver disease may alter the response to drugs in several ways:

- impaired drug metabolism may lead to increased toxicity
- hypoproteinaemia results in reduced protein binding and increased toxicity of highly protein-bound drugs, e.g. phenytoin
- reduced synthesis of clotting factors increases the sensitivity to oral anticoagulants
- hepatic encephalopathy may be precipitated by certain drugs, e.g. sedative drugs, opioid analgesics, diuretics, and drugs that cause constipation
- fluid overload (ascites, oedema) may be exacerbated by drugs that give rise to fluid retention, e.g. non-steroidals and corticosteroids
- hepatotoxicity is either dose related or unpredictable (idiosyncratic), and is more common in patients with liver disease.

Drugs to be used with caution

- For up-to-date guidance always consult your hospital pharmacist, the British National Formulary http://www.bnf.org/, or the electronic Medicines Compendium http://emc.medicines.org.uk/
- The following antimicrobials should be used with caution in liver disease:
    - antibacterials, e.g. ceftriaxone, chloramphenicol, co-amoxiclav, co-trimoxazole, daptomycin, flucloxacillin, fusidic acid, isoniazid, macrolides, ketolides, linezolid, meropenem, metronidazole, moxifloxacin, neomycin, ofloxacin, rifamycins, sodium fusidate, Synercid®, tetracyclines, tigecycline, tinidazole, Timentin®
    - antifungals, e.g. azoles, caspofungin, griseofulvin, terbinafine
    - antimalarials, e.g. mefloquine, pyrimethamine, Riamet®
    - antivirals, e.g. antiretrovirals, interferons, ribavirin, valaciclovir.

Duration of antibiotic therapy

There is limited evidence for the optimal duration of antimicrobial therapy for most conditions, and this subject is much debated. *Always follow your local antimicrobial guidelines*. Table 2.3 lists the typical duration of therapy given for some conditions in immunocompetent patients. This is only meant as a guide, and clearly response to treatment, e.g. symptoms, signs, and laboratory parameters should be continually reviewed.

Table 2.3 Guide to typical duration of antimicrobial therapy

Infection	Organism/condition	Duration
Urinary tract infections, p.716	Uncomplicated cystitis	3 days
	Pyelonephritis/catheter-associated UTI	7–14 days
	Asymptomatic bacteriuria in pregnancy	3–7 days
Intra-abdominal infections	Complicated intra-abdominal infections	4–7 days
Infectious diarrhoea, p.688	*Clostridium difficile*	10 days
	Salmonella spp. (age <6months or >65 years, severe)	5–7 days
	Shigella spp.	3 – 5 days
	Campylobacter spp.	5 days
	Giardia lamblia	10 days
Pharyngitis, p.632	Group A streptococcus	10 days
Community acquired pneumonia, p.654	Community-acquired pneumonia	5–7 days
	Hospital-acquired pneumonia	7–14 days
Skin and soft tissue infections, p.788	Impetigo	7–10 days
	Cellulitis/erysipelas	7–14 days
Bone and joint infections, p.802	Acute septic arthritis	3 weeks
	Acute haematogenous osteomyelitis	3 weeks
	Chronic osteomyelitis	12 weeks
Intravascular catheter-related infections, p.674	Coagulase-negative staphylococci	5 – 7 days[a]
	S. aureus	≥14 days[a]
	Gram-negative bacilli	7 – 14 days[a]
	Candida spp.	14 days[b]
Infective endocarditis, p.670	Viridans streptococci and *S. bovis* (native valve)	2–4 weeks
	S. pneumoniae and groups A, B, C, E and G, streptococci (native valve)	4 weeks
	S. aureus (native valve)	6 weeks
	S. aureus (prosthetic valve)	≥6 weeks
	Enterococci (native or prosthetic)	4–6 weeks
	HACEK organisms	4 weeks
	Culture negative	4–6 weeks

Table 2.3 Guide to typical duration of antimicrobial therapy (*Continued*)

Infection	Organism/condition	Duration
Bacterial meningitis p.754	*N. meningitidis*	7 days
	H. influenzae, type B	7 days
	S. pneumoniae	10–14 days
	Group B streptococci	14–21 days
	Gram-negative bacilli	21 days
	Listeria monocytogenes	≥21 days

[a]After catheter removal.

[b]After first negative blood culture.

References

IDSA guidelines for uncomplicated UTI *Clin Infect Dis* 1999;29:745–58.

IDSA guidelines for catheter-associated UTI in adults. *Clin Infect Dis* 2010;50:625–663.

IDSA guidelines for catheter-associated UTI in adults. *Clin Infect Dis* 2010;50:133–164.

IDSA guidelines for asymptomatic bacteriuria *Clin Infect Dis* 2005;40:643–54.

BTS guidelines for community acquired pneumonia *Thorax* 2001;56:1–64.

IDSA guidelines for community acquired pneumonia *Clin Infect Dis* 2007;44:S27–72.

IDSA guidelines for hospital-acquired pneumonia *Am J Resp Crit Care Med* 2005;171:388–416.

IDSA guidelines for skin and soft tissue infections *Clin Infect Dis* 2005;41:1373–406.

IDSA guidelines for infectious diarrhoea *Clin Infect Dis* 2001;32:331–50.

IDSA guidelines for intravenous catheter-related infections *Clin Infect Dis* 2009;49:1–45.

AHA guidelines for infective endocarditis *Circulation* 2005; 112: 2373.

IDSA guidelines for meningitis *Clin Infect Dis* 2004;39:1267–84.

Antimicrobial prophylaxis

Definitions

Prophylaxis is the administration of antibiotics to prevent the infection of a previously uninfected tissue. Primary prophylaxis aims to prevent initial infection or disease (e.g. to cover a surgical procedure), while secondary prophylaxis aims to prevent recurrent disease (e.g. giving penicillin to a patient who has had rheumatic fever). Surgical prophylaxis is usually primary and aims to target the operative period when the site may become contaminated.

Principles of antimicrobial prophylaxis

All hospitals should have a policy for prescribing antimicrobial prophylaxis for common procedures, which may be area/unit/surgeon specific. The following factors should be considered:

- before prescribing always consider:
  - does the patient have any known drug allergies?
  - has the patient received any recent antibiotics?
  - is the patient known to be colonized with resistant organisms?
- which drug? A bactericidal agent should be used which:
  - is active against the probable infecting organism(s)
  - penetrates the likely site of infection
  - has a favourable safety profile
- what dose? The aim is to maintain the drug concentration above the target MIC throughout the operative period. The number of doses usually depends on the length of procedure and likely blood loss
- which route? This depends on the nature of the procedure, whether or not the patient is nil by mouth, and the pharmacokinetics of the drug
- time of administration? Antimicrobial prophylaxis should be administered 0–2 h prior to the procedure in order to ensure adequate tissue levels
- duration? Prophylactic antibiotics should not usually be given for more than 24 h. If there is evidence of infection the patient should be carefully assessed and appropriate cultures sent. The organisms responsible for postoperative infections are unlikely to be sensitive to the prophylactic antibiotics. Seek advice on antimicrobial therapy from an infection specialist in light of the clinical picture and likely infecting organisms.

Risks of antimicrobial prophylaxis

While the benefit of prophylactic antibiotics is clear, there are also potential risks. These include:

- adverse effects associated with specific drug (e.g. penicillin anaphylaxis)
- selection of antibiotic-resistant organisms
- alteration of normal flora.

Before prescribing antimicrobial prophylaxis for a procedure, consider whether it is actually needed. Consult your local antibiotic policy or seek advice from an infection specialist if unsure.

Detailed indications for prophylaxis

For up-to-date information consult the British National Formulary (http://www.bnf.org). Antimicrobial prophylaxis is currently recommended in the following situations:

- prevention of recurrence of rheumatic fever
- prevention of a secondary case of group A streptococcal infection
- prevention of a secondary case of meningococcal infection
- prevention of a secondary case of *H. influenzae* type B disease
- prevention of a secondary case of diphtheria in a non-immune contact
- prevention of a secondary case of pertussis in a non-immune or partially immune contact
- prevention of pneumococcal infection in asplenia or sickle cell disease
- prevention of gas gangrene in high lower-limb amputations or following major trauma
- prevention of tuberculosis in susceptible close contacts or those that become tuberculin skin test-positive
- prevention of infection in gastrointestinal procedures
- prevention of infection in obstetric and gynaecological surgery
- prevention of infection in orthopaedic surgery
- prevention of infection in urological procedures
- prevention of infection in vascular surgery
- prevention of endocarditis (now only recommended for 'at risk' patients undergoing a GI or genitourinary (GU) procedure where infection is suspected http://www.nice.org.uk/guidance/index.jsp?action=byID&o=11938).

Other examples of antimicrobial prophylaxis

- Selective decontamination of digestive tract (SDD) – administration of antibiotics that are poorly absorbed when given orally to eliminate normal GI flora. Some intensive care units use it to prevent ventilator-associated pneumonia. This practice is controversial.
- Local infiltration into wound/incision line – current data suggest this can lead to higher rates of infection, unless combined with systemic administration. There is no additional reduction in wound infections if antibiotics are given by both routes simultaneously, compared with systemic antibiotics alone.
- Antibiotic-impregnated materials – gentamicin cement is used routinely in joint replacement; this is also controversial. Antibiotic-soaked Dacron vascular grafts are used in vascular surgery.

Penicillins

Penicillin was discovered by Alexander Fleming in 1928 but did not become widely available until the 1940s. The penicillins are closely related compounds comprising a β-lactam ring, a five-membered thiozolidine ring and a side chain (Fig. 2.1). The ring structures are essential for antibacterial activity, and the side chain determines the spectrum and pharmacological properties. Most penicillins in current use are semi-synthetic derivatives of 6-aminopenicillic acid. They inhibit bacterial cell wall synthesis and are thus bactericidal.

Classification

- Group 1 – benzylpenicillin and its long-acting parenteral forms
- Group 2 – orally absorbed penicillins, e.g. penicillin V
- Group 3 – antistaphylococcal penicillin, e.g. meticillin, flucloxacillin
- Group 4 – extended spectrum penicillins, e.g. amoxicillin
- Group 5 – antipseudomonal penicillins, e.g. ticarcillin, piperacillin
- Group 6 – β-lactamase-resistant penicillins

Mode of action

The penicillins inhibit cell wall synthesis by binding to PBPs and inhibiting transpeptidation of peptidoglycan.

Resistance

Bacteria may become resistant to penicillins by a number of mechanisms:

- destruction of the antibiotic by β-lactamases (β-lactamases, 📖 see p.68) this is the most common mechanism
- failure to penetrate the outer membrane of Gram-negative bacteria
- efflux across the outer membrane of Gram-negative bacteria
- low-affinity binding of antibiotic to target PBPs.

Some bacteria may display more than one resistance mechanism, e.g. in MRSA the *mecA* gene encodes for an additional PBP (i.e. altered target site), and most also produce a β-lactamase.

Clinical use

- Penicillin G is used in infections due to group A and group B streptococci; meningitis due to *S. pneumoniae* (if penicillin susceptible) and *N. meningitidis*, streptococcal and enterococcal endocarditis, neurosyphilis.
- Aminopenicillins are used in respiratory tract infections, endocarditis, meningitis, and urinary tract infections caused by susceptible organisms, and treatment of *H. pylori*.
- Extended-spectrum and antipseudomonal penicillins are used in infections due to resistant Gram-negative bacteria, usually in combination with an aminoglycoside.
- Penicillin V is also used prophylactically to prevent recurrent rheumatic fever, secondary cases in outbreaks of group A streptococcal disease, and pneumococcal and *H. influenzae* infections in asplenic patients

Pharmacology

- Penicillins differ markedly in their oral absorption (penicillin V 60%, amoxicillin 75%, antipseudomonal penicillins 0%).
- They vary in their degree of protein binding, and metabolism is minimal.
- They are rapidly excreted by the renal tubular cells; excretion may be blocked by probenecid. Dose modification may be required in renal failure.

Toxicity and side-effects

- Allergic reactions (skin rashes, serum sickness, delayed hypersensitivity) occur in <10% of those exposed. Anaphylactic reactions are rare (0.004–0.4%)
- Gastrointestinal – diarrhoea, enterocolitis (2–5%, usually ampicillin)
- Haematological – haemolytic anaemia, neutropenia; thrombocytopenia (1–4%)
- Laboratory – elevated transaminases (usually flucloxacillin), electrolyte abnormalities (hypernatraeamia, hypo or hyperkalaemia)
- Renal – interstitial nephritis, haemorrhagic cystitis
- Central nervous system – encephalopathy or seizures are rare but may occur in renal failure or if high prolonged doses of penicillin are used

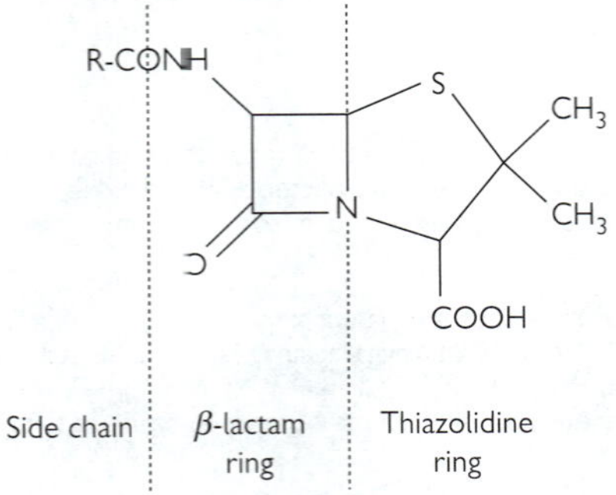

Fig. 2.1 Structure of pencillin.

Cephalosporins

Giuseppe Brotzu first demonstrated the antimicrobial activity of culture filtrates of the mould *Cephalosporium acremonium* in 1945. However, the cephalosporin class of antibiotics did not become widely used for another 20 years. Cephalosporins consist of a β-lactam ring, a six-membered dihydrothiazine ring, modified at certain positions to produce different compounds (Fig. 2.2). Most available cephalosporins are semi-synthetic derivatives of cephalosporin C.

Classification

The classification into 'generations' is the most commonly used:

- 1st generation – primarily active against Gram-positive bacteria, e.g. cefazolin, cefalothin, cefradine, cefalexin
- 2nd generation – enhanced activity against Gram-negative bacteria, with varying degrees of activity against Gram-positive bacteria e.g. cefuroxime, cefamandole, cefaclor. The cephamycin group (e.g. cefotetan and cefoxitin) have additional anaerobic activity against e.g. *B. fragilis*
- 3rd generation – markedly increased activity against Gram-negative bacteria, e.g. cefotaxime, ceftriaxone, ceftazidime, cefdinir, cefixime, cefpodoxime
- 4th generation – broad spectrum of activity against Gram-positive cocci, and gram-negative bacteria including *Pseudomonas* spp., e.g. cefepime, cefpirome.

Mode of action

They inhibit cell wall synthesis by binding to PBPs and inhibiting transpeptidation of peptidoglycan. They are bactericidal and exhibit significant post-antibiotic effect against Gram-positive (but not Gram-negative) bacteria.

Resistance

There are four resistance mechanisms:

- destruction of the antibiotic by β-lactamases (see 🕮 β-lactamases, p.68, and Box 2.6)
- reduced penetration through the outer membrane of Gram-negative bacteria
- enhanced efflux
- alteration in PBP target, resulting in reduced-affinity binding.

Clinical use

- 1st generation – staphylococcal and streptococcal skin and soft tissue infections, urinary tract infections
- 2nd generation – severe community-acquired pneumonia, otitis media, sinusitis, streptococcal pharyngitis, early Lyme disease
- Cephamycins – intra-abdominal, pelvic, and gynaecological infections, infected decubitus ulcers, diabetic foot infections, mixed aerobic–anaerobic soft tissue infections
- 3rd generation – penicillin-resistant pneumococci, meningitis, upper and lower respiratory tract infections, sinusitis, otitis media, nosocomial infections caused by Gram-negative bacilli, *N. gonorrheae*, chancroid, Lyme disease, typhoid, severe *Shigella* spp., and non-typhoidal salmonella infection, outpatient antibiotic therapy for endocarditis and osteomyelitis

- 4th generation – role not yet clear but effective in severe Gram-negative infections. Active against *P. aeruginosa*, *Enterobacter* spp., *Citrobacter* spp. and *Serratia* spp.

Pharmacology

May be given orally (PO) intravenously (IV) or intramuscularly (IM). Fourth generation drugs are all parenteral. Oral preparations have 80–95% bioavailability. Protein binding is variable (10–98%). Drugs are largely confined to the extra-cellular compartment. Poor CSF penetration unless meningeal inflammation. Cross placenta. Most drugs are not metabolized except cefotaxime and cealothin, which are metabolized in the liver. Most drugs are excreted by the kidneys. Ceftriaxone and cefoperazone are excreted by the biliary system.

Box 2.6 CAUTION!

The 2nd and 3rd generation cephalosporins are susceptible to inactivation by inducible β-lactamases (see 📖 β-lactamases, p.68). They should never be used to treat organisms that may have these enzymes, e.g. *Enterobacter* spp., *Serratia* spp., *Citrobacter freundii*, *Acinetobacter* spp., *Proteus vulgaris*, *Providencia* spp., *Morganella morganii* (ESCAPPM)

Toxicity and side-effects

- Hypersensitivity – rash (1–3%), urtcaria and serum sickness (<1%), anaphylaxis (0.01%)
- Gastrointestinal – diarrhoea (1–19%), nausea and vomiting (1–6%), transient hepatitis (1–7%), biliary sludging (ceftriaxone)
- Haematological – eosinophilia (1–10%), neutropenia, thrombocytopaenia, clotting abnormalities, platelet dysfunction, haemolytic anaemia
- Renal – interstitial nephritis
- CNS – seizures
- False-positive laboratory tests – Coombs' test, glycosuria, serum creatinine
- Other – drug fever, disulfiram-like reaction, phlebitis

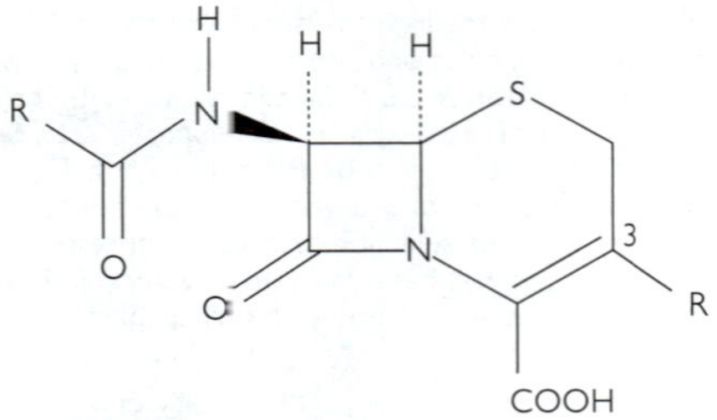

Fig. 2.2 Basic structure of cephalosporins.

β-lactamases

β-lactamases are enzymes that bind covalently to the β-lactam ring, hydrolyse it and make the antibiotic ineffective. Emergence of resistance to β-lactam antibiotics began even before penicillin was widely available, with the first β-lactamase (penicillinase) being described in *E. coli* in 1940. This was followed by the emergence of resistance in *S. aureus* due to plasmid-encoded penicillinase. Many genera of Gram-negative bacilli possess naturally occurring chromosomally mediated β-lactamases (AmpC); these enzymes are thought to have evolved from PBPs, to which they are very similar. The first plasmid-mediated β-lactamase in Gram-negative bacteria, TEM-1, was described in 1960. Within a few years it had spread worldwide and was found in many different species. Over the past 20 years, many antibiotics have been developed to be resistant to these β-lactamases. However, with each new class of drugs, new β-lactamases have emerged.

Classification

There are two classification systems for β-lactamases:
- Ambler classification is a molecular classification with four classes (A to D) based on the nucleotide/amino acid sequences of the enzymes:
  - classes A, C, and D are PBPs, without cell wall synthetic activity (serine β-lactamases)
  - class B are zinc-dependent enzymes (metallo-β-lactamases) that hydrolyse the β-lactam ring by a different mechanism
- Bush–Jacoby–Medeiros classification is a functional classification with four groups:
  - group 1 β-lactamases are cephalosporinases that are not inhibited by clavulanic acid. They correspond to Ambler group C
  - group 2 β-lactamases are penicillinases and/or cephalosporinases that are inhibited by clavulanic acid. This group corresponds to Ambler groups A and D, and includes the TEM and SHV enzymes
  - group 3 β-lactamases are zinc-dependent (metallo-β-lactamases) and are not inhibited by clavulanic acid. They correspond to Ambler group B
  - Group 4 β-lactamases are penicillinases that are not inhibited by clavulanic acid. They do not yet have a molecular class.

AmpC β-lactamases

These are chromosomally mediated β-lactamases that are active against 3rd generation cephalosporins and are not inhibited by clavulanic acid. They fall into the molecular group C/functional group 1. They are found in the ESCAPPM group of organisms, e.g. ***E****nterobacter* spp., ***S****erratia* spp., ***C****itrobacter freundii*, ***A****cinetobacter* spp., ***P****roteus vulgaris*, ***P****rovidencia* spp., ***M****organella morganii*. The use of 3rd generation cephalosporins to treat these infections results in the selection of stably derepressed mutants that hyperproduce AmpC, and has been associated with clinical failure. These infections are therefore usually treated with carbapenems.

Extended-spectrum β-lactamases (ESBLs)

These are β-lactamases which are capable of conferring bacterial resistance to the penicillins, 1st, 2nd and 3rd generation cephalosporins, and aztreonam (but not the cephamycins or carbapenems) by hydrolysis of these antibiotics, and which are inhibited by β-lactamase inhibitors such as clavulanic acid. They fall into functional groups 2be and 2d. ESBLs are most commonly found in *E. coli* and *K. pneumoniae* but have been described in many other Gram-negative bacilli. Most ESBLs are derivatives of the TEM and SHV enzymes (see below).

- TEM β-lactamases – TEM-1 is the most common β-lactamase in Gram-negative bacteria and is able to hydrolyse penicillins and early generation cephalosporins. TEM-2 has a similar spectrum. TEM-3 was the first ESBL, reported in 1989. Since then over 160 TEM enzymes have been described (see http://www.lahey.org/studies/temtable.htm). Most of these are inhibited by clavulanic acid but some inhibitor-resistant variants exist, particularly in Europe. TEM enzymes are most common in *E. coli* and *K. pneumoniae*, but are increasingly found in other species of Gram-negative bacilli.
- SHV β-lactamases – the SHV-1 β-lactamase is most commonly found in *K. pneumoniae* and accounts for ≤20% of ampicillin reistance in this species. Unlike TEM, there are relatively few SHV-1 derivatives.
- CTX-M β-lactamases – this family of plasmid-mediated β-lactamases preferentially hydrolyses cefotaxime. They have been found in *Salmonella enterica* serovar Typhimurium and *E. coli* as well as other enterobacteria. These enzymes are quite different to the TEM and SHV enzymes and show greater similarity to chromosomal AmpC enzyme of *Kluyvera ascorbata*, suggesting CTX-M may have originated from this species. CTX-M β-lactamases have previously been associated with outbreaks in Europe, South America, and Japan, although they are now reported worldwide.
- OXA β-lactamases – these are characterized by their high hydrolytic activity against oxacillin and cloxacillin and are poorly inhibited by clavulanic acid. They belong to molecular group D/functional group 2d. OXA-type ESBLs are mainly found in *P. aeruginosa*, but have been detected in other Gram-negative bacteria. More recently, non-ESBL OXA derivatives have been described.
- Other ESBLs – a number of ESBLs that are unrelated to the established families of ESBLs have been described, e.g. PER-1, PER-2, VEB-1, GES, BES, TLA, SFO, IBC.

ESBL detection methods

In general ESBL detection methods use a β-lactamase inhibitor (clavulanate) in combination with an oximino-cephalosporin, e.g. ceftazidime or cefotaxime. The clavulanate inhibits the ESBL, thereby reducing the level of resistance to the cephalosporin. A number of methods exist, e.g. Jarlier double disc method, Etest® for ESBLs (see 🕮 Breakpoints, p.43).

β-lactamase inhibitors

β-lactamase inhibitors are clavulanic acid and penicillanic acid sulfone derivatives. They have weak antibacterial activity but are potent inhibitors of many β-lactamases, e.g. penicillinases produced by *S. aureus*, *H. influenzae*, *M. catarrhalis*, and *Bacteroides* spp., and TEM and SHV β-lactamases produced by enterobacteriaceae. They can restore the antibacterial activity of certain antibiotics, e.g. amoxicillin, ampicillin, piperacillin, mezlocillin, and cefoperazone. Three β-lactamase inhibitors are in clinical use: clavulanic acid, sulbactam, and tazobactam. All are only available in combination with a β-lactam antibiotic; the antibiotic spectrum is determined by the companion antibiotic. Although there are minor differences in potency, activity, and pharmacology between the three compounds, they can be considered therapeutically equivalent (except for some *Klebsiella* spp. where clavulanate inhibits isolates resistant to sulbactam and tazobactam).

Co-amoxiclav (Augmentin®)

- Clavulanate is a potent inhibitor of many plasmid-mediated β-lactamases and a weak inducer of some chromosomal β-lactamases.
- It is available as a combination with amoxicillin and used for the treatment of a wide range of infection where β-lactamase-producing organisms may be present. Examples include otitis media, sinusitis, pneumonia, skin and soft tissue infections, diabetic foot infections, and bite infections.
- It is available as oral or parenteral formulations. In the oral formulation, the ratio of amoxillin to clavulanic acid is 2:1, e.g. 250 mg/125 mg. whereas in the intravenous formulation it is 5:1, e.g. 1000 mg/200 mg.
- Side-effects are similar to ampicillin. Cholestatic jaundice may occur during/after therapy, and is six times more common than with amoxicillin alone.

Ticarcillin-clavulanate (Timentin®)

- This combination is useful against infections caused by *Pseudomonas* spp., and *Proteus* spp.
- It has been used for the treatment of pneumonia, intra-abdominal infections, gynaecological infections, skin and soft tissue infections, and osteomyelitis.
- It is only available in parenteral form and is given intravenously.
- Side-effects are similar to those of other β-lactams. Cholestatic jaundice may also occur because of the clavulanic acid component.

Ampicillin-sulbactam (Unasyn®)

- Sulbactam is 6-desaminopenicillin sulfone. It has a broader spectrum of activity but is less potent than clavulanic acid.
- In the USA it is available as a combination with ampicillin and is given intravenously.
- It is used for the treatment of skin and soft tissue infections, intra-abdominal infections, and gynaecological infections caused by β-lactamase-producing bacteria. It has also recently been used to treat carbapenem-resistant *Acinetobacter baumanii* infections.
- Side-effects are similar to ampicillin.

Piperacillin-tazobactam (Tazocin®)

- Tazobactam is penicillanic acid sulfone β-lactamase inhibitor with a similar structure to sulbactam. Its spectrum of activity is similar to sulbactam but its potency is comparable to clavulanic acid.
- It is available as a combination with piperacillin (an antipseudomonal penicillin) and is given parenterally.
- It has a broad spectrum of activity and is used in the treatment of pneumonia (especially *P. aeruginosa*), skin and soft tissue infections, intra-abdominal infections, urinary tract infections, polymicrobial infections, bacteraemia, and febrile neutropenia (in combination with an aminoglycoside).
- Side-effects are similar to piperacillin.

Carbapenems

The carbapenems are β-lactam antibiotics derived from thienamycin, a compound produced by *Streptomyces cattleya.* Three carbapanems are licensed for use in the UK: imipenem, meropenem, and ertapenem. Other drugs in the same class include panipenem, doripenem, and faropenem.

Mode of action

These agents show high affinity to most high molecular weight PBPs of Gram-positive and Gram-negative bacteria. Carbapenems, particularly imipenem, traverse the outer membrane of Gram-negative bacteria through different outer membrane proteins (OprD) than those that are used by penicillins and cephalosporins (OmpC and OmpF). They also have excellent stability to β-lactamases. Consequently carbapenems have the broadest antibacterial spectrum of all the β-lactam antibiotics. Imipenem is slightly more active against Gram-positive bacteria, whereas meropenem and ertapenem are slightly more active against Gram-negative species. Meropenem is the most active against *P. aeruginosa.* Ertapenem has poor activity against *P. aeruginosa* and *Acinetobacter* spp.

Resistance

Resistance is due to one of four mechanisms:

- production of a low-affinity PBP target
- reduced outer membrane permeability due to absence of OprD in Gram-negative bacteria
- efflux of the drug in Gram-negative bacteria
- production of β-lactamases (📖 see β-lactamases, p.68) that hydrolyse carbapenems (carbapenemases). These enzymes fall into three groups. The molecular class A (functional group 2f) enzymes include SME, IMI, NMC, KPC and GES. The molecular class B (functional group 3) enzymes include IMP, VIM, GIM and SVM. The molecular group D (functional group 2d) include the OXA enzymes.

In Gram-negative bacteria, resistance is frequently due to a combination of impaired drug entry, drug efflux, and β-lactamase production

Clinical use

- Carbapenems may be used to treat a wide variety of severe infections, e.g. bacteraemia, pneumonia, intra-abdominal infections, obstetric and gynaecological infections, complicated urinary tract infections, and soft tissue and bone infections.
- Imipenem and meropenem are most appropriate for treatment of infections caused by the cephalosporin-resistant AmpC-producing organisms, e.g. ***E**nterobacter* spp., ***S**erratia* spp., ***C**itrobacter freundii,* ***A**cinetobacter* spp., ***P**roteus vulgaris,* ***P**rovidencia* spp., ***M**organella morganii* (the ESCAPPM group).
- Imipenem and meropenem are also used for the treatment of serious infections, e.g. patients with polymicrobial infections, febrile neutropenia, and nosocomial infections, such as those caused by *P. aeruginosa* and *Acinetobacter* spp.

- Meropenem is also licensed for the treatment of bacterial meningitis – imipenem should *not* be used because of its propensity to cause seizures.
- Ertapenem has similar uses to imipenem and meropenem but cannot be used in infections caused by *P. aeruginosa* and *Acinetobacter* spp. Its long plasma-half life means that it can be administered once daily, making it useful for outpatient antibiotic therapy (OPAT).

Pharmacology

- Imipenem, meropenem, and ertapenem have poor oral absorption and are given parenterally.
- Imipenem and meropenem are pharmacologically similar with a plasma half-life of 1 h, whereas ertapenem has a plasma half-life of 4 h, which permits once-daily dosing.
- All carbapenems are widely distributed and penetrate inflamed meninges.
- All are renally excreted and require dose modification in renal failure.
- Imipenem is a substrate for the renal dehydropeptidase-1 (DHP-1) enzyme and is therefore co-administered with cilastatin, a DHP-1 inhibitor.

Toxicity and side-effects

- Carbapenems are generally well tolerated.
- β-lactam allergic reactions are the most common side-effects, e.g. rash, urticaria, immediate hypersensitivity, cross-reactivity with penicillin.
- Imipenem causes nausea (if infused too quickly) and can cause seizures.

Monobactams

The monobactams are monocyclic β-lactam antibiotics produced by some bacteria (e.g. *Chromobacterium violaceum*). They are only active against Gram-negative bacteria.

Aztreonam

- Aztreonam is the only commercially available compound.
- It is active against most *Enterobacteriaeceae, H. influenzae,* and *Neisseria* spp. *S. maltophilia, B. cepacia*, and many *Acinetobacter* spp. are resistant. Some strains of *P. aeruginosa, E. cloacae*, and *C. freundii* are resistant.
- Aztreonam passes through the outer membrane and binds to PBP3 of Gram-negative bacteria. It is resistant to hydrolysis by most β-lactamases, apart from the AmpC β-lactamases.
- Aztreonam is not absorbed orally and is given intravenously or intramuscularly. It is widely distributed and penetrates inflamed meninges. It is mainly renally excreted and requires dose modification in renal failure.
- It is used for the treatment of a variety of infections, e.g. urinary tract infections, pneumonia, septicaemia, skin and soft tissue infections, intra-abdominal infections, gynaecological infections, wound and burn infections.
- Aztreonam should never be used alone as empiric therapy as it has no activity against Gram-positive organisms.
- Side-effects are similar to those of the other β-lactams except for hypersensitivity which does not occur.

Other cell wall agents

Bacitracin

Bacitracin binds to isoprenylphosphate and prevents dephosphorylation of the lipid carrier that transports the cell wall building block across the membrane. Without dephosphorylation, the native compound cannot be regenerated for another round of transfer. Similar reactions in eukaryotic cells may be why this agent is so toxic and it is therefore used topically. It is also used to identify group A streptococci (bacitracin resistant) in the diagnostic laboratory.

Fosfomycin

Fosfomycin inhibits pyruvyl transferase, and therefore formation of N-acetylglucosamine from N-acetylmuramic acid. It is a naturally occurring antibiotic with a fairly broad spectrum, particularly against Gram-negative rods. It is mainly used to treat urinary tract infections.

Cycloserine

This drug is often part of the 2nd-line regimen for drug-resistant tuberculosis. It is a structural analogue of D-alanine, and acts on alanine racemase and synthetase to inhibit the synthesis of the terminal D-alanyl-D-alanine. It thus prevents formation of the pentapeptide chain of muramic acid. See 🕮 Antituberculous agents-2nd line, p.112.

Isoniazid and ethambutol

These are 1st-line drugs used in the treatment of tuberculosis. They interfere with mycolic acid synthesis in mycobacterial cell walls. See 🕮 Antituberculous agents-1st line, p.110.

Glycopeptides

The glycopeptide antibiotics, vancomycin and teicoplanin are bactericidal against most Gram-positive bacteria. Vancomycin was first isolated from *Nocardia orientalis* and introduced into clinical practice in 1958. Teicoplanin was obtained from *Actinoplanes teichomyceticus* in 1978 and is available in Europe and Asia but not in the USA.

Mode of action

Glycopeptides inhibit cell synthesis by binding to D-alanyl-D-alanine tail of the muramylpentapeptide. This complex cannot be processed by the enzyme glycosyltransferase, inhibiting the incorporation of the murein monomers (*N*-acetylmuramic acid and *N*-acetylglucosamine) into the growing peptidoglycan chain.

Antimicrobial activity

Glycopeptides have broad activity against Gram-positive organisms e.g. staphylococci, *E. faecalis*, *S. pneumoniae*, groups A, B, C, and G streptococci, *S. bovis*, *S. mutans*, viridans group streptococci, *L. monocytogenes*, *Bacillus* spp., *Corynebacterium* spp., *Peptostreptococcus* spp., *Actinomyces* spp., *Propionibacterium* spp., and most *Clostridium* spp. Glycopeptides show no activity against Gram-negative species (except non-gonococcal *Neisseria* spp). The MICs of teicoplanin against coagulase-negative staphylococci are more variable than vancomycin.

Resistance

Vancomycin resistance may be intrinsic or acquired.

- Intrinsic vancomycin resistance occurs in *Leuconostoc*, *Pediococcus*, *Lactobacillus* and *Erysipelothrix rhusiopathiae*. Intrinsic teicoplanin resistance is seen in *S. haemolyticus*.
- Enterococci – six types of glycopeptide resistance have been described (VanA, VanB, VanC, VanD, VanE, and VanG), named on the basis of their ligase genes (*vanA*, *vanB* etc, see Table 2.4). These result in the formation of a peptidoglycan precursor with decreased affinity for glycopeptides. Resistance may be intrinsic (e.g. in *E. gallinarum*, *E. casseliflavus*) or acquired (e.g. in *E. faecium* and *E. faecalis*).
- *Staphylococcus aureus* – the first clinical isolate of *S. aureus* with diminished susceptibility to vancomycin was reported in Japan in 1997. This is referred to as a vancomycin-intermediate *S. aureus* (VISA) or glycopeptide intermediate *S. aureus* (GISA). VISA isolates have a thickened cell wall which may prevent glycopeptides from reaching their target sites. In 2002, two isolates of truly vancomycin-resistant *S. aureus* (VRSA) were reported, both of which carried the *vanA* gene, suggesting horizontal transfer of this gene from enterococci.
- *S. pneumoniae* – vancomycin tolerance has recently been reported.

Clinical use

Glycopeptides are used to the following conditions:

- severe infections caused by MRSA
- meningitis due to penicillin-resistant *S. pneumoniae*
- *C. difficile*-associated diarrhoea (oral vancomycin)

- febrile neutropenia
- continuous ambulatory peritoneal dialysis (CAPD) peritonitis
- endophthalmitis
- empiric treatment of intravascular catheter-related infections and cerebrospinal fluid (CSF) shunt infections.

Pharmacology

- Vancomycin is usually given intravenously but may also be given orally, intraperitoneally, intrathecally or intraocularly. It is widely distributed but has poor CSF penetration in the absence of meningeal inflammation. Vancomcyin is excreted unchanged in the kidneys, and dose reduction is required in renal impairment. Therapeutic drug monitoring of trough levels is performed after the 3rd dose (in patients with normal renal function) and should be 10–15 mg/L. Vancomycin shows time-dependent killing: if the trough level is too high, it is better to reduce the dose rather than increase the dosing internal.
- Teicoplanin is usually administered intravenously, intramuscularly, but may also be given intraperitoneally. It has a long plasma half-life (83–168 h), enabling daily dosing. Teicoplanin has better bone penetration than vancomcyin. It is excreted by the kidneys. Monitoring of teicoplanin levels is not needed for toxicity but is sometimes done in severe infections to ensure therapeutic levels.

Toxicity and side-effects

Toxicity is more common with vancomycin than teicoplanin.

- Ototoxocity is rare unless there is renal impairment.
- Nephrotoxocity occurs with high doses and is often associated with concomitant aminoglycoside usage.
- Infusion-related reactions can occur, e.g. 'red man syndrome' with rapid infusion of vancomcyin.
- Others, e.g. neutropenia, thrombocytopaenia, rashes, drug fever.

Table 2.4 Vancomycin resistance in enterococci and staphylococci

	Van A	Van B	Van C	Van D	Van E	Van G
Vanc MIC	64–>500	4–>500	2–32	64–128	16	12–16
Teic MIC	16–>500	0.5–2	0.5–2	4–64	0.5	0.5
Expression	Inducible	Inducible	Constitutive, inducible	Constitutive	Inducible	
Location	P, C	P, C	C	C	C	C
Species	*E. faecalis, E. faecium, S. aureus*	*E. faecalis, E. faecium*	*E. gallinarum, E. casselflavius, E. flavescens*	*E. faecium*	*E. faecalis*	*E. faecalis*

Vanc = vancomycin, Teic = teicoplanin, MIC = minimum inhibitory concentration (microgram/ml), P = plasmid, C = chromosome

Aminoglycosides

Streptomycin, produced by *Streptomyces griseus,* was the first aminoglycoside used in the initial treatment trials of tuberculosis in the 1940s. Today aminoglycosides remain an important part of the antibiotic arsenal. All aminoglycosides have an essential six-membered ring with amino group constituents (aminocyclitol). The term aminoglycoside results from glycosidic bonds between the aminocyclitol and two or more sugars. They are active against many Gram-negative and some Gram-positive organisms. In the UK the currently available aminoglycosides are: streptomycin, neomycin, kanamycin, paromomycin, gentamicin, tobramycin, amikacin, netilmicin, and spectinomycin. Other drugs (e.g. sisomicin, dibekacin, and isepamicin) are available in Japan and continental Europe.

Mode of action

Aminoglycosides bind to the A site of 30S ribosomal subunit, resulting in a conformational change that interferes with mRNA translation and translocation, and hence inhibit protein synthesis. Avidity of binding varies between aminoglycosides. The transport of aminoglycosides into the cell by energy-dependent mechanisms (EDP-I and EDP-II) results in accumulation of high concentrations of drug in the cell. The onset of cell death is coincident with the transition from EDP-I to EDP-II.

Resistance

Resistance to aminoglycosides may be intrinsic or acquired:

- intrinsic resistance may be non-enzymatic or enzymatic:
  - anaerobes are unable to generate a sufficient electrical potential difference across the membrane and are intrinsically resistant
  - mutations in the 16S ribosomal subunit can result in resistance to streptomycin in *M. tuberculosis*
  - methylating enzymes that modify the 16S rRNA may cause intrinsic resistance; this has not yet been seen in clinical isolates
- acquired resistance may occur by a variety of mechanisms:
  - reduced drug uptake
  - efflux pumps, e.g. activation of the Mex XY pump in *P. aeruginosa*
  - enzymatic modification of the drug may occur as a result of aminoglycoside-modifying enzymes (AMEs) that phosphorylate, acetylate, or adenylate exposed amino or hydroxyl groups. The enzymatically modified drugs bind poorly to ribosomes resulting in high levels of resistance.

Clinical use

- Empiric therapy – aminoglycosides may be given as empiric therapy for serious infections suspected to be due to Gram-negative bacteria. Depending on the clinical indication, they are usually combined with a β-lactam, vancomycin or an anaerobic agent.
- Specific therapy – once culture results are available, aminoglycosides may be useful for the specific treatment, e.g. infections due to *Pseudomonas* spp. or resistant Gram-negative species, endocarditis.

- Prophylaxis – aminoglycosides are sometimes used prophylactically, e.g. to prevent enterococcal endocarditis in 'at-risk' patients undergoing genitourinary or gastrointestinal procedures.
- Gentamicin is the most commonly used aminoglycoside in the UK. Its main use is in the empirical treatment of serious infections (e.g. septicaemia, febrile neutropenia, biliary sepsis, acute pyelonephritis, endocarditis. It is often incorporated into cement in orthopaedic procedures. Gentamicin drops are used in superficial eye infections and bacterial otitis externa.
- Amikacin is used in gentamicin-resistant infections, mycobacterial infections or nocardiosis.
- Tobramycin is slightly better for *P. aeruginosa* than gentamicin, and may be used in cystic fibrosis patients.
- Neomycin is given orally for bowel sterilization pre-surgery, or selective decontamination of the digestive tract (see 🕮 Antimicrobial prophylaxis, p.62).
- Netilmicin is used in Gram-negative infections that are resistant to gentamicin.
- Streptomycin is used to treat tuberculosis, particularly in the developing world. It is sometimes used synergistically in enterococcal endocarditis (if there is gentamicin resistance).
- Spectinomycin is used to treat gonococcal infections.
- Paromomycin is used to treat cryptosporidiosis.

Pharmacology

- The aminoglycosides share a number of important characteristics (see 🕮 Pharmacodynamics, p.48):
  - concentration-dependent bactericidal activity
  - significant post-antibiotic effect (PAE)
  - synergism particularly with cell-wall-active agents.
- Aminoglycosides have poor oral absorption and are usually administered intravenously or intramuscularly. They may also be administered orally (e.g. neomycin, paromomycin), topically, intrapleurally, intraperitoneally, or intrathecally.
- Aminoglycosides are highly soluble with low protein binding, resulting in distribution in the vascular and interstitial compartments. CSF penetration is poor, apart from in neonates. Aminoglycosides are excreted unchanged in the urine (99%).

- Aminoglycosides may be given once daily or in multiple daily doses. Once-daily dosing is simpler, as efficacious as multiple dosing, and may lower the risk of drug-induced toxicity. The usual suggested dose of gentamicin is 5–7 mg/kg/day. The dose is reduced in renal failure to 3 mg/kg/day. Exceptions: children, pregnancy, burns, endocarditis. If patients need to continue therapy beyond 48 h, trough drug levels should be monitored and the dosing interval adjusted according to the Hartford nomogram (Fig. 2.3).

Toxicity and side-effects

- Nephrotoxicity is the most common adverse effect (5–25%).
- Ototoxicity (cochlear and vestibular) may be irreversible.
- Neuromuscular blockade is rare.

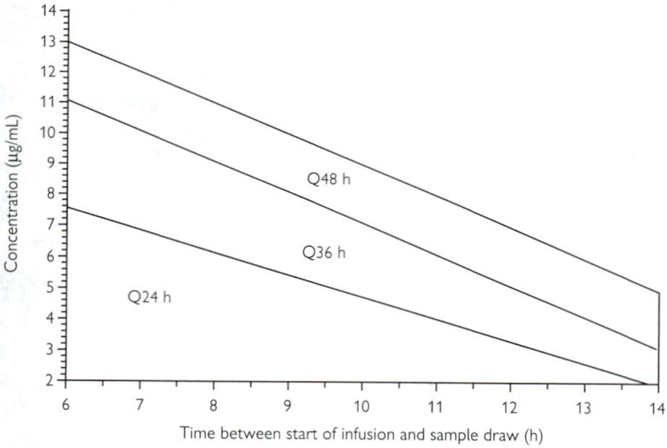

Fig. 2.3 Hartford nomogram for once-daily aminoglycosides. Reproduced with permission from Nicolau et al. (1995). Experience with a once daily aminoglycoside program administered to 2,184 patients, *Antimicrob Agents Chemother* **39**:650–5.

Macrolides

The macrolides (erythromycin, clarithromycin, azithromycin) and the lincosamides (lincomycin and clindamycin), although chemically unrelated have some similar properties such as antimicrobial activity, mechanisms of action, and resistance and pharmacology. The ketolides are a new class of antibiotics, derived from erythromycin, with activity against macrolide-resistant strains.

Erythromycin

- Erythromycin was derived from *Saccharopolyspora erythraea* in 1952. Erythromycin A is the active component. It consists of a 14-membered macrocyclic lactone ring attached to two sugars.
- Mode of action – inhibits RNA-dependent protein synthesis at the step of chain elongation by interacting with the peptidyl trasferase site. It also inhibits the formation of the 50S ribosomal subunit.
- Resistance – there are four resistance mechanisms:
  - decreased outer membrane permeability e.g. *Enterobacteriaceae*, *Pseudomonas* spp. *Acinetobacter* spp. are intrinsically resistant.
  - efflux pumps e.g. *msr(A)* gene of *S. aureus* and *mef(A)* gene of *S. pneumoniae* and *Group A. streptococci*
  - alterations of 23S rRNA by methylation of adenine. This confers resistance to macrolides, lincosamides, and streptogramins type B, and is referred to as the MLS_B phenotype. It is encoded by *erm* (erythromycin ribosomal methylase) genes.
  - enzymatic inactivation by phosphotransferases, mediated by *mph* genes. Hydrolysis of the macrocyclic lactone is encoded by esterase genes, *ere(A)* and *ere(B)*, on plasmids.
- Clinical use – community-acquired pneumonia, atypical pneumonia (e.g. *M. pneumoniae. C. pneumoniae, L. pneumophila*), *B. pertussis*, *Campylobacter* gastroenteritis.
- Pharmacology – given orally (stimulates gastrointestinal motility) or intravenously. Widely distributed in the tissues. Excreted in the bile and urine; some is inactivated in the liver.
- Toxicity and side-effects – gastrointestinal symptoms (nausea, vomiting, abdominal cramps, diarrhoea) are common; skin rash, fever, eosinophlia, cholestatic jaundice, transient hearing loss, QT prolongation, torsades de pointes, candidiasis, pseudomembranous colitis, infantile pyloric stenosis.

Clarithromycin

- Structure – 14-membered ring with methoxy group at postion 6
- Mode of action – same as erythromycin. More active than erythromycin against *S. pneumoniae, Group A. streptococci*, MRSA, *M. catarrhalis*, and *L. pneumophila*. Also active against *M. leprae, M. avium* complex (MAC) and *T. gondii*
- Resistance – similar to erythromycin
- Clinical use – similar to erythromycin. Treatment of MAC and other non-tuberculous mycobacterial infections, *H. pylori* eradication, Lyme disease
- Pharmacology– given orally or intravenously. Metabolized in the liver to active metabolites

Azithromycin

- Structure – 15-membered lactone ring (azalide)
- Mode of action – same as erythromycin. Greater activity against Gram-negative species than erythromycin and clarithromcyin. Also active against MAC and *T. gondii*
- Resistance – similar to erythromycin
- Clinical use – similar to erythromycin. Also use for treatment of trachoma, *B. microti, B. burgdorferi,* cryptosporidiosis
- Pharmacology – given orally but should be taken 1 h before or 2 h after food. Widely distributed in tissues with half-life of 2–4 days. Mostly not metabolized, and excreted in the bile
- Toxicity and side-effects – similar to erythromycin

Spiramycin

- Used in treatment of cryptosporidia and prevention of congenital toxoplasmosis

Ketolides

Ketolides are a new class of antibiotics derived from erythromycin A that have increased potency against bacteria that have become resistant to macrolides, e.g. *S. pneumoniae* and *S. pyogenes*. Telithromycin is currently the only available drug

Telithromycin

- Structure – 14-membered ring with ketone instead of L-cladinose at position 3 – this prevents induction of macrolide-lincosamide-streptogramin B (MLS_B) resistance (see Lincosamides, Box 2.7)
- Mode of action – similar to erythromycin
- Resistance – this is uncommon as ketolides are poor inducers of efflux pumps and MLS_B methylase genes. *S. aureus* strains with constitutive *erm* genes are resistant, whereas *S. pneumoniae* strains with constitutive *erm* genes remain sensitive
- Toxicity and side-effects – similar to clarithromycin and azithromycin. Reports of exacerbation of myasthenia gravis
- Pharmacology – good oral absorption and bioavailabilty. Metabolized in the liver by CYP3A4
- Clinical use – community-acquired pneumonia, acute exacerbation of chronic obstructive pulmonary disease (COPD), tonsillitis, pharyngitis, and sinusitis

Lincosamides

This group of antibiotics includes lincomycin (not available in the UK) and clindamycin. Lincomycin was isolated from *Streptomyces lincolnensis* in 1962. Clindamycin, which was produced by the chemical modification of lincomycin has better oral bioavailability and increased bacterial potency compared with lincomycin. Although chemically unrelated to erythromycin, many of the biological properties of lincosamides are similar to the macrolides.

Mode of action

Lincosamides inhibit protein synthesis by interacting with the peptidyl transferase site of the 50S ribosomal subunit. They also inhibit the formation of the 50S ribosomal subunit. Clindamycin is highly active against anaerobes (e.g. *B. fragilis*), pneumococci, group A streptococci, meticillin-sensitive *S. aureus* (MSSA), *T. gondii*, and *P. falciparum*

Resistance

There are several resistance mechanisms:
- alteration of 50S ribosomal proteins of the receptor site confers resistance to macrolides and lincosamides
- alteration in the 23S sunbunit by methylation of adenine results in the MLS_B phenotype (see Box 2.7) and confers resistance to macrolides, lincosamides and type B streptogramins. This MLS_B phenotype is encoded by *erm* (erythromycin ribosomal methylase) genes
- Inactivation by 3-lincomycin, 4-clinadamycin 0-nucleotide transferase. This is plasmid mediated and encoded by *linA* and *linA′* genes,
- Decreased membrane permeability in Gram-negative species, e.g. Enterobacteriaceae, *Pseudomonas* spp., *Acinetobacter* spp.

Clinical use

- Alternative to β-lactams in penillin-allergic patients with skin and soft tissue infections
- Staphylococcal bone and joint infections
- Severe group A streptococcal infections, e.g. necrotizing fasciitis, toxic shock syndrome
- Anaerobic infections e.g. intra-abdominal sepsis, anaerobic bronchopulmonary infections
- *Pneumocystis jirovecii* pneumonia (in combination with primaquine)
- *P. falciparum* malaria (in combination with quinine)

Pharmacology

- Clindamycin is given orally, intravenously or by deep intramuscular injection
- Well absorbed orally and widely distributed with good tissue penetration especially bone. CSF penetration is negligible
- Most of the drug is metabolized to products with variable antibacterial activity
- Excreted in the bile and urine – dose modification required in severe renal and liver disease

Toxicity and side-effects

- *C. difficile* colitis – discontinue clindamycin
- Allergic reactions – rashes, fever, erythema multiforme, anaphylaxis
- Laboratory abnormalities – transient hepatitis, neutropenia, thrombocytopaenia

Box 2.7 MLS_B resistance (also known as inducible resistance)

Macrolides, lincosamides and streptogramin type B (MLS_B) antibiotics bind to closely related sites on the 50S ribosome of bacteria. One consequence is that some bacteria (e.g. staphylococci, streptococci, and enterococci) with inducible resistance to erythromycin also become resistant to the other MLS_B agents, in the presence of erythromycin. The methylase enzyme involved is not induced by lincosamides or streptogramins, which therefore remain active in the absence of macrolides. Over 20 *erm* genes encode the MLS_B resistance, and it is becoming more common in group A streptococci and pneumococci.

Streptogramins

Streptogramins are a group of antibiotics derived from various *Streptomyces* spp. They consist of two macrocyclic lactone peptolide components referred to as streptogramin A and streptogramin B. A number of compounds exist:

- quinupristin-dalfopristin (Synercid®) is the only drug available in the UK and used for the treatment of resistant Gram-positive infections.
- pristinamycin (used for treatment of skin and soft tissue infections)
- virginiamycin (mainly used as an animal growth promoter)
- mikamycin

Mode of action

Streptogramins exert their action on the second or elongation stage of protein synthesis. The two components act synergistically as follows:

- streptogramin A molecules (e.g. dalfopristin) bind to the 50S ribosomal subunit and prevent aminoacyl-tRNA attaching to the catalytic site of the peptidyl transferase, thus inhibiting transfer of the growing peptide chain
- streptogramin B molecules (e.g. quinupristin) prevent the peptide bond forming, which leads to the premature release of incomplete polypeptides.

Quinupristin-dalfopristin is active against most Gram-positive organisms (except *E. faecalis*, which is intrinsically resistant).

Resistance

There are three mechanism of resistance:

- modification of the ribosomal target site (quinupristin). This results in resistance to macrolides, lincosamides, and streptogramin B (MLS_B phenotype) and is encoded by various *erm* genes (see 📖 Lincosamides, p.84–85, Box 2.7)
- enzymatic inactivation by acetyltransferases, encoded by *vat(A), vat(B), vat(C)* in staphylococci, and *vat(D)* in *E. faecium* (quinupristin and dalfopristin)
- active transport out of the cells by efflux pumps, encoded by *vga(A)* and *vga(B)* genes in staphylococci (quinupristin and dalfopristin).

Clinical use

- Vancomycin-resistant *E. faecium* (not active against *E. faecalis*)
- Skin and soft tissue infection caused by MSSA or group A streptococci
- Serious Gram-postive infections where there is no alternative antibiotic available

Pharmacology

- Quinupristin-dalfopristin is given intravenously, preferably into a central vein
- Exhibits significant post-antibiotic effect: 2.8 h for pneumococci, 4.7 h for staphylococci and 2.6–8.5 h for enterococci
- Wide volume of distribution but poor CSF penetration
- Metabolized in the liver and excreted in the faeces

Toxicity and side-effects

- Injection site reactions occur in >30%, so the drug should be given via a central vein
- Arthralgia and myalgia are common
- Nausea, vomiting, diarrhoea, skin rash, pruritis
- Laboratory abnormalities – hepatitis, hyperbilirubinaemia
- Inhibition of hepatic CYP3A4 resulting in increased levels of drugs metabolized by this enzyme

Lipopeptides

Daptomycin (Cubicin®), a fermentation product of *Streptomyces roseosporus*, was discovered in the 1980s. It is a 13-membered cyclic amino acid lipopeptide antibiotic with a lipophilic tail. It was approved in the UK in 2003 for the treatment of complicated skin and soft tissue infections.

Mode of action

The exact mechanism of action is unknown although it appears to bind to the cell membrane of Gram-positive bacteria in a calcium-dependent manner, disrupting the cell membrane potential. Daptomycin is active against Gram-positive organisms, e.g. staphylococci and streptococci, including those that are glycopetide resistant.

Resistance

Resistance to daptomycin is rare – strains with reduced susceptibility have been obtained after serial passage *in vitro*.

Clinical use

Daptomycin is used for complicated skin and soft tissue infections caused by Gram-positive bacteria.

Pharmacology

Daptomycin is given by intravenous infusion. The AUC/MIC profile and prolonged PAE enable once-daily dosing. Daptomycin is highly protein bound and is eliminated largely unchanged by the kidneys.

Toxicity and side-effects

- Common side-effects – nausea, vomiting diarrhoea, headache, rash, injection site reactions
- Muscle toxicity – myalgia, muscle weakness, and myositis are uncommon; rhabdomyolysis is rare. Serum creatinine kinase (CK) should be checked before starting treatment, and weekly during treatment. Stop treatment if symptoms develop
- Interference with prothrombin time/international normalized ratio (INR) assay – clotting sample should be taken just prior to administration of daptomycin

Oxazolidinones

The oxazolidinones are a new, purely synthetic class of antimicrobials with activity against staphylococcal and streptococcal species. At present linezolid (Zyvox®) is the the only oxazolidinone available in the UK. It was introduced in 2001. It is active against Gram-positive bacteria and is used for infections that are resistant to other antibiotics (e.g. MRSA and vancomycin-resistant enterococci (VRE)). Always involve an infection specialist when initiating therapy.

Mode of action

Oxazolidinones are protein synthesis inhibitors that are bacteriostatic against Gram-positive organanisms. They bind to the 50S ribosomal subunit at its interface with the 30S ribosomal subunit, preventing formation of the 70S initiation complex.

Resistance

Despite its recent introduction, resistance to linezolid among strains of MRSA and VRE has already been reported. The mechanism appears to be mutation in the 23S RNA domain V region. It is usually associated with long durations of therapy or prior exposure to linezolid.

Clinical use

Linezolid is approved for use in Gram-positive infections:

- pneumonia and complicated skin/soft tissue infections caused by Gram-positive bacteria
- serious infections due to resistant Gram-positive bacteria (e.g. MRSA, VRE and penicillin-resistant pneumococci).

Pharmacology

Linezolid may be given orally (100% bioavailability) or intravenously. Linezolid is widely distributed with good tissue and CSF penetration. It is metabolized by oxidation in the liver and excreted in the urine (85%) or faeces. No dose adjustment is required for renal or hepatic disease.

Toxicity and side-effects

Linezolid is generally well tolerated.

- Gastrointestinal symptoms, e.g. nausea, vomiting, diarrhoea are common.
- Myelosuppression – thrombocytopaenia, neutropenia, pancytopaenia have been reported. More common with prolonged therapy (>10 days) and usually reversible. Full blood count (FBC) should be monitored weekly in patients taking linezolid.
- Monoamine oxidase inhibition – linezolid is a monoamine oxidase inhibitor. Patients should be told to avoid tyramine-rich foods. Linezolid has been associated with serotonin syndrome in patients taking concomitant selective serotonin reuptake inhibitors (SSRIs).
- Optic neuropathy has been reported in patients taking >28 days' treatment. Patients should be told to report visual symptoms and referred to an ophthalmologist if necessary.
- Lactic acidosis has been associated with prolonged treatment.

Chloramphenicol

Chloramphenicol, initially called chloromycetin, was first isolated from *Streptomyces venezuelae* in 1947. It has a broad spectrum of activity against a wide range of bacteria, spirochaetes, rickettsiae, chlamydiae, and mycoplasmas. Soon after its introduction in 1949 reports of aplastic anaemia emerged, limiting its use. Furthermore, widespread use in the developing world has resulted in resistance, particularly in *S. typhi*. Despite this, chloramphenicol remains useful for the treatment of serious infections that are resistant to other antibiotics.

Mode of action

Chloramphenicol inhibits protein synthesis by binding to the 50S subunit of the 70S ribosome at a site that prevents the attachment of tRNA – this prevents association of the amino acid with peptidyltransferase and peptide band formation. This is a bacteriostatic effect in most organisms but is bactericidal in some meningeal pathogens e.g. *H. influenzae, S. pneumoniae*, and *N. meningitis*.

Resistance

There are several resistance mechanisms:

- reduced permeability or uptake
- ribososmal mutation
- production of acetyl transferase, an enzyme that acetylates the antibiotic into an inactive form. This mechanism also confers resistance to tetracyclines (see 📖 Tetracyclines, p.92), and is responsible for widespread epidemics of chloramphenicol resistance to *S. typhi* and *S. dysenteriae* seen in the developing world.

Clinical use

In the developed world chloramphenicol is rarely used (because of toxicity), but it remains a commonly used antibiotic in the developing world.

- Enteric fever due to *S. typhi* and *S. paratyphi* – high rates of drug resistance have been reported in India, Vietnam, and Central and South America
- Severe infections such as meningitis, septicaemia, epiglottitis due to *Haemophilus influenzae*
- Sometimes used in infective exacerbations COPD
- An alternative agent for infections in pregnancy, young children, or patients with immediate penicillin hypersensitivity
- Eye drops/ointment are widely used for superficial eye infections
- Ear drops are used for bacterial otitis externa

Pharmacology

Chloramphenicol may be administered orally, intravenously, intramuscularly, or topically. It has high lipid solubility and low protein binding, resulting in a wide volume of distribution in body fluids and tissues. CSF and ocular penetration is good. Chloramphenicol is metabolized in the liver by glucuronidation and excreted in the bile. Only 5–10% is excreted in the urine.

Toxicity and side-effects

- Bone marrow suppression is common, dose-related, and reversible. It is a direct pharmacological effect of the antibiotic, resulting from inhibition of mitochondrial protein synthesis. Manifestations include anaemia, reticulocytosis, leucopaenia, and thrombocytopaenia. Monitor FBC twice weekly during treatment.
- Aplastic anaemia is a rare, idiosyncratic, and often fatal complication, which may occur during or after completion of therapy. It occurs in 1 in 25–40,000 patients. The pathogenesis of this is condition is incompletely understood. Monitor FBC twice weekly during treatment and discontinue the drug if the white cell count (WCC) falls below 2.5×10^9/l.
- There are also reports of haemolytic anaemia in patients with glucose-6-phosphate dehydrogenase (G6PD) deficiency and childhood leukaemia after chloramphenicol therapy.
- Grey baby syndrome – high doses in neonates may result in grey baby syndrome (abdominal distension, vomiting, cyanosis, circulatory collapse) due to inability to metabolize and excrete the drug. If the drug is required in neonates the dose should be reduced and drug levels monitored.
- Other side-effects – rash, fevers, Jarisch–Herxheimer reactions, gastrointestinal symptoms, glossitis, stomatitis, optic neuritis, bleeding disorders, acute intermittent porphyria, interference with development of immunity after immunization.

Tetracyclines

Tetracyclines are a group of broad-spectrum bacteriostatic antibiotics active against Gram-positive, Gram-negative and intracellular organisms, e.g. *Chlamydia*, mycoplasmas, rickettsiae and protozoan parasites. The first tetracycline aureomycin (chlortetracycline) was isolated from *Streptomyces aureofaciens*, a soil organism. Since then, a number of other tetracyclines have been developed. Tetracyclines differ in their pharmacological properties rather than spectrum of cover, although minocycline has slightly broader spectrum.

Classification

- 1st generation – tetracycline, chlortetracycline, oxytetracycline, demeclocycline, lymecycline, and methacycline
- 2nd generation – doxcycline and minocycline
- 3rd generation (glycylcyclines) – tigecycline

Mode of action

- Tetracyclines inhibit bacterial protein synthesis by reversibly binding to the 30S ribosomal subunit. This blocks binding of aminoacyl–tRNA to the ribosomal 'A' site, preventing addition of new amino acids. As their binding is reversible these agents are mainly bacteriostatic.
- Tetracyclines also inhibit mitochondrial protein synthesis by binding to 70S ribosomal subunits in mitochondria in eukaryotic parasites. The mechanism of their antiprotozoal activity is unknown.

Resistance

The widespread use of tetracyclines has been accompanied by increasing drug resistance. This is mediated by acquisition of genes on mobile genetic elements (MGEs, see 📖 Molecular genetics of resistance, p.40). Many tetracycline-resistance genes have been identified; most belong to the *tet* family and some belong to the *otr* family. These genes confer resistance by the following mechanisms:

- efflux pumps – these membrane-associated proteins pump tetracyclines out of the cell. They confer resistance to 1st-generation tetracyclines
- ribosomal protection proteins are cytoplasmic proteins that release tetracyclines from their binding site by guanosine diphosphate (GDP)-dependent mechanisms. They protect the ribosome from 1st- and 2nd-generation tetracyclines
- enzymatic inactivation – this mechanism is seen in *B. fragilis*, where the *tet(X)* gene codes for a protein that modifies tetracyclines in the presence of nicotinamide adenine dinucleotide phosphate-oxidase (NADPH) and oxygen.

Clinical use

- Chlamydial infections – trachoma, psittacosis, salpingitis, urethritis, lymphogranuloma venereum
- Rickettsial infections
- Q-fever
- Brucellosis (doxycycline with either streptomycin or rifampicin)

- Lyme disease (*Borrelia burgdorferi*)
- *Mycoplasma* spp. infections
- Infective exacerbations of COPD (due to their activity against *H. influenzae*)
- Also used in acne, destructive (refractory) periodontal diseases, sinusitis, chronic prostatitis, pelvic inflammatory disease, melioidosis

Pharmacology

- Tetracyclines are usually given orally. Absorption of tetracycline and oxytetracycline is reduced by milk, antacids and some salts. Doxycycline and minocycline are highly bioavailable.
- They are sometimes divided into 3 groups on the basis of their half-lives: short-acting (tetracycline, oxytetracycline), intermediate acting (demeclocycline) and long-acting (doxycycline, minocycline).
- Tetracyclines are widely distributed and show good tissue penetration.
- Tetracycline is eliminated in the urine. Minocycline is metabolized in the liver. Doxycycline is mainly eliminated in the faeces.

Toxicity and side-effects

- Nausea, vomiting, diarrhoea, dysphagia, and oesophageal irritation are common.
- Photosensitivity reactions are common and appear to be toxic rather than allergic.
- Prolonged minocycline administration can cause skin, nail, and scleral pigmentation.
- Deposition occurs in growing bones and teeth so tetracyclines should not be given to children <12 years or pregnant/breast-feeding women.
- Hepatotoxicity due to fatty change may be fatal.
- Tetracyclines exacerbate renal impairment. All tetracyclines (except minocycline and doxycycline) should be avoided in renal failure. Demeclocycline causes nephrogenic diabetes insipidus and is used as a treatment for inappropriate antidiuretic hormone (ADH) secretion.
- Vertigo is unique to minocycline.
- Benign intracranial hypertension has been described with all tetracyclines.
- Superinfection – mucocutaenous candidiasis is common. *C. difficile* colitis may occur.
- Allergic reactions (rashes, urticaria, anaphylaxis) are uncommon.

Tigecycline (Tygacil®)

- A glycylcycline antibiotic, structurally related to the tetracyclines
- Active against Gram-positive and Gram-negative bacteria, including tetracycline-resistant organisms, and some anaerobes. It is also active against MRSA and VRE but not against *P. aeruginosa* and *Proteus* spp.
- Reserved for the treatment of complicated skin and soft-tissue infections and complicated abdominal infections caused by multi-drug-resistant organisms
- Side-effects are similar to those of the tetracyclines

Sulfonamides

The era of antimicrobial chemotherapy began in 1932 when Domagk reported the antimicrobial effects of Prontosil (a sulfachrysoidine dye) against murine streptococcal infections. This compound was found to exert its antibacterial effect through the release of sulfanilamide, which acted as a competitive inhibitor of dihydropteroate synthetase in the folate synthesis pathway (Figure 2.4). Although many sulfonamide drugs were developed, relatively few are in clinical use today, mainly because of their toxicity and increasing drug resistance. Those currently available in the UK include sulfamethoxazole, sulfadiazine, sulfadoxine, sulfasalazine, mafenide acetate, and sulfacetamide sodium.

Classification

The sulfonamides can be classified as follows:

- short- or medium-acting sulfonamides, e.g. sulfisazoxole, sulfamethoxazole, sulfadiazine, sulfamethizole, sulfadimidine, sulfacarbamide
- long-acting sulfonamides, e.g. sulfamethoxypyridazine, sulfameter, sulfadimethoxine, sulfadoxine
- sulfonamides limited to the GI tract, e.g. sulfaguanidine, sulfasuxudine, sulfathalidine, sulfasalazine
- topical sulfonamides, e.g. silver sulfadiazine, mafenide acetate, sulfacetamide sodium.

Mode of action

The sulfonamides are bacteriostatic and inhibit bacterial growth by interfering with folic acid synthesis. They are analogues of PABA and competitively inhibit the incorporation of PABA into tetrahydropteroic acid by the enzyme tetrahydropteroic acid synthetase. They are active against a broad spectrum of Gram-positive and Gram-negative bacteria, *Actinomyces, Chlamydia, Plasmodium* and *Toxoplasma* spp.

Resistance

Resistance to sulfonamides is widespread and increasingly common; cross-resistance between different sulfonamides occurs. Resistance may be due to:

- chromosomal mutations that result in:
    - overproduction of PABA, e.g. *S. aureus, N. gonorrheae*
    - alterations in dihydropteroate synthetase that result in reduced affinity for sulfonamides, e.g. *E. coli*
- plasmids that carry genes coding for:
    - production of drug-resistant enzymes
    - decreased bacterial permeability.

Plasmid-mediated sulfonamide resistance is common in enterobacteriaceae and has increased greatly in recent years, often in conjunction with trimethoprim resistance.

Clinical use

- Sulphamethoxazole is used in combination with trimethoprim for the treatment of a variety of infections (p.97).
- Sulfadiazine may be used for the prevention or rheumatic fever.
- Sulfadiazine is used combination with pyrimethamine (see Antiprotozoal drugs 3, p.160) for toxoplasmosis (unlicensed).
- Sulfadoxine is used in combination with pyrimethamine for treatment of falciparum malaria (see Miscellaneous antimalarials, p.154).
- Silver sulfadiazine (Flamazine®) is used topically to prevent/treat burn infections.

Pharmacology

- Usually administered orally. Sulfadiazine and sulfisoxazole are available as intravenous or subcutaneous preparations. Sulfacetamide is used in topically in eye drops. Silver sulfadiazine and mafenide acetate are used topically in burns.
- Oral sulfonamides are rapidly absorbed. Topical sulfonamides are also absorbed and may be detectable in the blood.
- Widely distributed with high concentrations in body fluids including CSF.
- Metabolized in the liver and excreted in the urine. Dose modification is required in renal impairment.

Toxicity and side-effects

- General – nausea, vomiting, diarrhoea, rash, fever, headache, depression, jaundice, hepatic necrosis, drug-induced lupus, serum sickness-like syndrome
- Haematological – acute haemolytic anaemia, aplastic anaemia, agranulocytosis, leucopaenia, thrombocytopaenia
- Hypersensitivity reactions – drug eruption, vasculitis, erythema nodosum, erythema multiforme, Stevens–Johnson syndrome, anaphylaxis
- Neonatal kernicterus if given in last month or pregnancy

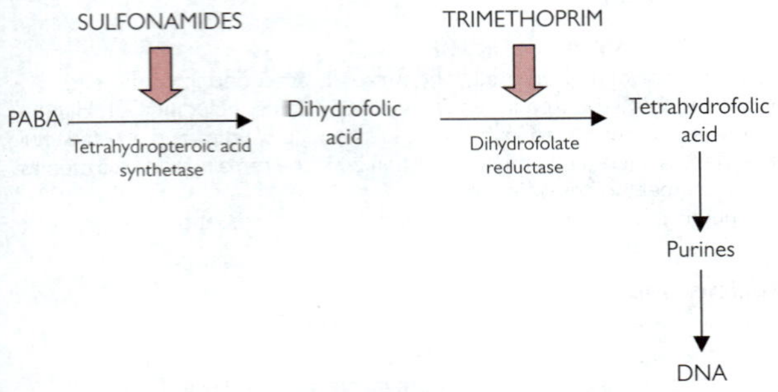

Fig. 2.4 Action of sulfonamides and trimethoprim on the bacterial folate synthesis pathway.

Trimethoprim

Trimethoprim is a diaminopyrimidine which was synthesized by Bushby and Hitchings in 1968. The other members of this class are pyrimethamine (an antiprotozoal), cycloguanil (a product of proguanil, see 📖 Miscellaneous antimalarials, p.154), and flucytosine (an antifungal). Trimethoprim has a fairly broad spectrum of activity against many Gram-positive bacteria and most Gram-negative rods except *P. aeruginosa* and *Bacteroides* spp.

Mode of action

Trimethoprim inhibits the bacterial enzyme dihydrofolate reductase (DHFR), preventing the conversion of dihydrofolate to tetrahydrofolate in the folate synthesis pathway (see 📖 p. 94–95, Fig. 2.4). It works on the same pathway as the sulfonamides but at a later point, resulting in synergistic activity. It is bactericidal or bacteriostatic, depending on the organism and drug concentration.

Resistance

Resistance is common in *Enterobacteriaceae*, particularly in developing countries. It may be chromosomal or plasmid mediated, and caused by:

- chromosomal mutations in the gene for DHFR (or its promoter) resulting in overproduction or modification of the target enzyme
- plasmid-encoded resistance (*dfr* genes in *enterobacteriaceae*), which may result in synthesis of an additional trimethoprim-resistant DHFR enzyme
- change in cell permeability/efflux pumps
- alterations in metabolic pathway.

More than one mechanism can occur in the same cell, resulting in higher resistance levels.

Clinical use

- Urinary tract infections (UTIs, 3 days for uncomplicated cystitis)
- Prophylaxis of recurrent UTIs
- Treatment of prostatitis and epididymo-orchitis
- An option for oral treatment of MRSA (in combination with rifampicin or fusidic acid)

Pharmacology

- Trimethoprim is given orally, and is rapidly absorbed from the gut.
- It is widely distributed in tissues and body fluids, including CSF. High concentrations are achieved in the kidney, lung, sputum, and prostatic fluid;
- 60–80% is excreted in the urine within 24 h; the remainder is excreted as urinary metabolites or in the bile.
- Synergism is seen with sulfamethoxazole, polymixins, and aminoglycosides.

Toxicity and side-effects

- Avoid in pregnancy, especially 1st trimester (antifolate)
- Contraindicated in blood dyscrasias
- Side-effects are similar to co-trimoxazole, but less severe and less frequent with trimethoprim alone
- Other side-effects include GI disturbance, pruritis, rashes, hyperkalaemia

Co-trimoxazole

Co-trimoxazole (Septrin®) is a synergistic combination of trimethoprim and sulfamethoxazole, in the ratio of 1:5.

Mode of action

Sequential inhibition of two enzymes (tetrahydropteroic acid synthetase and dihydrofolate reductase) in the bacterial folate synthesis pathway (see 📖 Sulfonamides, p.94).

Resistance

Resistance may be due to a variety of mechanisms (see 📖 Sulfonamides, p.94 and 📖 Trimethoprim, p.96 for details). Increasing drug resistance rates have been seen in *S. aureus*, many *Enterobacteriaceae* and *Pneumocystis jirovecii*.

Clinical use

- *Pneumocystis jirovecii* pneumonia (treatment and prophylasis)
- Toxoplasmosis (prophylaxis and 2nd-line therapy)
- Nocardiosis (2nd-line therapy)
- Multi-drug resistant organisms e.g. *Acinetobacter* spp.; *B. cepacia*; *S. maltophilia*; *M. marinum*; *M. kansasii*. Seek advice from an infection specialist
- Acute exacerbations of COPD (if organism sensitive and no other options)
- Urinary tract infections (if organism sensitive and no other options)
- Acute otitis media in children (if organism sensitive and no other options)

Pharmacology

Co-trimoxazole may be given orally or intravenously. Components have different volumes of distribution, so seek advice if treating complicated cases (e.g. at an unusual site).

Toxicity and side-effects

- Avoid in blood disorders, infants <6 weeks, hepatic impairment, renal impairment, pregnancy, and breast feeding.
- Side-effects are mainly due to sulfonamide component and may be more severe in the elderly
- Nausea, vomiting, diarrhoea, anorexia, and hypersensitivity most common
- Rashes including erythema multiforme, Stevens–Johnson syndrome, and toxic epidermal necrolysis are rare
- Haematological toxicity
- Renal dysfunction, interstitial nephritis, hyperkalaemia
- Drug-induced hepatitis, pancreatitis, hepatic failure

Quinolones

Nalidixic acid, the first quinolone antibiotic was identified in 1962. Development of the fluoroquinolones in the 1980s led to an expanded spectrum of activity and greater potency. Currently available quinolones include ciprofloxacin, gatifloxacin levofloxacin, moxifloxacin, nalidixic acid, norfloxacin, and ofloxacin. Several quinolones, e.g. temafloxacin, sparfloxacin, grepafloxacin, and trovafloaxin have been withdrawn from use because of toxicity.

Classification

- Group 1 – active against *Enterobacteriaceae*, e.g. nalidixic acid
- Group 2 – active against *Enterobacteriaceae*, *Pseudomonas* spp., and some Gram-positive cocci, e.g. ciprofloxacin, levofloxacin, norfloxacin, ofloxacin
- Group 3 – improved spectrum against *S. pneumoniae* and other Gram-positive cocci, e.g. gatifloxacin, sparfloxacin (neither available in the UK)
- Group 4 – enhanced activity against Gram-positive cocci, *S. pneumoniae* and anaerobes, e.g. gatifloxacin, moxifloxacin, trovafloxacin

Mode of action

The quinolones prevent bacterial nucleic acid synthesis by inhibiting two enzymes, DNA gyrase and topoisomerase IV. DNA gyrase consists of α- and β-subunits, which are encoded by the *gyrA* and *gyrB* genes respectively. Quinolones inhibit DNA supercoiling, mainly through action on the α-subunit of DNA gyrase. Topoisomerase IV also consists of two subunits, encoded by the *parC* and *parE* genes. Topoisomerase IV is involved in DNA relaxation and chromosomal segregation. DNA gyrase is the primary target for quinolones in Gram-negative bacteria, whereas topoisomerase IV is the main target in Gram-positive bacteria.

Resistance

Resistance is mainly due to spontaneous chromosomal mutations that:

- alter the target enzymes – exposure to quinolones results in the selection of drug-resistant mutants. Stepwise increasing resistance occurs by sequential mutations in *gyrA* (or *gyrB*) and *parC* (or *parE*)
- alter cell membrane permeability – this may be due to mutations that reduce entry through porin channels or increase efflux. In *P. aeruginosa*, resistance has been shown to be due to over-expression of genes that encode the MexAB–OprM efflux pump, comprising a membrane-fusion protein (MexA), inner-membrane efflux pumps (MexB) and an outer-membrane protein (OprM).
- Recently, plasmid-mediated quinolone resistance encoded by the *qnr* gene has been identified in *K. pneumoniae, E. coli* and others.

Clinical use

- Ciprofloxacin – UTIs, prostatitis, gonorrhea, pseudomonal infections in cystic fibrosis patients, prophylaxis of meningococcal meningitis (not licensed), anthrax
- Levofloxacin – sinusitis, COPD exacerbation, community-acquired pneumonia, UTI, chronic prostatitis, skin and soft tissue infections. Also active against *M. tuberculosis*
- Moxifloxacin – COPD exacerbation and community-acquired pneumonia (2nd line). Also active against *M. tuberculosis*
- Nalidixic acid – UTIs
- Norfloxacin – UTIs, chronic prostatitis
- Ofloxacin – UTIs, chronic prostatitis, lower respiratory tract infections, skin and soft tissue infections, gonorrhea, genital chlamydia, non-gonococcal urethritis, pelvic inflammatory disease

Pharmacology

- Well absorbed with bioavailability ranging from 50% to 100%. Oral bioavailability is reduced by co-administration of antacids.
- Moxifloxacin, nalidixic acid and norfloxacin are only given orally; ciprofloxacin, levofloxacin may also be given intravenously. Protein binding is low and volumes of distribution are high. Concentrations in prostate, lung, bile, and stool may exceed plasma concentrations.
- Ofloxacin, levofloxacin, and gatifloxacin are renally eliminated. Nalidixic acid and moxifloxacin, undergo hepatic metabolism. Most others are excreted by renal and non-renal pathways. Dose adjustments may be required in renal or liver disease – consult the *British National Formulary* (http://www.bnf.org) for details.

Toxicity and side-effects

- Gastrointestinal side-effects are common (3–17%), e.g. nausea, vomiting, abdominal discomfort, diarrhoea (risk factor for *C. difficile* disease).
- CNS symptoms occur in 0.9–11% of patients, e.g. headache, dizziness, insomnia, altered mood. Hallucinations, delirium, and seizures are rare. Quinolones may induce seizures and should not be given to epileptics or people who are prone to seizures.
- Allergic reactions – rash, photosensitivity, drug fever, urticaria, angioedema, vasculitis, serum sickness, interstitial nephritis.
- Arthropathy has been observed in animals and tendon rupture in adults. Quinolones are not recommended in young children (but may be used for specific indications under expert guidance) or in adults with a history of tendon disorders.
- QT prolongation may predispose to ventricular arrhythmias (grepafloxacin).
- Laboratory abnormalities – leucopaenia, eosinophilia, hepatitis (moxifloxacin), dysglycaemia (gatifloxacin).

Indications for consideration of higher dose ciprofloxacin

- Infections where antibiotic penetration is likely to be suboptimal e.g. septic arthritis; osteomyelitis; VAP; cystic fibrosis patients; meningitis; intra-ocular infections
- Serious pseudomonal infections

Nitroimidazoles

Metronidazole (Flagyl®)

Metronidazole is a nitroimidazole drug which was introduced for the treatment of *T. vaginalis* infections in 1959. It was subsequently found to be bactericidal for most anaerobic and facultatively anaerobic bacteria and protozoa.

Mode of action

Metronidazole is a prodrug that needs to be activated in susceptible organisms. It has a low molecular weight and enters the bacterial cell by passive diffusion. It is activated by reduction of its nitro group by a nitroreductase resulting in the formation of metronidazole radicals. These highly reactive compounds interact with nucleic acids and proteins causing breakage, destabilization and cell death.

Resistance

Resistance to metronidazole is rare and a combination of mechanisms is required. Both chromosomally mediated and plasmid-mediated resistance have been described. Reports of resistance in *Bacteroides* spp. have been attributed to the transferable genes *nimA* and *nimD*. Metronidazole resistance in *H. pylori* is associated with mutational inactivation of the *rdxA*, *frxA*, and *fdxB* genes. Metronidazole resistance in *T. vaginalis* and *Giardia* is probably multifactorial, with reduced activation of metronidazole and/or reduced transcription of the ferrodoxin gene.

Clinical use

- Parasitic infections, e.g. bacterial vaginosis, intestinal amoebiasis, giardiasis, amoebic liver abscess
- Anaerobic infections
- *C. difficile* colitis
- *H. pylori* eradication therapy
- Small bowel bacterial overgrowth
- Pouchitis (inflammatory bowel disease)
- Infected leg ulcers and pressure sores
- Pelvic inflammatory disease
- Acute ulcerative gingivitis
- Surgical prophylaxis

Pharmacology

- Metronidazole may be given orally, intravenously, per vagina, per rectum or topically. *It should never be taken with alcohol because of the risk of a disulfiram-like reaction* (see Toxicity and side-effects).
- When given orally it is absorbed rapidly and almost completely.
- Protein binding is low and the drug is widely distributed in fluids and tissues. Metronidazole shows excellent penetration into abscesses.
- It is metabolized in the liver by the CYP450 enzyme system.
- Metronidazole and its metabolites are primarily eliminated by the kidneys.

Toxicity and side-effects

Metronidazole is generally well tolerated.

- Abnormal metallic taste is commonly reported
- Gastrointestinal – nausea anorexia, epigastric discomfort, vomiting, diarrhoea, constipation
- Peripheral neuropathy occurs with prolonged treatment
- Disulfiram-like reaction with alcohol, e.g. nausea, vomiting, flushing, tachycardia, hypotension, acute confusion/psychosis, sudden death
- Genitourinary – transient darkening of the urine, dysuria, cystitis, incontinence
- Allergic reactions – rash, urticaria, flushing, bronchospasm, serum sickness
- CNS symptoms include headache, dizziness, syncope, vertigo, sleep disturbance, confusion, excitation, and depression. Cerebellar toxicity has been seen with high doses and/or prolonged therapy
- Other – fever, mucocutaneous candidiasis, neutropenia, thrombophlebitis with intravenous infusion

Antibiotics with Anaerobic Cover

If your patient is on one of the following drugs, seek advice about whether metronidazole should be continued

- Co-amoxiclav
- Imipenen or meropenem
- Clindamycin
- Piperacillin-tazobactam

Nitrofurans

The nitrofuran group of antibiotics comprises:
- nitrofurantoin (Furadantin®, Macrobid®, Macrodantin®)
- furazolidone (not available in UK)
- nitrofurazone (not available in UK).

Mode of action

The mechanism of action is poorly understood but requires enzymatic reduction within the bacterial cell (like metronidazole). The reduced derivatives bind to ribosomal proteins and block translation. They also appear to directly damage bacterial DNA (like quinolones) and inhibit DNA repair. Nitrofurans are bactericidal against urinary pathogens such as *E. coli, Citrobacter*, group B streptococci, *S. saprophyticus, E. faecalis*, and *E. faecium*. However, note that only a minority of *Enterobacter* spp. are sensitive, and most members of *Proteus, Providencia, Morganella, Serratia, Acinetobacter* and *Pseudomonas* spp. are resistant.

Resistance

Resistance is rare. In *E. coli*, resistance may be chromosomal or plasmid mediated and is associated with inhibition of nitrofuran reductase activity.

Clinical use

- Acute uncomplicated cystitis (not pyelonephritis) – 3 days
- Treatment of recurrent UTIs – 7 days
- Prophylaxis of recurrent UTIs

Pharmacokinetics

- Nitrofurantoin has good oral absorption, which is enhanced by food.
- Serum concentrations are low but urine concentrations are high.
- Renal elimination – dose reduction required in renal impairment.

Toxicity and side-effects

- Gastrointestinal – nausea, vomiting
- Pulmonary – acute hypersensitivity (fever, cough, dyspnoea, pulmonary infiltrates, myalgia, eosinophilia), chronic (pulmonary fibrosis, bronchiolitis obliterans organizing pneumonia)

Novobiocin

This was once commonly used as a reserve drug for staphylococcal infections, but is now out of favour because of problems with resistance and toxicity. It acts on the β-subunit of DNA gyrase (like quinolones). It is used in the diagnostic microbiology laboratory to identify *S. saprophyticus* (coagulase-negative, novobiocin resistant), a urinary pathogen.

Rifamycins

These are semisythetic derivatives of rifamycin B, a natural product originally isolated from *Streptomyces mediterranei*. Rifamycin B is poorly active but readily converted into rifamycin S, from which most active compounds are derived.

The rifamycins have a number of features in common:

- inhibit bacterial DNA-dependent RNA polymerase
- bactericidal effect against a variety of bacteria
- rapid emergence of resistance due to mutation in *rpoB* gene (encodes β-subunit of DNA-dependent RNA polymerase)
- used in combination with unrelated antibiotics to suppress emergence of resistance
- stimulate hepatic metabolism by CYP450 enzyme system
- predominantly biliary excretion.

Rifampicin

This is the most important rifamycin and is widely used for the treatment of tuberculosis and other bacterial infections.

- Mode of action – inhibits DNA-dependent RNA polymerase. Bactericidal against *S. aureus* (including MRSA), group A streptococcus, *S. pneumoniae*, *N. gonorrheae*, *N. meningitidis*, *H. influenzae*, *M. tuberculosis*, *M. kansasii*, *M. marinum*, *M. leprae*, *Legionella* spp., *L. monocytogenes*, *Brucella* spp.
- Resistance – mutations in *rpoB* gene
- Clinical use – tuberculosis, leprosy, serious or device-related infections with antibiotic-resistant staphylococci, pneumococci, legionella, elimination of nasopharyngeal carriage of *N. meningitidis* and *H. influenzae*
- Pharmacology – >90% oral absorption, widely distributed, low CSF penetration unless meningeal inflammation, metabolized in liver, predominantly excreted in bile
- Interactions – enhances its own metabolism and that of other drugs, e.g. warfarin, oral contraceptives, corticosteroids, protease inhibitors
- Toxicity and side-effect – orange discolouration of body fluids, skin rashes, GI upset, hepatitis, thrombocytopaenia (stop drug), 'rifampicin flu' (intermittent therapy)

Rifabutin

Rifabutin is mainly used to treat atypical mycobacterial infections.

- Mode of action – similar to rifampicin but more active against *M. avium* complex (MAC)
- Resistance – mutations in *rpoB* gene. Lower frequency of spontaneously resistant strains than rifampicin
- Clinical use – prophylaxis against MAC in AIDS patients, treatment of non-tuberculous mycobacterial disease, treatment of tuberculosis in those who cannot have rifampicin (unacceptable interactions, intolerance)
- Pharmacology – 12–20% oral absorption, widely distributed with concentrations in organs being higher than plasma

- Interactions – clarithromycin and ritonavir inhibit CYP450, increasing rifabutin levels
- Toxicity and side-effects – skin rashes, GI upset, hepatitis, neutropenia, uveitis, and arthralgia (with higher doses)

Rifapentine

- Unavailable in the UK, used in the USA.
- Mode of action – similar to rifampicin but more active against MAC. Also active against *L. monocytogenes* and *Brucella* spp.
- Clinical use – used once weekly in continuation phase of TB treatment in non-cavitatory, drug-suspectible, smear-negative (at 2 months) TB. Should not be given in HIV-infected patients as it has high treatment failure rates
- Pharmacology – 70% oral absorption; well distributed with tissue concentration exceeding plasma concentrations except CSF and bone
- Interactions – potent inducer of CYP450 resulting in reduced concentrations of co-administered drugs, e.g. protease inhibitors
- Toxicity and side-effects – neutropenia, hepatitis, animal evidence of teratogenicity and foetal toxicity (avoid in pregnancy)

Other rifamycins

- Rifamide – used in staphylococcal and biliary infections (limited availability)
- Rifamycin SV – can be given parenterally or topically (not available in UK)
- Rifaximin – used in traveller's diarrhoea, hepatic encephalopathy (not available in UK). May have a role in *C. difficile*-associated disease

Polymyxins

The polymyxins (polymyxin B and polymyxin E/colistin) were discovered in 1947 and were used parenterally until the 1960s when aminoglycosides entered common usage. In the 1980s polymyxins fell into disuse mainly because of nephrotoxicity. With the emergence of multidrug-resistant Gram-negative organisms, e.g. *Pseudomonas* spp. and *Acinetobacter* spp., injectable polymyxins are playing an increasing role.

Mode of action

Polymyxins are cyclic cationic polypeptide detergents. They penetrate cell membranes and interact with phospholipids to disrupt the membranes. They are rapidly bactericidal.

Clinical use

Polymyxin B is used for the treatment of severe infections caused by multi-drug resistant Gram-negative organisms. Colistin sulfate has been used for intestinal decontamination. Aerosolized colistemethate has been used to treat cystic fibrosis patients with pulmonary colonization or infection with multi-drug-resistant *Pseudomonas* spp. Intravenous colistemethate has been used to treat severe multi-drug-resistant Gram-negative infections, e.g. ventilator-associated pneumonia.

Pharmacology

Not absorbed orally. Good serum levels after IV administration but poor penetration of CSF, biliary tract, pleural fluid, and joint fluid. Renally excreted – reduce dose in renal impairment.

Polymyxin B is available topically and as a parenteral preparation that can be given intravenously or intramuscularly.

Colistin is available as colistin sulfate for topical use and colistimethate sodium for intramuscular or intravenous use.

Toxicity and side-effects

Dose-related nephrotoxicity and neurotoxicity (paraesthesia, peripheral neuropathy, and neuromuscular blockade).

Fusidic acid

Fusidic acid is a member of the fusidane class of antibiotics, derived from *Fusidium coccineum* and chemically related to cephalosporin P. The sodium salt of fusidic acid (fucidin) was introduced into clinical practice in 1962.

Mode of action

Bacteriostatic: inhibits protein synthesis by blocking elongation factor G. Fusidic acid also has *in vitro* and *in vivo* immunosuppressive effects.

Resistance

Occurs by chromosomal mutations in the *fusA* gene which codes for elongation factor, and by plasmid-mediated resistance resulting in reduced permeability to the drug. It is a particular concern with long-term monotherapy, so fusidic acid is often combined with another agent.

Clinical use

Fusidic acid is mainly used for the treatment of staphylococcal infections, e.g. skin and soft tissue infections, bacteraemia, septic arthritis, osteomyelitis, lower respiratory tract infections in cystic fibrosis patients. It has also been used to treat erythrasma due to *Corynebacterium minutissium* and lepromatous leprosy.

Pharmacology

Fusidic acid may be given orally, topically or intravenously. Oral absorption is rapid and almost complete. It is highly protein bound and widely distributed in most tissues. Metabolized in the liver by CYP450 and eliminated in the bile.

Toxicity and side-effects

Fusidic acid is generally well tolerated but the oral form may cause nausea, vomiting, and reversible jaundice (6%). The intravenous form is associated with thrombophlebitis and jaundice (17%). Ophthalmic preparations may cause itching or stinging. A drug-induced immune-mediated thrombocytopaenia has been decribed.

Mupirocin

Mupirocin is a pseudomonic acid, produced by *Pseudomonas fluorescens*, that is not related to any other antibiotic in clinical use.

Mode of action

Bacteriostatic: inhibits bacterial RNA and protein synthesis by binding to bacterial isoleucyl tRNA synthetase, preventing the incorporation of isoleucine into protein chains in the bacterial cell wall.

Resistance

Low-level resistance is due to spontaneous mutation resulting in altered access to binding sites in isoleucyl tRNA synthetase. High-level resistance is mediated on transferable plasmids by the *mupA* gene, which codes for a modified enzyme. Mupirocin resistance in MRSA has been associated with widespread use – prolonged use (>7 days) is discouraged.

Clinical use

Primarily used for skin infections, e.g. impetigo and folliculitis, and for nasal decolonization of *S. aureus* or MRSA carriage. Mupirocin has also been used for treatment of secondarily infected eczema, burns, lacerations, and ulcers.

Pharmacology

Given topically as a cream (Bactroban®) or nasal ointment (Bactroban nasal®).

Toxicity and side-effects

Local reactions such as pruritis, burning sensation, rash, urticaria may occur, particularly if used on broken skin.

Retapamulin

This is one of the pleuromutilins, which are a new class of agents with a unique binding site on the bacterial ribosome. They have a long post antibiotic effect, and it is said that resistance is slow to develop.

Retapamulin is only available topically, and is used for impetigo and other superficial bacterial skin infections caused by *S. aureus* and *S. pyogeues* that are resistant to first line topical agents.

Antituberculous agents – 1st line

Streptomycin was the first drug used in a randomized clinical trial in 1948, and was shown to reduce mortality from pulmonary tuberculosis. Since then a number of antituberculous agents have been developed. For treatment guidelines see 📖 M. tuberculosis, p.407.

Classification

- 1st-line drugs have superior efficacy and more acceptable toxicity, e.g. isoniazid, rifampicin, pyrazinamide, ethambutol, streptomycin.
- 2nd-line drugs may be less efficacious, more toxic, or more expensive, e.g. rifamycins, fluoroquinolones, aminoglycosides, para-aminosalicyclic acid, cycloserine, ethionamide, protionamide, thiocetazone, capreomycin, viomycin.

Modes of action

- Bactericidal drugs are active against replicating tubercle bacilli in cavities, e.g. isoniazid, rifampicin, ethambutol, streptomycin.
- Sterilizing drugs are active against slowly replicating organisms (persisters) or intracellular organisms, e.g. pyrazinamide, isoniazid.

Isoniazid (H)

- Isonicotinic acid hydrazide, synthetic agent
- Mode of action – inhibits mycolic acid synthesis. Rapidly bactericidal against actively replicating *M. tuberculosis* (MIC 0.01–2 mg/L). Most other mycobacteria are resistant
- Resistance – isoniazid resistance is one of the two most-frequent forms of resistance. Associated with mutations in *inhA* (mycolic acid synthesis), *katG* (catalase peroxidase) and *oryR-ahpC* genes
- Clinical usage – treatment of all forms of TB infection, chemoprophylaxis in contacts and highly susceptible patients
- Pharmacology – >95% oral absorption, widely distributed, good CSF penetration, metabolized in liver (*N*-acteyl transferase), excreted in urine. Patients may be fast or slow acetylators, depending on genetic polymorphism (no clinical significance). Dose reduction in renal failure
- Toxicity and side effects – neurotoxicity (reduced with pyridoxine), hepatitis, arthralgia, hypersensitivity, antinuclear antibody (ANA)-positive lupus-like syndrome increases plasma concentration of antiepileptic drugs

Rifampicin (R) (📖 Rifamycins, p.104)

- Semi-synthetic derivative of rifamycin B
- Mode of action – inhibits mycobacterial DNA-dependent RNA polymerase, so inhibiting RNA synthesis. Bactericidal against actively replicating *M. tuberculosis*, and other mycobacteria (*M. kansasii, M. marinum, M. leprae*). Also active against a variety of other bacteria (see Rifamycins)
- Resistance emerges rapidly during monotherapy. Caused by a point mutation in the *rpoB* gene (encodes β-subunit of RNA polymerase)
- Pharmacology – >90% oral absorption, widely distributed, low levels in CSF unless meningeal inflammation. Metabolized in liver (cytochrome P450 enzyme system), induces its own metabolism (autoinduction) and

that of corticosteroids. Excreted in bile and undergoes enterohepatic circulation; some excreted in urine. Dose reduction in liver failure
- Adverse effects – orange discolouration of body fluids, skin rash, GI upset, hepatitis, hypersensitivity, 'rifamipicin flu' (intermittent therapy), 'red man syndrome' (overdose)
- Clinical usage – treatment of all forms of TB infection, chemoprophylaxis of contacts and susceptible individuals (2nd line)

Pyrazinamide (Z)

- Pyrazinoic acid amide, synthetic nicotinamide analogue
- Mode of action – unknown. Activity requires conversion to pyrazinoic acid by mycobacterial pyrazinamidase. Active against replicating intracellular organisms and those in an acid pH environment, e.g. within necrotic inflammatory foci
- Resistance is uncommon and is due to mutations in the *pncA* (pyrazinamidase) gene
- Pharmacology – >90% oral absorption, widely distributed, good CSF penetration. Metabolized in liver; excreted by kidneys. Dose reduction in renal failure. Increase dose in dialysis patients
- Toxicity and side effects – GI upset, hepatotoxicity, gout (inhibits excretion of uric acid), arthralgia
- Clinical usage – essential component of short-course therapy – high relapse rates if not given in 1st 2 months of treatment

Ethambutol (E)

- Hydroxymethylpropylethylene diamine, synthetic compound
- Mechanism of action – inhibits arabinosyl transferase enzymes (synthesis of arabinogalactan and lipoarabinomannan)
- Bacteriostatic. Active against mycobacteria (*M. tuberculosis, M. kansasii, M. xenopi, M. malmoense*) and nocardia.
- Resistance is uncommon – caused by point mutations in the genes encoding arabinosyl transferase enzyme (*embA, embB*, and *embC*)
- Pharmacology – 75–80% oral absorption, widely distributed, 25–40% CSF penetration. Metabolized in liver, renal excretion. Dose modification in renal failure
- Adverse effects – optic neuritis (reversible), peripheral neuropathy, arthralgia, hyperuricaemia, rashes
- Clinical usage – 1st 2 months of tuberculosis treatment (if suspected or known drug resistance); other mycobacterial infections

Streptomycin (S) (📖 Aminoglycosides, p.78)

- Aminocyclitol antibiotic, derived from *Streptomyces griseus*
- Mode of action – binds to 16S rRNA, inhibits protein synthesis
- Bactericidal against *M. tuberculosis* and various other bacteria
- Resistance – mutation in *rpsL* gene (encodes ribosomal protein S12)
- Pharmacology – not absorbed from GI tract, administered intramuscularly. Widely distributed but poor penetration of CSF and abscesses, 99% renal excretion
- Adverse effects – injection site reactions, ototoxicity, hypersensitivity, neuromuscular blockade, peripheral neuritis, optic neuritis
- Clinical usage – treatment of tuberculosis, *M. kansasii* infections, plague, enterococcal endocarditis

Antituberculous agents – 2nd line

These drugs may be used to treat tuberculosis if there is:

- intolerance to 1st-line drugs
- resistance to one or more 1st-line drugs
- multi-drug resistance (resistance to isoniazid and rifamipcin, MDR-TB)
- Extensively drug-resistant TB (MDR plus resistance to a quinolone and an injectable agent)
- Co-infection with HIV.

Treatment in these situations may be complicated, and advice should be sought from a specialist experienced in treating such patients.

Rifamycins (see Rifamycins, p.104)

- Rifabutin induces CYP450 less than rifampicin. It is used to treat tuberculosis in HIV-infected patients on protease inhibitors or with atypical mycobacterial infections (see Rifamycins).
- Rifapentine is a greater inducer of CYP450 than rifampicin. It has been used to treat tuberculosis in HIV-infected patients, but is not recommended for intermittent therapy.

Fluoroquinolones (see Quinolones, p.98)

Ciprofloxacin, ofloxacin, levofloxacin, and moxifloxacin can be used to treat multi-drug-resistant tuberculosis (MDR-TB) or atypical mycobacterial infections (see Non-tuberculous mycobacteria, p.413).

Para-aminosalicyclic acid (PAS)

- Mode of action – interferes with the salicylate-dependent biosynthesis of iron-chelating compounds called mycobactins
- Clinical usage – limited use in developing countries because of low cost, poor compliance, primary resistance. Used for MDR-TB in developed countries
- Pharmacology – incomplete oral absorption, hepatic metabolism, renal excretion
- Adverse effects – GI upset, interferes with iodine metabolism

Capreomycin

- Cyclic polypeptide antibiotic, produced by *Streptomyces capreolus*
- Mode of action unknown
- Resistance – mechanism unknown, cross-resistance with viomycin and partial cross-resistance with other aminglycosides
- Clinical usage – 1st-line injectable agent in MDR-TB, particularly if streptomycin resistant.
- Pharmacology – administered IM, renal excretion
- Adverse effects – injection site reactions, ototoxic, nephrotoxic, neuromuscular blockade

Aminoglycosides

- Amikacin and kanamycin are used 3rd line after streptomycin and capreomycin for treatment of MDR-TB.
- Amikacin is ototoxic and nephrotoxic – monitor serum levels.
- Kanamycin offers no advantage over amikacin apart from cost.

Viomycin

- Cyclic polypeptide antibiotic, related to capreomycin, produced by several *Streptomyces* species
- Resistance – mutations in the *vicA, vic B* genes (encode 50S and 30S ribosomal subunits), *str* gene (streptomycin resistance), and *nek* gene (neomycin/amikacin resistance). Cross-resistance with capreomycin
- Clinical usage – limited availability
- Pharmacology – administered IM, renal excretion

Cycloserine

- Naturally occurring amino acid, derived from *Stretomyces orchidaceus*
- Mode – inhibits cell wall synthesis
- Broad spectrum of antibiotic activity – *S aureus*, streptococci, enterococci, enterobacteriaceae, *Nocardia* spp., *Chlamydia* spp., and mycobacteria
- Primary resistance is rare; secondary resistance develops slowly
- Clinical usage – re-treatment regimes, primary treatment of drug-resistant TB, non-tuberculous mycobacteria
- Pharmacology – well absorbed, widely distributed including CSF, 50% metabolized, 50% excreted unchanged in urine

Ethionamide

- Ethylthiosonicotinamide, isonicotinic acid derivative, introduced 1956
- Mode of action – inhibits mycolic acid synthesis. Bacteriostatic
- Resistance mechanism unknown
- Clinical usage – drug-resistant and MDR-TB, sometimes used in leprosy
- Pharmacology – well absorbed, widely distributed including CSF, metabolized in liver, 99% excreted as metabolites in urine
- Adverse effects – GI upset, hypersensitivity, hepatitis

Propylthioisonicotinamide

- Similar to ethionamide but better tolerated
- Rarely used for TB, sometimes used in leprosy

Thiacetazone

- Acetylaminobenzaldehyde thiomicarbazone
- Mode of action – poorly understood, inhibits mycolic acid synthesis
- Resistance – mechanism unknown. Primary and acquired resistance is common in developing countries where it has been extensively used
- Clinical usage – rarely used because of low efficacy and frequency of adverse effects. Should never be given to an HIV-infected patient
- Pharmacology – well absorbed, 20% eliminated in urine
- Adverse events – rash, exfoliative dermatitis, Stevens–Johnson syndrome (especially in HIV patients), GI upset, vertigo, conjunctivitis

Non-tuberculous mycobacteria

These organisms may be called atypical mycobacteria, environmental mycobacteria, opportunist mycobacteria, non-tuberculous mycobacteria (NTM), or mycobacteria other than tuberculosis (MOTT). They can cause a variety of diseases (pulmonary, skin, soft tissue, bone and joint infections, disseminated disease) particularly in the immune-deficient. Treatment depends on the causative organism and often requires a prolonged course of a combination of antibiotics. Surgical excision of lesions may also be required. The most commonly used drugs are:

- clarithromycin, azithromycin(see 📖 Macrolides, p.82)
- ethambutol (see Antituberculous agents-first line, p.110)
- rifampicin, rifabutin (see 📖 Rifamycins, p.104)
- clofazimine (see 📖 Antileprotics, p.116)
- ciprofloxacin, ofloxacin (see 📖 Quinolones, p.98)
- amikacin (see 📖 Aminoglycosides, p.78)
- minocycline, doxycycline (see 📖 Tetracyclines, p.92)
- co-trimoxazole (see 📖 Co-trimoxazole, p.97).

Treatment of atypical mycobacterial infections is summarized in Table 2.5.

Table 2.5 Treatment of atypical mycobacterial infections

Causative organism	British Thoracic Society (BTS) guidelines[a]	American Thoracic Society (ATS) guidelines[b]	Comments
M. avium complex (MAC), normal host	Pulmonary: rifampicin + ethambutol ± isoniazid for 24 months	Clarithromcyin (or azithromycin) + rifabutin (or rifampicin) + ethambutol (until culture negative for 1 year)	May add streptomycin for initial 2–3 months for severe disease; extrapulmonary disease: surgical excision.
M. avium complex, immuno-compromised	Rifabutin + ethambutol + clarithromcyin (or azithromycin)	Clarithromcyin (or azithromycin) + ethambutol + rifabutin. Amikacin or streptomycin initially for severe disease	Primary prophylaxis if CD4 count <50/mm^3. Lifelong treatment required
M. abscessus	Rifampicin + ethambutol + clarithromycin	Amikacin + cefoxitin for severe disease. Newer macrolides	Surgical excision.
M. chelonae	Pulmonary: rifampicin + ethambutol + clarithromycin; extrapulmonary: ciprofloxacin + aminoglycoside or imipenem ± clarithromycin	Tobramycin + cefoxitin or imipenem for severe disease. Clarithromycin or clofazimine orally.	Surgical excision.

Table 2.5 Treatment of atypical mycobacterial infections (*Continued*)

Causative organism	British Thoracic Society (BTS) guidelines[a]	American Thoracic Society (ATS) guidelines[b]	Comments
M. fortuitum	Pulmonary: rifampicin + ethambutol + clarithromycin; extrapulmonary: ciprofloxacin + aminoglycoside or imipenem ± clarithromycin	Pulmonary: 2 agents (macrolides, quinolones, doxycyline, minocycline); extrapulmonary: amikacin + cefoxitime or imipenem	Optimal regimen not defined; surgical excision
M. haemophilum	Pulmonary: rifampicin + ethambutol + clarithromycin	Ciprofloxacin + rifabutin + clarithromycin	Optimal regimen not defined; surgical excision
M. kansasii	Pulmonary: rifampicin + ethambutol (9 months)	Isoniazid + rifabutin + ethambutol (18 months, with 12 months smear negative)	
M. malmoense	Pulmonary: rifampicin + ethambutol ± isoniazid (24 months)	Rifampicin + isoniazid + ethambutol ± quinolones and macrolides	Extrapulmonary: excision
M. marinum	Rifampcin + ethambutol or co-trimoxazole or tetracycline	Clarithromycin or minocycline or doxycycline or co-trimoxazole or rifampicin + ethambutol (≥3 months)	Surgical excision
M. scrofulaceum		Clarithromycin + clofazimine ± ethambutol	Chemotherapy rarely indicated; surgical excision
M. ulcerans	Rifampicin + ethambutol + clarithromycin	Rifampicin + amikacin or ethambutol + cotrimoxazole (4–6 weeks)	Surgical excision and skin grafting
M. xenopi	Pulmonary: rifampicin + ethambutol ± isoniazid or 24 months	Macrolide + rifamycin + ethambutol ± streptomycin	Extrapulmonary: excision

[a] British Thoracic Society (BTS) guidelines. *Thorax* 2000;55:210–18.

[b] American Thoracic Society (ATS) guidelines, *Am J Respir Crit Care Med* 2007;175:367–416.

Antileprotics

Introduced by the WHO in 1982, multi-drug therapy (MDT) with a combination of dapsone, clofazimine, and rifampicin is the current treatment for infections with *M. leprae* (see Table 2.6). It has been very successful with a high cure rate, few side-effects, and low relapse rate.

Dapsone

Since first used to treat leprosy by Cochrane in 1947, dapsone has remained the cornerstone of treatment. The emergence of drug resistance in the 1960s has necessitated the addition of other drugs.

- Structure – diaminodiphenyl sulphone
- Mode of action – Dapsone is a diaminodipheny salphane which is active against many bacteria and some protozoa
- It inhibits dihydropterate synthetase, inhibiting synthesis of dihydrofolic acid. Bactersostatic and weakly bactericidal
- Resistance – resistance acquired by sequential mutations
- Pharmacology – >90% oral absorption; widely distributed but selectively retained in skin, kidneys, and liver; metabolized by oxidation and acetylation; mostly renally excreted
- Toxicity and side effects – GI upset, anorexia, headaches, dizziness, insomnia, 'dapsone syndrome' (fever, skin rash ± lymphadenopathy, jaundice, hepatomegaly), haemolysis (especially if G6PD deficiency), methaemoglobinaemia, sulfhaemoglobinaemia
- Clinical use – leprosy, malaria (treatment and prophylaxis) toxoplasmosis (prophylaxis), pneumocystis pneumonia (treatment and prophylaxis), dermatitis herpetiformis

Clofazimine

- Structure – iminophenazine dye
- Antibiotic activity – active against mycobacteria (*M. tuberculosis,*
- *M. scrofulaceum, M. leprae, M. avium intracellulare, M. fortuitum, M. chelonae*), *Actinomyces* spp., and *Nocardia* spp.
- Mechanism of action – unknown; has anti-inflammatory properties
- Resistance is rare (one case report)
- Pharmacology – well absorbed orally, taken up by adipose tissue and monyctes/macrophages, long half-life (10–70 days), excreted in urine and faeces
- Adverse effects – GI upset, skin discolouration (dose related, reversible), small bowel oedema/subacute obstruction (prolonged use)
- Clinical use – leprosy

Rifampicin

- Most effective antileprotic, rendering the patient non-infectious within days of starting therapy. Also used for treatment of TB and other bacterial infections (see 📖 Rifamycins, p.104)

Second-line therapies

One problem with MDT is the prolonged duration of treatment (up to 2 years in multibacillary leprosy). Recent research has focused on determining alternative regimens of shorter duration. Antibiotics that have been shown to have potent bactericidal activity against *M. leprae* are:

- minocycline(see Tetracyclines, p.92)
- ofloxacin (see Quinolones, p.98)
- clarithromycin (see Macrolides, p.82).

Single dose therapy

Although slightly less efficacious than conventional MDT, a single dose regimen (rifampicin, ofloxacin and minocycline) (Table 2.6) is available for single lesion paucibacillary leprosy

Adjunctive treatments

- Corticosteroids have been shown to be efficacious in the treatment of nerve damage.
- Reversal reactions (type 1 reactions) can be treated with aspirin (if mild) or prednisolone (if moderate to severe).
- Erythema nodosum leprosum (type 2 reactions) can be treated with aspirin (if mild), prednisolone, or thalidomide (if severe).

Table 2.6 MDT regimens for the treatment of *M. leprae* infections

Regimen	Drug	Duration
Paucibacillary leprosy	Dapsone 100 mg daily	6 months
	Rifampicin 600 mg monthly	
Multibacillary leprosy	Dapsone 100 mg daily	2 years
	Clofazimine 50 mg daily	
	Clofazimine 300 mg monthly	
	Rifampicin 600 mg monthly	

Antifungals

A number of antifungal agents are currently available:

- alkylamines inhibit ergosterol biosynthesis by inhibiting squalene epoxidase, e.g. terbinafine
- antimetabolites that interfere with DNA synthesis, e.g. flucytosine
- azoles inhibit ergosterol synthesis by blocking 14-α-demethylase, e.g. imidazoles and triazoles
- glucan synthesis inhibitors, e.g. echinocandins
- polyenes bind to the fungal cell membrane and cause it to leak electrolytes, e.g. nystatin, amphotericin
- miscellaneous agents, e.g. griseofulvin.

Polyenes

This group includes amphotericin and nystatin. Both drugs have the same mechanism of action, but very different clinical uses. Neither drug is absorbed when given by mouth. Nystatin is too toxic to be given parenterally, so is limited to topical treatment of mucosal *Candida* infections of the oropharynx, oesophagus, intestinal tract and vagina. However, lipid formulations of nystatin are in clinical trials for IV treatment of systemic infections. Amphotericin can be given topically, but is usually used IV for the systemic treatment of fungal infections.

Mode of action

Polyenes bind to ergosterol in the fungal cell membrane, resulting in increased membrane permeability, leakage of cell contents, and cell death.

Resistance

- Intrinsic resistance – the following organisms are intrinsically resistant to amphotericin: dermatophytes, *Aspergillus terreus, Fusarium* spp., *Pseudallescheria boydii*, *Scedosporium prolificans*, *Trichosporon beigelii*, some mucormycoses.
- Acquired resistance is rare apart from in AIDS patients with relapsing cryptococcal disease, and cancer patients with prolonged neutropenia and yeast infections.

Clinical use

- Conventional amphotericin B deoxycholate, given by IV infusion, is used to treat systemic fungal infections (including the dimorphic fungi). It is the drug of choice for aspergillosis. It is also commonly used for deep and disseminated candidiasis and cryptococcosis, either alone or with flucytosine (see 📖 Other antifungals, p.127).
- The lipid formulations of amphotericin B are recommended if toxicity or renal impairment preclude the use of conventional amphotericin. As the lipid formulations are very costly, hospitals usually have their own policy concerning their use. Other licensed indications include:
  - Abelcet® for systemic fungal infections not responding to conventional amphotericin or other antifungals (it can be given at a higher dose than the other lipid formulations)
  - AmBisome® for infections in febrile neutropenia unresponsive to broad-spectrum antibacterials, and visceral leishmaniasis.
- Nystatin and amphotericin pastilles are used for mucosal candidiasis.
- Amphotericin solution can be used for continuous bladder irrigation in mycotic infections.
- Nystatin cream or pessaries are used to treat vaginal candidiasis. It stains clothes yellow, and damages latex condoms and diaphragms.

Pharmacology

Amphotericin is usually given IV with a carrier (e.g. deoxycholate). It is highly protein bound, and penetrates the CSF and other body fluids poorly. Liver or renal impairment and dialysis have little effect on serum levels. The lipid formulations have widely diverse pharmacokinetics.

Toxicity and side-effects

- Despite the theoretical selective toxicity to fungal cell membranes compared to human cell membranes, conventional (micellar) amphotericin B is associated with infusion-related-reactions (chills, fever, headache, nausea, vomiting) and nephrotoxicity. A test dose is required (because of the risk of anaphylaxis) and close supervision is necessary (monitor FBC liver function tests (LFTs), renal function, and electrolytes). Prophylactic antipyretics or hydrocortisone may be tried in patients with previous acute adverse reactions, in whom ongoing treatment is essential. Toxicity has driven development of new lipid formulations (Table 2.7).
- Additional side-effects of IV amphotericin include GI (anorexia, nausea and vomiting, diarrhoea, epigastric pain), muscle and joint pain, anaemia, and other blood disorders, cardiovascular toxicity (including arrhythmias, especially if infused too quickly), neurological disorders, abnormal LFTS, rash and pain at the infusion site.

Table 2.7 Lipid formulations of amphotericin

Name	Formulation
Liposomal amphotericin (AmBisome®)	Drug encapsulated in phospholipid-containing liposomes
Amphotericin B colloidal dispersion (ABCD; Amphocil®)	Drug complexed with cholesterol sulphate to form small lipid discs
Amphotericin B lipid complex (ABLC; Abelcet®)	Drug is complexed with phospholipids to form ribbon-like structures

Imidazoles

Commonly used imidazoles include clotrimazole, miconazole, and ketoconazole. Other agents available (but rarely used) for the local treatment of vaginal candidiasis and for dermatophyte infections are econazole, fenticonazole, sulconazole, and tioconazole. The imidazoles are essentially fungistatic, but some may be fungicidal at high concentrations.

Mode of action

The imidazoles inhibit the synthesis of ergosterol, the main sterol in fungal cell membranes, by binding to the enzyme cytochrome P450 and interfering with demethylation of the 14-α-methyl sterol intermediates.

Resistance

Acquired ketoconazole resistance is rare, but there are case reports of ketoconazole resistance in patients treated for chronic mucocutaneous candidosis, and AIDS patients with oropharyngeal or oesophageal candidosis.

Clinical use

- Ketoconazole (Nizoral™) orally (PO) is used for the treatment of systemic mycoses, serious chronic resistant mucocutaneous candidiasis, serious resistant GI mycoses, chronic resistant vaginal candidiasis, and resistant dermatophyte infections of skin of fingernails (not toe nails). It is also suitable for non-life threatening infections with dimorphic fungi. It is also used for the prophylaxis of mycoses in immunosuppressed patients.
- Miconazole oral gel is used for the treatment and prophylaxis of oral and intestinal fungal infections. It should be held in the mouth near any localized lesions, after food.
- Clotrimazole or miconazole cream or pessaries are used to treat vaginal candidiasis. They can damage latex condoms and diaphragms.
- Clotrimazole or miconazole are used topically for many fungal skin infections, e.g. dermatophyte infections, pityriasis versicolour, and candidiasis.
- Clotrimazole solution is used for fungal otitis externa infections.

Pharmacology

- Ketoconazole is better absorbed by mouth than the other imidazoles, but levels are highly variable. It is only available orally, and co-administration of antacids should be avoided. Ketoconazole shows poor CSF penetration.
- Miconazole is available as an oral gel for mouth infections, but systemic absorption may result in significant drug interactions. Clotrimazole is used topically.

Toxicity and side-effects

Ketoconazole has been associated with fatal hepatotoxicity, so should not be used for superficial fungal infections. Always consult an expert before prescribing ketoconazole, and monitor liver function. Ketoconazole and miconazole are contraindicated in hepatic impairment, pregnancy, and breast feeding. Side-effects of these two drugs include GI (nausea, vomiting, abdominal pain), rashes, and headache. With topical preparations, avoid contact with eyes and mucous membranes, and discontinue if severe local irritation or hypersensitivity reactions occur.

Triazoles

The triazoles include fluconazole, itraconazole, voriconazole, posaconazole, and ravuconazole (currently in phase II trials).

Mode of action

Triazoles inhibit the synthesis of ergosterol (the main sterol in fungal cell membranes) by inhibiting cytochrome P450 14-α-demethylase (P45014DM). They are all essentially fungistatic, but some may be fungicidal at high concentrations.

Fluconazole (Diflucan®)

- Active against *Candida* spp. and *Cryptococcus* spp. Fluconozde is also active against *Coccidioides immitis* and has limited activity against *Histoplasma capsulatum*, *Blastomyces dermatitidis*, and *Sporothrix schenkii*. No meaningful activity against *Aspergillus* spp. or most other molds
- Resistance – *C. krusei* is intrinsically resistant to fluconazole *C. glabrata* has high fluconazole MICs; 10–15% are resistant. Acquired resistance to fluconazole has been reported in *C. albicans* in HIV patients
- Pharmacology – available as PO and IV preparations. Well absorbed orally. Renal excretion; dose reduction required in renal impairment
- Clinical use – treatment of oropharyngeal, vulvovaginal, and invasive candidiasis. Also used for prophylaxis in transplant patients
- Side-effects and toxicity – generally well tolerated. May cause abnormal LFTs

Itraconazole (Sporanox®)

- Active against yeasts, moulds, and dimorphic fungi, e.g. *Candida* spp., *Cryptococcus* spp., *Aspergillus* spp., *Pseudallescheria boydii, Sporothrix schenkii, Histoplasma capsulatum, Blastomyces dermatitidis, Coccidiodes immitis, Paracoccidioides braziliensis, Pencillium marneffei*, and dermatophytes. Active against fluconazole-resistant *C. krusei* and *C. glabrata*. Limited activity against *Fusarium* spp. and Zygomycetes.
- Resistance – detectable in most of the species above as well as others.
- Pharmacology – available as oral capsule, oral suspension (in cyclodextrin), or IV formulation (in hydroxyl-propyl-beta-cyclodextrin). Absorption of capsules in highly variable and unpredictable. Absorption and bioavailability of suspension is better. Gastric acidity and food affect absorption. Highly lipophilic; achieves high concentrations in fatty tissues and purulent exudates.
- Clinical use – used in the treatment of yeast and mold infections especially fluconazole resistant *Candida* spp. and *Aspergillus* spp.
- Side-effects and toxicity – side-effects are rare and similar to fluconazole. Hypertension, hypokalaemia, oedema, headache, and altered mental state have been reported. Seek advice before giving itraconazole to patients at risk of heart failure (e.g. elderly, those with cardiac disease or on negative inotropes, e.g. calcium channel blockers, or those on long courses/high doses of itraconazole).

Voriconazole (Vfend®)

- Structurally similar to fluconazole. Inhibits P45014DM to a greater extent than fluconazole. Licensed in 2002
- Active against a wide variety of fungi including *Candida* spp. (fungistatic), *Cryptococcus* spp. (fungistatic), *Aspergillus* spp. (fungicidal), *Blastomyces dermatitidis, Coccidioides immitis, Histoplasma capsulatum, Fusarium* spp., and *Penicillium marneffei.* Active against fluconazole-resistant *C. krusei, C. glabrata* and *C. guilliermondii.* Zygomycetes, e.g. *Mucor* spp. and *Rhizomucor* spp., have high MICs
- Pharmacology – available as PO or IV formulations
- Clinical use – licensed for treatment of invasive aspergillosis and invasive candidiasis. Also used in salvage therapy of *Scedosporium apiospermum* and *Fusarium* spp.
- Side-effects and toxicity – dose-related, transient visual disturbance, skin rash, abnormal LFTs

Posaconazole (Noxafil®)

- Excellent activity against *Candida* spp., including those that have reduced fluconazole susceptibility. Also active against *Aspergillus* spp., *Sporothrix schenkii, Histoplasma capsulatum, Blastomyces dermatitidis, Coccidiodes immitis, Paracoccidioides braziliensis, Pencillium marneffei*, and agents causing chromoblastomycosis, mycetoma, and phaeohyphomycosis, including *Scedosporium apiospermum* and species of *Exophiala, Alternaria,* and *Bipolaris.* It is the only azole with consistent activity against Zygomycetes
- Pharmacology – oral suspension available. IV formulation under development
- Clinical use – used in prevention of invasive antifungal infections in high-risk patients with neutropenia or graft versus host disease. Also used as salvage therapy in patients with invasive fungal infections that failed primary therapy (usually amphotericin B). Preliminary data suggest that posaconazole may be effective for zygomycosis unresponsive to amphotericin B
- Toxicity and side-effects – nausea, headache, rash, dry skin, taste disturbance, abdominal pain, dizziness, and flushing may occur

Ravuconazole

- Currently in phase II trials
- Structurally similar to fluconazole and voriconazole
- Active against *Candida* spp., *Cryptococcus neoformans, Aspergillus fumigatus*, and dermatophytes. Likely to be active against fluconazole-resistant *Candida* spp. Limited activity against *Sporothrix schenkii, Pseudallescheria boydii, Fusarium* spp. and Zygomycetes
- Pharmacology – oral formulation has been described
- Clinical use – under development
- Toxicity and side-effects – Ravuconazole is in phase II of clinical trials. No side-effects noted in rats and dogs treated with ravuconazole for 1 week

Echinocandins

This relatively new class of drugs inhibit glucan synthesis and are rapidly fungicidal for yeasts; their activity against moulds is more complex. Three drugs are currently licensed: caspofungin, micafungin, and anidulafungin.

Mode of action

Blocks the synthesis of 1–3-β-glucan (a fungal cell wall component) by inhibiting the enzyme 1–3-β-glucan synthase. This selective action results in fewer side-effects (as there is no mammalian target) and lack of cross-resistance with other antifungals.

Caspofungin (Cancidas®)

- Licensed in 2001
- Active against *Candida* spp. and *Aspergillus* spp. No activity against *Cryptococcus neoformans, Fusarium* spp, *Pseudallescheria* spp., or the Zygomycetes. Acquired resistance not yet described
- Pharmacology – available as an IV formulation. Protein binding is >90%. Widely distributed with high levels in the lungs, liver, spleen, and kidneys, and lower levels in CSF. Metabolized by the liver (not by cytochrome P450). Reduce dose in moderate hepatic impairment. Metabolites eliminated in the urine and faeces. No dose adjustment is required in renal impairment
- Clinical use – invasive *Aspergillus* infections unresponsive to amphotericin or itraconazole, or in patients who cannot tolerate amphotericin or itraconazole. It is also used for the treatment of invasive candidiasis and as empirical therapy in neutropaenic patients. Caspofungin is active against *Pneumocystis jiroveciii*, but not licensed for treatment
- Toxicity and side-effects – nausea, vomiting, abdominal pain, diarrhoea, flushing, fever, headache, and injection-site reactions. Transient LFT abnormalities occur in 11–24% of patients

Anidulafungin (Ecalta®)

- Active against *Candida* spp. Not active against *C. neoformans*. Limited activity against *Blastomyces dermatitidis, Histoplasma capsulatum, Sporothrix schenkii, Trichosporon beigelii, Acremonium strictum, Rhizopus arrhizus, Fusarium* spp., *Pseudallescheria boydii, Phialophora spp., Cladophialophora bantiana, Bipolaris* spp.
- Pharmacology – available as an IV formulation
- Clinical use – oesophageal candida, candidaemia
- Toxicity and side-effects – abnormal LFTs

Micafungin (Mycamine®)

- Active against *Candida* spp. and *Aspergillus* spp.
- Available as an IV formulation
- Clinical use – oesophageal candidiasis, candidaemia, invasive candidiasis, prevention of *Candida* infections post bone marrow transplantation
- Toxicity and side-effects – fever, electrolyte disturbances, abnormal LFTs

Other antifungals

Flucytosine (Ancotil®)

A fluorine analogue of cytosine (pyrimidine) that inhibits DNA synthesis.

- Mode of action – converted to metabolites which either inhibit thymidylate synthetase (interfering with DNA synthesis) or cause aberrant transcription of RNA
- Resistance – may be due to loss of cytosine permease (that permits entry of the drug) or loss of enzymes that convert it into its active metabolites. Resistance emerges rapidly with monotherapy
- Clinical use – cryptococcosis, candidiasis, chromoblastomycosis, usually used as combination therapy, e.g. with amphotericin.
- Pharmacology – given IV. Rapidly and almost completely absorbed. Low protein binding. CSF concentrations are 74% of plasma concentrations; 90% is excreted unchanged in the urine – dose reduction required in renal impairment
- Toxicity and side-effects – rash, diarrhea, and abnormal LFTs. Leucopaenia, thrombocytopaenia, and enterocolitis may occur in patients with renal impairment – monitor FBC, renal, LFTs and serum flucytosine concentrations weekly during treatment. Flucytosine is teratogenic in rats and contraindicated in pregnancy

Griseofulvin

Griseofulvin was previously used for fungal nail infections, but poor response rates and significant relapse rates have limited its use. Used in children for tinea capitis.

- Mode of action – disruption of fungal cellular microtubules
- Clinical use – tinea capitis in children. Dermatophyte infections of skin, scalp, nails, and hair where topical therapy has failed or is inappropriate
- Pharmacokinetics – given PO
- Toxicity and side-effects – impaired performance of skilled tasks, enhancement of effects of alcohol, headache, nausea, vomiting, and rashes. May diminish the anticoagulant effect of warfarin. Avoid in pregnancy, breast -feeding, systemic lupus erythematosus (risk of exacerbation), and liver disease

Terbinafine (Lamisil®)

- Mode of action – acts on the enzyme squalene epoxidase, blocking the transformation of squalene to lanosterol and thus inhibiting ergosterol synthesis. The intracellular accumulation of squalene also results in disruption of fungal cell membranes
- Resistance – not yet reported
- Clinical use – dermatophyte and ringworm infections (including tinea pedis, cruris, and corporis) where oral therapy is appropriate. Fingernail infections need a 6-week course, toenails usually 12 weeks. Also available topically to treat fungal skin infections
- Pharmacology – given PO or topically. Metabolized by cytochrome P450 enzymes
- Toxicity and side-effects – abdominal discomfort, anorexia, nausea, diarrhoea, headache, rash, and urticaria. Rare events include liver toxicity and serious skin reactions (e.g. Stevens–Johnson syndrome)

Aciclovir and valaciclovir

Aciclovir was the first agent with demonstrated effectiveness against herpes simplex virus (HSV) infection. Valaciclovir is the l-valyl ester prodrug of aciclovir and is preferred due to its better oral bioavailability. It is rapidly and nearly completely converted to aciclovir in first-pass enzymatic hydrolysis in the liver. Penciclovir and its prodrug, famciclovir, are similar to aciclovir.

Mode of action

- An acyclic nucleoside analogue, substrate for HSV-specific thymidine kinase (TK).
- Phophorylated in HSV-infected cells to form aciclovir-triphosphate, a competitive inhibitor of viral DNA polymerase.
- Incorporated into viral DNA chain. Causes chain termination. Chain–enzyme complex formation may irreversibly inactivate the polymerase.

Resistance

- Most commonly associated with reduced TK activity
- Cross-resistance to other agents activated by TK (e.g. ganciclovir)
- Prevalence – aciclovir-resistant HSV is <1% immunocompetent patients, 6–8% immunocompromised, and up to 17% in patients with AIDS and transplantation patients receiving over 2 weeks of aciclovir therapy
- Aciclovir-resistant varicella zoster virus (VZV) is less common – increased risk in chronic suppressive therapy with subtherapeutic doses

Pharmacology

- Oral bioavailability: aciclovir 15–20%, valaciclovir up to 54%
- Oral valaciclovir achieves similar total aciclovir exposure to that of IV aciclovir but with lower peak plasma concentrations
- 60–90% renally excreted, remainder metabolized. Removed by haemodialysis. Reduce dose in renal impairment
- Topical absorption low
- Crosses placenta, high levels in breast milk. Not known to be harmful in pregnancy but manufacturers advise caution

Interactions

- Zidovudine – lethargy
- Ciclosporin and other nephrotoxic agents – renal impairment
- Probenecid – decreased clearance and prolonged half-life
- May decrease the clearance of other drugs eliminated by active renal secretion (e.g. methotrexate)

Side-effects and toxicity

- Topical may cause skin irritation
- Nausea, rash, headache
- Neurotoxicity in 1–4% of those receiving IV aciclovir, increased in renal impairment
- Reversible renal impairment in up to 5% receiving IV aciclovir
- High-dose valaciclovir may cause GI disturbance, confusion, and hallucinations

Antivirals for HSV and VZV

Table 2.8 Antiviral therapy for HSV and VZV

Drug	Preparations	Use
Aciclovir	Oral	Treatment and prophylaxis for HSV, treatment of VZV
	Intravenous	Treatment of HSV in the immunocompromised, treatment of severe genital herpes, VZV, and HSV encephalitis
	Topical	Skin and eye infections (e.g. dendritic corneal ulcer)
Valaciclovir	Oral	Treatment of herpes zoster and HSV infections of skin and mucous membranes
Penciclovir (acyclic guanosine analogue)	Topical	Similar to aciclovir in potency and spectrum of activity. Cross-resistance common. Labial HSV treatment
Famciclovir (penciclovir prodrug)	Oral	Treatment of herpes zoster, acute and recurrent genital HSV
Cidofovir(see 📖 Antivirals for CMV, p.130)	Intravenous	Aciclovir-resistant HSV strains as activation is not dependent on virus-specified enzymes

Antivirals for cytomegalovirus

Ganciclovir, foscarnet, and cidofovir all act by inhibiting viral DNA polymerase and are effective in treating cytomegalovirus (CMV) end-organ disease. Valganciclovir is an ester of ganciclorir which may be given orally-it has an identical toxicity profile.

Ganciclovir and valganciclovir

Inhibitory activity against herpesviruses. Particularly potent inhibition of CMV replication; ~33% of patients receiving IV therapy interrupt or prematurely stop therapy due to marrow or CNS toxicity. Due to its side-effects it is restricted to life-threatening or sight-threatening CMV infections in immunocompromised patients, or to prevention of CMV disease during immunosuppressive therapy after organ transplantation. Valganciclovir is licensed for induction and maintenance of CMV retinitis in AIDS and prevention of transplant associated CMV disease.

- Mechanism of action – nucleoside analogue of guanosine – phosphorylated by a CMV viral protein kinase (not TK as in HSV). Inhibits viral DNA chain elongation but does not necessarily cause chain termination (unlike aciclovir), leading to accumulation of short non-infectious viral DNA fragments
- Resistance – due to mutations in the protein kinase or DNA polymerase. High-level resistance seen in prolonged therapy for those with AIDS, or transplantation-related disease
- Pharmacology – oral bioavailability 5–10%. Aqueous, vitreous, and sub-retinal levels similar to serum. Most eliminated unaltered renally. Removed by haemodialysis. Available as intravitreal implant. Usually administered IV
- Interactions – increases didanosine levels and may increase cyclosporin levels. Zidovudine and other cytotoxic agents – myelosuppression significantly increased, they should not normally be given together. Probenecid – decreased clearance and prolonged half-life. Renal dysfunction with concurrent ciclosporin or amphotericin B
- Adverse effects – myelosuppression, e.g. neutropenia and thrombocytopenia occur in 15–20% of AIDS patients receiving IV therapy (less in transplantation patients) Usually seen in 2nd week of therapy and reversible within 1 week of cessation in most cases. Recombinant granulocyte-colony-stimulating factor (G-CSF) may be useful; 5–15% have CNS effects – headache to confusion and convulsions. Others – renal impairment, LFT abnormalities, rash, fever, phlebitis at IV site
- Contraindications – pregnancy (ensure effective contraception up to 90 days after therapy), breast feeding, low haemoglobin, neutrophil counts, or platelets
- Monitoring – FBC and renal function during treatment

Foscarnet

Used in CMV retinitis in AIDS patients and aciclovir-resistant mucocutaneous HSV. Ninety per cent of retinitis patients experience clinical stabilization. In those with persistent or relapsed retinitis, combined foscarnet/ganciclovir delays progression longer than high doses of the individual agents. Foscarnet is used in AIDS-related CMV GI and pulmonary infections.

- Mode of action – inorganic pyrophosphate analogue inhibitory for herpesviruses and HIV. No intracellular metabolism. Directly inhibits viral DNA polymerase or reverse transcriptase (100-fold greater effect compared with cellular DNA polymerase). Active against most ganciclovir-resistant CMV strains (and aciclovir-resistant HSV/VZV)
- Resistance – due to point mutations in DNA polymerase of HSV/CMV or reverse transcriptase of HIV. Occurs in <5% of patients
- Pharmacology – oral bioavailability 8%. Vitreous concentrations 1.4 times higher than plasma. Renal elimination – most unaltered. Removed by haemodialysis. Prolonged terminal half-life due to bone deposition
- Interactions – hypocalcaemia with concomitant IV pentamidine. Renal dysfunction with concurrent ciclosporin, amphotericin B, and other nephrotoxic agents
- Side-effects and toxicity – nephrotoxicity, proteinuria, sometimes acute tubular necrosis (one-third develop significant renal impairment – reversible within 3–4 weeks of cessation of therapy). Saline loading may reduce incidence. Metabolic – hypo- and hypercalcaemia, hypo- and hyperphosphataemia, hypomagnesaemia, hypokalaemia. CNS – secondary to hypocalcaemia, and direct effects: seizures, hallucinations. Nausea, rash, abnormal LFTs, heart block
- Contraindications – pregnancy and breast feeding
- Monitoring – electrolytes, calcium, magnesium, phosphate, and renal function

Cidofovir

Used in CMV retinitis in AIDS patients, particularly where ganciclovir or foscarnet therapy not tolerated or failed. Topical gel used in mucocutaneous lesions. It has been used intravitreally for CMV retinitis but very toxic. Anecdotal (and conflicting) evidence for its use in the treatment of progressive multifocal leucoencephalopathy (caused by reactivation of JC virus).

- Mode of action – acyclic phosphonate nucleotide analogue of deoxycytidine monophosphate active against herpesviruses and other DNA viruses. Activation doesn't require virus-specific enzymes, therefore inhibitory for certain aciclovir- and ganciclovir-resistant HSV and CMV strains. Cellular enzymes metabolize it to the active form – competitively inhibits viral DNA polymerase
- Resistance – some ganciclovir and foscarnet cross-resistance. Development of resistance secondary to cidofovir therapy uncommon
- Pharmacology – oral bioavailability <5%. Renal elimination – most unaltered
- Interactions – nephrotoxic drugs
- Side-effects and toxicity – dose nephrotoxicity – concomitant probenecid and saline prehydration reduce incidence. Neutropenia 20%
- Dose-related topical reactions and intravitreal dosing can cause iritis, vitreitis, and visual loss
- Contraindications – renal impairment (creatinine clearance <55 mL/min). Other nephrotoxic agents. Pregnancy and breastfeeding. Men should not father a child during or within 3 months of finishing treatment

Antivirals for influenza

Vaccination is the most effective way of preventing influenza. Antiviral agents are not a substitute. In the UK the National Institute for Health and Clinical Excellence (NICE; www.nice.org.uk) issues evidence-based guidance on the use of these agents (see 📖 Influenza-treatment and prevention, p.423). Osetalmavir and zanamivir are licensed for the treatment and prophylaxis of influenza A and B in 'at-risk' groups. Amantadine and rimantadine are licensed but no longer recommended for the treatment and prophylaxis of influenza A.

Zanamavir

This is effective against both influenza A and B.

- Mode of action – selective inhibitor of the viral neuraminidase and is administered by inhalation. It is licensed for the treatment of influenza A and B in those over 12 years old if given within 48 h of symptom onset
- Resistance – some reported but the impact is not yet clear
- Pharmacology – poor oral bioavailability, thus given by inhalation. Levels far above those required for viral inhibition achieved at respiratory mucosa after inhaling
- Side-effects and toxicity – gastrointestinal disturbance. Rare – angioedema, rash, bronchospasm (use with care in those with asthma and COPD – bronchodilators should be available)
- Contraindications – breast feeding, caution in pregnancy, severe asthma

Oseltamavir

- A neuraminidase inhibitor effective against both influenza A and B.
- Resistance has developed rapidly in H1 strain, the vast majority of H1N1 now being resitant. H3 strains are mostly susceptible at the time of writing. Resistance described in human cases of H5N1 virus infection in Vietnam. *In vitro* reports of cross-resistance with zanamavir
- Pharmacology – prodrug which is hydrolysed in the liver to produce the active agent. Excreted in the urine
- Side-effects and toxicity – nausea, vomiting, abdominal pain, headache, fatigue, and insomnia. Rare – rashes, hypersensitivity, Stevens–Johnson syndrome
- Contraindications – use only if benefit outweighs risks in pregnancy and breast feeding. Avoid in severe renal impairment

Amantadine and rimantadine

Symmetric tricyclic amines that specifically inhibit replication of influenza A at low concentrations. Initiating therapy within 2 days of symptoms reduces the duration of illness by 1–2 days. Its use is limited by its side-effect profile, limited antiviral spectrum, and the rapid development of resistance.

- Mode of action – at low concentrations interact with the M2 protein of susceptible viruses preventing viral uncoating in endosomes. Higher concentrations increase lysosomal pH and inhibit virus-induced membrane fusion
- Resistance – just over 2% of circulating UK influenza A has a degree of resistance; 30% of drug-treated patients shed resistant virus by 5th day

of therapy. Failure of drug prophylaxis due to transmission of resistant virus from rimantadine-treated index case to household contacts has been seen (thus some advocate the use of different agents for treatment and prophylaxis within the same household)

- Pharmacology – well absorbed orally. Amantadine excreted unchanged by kidney (decrease dose in elderly and renally impaired). Rimantadine extensively metabolized by liver before renal excretion (reduce doses in those with severe liver or renal dysfunction)
- Interactions – increased CNS side-effects with concomitant antihistamines or anticholinergics
- Side-effects and toxicity – diarrhoea, nausea, difficulty concentrating.
- Neurotoxicity at high doses or in those with renal impairment e.g. tremor, seizures, coma. Anticholinergic symptoms with high doses of amantadine anticholinergic features, e.g. dry mouth, papillary dilation, toxic psychosis, and cardiac arrhythmias. Psychiatric symptoms in those with Parkinson's disease or schizophrenia have been associated with amantadine
- Contraindications – epilepsy, pregnancy and breast feeding, gastric ulceration, severe renal impairment. Use with caution with congestive cardiac failure. Not licensed for use in children <10 years

Antivirals for respiratory syncytial virus

Ribavirin

Broad-spectrum antiviral, used in the treatment of respiratory syncytial virus (RSV), hepatitis C, influenza, and Lassa fever. Studies assessing its role in the treatment of RSV infection in children have been small and although they show no reduction in hospital stay there is a beneficial effect on pulmonary function at 1 year. Currently used in the treatment of RSV bronchiolitis and pneumonia in hospitalized children – especially those with complicated or severe disease.

- Mode of action – guanosine analogue: interferes with nucleic acid synthesis and may also block production of viral mRNA
- Resistance – not demonstrated in RSV
- Pharmacology – excreted renally (around 40%), in faeces (15%), and metabolized by liver. Some is retained in tissues (especially red blood cells). Given by nebulizer in bronchiolitis, orally or IV in other conditions
- Side-effects and toxicity – dose-related anaemia (extravascular haemolysis) and marrow suppression at high doses. Itch, nausea, depression, and cough. With aerosolized preparations – conjunctivits, rash, bronchospasm
- Contraindications – pregnancy (teratogenic risk – contraception for 6 months after therapy in both men, and women and condoms must be used if partner of male patient is pregnant as ribavirin is present in semen) and breast feeding. Avoid oral therapy in: severe cardiac disease, haemoglobinopathies, severe hepatic dysfunction, autoimmune disease
- Monitoring – FBC and biochemistry on initiation of therapy and at weeks 2 and 4. ECG if history of cardiac disease

Other therapies for RSV

These are discussed in more detail later (see 🕮 Bronchiolitis, p.652). Some authorities advocate the use of ribavirin in combination with RSV immunoglobulin – particularly in the immunosuppressed.

Antivirals for chronic viral hepatitis

A number of drugs are licensed for the treatment of viral hepatitis. These are generally used for the treatment of chronic hepatitis B (HBV, see 📖 Hepatitis B virus, p.459) and hepatitis C (HCV, see 📖 Hepatitis C virus, p. 462) virus infections. They have rarely been used in fulminant acute viral hepatitis but this approach is experimental and referral to a transplant centre is recommended.

Indications for therapy

- Chronic HBV infection with ongoing viral replication for at least 6 months: HbsAg, HbeAg and HBV DNA-positive (wild-type infection) or HbsAg, anti-Hbe and HBV DNA-positive (anti-HBe variant)
- Chronic HCV infection with moderate to severe disease, defined as significant fibrosis and/or significant necrotic inflammation

Interferon-α

- Synthetic analogue of a natural compound with a number of antiviral and immunomodulatory effects
- Indications – treatment of chronic HBV and HCV infections
- Pharmacology – given by IM or s/c injection. The usual dose is 5–10 MU 3 times weekly for 3–6 months. Pegylated interferon-α has a longer half-life and can be given weekly
- Duration of treatment – 24 weeks for HCV genotype 2 or 3 infections; 12 weeks initially for genotypes 1, 4, 5, and 6; continued to 48 weeks if viral load suppressed to <1% of baseline level (2 log reduction)
- Side-effects – fever, chills, fatigue, myalgia, myelotoxicity (monitor FBC), impaired concentration, altered mood, exacerbation or development of autoimmune thyroid diseases, alopecia, arthralgia, hypersensitivity (rare), pulmonary infiltrates (rare)
- In HbeAg-positive chronic HBV the response rate to 4–6 months' therapy with interferon-α is 30%. In HbeAg-negative chronic HBV, the response rate to pegylated interferon-α is higher
- In chronic HCV infection the sustained virological response rate (HCV RNA undetectable 6 months after stopping treatment) is 38–43% for interferon-α + ribavirin and 54–56% for pegylated interferon + ribavirin

Ribavirin

A guanosine analogue with broad-spectrum antiviral activity.

- Indications – used in combination with interferon-α or pegylated interferon-α for treatment of chronic HCV infection
- Mechanism – inhibits inosine monophosphate dehydrogenase
- Resistance – not reported
- Pharmacology – rapidly absorbed orally, rapidly metabolized in liver; excreted in urine (50%), faeces (15%), or retained in tissues, principally in red blood cells
- Side-effects – haemolytic anaemia (monitor FBC), pruritis, rash, hypersensitivity (rare), teratogenicity (avoid conception)
- Outcome – combination therapy with interferon-α shows sustained response rates of 30–45%. Combination with pegylated interferon further increases sustained response rates to 54–56%. Response rates are lower with viral genotype 1 compared with viral genotypes 2 and 3.

Lamivudine

- Nucleoside analogue that is widely used for the treatment of HIV and HBV infection
- Indications – end-stage HBV disease pre-, peri- and post-transplant; patients who have failed interferon therapy; patients for whom interferon is contraindicated
- Mechanism – inhibits viral reverse transcriptase enzyme, inhibiting viral replication
- Resistance emerges with monotherapy, e.g. the YMDD variant. Resistance rates 50% at 2 years, 90% at 4 years
- Pharmacology – given orally, usual dose 100 mg once daily, well absorbed, renal excretion (reduce dose in renal impairment)
- Side-effects – minimal
- Outcome – response rate increases with time; 17% at 1 year, 27% at 2 years. One study showed 66% response rate at 4 years

Adefovir dipivoxil

- Prodrug of the nucleotide analogue adefovir, with activity against herpes viruses and HBV
- Lamivudine-resistant strains have been shown to be sensitive to adefovir
- Treatment with 10 mg daily decreases HBV DNA levels, improves hepatic histology scores, and induces loss of HbeAg
- Concerns about renal toxicity when used as HIV therapy

Entecavir

- Nucleoside analogue – inhibits HBV DNA polymerase
- Indications – treatment of chronic HBV, active against laminudine resistant strains.
- Pharmocology – given orally
- Side effects – GI symptoms, mused amylose, and CNS symptoms.

Tenofovir disoproxil fumarate

- A synthetic nucleotide analogue with activity against HIV and HBV
- Recent studies have shown benefit in HBV infection
- Pharmacology – 25% oral absorption, not metabolized, renal excretion
- Side-effects – GI symptoms, renal impairment, lactic acidosis, and hepatic steatosis (rare)

Immunomodulatory agents

A number of strategies have been tried in HBV but these have generally showed no benefit, e.g. corticosteroids ± interferon, thymic hormones, levimasole, and inosine pranobex. Trials are currently under way with lamivudine + corticosteroids.

Future therapies

Combination therapy with interferons and nucleotide analogues is likely to become standard for HBV. HCV has serine protease and helicase activity to which specific inhibitors may be developed in the future.

Principles of HIV treatment

Before 1988, the treatment of HIV and AIDS was primarily the treatment and prevention of opportunistic infections. Early studies of zidovudine monotherapy showed short-term benefits in delaying disease progression in symptomatic patients, but this finding was later refuted by the Concorde trial. In 1995, several studies proved that dual nucleoside analogue regimens were superior to monotherapy. In 1996, introduction of the protease inhibitors allowed the development of potent triple drug combinations and the era of highly active antiretroviral therapy (HAART) began. Although HAART has dramatically decreased HIV-related morbidity and mortality in the developed world, a number of problems remain: ongoing viral replication, emergence of drug resistance, high treatment failure rates, and concerns about the long-term metabolic complications of antiretroviral therapy.

Antiretroviral drugs

There are five main classes of antiretroviral drugs:

- nucleoside/nucleotide reverse transcriptase inhibitors = NRTIs (see 🕮 Nucleoside/nucleotide analogue reverse transcriptase inhibitors (NRTIs), p.140)
- protease inhibitors = PIs (see 🕮 Protease inhibitors, p.144)
- non-nucleoside reverse transcriptase inhibitors = NNRTIs (see 🕮 Non-nucleoside reverse transcriptase inhibitors (NNRTIs), p.142)
- entry inhibitors (see 🕮 Other HIV therapies, p.146)
- integrase inhibitors (see 🕮 Other HIV therapies, p.146).

Considerations when starting antiretroviral therapy (Table 2.9)

The aim of antiretroviral therapy is to prolong and improve quality of life by maintaining viral load suppression for as long as possible. Factors determining when to start and choice of therapy are:

- risk of disease progression (CD4 count, HIV viral load)
- willingness of patient to start therapy
- clinical effectiveness of combination regimen
- ability of patient to adhere to therapy
- comorbidity, e.g. tuberculosis, liver disease, cardiovascular disease, psychiatric conditions
- pill burden, dosing schedule, food considerations
- adverse effects
- drug–drug interactions
- pregnancy potential
- gender and CD4 count (if considering nevirapine)
- drug resistance potential/results of genotypic testing
- future therapeutic options.

Primary HIV infection

The rationale for starting treatment during or shortly after infection is to attempt to maintain specific and robust CD4 T-cell responses, which are generally lost in chronic HIV infection. This should be balanced against the risks of toxicity, development of drug resistance, and difficulties of long-term adherence. This remains an area of active research.

Established HIV infection

Any patient who is symptomatic or has a CD4 count <350 cells/mm^3 should be offered therapy. The data supporting this recommendation are strongest for patients with a CD4 count <200 cells/mm^{3}or an AIDS defining condition.

Special circumstances

Antiretroviral therapy should also be started in the following patients, regardless of CD4 cell count:

- pregnant women
- HIV-associated nephropathy
- HBV infection requiring treatment.

Initial regimen in treatment naive patients

Any HAART regimen should be individualized to achieve the best potency, adherence and tolerability and to minimize drug interactions and toxicity. Although baseline drug resistance testing is recommended prior to starting antiretroviral therapy, this is unavailable in developing countries

Recommended initial combinations are:

- NNRTI based – NNRTI plus two NRTIs
- PI based – 1 to 2 PIs plus two NRTIs.

Regimens that are not recommended are:

- monotherapy (rapid development of resistance, inferior efficacy)
- dual therapy (inferior efficacy)
- triple NRTI combinations (inferior efficacy).

Table 2.9 Recommendations for starting antiretroviral therapy in adults

Disease stage	UK guidelines[a]	US guidelines[b]
Primary HIV infection	Only in clinical trial	Consider treatment
Established infection, CD4 < 200 cells/mm^3	Treat	Treat
Established infection, CD4 201–350 cells/mm^3	Treat	Treat
Established infection, CD4 > 350 cells/mm^3	Defer	May be considered
Symptomatic disease or AIDS	Treat	Treat
Special situations		Pregnant women, HIV nephropathy, HBV infection requiring treatment

CD4 = CD4 T-cell count, VL = HIV RNA level.

[a]British HIV Association guidelines (2008) http://www.bhiva.org;

[b]Guidelines for the use of antiretroviral agents in HIV-1-infected adults and adolescents. Department of Health and Human Services. December 1, 2009; 1–161. http://www.aidsinfo.nih.gov/ContentFiles/AdultandAdolescentGL.pdf

Nucleoside/nucleotide analogue reverse transcriptase inhibitors

HIV reverse transcriptase is an RNA-dependent DNA polymerase that is essential for virus replication. It enables transcription of virus RNA into a DNA copy, which is then integrated into the host genome. This enzyme can be inhibited by:

- nucleoside or nucleotide reverse transcriptase inhibitors (NRTIs)
- non-nucleoside reverse transcriptase inhibitors (NNRTIs, (see 📖 Non-nucleoside reverse transcriptase inhibitors, p.142).

NRTIs (Table 2.10) are commonly used as the backbone of therapy for chronic HIV infection in combination with protease inhibitors (PIs, see 📖 Protease inhibitors, p.144) or NNRTIs. For treatment guidelines see http://www.bhiva.org and http://AIDSinfo.nih.gov

Specific drugs

- Abacavir (ABC, Ziagen®)
- Didanosine (ddI, Videx®, Videx EC®)
- Emtricitabine (FTC, Emtriva®)
- Lamivudine (3TC, Epivir®)
- Stavudine (d4T, Zerit®)
- Tenofovir disoproxil fumarate (TDF, Viread®)
- Zidovudine (AZT, ZDV, Retrovir®)

Structure

- Nucleoside or nucleotide analogues

Mechanism of action

- Inhibit HIV reverse transcriptase enzyme
- Also inhibit mitochondrial DNA polymerases, resulting in toxicity

Resistance

- Caused by one or more mutations in reverse transcriptase gene
- Some specific mutations confer cross-resistance to other NRTIs

Pharmacology

- Variable oral absorption (25–93%), reduced by food
- Widely distributed
- Variable CSF penetration (12–50%)
- Cross the placenta, secreted in breast milk
- Some metabolized in liver
- Renal excretion – some need dose adjustment in renal impairment

Adverse effects

- All may cause gastrointestinal symptoms, lipodystrophy, neurological symptoms, and lactic acidosis with hepatic steatosis (rare but may be fatal).
- Some also cause myelosuppression (AZT), peripheral neuropathy (d4T, ddI, ddC), pancreatitis (d4T, ddI, ddC), and hypersensitivity (ABC).

Table 2.10 Characteristics of NRTIs

Drug	Pharmacology	Adverse effects
Abacavir	83% oral bioavailabilty; metabolized in liver, metabolites excreted in urine	Hypersensitivity reaction (can be fatal)
Didanosine	Take ½ h before or 2 h after meals; 30–40% oral bioavailability; renal excretion 50% – adjust dose in renal impairment	Pancreatitis, peripheral neuropathy. Coadministration with tenofovir increases didanosine levels
Emtricitabine	93% oral bioavailability; renal excretion – adjust dose in renal impairment	Minimal toxicity
Lamivudine	86% oral bioavailability; renal excretion – adjust dose in renal impairment	Minimal toxicity
Stavudine	86% oral bioavailability; renal excretion 50% – adjust dose in renal impairment	Peripheral neuropathy, pancreatitis, lipo-dystrophy, rapidly progressive ascending neuromuscular weakness (rare)
Tenofovir	Bioavailability: 25% if fasting, 39% with high-fat meal; renal excretion – adjust dose in renal impairment	Rare reports of renal insufficiency. Coadministration with didanosine increases didanosine levels
Zidovudine	60% oral bioavailability; metabolized to glucuronide, renal excretion of metabolite	Bone marrow suppression

Non-nucleoside reverse transcriptase inhibitors

This is the 3rd class of commonly used antiretrovirals. Based on clinical trial data, efavirenz-based regimens are considered superior to other regimens in terms of antiviral potency, durability, and safety. However, because of teratogenicity in primates, efavirenz is not recommended in pregnancy and women of childbearing potential. Nevirapine-containing regimens are also potent, but the data for antiviral activity compared to other regimens is less consistent. Nevirapine also has a higher incidence of serious adverse toxicities (hepatitis, Stevens–Johnson syndrome), making it more appropriate as an alternative to efavirenz in treatment-naive patients. Delaviridine is the least potent and is not recommended as part of an initial antiretroviral regimen. For treatment guidelines see http://www.bhiva.org and http://AIDSinfo.nih.gov

Specific drugs

Four NNRTIs (Table 2.11) are currently marketed for use:
- delaviridine (Rescriptor®)
- efavirenz (EFV, Sustiva® or Stocrin®)
- nevirapine (NVP, Viramune®)
- etravirine (TMC-125, INTELENCE™).

Mechanism of action

- Inhibit HIV reverse transcriptase

Resistance

- Associated with mutation in reverse transcriptase gene
- Single mutation confers resistance to all drugs in class

Pharmacology

- Well absorbed orally
- Metabolized by cytochrome P450 enzyme system
- Excretion of metabolites in urine
- Unchanged drug excreted in faeces
- Long half-life – this is a particular problem when used in pregnancy as monotherapy, as it results in development of resistance which may affect future treatment options. Also an issue when stopping drug regimens because of the mismatch in half-life of nevirapine compared with NRTIs and PIs

Adverse effects

- Rash common
- Raised transaminases
- CNS symptoms common with efavirenz (dizziness, insomnia, somnolence, abnormal dreams, confusion, impaired concentration, amnesia, agitation, hallucinations, euphoria). Take tablets at night. Symptoms usually resolve in 2–4 weeks

- Hepatitis (most common with nevirapine, especially in women with high CD4 counts
- Stevens–Johnson syndrome (rare, most common with nevirapine)
- Teratogenicity in monkeys (efavirenz, no data on other NNRTIs)

Table 2.11 Characteristics of NNRTIs

Drug	Pharmacology	Adverse effects
Delaviridine	85% oral bioavailability; CYP3A4 inhibitor. Excretion: 51% in urine, 44% in faeces	Rash, increased transminases, headaches
Efavirenz	Take on an empty stomach; oral bioavailability not known; mixed CYP3A4 inducer and inhibitor; excretion: 14–34% in urine, 16–61% in faeces	Rash, CNS symptoms (52%), increased transaminases, false-positive cannabinoid test, teratogenic in monkeys
Nevirapine	>90% oral bioavailability; CYP3A4 inducer; excretion: 80% in urine, 10% in faeces	Rash, hepatitis, Stevens–Johnson syndrome, toxic epidermal necrolysis, DRESS syndrome (drug rash with eosinophilia and systemic symptoms)
Etravirine	Take with food; oral bioavailability not known; 99% protein bound; metabolized by CYP3A4; eliminated in faeces and urine	Rash, GI symptoms, fatigue, peripheral neuropathy, headache, hypertension

Protease inhibitors

HIV protease inhibitors (PIs, Table 2.12) are a group of structurally related molecules that represent synthetic analogues of one of the HIV *gag-pol* cleavage sites. The introduction of PIs in 1996 heralded the era of HAART, enabling the combination of drugs with two different enzymatic sites of action in the virus life cycle. They are used in combination with NRTIs for treatment and post-exposure prophylaxis of HIV infection. For treatment guidelines see http://www.bhiva.org and http://AIDSinfo.nih.gov

Specific drugs

- Amprenavir (Agenerase®)
- Atazanavir (ATV, Reyataz™)
- Darunavir (DRV, Prezista™)
- Fosamprenavir (FPV, Telzir®, Lexiva™)
- Indinavir (Crixivan®) – no longer recommended
- Lopinavir + ritonavir (LPV/r, Kaletra®)
- Nelfinavir (NFV, Viracept®) – no longer recommended
- Ritonavir (RTV, Norvir®)
- Saquinavir (SQV) hard gel capsule (Invirase®)
- Tipranavir (TPV, Aptivus™)

Mechanisms of action and resistance

- Inhibit HIV protease, so inhibiting cleavage of viral polypeptides to form structural proteins and viral enzymes.
- Mutations in HIV protease gene – many mutations confer cross-resistance to other PIs.

Pharmacology

- Oral absorption variable (13–78%), not known for some drugs
- Metabolized by cytochrome P450 enzyme system
- Distribution unknown for some drugs, CNS penetration poor
- Eliminated primarily in faeces
- Low-dose ritonavir is used in combination with other PIs to decrease metabolism and increase plasma levels of the other PI. This is known as 'ritonavir-boosted therapy'

Side-effects and toxicity

- Gastrointestinal symptoms (nausea, vomiting, diarrhoea)
- Neurological symptoms (headache, paraesthesias)
- Skin rash
- Fat redistribution
- Abnormal liver function tests, hyperglycaemia, hyperlipidaemia

Interactions

Drugs that should *not* be administered with PIs include:

- antiarrhythmics – amiodarone, IV lidocaine, quinidine
- antihistamines – astemizole, terfenadine
- ergot derivatives – ergotamine, dihydroergotamine
- motility agents – cisapride
- neuroleptics – pimozide

- sedatives – midazolam, triazolam
- antimycobacterials – rifampicin
- statins – simvastatin, lovastatin
- phosphodiesterase inhibitors – sildenafil
- herbal products – St John's wort

Table 2.12 Characteristics of PIs

Drug	Pharmacology	Adverse effects
Amprenavir	Avoid high-fat meal; oral bioavailability not determined; CYP3A4 inhibitor	GI symptoms, rash, oral paraesthesia, raised transaminases, hyperglycaemia, hyperlipidaemia, fat redistribution
Atazanavir	Take with food; oral bioavailability not determined; CYP3A4 inhibitor and substrate	Hyperbilirubinaemia, prolonged PR interval, hyperglycaemia, fat redistribution
Darunavir	Adminster with food; oral bioavailability 37%; CYP3A4 inhibitor and substrate	Skin rash, diarrhoea, nausea, hyperlipidaemia, transaminitis, hyperglycaemia, fat redistribution, possible increased bleeding in haemophiliacs
Fosamprenavir	No food restrictions; oral bioavailability not determined; CYP3A4 inhibitor, inducer and substrate	Skin rash, GI symptoms, headache, raised transaminases, hyperglycaemia, hyperlipidaemia, fat redistribution
Indinavir	Take 1 h before or 2 h after meals; 65% oral bioavailability; CYP3A4 inhibitor	Nephrolithiasis, GI symptoms, headache, asthenia, blurred vision, dizziness, rash, metallic taste, hyperglycaemia, hyperlipidaemia, fat redistribution
Lopinavir + ritonavir	Take with food; oral bioavailability not determined; CYP3A4 inhibitor; refrigerate.	GI symptoms, asthenia, raised transaminases, hyperglycaemia, hyperlipidaemia, fat redistribution
Nelfinavir	Take with food; oral bioavailability 20–80%; CYP3A4 inhibitor	Diarrhoea, hyperglycaemia, hyperlipidaemia, fat redistribution
Ritonavir	Take with food; oral bioavailability not determined; inhibits CYP3A4 > CYP2D6; refrigerate	GI symptoms, parasthesias, hepatitis, pancreatitis, asthenia, taste disturbance, hyperglycaemia, hyperlipidaemia, raised transaminases, uric acid, CK
Saquinavir	No food effect. Bbioavailability: 4% (hard gel). CYP3A4 inhibitor.	GI symptoms, headache, raised transaminases, hyperglycaemia, hyperlipidaemia, fat redistribution
Tipranavir	Take with food; CYP3A4 inhibitor and substrate; refrigerate capsules	Hepatotoxicity, skin rash, hyperlipidaemia, hyperglycaemia, fat reditribution, intracranial haemorrhage (rare)

Other HIV therapies

These include:

- fusion entry inhibitors, e.g. T20
- chemokine receptor antagonists, e.g. maraviroc, vicriviroc
- integrase inhibitors, e.g. raltegravir
- immunotherapies, e.g. interleukin-2 (IL-2), therapeutic vaccines.

Enfuvirtide (T20, Fuzeon®)

- The first of a new class of drugs called the entry inhibitors
- Binds to the HIV surface protein gp41 so preventing binding of gp41 to CD4 cells
- Reserved for use in treatment-experienced patients
- Given twice daily by subcutaneous infection
- Side-effects include injection site reactions, increased risk of bacterial pneumonia, hypersensitivity reactions (<1%)

Maraviroc (MVC, Selzentry™, Celsentri®)

- A chemokine receptor antagonist that prevents the binding of HIV to the CCR5 receptor on CD4 T cells
- Used in highly treatment-experienced patients
- Requires a tropism assay prior to initiating treatment to determine if the virus is CCR5 (susceptible) or CXCR4 (not susceptible)
- Given orally (no food effect)
- Metabolized in the liver by CYP3A
- Side-effects – abdominal pain, cough, dizziness, musculoskeletal symptoms, pyrexia, rash, upper respiratory tract infections, hepatotoxicity, postural hypotension

Vicriviroc (SCH-D)

- A chemokine receptor antagonist that prevents the binding of HIV to the CCR5 receptor on CD4 T cells
- Investigational drug undergoing clinical trials – not yet Food and Drug Administration (FDA) approved
- Side-effects – headache, fatigue, nausea, dizziness, abdominal pain

Raltegravir (RAL, Isentress®)

- Novel agent that inhibits integration of viral genetic material into the human chromosome (strand transfer inhibitor)
- Used in highly treatment-experienced patients
- Given orally (no food restrictions)
- Metabolized by glucuronidation
- Side-effects – nausea, diarrhoea, pyrexia, CK elevation

Interleukin-2 (IL-2)

- Aldesleukin is a synthetic IL-2 derivative with antineoplastic and immunomodulating activities
- Previous studies have shown that IL-2 in combination with antiretroviral therapy significantly increases CD4 counts

- Studies are now underway to see if IL-2 in combination with approved and investigational HIV treatments can improve the immune system and delay disease progression

Therapeutic HIV vaccines

- Therapeutic HIV vaccines are designed to boost the host immune response to HIV in order to control the infection.
- There are currently no licensed HIV therapeutic vaccines although clinical trials are ongoing.

Antimalarials

Historical aspects

- Quinolines have been the mainstay of antimalarial therapy since the 17th century when Peruvian Indians used the bark of the cinchona tree. Despite its longevity, quinine remains an important drug in the treatment of malaria, although resistance to it is gradually increasing. Another quinoline, chloroquine, was widely used but has succumbed to global resistance in *P. falciparum* and increasing resistance in *P. vivax*. Newer quinolines include mefloquine, amodiaquine, and tafenoquine.
- The antimalarial activity of the biguanides (proguanil) was discovered during the 2nd world war. These, along with the diaminopyrimidines (pyrimethamine), are referred to as the antifolate drugs. They are used in combination with other drugs for prophylaxis (e.g. chloroquine + proguanil or atovaquone + proguanil) or treatment (e.g. sulfadoxine-pryrimethamine).
- Qinghao derived from *Artemisia annua* has been used by the Chinese to treat fever for over 2000 years. Artemisin and its derivatives (dihydroartemisin, artemether, arteether, and artesunate) have been found to have potent antimalarial activity and are widely used for the treatment of malaria in southeast Asia. In uncomplicated malaria these drugs are used in combination with other drugs, e.g. mefloquine-artesunate, arthemether-lumefantrine (Riamet®) and dihydroartemisinin-piperaquine (Artekin®).

Resistance mechanisms

Antimalarial drug resistance arises through the selection of rare naturally arising mutants with reduced drug susceptibility. Unlike bacteria, plasmodia do not have transferable resistance mechanisms but they are eukaryotes and can acquire or lose polygenic resistance mechanisms during meiosis. Resistance arises readily to drugs such as the antifolates or atovaquone because a single point mutation confers resistance and per-parasite mutation frequencies are relatively high.

- Resistance to antifolates is associated with progressive acquisition of mutations in the dihydrofolate reductase (DHFR) gene or dihydropterate synthetase (DHPS) gene respectively.
- The mechanism of resistance in quinoline antimalarials remains controversial but has been associated with mutations in *pfmdr1* (an efflux mechanism similar to that found in multi-drug-resistant mammalian tumour cells) and *pfCRT* (a food vacuolar membrane protein with transporter function).
- Atovaquone resistance arises from point mutations in the gene encoding cytochrome bc_1 mitochondrial complex.
- Although resistance to artemisinin and its derivatives occurs *in vitro*, clinical resistance is yet to be reported.

Rationale for combination therapy

The simplest reason to combine antimalarials is to increase efficacy. Drug combinations can also shorten duration of treatment, hence increasing compliance and decreasing the risk of resistant parasites arising by mutation during therapy. As with other infectious diseases such as TB and HIV, the emergence of resistance can be prevented by the use of drugs with different mechanisms of action. If two drugs with different mechanisms of action are combined, then the per-parasite probability of developing resistance is the product of their individual per-parasite probabilities. There are some limitations, however:

- if some patients receive only one component of the combination or there is already high resistance to one of the components, then resistance can arise to the other
- combinations are more expensive than single drugs (but the increased cost is outweighed by longer-term benefits)
- although artemisinin combinations appear to be the most promising, the best combinations have yet to be determined.

Guidelines for the prevention and treatment of malaria are given in Box 2.8.

Box 2.8 Guidelines for prevention and treatment of malaria

Prophylaxis

- UK guidelines: Chiodini P. et al. Guidelines for malaria prevention in travellers from the United Kingdom. London, Health Protection Agency, January 2007 http://www.hpa.org.uk/webw/HPAweb&HPAwebStandard/HPAweb_C/1203496943315?p=1249920576136
- US guidelines: National Center for Infectious Diseases Travelers' Health. *The Yellow Book – Health Information for International travel 2003–2004*. Atlanta: Centers for Disease Control. http://wwwn.cdc.gov/travel/ybToc.aspx
- WHO guidelines: World Health Organization. *International Travel and Health: vaccination requirements and health advice*. Geneva: WHO, 2004. www.who.int/ith

Treatment

- UK malaria treatment guidelines: Lalloo D et al. *Journal of Infection* 2007;54(2):111–21. See also section on treatment of malaria in *British National Formulary* www.bnf.org
- US guidelines: CDC Treatment of malaria (Guidelines for Clinicians) 2009. http://www.cdc.gov/malaria/diagnosis_treatment/treatment.html
- WHO guidelines *for the Treatment of Malaria* (2006). http://www.who.int/malaria/publications/atoz/9241546948/en/index.html

Quinine and related compounds

Quinine

- A quinolinemethanol derived from the bark of the cinchona tree. Inhibits eyrthrocytic stages of human malaria parasites but not all liver stages. Also active against gametocytes of *P. vivax*, *P. ovale*, and *P. malariae*
- Resistance is widespread in southeast Asia where some strains are also resistant to chloroquine, mefloquine, and sulfadoxine-pyrimethamine. Cross resistance with mefloquine in Central Africa
- Pharmacology – well absorbed orally; IM administration more predictable than IV administration. Extensive hepatic metabolism. Urinary clearance <20%
- Adverse effects – cinchonism (tinnitus, vomiting, diarrhoea, headache), hypogylcaemia (monitor blood glucose), hypotension, cardiac arrhythmias, haemolytic anaemia (blackwater fever)
- Use – treatment of falciparum malaria. Quinidine, an antiarrhythmic drug with greater toxicity is used more commonly in the United States

Chloroquine (Avloclor®, Nivaquine®)

- A synthetic 4-aminoquinoline, active against erythrocytic stages of all four human malaria species and gametocytes of *P. vivax*, *P. ovale*, and *P. malariae*
- Mode of action – inhibits parasite haem detoxification
- Resistance of *P. falciparum* to chloroquine is widespread and is due to increased drug efflux (mutation in *pfmdr1* gene) and/or decreased drug uptake (mutation in *pfCRT* gene)
- Pharmacology – 80–90% oral absorption. Widely distributed. Extensive tissue binding with high affinity for melanin-containing tissues. Extensive metabolism to active metabolite. 50% renal excretion
- Adverse effects – dizziness, headache, rashes, nausea, diarrhea, pruritis. Long-term treatment may cause CNS effects and retinopathy. Rarely, photosensitivity, tinnitus, and deafness may occur
- Use – prophylaxis and treatment of chloroquine-sensitive malaria

Mefloquine (Lariam®)

- A synthetic 4-quinolinemethanol, formulated as hydrochloride. Active against erythrocytic stages of *Plasmodium* spp. Effective against strains of *P. falciparum* that are resistant to chloroquine, sulfonamides and pyrimethamine. Also active against bacteria (e.g. MRSA) and some fungi
- Resistance is increasing – 15% high-grade resistance and 50% low-grade resistance in southeast Asia. Cross-resistance with quinine and halofantrine and inverse relationship with chloroquine resistance
- Pharmacology – well absorbed orally, concentrated in erythrocytes, metabolites not active, predominantly excreted in bile
- Adverse effects – nausea, dizziness, fatigue, confusion, sleep disturbance. Psychosis, encephalopathy and convulsions in 1/1200–1700 patients; 1/10,000 risk of serious toxicity in prophylaxis

- Use – prophylaxis in areas of chloroquine resistance. Treatment of uncomplicated multi-drug-resistant malaria

Amodiaquine

- A 4-aminoquinoline, active against *P. falciparum* and *P. vivax*
- Pharmacologically similar to chloroquine. Rapidly and extensively metabolized to desethylamodiaquine (active metabolite) before undergoing transformation to a quinone amine. Elimination half-life of 1–3 weeks
- Adverse effects appear to be related to immunogenic properties of quinone amine rather than direct toxicity of parent compound or metabolite. Severe/life-threatening adverse events have only been associated with prophylaxis, e.g. agranulocytosis, hepatotoxicity, aplastic anaemia
- Use – treatment of falciparum malaria

Piperaquine

- 4-aminoquinoline; structurally related to chloroquine, but active against chloroquine-resistant *P. falciparum*
- Pharmacology – well absorbed orally; elimination half-life 17 days
- Adverse effects – similar to chloroquine but pruritis uncommon
- Has been used since 1978 to treat malaria in China
- Combination with dihydroartemisinin has shown excellent tolerability and high cure rates of multi-drug-resistant falciparum malaria

Primaquine

- A synthetic 8-aminoquinoline, formulated as diphosphate. Active against hepatic stages of *Plasmodia* spp., including the hypnozoite stage of *P. vivax*. Poor activity against erythrocytic stages, but active against gametocytes. Also active against *Pneumocystis* spp., *Babesia* spp., *Leishmania* spp., *Trypanosoma cruzi*
- Resistance – failure rates of up to 35% reported in southeast Asia in patients treated for *P. vivax*
- Pharmacology – well absorbed orally, extensive tissue distribution, metabolized to carboxyprimaquine, methoxy and hydroxy metabolites, <4% excreted unchanged in urine
- Adverse effects are generally mild – abdominal cramps, anaemia, leucocytosis, methaemoglobinaemia. Haemolysis occurs in G6PD deficiency (check G6PD levels first)
- Use – treatment of *P. vivax* and *P. ovale*. 2nd line for treatment of *Pneumocystis jirovecii* (in combination with clindamycin)

Tafenoquine

- 8-aminoquinoline with activity against *P. falciparum* and hepatic stages of *P. vivax* and *P. ovale*
- Pharmacology – similar to primaquine but more potent, less toxic, and longer half-life (14 days), enabling weekly or monthly administration
- Adverse effects – methaemoglobinaemia, haemolysis in G6PD deficiency
- Use – prophylaxis and treatment of malaria

Artemisinin and its derivatives

Artemisinin (quinghao) is derived from *Artemisia annua* (sweet wormwood). Artemisinin and its derivatives act by inhibiting *P. falciparum*-encoded sarcoplasmic-endoplasmic reticulum calcium ATPase. Most clinically important artemisinins are metabolized to dihydroartemisinin, in which form they have comparable antimalarial activity. They are highly effective alternatives to quinine for the treatment of chloroquine-resistant falciparum malaria. In both uncomplicated and severe infections they have shown faster fever and parasite clearance times and have proved effective in cerebral malaria. They are widely used in the developing world to treat malaria. Their use in the developed world is hampered by a lack of availability of GMP (good manufacturing practice) products.

Artemisinin

- A sesquiterpene peroxide active against erythrocytic and gametocyte stages of *Plasmodium* spp. Also active against *Toxoplasma gondii, Leishmania major*, and *Schistosoma mansoni* in experimental models
- Resistance develops in experimental models but as yet there are no clinical reports of resistance
- Pharmacology – incomplete oral absorption. Concentrated in erythrocytes and hydrolysed to dihydroartemisinin. Metabolized by hepatic cytochromes. Peak concentrations after 1–3 h. Elimination half-life <30 min. Can be given rectally (PR)
- Adverse effects – drug-induced fever; reversible decrease in reticulocytes. In animal models, high doses of dihydroartemisinin associated with neurotoxicity. No clinical reports of neurotoxicity
- Use – treatment of malaria

Artemether and arthether

- Methyl and ethyl esters of dihydroartemisin.
- Oil-based preparations; can be given PO, PR, or IM
- Absorbed slowly and erratically– not suitable for severely ill patients

Artesunate

- Water-soluble hemisuccinate of dihydroartemisinin
- Can be given PO, PR, or IV
- Widely used in treatment of malaria in the developing world but not licensed in the UK
- Shown to be superior to quinine in efficacy (mortality 15% versus 24%) and tolerability for the treatment of severe malaria
- Unfortunately since 1988 counterfeit artesunate (containing little or no active drug) has been sold in southeast Asia. This not only threatens successful treatment of malaria but may also result in the emergence of drug resistance.

Dihydroartemisinin

- Active metabolite of artemisinin
- Can be given orally
- Used in treatment of uncomplicated malaria

Artemisin combination therapies

These are generally accepted as the best treatments for uncomplicated falciparum malaria. They are rapidly and reliably effective; their efficacy is determined by the drug partnering the artemisinin derivative and is usually >95%. The use of certain combinations is limited by resistance in the partner drug. They are safe and well tolerated, although hypersensitivity may occasionally occur. The adverse effect profiles are determined by the partner drug. They should not be used in the 1st trimester of pregnancy (safety not established) unless there is no alternative.

Artesunate – mefloquine

This combination is safe, well tolerated and highly effective and has been used in Thailand, South America, and Africa. Disadvantages include its price and the pharmacokinetic mismatch of its components. However, in Thailand resistance to mefloquine has actually decreased since its introduction.

Artesunate – sulfadoxine-pyrimethamine

Initially gave promising results in African children but subsequent studies have been disappointing.

Artesunate – amodiaquine

One African multicentre trial showed better overall efficacy than amodiaquine alone but 6% of patients developed neutropenia. The pharmacokinetic mismatch of the components raises concerns about prolonged exposure of parasites to amodiaquine. There is increasing resistance in eastern and southern Africa.

Artemether – lumefantrine (Riamet®, Coartem®)

First fixed dose combination of artemisinin derivative and another drug. It has become increasingly available in tropical countries and well tolerated and affordable ($1). It has been associated with irreversible hearing loss. Pharmacokinetic mismatch of components necessitates a complex 3-day dosing regimen.

Dihydroartemisin-piperaquine (Artekin®)

This combination that has good tolerability and has shown high cure rates in multi-drug-resistant falciparum malaria in Cambodia, Vietnam and Thailand. The slow elimination of piperaquine determines parasitological efficacy and post-treatment prophylactic effect.

Further reading

1 International Artemisinin Study Group. Artesunate combinations for treatment of malaria: meta-analysis. *Lancet* 2004;**363**:9–17.

2 Kremsner PG, S Krishna S. Antimalarial combinations. *Lancet* 2004;**364**:285–94.

3 Dondorp A, Nosten F, Stepniewska K et al. Artsunate versus quinine for the treatment of severe malaria. *Lancet* 2005;**366**:717–25.

4 Nosten F, White NJ. Artemisinin based combination therapies for falciparum malaria *American Journal of Tropical Medicine and Hygiene* 2007;**77**:181–92.

Miscellaneous antimalarials

Proguanil (Paludrine®)

- A synthetic arylbiguanide. Its metabolite cycloguanil inhibits the erythrocytic stages of all four *Plasmodium* species and the hepatic stage of *P. falciparum*. Acts synergistically with atovaquone
- Worldwide resistance of *P. falciparum* associated with point mutations in *DHFR* gene. Resistance of *P. vivax* and *P. malariae* reported in southeast Asia
- Pharmacology – >90% oral absorption; 75% protein bound; concentrated in erythrocytes; 20% metabolized by CYP450 enzyme system to cycloguanil (active metabolite). Non-metabolizers occur in Japan and Kenya, leading to resistance; 60% excreted in urine
- Adverse effects – GI and renal effects at high doses (>600 mg/day)
- Use – antimalarial prophylaxis (with chloroquine); treatment and prophylaxis of drug-resistant falciparum malaria (with atovaquone)

Pyrimethamine

- A synthetic diaminopyrimidine. Active against *Plasmodium* spp., *Toxoplasma gondii*, and *Pneumocystis jirovecii*
- Available as a single agent (Daraprim®) or in combination with sulfadoxine (Fansidar®), dapsone (Maloprim®), or mefloquine and sulfadoxine (Fansimef®)
- Resistance is due to point mutations on the *DHFR* gene and is widespread in Africa and Latin America. Sequence variations in *P. falciparum* and mutations in *pfmdr1* gene may also be important
- Pharmacology – well absorbed orally. Long plasma half-life (111 h). CSF levels 10–25% plasma levels. Secreted in breast milk and crosses the placenta. Hepatic metabolism
- Adverse effects – megaloblastic anaemia, leucopaenia, thrombocytopaenia, pancytopaenia. Very large doses in children have caused vomiting, convulsions, respiratory failure, and death. Aggravation of subclinical folate deficiency. Teratogenic in animals
- Use – treatment of malaria in combination with other drugs, treatment of toxoplasmosis and *Pneumocystis* pneumonia

Atovaquone-proguanil (Malarone®)

- Atovaquone is a hydroxynaphthoquinone that is more active than standard antimalarials against all stages of *P. falciparum*. It is also active against *Babesia* spp., *Toxoplasma gondii*, and *Pneumocystis jirovecii*.
- Resistance is due to point mutations in the parasite's cyctochrome bc_1 gene. Resistance emerges rapidly and it is therefore used in combination with proguanil, with which it appears to have synergistic activity.
- Pharmacology – poor oral absorption, improved when given with meals; 99% protein bound; poor CSF penetration (<1%). Not metabolized. Elimination half-life 73 h
- Adverse effects – fever, nausea, diarrhea, and rash
- Use – prophylaxis and treatment of malaria; treatment of *Pneumocystis* pneumonia (atovaquone alone)

Halofantrine

- A phenathrine methanol with inhibits erythrocytic stages of *Plasmodium* spp.
- Resistance reported in central and West Africa and Thailand
- Pharmacology – widely variable absorption, bioavailability increased by a fatty meal. Not concentrated in erythrocytes; 20–30% metabolized by CYP450 enzyme system. Little excreted in urine
- Adverse effects – abdominal pain, diarrhoea, pruritis. High doses are cardiotoxic with prologation of PR and QT intervals; mefloquine enhances these effects and sequential use is contraindicated
- Use – treatment of multi-drug-resistant falciparum malaria

Chlorproguanil-dapsone (LapDap)

- Dapsone (see Antileprotics, p.116) is a sulfa drug which has been combined with chloproguanil and developed as an affordable combination treatment for African children.
- Although well tolerated and efficacious, it is associated with a higher frequency of serious haematological toxicity than sulfadoxine-pyrimethamine.
- There are also concerns about the potential emergence of cross-resistance to this combination in areas where resistance to sulfadoxine-pyrimethamine is already high.

Tetracycline and doxycycline

- Tetracycline and doxycycline (see Tetracyclines, p.92) are both protein synthesis inhibitors. They are well absorbed orally with elimination half-lives of 3 h (tetracycline) and 20 h (doxycycline).
- Quinine + tetracycline has been used for treatment of multi-drug-resistant falciparum malaria in Thailand for years.
- Quinine + doxycycline is recommended for treatment of falciparum malaria in the UK. Doxycycline also used in prophylaxis of malaria (in regions with chloroquine or mefloquine resistance).
- Main limitations – contraindicated in children and pregnant women; photosensitivity with doxycycline; emergence of parasite resistance to tetracycline in areas where it has been extensively used.

Clindamycin

- Clindamycin (see Lincosamides, p.84) is a lincosamide antibiotic that also acts on the malaria parasite's apicoplast.
- Although numerous studies of quinine + clindamycin have shown good efficacy and safety profiles in various populations, it has never been widely used.

Fosmidomycin

- A new antimalarial which also acts on the malaria parasite's apicoplast.
- Fosmidomycin + clindamycin showed rapid parasite clearance and 100% cure rates at 28 days in one study of Gabonese children.

Antiprotozoal drugs (1)

Albendazole

- A benzimidazole derivative, active against a variety of parasites
- Use – giardiasis, microsporidiosis, intestinal worm infections, trichinosis, cutaneous larva migrans, hydatid disease, neurocysticercosis, lymphatic filiariasis

Amphotericin B (see 📖 Polyenes, p.120)

- A polyene antifungal active against a variety of fungi and some protozoa, e.g. *Leishmania* spp., *Naegleria*, and *Hartmanella*
- Use – treatment of leishmaniasis

Antimony compounds

- Examples – sodium stibogluconate, meglumine antimonite
- Active against amastigotes of *Leishmania* spp. within macrophages, with variation in sensitivity of different species
- Acquired resistance results in poor response and high resistance rates. Relapse also common in immunosuppressed or HIV patients
- Pharmacology – given IM or IV with peak concentrations occurring after 1 h. Slow accumulation in CNS and tissues. Excreted in urine.
- Adverse effects – cough/vomiting (if infused too quickly), arthralgia, myalgia, bradycardia, abdominal cramps, diarrhoea, rash, pruritis, ↑LFTs, ↑creatinine, ↑amylase.
- Use – treatment of leishmaniasis

Atovaquone (see 📖 Miscellaneous antimalarials p.154)

- Antimalarial with activity against other protozoa, e.g. *Babesia* spp., *Toxoplasma gondii*, and *Pneumocystis jirovecii*
- Use – treatment of *Pneumocystis jirovecii* pneumonia. Has also been used in cerebral toxoplasmosis and babesiosis but further studies required

Benznidazole

- A synthetic 2-nitroimidazole active against *Trypanosoma cruzi*
- Pharmacology – well absorbed orally
- Adverse effects – photosensitivity (50%), anorexia, nausea, vomiting, abdominal pain, disorientation, insomnia, paraesthesias, polyneuritis, seizures,
- Use – treatment of Chagas' disease

Ciprofloxacin (see 📖 Quinolones, p.98)

- A fluoroquinolone antibiotic, active against a wide variety of bacteria, legionella, mycoplasma, chlamydia, mycobacteria, *Cyclospora cayetanensis*, and *Isospora belli*
- Use – 2nd-line treatment of cyclospora and isospora

Clarithromycin (see 📖 Macrolides, p.82)

- A macrolide antibiotic, active against a variety of bacteria, *Legionella* sp., *Mycoplasma pneumoniae*, *Chlamydia* spp., atypical mycobacteria, and *Toxoplasma gondii*
- Use – 2nd-line treatment of toxoplasmosis

Clindamycin (see 📖 Lincosamides, p.84)

- A lincosamide antibiotic, active against Gram-positive and anaerobic bacteria and some protozoa, e.g. *P falciparum*, *Babesia* spp., *Toxoplasma gondii*, and *Pneumocystis jirovecii*
- Use – 2nd-line treatment of *Pneumocystis jirovecii* pneumonia; 2nd-line treatment and 2° prophylaxis of cerebral toxoplasmosis

Co-trimoxazole (see 📖 Co-trimoxazole, p.97)

- A diaminopyrimidine-sulfonamide antibiotic, active against a wide variety of bacteria, atypical mycobacteria, *Nocardia* spp., *Pneumocystis jirovecii, Toxoplasma gondii*, *Cyclospora cayetanensis*, and *Isospora belli*
- Use – treatment and prophylaxis of *Pneumocystis jirovecii* pneumonia and cerebral toxoplasmosis

Dapsone (see 📖 Antileprotics, p.116)

- Sulfonamide derivative, active against *Mycobacterium leprae, Plasmodium spp., Toxoplasma gondii, Pneumocystis jirovecii*
- Use – treatment and prophylaxis of *Pneumocystis jirovecii* pneumonia and cerebral toxoplasmosis

Diloxanide furoate

- Dichloromethylacetamide, active against *Entamoeba histolytica*
- Resistance – none reported
- Pharmacology – limited human data. Animal data show rapid oral absorption, hydrolysed in gut, 75% excreted via kidneys within 24 h
- Adverse effects – nausea, vomiting, abdominal distension, flatulence, pruritis, urticaria
- Use – treatment of intestinal amoebiasis, eradication of cysts after acute amoebiasis

Eflornithine

- α-difluoromethylornithine, active against *Trypanosoma brucei gambiense*, *P falciparum* (experimental models), *Leishmania* promastigotes and *Giardia lamblia* (*in vitro*)
- Pharmacology – given IV, plasma half-life 3 h, good CNS penetration, rapid renal excretion
- Adverse effects – osmotic diarrhoea, bone marrow suppression, convulsions
- Use – late-stage *T brucei gambiense* infections. Has also been used speculatively in *Pneumocystis* pneumonia

Fluconazole (see 📖 Triazoles, p.124)

- A triazole antifungal, active against a wide variety of fungi and *Leishmania* spp. Used to treat cutaenous leishmaniasis

Fumagillin

- Antibiotic derived from *Aspergillus fumigatus*. Given orally. Adverse effects – neutropenia, ↓platelets
- Use – 2nd-line treatment of microsporidiosis

Antiprotozoal drugs (2)

Furazolidone (see 📖 Nitrofurans, p.102)

- A nitrofuran antibiotic, active against a variety of bacteria, *Giardia lamblia* and *Trichomonas vaginalis*
- Adverse effects – disulfiram reaction with alcohol. Occasionally: fever, urticaria, hypotension, arthralgia, nausea, vomiting, headache, haemolysis (G6PD deficiency)
- Use – treatment of giardiasis

Iodoquinol

- An 8-aminoquinoline, active against *Entamoeba histolytica* and *Dietamoeba fragilis*
- Pharmacology – slowly and incompletely absorbed (<10%), hepatic metabolism and renal excretion
- Adverse effects – nausea, abdominal cramps, rash, acne. Conraindicated if allergic to iodine
- Use – asymptomatic or mild intestinal amoebiasis

Melarsoprol

- An arsenical compound, active against *Trypanosoma brucei* spp.
- Resistance – due to reduced uptake by trypanosomes
- Pharmacology – given IV, rapidly metabolized to melarsen oxide which crosses the blood–brain barrier, biphasic elimination
- Adverse effects – fever, abdominal pain, vomiting, peripheral neuropathy; 10% risk of post-treatment encephalopathy (may be reduced by perdnisolone), 2–4% risk of death secondary to treatment
- Use – late-stage African sleeping sickness

Mepacrine (quinacrine)

- A synthetic acridine derivative with broad antiprotozoal activity: *Blastocystis hominis*, *Entamoeba histolytica*, *Giardia lamblia*, *Leishmania* spp., *Plasmodium* spp., *Trichomonas vaginalis*, *Trypanosoma cruzi*. Also active against tapeworms
- Pharmacology – well absorbed orally, extensive tissue binding, concentrated in leucocytes, 10% daily dose excreted in urine
- Adverse effects – yellow staining of skin, dizziness, headache, vomiting, psychosis, haemolytic anaemia, ↓WCC, ↓platelets, urticaria, fever, rash
- Use – giardiasis, prophylaxis of malaria, treatment of tapeworm, cutaneous leishmaniasis

Metronidazole (see 📖 Nitroimidazoles, p.100)

- A nitroimidazole antibiotic, active against anaerobic bacteria and protozoa, e.g. *Trichomonas vaginalis*, *Giardia lamblia*, *Entamoeba histolytica*, *Balantidium coli*, *Blastocystis hominis*
- Use – anaerobic bacterial infections, bacterial vaginosis, giardiasis, intestinal and amoebiasis, amoebic liver abscess

Miltefosine

- Hexadecylphosphocholine, active against *Leishmania* spp., *Trypanosoma* spp., and *Entamoeba histolytica*
- Pharmacology – well absorbed and widely distributed
- Adverse effects – vomiting, diarrhoea, teratogenic (avoid in pregnancy)
- Use – treatment of leishmaniasis

Nifurtimox

- A nitrofuran antibiotic, active against a variety of bacteria and *Trypanosoma cruzi*
- Adverse effects – GI symptoms (40–70%), CNS symptoms (33%), skin rash, haemolysis (G6PD deficiency)
- Use – treatment of Chagas' disease

Nitazoxanide

- A broad-spectrum antiparasitic heterocycle, active against *Cryptosporidium parvum* and *Enterocytozoon bieneusi*
- Use – HIV-associated cryptosporidiosis (if HAART fails)

Ornidazole (see Nitroimidazoles, p.100)

- A nitroimidazole antibiotic, similar to metronidazole
- Use: Anaerobic bacterial infections, bacterial vaginosis, giardiasis, intestinal and amoebiasis, amoebic liver abscess

Paromomycin (see Aminoglycosides, p.78)

- Aminoglycoside antibiotic, similar to neomycin
- Use – intestinal amoebiasis (oral), cutaneous leishmaniasis (topical), nirtoimidazole-resistant trichomiasis (topical)

Pentamidine isetionate

- A synthetic diamidine, active against *Plasmodium falciparum*, *Toxoplasma gondii*, *Leishmania* spp., *Trypanosoma* spp., *Babesia* spp., and *Pneumocystis jirovecii*
- Resistance – relapse rates of 7–16% reported in the treatment of *Trypanosoma brucei gambiense*
- Pharmacology – negligible oral absorption, given IV or by nebulizer, hepatic metabolism, 15–20% excreted in urine. Poor CSF penetration (<1%). Retained in tissues, e.g. liver, kidneys, adrenals, spleen, and lungs resulting in long terminal half-life (>12 days)
- Adverse effects – phlebitis, injection site abscess, GI symptoms, hypotension, hypoglycaemia/hyperglycaemia, hypocalacaemia, neutropenia, ↓platelets, ↑creatinine, ↑LFTs, pancreatitis, rash
- Use – treatment of African trypanosomiasis (early stage), prophylaxis and treatment of *Pneumocystis* pneumonia, treatment of leishmaniasis (antimony-resistant)

Antiprotozoal drugs (3)

Primaquine (see 📖 Quinine and related compounds, p.150)

- A synthetic 8-aminoquinoline, active against hepatic stages of *Plasmodium vivax* and *ovale*, *Pneumocystis jirovecii*, *Babesia* spp., *Leishmania* spp., and *Trypansoma cruzi*
- Use – treatment of *P vivax* or *P ovale* malaria, treatment of pneumocystis pneumonia (with clindamycin)

Pyrimethamine (see 📖 Miscellaneous antimalarials, p.154)

- A synthetic diaminopyrimidine, active against *Plasmodium* spp., *Toxoplasma gondii*, *Pneumocystis jirovecii*
- Use – treatment of malaria (in combination with sulfadoxine or dapsone), toxoplasmosis (with sulfadiazine), and pneumocystis pneumonia (with dapsone)

Spiramycin (see 📖 Macrolides, p.82)

- A macrolide antibiotic, active against a variety of bacteria and *Toxoplasma gondii*
- Use – toxoplasmosis (especially in pregnancy)

Sulfadiazine (see 📖 Sulfonamides, p.94)

- A sulfonamide antibiotic, active against a variety of bacteria and *Toxoplasma gondii*
- Use – toxoplasmosis (in combination with pyrimethamine)

Suramin

- A sulfated naphthylamine, active against *Trypanosoma brucei* spp.
- Mode of action – binds to plasma proteins and taken up into trypanomes by endocytosis. Acts synergistically with nitroimidazoles and enflornithine
- Resistance – clinical relapse rates of 30–50% reported in East Africa
- Pharmacology – poor oral absorption, given by slow IV infusion, >99% protein bound, poor CSF penetration, plasma half-life >40 days, not metabolized, high tissue distribution (liver, kidney, adrenal glands), renal excretion
- Adverse effects – highly toxic especially in malnourished patients. Immediate reactions (nausea, vomiting, cardiovascular collapse) can be avoided by slow IV injection. May be followed by fever, urticaria. Anaphylaxis is rare (<1 in 2000). Delayed reactions include exfoliative dermatitis, anaemia, leucopaenia, jaundice, diarrhoea
- Use – African sleeping sickness (early stage), onchocerciasis

Tetracycline (see 📖 Tetracyclines, p.92)

- A tetracycline antibiotic, active against a variety of bacteria, chlamydia, rickettsia, spirochaetes, *Plasmodium falciparum, Balantidium coli, Dientamoeba fragilis*

- Use – treatment and prophylaxis of drug-resistant falciparum malaria, treatment of *Balantidium coli* and *Dientamoeba fragilis*

Tinidazole (see 🕮 Nitroimidazoles, p.100)

- A nitroimidazole antibiotic, similar activity to metronidazole
- Use – amoebiasis, giardiasis, trichomoniasis

Antihelminthic drugs (1)

Most antihelminthics were discovered and developed for use in veterinary medicine. Although no new antihelminthics have come to the market in recent years, satisfactory results can be achieved with current drugs. The exceptions to this are the treatment of larval cestodes, disseminated strongyloidiasis, and guinea worm.

Generic properties

- Mode of action – cause degenerative alterations in the tegument and intestinal cells of the worm by inhibiting its polymerization into microtubules. This results in inability to uptake glucose by the larva and adult stage.
- Side-effects – the main side-effects are GI and neurological symptoms, although allergic/anaphylactic reactions may rarely occur as a result of the death of large numbers of worms.
- Drug resistance – clinical drug resistance is rare.

Albendazole

- A benzimidazole carbamate, active against *Enterobius vermicularis, Ascaris lumbricoides, Ancylostoma duodenale, Necator americanus, Strongyloides stercoralis, Trichuris trichura, Trichinella spiralis*, animal hookworms, microfilaria, *Echinococcus* spp. Also active against *Giardia lamblia,* microsporidia
- Pharmacology – well absorbed orally, metabolized to albendazole sulfoxide (active metabolite), half-life 8 h, renal excretion. Topical albendazole is also available and can be used for some helminths (e.g. cutaeous larva migrans)
- Adverse effects – GI symptoms common, leucopaenia and ↑LFTs with prolonged use. Rarely, neutropenia, pancytopaenia, agranulocytosis, thormbocytopaenia. Avoid in pregnancy (teratogenic)
- Use – intestinal worm infections, trichinosis, cutaneous larva migrans, hydatid disease (±surgery), neurocysticercosis, lymphatic filiariasis (±ivermectin), giardiasis, microsporidiosis

Diethylcarbamazine

- A carbamyl derivative of piperazine, active against filarial worms: *Loa loa, Brugia malayi, Wuchereria bancrofti, Onchocerca volvulus*
- Pharmacology – >90% oral absorption, 50% metabolized and excreted by faeces, 50% excreted unchanged in urine
- Adverse effects – drug-induced death of microfilariae may trigger Mazotti reactions, e.g. skin rash, fever, headache, malaise, nausea, myalgia, arthralgia, vertigo, tachycardia, hypotension, cough, respiratory distress, ocular signs, neurological problems
- Use – filiariasis

Flubendazole

- A benzimidazole carbamate used in some countries instead of albendazole for treatment of ascariasis. Less well-absorbed orally

Ivermectin

- A mixture of two semisynthetic derivatives of avermectins, a complex of macrocyclic lactone antibiotics produced by *Streptomyces avertimilis*. Active against *Ascaris lumbricoides, Strongyloides stercoralis, Onchocerca volvulus, Loa loa*, and *Sarcoptes scabiei*
- Pharmacology – 60% oral absorption, rapidly metabolized by liver, highest concentrations in liver and fat, excreted in faeces
- Adverse effects – mazotti reactions, GI symptoms, neurological symptoms
- Use – onchocerciasis, non-disseminated strongyloidiasis, lymphatic filiariasis (with albendazole), scabies

Levamisole

- l-isomer of tetramisole, active against *Ascaris lumbricoides* and hookworms
- Mode of action – worms are paralysed and passed out in the faeces
- Pharmacology – 90% oral absorption, extensively metabolized, excreted in urine
- Adverse effects – GI symptoms, mild neurological symptoms
- Use – ascariasis, hookworm infection (rarely used)

Mebendazole

- A benzimidazole carbamate, active against *Enterobius vermicularis, Ascaris lumbricoides, Ancylostoma duodenale, Necator americanus, Strongyloides stercoralis, Trichuris trichura*
- Pharmacology – poor oral absorption, most of the drug and its metabolites (inactive) are retained in GI tract and excreted in faeces, <2% excreted by kidneys
- Adverse effects – GI symptoms. Avoid in pregnancy
- Use – intestinal worm infections, trichinosis

Metrifonate

- An organophosphate compound, active against *Schistosoma haematobium*
- Pharmacology – rapidly absorbed, undergoes chemical transformation to dichlorvos (active), rapidly and extensively metabolized, excreted in urine
- Adverse effects – GI symptoms, vertigo, ↓erythrocytes
- Use – urinary schistosomiasis (especially mass chemotherapy programmes)

Niclosamide

- A synthetic chlorinated nitrosalicylanide, active against *Taenia saginata, Taenia solium, Diphyllobothrium latum, Hymenolepis nana*
- Pharmacology – level of absorption uncertain, metabolized in liver, passed in urine and faeces (stains them yellow)
- Adverse effects – GI symptoms, dizziness
- Use – intestinal tapeworm infections

Antihelminthic drugs (2)

Oxaminiquine

- A synthetic quinolinemethanol, active against *Schistosoma mansoni*
- Resistance – some strains in Egypt and southerm Africa require higher doses
- Pharmacology – well absorbed orally. Can be given IM. Extensively metabolized to inactive metabolites that are excreted in the urine
- Adverse effects – dizziness, sleepiness, nausea and headache
- Use – treatment of *S. mansoni* infections

Piperazine

- A synthetic agent, active against *Enterobius vermicularis* and *Ascaris lumbricoides*
- Pharmacology – variable absorption and half-life, excreted in urine
- Adverse effects – transient mild GI or neurological symptoms; hypersensitivity
- Use – treatment of pinworm and ascariasis

Praziquantel

- A synthetic pyrazinoquinolone, active against schistosomes, larval tapeworms, *Fasciolopsis buski*, *Metagonimus yokogawi*, *Heterophyes heterophyes*, *Nanophytes salmincola*, *Clorchis* spp., *Opisthorcis* spp., *Paragonimus* spp., *Fasciola hepatica* (variable activity)
- Resistance is emerging in schistosomes
- Pharmacology – >80% oral absorption, undergoes rapid 1st-pass metabolism to inactive metabolites, low plasma levels, 90% excreted in urine by 24 h
- Adverse effects – GI symptoms, mild neurological effects during treatment of schistosmiasis; cerebral inflammation and oedema during treatment of neurocysticercosis; risk of visual impairment with ocular cysticercosis
- Use – schistosomiasis, tapeworm infections, trematode infections (except *Fasciola hepatica*)

Pyrantel

- A tetrahydropyrimidine, active against *Enterobius vermicularis*, *Ascaris lumbricoides*, *Ancylostoma duodenale*, *Necator americanus*
- Pharmacology – <5% absorbed orally, metabolized, and excreted in urine; the rest passes unchanged in the faeces
- Adverse effects – GI symptoms and neurological effects (rare). Antagonistic with piperazine (do not use together)
- Use – pinworm, ascariasis, hookworm (especially *Ancylostoma duodenale*), trichostrongyliasis

Tiabendazole

- A thiazoloyl benzimidazole, active against most common intestinal nematodes. Also effective against strongyloidiasis, trichinosis, cutaneous and visceral larva migrans
- Pharmacology – well absorbed orally, peak plasma levels after 1–2 h, metabolized to 5-hydroxyderivative (inactive), 90% excreted in urine within 24 h, remainder excreted in faeces
- Adverse effects – GI symptoms, fever, neurological symptoms
- Use – intestinal worms, strongyloidiasis, cutaneous larva migrans

Chapter 3

Infection control

Introduction to infection control

At any one time in the UK, it is estimated that 10% of hospital inpatients are suffering from a nosocomial infection.[1] The top five most common nosocomial infections are as follows:

- intravascular device related bacteraemia
- urinary tract infection
- lower respiratory tract infection
- surgical wound infection
- skin infection

The socioeconomic impact in terms of increased length of stay and financial cost is immense, as is the personal cost to each individual who acquires an essentially preventable infection. Hospital-acquired infections (HAIs) have triggered considerable political interest over recent years, resulting in re-organization of hospital and community infection control services and the publication of numerous documents. It is likely that infection control will maintain its high political profile.

References to relevant topics in other chapters

- 📖 Management of rash contact in pregnancy, p.828
- 📖 Bioterrorism, p.876
- 📖 Glycopeptide resistance in *S. aureus*, p.278

Box 3.1 Important definitions in infection control

Infection – the deposition and multiplication of organisms in tissues or on body surfaces, which usually causes adverse effects.
Colonization – organisms are present but cause no host response.
Carrier – organisms remain in the individual (e.g. tissues/body cavity) after a clinical infection, but cause no symptoms. An immunological response may remain. Typhoid Mary was one of the most famous carriers of all time (see 📖 Enteric fever, p.690).
Nosocomial infection – hospital-/healthcare-acquired infection that was not present or incubating at the time of admission. This includes infections that only appear after discharge, e.g. postoperative wound infection. It also includes occupational infections among healthcare staff.
Community infection – infection in the community. Be careful to differentiate between community-onset (may include patients recently discharged from hospitals), and community-acquired (community patients with no history of direct or indirect contact with healthcare) infections.
Decontamination – a process or treatment that renders a medical device, instrument, or environmental surface safe to handle.
Disinfection – the destruction of pathogenic and other microorganisms by physical or chemical means. Disinfection is less lethal than sterilization, because it destroys most recognized pathogenic microorganisms, but not necessarily all microbial forms, such as bacterial spores (see p.224).
Sterilization – a physical or chemical procedure that destroys all organisms, including large numbers of resistant bacterial spores (see p.227).

Useful infection control websites

- Health Protection Agency – www.hpa.org.uk
- Hospital Infection Society – www.his.org.uk
- Department of Health Healthcare Associated Infections – www.doh.gov.uk/hai
- World Health Organization – www.who.int/csr/resources
- CDC Hospital Infection Program – www.cdc.gov/ncidod/dhqp/index.html
- Infection Control Nurses Association – www.icna.co.uk
- Society for Healthcare Epidemiology of America – www.shea.org
- Medical Devices Agency – www.medical-devices.gov.uk
- EPIC Guidelines (Thames Valley University) – www.epic.tvu.ac.uk
- World Alliance for Patient Safety – www.who.int/patientsafety/worldalliance/en
- National electronic Library for Infection: this includes the National Resource for Infection Control – www.NeLi.org.uk
- European Centre for Disease Prevention and Control – http://ecdc.europa.eu/

Box 3.2 Infection control abbreviations used in the UK

- **ACOP** – Approved Code of Practice, e.g. for legionella
- **CCDC** – Consultant in Communicable Disease Control
- **CDC** – Centers for Disease Control (Atlanta, USA)
- **CDSC** – HPA Communicable Disease Surveillance Centre: now Cfi
- **Cfi** – HPA centre for infections at Colindale, has replaced CDSC
- **CLASSP** – Coordinated Local Authorities Sentinel Surveillance of Pathogens
- **CSSD** – Centre for Surgical Sterilization and Disinfection also called **TSU** (Theatre Supplies Unit)
- **DH** – Department of Health
- **DIPC** – Director of Infection Prevention and Control
- **ECDC** – European Centre for Disease Prevention and Control (Stockholm, Sweden)
- **HACCP** – Hazard Analysis Critical Control Points
- **HBN** – Health Building Note
- **HGM** – Health Guidance Memorandum
- **HICPAC** – Hospital Infection Control Practices Advisory Committee (U.S.A.)
- **HII** – High Impact Intervention: from 'Saving Lives' programme
- **HSE** – Health and Safety Executive
- **HTM** – health technical memorandum
- **ICC** – infection control committee
- **ICD** – infection control doctor
- **ICN** – infection control nurse
- **ICT** – infection control team
- **LACORS** – Local Authorities Co-ordinators of Regulatory Services
- **MDA** – Medical Devices Agency
- **RIDDOR** – Reporting of Injuries, Diseases and Dangerous Occurrences Regulations
- **RM** – regional microbiologist
- **RMN** – regional microbiology network

National agencies that support the NHS in reducing HAIs include:

- **HELICS** – Hospital in Europe Link for Infection Control through Surveillance
- **HCC** – Health Care Commission
- **HPA** – Health Protection Agency
- **NHS** Institute for Improvement and Innovation
- **NPSA** – National Patient Safety Agency

Table 3.1 Some of the significant UK documents relating to infection control (published since 2000)

Year	Organization	Title
2000	National Audit Office Report	*The Management and Control of HAI in acute NHS Trusts in England.* http://www.nao.org.uk/publications/9900/hospital_acquired_infection.aspx
2001	EPIC group, Thames Valley University	The EPIC project: developing national evidence-based guidelines for preventing HAI. Phase 1 guidelines for preventing HAI. *J Hosp Infect* 2001;47:S3–S82. www.epic.tvu.ac.uk
2002	National Patient Safety Agency	*Ready Steady Go! The full guide to implementing the Clean***your***hands campaign in your trust.* http://www.npsa.nhs.uk/cleanyourhands
2002	Department of Health (DH)	*Getting Ahead of the Curve – a strategy for combating infectious diseases.* http://www.dh.gov.uk/en/Consultations/Closedconsultations/DH_4016942
2003	Department of Health (DH)	*Winning Ways: working together to reduce HCAI in England.* www.dh.gov.uk/publications (ref: 34152)
2003	NICE	*Prevention of HAI in Primary and Community Care.* http://www.nice.org.uk
2004	National Audit Office	*Improving Patient Care by Reducing the Risk of HAI: a progress report.* http://www.nao.org.uk/publications/0304/improving_patient_care.aspx
2004	DH	*Towards Cleaner Hospitals and Lower Rates of Infection – a campaign for action.* http://www.dh.gov.uk/en/Publicationsandstatistics/Publications/PublicationsPolicyAndGuidance/DH_4085649
2005	DH	*Saving Lives: reducing infection, delivering clean and safe care.* http://www.dh.gov.uk/en/Publicationsandstatistics/Publications/PublicationsPolicyAndGuidance/DH_4008701
2006	DH	*Code of Practice for Prevention and Control of HCAI*: Under the 2006 Health Bill, trusts have a duty of care to comply with this code of practice, which will be overseen by the Health Care Commission. The code is divided into three areas: 1 Management, organization, and the environment; 2 Clinical care protocols; 3. Healthcare workers. www.dh.gov.uk/publications (ref: 277472)

Box 3.3 Current emphasis on infection control

Implementation of the Code of Practice for prevention and control of HAI (Health Act 2006) is a legal requirement for hospitals. This states that "effective prevention and control of HAI has to be embedded into everyday practice and applied consistently to everyone". **Saving Lives** provides the tools and resources for hospitals to achieve this.

- **High Impact Interventions 1–7** have been published by the DH, based on the care bundle approach. These are simple evidence-based tools, which reinforce the practical actions that clinical staff need to undertake to significantly reduce HAI. See www.clean-safe-care.nhs.uk. The seven High Impact Interventions focus on:
  1. central venous catheter care (see Box 3.29, p.242)
  2. peripheral intravenous canula care (see Box 3.31, p.254)
  3. renal dialysis catheter care
  4. prevention of surgical site infection (see Box 3.28, p.241)
  5. care for ventilated patients (or tracheostomies where appropriate) (see Box p.249)
  6. urinary catheter care (see Box 3.30, p.245)
  7. reducing the risk of *C. difficile* (see Box 3.32, p.259)
- **The revised Saving Lives** tools (June 2007) include:
  - screening for MRSA colonization (see p.218)
  - best practice for taking blood cultures (see p.243)
  - summary of best practice on antimicrobial prescribing (see Box 3.18, p.211)
  - isolating patients with HAI: summary of best practice (see Box 3.16, p.207)
- Other recent approaches include the following:
  - care bundles – these involve multiple discrete steps in the prevention of infection, and should be implemented and monitored by multidisciplinary teams. See www.ihi.org/IHIPrograms/Campaign/100kCampaign
  - benchmarking is inadequate: a culture of *zero-tolerance* is required
  - moving towards a culture of *accountability*, with adequate *administrative support*

Box 3.4 Clinical trials in infection control/hospital epidemiology

Historically, there have been few controlled trials in infection control, and the evidence base for many procedures is sparse. There have been several initiatives to rectify this in recent years, including formation of ORION (Outbreak Reports and Intervention Studies of Nosocomial Infection). They have published a CONSORT (Consolidated Standards of Reporting Trials) equivalent statement, in order to raise the standards of research and publication. It consists of a 22-item checklist, and a summary table. The emphasis is on transparency to improve the quality of reporting and on the use of appropriate statistical techniques, so that the work is robust enough to influence policy and practice.[2]

References

1 Plowman R, Graves N, Griffin MA et al. The rate and cost of hospital acquired infections occuring in patients admitted to selected specialties of a district general hospital in England and the national burden imposed. *JHI* 2001; **47**: 198–209.

2 Stane S, Cooper B, Kibbler C, et al. The ORION statement: guidelines for transparent reporting of outbreak reports and intervention studies of nosocomical infection. *Lancet Infect Diseases* 2007; **7**(4):282–88.

Basic epidemiology of infection (1)

In order to introduce effective infection control measures, the basic epidemiology of an infection (route of transmission, host risk factors etc) must be considered. The incidence and nature of a HAI (as with any infection) depend on:

- the organism
- the host (patients and staff)
- the environment.

The organism

The organisms responsible for common nosocomial infections are listed in Table 3.2. These may be acquired endogenously or exogenously.

Endogenous infection

Endogenous infection, also called auto-infection or self-infection, occurs when the infection is caused by organisms from the patient's own body. These may be part of their normal flora, or acquired in hospital. Antibiotics change the normal flora of the host, and may select for resistant organisms. Risk of infection may be reduced by protecting any potential sites of entry, e.g. intravascular lines.

Exogenous infection

Exogenous infection refers to an infection when the causal organism originates outside the patient. This is usually from the environment, and organisms are generally acquired by the airborne, contact, or percutaneous routes. Within the hospital setting, environmental infection is usually due to a contaminated item of equipment, and can be minimized by implementing the correct decontamination, sterilization, and infection control procedures. Cross-infection refers to infection acquired in hospital from another person, either patients or staff. Risks can be reduced by focusing on measures to interrupt transmission, e.g. handwashing.

The host

Patient risk factors that result in increased likelihood of acquiring an infection in hospital include:

- degree of underlying illness – more-severely ill patients are more vulnerable to acquiring an infection, and more likely to become sicker
- use of medical devices – these breach host defences and provide possible portals of entry for organisms
- extremes of age – the elderly and very young are at higher risk of infection
- immunosuppression.

Staff risk factors include:

- immunosuppression, e.g. HIV; pregnancy
- staff who perform exposure-prone procedures are more likely to be exposed to blood-borne viral infections
- skin conditions (e.g. eczema) increase prolonged carriage of organisms such as MRSA.

Table 3.2 Organisms commonly involved in hospital-acquired infections

Infection	Organism(s) involved
Urinary tract infections	Gram-negative bacteria, e.g. *E. coli*, *Proteus* spp., *Klebsiella* spp., *Serratia* spp., *P. aeruginosa*
	Gram-positive bacteria less common, e.g. *Enterococcus* spp.; Fungi less common, e.g. *C. albicans*
Respiratory infections	Bacteria, e.g. *H. influenzae*, *S. pneumoniae*, *S. aureus*, Enterobacteriaceae
	Viruses: respiratory viruses
	Fungi e.g. *Candida* spp., *Aspergillus* spp
Wounds and skin sepsis	Bacteria e.g. *S. aureus; S. pyogenes; E. coli; Proteus* spp.; anaerobes; *Enterococcus* spp.;
Bloodstream infection (BSI)	Gram-positive bacteria, e.g. *S. aureus*, meticillin-resistant S. aureus (MRSA), *Enterococcus* spp., coagulase-negative staphylococci
	Gram-negative bacteria, e.g. *E. coli*, *Proteus* spp., *Klebsiella* spp., *Serratia* spp., *P.aeruginosa*
	Fungi, e.g. Candida spp.
Gastrointestinal infections	Bacteria, e.g. *C. difficile*
	Viruses, e.g. norovirus

Basic epidemiology of infection (2)

The environment

The hospital environment includes all of the physical surroundings of the hospital patients and staff, i.e. the building, fittings, fixtures, furnishings, equipment, and supplies. The following are important environmental issues in the control of infection:

- environmental cleaning – see Box 3.5
- environmental disinfection
- decontamination of equipment
- building and refurbishment, including air-handling systems
- clinical waste management
- pest control
- food services/food hygiene
- isolation facilities/ability to cohort patients.

Box 3.5 Hospital cleaning

'Dirty hospitals' are frequently reported by the media, with attention drawn to the lack of investment and poor support for hospital cleaning. Although there is to known correlation between the cleanliness of wards and infection rates, resources should be directed towards education and training of medical and nursing staff, and employment of adequate cleaning services. The physical removal of dirt, fomites, dust and human body fluids is most important. New technologies, such as water-based cleaning using ultra-microfibres (UMF) and dry steam vapour look promising. The HACCP process seeks to identify hotspots in the environment that require more cleaning.

Routes of transmission

The isolation precautions required depend on the likely route of transmission of the organism. The main routes are:

- airborne – this is when the infection usually occurs by the respiratory route, with the agent being carried in aerosols (<5 micrometre diameter)
- droplet – large droplets carry the infectious agent (>5 micrometre diameter)
- direct contact – infection occurs through direct contact between the source of infection and the recipient, i.e. person-to-person spread
- indirect contact – infection occurs through 'indirect contact', i.e. via equipment contaminated with body fluids such as urine, faeces, and wound exudates. This route also includes contact via an environmental source, e.g. an outbreak of gastroenteritis transmitted by food
- inoculation – infection occurs through direct inoculation, e.g. needlestick injury. Other routes include via blood products (hepatitis A, *Yersinia enterocaritica, seratia*), TPN and other fluids (*Enterobacter*; *B. cepacia*). Multi-dose vials should be avoided.

Infection control committee

The National Audit Office report (July 2004) stated that all NHS trusts are responsible for clarifying and explaining accountabilities, including the role, membership, and responsibilities, of the hospital infection control committee (ICC). Thus the chief executive and trust board are responsible for ensuring effective arrangements for infection control within the trust.

Director of infection prevention and control (DIPC)

In 2003 it became mandatory for every trust to appoint a DIPC, to lead the ICC. This was the first time an infection-related post was created at board level, with direct reporting to the chief executive. Clear competencies of the DIPC have been defined.[1]

Responsibilities of the Hospital ICC

- Endorsing all infection control policies, procedures and guidelines
- Providing advice and support on the implementation of policies
- Collaborating with the ICT to develop the annual infection control programme and monitor its progress

Membership of the Hospital ICC may include:

- DIPC
- the ICT (see below)
- chief executive or representative
- occupational health physician and nurse
- senior clinical representatives
- nurse executive director or representative
- consultant in communicable disease control (CCDC).

The infection control team (ICT)

The ICT should be led by the DIPC and include the infection control doctor(s) and nurse(s). The ICT is mainly accountable to the trust chief executive and trust board, and the ICT responsibilities include the following:

- ensuring advice on infection control is available on a 24 h basis
- producing the annual infection control programme, in consultation with the ICC, health professionals, and senior managers. This programme will include surveillance of infection, and an audit of the implementation and compliances with selected policies
- providing education and training on the prevention and control of HAI to all grades of hospital staff.

Management of infection control in the community

The CCDC has a key role to play in collaborating with the ICT on the management of hospital and community outbreaks. They are responsible for advising health authorities and primary care organizations. They provide epidemiological advice and have overall responsibility for the surveillance, prevention, and control of communicable diseases and infections in the community (p.262).

References

1 Competencies for directors of infection prevention and control. Department of Health (2004). http://www.dh.gov.uk/en/publicationsandstatistics/lettersandcirculars/Dearcolleagueletters/DH_4083982.

Overview of surveillance

Definitions

Surveillance comes from the French 'to watch over' and means 'vigilant supervision' or 'ongoing scrutiny'. Langmuir, the founder of CDC, Atlanta described surveillance as 'the continued watchfulness over the distribution and trends of diseases through systematic collection, consolidation, and evaluation of data'. Often, in the real world, total accuracy of the data must be sacrificed in favour of the practicability, cost-effectiveness, uniformity, and timeliness in which the information can be collected. An important point to remember is there is no point just collecting interesting data, you need to do something with it – as emphasized by an alternative definition of surveillance: 'information for action'.

Surveillance may be active or passive. **Active** surveillance is when a special effort is made to collect the data, e.g. investigation of an outbreak or a survey of a particular disease. **Passive** surveillance uses routine data that have already been collected, e.g. the UK national census.

Aims of surveillance

The main aims of surveillance of hospital infection are to keep infection to a minimum level, and prevent and control outbreaks. Every member of medical and nursing staff and every department has a responsibility to collect relevant data on infection rates and to perform regular audits. The infection control team advises on this. In addition, routine surveillance from the laboratory can identify new antibiotic resistant or 'alert' organisms.

Methods of surveillance employed in HAI usually involve a combination of active and passive surveillance systems. Practice varies within and between hospitals and is largely dependent on availability of staff and resources. In practice, a number of surveillance methods are used, as each provides different information. These include:

- each department collects data on their patient risk factors and infection rates. This is particularly relevant in areas such as ICU and surgical wards.
- laboratory-based ward liaison, i.e. the ICN reviews patients with 'significant' microbiology cultures, and also visits each ward to discuss all patients with the ward staff/infection control link nurse.
- others include infection control link nurses, review of patients on antibiotics, with known risk factors or indwelling devices, and principles of surveillance (see Box 3.6).

Box 3.6 Principles of surveillance

- 'Surveillance is a circular process'

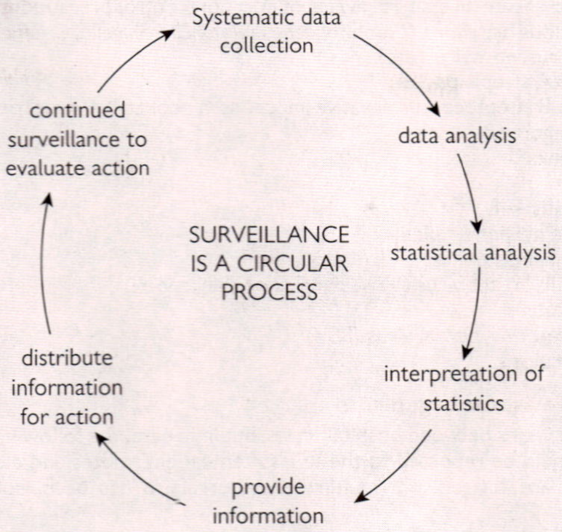

Evaluation of a surveillance system

- Define the event to be measured ('case definition')
- Define the population under surveillance
- List the objectives of system
- Define the public health importance of the health event (e.g. number of deaths, case fatality ratio, morbidity, economics, preventability)

Some characteristics of a surveillance system are listed in Box 3.7 (p.183).

Does surveillance work?

There is good evidence that surveillance in hospitals actually reduces infection rates. The SENIC study (the Study on Efficacy of Nosocomial Infection Control) was carried out over two decades ago in the USA.[1] One of the main objectives was to determine whether surveillance and infection control programmes lowered the rate of nosocomial infection. SENIC demonstrated that hospitals with active surveillance and infection-control programmes could reduce the incidence of infection by 32%. Another important finding of the SENIC study was the need to close the loop and present data back to the clinicians. Clinical audit was also found to be important.

References

1 Haley RW, Morgan WM, Culver DH. Study on Efficacy of Nosocomial Infection Control. *An J Infect Control* 1985 June; 13(3); 97–108

Surveillance of alert organisms

This may vary between institutions, but in most UK hospitals the ICT should be, informed of patients suffering from the following conditions, in order to ensure all staff are aware of infection control precautions:

- infectious diarrhoea (*Campylobacter, Salmonella, Shigella, C. difficile*, rotavirus, norovirus, etc)
- group A streptococcus
- group B streptococcus (invasive infections in neonatal or maternity units)
- meningococcal disease
- influenza
- HIV
- hepatitis A–E
- severe herpes simplex
- legionnaire's disease
- varicella zoster virus (VZV; shingles or chickenpox)
- lice/scabies
- respiratory syncytial virus (RSV)
- tuberculosis.

Multi-resistant organisms

Practice varies between hospital trusts, but in general the following organisms should be reported to the ICT. Often patient's notes and electronic records are 'flagged' if a multi-resistant organism has been isolated in the past:

- MRSA
- vancomycin-resistant enterococci (VRE)
- penicillin-resistant/non-susceptible pneumococcus
- multi-resistant Gram-negative organisms – the definition of this varies between hospitals. Originally aminoglycoside-resistant organisms were usually flagged; however, these have become so widespread that in many hospitals this would not be practical. Some hospitals flag isolates producing extended spectrum β-lactamases (ESBLs).

For national surveillance (the compulsory reporting of MRSA bacteraemia and *C. difficile rates*), see Table 3.3, p.184.

Box 3.7 What makes a good surveillance system?

For surveillance to succeed, it should have the following attributes:

- simple (well-designed reporting forms make all the difference!)
- flexible
- acceptable to the population studied
- sensitive
- representative
- timely
- reasonable cost

The quality of the data provided can be evaluated in terms of:

- Sensitivity and specificity
- Predictive value (positive and negative)
- Usefulness, in relation to the goals of the surveillance (quality indicators)

It is sometimes more valuable (although usually more difficult) to focus on the outcome (e.g. the number of cases of polio) rather than the process (e.g. the number of polio vaccinations).

Ownership is important – feeding the data back to those who collected it will make them more willing to help again in the future!

Surveillance of hospital-acquired infection

Definition

Hospital-acquired infection (HAI) is an infection that was neither present nor incubating at the time of hospital admission, which normally manifests itself more than 48h later. The term healthcare-associated infection (HCAI) encompasses any infection by any infectious agent acquired as a consequence of a person's treatment by the NHS or which is acquired by a healthcare worker in the course of their NHS duties (Health Act 2006).

Criteria for HAI surveillance schemes

In order to obtain accurate results that can be compared to other hospitals, the following criteria are important:

- agreed definitions of infection
- accurate denominator data
- correction of infection rates for risk factors, e.g. pre-existing diseases
- identification of HAI post-discharge.

Brief history of national surveillance schemes (Table 3.3)

The Nosocomial Infection National Surveillance Scheme (NINSS), launched in the UK in 1996, was set up to monitor the rate of HAI. Trusts enrolled for 3-month modules, and were able to compare their performance to others. This was replaced by new mandatory laboratory-based MRSA bacteraemia surveillance in 2001. The Health Protection Agency (HPA) is now responsible for taking forward surveillance under a service level agreement with the DH.

Table 3.3 National surveillance schemes in the UK

Surveillance activity	Date	Notes
S. aureus (including MRSA) bacteraemia	2001	6-monthy reports, now quarterly
Gycopeptide-resistant enterococcal (GRE) bacteraemia	2003	Annual reports initially, now quarterly
Clostridium difficile-associated diarrhoea (CDAD)	2004	Annual reports initially, now quarterly
Orthopaedic surgical site infection	2004	Annual reports
MESS: MRSA Enhanced Surveillance Scheme – additional details of cases of MRSA bacteraemias	2005	Enhanced surveillance data

International surveillance schemes

Most other countries have established surveillance programmes, although data are often not comparable between countries/continents because of variations in definitions and methods.

The nature of risk and its assessment

Risk management is something we do every day in our personal lives – such as when crossing the road – but has recently evolved as an important science in healthcare settings. In combination with total quality management (TQM), it aims to integrate and coordinate all quality-assessment activities and focus on the identification and correction of any problems, with the ultimate goal of protecting patients.

Key definitions

- **Risk management** is a systematic process of risk identification, analysis, treatment, and evaluation of potential and actual risks.
- **Hazard versus risk** – a hazard is the potential to cause harm while a risk is the likelihood of harm (in defined circumstances, and usually qualified by some statement of the severity of the harm).
- **Total quality management** is a management strategy aimed at embedding awareness of quality in all organizational processes.
- **Controls assurance** – the need to be seen to be doing our 'reasonable best' to reduce risk by using resources effectively.
- **Clinical governance** – a framework through which NHS organizations are accountable for continually improving the quality of their services and safeguarding high standards of care, by creating an environment in which excellence in clinical care will flourish

Risk management and economics

Health economics and cost-effectiveness play a big part in risk management. It is particularly important today because of rising healthcare costs, governmental cost containment and adverse claims experience.

Goals of risk management

- Survival of the organization (the NHS)
- Enhanced quality and standard of care
- Minimization of risk of medical or accidental injuries and losses
- Improvement of programme effectiveness and efficiency through administrative direction and control
- Coordination and integration of current policies, functions, programmes, committees, and other aspects relative to the risk management process
- Avoidance of adverse publicity
- Minimization of cost of risk transfer (insurance)

The risk-management cycle is outlined in Box 3.8.

Box 3.8 The risk-management cycle

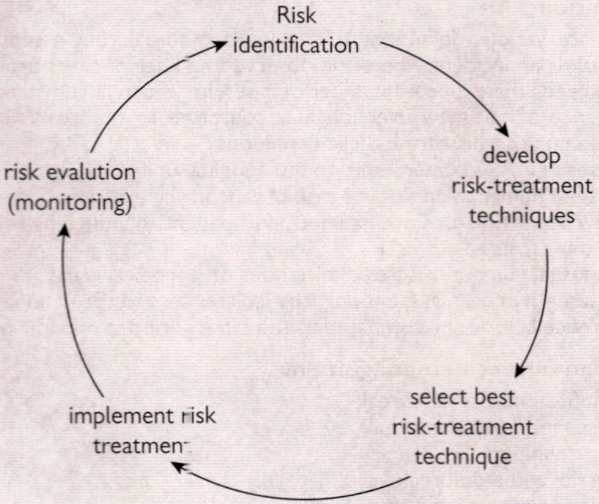

Management of outbreaks

Definition

There are various definitions of the word outbreak, but essentially an outbreak is an incident where the observed number of cases unaccountably exceeds the expected number of cases. In certain circumstances, the emergence of one new infection may constitute an outbreak. Here are some examples to illustrate these definitions:

- 2 cases of MRSA bacteraemia on the neonatal unit
- 1 case of human avian flu or 1 case of vancomycin-resistant *S. aureus*
- 2 new hepatitis C-positive haemodialysis patient, identified by regular screening tests
- an unusual number of cases of diarrhoea on a medical ward
- increased numbers of *Campylobacter* isolates sent to the national reference laboratory, compared to the same period in previous years.

Steps in outbreak management

1 Identification of an outbreak
2 Investigation of an outbreak
3 Case definition
4 Describing the outbreak
5 Propose and test hypothesis
6 Control measures and follow-up
7 Communication

Management of hospital outbreaks

Outbreaks of nosocomial infections must be identified and thoroughly investigated, because of their importance in terms of morbidity and mortality, the need to identify any breakdown in a process and to prevent spread or future recurrences. The transfer of patients between clinical areas and different wards (and different hospitals) can make contacts difficult to identify, and outbreak investigation is often time consuming. In addition to short-term benefits, effective outbreak investigations often lead to sustained improvement in patient care.

Management of risks to healthcare workers from patients

Practices are in place to reduce the risk of healthcare workers (HCWs) acquiring infections from their patients. However, HCWs (including laboratory staff) do become infected at work, with organisms ranging from the relatively benign norovirus to life-threatening conditions. Infection control procedures for patients infected with HIV, hepatitis B or hepatitis C may be considered together because routes of acquisition are similar. The main risk of transmission is from the accidental inoculation of blood (see 📖 Management of risk from sharps, p.192). Clearly, it is impossible to identify all carriers of blood-borne viruses, so universal infection control precautions (see 📖 p.208) are required for all patients.

Infectivity of body fluids from patients with HIV and hepatitis B

The following fluids are potentially infectious:

- blood and any fluid from the list* visibly contaminated with blood
- cerebrospinal fluid (CSF), peritoneal fluid, pleural fluid, pericardial fluid, synovial fluid, amniotic fluid, semen, vaginal secretions, saliva in the context of dentistry.

*The following are *not* regarded as infectious, unless visibly contaminated with blood: faeces, nasal secretions, saliva (except in dentistry), sputum, sweat, tears, urine, and vomit.

Care of patients with HIV or hepatitis B or C

Patients known to be positive for these viruses (or likely to be positive as a result of certain risk factors) should have a risk assessment performed. The risk assessment should fully respect patient confidentiality. The likelihood of exposure of staff or other patients to the patient's blood/body fluids should be considered. Medical and nursing care of infected patients depends on local hospital policy, but here are a few practical suggestions:

- single room isolation – this is only usually required if uncontrolled bleeding or loss of other body fluids is likely, or if the patient has another condition requiring isolation (e.g. diarrhoea; open pulmonary tuberculosis; salmonellosis; herpes zoster). On vacation of the room, terminal cleaning is necessary
- disposal of sharps – if the patient needs single room isolation, sharps must be disposed of in a sharps bin *inside* the room
- spills – any spills of blood/body fluids should be covered immediately with chlorine-releasing granules or strong (1%) hypochlorite
- equipment disinfection and sterilization – use single-use/disposable items whenever possible
- protective clothing – depends on the likelihood and degree of exposure to the patients blood and body fluids. Gloves, mask and protective eyeware are required wherever there is a possibility of blood contact or aerosolization
- linen from patients in isolation or contaminated with blood or body fluids should be regarded as 'infected linen' (see 📖 Laundry, p.231)

- toilet/bathroom facilities – patients may use the ward facilities except if there is bleeding (or risk of bleeding) when toilet/bathroom facilities must be reserved for this patient only (or use a commode)
- crockery – disposable crockery must be used if there is bleeding from the mouth, otherwise no special precautions are needed
- waste – clinical waste and disposable items must be placed in yellow bags (see 📖 Waste, p.232), with double-bagging if leakage of body fluids is possible
- specimen transport – label specimens and accompanying forms with 'Danger of Infection' stickers.

Other infections acquired by HCWs from patients

Many infections may be transmitted to hospital staff, but the following are recognized more frequently or have more significant consequences:

- viral gastroenteritis
- *Neisseria meningitidis* – HCWs who intubate the patient or perform mouth-to-mouth resuscitation.
- *Mycobacterium tuberculosis* (MTB)
- varicella zoster
- influenza
- others include pertussis, diphtheria, rabies, viral haemorrhagic fevers.

Box 3.9 Education in infection control procedures works!

During the SARS epidemic (2004), the proportion of infected HCWs varied from 20–60% of cases worldwide, with notable differences between hospitals. The better the education in infection control, the lower the risk of acquiring the infection.[1]

Further information

- HPA guidelines on needlestick injuries www.hpa.org.uk
- Health and Safety Executive advice on needle stick injuries www.hse.gov.uk/healthservices/needlesticks
- Banatvala JE et al. *HIV Infection: hazards of transmission to patients and HCWs during invasive procedures – report of a working group of the Royal College of Pathologists.* 1992
- Joint working party of Hospital Infection Society and Surgical Infection Study Group. Risks to surgeons and patients from HIV and hepatitis. *BMJ* 1992;305:1337–43.

Reference

1 McDonald L, Simor AE, Su IJ, et al. SARS in healthcare facilities, Toronto and Taiwan. *Emerg Infect Dis* 2004 May; 10(5)777–81.

Management of risk from sharps

By following universal infection control precautions all healthcare workers can minimize risks of infection associated with blood and body fluids The main risk associated with needlestick and other sharps injuries is transmission of blood-borne viruses; with a significant innoculation of blood through a hollow bore needle the rate of transmission to the recipient is as follows

- approx 30% if the donor is hepatitis B surface antigen-positive (the risk is as high as 30% if the patient is hepatitis B e antigen-positive, while it decreases to 3% if hepatitis B e antibody-positive);
- approx 3% if the donor is hepatitis C antibody-positive;
- approx 0.3% if the donor is HIV-positive.
- There is also the risk of emerging or unknown agents.

Prevention of sharps injuries

The National Audit Office report (2000) noted poor practice in sharps disposal.[1] Good clinical practice is important in preventing needlestick injuries, e.g.:

- handle needles and other sharps carefully
- never re-sheath needles
- staff using needles or other sharps are responsible for their safe disposal
- use sharps bins as indicated.

Action in event of sharps injury

- Immediate first aid – encourage bleeding of the area, wash with soap under running water, and cover with waterproof dressing. If eye splash, irrigate well with running water. If splash into mouth, do not swallow and rinse out mouth several times with cold water.
- Report the incident to the senior person in that area. Complete an incident form/accident book.
- Contact occupational health or another expert about hospital guidelines (e.g. microbiology/infectious diseases physician).
- Risk assessment – consider the hazard (potential to cause harm) and risk (likelihood that harm will occur), based on the injury, the source and the recipient (see Box 3.10).

Management of a 'needlestick' injury (Box 3.12)

- Always seek expert advice, usually from occupational health, microbiology or infectious diseases.
- Refer the recipient to occupational health/microbiology/infectious diseases immediately. They will make a risk assessment of the severity of the injury.
- If this was a significant injury then a member of the team looking after the donor (but not the recipient of the injury) should counsel the donor. Blood should be requested for HIV, HBs Ag, HCV antibody. See Table 3.5 p.194 for management of sharps injury from infected donor.
- If pre-test counselling is required, involve trained personnel.

1 National Audit Office (2000). www.nao.org.uk/publications/9900/hospital_acquired_infection.aspx

- Note that the source patient may decline to be tested, and this will not affect their medical care.
- Provide patient information leaflets as appropriate.
- General Medical Council (GMC) guidance offers advice on difficult situations, e.g. if there is conflict between the needs of the recipient and the wishes of the source, or if the source patient is unconscious or has died. Decisions about testing an incapacitated patient must take account of the current legal framework governing capacity issues and the use of human tissue. Consult local occupational health services and infection experts. Currently, blood should not be tested for blood borne viruses from an unconscious patient if it is not in the medical interests of that patient.
- If the injury was negligible, blood will be taken for storage and to check anti-HBsAg if unknown. This is an opportunity for hepatitis B vaccination.
- If the injury was significant then blood will be taken for HBsAg and long term storage. Hepatatis B vaccination (booster) will be considered. With a high risk donor, post exposure prophylaxis (PEP) and/or Hepatitis B immunoglobulin may be considered.

Box 3.10 Risk assessment –

Consider the injury, the source, and the recipient:

- Injury
  - Extent and depth of injury
  - Size and type of needle
  - Visible contamination with blood
  - Site
- Source (usually the patient undergoing the procedure, but maybe unknown)
  - Hepatitis B and C, HIV status (if known)
  - Viral load/stage of illness if HIV-positive
- Recipient
  - Hepatitis B vaccine status/antibody level
  - Any conditions that may affect treatment, e.g. pregnancy

Reporting

All injuries should be reported to the senior on duty, and recorded on an incident form/accident book. If the injury was with a used needle or instrument, advice should be sought from infection experts/occupational health. A review of equipment/procedures should occur, led by a senior member of the department.

Box 3.11 Grading of exposure

Risk associated with type of exposure is graded as follows:

- negligible risk
  - intact skin visibly contaminated with blood or body fluids, i.e. no actual injury occurred
- low risk
  - mucous membranes or conjunctival contact with blood or body fluids
  - intradermal injury associated with needle or instrument contaminated by blood or body fluids (possible parenteral exposure)
- medium risk
  - skin penetrating needle contaminated with blood or body fluids or wound which causes bleeding and is produced by an instrument that is visibly contaminated (definite parenteral exposure)
- high risk
  - significant exposure to blood or body fluids from source known to be hepatitis B/hepatitis C/HIV-positive.

Table 3.5 Management of sharps injury from infected donor

Donor known positive or likely positive	Action
HIV	*URGENT.* Discuss with an infection expert who will consider PEP. In general PEP is not recommended where HIV status is unknown, due to the significant morbidity and mortality associated with the drug regimen. The risks associated with PEP versus the risk of acquiring HIV will be explained to the recipient. PEP usually consists of a 3-day starter pack of antiviral therapy, with plans for ongoing counselling/treatment depending on results. Further information available from—HIV post-exposure prophylaxis: Guidance from the UK Chief Medical Officer's Expert Advisory Group on AIDS (Department of Health, 2008) http://www.dh.gov.uk/en/PublicationsandStatistics/Publications/PublicationsPolicyandGuidance/DH_088185.
Hepatitis B (HBV)	*URGENT.* Discuss with an infection expert who will consider hepatitis B immune globulin (HBIG) and accelerated vaccination or booster(s) as in Table 3.6.
Hepatitis C (HCV)	No immediate intervention. If donor anti-HCV positive, HCV RNA will be tested in the recipient by PCR at 6 and 12 weeks, and anti-HCV will be tested at 12 and 24 weeks. If signs of infection in the recipient, the test will be confirmed and the patient may be referred to further experts to consider HCV chemotherapy.

Table 3.6 Management of significant exposure to hepatitis B virus (HBV); In addition to percutaneous inoculation (needlestick, scratch, bite, etc) this may result from contamination of mucous membranes (e.g. spillage into eyes or mouth) or contamination of non-intact skin (open wounds, dermatitis eczema). For management of HBV exposure in newborn infants and in sexual contacts see 'The Green Book' (Immunication against infections disease www.doh.gov.uk/en/publichealth/healthprotection/Immunization/greenbook/DH–4097254)

HBV status of person exposed	**Significant exposure**	
	HBsAg-positive source	**Unknown source**
1 dose or less of HBV vaccine pre-exposure	Accelerated course of HBV vaccine (doses at 0, 1, and 2 months. May need booster dose at 12 months if continuing risk of exposure) Also give HBIG x1	Accelerated course of HBV vaccine (doses at 0, 1, and 2 months. May need booster dose at 12 months if continuing risk of exposure)
2 doses or more of HBV vaccine pre-exposure (anti-HBS not known)	1 dose of HBV vaccine followed by 2nd dose 1 month later	1 dose of HBV vaccine
Known responder to HBV vaccine (anti-HBS >10 microunits/mL)	Consider booster dose of HBV vaccine	Consider booster dose of HBV vaccine
Known non-responder to HBV vaccine (anti-HBS <10 microunits/mL) 2–4months post immunization	HBIG x1	HBIG x1
	Consider booster dose of HBV vaccine	Consider booster dose of HBV vaccine

Additional notes regarding management of significant exposure to HIV

Factors known to increase the risk of transmission of HIV

- Deep and penetrating injury
- Visibly blood-stained device
- Needle involved has been in the source patient's artery or vein
- Source has terminal HIV disease/high viral load

Box 3.12 Summary flowchart for needlestick injury

Check basic first aid has been performed

↓

Discuss with occupational health/infectious diseases/microbiology
Risk assessment – significant injury?

No

↓

- occupational health referral for routine assessment
- recipient sample for LTS (Long Term Storage) and antiHBsAg if unknown
- encourage HBV vaccination if not done

Yes

↓

- recipient sample for LTS and antiHBsAg if unknown
- request source sample for HbsAg; anti-HCV; anti-HIV testing, LTS
- unknown source: gather as much information as possible to make a risk assessment. HIV PEP unlikely to be given

↓

Further risk assessment:

- known or highly likely to be HIV+ve – URGENT: Infection expert will consider PEP
- known or highly likely to be HBV+ve – URGENT: Infection expert will consider HBIG and accelerated vaccine/booster
- known or highly likely to be HCV+ve – no immediate intervention. Infection expert will check HCV RNA on source if known anti-HCV +ve donor. Follow up recipient at 6 and 12 weeks (PCR), 12 and 24 weeks (anti-HCV)

Risk from tissues for transplantation

The number of organs and tissues that can be successfully tranplanted is increasing. The current list of organ transplants includes heart, kidneys, liver, lungs, pancreas, and intestine. Tissues include bones, tendons, cornea, heart valves, veins, and skin. Rejection, and infection arising from anti-rejection therapy are the main risks. Haematology and oncology patients who have received bone marrow transplants are at particularly high risk of infection.

Infection in a transplant recipient may either be transmitted from the donor with the organ, or arise due to the immunosuppressed state of the recipient. Certain infectious agents are particularly recognized as causing infections post transplant. Depending on which organ has been transplanted, different infections agents are commonly implicated at different time periods. Box 3.13 describes typical problems after a renal transplant.

Box 3.13 Example – infections post renal transplant

Minor infections are common after a kidney transplant. Urine infections affect ~50% of transplant recipients, especially if the patient has reflux nephropathy or diabetes. More serious infections in the first 6 months post transplant include pneumonia (e.g. *Pneumocystis* pneumonia (PCP), pneumococcus), cytomegalovirus (CMV), chickenpox, BK virus, and disseminated fungal infection. Each transplant unit will have guidelines for prophylaxis, which may include:

- co-trimoxazole (PCP)
- amphotericin (oral candida)
- isoniazid for those at risk of TB
- antibiotics (if urinary tract infections are common)
- valganciclovir or valaciclovir if CMV +ve donor into CMV –ve recipient
- vaccination e.g. influenza, Pneumovax®.

Surveillance of infections among tissue donors

A tissue donation and banking programme is operated by The National Blood Service (NBS) 'Tissue Services'. Donations may come from living and/or cadaveric donors, and include surgical bone (mainly femoral heads), tendons, skin, and heart valves. The NBS also operate the National Bone Marrow Registry and a cord blood bank.

All tissue donors (including stem cell and cord blood donors) are routinely tested for HIV, HCV, HBV, human T cell lymphotropic virus (HTLV), and syphilis infections. Data concerning rates of infection are collated by the HPA Center for Infections and the NBS.

Surveillance of infections in blood donors

Every blood donation is tested for markers of HIV, HCV, HBV, HTLV and syphilis and only used if all tests are negative. If an infection is detected, the donor is invited to return to the blood centre, when they will be told about their test results, asked for a repeat sample, asked to stop donating blood, and referred to a specialist. In addition to ensuring our blood supply is safe, these data improve our understanding of the epidemiology of blood-borne infections. The National Blood Service and the HPA manage a series of schemes which monitor infections in blood and tissue donor and transfusion recipients. This benefits patient safety and public health. See www.hpa.org.uk for details on 'BloodBorne Infections in blood donors'.

SHOT (Serious Hazards of Transfusion) is a scheme for reporting investigations into infections in transfusion recipients (http://www.shotuk.org)

Epidemiological data are also available by testing all pregnant women for blood-borne infections.

Management of risk from virally infected healthcare workers

Virally infected staff are usually managed by the occupational health department. Please note that up to date guidance from the Department of Health should be consulted. The following is a guide to current recommendations at the time of this publication.

Hepatitis B-infected HCWs

- Health Service Guidelines HSG (93)40: *Protecting Health Care Workers and Patients from Hepatitis*
  http://www.dh.gov.uk/en/Publicationsandstatistics/Lettersandcirculars/Healthserviceguidelines/DH_4084234

This guidance recommends that HBV carriers who are e-antigen-positive must not carry out procedures where there is a risk that injury to themselves will result in their blood contaminating a patient's open tissues (i.e. should not perform exposure-prone procedures (EPPs)).

- Health Service Circular (HSC) 2000/020: *Hepatitis B Infected Health Workers*
  http://www.dh.gov.uk/en/Publicationsandstatistics/Lettersandcirculars/Healthservicecirculars/DH_4004553

This circular recommends testing HBV-infected HCWs who are e-antigen (HBeAg)-negative and perform EPPs. Those with higher viral loads (>1000 genome equivalents per mL) should have their working practices restricted.

Hepatitis C-infected HCWs

- Health Service Circular HSC 2002/010: *Hepatitis C Infected Health Care Workers*
  http://www.dh.gov.uk/en/publicationsandstatistics/lettersandcirculars/healthservicecirculars/DH_4004561

This circular builds upon previous advice from the Advisory Group on Hepatitis and recommends that HCWs who know they are carrying HCV should not perform EPPs.

HIV-infected HCWs

- Best practice guidance. HIV Infected Health Care Workers: guidance on management and patient notification (DH, 2005)
  http://www.dh.gov.uk/en/Publicationsandstatistics/Publications/PublicationsPolicyAndGuidance/DH_073132

This guidance replaces the previous version published in 1998. It advises that it is no longer necessary to notify every patient who has undergone an EPP by an HIV-infected HCW, because of the low risk of transmission and the anxiety caused to patients and the wider public. However, the long-standing restriction on HIV-infected HCWs carrying out EPPs remains. The decision on whether a patient notification exercise should be undertaken should be assessed on a case-by-case basis using a criteria-based framework, as set out in this document. In line with *Shifting the Balance of Power* (Department of Health), directors of public health and primary care trusts will be responsible for deciding whether patient notification is necessary. UKAP, the UK Advisory panel for healthcare workers infected with bloodborne viruses, will be available to provide advice (Box 3.14).

- See also Expert Advisory Group on AIDS (EAGA) www.advisorybodies.doh.gov.uk/eaga

Box 3.14 Further advice from UKAP (UK Advisory Panel for healthcare workers infected with blood-borne viruses)

Although this was originally set up to consider individual cases of HIV-infected HCWs, the UKAP now considers other blood-borne viruses, in particular hepatitis B and, more recently, hepatitis C. They can provide advice for specific situtions, as well as general policies, and can be contacted at ukap@hpa.org.uk.

Hazard groups

In 1995, the Advisory Committee on Dangerous Pathogens (www.advisorybodies.doh.gov.uk/acdp) classified all microorganisms into hazard groups 1–4, based on their hazard and containment level (Table 3.7). The framework is based on their pathogenic potential route of transmission epidemiological consequences of escape host susceptibility. The hazard groups 1–4 are thus handled at different containment levels (CLs) in the laboratory, defined by Control of Substances Hazardous to Health (COSHH). ACOP category 3 organisms are listed in Box 3.15. See http://www.hse.gov.uk/pubns/misc208.pdf for the complete approved list of biological agents for all categories.

Table 3.7 ACDP hazard groups 1–4

Hazard group	CL (p204)	Examples
Group 1 Unlikely to cause human disease	CL-1	Coagulase-negative staphylococcus
Group 2 May cause human disease May be a hazard to laboratory workers Unlikely to spread in the community Treatment or prophylaxis available	CL-2	*Staphylococcus aureus* *Salmonella enteritidis*
Group 2+ organisms		*N. meningitidis*; *Legionella*
Group 3 May cause severe human disease Serious hazard to laboratory workers May spread in the community Treatment or prophylaxis available	CL-3	See full list in Box 3.16 Examples include: *S. Typhi*; *Mycobacterium tuberculosis*; *Brucella* spp.; HIV; hepatitis B, C, D
Group 4 Severe disease Serious hazard to laboratory workers High risk of spread in the community No treatment or prophylaxis		Haemorrhagic fevers, e.g. Lassa fever Note that this group only contains viruses, not bacteria

Box 3.15 List of ACDP category 3 organisms

- Bacteria
  - *Bacillus anthracis*
  - *Pseudomonas mallei*
  - *Pseudomonas pseudomallei*
  - *Chlamydia psittaci*
  - *Coxiella burnetii*
  - *E.coli 0157 and* verocytotoxigenic strains
  - *Francisella tularensis*
  - *Salmonella typhi*
  - *Salmonella paratyphi* A,B,C
  - *Mycobacterium tuberculosis*
  - *Mycobacterium* (certain species)
  - *Brucella*
  - *Rickettsia* spp.
  - *Shigella dysenteriae* type 1
  - *Yersinia pestis*
- Viruses (certain viruses from the following groups)
  - HIV and other retroviruses
  - hepatitis B, C ,D, E
  - arenaviridae
  - bunyaviridae
  - hantavirsues
  - phleboviruses
  - nairoviruses
  - calicicviruses
  - togaviruses
  - flaviviruses, e.g. dengue
  - tick-borne virus group
  - poxviruses
  - rhabdoviruses, e.g. rabies
  - SARS coronavirus
- Parasites (certain parasites from the following groups)
  - *Echinococcus* spp
  - *Leishmania* spp
  - *Plasmodium falciparum*
  - *Naegleria fowleri*
  - *Taenia solium*
  - *Trypanosoma cruzi* and *Trypanosoma brucei rhodesiense*
- Fungi
  - *Blastomyces dermatitidis*
  - *Coccidioides immitis*
  - *Paracoccidioides brasiliersis*
  - *Histoplasma capsulatum*
  - *Penicillium marneffei*
  - *Cladophialophora bantiana*
- Others
  - Creutzfeldt–Jakob disease (CJD) and variant CJD (vCJD)
  - Kuru
  - Gerstmann–Sträussler–Scheinker disease (GSS) and other transmissible spongiform encephalopathies (TSEs)

Containment levels

The containment level (CL) refers to the physical requirements necessary for working with organisms of different pathogenicity, and includes guidance about the facilities, working environment, safety equipment and procedures (e.g. staff training). There are four different levels (CL1–CL4), and the CL of an organism usually corresponds with its categorization, e.g. all group 3 organisms must be handled at CL3.

Summary of requirements

Containment level 1 (CL1), i.e. low individual and community risk

- No special facilities, equipment or procedures are required. Standard well-designed laboratory facilities and basic safe laboratory practices suffice.
- Handwashing facilities must be provided.
- Disinfectants must be properly used.

Containment level 2 (CL2), i.e. moderate individual risk, limited community risk

- Laboratory should be separated from other activities, biohazard sign, room surfaces impervious and readily cleanable.
- Equipment should include an autoclave, certified high-efficiency particulate air (HEPA)-filtered class I or II biological safety cabinet for organism manipulations, and personal protective equipment to include laboratory coats worn only in the laboratory.
- All contaminated material should be properly decontaminated.

Containment level 3 (CL3), i.e. high individual risk, low community risk

- Specialized design and construction of laboratories, with controlled access double door entry and body shower. All wall penetrations must be sealed. Ventilation system design must ensure that air pressure is negative to surrounding areas at all times, with no recirculation of air; air should be exhausted through a dedicated exhaust or HEPA filtration system. Minimum furnishings, all readily cleanable and sterilizable (fumigation). Laboratory windows sealed and unbreakable. Backup power available.
- Equipment must include an autoclave, certified HEPA-filtered class II biological safety cabinet for organism manipulations, and a dedicated handwashing sink with foot, knee, or automatic controls, located near the exit. Personal protective equipment should include solid front laboratory clothing worn only in the laboratory, head covers and dedicated footwear, gloves, and appropriate respiratory protection, depending on the infectious agents in use.
- All activities involving infectious materials to be conducted in biological safety cabinets or other appropriate combinations of personal protective and physical containment devices.
- Laboratory staff must be fully trained in the handling of pathogenic and other hazardous material, in the use of safety equipment, disposal techniques, handling of contaminated waste, and emergency response. Standard operating procedures must be provided and posted within the laboratory, outlining operational protocols, waste disposal, disinfection

procedures, and emergency response. The facility must have a medical surveillance programme appropriate to the agents used.

Containment level 4 (CL4), i.e. high individual risk, high community risk

- CL4 is the highest level of containment and represents an isolated unit that is completely self-contained to function independently. Facilities are highly specialized and secure, with an air lock for entry and exit, class III biological safety cabinets, or positive pressure-ventilated suits, and a separate ventilation system with full controls to contain contamination.
- Only fully trained and authorized personnel may enter the CL4 containment laboratory. On exit from the area, personnel will shower and redress in street clothing. All manipulations with agents must be performed in class III biological safety cabinets or in conjunction with one-piece, positive-pressure-ventilated suits.

Patient isolation

The use of universal infection control precautions (UICP p.208) should minimize the need for isolation of most patients. In practice, isolation depends on a risk assessment for each patient, and the side-rooms/facilities available in each trust. Always act on the patient's clinical presentation, and do not wait for laboratory results to be available as it may be too late. Involve your Infection control team early, and consult the DH guidance for further advice (Box 3.16) early.

Effective isolation relies on all staff following the necessary procedures, to make sure that none of the transmission barriers are breached. The simplest solution is to use single-rooms, but in an outbreak multi-bedded bays or even whole wards may be used.

Patients are isolated for two reasons:

- **source isolation** – to minimize the chance of infecting other patients, e.g. patient with open TB. The air inside the isolation room should be at negative pressure (exhaust-ventilated) compared to the corridor. Patients with highly contagious infections such as the viral haemorrhagic fevers must be nursed in a high-security isolation unit.
- **protective isolation** – to minimize that patient becoming infected, e.g. to protect susceptible or immunosuppressed patients. The air inside the isolation room should be at positive-pressure (pressure-ventilated) compared to the corridor. In some critical situations such as bone marrow transplant units, where airborne contamination with fungal spores is a problem, the efficiency of air filtration may be increased and laminar flow maintained as a barrier around the patient.

Source isolation

The following measures apply to patients in source isolation:

- limit transport to other departments (e.g. x-ray) to essential investigations only
- if a patient does need to go to another department, brief the porters and other staff what precautions are necessary
- do not transfer the patient to another ward or healthcare institution without discussion with the ICT
- if the patient is well enough, consider sending them home
- keep staff caring for infected patients to a minimum. Try not to let these staff work elsewhere in the hospital
- after death of an infected patient, maintain infection control precautions and consult your trust policy for dealing with the body.

Box 3.16 Isolating patients with HAI

Guidance and summary of best practice on isolating patients with health care acquired infection has been published by the DH (www.clean-safe-care.nhs.uk)

Recommendations cover the following:

1. Single room nursing
2. Cohort nursing
3. Management of the patient once isolated
    - Hand hygiene and personal protective equipment (PPE)
    - Cleaning and decontamination
    - Movement of patient

Universal infection control precautions and barrier nursing

Universal infection control precautions (UICP) involve following simple infection control precautions for *all* patients. It is difficult to tell which patients are infected and which are not, so all patients should be regarded as 'potentially infected'. Adherence to UICP for all patients should minimize the transmission of HIV, hepatitis B and C, and other infectious agents. It also eliminates confusion among staff as to which patients are to be treated as 'infected', and should also prevent any breach of confidentiality.

The main components of UICP

- Handwashing (p.212)
- Protective clothing
- Gloves and aprons should be worn if the episode of patient contact involves blood or body fluid, but the risk of splashing is low. If the risk of splashing is high, a waterproof gown, mask and eye protection should also be worn
- Disposal of linen and waste (p.231–233)
- Broken skin:
  - clinical staff should cover all skin lesions with a waterproof dressing
- Sharps (p.192 and Box 3.17):
  - never resheath, bend, or break a needle or any other sharp
  - dispose of all sharps as a single unit, in a suitable sharps bin
  - never attempt to retrieve anything from a sharps bin
  - only fill a sharps bin to ¾ full, and secure the lid before disposing of it according to local policy
- Spills:
  - any spill of blood or other body fluid that contains blood should be treated with chlorine-releasing granules, and left in place for 2 min. Afterwards this should be cleared up with paper towels, wearing gloves and aprons. The area should then be washed with hot water and detergent
  - if granules are not available, a solution of hypochlorite diluted to 10,000 ppm (1%) should be used in the same way
  - any spill of urine should be dealt with immediately using hot water and detergent

Box 3.17 Preventing the risk of microbial contamination from Saving Lives High Impact Intervention No. 1. Central Venous Catheter Care

The elements of the care process listed below form the basis of reducing the risk of bacterial contamination. This underpins all other High Impact Intervention (HII), and should be recorded at the time of the procedure. The three elements are as follows:

Hand hygiene

- Decontaminate hands before and after each patient contact
- Use correct hand hygiene procedure
- Personal protective equipment (PPE)
- Wear examination gloves if there is risk of exposure to body fluids
- Gloves are single-use items
- Gowns, aprons, eye/face protection may be indicated if there is a risk of splashing with blood or body fluids

Aseptic technique

- Gown, gloves and drapes as indicated should be used for the insertion of invasive devices

Sharps

- Safe disposal of sharps
- Sharps container available at point of use
- No disassembling of needle and syringe
- Sharps should not be passed from hand to hand
- The container should not be overfilled

Antibiotic policies

An antibiotic policy consists of written guidance that recommends antibiotics and their dose for treating and preventing specific infections. In general, a hospital antibiotic policy covers empirical treatment (when the initial diagnosis is made and before a causative organism is isolated) and specific treatment (when causative organism is known) and agents for prophylaxis.

Key objectives of an antibiotic policy

- Best possible treatment for individual
- Tailored to local units with their specific microbiological issues
- Reduce antibiotic resistance
- Clear and easy to follow. Information on practical aspects, e.g. dose, duration, should be included
- Minimal side-effects and toxicity (but should include information about monitoring levels etc)
- Low cost
- Debate re comprehensiveness – some prefer to only include common and important infections, so junior doctors will seek expert advice from an infection specialist in more difficult cases

Advantages

- Ensure patients receive effective treatment
- Minimal side-effects and complications, e.g. *C. difficile*
- Reduce the overuse of broad spectrum agents
- Reduce emergence of resistance
- Limit unnecessary treatment
- Education

Disadvantages

- Poor compliance amongst clinicians
- May lead to overuse of antibiotics, if clinical decision making is poor
- No control for quality of diagnosis

Box 3.18 DH strategy: Antimicrobial Prescribing – a summary of best practice

This strategy aims to reduce the risk of infections from MRSA, other resistant bacteria, and clostridium difficile associated diarrhoea (CDAD), and maintain the effectiveness of antibiotics by reducing antibiotic resistance. See 'Saving Lives' document for details, but here are the summary recommendations for overall antimicrobial management:

- every trust should have an antimicrobial prescribing and management group and policy with a strategy for implementation
- every trust should have an antimicrobial formulary and guidelines for antimicrobial treatment and prophylaxis.

This document also includes good practice recommendations for appropriate prescribing of antimicrobials, as follows:

- the decision to prescribe should always be clinically justified
- IV therapy should only be used for severe infections and/or those who cannot take oral medications. Generally, IV therapy should only be prescribed for 2 days, and oral switch considered at 48 h.
- antimicrobial treatment should be reviewed at least daily
- broad-spectrum antimicrobials should be minimized
- use of single dose for surgical prophylaxis

Reference

www.dh.gov.uk/publichealth/healthprotection/Healthcareacquired infection

Box 3.19 Resources with information on antibiotic policies and antimicrobial resistance

- Alliance for the Prudent Use of Antibiotics (APUA) (Tufts University, Boston, USA) – http:www.tufts.edu/med/apua/home.html
- Bugs and Drugs on the Web (National Electronic Library of Infection (NeLI)) – http:www.antibioticresistance.org.uk
- Drug Resistance (World Health Organization (WHO)) – http:www.who.int/drugresistance/en/
- Antibiotic Resistance (US Food and Drug Administration (FDA)) – http:www.fda.gov/oc/opacom/hottopics/anti_resist.html
- Drug Resistance (CDC Atlanta, USA) – http://www.cdc.gov/drugresistance/healthcare/default.htm
- Antibiotic Resistance Control and Prevention (European Commission) – http://www.abdn.ac.uk/arpac/index
- Health Protection Agency, UK – http://www.hpa.org.uk
- Department for Rural Affairs, UK – http://defra.gov.uk
- National Antimicrobial Resistance Monitoring System (USA) – http://www.cdc.gov/narms/
- The Alexander project – Felmingham D, White AR, Jacobs MR. The Alexander Project : the benefits from a decade of surveillance. JAC 2005, 56 S1, ii 3–ii 21(-)
- The Mystic project – www.ann-clinmicrob.com/pubmed/10969060

Handwashing

The importance of handwashing has been recognized since the 19th century when Semmelweiss encouraged medical students in Vienna to wash their hands in chlorinated lime solution on the delivery unit. The maternal mortality rate from puerperal fever in patients attended to by medical students was far lower than those attended to by midwives who did not wash their hands.

Handwashing practices these days are not always ideal.
- The National Audit Office report (2000) noted a lack of adherence to handwashing procedures.[1]
- Average compliance of healthcare workers (HCWs) with handwashing is <50%, and technique is often poor and rushed.
- Research has begun to focus on how to change the culture and behaviour on the ward, and improve adherence to policies.
- There is no doubt that good handwashing practice reduces transmission of infections, but to be effective all HCWs must comply all of the time.

Skin flora can be divided into two types:
- transient organisms – these are not normally part of the normal flora, and can be picked up from the patient or their environment. Examples include *E.coli, S. aureus, Klebsiella* spp., *Pseudmomonas* spp. They are usually removed by a 'social' handwash (Box 3.20), with soap and water. 'Hygienic hand disinfection' (Box 3.20), e.g. with alcohol gel, aims to remove or destroy all transient flora, and there may be a prolonged effect
- resident organisms – these organisms are usually found deep in the dermis. They do not usually cause infection, except if introduced during invasive procedures, e.g. line insertion or surgical procedures. Examples include coagulase-negative staphylococci, and aerobic and anaerobic diphtheroids. They are not usually removed by a single handwashing procedure.

How to wash?

- Using a good handwashing technique (Fig. 3.1) will clean areas that are often missed (e.g. between the fingers, thumbs, fingertips, areas of palms, and backs of the hands). This should only take 15–30 s. Make sure hands are wet before applying soap, and rinse thoroughly before drying. If using a gel, effective decontamination only occurs when the alcohol is rubbed in until the skin is dry.

When to wash?

- Wash hands after any process that contaminates the skin, and before food preparation, patient contact or any clinical procedure. Examples include after leaving a source isolation room, or before entering a protective isolation room.

Box 3.20 Common handwashing terms[2]

Social hand wash
- Cleaning of hands with plain non-medicated bar or liquid soap and water, for the removal of dirt, soil and various organic substances

Hygienic hand wash
- Cleaning of hands with antimicrobial or medicated soap and water. Most antimicrobial soaps contain a single active agent and are usually available as liquid preparations

Hygienic hand disinfection

- Normally consists of the application of an alcohol-based hand rub onto dry hands without water

Surgical scrub

This procedure aims to remove or destroy all transient flora, and reduce resident flora. There must also be a prolonged effect. Chlorhexidine, povidone-iodine, or alcohol are usually used.

Measuring compliance with handwashing

The gold standard for measuring compliance with handwashing is direct observation. This may, however, be subject to the Hawthorne effect in that behaviour tends to improve when an individual is being watched. Newer methods include devices to electronically monitor use of soap and hand-wash dispensers.

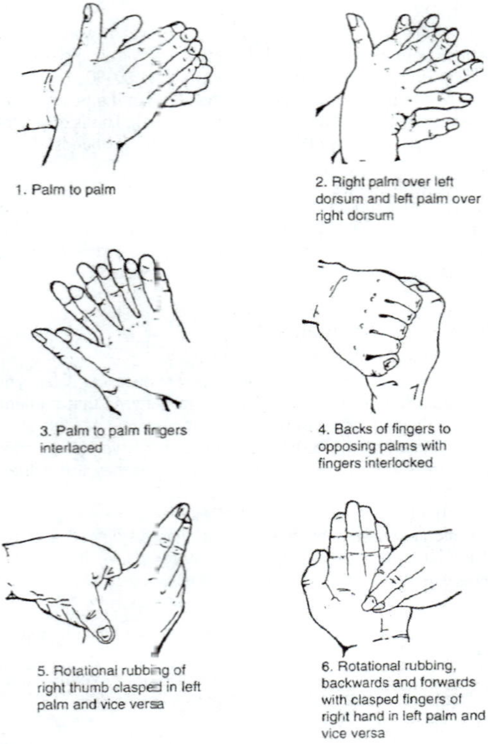

Fig. 3.1 Hand decontamination. Techique based on the procedure described by Aycliffe et al in *J Clin Pathol* 1978;31:923. Reproduced with permission from BMJ Publishing Group.

What to use?

- In most clinical situations, soap and water or an alcohol rub are adequate. The length of time and handwashing technique are more important than which soap is used. There are specifications for sinks in clinical areas. Hand lotions and creams may be used after handwashing to prevent soreness.

Handcare in general

- Nailcare – keep nails short and clean. False nails have been the source of HAI including endocarditis, so should not be worn.
- Jewellery and watches may harbour bacteria and hinder handwashing. Trusts often limit jewellery to a plain wedding band.
- If hands get dry and sore, as often occurs after repeated handwashing, then transient flora may become resident in skin cracks. HCW should consult occupational health if they are concerned.

Changing the culture

The main factors preventing compliance with hand hygiene are time and system constraints. Many feel that full compliance with complete guidance is unrealistic. Washing with soap and water can take 60–90 s; therefore many institutions have moved to alcohol-based hand rub at the point of care (i.e. at the bedside rather than at the entrance to the ward). This is easier, takes only 15–20 s, is microbiologically efficacious, and is better for hands.

Multifaceted approach

Evidence suggests a multipronged approach is the only way to bring about change. For example, key parameters in compliance with alcohol-based hand rubs include education of HCWs, monitoring and feedback to HCW, good administrative support, and introducing a system change (i.e. putting alcohol gel by each patient).

Cleanyourhands Campaign www.npsa.nhs.uk/cleanyourhands

The National Patient Safety Agency has produced a Cleanyourhands toolkit available free in the UK to NHS organizations that implement the campaign. The key elements of the campaign are:

- place disinfectant handrubs near to where staff have patient contact
- display posters and promotional material where they will influence staff and patients
- involve patients in improving hand hygiene

An economic assessment by the National Patient Safety Agency on the costs of implementing the Cleanyourhands campaign has found it to be cost saving, even if the reduction in HAI rates were as low as 0.15.

References

1 National Audit Office (2000) http://www.nao.org.uk/publications/9900/hospital_acquired_infection.aspx

2 Kampf G & Kramer E, Epidemiological background of hand hygiene and evaluation of the most important agents for scrubs and rubs. *Clin Microbial Rev* 2004; 17(4): 863-93.

Table 3.8 Best practice guidelines for handwashing

	Body	Reference
EPIC guidelines	Thames Valley University	http://www.epic.tvu.ac.uk/epicphase/1.html
Cleanyourhands	National Patient Safety Agency	Implementation of guidelines and results of pilot campaign available at http://www.npsa.nhs.uk/cleanyourhands
ICNA audit tool	Infection Control Nurses Association	http://www.icna.co.uk
HICPAC	American Healthcare Infection Control Practices Advisory Committee	http://www.cdc.gov/mmwr/pdf/rr/rr5303.pdf Also in Morbidity and Mortality Weekly Report 2002;51:1–44.
WHO Patient Safety	WHO Patient Safety	www.who.int/patientsafety – These hand hygiene guidelines on the WHO site should be applicable to all healthcare settings.

Management of antibiotic-resistant organisms

In the following section, management of antibiotic-resistant organisms will be discussed at the population level. For treatment of individuals see MRSA (📖 p.276), vancomycin-intermediate *S. aureus* (VISA) (📖 p.278), Gram-negative bacteria (📖 p.317).

Control of outbreaks of antibiotic-resistant organisms

The main steps in controlling an outbreak of antibiotic-resistant organisms are as follows

- Identify reservoirs of resistant organisms
  - colonized and infected patients
  - environmental contamination
- Halt transmission
  - improve handwashing and asepsis
  - isolate colonized and infected patients
  - eliminate any common source and disinfect environment
  - separate susceptible from infected and colonized patients
  - consider closing unit to new admission
- Modify host risk
  - discontinue compromising factors if possible
  - control antibiotic use (consider rotation, restriction, or discontinuing antibiotics)

Control of endemic antibiotic resistance

- Appropriate use of antibiotics – this should include optimal choice of agent, the dose and its duration, based on defined antibiotic policies for both treatment and prophylaxis. The use of topical antibiotics should be limited.
- Infection control – institute guidelines for intensive infection control procedures and provide adequate facilities and resources, especially for handwashing, barrier precautions (isolation), and environmental control measures.
- Improve antimicrobial prescribing practices through educational and administrative methods.
- Monitor local antibiotic resistance rates, and ensure antimicrobial guidelines are up to date.

UK MRSA prevention guidelines

Until 1998, the UK largely operated an MRSA search and destroy policy. Increasing colonization rates led to the revision of UK guidelines in 1998, and patients were stratified in terms of risk , and prevention measures targeted accordingly. In recent years various guidelines covering management of MRSA in the UK have been published (Table 3.9). The evidence for treatment of MRSA nasal/extranasal colonization has been evaluated (www.clinicalevidence.com) Box 3.21.

Table 3.9 UK guidelines for management of MRSA

Publication	Ref	Notes
Guidelines for the prophylaxis and treatment of MRSA infections in the UK	*Gemmel CG: J Antimicrob Chemother* 2006;57:589–608	Prevention and treatment with antibiotics
Guidelines for the laboratory diagnosis and susceptibility testing of MRSA	*Brown DFJ: J Antimicrob Chemother* 2005;56:1000–18	Reliable detection of MRSA colonization
Guidelines for the control and prevention of MRSA in healthcare facilities	*Coia JE: J Hosp Infect* 2006; 63S:S1–S44	Infection control guidance and strategies for preventing spread of MRSA or infection with MRSA

Box 3.21 MRSA treatment of nasal/extranasal colonization (www.clinicalevidence.com)

- Likely to be beneficial:
  - mupirocin nasal ointment
- Unknown effectiveness:
  - antiseptic body washes
  - chlorhexidine-neomycin nasal cream
  - mupirocin nasal ointment for 5 days (compared with >5 days)
  - systemic antimicrobials
- Unlikely to be beneficial:
  - tea tree preparations

Box 3.22 Treatment of MRSA – infection of any body site (www.clinicalevidence.com)

- Trade-off between benefits and harms:
  - linezolid (compared with glycopeptides)
  - glycopeptides compared with linezolid, quinupristin-dalfopristin, or trimethoprim-sulphamethoxazole
- Unknown effectiveness:
  - azithromycin, clarithromycin, erythromycin
  - ciprofloxacin, levofloxacin, moxifloxacin
  - clindamycin
  - daptomycin
  - doxycycline, minocycline, oxytetracycline
  - fusidic acid
  - quinupristin-dalfopristin
  - rifampicin
  - trimethoprim

Screening for MRSA colonization Department of Health 'Saving Lives'

This strategy aims to reduce MRSA infection by screening patients identified as 'at risk' from MRSA colonization[1]. All trusts should have screening and decolonization policies in place. Screening should be considered in the following groups:

- pre-operative patients in certain surgical specialties
- emergency orthopaedic and trauma admissions
- critical care (including intensive care units (ICU) and high-dependency units)
- renal medicine
- other specific patient groups (depending on local risk assessment and practicalities) – all patients previously known to be MRSA-positive; all elective surgical patients; oncology/chemotherapy inpatients; patients admitted from high-risk settings; all emergency admissions.

This document also discusses screening and testing methods, decolonization and isolation of patients.

References

1 Saving Lives Screening for meticillin-resistant staphylococcus aureus (MRSA) colonisation. A strategy for NHS trusts: a summary of best practice. Department of Health (2007). http//www. dh.gov.uk/en/Publichealth/Healthprotection/Healthcareacquiredinfection/Healthcareacquiredgeneralinformation/ThedeliveryprogrammetoreducehealthcareassociatedinfectionsHCAincludingMRSA/index.htm

Control of TB in hospitals

While most TB infections are acquired in the community, the risk of healthcare-associated TB remains for patients and HCWs. Most health-care-acquired TB cases result from delayed diagnosis of TB, inadequate treatment of latent TB, and lack of isolation facilties. The successful control and prevention of TB in hospitals may be achieved through three approaches:[1]

- administrative – i.e. early investigation and diagnosis of those suspected to have TB
- environment/engineering – e.g. single rooms until patient has received appropriate treatment for 2 weeks. Negative pressure isolation rooms are only recommended for patients with suspected/proven multi-drug-resistant (MDR) TB[2]
- personal respiratory protection – barrier nursing; gowns; filtered masks ('duck masks' or FFP3). In the UK, filtered masks are only indicated for patients with suspected/proven MDR TB or for aerosol-generating procedures (e.g. bronchoscopy). All masks must be correctly fitted.

Transmission of TB from patients

Stopping TB in England: an action plan from the Chief Medical Officer was published by the Department of Health in October 2004.[3] Action 5 of the five-point plan for providing well-organized and coordinated patient services is to identify, facilitate access to, and ensure staff are aware of the appropriate isolation facilities and infection control precautions to be taken for patients with infectious, or potentially infectious TB, or who have drug-resistant TB.

Transmission of TB from HCWs

When a HCW develops TB, this may lead to expensive, time-consuming, large-scale contact investigations to determine the extent of transmission and prevent further spread. The incidence of acute and latent TB is higher among foreign-born HCWs. Difficulties arise in interpretation of tuberculin skin tests in this group. Management of HCW with TB is usually shared by occupational health and infection control.

References

1 Humphreys H. Control and prevention of healthcare-associated tuberculosis: the role of respiratory isolation and personal respiratory protection. *J Hosp Infect* 2008; 69(1), 91–92.

2 NICE guidelines March 2006. Clinical Guideline 33. Tuberculosis: clinical diagnosis and management of tuberculosis, and measures for its prevention and control. http://www.nice.org.uk/guidance/CG33/niceguidance/pdf/English

3 *Stopping TB in England: an action plan from the Chief Medical Officer*. London: Department of Health, 2004. http://www.dh.gov.uk/en/Publicationsandstatistics/Publications/PublicationsPolicyAndGuidance/DH_4090417 (accessed 28 July 2008).

Control of CJD and transmissible spongiform encephalopathies

The transmissible spongiform encephalopathies (TSEs) include Creutzfeld–Jakob disease (CJD) (which may be sporadic, familial, iatrogenic, or variant CJD (vCJD)) and Gerstmann–Sträussler–Scheinker syndrome (GSS) (p.514). These conditions are characterized by neurodegeneration, with spongiform changes in the brain and central nervous system (CNS). The prion proteins associated with TSEs are unusually resistant to inactivation by heat and chemicals, so special decontamination procedures are required. Cases of iatrogenic CJD have been transmitted by contaminated pituitary-derived hormones, dura mater grafts, neurosurgical instruments, corneal transplantation, organ transplantation, and blood transfusion. Current challenges faced are in the early detection and diagnosis of vCJD, therapy and support for those affected, and improved understanding of transmission risks.

Variant CJD/new variant CJD (vCJD)

This emerged in the late 1990s, and there have been 164 definite and probable cases in the UK. It has a distinct clinical presentation (p.512) and tends to infect a younger age group than classical CJD. The number of new cases of vCJD is falling, but the total number of predicted cases is still a topic of debate. In vCJD, prion proteins have been detected in systemic lymphoid tissue before the patient is symptomatic, and the infectious agents are more resistant to inactivation than previously observed. These issues have had widespread consequences in infection control procedures, such as increased use of single-use devices, traceability of endoscopes, and quarantining of surgical instruments.

Risk assessments

All patients admitted for surgery should have a risk assessment performed. The potential for transmission depends on factors including:

- prevalence of disease in the population
- type and frequency of invasive procedures involving potentially infectious tissues
- effectiveness of strategies for instrument decontamination and management
- donor screening and selection (e.g. in transplantation).

Classification of risk

Risk for each individual patient should be classified as follows:

- low risk i.e. no identified risk
- asymptomatic but potentially at risk of CJD (e.g. recipients of pituitary hormones or dura mater grafts, or relatives with familial CJD)
- symptomatic (definite, probable, or possible CJD or vCJD, or diagnosis actively under consideration).

Prevention of transmission of CJD

Always consult Trust policy and involve the infection control team. Some general principles are advised for the processing of instruments (Box 3.23) and in blood transfusion (Box 3.24) in order to minimize transmission of CJD. There have been four cases of blood transfusion-associated cases of CJD in the UK. The blood products were all donated when the donor was in the preclinical stage of the disease, and none were leucodepleted. All clinical specimens from known, suspect or at-risk patients should be handled at CL3 in the Microbiology laboratory, as the agent of CJD is in hazard group 3.

Box 3.23 Precautions regarding processing of instruments to minimize transmission of CJD

- Staff should practice universal precautions.
- Clean all surgical instruments to remove organic matter before sterilization.
- Consider using single-use instruments whenever possible. Never reprocess single-use kits – throw them away immediately. Use single-use kits for all lumbar punctures.
- The Spongiform Encephalopathy Advisory Committee (SEAC) recommends high-vacuum porous-load autoclaving at 134–137°C for 18 min. The following methods are ineffective for prions: autoclaving at 134°C for 3 min; alcohol; ethylene oxide; glutaraldeyde; formalin.
- Record the unique identification number of all flexible endoscopes every time they are used, and ensure all instruments are traceable through the audit trail.
- Recent research in mice has suggested that dental tissue may be infective, thus instruments used for root canal work should be single-use.

Box 3.24 Precautions regarding blood transfusion to minimize transmission of CJD

- Blood donors who have themselves received a blood transfusion or tissue-donation are now excluded (2005).
- Leucodeplete all blood donations – this will reduce infectivity, but not eliminate it.
- Plasma for use in those <16 years old is purchased outside the UK.
- Note that screening tests are being developed to detect the prion protein in blood products but none are currently in use.

Box 3.25 Care of symptomatic or at-risk patients

All patients should be subject to universal precautions. Patients known to have CJD do not require special nursing precautions or special precautions for management of sharp injuries, exposure to blood and body fluids, used and infected linen, and the disposal of clinical waste. However, particular care must be taken to adhere to trust policy for the following:

- collection, labeling, and transport of clinical specimens – use 'Danger of Infection' stickers and provide adequate clinical information for laboratory staff to undertake a risk assessment
- CNS and lymphoid tissue biopsies – these procedures should be performed by experienced staff, using disposable equipment. Gloves, goggles, and aprons should be worn and any contaminated objects should be incinerated
- surgical procedures on symptomatic or at-risk patients – perform a risk assessment considering the likelihood the patient is carrying the infectious agent and the chance that the agent could be transmitted by the specific procedure. Consult trust policy (or SEAC website), for details about single-use protective clothing, single use instruments, incineration, cleaning of theatres, and other precautions.

Further information is available from:

- Advisory Committee on Dangerous Pathogens – TSE Agents. www.archive.official-documents.co.uk/document/doh/spongifm/part-4b.htm
- Spongiform Encephalopathy Advisory Committee (SEAC) – www.seac.gov.uk/
- Department of Health. HSC 1999/178: Variant Creutzfeldt–Jakob disease (vCJD): minimising the risk of transmission. London: 1999. http://www.dh.gov.uk/en/Publicationsandstatistics/Lettersandcirculars/Healthservicecirculars/DH_4004969

Disinfection

Disinfection is the process of killing of microorganisms by physical or chemical means, to render the object 'safe'. It does not imply complete inactivation of all viruses or removal of bacterial spores (as occurs in sterilization).

A disinfectant is defined as a chemical used to destroy microorganisms. These agents only act on surfaces (environmental surfaces, equipment, or body surfaces), and do not penetrate layers of dirt or grease. Thus disinfection is not a substitute for cleaning. Disinfectants do not usually have a persistent effect.

Use of disinfectants

The environmental use of disinfectants should be restricted to accidental spills or build-up of infected material in areas where this may be a hazard to patients or HCWs.

Different disinfectants have different properties: many are corrosive and toxic and the speed of action is highly variable. Some, if used correctly, can kill all germs (i.e. sterilize), but most are highly selective and only kill a limited range of organisms. They fall into two main groups:

- environmental disinfectants – often too toxic for use on skin, may require protective clothing. Hypochlorite is most commonly used.
- skin disinfectants (also called antiseptics) – often have limited range of action so are inappropriate for environmental disinfection, and usually relatively expensive. Chlorhexidine is the preferred agent. Alternatives are alcohol (inferior)and iodine(irritant)

Other disinfectants may be used to sterilize instruments, but heat treatment is usually preferred (p.227).

Selection of a disinfectant

Considerations include the following:

- which organisms do you want to destroy?
- what is the construction of the object to be disinfected?
- does the object need cleaning first?
- when an agent has been chosen, what concentration of disinfectant is required?

Table 3.10 Environmental and skin disinfectants

Class	Examples	Use	Notes
Environmental disinfectants			
Hypochlorite (bleach)	Hypochlorite powder, e.g. Titan sanitiser; detergent hypochlorite (liquid or tablet), e.g. Domestos, Presept	Best general purpose disinfectant available. However, not suitable for particularly dirty situations. Generally use a solution of 1000 ppm (parts per mL) available chlorine, but increase concentration to 1 ppm if need to destroy hepatitis viruses (e.g. dialysis units). Use hypochlorite granules for spillage of body fluids except urine	Sodium hypochlorite acts by the release of chlorine on contact with organic matter, so rapidly destroys all bacteria and viruses. However some agents are unstable, and disinfecting properties may be lost by the rapid release of chlorine on contact with blood, faeces, or textiles. Strong solutions are corrosive to aluminium and other metals
Phenolics	Black fluids, e.g. 'Jeyes fluid.' White fluids, e.g. 'Izal.' Clear phenolics, e.g. 'Stericol'	Not for routine environmental cleansing or disinfection; used in laboratory and post-mortem rooms.	Derived from coal-tar, and in common use in hospitals for over a century. Reasonable in visibly dirty situations. However, many bacteria and viruses are resistant and prolonged exposure is needed for effective action. Also toxic so handle with special precautions
Chlorxylenols	'Dettol'	Household disinfectant	Said to combine some of the properties of the phenolics with hypochlorites. Less effective at killing Gram-negatives than the phenolics, and expensive

Table 3.10 Environmental and skin disinfectants (*Continued*)

Class	Examples	Use	Notes
Skin disinfectants			
Alcohols	Ethyl alcohol (available as industrial methylated spirit (IMS)); isopropyl alcohol (e.g. 'Mediswabs')	Often used as a base for other skin disinfectants e.g.iodine or chlorhexidine.	Ethyl alcohol (70%) effectively kills organisms on the skin, but this effect ceases after evaporation. Isopropyl alcohol evaporates less rapidly, but is thought to be less effective against some viruses
Iodine	Huge variety of products, e.g. Betadine	Surgical scrubs, shampoos, etc	Iodine dissolved in 70% alcohol is less popular nowadays as it is messy and may be an irritant
Chlorhexidine	'Hibitane' 'Hibiscrub'	Surgical scrubs, popular disinfectant in hospitals and laboratories	Often marketed with other disinfecting agents and in alcoholic or aqueous solution. Some hospital organisms show resistance to it, but this is not a problem when using the alcoholic solution
QAC (quaternary ammonium compounds)	Cetrimide	Trauma wounds and other special situations, not used in routine wound cleaning	Ineffective against some hospital organisms. e.g. *Pseudomonas* spp.

Sterilization

Sterilization is the process by which transmissible agents are killed or eliminated. This includes fungi, bacteria, viruses and spores, but not prions. There are two main types of sterilization:

- Physical sterilization: this includes heat sterilization and radiation (e.g. electron beams, X-rays, gamma rays)
- Chemical sterilization: this includes *ethylene oxide; ozone; chlorine bleach; glutaraldehyde; formaldehyde; hydrogen peroxide; peracetic acid*

Cleaning of instruments

Whichever sterilization process is deemed most appropriate, thorough cleaning is required, otherwise any dirt or biological matter may shield any organisms present. Physical scrubbing with detergent and water is recommended. Cool water is needed to clean organic matter from instruments as warm or hot water may cause coagulation of organic debris. Alternative cleaning methods include ultrasound or pulsed air.

Physical sterilization: heat

This can be either dry heat or moist heat sterilization.

- Dry heat sterilization uses hot air which is free (or almost free) from water vapour, so any moisture plays no role in the process of sterilization. Methods include the hot air oven, radiation, and microwave. The dry heat coagulates the proteins in any organism, and causes oxidative free radical damage and drying of cells.
- Moist heat sterilization uses hot air heavily laden with water vapour. Moist heat coagulates the proteins in any organism, which is helped by the water vapour that has a very high penetrating property, and also causes oxidative free radical damage. Methods include autoclaving, pressure cooking, pasteurisation of milk, boiling, and steam sterilizing (Steam at *atmospheric pressure* for 90 minutes).

Autoclaves: Steam sterilization using autoclaves is commonly used in hospitals, and provides an inexpensive means of sterilizing large numbers of surgical instruments. To achieve sterility, a holding time of >15 minutes at 121°C, or 3 minutes at 134°C is required. Liquids and instruments packed in cloth may take longer to reach the specified temperature, so usually need more time. Autoclave treatment inactivates all bacteria, fungi, viruses and spores. Certain prions may be eradicated by autoclaving at 121–132°C for 60 minutes or 134°C for >18 minutes, but this process is not 100% reliable for CJD (see Box 3.24). Monitoring an autoclave cycle is important, and involves recording temperature and pressure over time. To ensure adequate conditions have been met, most hospitals use indicator tape which changes colour. Bioindicators (e.g. based on the spores of *B. sterothermophilus*) are also used to independently confirm autoclave performance. These indicators should be positioned to ensure that steam penetrates the most difficult places. Note that autoclaving is often used to sterilize medical waste prior to disposal.

Chemical sterilization

Chemical sterilization is generally used when heat methods are inappropriate, e.g for sterilizing heat-sensitive materials such as plastics, paper, biological materials, fibre-optics, and electronics. Options include:

- Ethylene Oxide (EO) sterilization is very common, particularly for disposable medical devices. Sterilization is usually carried out between 30°C–60°C, for objects that are sensitive to higher temperatures e.g. plastics and optics. EO gas penetrates well and is highly effective, killing all viruses, bacteria, fungi and spores. However it is highly flammable, takes longer than any heat treatment, and produces toxic residues. *B. subtilis* spores are used as a rapid biological indicator for EO sterilizers.
- Ozone is used to sterilize water and air in industrial settings, and as a surface disinfectant. It can oxidize most organic matter, but may be impractical because it is toxic and unstable and must be produced on-site.
- Chlorine bleach will kill bacteria, fungi and viruses and most spores. Household bleach (5.25% sodium hypochlorite) is usually diluted to 1/10 before use. To kill MTB it should only be diluted 1/5, and to inactivate prions it should be 1/2.5 (1 part bleach and 1.5 parts water). For full sterilization, bleach should be allowed to react for 20 minutes. It is highly corrosive (including some stainless steel surgical instruments).
- Glutaraldehyde and formaldehyde are volatile liquids, which are only effective if the immersion time is long enough. It can take up to 12 hours to kill all spores in a clear liquid with glutaraldehyde, and even longer with formaldehyde. Both liquids are toxic if inhaled or if they come into contact with skin. Glutaraldehyde is expensive and has a shelf life of <2 weeks. Formaldehyde is cheaper, but much more volatile (in fact it may be used as a gaseous sterilizing agent).
- Ortho-phthalaldehyde has many advantages over glutaraldehyde: shows better myco-bactericidal activity, kills glutaraldehyde-resistant spores, is more stable, less volatile, less irritating and acts faster. However it is more expensive, and stains skin and proteins grey.

Hydrogen peroxide is a non-toxic chemical at low concentrations and leaves no residue. It can be used to sterilize endoscopes, either in Low Temperature Plasma sterilization chambers or mixed with formic acid. It can also be used at a concentration of 30-35% under low pressure conditions in the dry sterilization process (DSP). This process achieves bacterial reduction of 10^{-6}–10^{-8} in approx 6 seconds, and the surface temperature is increased only 10–15°C.

Ventilation in healthcare premises

Definitions of different ventilation systems

Positive-pressure ventilation

Positive pressure isolation rooms are used to prevent the entry of microorganisms, if patients are susceptible to infections. The air is filtered before entering a sealed room, with a High Efficiency Particulate Air (HEPA) filter, and air is pumped into the room at a greater rate than it is expelled. This forces air out of the isolation room, keeping the room free of microorganisms. There is usually an anteroom to facilitate the donning of protective clothing and airflows of at least 12 air changes per hour.

Negative-pressure ventilation

Negative pressure isolation rooms are used to prevent pathogens (e.g. TB) from an infected patient infecting other patients or HCW in the hospital. It is usually a sealed room except for a small gap under the door, through which air enters. Direction of air flow can be confirmed by a smoke test (hold a smoke tube ~5 cm in front of the bottom of the door, and if the room is at negative pressure, the smoke will travel under the door and into the room).

HEPA filter

HEPA filters can remove almost all airborne particles 0.3 micrometres in diameter. e.g. aspergillus spores.

Plenum ventilation

This is the most frequently used system in general-purpose operating theatres. Atmospheric air is filtered in two stages:

- coarse filter to remove dust and debris
- bacterial filter of ~2 micrometres pore size with 95 % efficiency is used inside the inlet grill.

Some air may be recirculated within the suite. An exhaust system removes the air to the outside. There are ~20 air changes/hour.

Laminar flow

Laminar flow is used in orthopaedic theatres to reduce the number of microorganisms present. This is of particular value in preventing prosthetic joint infections. A continuous flow of filtered air is recirculated under positive pressure into the operating field, and any air contaminants generated under surgery are removed from the site. There are approximately 300 air changes/hour, which should result in <10 cfu/m^3 (colony forming units). Different systems include introduction of air horizontally or vertically, in an enclosed, semi-enclosed or open manner.

Operating theatres (OTs)

HTM2025 (Specialist Ventilation in Healthcare premises) has been revised as HTM03-01 (Part A covers design and validation and Part B covers operational management and performance ventilation). This is a new set of standard schemes for ventilation of conventional and ultraclear ventilation UCV operating theatres. There are four main sections:

- management policy – management responsibilities/legal issues
- design requirements
- validation/verification – commissioning (i.e. when a new theatre is built or after major constructional changes), performance tests, handover

- operative management – day-to-day issues such as minimum standards, maintaining performance, routine maintenance etc.

See Box 3.26 for a discussion of practices in OTs.

Air sampling

Microbiological tests are needed to complement the physical monitoring systems, although few studies have demonstrated a link between microbiological air quality and wound infections. For details of performing air sampling and testing infection control rooms see Walker et al[1] and for operating theatres see HTM03-01(Department of Health). Separate guidance is given for empty and in-use theatres.

- If an empty theatre fails the air sampling tests, check the technique and repeat sampling. If it fails again, discuss the findings with the ICT, engineers or other experts. Consider testing the particle penetration of filters and/or the air velocity.
- If an in-use theatre fails the air sampling tests, check the technique and repeat sampling when the theatre is empty.

Box 3.26 Rituals and behaviours in operating theatres

All OTs should have their own up-to-date infection control policy. This should include standard precautions for every invasive procedure, and outline the need for an additional risk assessment for each patient to see if other specific precautions are required. There are many 'rituals and behaviours' that have crept into 'standard practice' in many OTs – some of which are beneficial, some are harmful. The standard of evidence varies, but a few practical pointers are listed below. See Woodhead K et al for further discussion.[2]

- Patients' clothes – it may not be necessary to change, e.g. for cataract surgery. Jewellery only needs to be removed (for infection control purposes) if near the site of operation.
- Shaving should be avoided. Depilatory cream the day before surgery is preferable, or clippers in the anaesthetic room immediately pre-op.
- Hand hygiene – Scrubbing brushes should not be used on the skin.
- Drapes – there is no evidence for adhesives around the edge of wounds.
- Gloves – needle puncture is not an indication to change gloves; if necessary, a second pair should be worn on top.
- Masks – there is no good evidence that masks reduce infection rates; however, they are recommended to protect the surgeon. The scrub team should wear masks and hats for implants, and the mask should be changed for each procedure. Non-scrubbed staff in plenum ventilated theatres do not require masks or hats.
- Linen – should be waterproof and disposable, in accordance with European standards

References

1 Walker JT, Hoffman P, Bennett AM, et al. Hospital and community acquired infection and the built environment-design and testing of infection control rooms. *J Hosp Infect* 2007;65 S2:43–9.

2 Woodhead K, Taylor EW, Bannister G. Behaviours and rituals in the operating theatre. A report from the HIS Working Party. *J Hosp Infect* 2002;51(4):241–55.

Laundry

Hospital linen should be processed so it is not an infection risk to future users. The laundry should remove evidence of previous use, including organisms, but cannot be expected to kill bacterial spores. Department of Health guidance (HSG(95)18) differentiates used (soiled or fouled) and infected linen, and describes a framework to reduce risk to all staff concerned (porters, laundry staff etc).[1] Table 3.11 defines soiled, fouled, and infected linen.

Washing temperature requirements vary between countries, but in the UK the temperature of the cycle should reach 65°C for at least 10 minutes or 71°C for at least 3 minutes. Heat-labile laundry that would be damaged at high temperatures, can be washed at 40°C with sodium hypochlorite to give a final concentration of 150 ppm available chlorine to the final rinse.

Monitoring of critical points is important under HACCP. These include temperature and exposure times, treatment of rinse water (if potentially contaminated), in-use detergent concentrations and drying temperatures.

Curtains should be laundered as follows:

- if visibly soiled
- after an outbreak of viral gastroenteritis as part of the terminal clean
- if they are around the bed of a patient who has been barrier nursed (e.g. with MRSA, *C. difficile*), they should be washed on patient discharge
- routinely (3 months minimum).

Table 3.11 Categories of Linen

Linen category	Definition	Bag colour	Notes
Soiled	All used linen that is not fouled or infected	White nylon bag	
Fouled	Contaminated with any bodily fluid	White or clear plastic bag, then white linen outer bag	Treat as potentially infected, so wear gloves and aprons when handling it
Infected	Linen from patient with or suspected to have infection with enteric organism, e.g. diarrhoea and/or vomiting, *campylobacter*, viral gastroenteritis, *salmonella*, *shigella*, hepatitis A etc	Red alginate bag, then red nylon outer bag, labelled with point of origin	
	Blood-stained linen from patient with HIV/hepatitis B or C		
	Linen visibly contaminated with sputum from patient with open TB		

References

1 HSG (95)18. www.dh.gov.uk/en/publicationandstatistics/lettersandcirculars/Healthserviceguidelines/DH_4017865

Waste

Disposal of waste is subject to strict legislation set out by the Department of Health[1], Department of the Environment, the Health and Safety Commission Advisory Group, and the amended European Communities Framework Directive on Waste. This legislation places a duty of care on anyone handling clinical waste, including porters and incineration staff. Thus hospital guidelines apply not only to clinical areas, but throughout the hospital. There is a move towards a unified approach for disposal of infectious and medicinal waste, and to operate the same category codes throughout Europe.

Clinical waste

Clinical waste is always incinerated and is defined as follows:

- all human tissue and body fluids and related items (e.g. dressings, incontinence pads, stoma bags, and urine containers)
- sharps and contaminated sharp items, e.g. glass
- certain pharmaceutical products and their containers
- potentially infected laboratory waste.

Table 3.12 summarizes other types of waste, the national colour-coded system, and recommendations for disposal.

References

1 Environment and Sustainability Health Technical Memorandum 07-01: Safe management of health care waste, Department of Health.

Table 3.12 National colour-coded system for waste disposal containers

Type of waste	Container	Notes
Clinical waste	Yellow plastic bags (Orange/Red in some hospitals)	Make sure bags are only two-thirds full. Tie them securely and label each bag with ward of origin before sending for incineration. If there is a chance that body fluids may leak from a single yellow bag, then use double-bagging.
Non-clinical waste	Black plastic bags	
Sharps	Yellow rigid plastic boxes	Do not overfill these boxes. Never resheath needles or separate needles from a syringe unless using an approved safe method. Put IV giving sets into these boxes too.
Glass (larger items)	Black dustbins	
CSSD	Brown bags marked CSSD	
Confidential or cytotoxic waste	Local policies apply	

Introduction to prevention of HAI

In the year 2000, it was estimated that 9% of hospital inpatients had a HAI during their stay. Patients on intensive care units are at increased risk, and estimates suggest that 15–40% of patients will have at least one HAI. This amounts to at least 100,000 HAIs a year in the UK alone. The effects varied from increased length of stay and discomfort, to prolonged and permanent disability, and to at least 5000 deaths a year. The estimated cost to the NHS is £1 billion annually.[1]

It is estimated that around 15% of HAIs are preventable through better application of 'good practice'.[2] It is difficult to calculate the costs of introducing prevention methods on a trust basis, but all research so far suggests that it is cheaper to focus on prevention rather than pay for costs of treating HAI. The NAO report noted that a reduction in HAI by 15% could save £150 million.[1]

Prevention is everyone's business . . . not just the ICT!

There is clearly a need to change the culture and staff behavior. Currently this has been highlighted as one of biggest obstacles. Evidence suggests that a variety of approaches are required so that the individual HCW accepts personal responsibility. Training and education must be continual, and constant reminders, e.g. poster campaigns, handwashing publicity, and infection awareness days, are effective. Named individuals acting as liaison representatives and role models in each specialty are beneficial. Feedback of infection rates at ward/team level is vital to engage staff, encourage a sense of ownership, and encourage continual review of practice.

Focus points re prevention

- Education and training for healthcare staff, especially doctors
- Better compliance with hand hygiene, care of indwelling lines, catheter care, and aseptic technique
- Good antibiotic prescribing
- Hospital cleanliness
- Consultation with ICT on wider issues, e.g. new-build projects

References

1 NAO, 2004. – http://www.nao.org.uk/Publications/0304/improving_patient_care.aspx

2 NAO, 2000. http://www.nao.org.uk/publications/9900/hospital_acquired_infection.aspx

Infections in intensive care

Nosocomial infections complicate 25–40% of all ICU admissions. Although ICUs represent <5% of hospital beds, nosocomial infections in the ICU consume a significant amount of hospital resources.

Patients on ICU are exposed to more broad-spectrum antibiotics, and medical devices, and more procedures than those on normal hospital wards. Hand hygiene, barrier precautions, cohorting of personnel, and antibiotic policies are particularly important in controlling infection.

Organisms

- The causal organism(s) isolated depend on length of ICU stay (Table 3.13).
- Recent shift from a preponderance of Gram-negative infections to Gram-positives, probably due to increased line and device-associated infections.
- Increasing prevalence of candida infections including non-albicans which may be more drug resistant.
- More infections with antibiotic resistant organisms – MRSA, VRE, multi-resistant Gram-negative species, such as *E. coli*, *Klebsiella* spp., *Serratia* spp., *Acinetobacter* spp., *S. maltophilia*, *Enterobacter* spp.

Table 3.13 Organisms isolated depend on the length of ICU stay

Early infection (≤4 days)	Late infection (>4 days)
S. pneumoniae	*Enterobacter* spp.
H. influenzae	*Serratia* spp.
Enterobacteriaceae	*P. aeruginosa*
S. aureus – MSSA	*Acinetobacter* spp.
Streptococcus spp.	*S. aureus* – MRSA
Anaerobes	*Enterococcus* spp.
	Fungi

MSSA: meticillin-sensitive *S. aureus*.

Patients

- Increasing population of immunosuppressed patients, e.g. HIV, bone marrow transplant (BMT), solid organ transplant patients
- Increasing use of devices, e.g. lines, balloon pumps, pacing wires, endotracheal tubes
- More-invasive procedures, e.g. ventilator, drains,

ICU environment

- Isolating patients with resistant organisms is the aim. Ideally there should be at least one side room for every six beds. There should also be sufficient space around each bed (20 m^2), wash hand basins between every other bed, adequate ventilation, and sufficient storage space and utility space.

How to minimize infections on ICU

- Follow evidence-based guidelines and policies
- Good infection control measures, e.g. handwashing
- Good 'antibiotic control' e.g. specific policy based on local knowledge
- Close liaison with infection services, pharmacy, engineers, estates, etc
- Feed back results of surveillance of resistant organisms

Box 3.27 Studies of HAIs on ICU

- The European Prevalence of Infection in Intensive Care (EPIC) study, 1992 was a 1 day point prevalence study looking at >10,000 patients from 17 countries on all ICUs except paediatrics and coronary care unit.[1] The infection data were linked to the patients' APACHE score and 6 week outcome, presence of lines, specific interventions, demographics. Over all 45% of patients had some sort of an infection, and 20% had at least one infection acquired on ICU. The most common were pneumonia (47%), lower respiratory tract infection (18%), UTI (18%), bacteraemia (12%), and wound infection (7%). Organisms were split 50/50 Gram-positive and Gram-negative, the most common being *S. aureus*, *P. aeruginosa*, coagulase-negative staphylococci and enterococcus.
- The Study on the Efficacy of Nosocomial Infection Control (SENIC) looked at the relative change in nosocomial infection over a 5-year period.[2] Overall, when infection control measures were introduced, nosocomial infections were reduced by 32%.

Selective decontamination of the digestive tract (SDD)

This is a prophylactic technique, which remains controversial. It aims to eradiacate aerobic Gram-negative rods (GNR) from the oropharynx, and consists of four components:

- oral antimicrobial applied topically to the mouth four times daily
- a liquid suspension containing the same antimicrobials given via nasogastric tube
- intravenous antimicrobials for 3 days
- stringent infection control measures.

A systematic review and meta-analysis of >50 randomized controlled trials showed that SDD resulted in a significant decrease in levels of overall bloodstream infections (BSIs), Gram-negative BSIs and overall mortality, but had no effect on Gram-positive BSIs.[3] There are concerns, however, about the generation of resistant organisms.

DH Guidance: Infection Prevention and Control in adult critical care

This contains recommendations for reducing the risk of infection through best practice, and sustaining this reduction.

- Sustainable reductions in HCAI require the engagement and active involvement of all staff working in ICU supported by the ICT and clinical champions.
- No single action will produce effective infection prevention and control practice. This is achieved by sustained and close adherence to best practice by every member of the ICU team.

- All individuals who come into contact with ICU patients have a responsibility to ensure effective infection prevention and control afforded to them.

References

1 Vincent JL, Bihari DJ, Suter PM. EPIC study (1992). *JAMA* 1995; 274: 639–44.

2 Haley RW, Morgan WM, Culver DH. Study on Efficiancy of Nosocomial Infection Control. *Am J Infect Control* 1985; 13(3):97–108.

3 Silvestri L, van Saene HK, Milanese M, Gregori D, Gulls A. Selective decontamination of the digestive tract reduces bacterial bloodstream infection and mortality in critically ill patients. Systematic review of randomized, controlled trials. *J Hosp Infect:* 2007; 65(3):187–203.

Surgical site infection

Introduction

Surgical site infections(SSIs) are seen in at least 5% of patients undergoing a surgical procedure. They range from a minor wound discharge through to sterile osteomyelitis. SSIs make up almost 20% of all HAIs, and cause significant morbidity, increased length of stay and increased costs. Most infections are endogenous (i.e. result from contamination of the incision by the patients own microbes during surgery). The other route of acquisition is exogenous from the environment or other people. Factors associated with SSI have been well defined (Table 3.14), and multiple guidelines and recommendations have been published to aim to prevent SSI occurring. This includes the High Impact Intervention as part of the DH Saving Lives Program (Box 3.28).

National surveillance of surgical site infection

Mandatory reporting of SSI in orthopaedic surgery has shown that rates are highest in hip hemiarthroplasty – partly due to the increased risk of infection in these patients and also due to increased detection of infection as the patients tend to have a long hospital stay. Most of the SSIs reported affected the superficial layers of the wound, but approx 25% involved deeper tissues. *S. aureus* accounts for about half of all infections, and nearly one-third of SSIs are due to MRSA. A study from the USA found that compared to patients with SSI due to MSSA, the group with MRSA SSI had increased mortality and higher hospital costs.[1]

Table 3.14 Factors associated with SSI[2]

Patient related	Procedure related
Colonization with *S. aureus/roti*	Antimicrobial prophylaxis
Corticosteroid use	Duration of procedure
Diabetes mellitus	Duration of surgical scrub
Extremes of age	Foreign material
Immunosuppression	Sterilization of instruments
Longer hospital stay	Operating room ventilation
Malnutrition	Pre-operative shaving
Obesity	Pre-operative skin preparation
Remote infection	Skin antisepsis
Smoking	

Non-pharmacological measures for reducing SSI[2]

- Appropriate hair removal
- Appropriate operating room air exchanges
- Appropriate surgical attire
- Glycaemic control
- Maintenance of good oxygenation

- Limit in-and-out traffic
- Maintain normothermia
- Proper preparation of the surgical field

Box 3.28 Care Bundle for Preventing SSI: 'Saving Lives' High Impact Intervention No. 4

Pre-operative

- MRSA screening:
  - all patients undergoing implant, cardiothoracic, and neurosurgery
  - other patients according to local trust policy, e.g. orthopaedic, vascular
- MRSA decontamination – see recommendations technique on Hospital Infection Society website (www.his.org.uk)
- Hair removal – use a clipper with a disposable head. Shaving with a razor is not recommended.

Perioperative

- Prophylactic antimicrobial – where indicated, at correct timing, with appropriate antimicrobial. Remember repeat dosing in longer procedures.
- Glucose control – maintaining a blood glucose <11 mmol/L has been shown to reduce would infection in diabetic patients
- Normothermia – maintaining a body temperature above 36°C in the peri-operative period has been shown to reduce infection rates

Other useful resources

- American Healthcare Infection Control Practices Advisory Committee (HICPAC) guidelines for prevention of SSI (1999)[1,2]
- NICE guidelines: prevention and treatment of surgical site infection October 2008. www.nice.org.uk/CG74
- Specific recommendations for MRSA screening and decontamination published by a working party of the British Society for Antimicrobial Chemotherapy, the Hospital Infection Society, and the Infection Control Nurses Association (www.his.org.uk/_dh/documents/MRSA_guidelines_pdf.pdf)
- Advice on peri-operative prophylactic antibiotics can be found from the Health Technology Assessment programme; Drug and Therapeutics Bulletins and the Scottish Intercollegiate Guidelines Network (SIGN), all referenced in Saving Lives High Impact Intervention No. 4.

References

1 Engemann JJ, Carmeli Y, Cosgrove YE, et al. Adverse clinical and economic outcomes attributate to meticillin resistance among patients with Staphylococcus aureus surgical site infection. *Clin Infect Dis* 2003;36:592–98.

2 Mangram AJ. Horan TC, Pearson ML, et al. Guideline for prevention of surgical site infection. Centers for Disease Control and Prevention (CDC) Hospital Infection Control Practices Advisory Committee. *Am J Infect Control* 1999; 27(2):97–132.

Bloodstream infection (BSI)

Definitions

Primary bacteraemia is commonly defined as organisms cultured from the blood, without a documented distal source of infection, but including those resulting from an intravenous or arterial line infection. Over 95% of BSI on ICUs are primary bacteraemias, most of which are line related.[1] Secondary bacteraemia is when organisms are cultured from the blood, and are related to a documented focus of infection, e.g. infected leg ulcer.

Catheter-related bloodstream infections (CR-BSI)

Peripheral venous catheters are more commonly used for vascular access, but the risk of bloodstream infection is low (📖 p.258).

Bloodstream infections are more commonly associated with central venous catheter (CVC) insertion and are a significant cause of morbidity. Estimates suggest that up to 6000 patients a year in England may acquire a CR-BSI.[2] In 2000, the NAO estimated the additional cost of a bloodstream infection to be over £6000 per patient.

Prevention of CR-BSI

The combination of a CVC insertion guideline and a monitoring tool has been shown to significantly reduce the incidence of CR-BSI in an ICU.[3] Coated catheters and antibiotic-containing locks are helpful (📖 p.253).

Box 3.29 'Saving Lives' High Impact Intervention Number 1: Central Venous Catheter care

On insertion

- Catheter type – single lumen unless indicated otherwise for patient care; Consider antimicrobial impregnated line if 1-3 wks duration likely, and high risk BSI.
- Insertion site – subclavian or internal jugular
- Use alcoholic chlorhexidine gluconate for skin preparation, and allow to dry
- Prevent microbial contamination – hand hygiene, aseptic technique
- Sterile, transparent, semi-permeable dressing

Continuing care

- Full documentation
- Regular observation of line insertion site – at least daily
- Catheter site care – intact, clean dressing
- Catheter access
- Aseptic techniques when accessing catheter ports
- No routine catheter replacement

For further discussion of line-related sepsis including the care of peripheral lines and VIP score, see 📖 Line-related sepsis, p.252.

DH Guidance – *Taking Blood Cultures: a summary of best practice*

These recommendations aim to ensure that blood cultures are taken for the correct indication at the correct time using the correct technique. See 'Saving lives' document for details,[4] but here are the summary points:

- only take blood for culture when there is a clinical need to do so and not as routine
- members of staff taking blood cultures should be trained and competent in the procedure
- always make a fresh stab
- thoroughly disinfect the skin before inserting the needle
- once disinfected don't touch the skin again (no touch technique)
- disinfect the culture bottle cap before transferring the sample.

Other useful resources

- American Healthcare Infection Control Practices Advisory Committee (HICPAC) guidelines (2002)
- The Infection Control Nurses Association (www.icna.co.uk) audit tool for managing central lines
- EPIC guidelines www.epic.tvu.ac.uk

References

1 Richards MJ, Edwards JR, Culver DH, Gaynes RP. Nosocomial infections in medical intensive care units in the United States. National Nosocomial Infections Surveillance System. *Crit Care Med* 1999; 25(5):887-92.

2 NHS. www.clean-safe-care.nhs.uk

3 Berenholtz SM, Pronovost PJ, Lipsett PA, etal. Eliminating Catheter-related bloodstream infections in the intensive care unit. *Crit Care Med* 2004;32:2014–20.

4 NHS. www.clean-safe-care.nhs.uk

Urinary catheter-associated infection

Urinary tract infections are the largest single group of HAIs, accounting for 23%.[1] The presence of a urinary catheter and the length of time it is in place are contributory factors. Estimates from The National Audit Office report (2000) put the extra financial cost of urinary infection at £1122 per patient.[2] This report also suggested that revised urinary catheter management policies could reduce the number of UTIs. The DH Saving Lives programme includes a care bundle for insertion and ongoing care of urinary catheters (Box 3.30)[3]

Risk factors

- Presence of urinary catheter/conveen
- Duration of catheter
- Advanced age/diabetes/immunosuppression

Prevention

Before inserting a urinary catheter

- Is it really necessary? Review the indication for inserting a catheter in this particular patient at this particular time. Only use an indwelling urethral catheter after considering alternative options (penile sheath, incontinence pads). Suprapubic catheters, commonly used for acute retention, have a lower risk of infection.
- Choose the correct catheter type, catheter size, and drainage system. By selecting the optimum equipment, the risk of infection from re-catheterization can be reduced. Use the smallest catheter possible which allows adequate drainage, and make sure the length is appropriate for male/female patients. In general, a catheter with a 10 mL balloon capacity should be used, except for specific urology cases.
- Document the date of insertion and the type and size of catheter.

Insertion of the catheter

- Use sterile equipment and an aseptic technique. Clean the urethral meatus prior to insertion, using soap and water (antiseptic preparations are not necessary). Using a sterile lubricant in both male and female patients should reduce urethral trauma, thus decreasing the risk of infection.
- Antibiotic prophylaxis is *NOT* indicated in most patients. However, in some individual cases it may be beneficial.

Ongoing management of a catheterized patient

- Review the need for the catheter daily. Remove it as soon as possible.
- Empty the urinary drainage system frequently, to ensure adequate flow and prevent reflux. Use a separate container for each patient, and avoid contact between the drainage tap and container. The drainage bag should only be changed when necessary, according to the manufacturer's instructions.

- Management of the drainage bag requires universal precautions. Wash your hands and wear a new pair of gloves before manipulating the catheter. Always position the drainage bag below the level of the bladder (to prevent backflow). If this is not possible, e.g. when the patient is being moved, clamp the drainage tube, and ensure that the clamp is removed as soon as dependent drainage can be resumed.
- Clean the catheter urethral meatus junction daily with soap and water. Do not use antiseptic creams as these may increase infection. Advise the patient to have a shower rather than a bath.
- Maintain the connection between the urinary catheter and the drainage system, and only break it for good clinical reasons.
- Only flush a drainage bag if there is a clear indication (e.g. after some surgical procedures, or to manage obstructive problems).
- Do not change a catheter routinely – assess each patient's needs.
- Record ongoing management in the care plan/nursing notes.

Obtaining a urine sample from a catheterized patient

Clean the sampling port with an alcohol swab, then use sterile equipment and an aseptic no-touch technique. If there is no sampling port available, send a sample from the drainage bag (and label it as such).

Box 3.30 Urinary catheter care bundle: High Impact Intervention 6

At insertion

- Assess need for catheterization – avoid if possible
- Clean the urethral meatus prior to catheter insertion
- Sterile, closed drainage systems are recommended
- Correct hand hygiene, aseptic technique and personal protective equipment

Continuing care

- Sterile sampling of urine – perform aseptically via designated catheter port
- Drainage bag position – above floor but below bladder level to prevent reflux or contamination
- Examination gloves – should be worn to manipulate a catheter, preceded and followed by hand decontamination
- Correct hand hygiene
- Clean catheter site regularly
- Remove catheter as soon as possible.

Other useful resources

- EPIC guidelines for urinary catheter management, including insertion and management of short term indwelling urinary catheters in acute care (www.epic.tvu.ac.uk/epicphase/1.html)

- Infection Control Nurses Association (ICNA) audit tool for managing urethral catheters (www.icna.co.uk)

References

1 Emmerson AM et al. The second national prevalence survey of infection in hospitals. *J Hosp Infect* 1996;32:175–90.

2 National Audit Office 2000 – www.nao.org.uk/publications/9900/hospital_acquired_infection.aspx

3 www.clean-safe-care.nhs.uk

Ventilator-associated pneumonia

Hospital-acquired pneumonia

Respiratory infections are the second largest contributor to HAI in England.[1] Approximately 1% of hospital inpatients suffer from hospital-acquired pneumonia, which results in increased length of stay (7–9 days), increased morbidity, and increased health complications. The causes of hospital-acquired pneumonia are divided into those causing early onset (<5 days after admission) and late onset (>5 days) infections (Table 3.15). The British Society of Antimicrobial Chemotherapy (BSAC) guidelines discuss prevention, diagnosis, and treatment of hospital-acquired.[2] *Pneumonia* in more detail. See also 📖 Hospital-acquired pneumonia, p.660.

Table 3.15 Microbiology of hospital-acquired pneumonia

Early onset <5 days	Late onset >5 days	Others based on specific risks
S. pneumoniae	*P. aeruginosa*	Anaerobic bacteria
H. influenzae	*Enterobacter* spp.	*Legionella pneumophila*
S. aureus	*Acinetobacter* spp.	Viruses: influenza A and B; RSV
Enterobacter spp.	*Klebsiella* spp.	Fungi
	S. marcescens	
	E. coli	
	Other GNRs	
	S. aureus/MRSA	

Ventilator-associated pneumonia (VAP)

Pneumonia occurring during mechanical ventilation is the most common infection in intensive care units and a leading cause of deaths. In the European Prevalence of Infection in Intensive Care study,[3] VAP contributed to 45% of all infections in ICUs in Europe. Its incidence can vary between 9% and 68% in mechanically ventilated patients. VAP may be due to micro-aspiration of oropahryngeal secretions, aspiration of gastric contents, inhalation of infected aerosols, haematogenous spread from a distant site, and direct inoculation from staff (cross-infection). Predisposing factors include impaired conscious level, presence of endotracheal or nasogastric tubes, replacement of normal flora due to prior antibiotic treatment, severely ill and immunocompromised patients. About 50% is defined as early VAP, i.e. within the first 5 days. VAP has significant consequences at the individual and population level:

- increased duration of ventilation
- increased length of ICU stay and hospital stay
- increased cost (estimated at almost $12000 US per patient)
- possible increased mortality.

Emphasis here is on prevention of VAP; for further discussion of pathogenesis, clinical features, diagnosis and treatment of VAP see 📖 Ventilator-associated pneumonia, p.660.

Prevention of VAP

Recommendations to prevent VAP in the ICU include:

- appropriate disinfection and care of tubing, ventilators and humidifiers to limit contamination (although recommended by the WHO, a recent analysis of randomized controlled trials concluded that active or passive humidifiers were *NOT* beneficial in preventing VAP[4])
- no routine changes of ventilator tubing
- avoid antacids and H_2 blockers
- sterile tracheal suctioning
- nurse in head-up position
- Selective Decontamination of Digestive Tract (📖 p.237) is controversial

The *Ventilator Care Bundle* was initially introduced as part of the 100,000 Lives Campaign in the USA[5]. Its success in preventing VAP depends on all five individual steps of the bundle being performed. These five steps are:

- elevation of head of bed to 30–45 degrees
- daily 'sedation vacations' or gradually lightening the use of sedatives
- daily assessment of readiness to extubate or wean from the ventilator
- peptic ulcer prophylaxis
- deep venous thrombosis prophylaxis.

Additional elements, from 'Saving Lives' HII No. 5

- Appropriate humidification of inspired gas (to prevent inspissation of secretions)
- Tubing management – only replace when visibly soiled or mechanically malfunctioning
- Continuing care – suctioning of respiratory secretions; wear gloves and decontaminate hands before and after the suction procedure

Impact of Ventilator Care Bundle

A study from Alabama reported no cases of VAP for 255 days after implementing the Ventilator Bundle.[5] Similarly, introduction of ventilator bundles in 35 ICUs in the USA resulted in an average 44.5% reduction of VAP.[6] One year after introducing a ventilator bundle to a general ICU in the UK, the mean ICU length of stay and the mean number of ventilator days were reduced.[7] Other benefits associated with reduced VAP include better patient outcome, shorter hospital stay, lower costs, and improved staff morale.

Other useful resources

- Centers for Disease Control Guidelines: Mortality Morbidity Weekly Report 2004;53:1–36 available from www.cdc.gov
- Canadian Critical Care Society Guidelines (*Ann Intern Med* 2004;141:305–13)
- American Healthcare Infection Control Practices Advisory Committee (HICPAC) guidelines (2003)

References

1 Emmerson AM, Enstane JE, Griffin M, etal. The Second National Prevalence Surgery of infections in hospitals – Overview of the results. *J Hosp Infect* 1996;32:175–90.

2 BSAC guidelines 2006, available at www.bsac.org.uk

3 Vincent JL, Bihari DJ, Suter PM. EPIC study (1992). *JAMA* 1995; 274: 639–44.

4 *J Hosp Infect* April 2007 – Niel-Weise BS. Wille JC, Van der B Foek PJ. Humidification polices for mechanically ventilate intensive care patients and prevention of ventilator associated pneumonia: a systematic review of randomized controlled trials. *JHI* 2007; 65(4): 285.

5 www.ihi.org/IHI/Programs/Campaign/ (ref = 100k lives)

6 Resar R, Pronovost P, Haraden C et al. Using a bundle approach to improve ventilation care process and reduce ventilator-associated *Pneumonia Jt Comm J Qual Patient Saf 2005;* 31(5):243-8.

7 Cruden E, Boyce C, Woodham H, Bray B. An evaluation of the impact of the Ventilator Care bundle. *Nurse Unit Care* 2005; 10(5): 242-6.

Line-related sepsis

Intravascular devices may be complicated by local infections (e.g. phlebitis) or systemic infections (e.g. BSI, endocarditis, osteomyelitis). The most common organisms that cause line-related sepsis are coagulase-negative staphylococci; *S. aureus* (including MRSA); enterococci; Enterobacteriaecae; *Pseudomonas* spp. and *Candida* spp. Infection may arise in numerous ways. Usually lines become contaminated by the patient's skin flora at the insertion site, or by the introduction of other organisms via the cannula hub or injection port.

Always consider whether a line is absolutely necessary, or whether an alternative route of administration may suffice (e.g. nasogastric, rectal, subcutaneous). Review the continued need for a line daily.

Peripheral venous catheters (PVCs)

PVCs or 'venflons' are used most frequently for vascular access. Although they have a low risk of systemic complications, the overall total morbidity is high because they are so widely used. Almost all systemic infections are preceded by a visible phlebitis, which should act as a trigger for their removal (see Table 3.16). The care bundle approach for minimizing peripheral line infections is summarized in Box 3.31.

Central vascular catheters (CVCs)

CVCs (non-tunelled) or 'central lines' have the highest rates of catheter-related BSI. These are less easy to prevent than PVC infections. For full discussion of advances in diagnosis, prevention and management see Raad et al and High Impact Intervention no.1 (Box 3.30) for care of central venous catheters.[1,4]

Risk factors

Risk factors associated with line infection include:

- patient characteristics – age, underlying illness, immunosuppression etc.
- catheter characteristics – material, type, size, coating/impregnation
- infusate and dressing type
- experience of person inserting the line, site preparation, anatomical insertion site, and duration of insertion
- standard of daily line care.

Minimizing line infections

- Hand decontamination – wash hands thoroughly first (see 📖 Handwashing, p.212)
- Aseptic technique – maintain a strict non-touch aseptic technique when manipulating any part of the line or cannula
- Cannula selection – chose the smallest possible lumen for the fluid to be infused; use a single-lumen CVC unless multiple ports are required; consider antimicrobial-impregnated or coated CVCs if line is needed for >3 days. PICC lines (peripherally inserted central catheters) may be considered for patients anticipated to need vascular access for a longer time period.

- Insertion site – for PVC look on the distal arm, away from previous sites and joint areas. For non-tunelled CVC, consider each case carefully as choice of site is important is minimizing infection. The subclavian has the lowest risk of infection
- Skin preparation – for PVC use a 70% alcohol swab; for CVC use alcoholic chlorhexidine gluconate
- Dressing – a transparent film or sterile gauze is ideal. Write the date of insertion on the dressing. Always replace the dressing after inspecting the insertion site, or if it becomes damp, loosened or soiled.
- Observation – at least daily, or whenever the line is manipulated. The visual infusion phlebitis (VIP) score may be useful (Table 3.16)[6]
- Catheter removal – replace any lines inserted as an emergency within 24h. Remove any catheter after 72h (PVC) or 7 days (CVC) or if there are signs of infection e.g. VIP score ≥2.
- Training and audit play an integral part

Table 3.16 Visual infusion phlebitis (VIP) score[6]

Score	Description
0	Site looks healthy
1	Mild pain or redness near site
2	Two of the following evident at site: redness, pain, swelling
3	All of the following evident: redness, pain along cannula site, swelling
4	All of the following evident and extensive: redness, pain along path of cannula, swelling, palpable venous cord
5	All of the following evident and extensive: redness, pain along path of cannula, swelling, palpable venous cord and pyrexia

Antimicrobial coated or impregnated catheters

There has been increased interest in using antimicrobial coated or impregnated catheters. Agents include antiseptics (e.g. chlorhexidine and silver sulfadiazine, silver, quaternary ammonium compounds) or antibiotics (e.g. minocycline and rifampicin). Ideally compounds should be active on internal and external surfaces, and mainly target Gram-positive organisms. Evidence supports their use, but some trials were poorly designed and cost-effectiveness has been queried. Follow your hospital guidelines. One approach is to use coated or impregnated catheters if the line is likely to be in place for more than 3 days.

Antibiotic-containing locks

A meta-analysis has shown that vancomycin lock solutions in high-risk patients being treated with long term central intravascular devices reduces the risk of BSI.[7] There may be concerns re increasing resistance, e.g. VRE.

Box 3.31 Peripheral line care: High Impact Intervention No.2

On insertion

- Asepsis – prevent microbial contamination by correct hand hygiene and personal protective equipment
- Skin preparation – use (2%) alcoholic chlorhexidine gluconate, allow to dry for maximal effect
- Dressing – a sterile, semi-permeable, transparent dressing to allow observation of insertion site
- Documentation – date and site of insertion recorded in notes

Continuing care

- Continuing clinical indication – ensure all lines and associated devices are still indicated. If there is no indication then the lines or devices should be removed
- Line insertion site – regular observation for signs of infection, at least daily
- Dressing – an intact, dry, adherent transparent dressing is present
- Line access – use aseptic techniques and swab ports or hub with alcohol prior to accessing the line or administering fluids or injections
- Administration set replacement – immediately after administration of blood, blood products or lipid feeds. Replace all other fluid sets after 72 h
- Routine line replacement – replace in a new site after 72 h, or earlier if indicated clinically

See 📖 Bloodstream infection, p.242 (Box 3.29) for 'Saving Lives' High Impact Intervention No. 1: Central venous catheter care.

Other useful resources

- American Healthcare Infection Control Practices Advisory Committee (HICPAC) guidelines (2002)
- The ICNA audit tool for managing central and peripheral lines
- Canadian intravascular access devices infection control guidelines
- EPIC guidelines (www.epic.tvu.ac.uk)

References

1 Raad I, Hanneit, Maki D. Intravascular catheters-related infections: advances in diagnosis, prevention and management. *Lancet Inf Dis* 2007; 7:645–57.

2 Jackson A. Infection Control-a battle in vein: infusion phlebitis. *Nurs Times* 1998 94;68, 71.

3 Safdar N. Maki DE. Use of vancomycin containing lock or flush solution for prevention of BSI associated with central venous access devices: a meta-analysis of prospective randomized trials. *Clin Infect Dis* 2006;43:474–84.

4 www.clean-safe-care.nhs.uk

Hospital epidemics of diarrhoea and vomiting

Most outbreaks of diarrhoea and vomiting in hospitals are caused by viruses (norovirus/SRSV/winter vomiting virus). However, remember to exclude other important causes such as *C. difficile*, *Salmonella* spp. and *Shigella* spp., although these bacteria predominantly cause diarrhoea, rather than vomiting.

Enteric precautions

In cases of viral diarrhoea, these precautions apply from when the diarrhoea first starts until 48 h after symptoms have settled. If an alternative cause for the diarrhoea is found, enteric precautions are required for different lengths of time, so seek advice from the ICT.

- As soon as the diarrhoea starts, move patients to single rooms, if available. Do not wait for the stool culture result to come back. Ideally each patient with diarrhoea or vomiting should have their own toilet, commode, or bedpan. If isolation is not feasible, clean equipment after use with a detergent hypochlorite solution (1000ppm).
- Clean the bed space the patient has moved from with a detergent hypochlorite solution (1000ppm).
- Staff & visitors should wear gloves and aprons when entering the room.
- Careful hand washing is vital, after each contact with the patient. Use soap and water followed by alcohol gel.
- Clear up diarrhoea or vomit immediately, and clean the area with a detergent hypochlorite solution (1000ppm). Aerosols from vomiting are an important route of transmission in viral gastroenteritis
- During an outbreak, clean all toilets on the ward with a detergent hypochlorite solution (1000ppm) at least twice a day. Pay particular attention to toilet flush handles, toilet seats, and door handles.
- Dispose of used linen as infected laundry.
- No special treatment is needed for washing crockery or cutlery.

Hospital outbreaks of diarrhoea and vomiting

Patients

- Involve the ICT, and consider holding an outbreak meeting. Notify the ICT of any new cases immediately.
- Patients may need to be cohort nursed in bays, if there is a lack of side rooms.
- For each patient, send one sample of vomit for virology, and one stool sample for virology, culture and *C. difficile* toxin.
- Involve the HPA regional laboratory. Note that only six samples are required to confirm an outbreak of norovirus by polymerase chain reaction (PCR). If positive for norovirus, one sample is sent for genogrouping.
- Patients must be asymptomatic for 48 h before they can be transferred to another healthcare setting.

- Patients who have not yet developed symptoms should not be transferred elsewhere without consulting the ICT, as they may be incubating the infection.
- Patients may be sent home at any time, even if they still have symptoms.

Staff

- Staff should pay particular attention to handwashing.
- No food or drink should be consumed in clinical areas.
- If symptoms develop, staff should report to occupational health immediately. A sample of stool or vomit should be submitted as soon as possible, and the member of staff should stop work immediately. They can return to work 24 h after symptoms have settled.
- Nursing staff must not work in any other clinical area without consulting the ICT.
- Key areas of the ward should be cleaned at least twice daily with a detergent hypochlorite solution at 1000ppm available chlorine. This includes the environment around symptomatic patients, the toilets/ commodes/bedpans, bathrooms and showers and the sluice (especially the macerator/bedpan washer).
- Terminal cleaning/environmental decontamination is important.
- The use of bank staff should be kept to a minimum, and they should not look after infected patients.

Clostridium difficile-associated diarrhoea

C. difficile was originally defined as the cause of antibiotic-associated colitis in 1978, and continues to have a significant impact on patient morbidity and mortality. On average, patients with *C. difficile*-associated diarrhea (CDAD) have an increased length of stay of 21 days, and the total cost per case is estimated at over £4000. Figures suggest that a 10% reduction in *C. difficile* infection could save 84,000 bed days and £16million per year.[1] The political profile is high, with compulsory reporting of *C. difficile* rates by acute trusts and the newly recognized ribotype 027/NAP-1 strain.

See 🕮 Antibiotic-associated colitis, p.694 for epidemiology, clinical features, pathogenesis, diagnosis and management of *C.difficile* associated diarrhoea.

Diagnosis

Current tests in use in the UK include enzyme-linked immunosorbent assays (ELISAs) for toxins A/B, vero-cell culture for cytopathic effect, and examination under fluorescent light after culture on specific agar (CCFA: cefoxitin cycloserine fructose agar or CCEY: cefoxitin cycloserine egg yolk). Rapid tests and molecular test are being developed.

Policies for testing samples vary: in some hospitals, tests for *C. difficile* must be specifically requested, while elsewhere all unformed stools will be tested. The three-day rule (i.e. patients must have been admitted for >3 days) misses community cases of CDAD, which are increasing in number. It is usual practice not to repeat the test once a patient has a positive result. In general, three negative results are needed to exclude CDAD.

Typing

Various typing techniques have recently been placed in order of decreasing discriminatory ability. These are (most discriminatory first): multiple loci variable-number tandem-repeat analysis (MLVA); restriction endonuclease analysis (REA), pulsed-field gel electrophoresis (PFGE), surface layer protein A gene sequence typing (slpAST), PCR-ribotyping, multi locus sequence typing (MLST) and amplified fragment length polymorphism (AFLP).[2] The most common technique used in the UK is PCR-ribotyping.

Box 3.32 Reducing the risk of infection from *C. difficile*: 'Saving Lives' High Impact Intervention No. 7

- Prudent antibiotic prescribing, as per local policy. Minimize broad spectrum agents, review prescription daily include stop dates
- Hand hygiene – wash hands with soap and water before and after each patient contact; implement cleanyourhands campaign trust wide
- Enhanced environmental cleaning – use chlorine-based disinfectants to reduce environmental contamination with *C. difficile* spores as per local policy. Deep clean and decontaminate a room after a CDAD patient has been discharged
- Isolation – always use a single room if available; cohort patient care should be applied if a single room is not available
- Personal protective equipment – always use disposable gloves and apron when handling body fluids and when caring for CDAD-infected patients

Control of infection

CDAD is transmitted by clostridial spores, which are shed in large numbers by infected patients and are capable of surviving for long periods in the environment. Enteric precautions should be followed, as outlined in 📖 Hospital epidemics of diarrhoea and vomiting, p.256. Specific issues relating to the management of patients with CDAD include:

Management of the patient

- Review the individual's antibiotic prescription. Consider stopping antimicrobial agents and protein pump inhibitors. Seek advice from an infection expert if in doubt.
- Most trusts recommend metronidazole as first line, as it is cheaper and results in less VRE than using oral vancomycin. However, oral vancomycin may be indicated in cases of severe or persistent disease. There is a suggestion that metronidazole may be less efficacious for ribotype 027. Consult your local trust policy. There is no benefit of adding rifampicin to metronidazole.
- Enteric precautions must continue until the diarrhoea settles – if in doubt, seek advice from the ICT. Note that stool may remain positive for CDT for considerable time afterwards, so microbiological clearance and repeat specimens are not required.
- Management of recurrence (which can be as high as 25%) may include probiotics, faecal implants, anion-exchange resins to absorb toxins, intravenous immunoglobulin, *Saccharomyces boulardii*, and vancomycin with tapering doses or pulse doses.
- Areas of research/development include tolevamer (a polymer that binds toxin and reduces the recurrence rate), rifaximin, monoclonal antibody, and nitazoxanide (antiparasitic agent).

Prevention of infection

- *C. difficile* spores survive in the environment. Disinfect all furniture and horizontal surfaces with dilute hypochlorite solution (1000ppm).
- When enteric precautions are discontinued, the curtains around the patient's bed must be laundered.
- Handwashing – By washing with soap and water, the dilutional effect of the water and friction through rubbing the hands may help to remove some of the spores. The only way to reduce transmission is to wear gloves, but overall the environment is more important in transmission.
- Probiotics/yoghurt drink is used in some hospitals to try to prevent *C. difficile*, but firm evidence is lacking.
- *C. difficile* vaccines and anti-toxin immunoglobulins are still in research/development stages.
- There is no evidence that giving 'pre-emptive' metronidazole or vancomycin when a patient starts broad-spectrum antibiotics is beneficial.

Other useful resources

- *C difficile* infections – *Prevention and Management: a report by a Department of Health/PHLS joint working group*. Manchester: BAPS Health Publications Unit, 1994
- National *C. difficile* Standards Group. *Report to the Department of Health*. London: The Stationery Office, 2003 (presents evidence about the diagnosis and control of CDAD, and mandatory reporting of CDAD in patients >65 years) (www.dh.gov.uk/hcai)

C. difficile 027/NAP1

This new epidemic strain causes disease that is more frequent, more severe and more refractory to normal treatment than previous types. Ribotype 027 produces more toxin A and B *in vitro* (due to an 18 bp deletion in the *tcdC* gene) and also more binary toxin (significance uncertain). Ribotype 027 is resistant to fluoroquinolones, and has higher MICs to metronidazole (but is still sensitive to this agent).

References

1 www.clean-safe-care.nhs.uk

2 Killgore G, Thompson A, Johnson S et al. Comparison of seven techniques for typing international epidemic strains of Clostridium difficile: restriction endonuclease analysis, pulsed-field gel electrophoresis, PCR-ribotyping, multilocus sequence typing, multilocus variable-number tandem-repeat analysis, amplified fragment polymorphism, and surface layer protein. A gene sequence typing. *J Clin Microbiol* 2008;46(2):431–7.

Infection control in the community

The burden of HAI outside hospitals is unknown. As more patients continually move between hospitals and the community, particularly the older and more dependent patients, it is envisaged that in the future HAI in hospitals and the community will be managed as one.

At the present time, infection control in the community is defined as the infection control service provided outside acute and major hospitals to those in another care setting. This covers a wide group, including nursing home residents, renal patients on home dialysis, and outbreaks of diseases in school and places of work. The nurses responsible for infection control in the community have many other remits, which may include contact tracing for TB.

Until recently, this has been an underdeveloped area with considerable uncertainty about local arrangements and no single pattern of service provision. Therefore a questionnaire-based survey was carried out by the CDSC (Communicable Disease Surveillance Centre). The resulting 'Infection Control in the Community Study' was published in June 2002 and recommended five indicative standards (Box 3.33).

Community *C. difficile* – CDAD is being increasingly recognized in the community. It is generally defined as patients who are diagnosed with CDAD in the community, or within 48 h of hospital admission, and research suggests most patients had been in a healthcare setting recently, and received antibiotics.

Community acquired MRSA. This is genetically distinct for hospital acquired MRSA, and is becoming more common (📖 p.276).

Control of antimicrobial resistance in the community

Some countries have launched campaigns to educate doctors and patients about antibiotic misuse and the threat of drug resistance. In the USA the 'Get Smart' campaign has been driven by the Centers for Disease Control and Prevention (http://www.cdc.gov/drugresistance/community) and Health Canada is behind the 'Do Bugs need Drugs' initiative (http://www.dobugsneeddrugs.org). In the UK, Andy Biotic is a cartoon character who encourages the public to use antibiotics responsibly.

Box 3.33 Recommended standards for infection control in the community

- The infection control nurses should be competent for this role and maintain their competence through appropriate professional development.
- There should be sufficient staffing resource to enable the function to be carried out satisfactorily and in accordance with accepted national standards (e.g. decontamination, infection control).
- There should be sufficient support for the team that undertakes the infection control in the community function for a defined population.
- Primary care trusts must have robust arrangements to fulfil their infection control functions.
- There should be a formally agreed infection control guidance for community healthcare settings and other establishments that have particular infection control needs

Chapter 4

Systematic microbiology

Bacteria

Identification of bacteria

Identification of bacteria in the diagnostic laboratory is based on phenotypic characteristics such as:

- Microscopic appearance
- Growth requirements
- Colonial morphology
- Haemolysis pattern
- Biochemical tests
- Antimicrobial susceptibility patterns

Many laboratory technicians are able to make a preliminary identification to genus level based on clinical data, cultural characteristics and a limited range of tests. Commercial identification systems such as the API (Analytical Profile Index) system (Biomerieux) contain a battery of biochemical tests that can identify the organism to species level.

Microscopy

Staining and microscopic examination of samples or cultures reveals the size, shape and arrangement of bacteria and the presence of inclusions e.g. spores. The following stains are commonly used:

- **Gram stain** – A fixed slide is flooded with 0.5% methyl / crystal violet (30 secs), followed by Lugol's iodine (30secs) , followed by rinsing with 95-100% ethanol or acetone, followed by counterstaining with 0.1% neutral red, safranin or carbol fuschin (2 mins). Gram positive organisms stain deep blue / purple and Gram-negative organisms stain pink / red.
- **Acridine orange stain** – This is used to identify *Trichomonas vaginalis* in vaginal specimens. The slide is stained with acridine orange (5-10 secs), decolourised with alcoholic saline (5-10 secs) and rinsed with normal saline. Once dry a drop of saline or distilled water and a coverslip are added and the slide is examined under fluorescence microscopy. Trophozoites of *T. vaginalis* stain brick red with green nuclei.
- **Auramine stain** – This is used to identify mycobacteria in clinical specimens. It is considered more sensitive than the Ziehl Neelsen stain. A heat-fixed slide is flooded with auramine-phenol (1:10 v/v) for 10 mins. It is rinsed with water and then decolourised with 1% acid-alcohol for 3-5 mins (until no further stain seeps from the film). It is rinsed and stained with 0.1% potassium permanganate for 15 secs. It is rinsed and allowed to air dry before examination under fluorescence microscopy. Acid-fast bacilli appear bright yellow/green against a dark background.
- **Ziehl Neelsen (ZN) stain** – This is used to identify mycobacteria in cultures and provides better morphological detail than an auramine stain. A heat fixed slide is flooded with strong carbol fuschin and heated gently until it is just steaming. It is left to cool (3-5 mins), rinsed with water, and decolourised with a 3% v/v acid-alcohol solution (5-7 mins, until the slide is faintly pink). The slide is rinsed with water and counterstained with 1% v/v methylene blue or malachite green (30 secs). It is allowed to air dry before examination under oil immersion light microscopy. Acid-fast bacilli appear red on a blue or green background. A modified ZN stain is used for identification of *Nocardia* spp. and cryptosporidia.

- **Nigrosin (India ink) stain** – This stain is used to identify *Cryptococcus neoformans* in clinical specimens. A drop of India ink is put on the slide followed by a drop of the specimen and mixed together. A coverslip is applied and the slide is examined under light microscopy. *Cryptococcus neoformans* is identified by a clear zone (capsule) around the organism.

Growth requirements

These can vary considerably and include:

- **Atmosphere** – Organisms can be divided into categories according to their atmospheric requirements:
  - Strict aerobes grow only in the presence of oxygen
  - Strict anaerobes grow only in the absence of oxygen
  - Facultative organisms grow aerobically or anaerobically
  - Microaerophilic organisms grow best in atmospheres with reduced oxygen concentration (e.g. 5-10% CO_2).
  - Capnophilic organisms require additional CO_2 for growth
- **Temperature** – Organisms can also be differentiated by their temperature requirements:
  - Psychrophilic organisms grow at temperatures of 10-30°C
  - Mesophilic organisms grow at temperatures of 30-40°C
  - Thermophilic organisms grow at temperatures of 50-60°C

  Most clinically encountered organisms are mesophilic.
- **Nutrition** – Some organisms grow readily on ordinary nutrient media whereas others have particular nutritional requirements e.g. *Haemophilus influenzae* requires the specific growth factors such as factor X (haemin) and factor V (NAD).

Colonial morphology

Bacterial colonies of a single species, when grown on specific media under controlled conditions, are described by their characteristic size, shape, texture and colour. Colonies may be flat or raised, smooth or irregular and pigmented (e.g. *Pseudomonas aeruginosa* is green/blue or *Serratia marcescens* is pink) or non-pigmented. Experienced laboratory technicians can often provisionally identify an organism using colonial appearance alone.

Haemolysis

Some organisms produce haemolysins which cause lysis of red blood cells in blood containing media. This haemolysis may be:

- β-haemolytic – a clear zone of complete haemolysis around the colony
- α-haemolytic – a green zone of incomplete haemolysis
- non-haemolytic

This feature is often used in the initial identification of streptococci.

Further information

For information on the identification of bacteria see the National Standard Methods which are available from the Health Protection Agency at http://www.hpa-standardmethods.org.uk/

Biochemical tests

A variety of biochemical tests may be used for the identification of bacteria in the diagnostic laboratory.

Catalase test

Many aerobic and facultatively anaerobic organisms are catalase positive whereas streptococci and enterococci are catalase negative. Hydrogen peroxide solution is drawn up into a capillary tube and the tip is then touched onto a colony. Vigorous bubbling indicates the presence of catalase. NB media containing blood may produce a false positive result.

Coagulase test

This test is used to differentiate the staphyloocci. Coagulase exists in two forms: bound coagulase/clumping factor (detected by the slide coagulase test) and free coagulase (detected by the tube coagulase test).

- **Slide coagulase test** – A colony is emulsified in a drop of distilled water on a slide. A loop or wire is dipped into plasma and then mixed into the bacterial suspension. A positive test result occurs if agglutination is seen within 10 secs
- **Tube coagulase test** – A colony is emulsified in a tube containing plasma and incubated at 37°C for four hours. A visible clot indicates a positive result. If negative at 4 hours the tube should be reincubated overnight. NB Some species e.g. MRSA may give a negative result at 4 hours.

Deoxyribonuclease (DNase) test

This test is used to identify pathogenic staphylococci (e.g. *S. aureus* and *S. schleiferi*) which produce large quantities of extracellular DNase. A colony is streaked onto a DNase plate and incubated at 37°C for 18-24 hours. The following day the plate is flooded with hydrochloric acid – unhydrolysed DNA is precipitated producing a white opacity in the agar. Cultures surrounded by a clear zone (hydrolysed DNA) are DNase positive. NB some strains of MRSA are DNase negative and *S. epidermidis* may be weakly positive.

Optochin test

This test is used to differentiate *S. pneumoniae* (optochin sensitive) from other α-haemolytic streptococci (optochin resistant). Optochin (ethylhydrocupreine hydrochloride) is a chemical that causes lysis of the cell wall of *S. pneumoniae*. An optochin disc is placed in the centre of the bacterial inoculum and incubated at 37°C in 5% $C0_2$ for 18-24 hours. A zone of inhibition ≥5mm indicates a positive result.

Aesculin hydrolysis test

This test is used to differentiate enterococci (aesculin positive) from streptococci (aesculin negative). It tests the ability of the organism to hydrolyse aesculin to aesculetin and glucose in the presence of 10-40% bile. The aesculetin combines with ferric ions in the medium to form a black complex. The organism is inoculated onto a bile aesculin plate or slope and incubated at 37°C for 24 hours. Presence of a dark brown or black halo indicates a positive result.

Indole test

The indole test is used to differentiate the *Enterobacteriaceae*. It detects the ability of an organism to produce indole from the amino acid tryptophan. A coloured product is obtained when indole is combined with certain aldehydes. There are 2 methods:

- Spot indole test – A piece of filter paper is moistened with the indole reagent and a colony is smeared onto the surface. A green / blue colour indicates a positive result
- Tube indole test – The organism is emulsified in a peptone broth and incubated at 37°C for 24 hours. 0.5mL of Kovac's reagent is added; a pink colour in the top layer indicates a positive result.

ONPG (β-galactosidase) test

This test is used as an aid to differentiate the *Enterobacteriaceae*. Two enzymes, permease and β-galactosidase are required for lactose fermentation. Late lactose fermenters do not possess permease but do have β-galactosidase. Tubes containing o-nitrophenyl-ß-D-galactopyranoside (ONPG) are inoculated with the organism and incubated at 37°C for 24 hours. If present, β-galactosidase hydrolyses ONPG to produce galactose and o-nitrophenol, a yellow compound.

Urease test

This test is used to differentiate urease-positive *Proteus* spp. from the other Enterobacteriaceae Some strains of *Enterobacter* and *Klebsiella* spp. are urease positive. Inoculate a slope of Christensen's medium with the test organism and at 37°C for 24 hours. A pink / purple colour indicates a positive result.

Oxidase test

This test determines if an organism has the cytochrome oxidase enzyme and is used as an aid in the differentiation of *Pseudomonas, Neisseria, Moraxella, Campylobacter* and *Pasteurella* spp (oxidase positive). Cytochrome oxidase catalyses the transport of electrons from donor compounds (e.g. NADH) to electron acceptors (usually oxygen). The test reagent, *N, N, N', N'-tetra-methyl-p-pherylenediamine dihydrochloride* acts as an artificial electron acceptor for the enzyme oxidase. The oxidised reagent forms the coloured compound indophenol blue.

Further information

For information on the identification of bacteria see the National Standard Methods which are available from the Health Protection Agency at http://www.hpa-standardmethods.org.uk/

Overview of Gram-positive cocci

Gram-positive cocci are commonly isolated from clinical specimens. They are widely distributed in the environment and are found as commensals of the skin, mucous membranes and other body sites. Because of their ubiquitous nature, recovery of these organisms from specimens should always be interpreted in the context of the clinical presentation. The two medically most important groups are the staphylococci and the streptococci.

Classification

The classification of Gram-positive cocci is based on a combination of phenotypic and genotypic characteristics and has changed over time as a result of genetic sequencing data. The table below summarises the current classification of some medically important Gram-positive cocci:

- Family *Staphylococcaceae*, genus *Staphylococcus*
- Family *Streptococcaceae*, genus *Streptococcus*
- Family *Enterococcaceae*, genus *Enterococcus*
- Family *Leuconostocaceae*, genus *Leuconostoc*
- Family *Micrococcaceae*, genus *Micrococcus* and genus *Rothia*
- Family *Lactobacillaceae*, genus *Pediococcus*
- Family *Aerococcaceae*, genus *Aerococcus* and genus *Abiotrophia*
- Family *Carnobacteriaceae*, genus *Alloiococcus*
- Family *Peptococcaceae*, genus *Peptococcus*
- Family *Peptostreptococcaceae*, genus *Peptostreptococcus*
- Other anaerobic cocci e.g. genus *Anaerococcus*, genus *Finegoldia*, genus *Parvimonas*, and genus *Peptoniphilus*

Staphylocci

- Staphylococci are non-motile, non-spore-forming, catalase-positive, Gram-positive cocci.
- They occur as single cells, pairs, tetrads or grape-like clusters (most common).
- Most species are facultative anaerobes, except *S. aureus* subsp. *anaerobius* and *S. saccharolyticus*, which are anaerobic.
- Staphylocci are normally found on the skin and mucous membranes of animals. In some cases their location may be very specific e.g. *S. capitis* subsp. *capitis* on the scalp or *S. auricularis* in the external auditory canal.
- *S. aureus*, *S. epidermidis* and *S. saprophyticus* are the main human pathogens.
- They may be differentiated on the basis of the coagulase test (see Box 4.1) into coagulase-positive (e.g. *S. aureus*) and coagulase-negative (CoNS, e.g. *S. epidermidis*).
- In addition, *S. aureus* produces DNase (deoxyribonuclease) while other staphylococci are usually DNase-negative.
- There are 30 or so species of CoNS but it is rarely necessary to identify them at the species level.

Streptococci

- The streptococci are non-sporing, non-motile organisms, catalase-negative Gram-positive cocci that grow in pairs and chains.
- Some species are capsulated.
- They are facultative anaerobes and often require enriched media to grow.
- The streptococci are subdivided on the basis of their 'classic' appearance on horse blood agar into α-, β- and non-haemolytic streptococci.
- The α-haemolytic streptococci (incomplete haemolysis on blood agar, resulting in a greenish tinge) include *Streptococcus pneumoniae* (📖 p.283), and the viridans streptococci.
- The β-haemolytic streptococci (complete haemolysis/clear zone on blood agar) are grouped on the basis of their Lancefield carbohydrate antigens. The medically important ones are Groups A, B, C, F, and G.
- The enterococci (*E. faecalis* and others, 📖 p.285) were originally called *Streptococcus faecalis*, and often react with group D antisera, but are now a separate genus.
- Non-haemolytic streptococci make up the remainder and include the viridans streptococci (*S. mutans, S. salivarus, S. anginosus, S mitis*, and *S. sanguinis* groups), the anaerobic and the nutritionally variant streptococci.
- For a comprehensive review of taxonomic and nomenclature changes of the streptococci see Facklam 2002).[1]

Other Gram-positive cocci

These include: *Peptococcus, Peptostreptococcus, Leuconostoc, Pedioccoccus, Abiotrophia, Micrococcus*, and *Stomatococcus*.

Reference

1 Facklam R. What happened to the Streptococci: overview of taxonomic and nomenclature changes. *Clin Micro Reviews* October 2002; 15(4): 613-630

Staphylococcus aureus

S. aureus is a facultatively anaerobic, non-motile, non-spore-forming catalase-positive, coagulase-positive Gram-positive coccus. It is a major human pathogen and can cause a wide variety of infections ranging from superficial skin infections to severe life-threatening conditions, e.g. toxic shock syndrome

Epidemiology

S. aureus is a skin colonizer and is found in the anterior nares of 10–40% of people. Chronic carriage is associated with an increased risk of infection, e.g. in haemodialysis patients. Nasal carriage has contributed to the persistence and spread of meticillin-resistant *S. aureus* (MRSA)

Pathogenesis

S. aureus possesses a wide array of virulence factors including:

- **biofilm** – this is an extracellular polysaccharide network produced by staphylococci (and other organisms) that result in colonization and persistence on prosthetic material. Polysaccharide intracellular adhesin (PIA) is synthesized by the *ica* operon
- **capsule** – more than 90% of *S. aureus* isolates have a capsule, with 11 serotypes reported
- **surface adhesins** also know as microbial surface components reacting with microbial surface components recognizing adherence matrix molecules or MSCRAMMs. These include protein A, clumping factor A and B, collagen-binding protein, fibronectin-binding protein, serine aspartate repeat protein, plasmin-sensitive protein, and surface proteins A to K
- **techoic and lipotechoic acids** – these are components of the cell wall. Lipotechoic acids trigger release of cytokines by macrophages
- **peptidoglcan** is the scaffold for anchoring the MSCRAMMs. It also triggers release of cytokines. Modification of peptidoglycan synthesis is associated with antimicrobial resistance
- **haemolysins** – *S. aureus* possesses four haemolysins (α, β, γ, and δ).
- **Panton–Valentine leucocidin** (PVL) – this is a haemolysin encoded by two genes (*lukS* and *lukF*) which are carried on a mobile phage (φSLT). PVL-producing strains are associated with furunculosis, severe haemorrhagic pneumonia, and clusters of MRSA skin infections
- **exfoliative toxins** – ETA and ETB are encoded by the *eta* and *etb* genes respectively. They cause staphylcococcal scalded skin syndrome
- **superantigens** – this group includes the toxic shock syndrome toxin (TSST-1) and the staphylococcal enterotoxins (SEs). TSST-1 is associated with toxic shock syndrome, whereas the SEs are associated with food poisoning
- **pathogenicity (genomic) islands** – these are structures that vary in size from 15 to 70 kB and harbour virulence and drug resistance genes, e.g. SaPI1 and SaPI2 carry the gene for TSST-1
- **resistance islands** – MRSA contains a resistance island called SCC*mec* which confers resistance to meticillin.

Clinical features

S. aureus can cause a wide spectrum of clinical infections including:

- skin and soft tissue infections, e.g. impetigo, folliculitis, furuncles, and carbuncles, hidradenitis suppuritiva, mastitis, wound infections, erysipelas, cellulitis, pyomyositis, necrotizing fasciitis
- pneumonia
- bone and joint infections, e.g. septic arthritis, ostemyelitis
- systemic infections, e.g. bacteraemia, endocarditis, meningitis
- prosthetic device-related infections, e.g. intravascular catheter associated, pacemaker infections, prosthetic joint infections etc
- toxin-mediated diseases, e.g. scalded skin syndrome, toxic shock syndrome.

Diagnosis

In some cases, e.g. skin and soft tissue infections, the diagnosis is clinical. In others appropriate samples e.g. pus, tissue, or blood should be taken and submitted to the laboratory for microscopy, culture and identification:

- Gram stain – Gram-positive cocci in clusters
- culture on blood agar or liquid media – growth usually occurs within 18–24 h. Prolonged incubation detects small colony variants
- biochemical tests – catalase-positive, coagulase-positive, DNase positive.
- identification – API Staph (Biomerieux)
- typing methods include PFGE, toxin typing, SCCmec typing and spa typing
- molecular diagnosis, e.g. 16S or 23S RNA polymerase chain reaction (PCR) or *mecA* gene PCR (for meticillin resistance)

Treatment

- Treatment depends on the type of infection and the drug susceptibility of the organism.
- Flucloxacllin PO (orally) or IV (intravenous) is used for meticillin-sensitive *S. aureus* isolates.
- Vancomycin IV is used in suspected *S. aureus* infections where MRSA is a possibility, e.g. hospitalized patient with intravascular catheter.
- Aminoglycosides exhibit synergism and are used in endocarditis to help sterilize the blood cultures.
- Other agents active against *S. aureus* include clindamycin, teicoplanin, and linezolid.
- Duration of treatment depends on the cause – 7 days for skin and soft tissue infection and up to 4 weeks for endocarditis. In bacteraemia if the source is removeable, e.g. intravascular catheter, 2 weeks of treatment is adequate

Prevention

- Prevention of *S. aureus* infections is based on bacterial decolonization of carriers with local antiseptics, e.g. nasal mupirocin and chlorhexidine soap. This is usually only done for MRSA carriers.
- Vaccines – a conjungate vaccine has been shown to reduce *S. aureus* bacteraemia rates in haemodialysis patients.[1]

Reference

1 Shinerfield H et al. Use of Staphylococcus conjugate vaccine in patients receiving haemodialysis. *N Engl J Med* 2002;346:491–6.

Meticillin-resistant *S. aureus*

Meticillin-resistant *S. aureus* (MRSA) was first detected in 1961, a few months after meticillin was introduced into clinical practice. However, it was not until the 1980s that endemic strains of MRSA with multi-drug resistance became a global nosocomial problem. The epidemic strains of MRSA have been classified as E-MRSA 1 to 17, and the common ones currently circulating in the UK are E-MRSA-15 and E-MRSA–16. These strains have different genetics compared to the other epidemic strains and produce different toxins. They are also resistant to the macrolides, clindamycin, and ciprofloxacin ± other agents.

Mechanism of resistance

MRSA strains are resistant to all β-lactams due to alteration in the penicillin-binding protein PBP2′ and consequently the structure of the cell wall. Meticillin resistance is due to the *mecA* gene which codes for the low-affinity penicillin-binding protein PBP2′. *mecA* is usually located on a mobile genetic element called SCC-*mec* (staphylococcal cassette chromosome). There are five SCC-*mec* elements, defined by class of *mecA* gene and type of *ccr* complex (cassette chromosome recombinase). SCC-*mec* types I to III are usually found in hospital strains, while type IV is more common in the community strains.

Epidemiology

- Risk factors for MRSA include increasing age, prior antibiotics, indwelling catheters, severe underlying disease, intensive care unit (ICU) stay.
- MRSA rates vary in different parts of the world. Countries like Finland, Denmark and the Netherlands with very low levels of MRSA (<5%) have strictly enforced contact precautions, take surveillance cultures of patients and personnel, and limit the use of broad-spectrum antibiotics. By contrast, some Asian countries (e.g. Japan, China) have high MRSA rates, probably because of antibiotic overuse. Many middle-income (e.g. Turkey) and some high-income countries (e.g. UK, USA) have hyperendemic MRSA and usually focus available resources on high-risk patients. Some countries with endemic MRSA (e.g. Australia, France, Belgium) have managed to stabilize or even lower MRSA prevalence in defined areas.
- Recent changes in the epidemiology of MRSA include:
  - the increase in MRSA bacteraemia rates – in the early 1990s, 2% of *S. aureus* bacteramias in the UK were due to MRSA; the mean figure now is 45%. Reporting of MRSA bacteraemia rates is now mandatory in the UK
  - the emergence of new community-associated MRSA strains, which are genetically different from previous healthcare-associated strains, and tend to be more virulent.

Clinical features

- MRSA causes similar infections to meticillin-sensitive strains of *S. aureus* (see 📖 *Staphylococcus aureus*, p.274). Community-acquired MRSA (CA-MRSA) was originally seen in people with a previous history of hospitalization or who were related to healthcare workers. Since the late 1990s, however, serious infections of caused by CA-MRSA have been reported in previously healthy individuals. CA-MRSA is several times more likely to cause skin and soft tissue infections, often complicated by deep abscesses or necrotizing fasciitis.[1] There have also been community outbreaks of severe CA-MRSA pneumonia.[2] CA-MRSA is defined genetically (type IV SCC*mec*; distinct PFGE profile) and has evolved from community MSSA, rather than hospital MRSA. This genotype is often sensitive to non-beta-lactam antibiotics (e.g. ciprofloxacin) and Panton–Valentine leococidin toxin-positive. The highly successful USA-300 clone has caused considerable morbidity and mortality in the USA,[3] but is not yet a problem in Europe.

Laboratory diagnosis of MRSA

See National Standard Method.[4]

- Criteria for meticillin resistance – presence of *mecA* gene or oxacillin minimum inhibition concentration (MIC) >2 mg/L or meticillin MIC >4 mg/L or cefoxitin MIC >4 mg/L.
- Molecular detection of the *mecA* gene is the diagnostic gold standard but is not available in many routine laboratories.
- Conventional methods for detecting MRSA rely on the use of selective media (mannitol salt agar with 7% NaCl, incubated at 37°C for 18–48 h). Broth enrichment in 7% NaCl prior to plating on selective media increases the diagnostic rate but increases the time for diagnosis by 24 h.
- Chromogenic agars (e.g. MRSA-ID) look promising and are recommended by the UK Health protection Agency (HPA) for MRSA screening.
- Isolates from patients with CA-MRSA or suspected toxin-mediated diseases should be submitted to the Laboratory for Healthcare Associated infection (LHCAI) at Colindale.

Treatment of MRSA

- All UK isolates are so far susceptible to the glycopeptides (e.g. vancomycin and teicoplanin), which are the treatment of choice. Vancomycin levels should be monitored, and current practice is tending towards greater serum antibiotic levels (e.g. trough of >15 mg/L).
- There is variable susceptibility to trimethoprim, rifampicin, tetracycline, doxycycline, fusidic acid, aminoglycosides (e.g. gentamicin), and nitrofurantoin (treatment of urinary tract infections (UTIs) only). These may provide alternative treatment choices if oral therapy or a second agent is required.
- Newer agents active against MRSA include linezolid, quinupristin with dalfopristin (Synercid®) and daptomycin. These are all licensed for treating skin and soft tissue infections. In addition, linezolid and synercid may be used for pneumonia.

Infection control issues

The main strategies to control MRSA are isolation/cohorting of patients, appropriate hand hygiene by healthcare workers, and effective cleaning of shared equipment.

References

1 Miller LG. Necrotizing faciitis caused by community-associated methicillin-resistant Staphylococcus aureus in Los Angeles. *N Engl J Med*; 352: 1445–53.

2 Francis S et al. Severe community-onset pneumonia in healthy adults caused by methicillin-resistant Staphylococcus aureus carrying the Pantone-Valentine leukocidin genes. *Clin Infect Dis* 2005; 40: 100–107.

3 GJ Moran et al. Methicillin-resistant S. aureus infections among patients in the emergency department. *New Engl J Med* 2006; 355:666–74.

4 Investigation of specimens for screening of MRSA, *BSOP* 29. Available from http://www.hpastandardmethods.org.uk/pdf-sops. (accessed 23 July 2008).

Glycopeptide resistance in *S. aureus*

Mechanism of resistance

The mechanisms of vancomycin resistance in *S. aureus* are:

- an increase in cell wall turnover that leads to an increase of non-cross-linked D-alanyl-D-alanine side chains that bind vancomycin outside the cell wall and inhibit binding to target peptides
- transfer of the enterococcal *vanA* determinant from *E. faecium* to *S. aureus*,

Intermediate resistance

Strains of *S. aureus* with homogenously reduced susceptibility to glycopeptides (vancomycin MIC ≥8mg/L) have been reported in several countries, e.g. Japan, France, and the USA. These organisms have been termed vancomycin-intermediate *S. aureus* (VISA), or glycopeptide-intermediate sensitivity *S. aureus* (GISA) based on the US Clinical Laboratory Standards Institute (CLSI) vancomycin MIC breakpoints:

- susceptible ≤4 mg/L
- intermediate 8–16 mg/L (VISA)
- resistant ≥32 mg/L (VRSA).

They would be considered resistant to vancomycin, according to the British Society of Antimicrobial Chemotherapy (BSAC) MIC breakpoints (susceptible ≤4 mg/L and resistant ≥8 mg/L).

Heteroresistance

A much more common situation is for a strain to yield a small proportion of daughter cells (1 in 10^5) able to grow in the presence of 8 mg/L of vancomycin. Such heterogeneously resistant strains are called hetero-VISA and there is considerable debate about their clinical significance.

Vancomycin resistance.

The first clinical VRSA infection was reported in the USA in 2002. VRSA (vancomycin MIC >128 mg/L, teicoplanin MIC 32 mg/L) was isolated from a haemodialysis catheter tip and a chronic foot ulcer of a patient in Michigan. Vancomycin-resistant *E. faecalis* was also isolated from the ulcer, raising the possibility of transfer of the *vanA* determinant.

Clinical and epidemiological characteristics

- The first GISA infection occurred in 1995 in France in a child with leukaemia and catheter-associated MRSA bacteraemia.
- The second GISA infection occurred in 1996 in Japan in a 4-month-old infant with an MRSA sternal wound infection.
- Seven VISA infections were reported in the USA between 1997 and 2000. These all had several features in common:
  - prolonged vancomycin exposure (3–18 weeks)
  - prior MRSA infection
  - underlying disease, especially renal failure
  - prosthetic devices e.g. intravascular catheters, peritoneal dialysis (PD) catheter
- The first VRSA infection was reported in the USA in 2000 and since then six further VRSA infections have been reported in the USA.

Laboratory detection

The detection of glycopeptide resistance in *S. aureus* is problematic as both VISA and hetero-VISA isolates appear susceptible to vancomycin by routine disc diffusion tests. Furthermore, there have been conflicting recommendations regarding methods of detection:

- US guidelines – all laboratories should have an algorithm by which they can identify strains of *S. aureus* that may need additional testing. Laboratories should used acceptable confirmatory testing methods, e.g. 24-h incubation and MIC susceptibility testing method. Any *S. aureus* with a vancomycin MIC ≥4 mg/L should be referred to the Centers for Disease Control (CDC) for confirmatory testing
- UK recommendations – there are no specific guidelines for the detection of VISA/VRSA in the UK. However, the HPA recommends that reduced vancomycin susceptibility should be confirmed by E test. All vancomycin-resistant strains should be referred to the reference laboratory for confirmation.

Infection control issues

MRSA is known to be highly transmissible in healthcare settings and it seems reasonable to assume that VISA and VRSA will be likewise highly transmissible. Although infection control experience with VRSA is limited, implementation of rigorous infection control procedures is crucial for containing an outbreak of VRSA in a hospital setting. The US Hospital Infection Control Practices Advisory Committee (HICPAC) has published infection control guidelines for all staphylococci with a vancomycin MIC ≥8 mg/L:

- use contact precautions as recommended for multi-drug-resistant organisms (handwashing, gloves, gowns ± masks). Monitor and enforce compliance with contact precautions
- isolate patient in private room. Minimize number of people in contact with/caring for patient. Begin one-to-one care with specified personnel
- initiate epidemiological and laboratory investigations with the help of the state health department and CDC. Determine the extent of transmission within the facility. Assess the efficacy of precautions by monitoring the acquisition of VISA/VRSA by personnel
- educate all healthcare personnel about the epidemiology of VISA/VRSA and appropriate infection control precautions
- consult with state health departments and CDC before transferring or discharging the patient
- inform appropriate personnel about the presence of VISA/VRSA, e.g. emergency department personnel, admitting medical team.

Prevention

The appropriate use of antimicrobials, especially vancomycin, is paramount in preventing the continued emergence of VISA and VRSA. Several studies have shown that vancomycin is frequently used for inappropriate reasons. Strategies to reduce inappropriate vancomycin use are essential, e.g. minimize use of temporary central venous catheters, use diagnostic techniques to avoid prolonged empiric use of vancomycin, prompt removal of *S. aureus*-infected prosthetic devices.

Coagulase-negative staphylococci

The coagulase-negative staphylococci (CoNS) may present as culture contaminants or be true pathogens. Infection is often associated with the presence of prosthetic material, e.g. intravascular catheters, cardiac valves, joint implants. Infections are often indolent but treatment may require removal of the foreign material. These organisms are often resistant to multiple antibiotics, which can make therapy difficult.

Epidemiology

CoNS are ubiquitous and are natural inhabitants of the skin. *S. epidermidis* is the most common species, accounting for 65–90% of all isolates, followed by *S. hominis*. *S. saprophyticus* is a urinary pathogen. *S. saccharolyticus* is the only strict anaerobe. Other less-frequent species include *S. haemolyticus*, *S. warneri*, *S. xylosus*, *S. cohnii*, *S. simulans*, *S. capitis*, *S. auricularis*, *S. lugdunensis* (coagulase-positive), *S. schleiferi* (coagulase-positive).

Pathogenesis

Plasmid DNA is abundant in all species of CoNS but only a few of the-plasmid-encoded genes have been identified. Plasmid-mediated antibiotic resistance to a wide variety of antibiotics is known to occur and may be transferred by conjugation with other organisms. CoNS also produce polysaccharide intracellular adhesin (PIA) resulting in biofilm formation, particularly on prosthetic devices. This biofilm protects the organsisms from antibiotics and host defence mechansisms. *S. saprophyticus* produces a number of substances that enable it to attach and invade the uroepithelium.

Clinical features

- Nosocomial bacteraemia (most common cause)
- Endocarditis
- Intravascular catheter-related infections e.g. lines, pacemaker wires
- Cerebrospinal fluid (CSF) shunt infections
- Peritoneal dialysis catheter-associated peritionitis
- Urinary tract infections (*S. saprophyticus*)
- Bacteraemia in immunocompromised patients
- Sternal osteomyelitis (post cardiothoracic surgery)
- Prosthetic joint infections
- Vascular graft infections
- Neonatal nosocomial bacteraemias
- Endophthalmitis (after surgery or trauma)

Diagnosis

Appropriate samples, e.g. blood, pus, tissues should be taken and submitted to the laboratory for microbiological examination. The following tests may be performed:

- Gram stain – Gram-positive cocci in clusters
- culture on blood agar or liquid media – growth usually occurs within 18–24 h
- catalase-positive
- coagulase-negative (exceptions: *S. lugdunensis*, *S. schleiferi*)
- DNase test-negative, or weakly positive
- antimicrobial susceptility testing

- biochemical tests, e.g. API Staph
- typing, e.g. pulsed field gel electrophoresis
- molecular diagnosis e.g 16S or 23S RNA PCR or *mecA* gene PCR (for meticillin resistance)

Treatment

Infections usually require the removal of prosthetic material, if present. CoNS are often resistant to multiple antibiotics; >80% are resistant to meticillin. Most CoNS are sensitive to vancomycin, linezolid, quinupristin/dalfopristin, and daptomycin. Sensitivity to teicoplanin is variable and a teicoplanin MIC must be checked before using this antibiotic. *S. saprophyticus* urinary tract infections may be treated with trimethoprim, nitrofurantoin or a fluoroquinolone.

Streptococci – overview

For an introduction to streptococci see Overview of Gram-positive cocci, p.272.

Table 4.1 Classification of streptococci

Haemolysis	Lancefield group	Species name	Clinical syndromes
α		*S. pneumoniae*	Pneumococcal pneumonia, bacteraemia, meningitis, otitis media, sinusitis
α		*S. sanguis* group, *S. mutans* group, *S. mitis* group and *S. salivarius* group *Viridans Streptococci*	Dental caries, endocarditis
β	A	*S. pyogenes*	Invasive (necrotising fasciitis, GAS toxic shock syndrome, bacteraemia etc); tonsillitis, skin infections etc
β	B	*S. agalactiae*	Neonatal meningitis and bactermaemia
β	C	*S. dysgalactiae subsp. dysgalactiae* *S. dysgalactiae subsp. equisimilis* *S. equi subsp. equi* *S.equi subsp. zooepidemicus*	Sore throat and bacteraemia, cellulitis
β	D	*S. bovis* and others	*S. bovis* bacteraemia endocarditis, *S. suis* bacteraemia and meningitis
β	G		sore throat and bacteraemia cellulitis
β or non	A, C, F, G	*S. milleri* (reclassified as *S. constellatus*, *S. intermedius* and *S. anginosus*).	Infective endocarditis, Abscesses
Non; anaerobe		*Peptostreptococcus*	Abscesses
Genera closely related to streptococci		*Enterococcus* *Leuconostoc*; *Pediococcus*; *Abiotrophia*; *Gemella*; *Aerococcus*	

Streptococcus pneumoniae

S. pneumoniae was first isolated in 1881 by Sternberg in the USA and Louis Pasteur in France. It became recognized as the most common cause of lobar pneumonia and was given the name pneumococcus. *S. pneumoniae* is an important bacterial pathogen of humans causing meningitis, sinusitis, otitis media, endocarditis, septic arthritis, peritonitis, and a number of other infections. It is a Gram-positive coccus that grows in pairs (diplococci) or chains. It produces pneumolysin which causes α-haemolysis (green discolouration due to breakdown of haemoglobin) of blood agar. It is catalase-negative, inhibited by ethyl hydrocupreine (optochin sensitive), and lysed by bile salts.

Epidemiology

- *S. pneumoniae* colonizes the nasopharynx of 5–10% of healthy adults and 20–40% of healthy children. The rate of colonization is seasonal, with an increase in winter.
- The rate of invasive pneumococcal disease is 15/100,000 persons/year. The incidence is up to 10-fold higher in certain populations, e.g. African-Americans, Alaskans, and Australia aboriginals. Invasive pneumococcal disease is more common at the extremes of age (age <2 years or >65 years).
- Risk factors for pneumococcal infection include antibody deficiencies, complement deficiency, neutropenia or impaired neutrophil function, asplenia, corticosteroids, malnutrition, alcoholism, chronic diseases (liver, renal, diabetes, asthma, chronic obstructive pulmonary disease (COPD), overcrowding).
- Antimicrobial resistance is increasing. Rates are high in European countries, e.g. Spain, Hungary, and in Asia, e.g. Thailand, Hong Kong, Vietnam and Korea. The major source of resistance is the worldwide geographic spread of a few clones that harbour resistance determinants.

Pathogenesis

A number of virulence factors have been identified:

- capsular polysaccharide >90 serotypes (prevents phagocytosis, activates complement)
- cell wall polysaccharide (activates complement and cytokine release)
- pneumolysin (activates complement and cytokines)
- PspA (inhibits phagocytosis by blocking activation and deposition of complement)
- PspC (inhibits phagocytosis by binding complement factor H)
- PsaA (mediates adherence)
- autolysin (causes release of bacterial components)
- neuraminidase (possible mediates adherence).

Antimicrobial resistance mechanisms

- Penicillin resistance mediated by alterations in PBP2A (low-level resistance) and mutations in PBP2X (high-level resistance).
- Macrolide resistance is mediated by acquisition of *ermB* (ribosomal methylase) and *mefA* (efflux pump) genes.

Clinical features

S. pneumoniae may cause infection either by direct spread of the organism from the nasopharynx to contiguous structures (e.g. middle ear, sinuses, and lungs), or by haematogenous spread (to the central nervous system (CNS), heart valves, bones, joints, peritoneum). Clinical syndromes include:

- otitis media
- sinusitis
- meningitis
- exacerbation of chronic bronchitis
- pneumonia
- meningitis
- endocarditis
- septic arthritis
- osteomyelitis
- peritonitis
- others – percarditis, epidural abscess, cerebral abscess, skin and soft tissue infections. Unusual infections in young people should prompt investigation for HIV.

Diagnosis

- Grows on routine media – causes α-haemolysis of blood agar
- Gram-positive lanceolate diplococci, often with visible capsule
- Identification – catalase-negative, optochin sensitive, soluble in 10% bile salts. Commercial identification tests (e.g. API strep, latex agglutination tests, and serotyping tests) are available
- Penicillin MIC should be determined for invasive isolates

Treatment

- Depends on the nature and severity of the presenting infection and drug susceptibility results
- Penicillin MIC <0.1mg/L penicillin or amicillin
- Penicillin MIC $> 0.1 \leq 1.0$ mg/L – ceftriaxone or cefotaxime for meningitis. High-dose penicillin or ampicillin is likely to be effective for non-meningeal sites of infection, e.g. pneumonia.
- Penicillin MIC ≥ 2.0 mg/L – vancomycin ± rifampicin. If non-meningeal site, also consider ceftriaxone or cefotaxime, high-dose ampicillin, carbapenem, active fluoroquinolone, e.g. moxifloxacin

Prevention

- **Immunization** – the 7-valent pneumococcal conjugate vaccine (PCV) was introduced into the UK childhood immunization schedule in 2006, and is given to all children >2 months old in the UK. The 23-valent unconjugated polysaccharide vaccine is given to 'at-risk' groups, e.g. adults >65 years old, homozygous sickle cell disease, asplenia/severe splenic dysfunction, chronic renal disease or nephrotic syndrome, celiac disease, immunodeficiency or immunosuppression due to disease or treatment, including HIV infection, chronic diseases (cardiac, respiratory, liver, renal), diabetes mellitus, patients with cochlear implants.
- **Antimicrobial prophylaxis** – oral penicillin V is recommended for the prevention of pneumococcal disease in asplenic patients.

Enterococci

Enterococci are environmental organisms that are found in the soil, water, food, and the gastrointestinal tract of animals. They are Gram-positive cocci that occur singly, in pairs or in chains and thus resemble streptococci. Until fairly recently they were classified among the Lancefield group D streptococci. In the 1980s they were reclassified as a separate genus, *Enteroccoccus*, because of different pathogenic, biochemical and serological profiles. At least 12 different species exist. *E. faecalis* is the most common clinical isolate (80–90%), followed by *E. faecium* (5–10%). Others include *E. avium, E. casseliflavus, E. durans, E. gallinarum, E. hirae*, and *E. raffinosus*

Epidemiology

Enterococci are part of the normal gut flora and can cause endogenous or exogenous infections, both in and out of hospital. In the hospital setting, enterococci are readily transmissible between patients and institutions. In the USA enterococci are a common cause of nosocomial infections. Risk factors for nosocomial enterococcal infections include GI colonization, severe underlying disease prolonged hospitalization, prior surgery, renal failure, neutropenia, transplantation, urinary or vascular catheters, intensive care unit (ICU) admission.

Pathogenesis

- Enterococci are less intrinsically virulent than organisms such as *S. aureus* and group A streptococci. They do not have classical virulence factors but are able to adhere to heart valves and renal epithelial cells. Several extracellular molecules play an important role in colonization and adherence, e.g. aggregation factor and extracellular surface protein). Other virulence factors include extracellular serine protease and gelatinase (GelE) and haemolysins.
- Enterococci are frequently found in cultures of intra-abdominal and pelvic infections – their role in this setting has not been clearly defined.
- Enterococcal bacteraemia carries a high mortality (42–68%) but it is not clear whether this is due to the organism itself or a marker of severe debilitation. However, epidemiological studies have calculated an attributed mortality of 31–37% in patients with enterococcal bacteraemia.
- The intrinsic resistance of enterococci to many antibiotics enables them to survive and multiply in patients receiving broad-spectrum agents, and accounts for their ability to cause superinfections.

Clinical features

- Urinary tract infections (most common)
- Bacteraemia and endocarditis
- Intra-abdominal and pelvic infections
- Skin, wound, and soft tissue infections
- Meningitis (associated with anatomical defects, trauma, or surgery)
- Respiratory infections (rare)
- Neonatal sepsis

Diagnosis

- Gram stain – elongated Gram-positive cocci ('cigar-shaped'), often in pairs and short chains
- Culture – facultative anaerobes that can grow under extreme conditions, e.g. 6.5% NaCl, pH 9.6, temperatures of 10°C to 45°C
- Biochemical tests – enterococci hydrolyse aesculin and L-pyrrolidonyl-β-naphthylamide (PYR)
- They usually agglutinate with Group D in streptococcal grouping kits.
- Intrinsically resistant to aminoglycosides (low levels), β-lactams (high MICs), lincosamides (low level), co-trimoxazole (*in vivo*), and quinupristin/dalfopristin (*E. faecalis*)
- All isolates should be tested for susceptibility to ampicillin, gentamicin, vancomycin, teicoplanin, linezolid, Synercid, chloramphenicol and nitrofurantoin (UTIs)

Treatment

- Enterococci are intrinsically resistant to many agents (e.g. cephalosporins, ciprofloxacin), and readily acquire new resistance mechanisms.
- Ampicillin is usual first-line agent for *E. faecalis* infections, with vancomycin as an alternative. *E. faecium* is usually resistant to ampicillin.
- When bactericidal therapy is needed, (e.g. endocarditis, meningitis), combination synergistic therapy of a cell-wall agent plus aminoglycoside is standard.
- Ciprofloxacin may be active *in vitro*, but is not usually recommended clinically (apart from occasionally for UTIs). Newer fluoroquinolones are said to be more active against the enterococci, but not against ciprofloxacin-resistant strains, which may preclude their usefulness.
- *E. gallinarum* and *E. casseliflavus* are intrinsically resistant to glycopeptides (pentapeptide terminates D-alanine-D-serine).
- High-level resistance to aminoglycosides and vancomycin resistance (VRE) are increasing problems, particularly on renal units.
- VRE bacteraemia has a worse prognosis than vancomycin-sensitive enterococcal bacteraemia, but this may be related to comorbidity and delay in receiving appropriate antibiotic therapy.

Streptococcus bovis

S. bovis bacteraemia and endocarditis are associated with gastrointestinal disease (primarily colonic malignancy). There are two biotypes of *S. bovis*: *S. bovis* biotype 1 bacteraemia has a higher correlation with underlying GI malignancy and endocarditis (71% and 94% respectively, in one study) than *S. bovis* biotype 2.

Pathogenesis

It is not clear whether *S. bovis* is a marker for malignancy or has an aetiological role. In some cases, *S. bovis* bacteraemia is the only pointer to the GI disease. There are also reports of the malignancy being found up to 2 years after the initial *S. bovis* infection. There seems to be an increase in stool carriage of *S. bovis* in patients with malignancy or pre-malignancy compared to healthy subjects. Some investigators have suggested biotype I has a type-specific adherence mechanism, which enables adherence to both cardiac valves and abnormal colonic mucosa.

Clinical features

The main clinical infections due to *S. bovis* are bacteraemia and endocarditis. Occasionally, *S. bovis* causes other infections such UTIs, meningitis or neonatal sepsis. The GI tract is the usual portal of entry for bacteraemia, and there is a strong association of bacteraemia with endocarditis. Most patients with endocarditis have an underlying valve abnormality or prosthetic valve. They tend to have a subacute course, indistinguishable clinically from endocarditis due to the *Streptococcus viridans* group, but studies suggest *S. bovis* endocarditis has a higher mortality rate (45%) compared to non-*S. bovis* endocarditis (25%).

Diagnosis

- *S. bovis* may be misidentified as enterococci or viridans streptococci (notably *S. salivarus*).
- Biochemical tests – *S. bovis* shares a number of properties with enterococci, e.g. they agglutinate with group D antisera, hydrolyse aesculin, and are bile tolerant. However, they differ from enterococci by growing in 6.5% salt and in the results of PYR test.
- Identification – the 'API Rapid Strep' reliably identifies *S. bovis*, and differentiates it to the biotype level, which is important for association with malignancy and endocarditis. Generally, *S. bovis* biotype 1 strains produce extracellular glucan from sucrose, hydrolyse starch, and ferment mannitol: *S. bovis* biotype 2 strains are usually negative for these tests. A PCR to differentiate the biotypes has been developed.

Treatment

- *S. bovis* is highly susceptible to penicillin (MICs 0.01–0.12 microgram/mL). It is also susceptible to ampicillin, the antipseudomonal penicillins, erythromycin, clindamycin, and vancomycin.

- Penicillin is the treatment of choice for *S. bovis* infections. Vancomycin is an alternative in β-lactam-allergic patients.
- Although penicillin/aminoglycoside combinations show synergy against *S. bovis*, combination therapy is no more effective than penicillin alone for treatment of endocarditis.

All patients with *S. bovis* bacteraemia should have a comprehensive work-up to exclude colonic malignancy and endocarditis.

Viridans streptococci

The viridans streptococci, sometimes known as the oral streptococci, are important in dental caries and endocarditis, bacteraemia, and deep-seated infections. They include *S. sanguis*, *S. mutans*, *S. mitis* and *S. salivarus*. This heterogenous group has been reclassified in to five distinct groups, on the basis of 16S rDNA analysis:

- the *S. mutans* group is now divided into seven species and collectively known as the 'mutans streptococci'. The most common are *S. mutans* and *S. sobrinus*
- the *S. sanguinis* group is now divided into *S. sanguinis*, *S. gordonii*, *S. parasanguis*, and *S. crista*
- the *S. milleri* group is now divided into three species: *S. constellatus*; *S. intermedius*, and *S. anginosus* – now called the *S. anginosus* group/ group F
- the *S. mitis* group includes *S. mitis*, *S. mitior*, and *S. oralis*.
- the *S. salivarus* group includes *S. salivarus* and *S. vestibularis*

Epidemiology

The viridans streptococci are commensals of the human upper respiratory tract, femal genital tract, and gastrointestinal tract, with large numbers present in the mouth. Each species has its own particular ecological niche.

Pathogenesis

- These organisms seem to possess few virulence factors.
- The ability to produce acid, especially by *S. mutans*, is thought to be important in dental caries.
- Production of various carbohydrates, which aid adherence to tooth enamel and gums, is important in the establishment and maintenance of colonization.
- Extracellular dextran production is important in the adherence of organisms to heart valves and in resistance to antimicrobial therapy.
- Fibronectin production also mediates adherence to heart valves.

Clinical features

- Endocarditis (common cause in patients with abnormal valves)
- Bacteraemia (especially in neutropaenic patients)
- Meningitis
- Pneumonia
- Other infections – abscesses, pericarditis, peritonitis, sialadenitis, odontogenic infections, endophthalmitis

Diagnosis

- Facultatively anaerobic Gram-positive cocci, catalase-negative
- Most are α-haemolytic on blood agar; some are non-haemolytic
- Resistant to optochin and lack bile solubility (unlike pneumococci)
- Unable to grow in 6.5% NaCl (unlike enterococci)
- Can be identified by biochemical tests or API STREP

Treatment

- Community-acquired infections are usually sensitive to penicillin which is the treatment of choice.
- Other β-lactams, e.g. ceftriaxone also have good in vitro acitivity against viridans streptococci.
- Nosocomial infections are associated with increased resistance to pencillin and other β-lactams.
- Some strains, e.g. *S. sanguis and S. gordonii* exhibit tolerance – inhibited at low concentrations of antibiotic but high levels required for bactericidal activity.
- Often resistant to aminoglyocides (when traditional breakpoints are applied) but exhibit synergy in combination with B-lactam antibiotics. This principle underlies combination treatment for bacterial endocraditis.
- Vancomycin is used in penicillin-allergic patients and penicillin-resistant infections.

Group A *Streptococcus*

Group A *Streptococcus* (GAS), also known as *Streptococcus pyogenes*, is responsible for a variety of conditions, ranging from sore throat to severe invasive infections, such as necrotizing fasciitis, which have a mortality approaching 10%. There are also a number of post-infectious 'immunological' conditions such as post-streptococcal glomerulonephritis and rheumatic fever.

Epidemiology

GAS are upper respiratory tract commensals in 3–5% of adults and up to 10% of children. Transmission is mainly via droplet spread. Some people develop pharyngitis/tonsillitis, others are asymptomatic, and a handful will become carriers of GAS in the throat. In the 1990s, the number of reports of invasive GAS increased globally, probably due to a re-emergence of more virulent strains. Risk factors for sporadic disease include people >65 years old, those with recent varicella zoster (VZV) infection, HIV-positive individuals, those with diabetes, heart disease, cancer, injecting drug use, or those on high-dose steroids. Over time, the epidemiology of GAS infection in terms of clinical manifestation of disease has changed, e.g. scarlet fever and acute rheumatic fever have become less common and toxic shock more common over the last few decades.

Pathogenesis

Group A streptococci possess a number of virulence factors:

- somatic constitutents – hyaluronic capsule, M protein, serum opacity factor, lipotechoic acid, fibronectin-binding proteins
- extracellular products – streptolysin O, streptolysin S, DNases A to D, hyaluronidase, streptokinase, streptococcal pyrogenic exotoxins (SpeA, SpeB, SpeC, SpeF), C5a peptidase, and streptococcal superantigens (SSA).

Clinical features

- Pharyngitis – most common infection. Suppurative complications include tonsillitis, peritonsillar abscess, retropharngeal abscess, suppurative cervical lymphadenitis, mastoiditis, sinusitis, otitis media
- Scarlet fever – notifiable disease. Similar to pharyngitis but associated with scarlatinal rash due to erythrogenic toxin production
- Rheumatic fever – may occur 1–5 weeks after pharyngitis. Relapses may occur
- Post-streptococcal glomerulonephritis – may occur after throat infections (commonly M types 12, 1, 25, 4, and 3) and skin infections (commonly M types 49, 52, 53–55, and 57–61), and is due to immunological cross-reactions between components of the glomerular basement membrane and cell membranes of nephritogenic streptococci
- Impetigo, erysipelas, cellulitis, necrotizing fasciitis, pyomyositis
- Bacteraemia – recent increase in group A streptococcal bacteraemia in previously healthy adults. Also associated with intravenous drug users (IVDUs)
- Puerperal sepsis – historically associated with group A streptococci

- Streptococcal toxic shock syndrome – fulminant disease with a high mortality is mainly associated with types M1 and 3, but types 12 and 28 are also involved. It is differentiated from the other types of invasive disease by the occurrence of shock and multi-organ failure early in the course of the infection
- Others – meningitis, osteomyelitis, and septic arthritis

Diagnosis

- GAS are facultative anaerobic, catalase-positive Gram-positive cocci, which tend to form long chains. They are non-sporing, non-motile and usually non-capsulate.
- Culture on blood agar produces smooth, circular colonies of 2–3 mm diameter, which are usually β-haemolytic. Strains that produce haemolysin O and not haemolysin S will only demonstrate β-haemolysis when cultured anaerobically.
- Lancefield grouping will reliably and accurately identify GAS.
- Most GAS are sensitive to bacitracin.
- Serology – used to diagnose immunological complications, e.g rheumatic fever rather than in caute disease. A rise in anti-streptolysin O titre (ASOT) confirms recent group A streptococcal disease. ASOT is reliable in the throat-associated disease, while antiDNAase B is higher and more frequently raised in pyoderma-associated disease.

Treatment

- The treatment of choice is oral phenoxymethylpenicillin (mild infections) or IV benzylpenicillin (severe infections). Pharyngitis is treated for 10 days.
- In penicillin-allergic patients, options include azithromycin (which has comparative clinical and bacteriological response rates to phenoxymethylpenicillin, but higher GI side-effects) or erythromycin. In 2003 3–4% of GAS isolates in the UK were resistant to macrolides so sensitivity testing is required.
- Treatment of more-severe infections, e.g. toxic shock syndrome usually requires addition of a second agent to the penicillin – options include clindamycin (prevents toxin secretion).
- Urgent surgical debridement is required in necrotizing fasciitis.

Prevention

- Infection control – GAS can spread from infected patients to close contacts, so isolate patients with invasive disease and involve the infection control team early.
- Antimicrobial prophylaxis – The available evidence suggests that routine administration of prophylactic antibiotics for close contacts of invasive disease is not justified.[1] All household contacts should be informed of clinical manifestations of invasive disease and instructed to seek medical attention immediately if they develop any symptoms. Antibiotics are only given to certain 'high-risk' groups.

Reference

1 Health Protection Agency. Interium UK guidelines for management of close community contacts of invasive group A streptococcal disease. *Comm Dis Public Health* 2004;7(4):354–61.

Group B *Streptococcus*

Group B streptococci (GBS, *Streptococcus agalactiae*) were first reported as causes of puerperal sepsis in 1938. By the 1970s, group B streptococci have become the main cause of neonatal sepsis in infants aged <3 months.

Epidemiology

- 5–40% of women are colonized with GBS (genital tract or lower GI tract). Colonization of neonates usually occurs via the mother's genital tract. Risk factors – African-American, diabetes
- Early-onset neonatal GBS disease (≤7 days) – risk factors include GBS bacteriruria, premature rupture of membranes, delivery <37 weeks, intra-partum fever or amnionitis, prolonged rupture of membranes
- Late-onset neonatal GBS disease (7–90 days) – risk factors include overcrowding, poor hand hygiene, increased length of stay
- Over the past 20 years there has been an increase in invasive group B streptococcal disease in non-pregnant adults, most of whom had underlying medical conditions. Risk factors include diabetes, chronic diseases (liver, renal, cardiovascular, pulmonary, GI, urological), neurologic impairment, malignancy, HIV, corticosteroids, splenectomy

Pathogenesis

Bacterial virulence factors that influence the outcome between exposure and development of colonization/invasive disease include the polysaccharide capsule (in particular high amounts of sialic-acid and type III virulent strains).

Clinical features

- Early-onset neonatal disease (defined as systemic infection in the first 6 days of life; mean age of onset = 12 h, ± pneumonia or meningitis), tends to results from vertical transmission *in utero* or at the time of delivery.
- Late-onset neonatal disease (onset 7 days to 3 months of age, mean = 24 days) arises from either horizontal transmission (often nosocomial, due to suboptimal nursery conditions) or vertical transmission.
- GBS infection in adults and older children, especially those with underlying disease includes bacteraemia, postpartum infections, pneumonia, endocarditis, meningitis, arthritis, osteomyelitis, otitis media, conjunctivitis, UTI, skin and soft tissue infections, and meningitis.

Diagnosis

- GBS are facultative anaerobic, catalase-negative Gram-positive cocci. They are non-sporing, non-motile and usually capsulate.
- Culture on blood agar produces smooth, circular colonies of 2–3 mm diameter, which are usually surrounded by a very small zone of β-haemolysis.
- Selective media containing Todd Hewitt broth and antimicrobials are used to enhance recovery of group B streptococci.
- Identification – Lancefield group B, resistant to bacitracin, hydrolyse sodium hippurate, do not hydrolyse aesculin hydrolysis, production of CAMP factor (results in synergistic haemolysis with the β-lysin of *S. aureus* on sheep blood agar plate).
- Typing – GBS may be classified as serotypes I to VIII, based on the basis of capsular polysaccharide and surface protein antigens. Other typing methods: multi locus sequence typing (MLST), pulsed field gel elctrophoresis (PFGE).

Treatment

- Neonatal infections are usually treated with IV ampicillin + gentamicin initially, then penicillin G.
- Adults usually receive 10–14 days of IV penicillin G (+2 weeks gentamicin for endocarditis); vancomycin if penicillin-allergic.

Prevention[1]

- Routine screening for antenatal carriage not recommended
- Antibiotic treatment of GBS carriage not recommended
- Newborns with signs of sepsis should be treated with broad-spectrum antibiotics that cover GBS
- Infant whose mother is colonized with GBS or has had a previous infant with GBS should be monitored for at least 12 h
- Consider intrapartum antibiotics if two or more risk factors for early-onset GBS disease
- Give intrapartum antibiotics if previous child had GBS disease
- If chorioamnionitis suspected treat with broad-spectrum antibiotics active against GBS
- Vaccines – capsular polysaccharide vaccines are under development, including a vaccine conjugated with tetanus toxoid

Reference

1 Green-top guideline no. 36. *Prevention of Early Onset Neonatal Group B Streptococcal Disease* – available from http://www.rcog.org.uk/index.asp?pageID=520

Other β-haemolytic streptococci

- Group C streptococci – there are four species in this group: *S.dysgalactiae* subsp. *dysgalactiae, S. dysgalactiae* subsp. *equisimilis, S.equi* subsp. *equi, S. equi* subsp. *zooepidemicus.* They are primarily animal pathogens, but *S. equisimilis* and *S. zooepidemicus* can cause a range of infections in humans. The most common problem in humans is outbreaks of tonsillitis, especially in schools and institutions. The group C streptococci can cause syndromes similar to group A streptococci such as postpartum sepsis, septicaemia, meningitis, pneumonia, skin, and wound infections – but group C infections are usually less severe. Group C streptococci are usually sensitive to the penicillins.
- Group F streptococci – these were formerly known as *Streptococci milleri,* which has a characteristic caramel odour when cultured in the laboratory. However, *S. milleri* has been reclassified within the viridans streptococcus group into *S. constellatus, S. intermedius,* and *S. anginosus.*
- Group G streptococci – these produce infections similar to group A and C streptococci, such as sore throat, erysipelas, cellulitis, bone and joint infection, pneumonia, and septicaemia. Occasionally group G streptococci bacteraemia is associated with underlying malignancy.

Other Gram-positive cocci

Leuconostoc

Leuconostoc are catalase-negative Gram-positive cocci or coccobacilli, which occasionally cause opportunistic infections. They are usually found on plants and vegetables, or rarely in dairy products and wine. There are only a few case reports of human infections, including bacteraemia (± indwelling line infection), meningitis, and dental abscess.

Note that *Leuconostoc* are intrinsically resistant to the glycopeptides, because the pentapeptide cell wall precursors terminate in D-alanine-D-lactate. The usual agent of choice for these infections is penicillin or ampicillin, but they are generally susceptible to most agents with activity against streptococci.

Abiotrophia

Abiotrophia is the new name for the nutritionally variant streptococci (NVS). These organisms have been classified in various ways, but 16S rRNA sequencing defined the new genus *Abiotrophia* to be distinct from the streptococci. NVS are defined by the need for pyridoxal or thiol group supplementation for growth, and thus appear as satellite colonies around bacteria such as *S. aureus*. Gram staining tends to show pleomorphic variable-staining cells. The two main species, *A. defectiva* and *A. adiacens*, are resistant to optochin and susceptible to vancomycin. However, because they grow poorly on solid media, they are easily overlooked if not grown in broth or subcultured appropriately.

Abiotrophia are normal flora in the upper respiratory, urogenital, and GI tract, and are clinically important as they cause approx 5% of cases of endocarditis. Abiotrophia endocarditis responds less well to antibiotics, and has higher morbidity and mortality compared to endocarditis due to other streptococci. Correlation of *in vitro* antibiotic susceptibility testing and clinical outcome is a specialist field, and the general recommendation is for long-term combination therapy (e.g. penicillin and gentamicin for 4–6 weeks). Bacteriological failure and relapse rates are high.

Anaerobic Gram-positive cocci

The anaerobic Gram-positive cocci have recently undergone multiple taxonomic changes. Genetic sequencing has resulted in the reclassification of most organisms in the genus *Peptostreptococcus*.

The anaerobic Gram-positive cocci are part of the normal flora of the mouth, upper respiratory tract, GI tract, vagina and skin. The most commonly isolated species are *Anaerococcus* spp., *Finegoldia magna, Parvimonas micra, Peptostreptococcus* spp., and *Peptinophilus* spp. They cause abscesses (e.g. brain abscess, often associated with otitis media, mastoiditis, chronic sinusitis, and pleuropulmonary infections), anaerobic pleuropulmonary disease, and bacteraemia (notably due to oropharyngeal, pulmonary, and female genital tract sources). When mixed with other bacteria, they may be involved with serious soft tissues infections such as necrotizing fasciitis. Anaerobic Gram-positive cocci cause anaerobic osteomyelitis and arthritis at all sites, including bites and cranial infections. Little is known about virulence factors or pathogenesis of infection. Regarding treatment, anaerobic Gram-negative cocci are often mixed with aerobes and anaerobes on culture plates. Obtaining appropriate specimens may be difficult, culture can be prolonged, and anaerobic sensitivity testing can also be challenging. Usually a combination of surgery (e.g. drainage/debridement) and antibiotic therapy is required. Most anaerobic Gram-positive cocci are sensitive to metronidazole, penicillin, and clindamycin.

Aerococcus

Aerococcus viridans, and the recently described *Aerococcus urinae*, are catalase-negative, Gram-positive cocci. They tend to form tetrads and may resemble staphylococci on Gram stain, but their biochemical and growth characteristics are more characteristic of α-haemolytic streptococci.

Aerococcus viridans is generally considered a contaminant on culture, but occasionally may be implicated in bacteraemia and endocarditis. It is a low-virulent organism and only causes systemic infections in the immunocompromised. Optimal treatment of such cases is unclear, so consult an infection specialist.

Aerococcus urinae, first reported in 1989, has been implicated as a cause of approx. 0.5% of UTIs. Most patients were elderly with predisposing conditions. It has also been found in patients with urogenic bacteraemia/septicaemia with or without endocarditis. *A. urinae* is usually susceptible to penicillin and resistant to sulphonamides and aminoglycosides.

Overview of Gram-positive rods

The Gram-positive rods can be divided into three groups based on the growth characteristics and/or morphology:

- Aerobic Gram-positive rods e.g. *Bacillus, Lactobacillus, Corynebacterium, Arcanobacterium, Listeria, Erysipelothrix rhusiopathiae, Rhodococcus equi*
- Anaerobic Gram-positive rods e.g *Clostridium, Propionibacterium*
- Branching Gram-positive rods e.g. *Actinomyces, Nocardia, Actinomadura, Streptomyces*

Table 4.2 Classification of Gram-positive rods

Group	Examples
Aerobic	*Bacillus* spp.
	Corynebacterium spp.
	Listeria spp.
	Erysipelothrix rhusiopathiae
	Rhodococcus equi
Anaerobic	*Clostridium* spp.
	Propionobacterium spp.
Branching	*Actinomyces*
	Nocardia spp.
	Actinomadura
	Streptomyces

Bacillus species

Bacillus spp. are environmental saprophytes that are found in water, vegetation, and soil. They are Gram-positive (or Gram-variable) aerobic or facultatively anaerobic rod-shaped bacilli with rounded or square ends. They form endospores that tolerate extremes of temperature and moisture. The ubiquitous nature of *Bacillus* spp. means that isolation from clinical specimens may represent contamination. Members of the group include:

- *B. anthracis* (see 🕮 *Bacillus anthracis*, p.298)
- *B. cereus*
- *B. circulans*
- *B. licheniformis*
- *B. megaterium*
- *B. pumilis*
- *B. sphaericus*
- *B. subtilis*
- *B. stearothermophilus.*

Clinical features

- Food-poisoning – *B. cereus* is the most common cause; may also be caused by *B. licheniformis* and *B. pumilis*. Occurs within 24 h of ingestion of the preformed toxin in food. The emetic form presents after 1–5 h, with nausea, vomiting, and abdominal cramps. The diarrhoeal form occurs 8–24 h after ingestion of food. Production of a heat-labile toxin results in profuse diarrhoea and abdominal cramps (fever and vomiting are rare). Symptoms usually resolve in 24 h.
- Bacteraemia is the most common systemic infection and is often associated with the presence of an intravascular catheter. *B. cereus* is the most common isolate but other species, e.g. *B. licheniformis* have been reported. Bacteraemia or endorcarditis may occur in injecting drug users.
- Disseminated infection has been reported in neonates and young children. Neonatal infection is acquired perinatally. Multisystem involvement may occur. Immunocompromise, e.g. neutropenia is associated with severe and sometimes fatal infections.
- CNS infections may occur following trauma or neurosurgery, or in association with a CSF shunt. Removal of hardware is required. Lumbar puncture may result in *Bacillus* spp. meningitis.
- Eye infections – endophthalmitis may occur following trauma, eye surgery, or haematogenous dissemination. *B. cereus* is the most common cause. Keratitis may occur after corneal trauma.
- Soft tissue and muscle infections may occur after injuries or wounds, e.g. road traffic accidents or after orthopaedic surgery.

Diagnosis

Bacillus spp. grow readily on ordinary culture media at environmental temperatures (25–37°C). All species may form spores but they vary in their colonial morphology, motility, and nutritional requirements. Microscopically they are large bacteria and are usually Gram-positive (older cultures may be Gram-variable or Gram-negative). Colonies are described as anthracoid as they resemble *B. anthracis*. However, most *Bacillus* spp. are β-haemolytic and motile (unlike *B. anthracis*). They also lack the glutamic acid capsule (thus negative McFadyean's stain).

Treatment

- There is no specific treatment for food poisoning syndromes and most cases settle in 24 h.
- For intravascular catheter- or prosthetic device-related infections, removal of the catheter or device is required for cure.
- Most *Bacillus* spp. isolates are susceptible to vancomycin, clindamycin, fluoroquinolones, aminoglycosides, and carbapenems.
- Serious infections are usually treated with vancomcyin or clindamycin ± an aminoglycoside.

Bacillus anthracis

The name anthrax is derived from a Greek word for coal and refers to the eschar seen in cutaneous anthrax. Anthrax occurs most commonly in wild and domestic animals in Asia, Africa, South and Central America, and parts of Europe. Humans are rarely infected and the most common form of infection is cutaneous anthrax, which is associated with occupational exposure to animal products, e.g. wool, hair, meat, bones, and hides. Anthrax was used as an agent of bioterrorism in the United States in 2001 when *B. anthracis* spores were sent in contaminated letters.

Pathogenesis

B. anthracis has a number of virulence factors:

- **capsule** – under anaerobic conditions a polypeptide capsule consisting of poly-D-glutamic acid is produced. Synthesis of the capsule is by three enzymes encoded by the *capA, capB*, and *capC* genes on the pX-02 plasmid. A fourth protein, encoded by the *dep* gene, catalysese the formation of low molecular weight polyglutamates that inhibit phagocytosis
- **toxin** – two binary toxins (o)edema factor (EF) and lethal factor (LF) bind a third toxin component, protective antigen (PA) before entering the target cell. The three toxin components are also encoded on a plasmid pX-01. The cellular receptor for PA, the anthrax toxin receptor was identified in 2001. LF is a zinc-dependent metallopeptidase that inhibits dendritic cell function. EF converts adenosine monophoshpate (AMP) to cyclic AMP (cAMP), resulting in dysregulation of water and ions.

Clinical features

There are four forms of human disease:

- **cutaneous anthrax** – >95% of cases, usually acquired by direct contact with infected animals. The incubation period is 1–12 days and the initial lesion is a pruritic papule, which becomes a vesicular or bullous lesion surrounded by non-pitting oedema. The central part becomes necrotic and haemorrhagic and may develop satellite vesicles. Finally there is a classic black eschar which falls off in 1–2 weeks, unless systemic disease ensues
- **gastrointestinal anthrax** – accounts for <5% of case. Oropharyngeal anthrax presents with febrile neck swelling due to cervical adenopathy and soft tissue oedema after ingestion of contaminated meat. Intestinal anthrax is more common and presents with fever, syncope and malaise followed by abdominal pain, nausea and vomiting. Examination shows abdominal distension and a mass in the right iliac fossa or periumbilical area. The third phase is characterized by paroxysmal abdominal pain, ascites, facial flushing, red conjunctivae, and shock
- **inhalational anthrax** – very rare. Occurs after inhalation of spores. The incubation period is <1 week. It presents as a flu-like illness with non-productive cough, haemorrhagic mediastinal lymphadenopathy, and multilobar pneumonia ± pleural effusions and bacteraemia. Chest x-ray (CXR) typically shows a widened mediastinum. High mortality rate (45–85%)

- **CNS disease** – very rare. Presents with a haemorrhagic meningoencephalitis; 95% mortality.

Diagnosis

- Specimens – *B. anthracis* may be isolated from wound swabs (if cutaneous disease); nasal swabs and blood cultures.
- Microscopy – Gram-positive rods 4 x 1 micrometre in 'box car'- or cigar-shaped chains. The spore is oval shaped and central or subterminal. McFadyean's stain shows capsulated, dark, square-ended bacilli in short chains.
- Culture – *B. anthracis* grows readily on ordinary media (optimal incubation temperature 35°C), after 2–5 days incubation. Colonies are white or grey-white with a characteristic 'medusa head' appearance. In contrast to most other *Bacillus* spp., *B. anthracis* is non-haemolytic and non-motile.
- Identification – *B. anthracis* can be identified by PCR or phage lysis.
- For flowcharts outlining the clinical evaluation and management of possible anthrax, see www.hpa.org.uk

Treatment

- Cutaneous anthrax – ciprofloxacin or doxycycline for 60 days
- Inhalational anthrax – initial therapy: intravenous ciprofloxacin or doxycycline and one or two additional antimicrobials (e.g. rifampicin, vancomycin, penicillin, ampicillin, chloramphenicol, imipenem, clindamycin, clarithromycin). This is followed by ciprofloxacin or doxycycline until day 60.

Prevention

- Vaccines – human and animal vaccines are available to prevent anthrax. Vaccination is recommended for workers at risk of cutaneous anthrax. Vaccination is also recommended post-exposure to inhalational anthrax.
- Antibiotic prophylaxis – oral ciprofloxacin or doxycycline are indicated for post-exposure prophylaxis of inhalational anthrax.

Corynebacterium diphtheriae

The name for *C. diphtheriae* is derived from the Greek 'korynee' meaning club and 'diphtheria' meaning leather hide (for the leathery pharyngeal membrane it provokes). Diphtheria is rare in the UK but remains a common problem in developing countries and the former Russian states. The organism spreads via nasopharyngeal secretions, and can survive for months in dust and contaminated dry fomites. Incidence is highest in young children (>3–6 months old), when protective maternal antibodies wane.

Pathogenesis

C. diphtheriae exerts its effects by production of a potent exotoxin which inhbits protein synthesis in mammalian cells. It consists of two fragments: fragment A (which inhibits polypeptide chain elongation at the ribosome) and fragment B (which helps transport fragment A into the cell). Inhibition of protein synthesis probably accounts for the toxin's necrotic and neurotoxic effects, which are mainly on the heart, nerves and kidneys.

Clinical features

- **Respiratory tract** – asymptomatic upper respiratory tract carriage is common in countries where diphtheria is endemic and is an important reservoir of infection. Anterior nasal infection presents with a serosanguinous or seropurulent nasal discharge often associated with a whitish membrane. Faucial infection is the most common site for clinical diphtheria. Clinical features include fever, malaise, sore throat, pharyngeal injection, development of a pseudomembrane which is initially white, then grey with patches or green or black necrosis. Cervical lymphadenopathy may result in a characteristic 'bull neck' and inspiratory stridor.
- **Cardiac disease** – myocarditis occurs after 1–2 weeks, usually as the oropharngeal disease is improving. Patients should be monitored by electrocardiogram (ECG) which may show ST segment changes, heart block, and arrhythmias. Clinical features include dyspnoea, cardiac failure, arrhythmias and circulatory collapse.
- **Neurological disease** – local paralysis of the soft palate and posterior pharyax lead to nasal regurgitation of fluids. Cranial nerve palsies and ciliary muscle paralysis may follow. Peripheral neuritis occurs 10–90 days after onset of pharyngeal disease and presents with motor deficits
- **Skin infections** – in the tropics chronic non-healing ulcers with grey membranes may be due to *C. diphtheriae*. Outbreaks have been described in homeless alcoholics in the USA.
- **Invasive disease** – endocarditis, mycotic aneurysms, septic arthritis, and osteomyelitis have been described, caused by non-toxigenic strains.

Diagnosis

- Culture – nasopharyngeal, throat, or skin swabs should be immediately transported to the laboratory and cultured on suitable culture media (e.g. Loeffler's, Hoyle's tellurite, Tinsdale media). The colonies are black on tellurite media. *C. diphtheriae* shows a halo effect on Tinsdale's agar.
- Microscopy – Gram staining of *C. diphtheriae* shows characteristic palisades, resembling Chinese letters. The beaded appearance obtained

by Neisser or Albert stains, whereby the volutin/metachromatic granules are dark purple compared to brown/green counterstain, is characteristic.
- Identification – *C. diphtheriae* is a non-motile, non-sporing and non-capsulate Gram-positive rod. It is catalase-positive, urease-negative, nitrate-positive, pyrazinamidase-negative and cystinase-negative. It can reliably be identified with the API Coryne. Isolates should be submitted to the HPA reference laboratory at Colindale for toxigenicity testing. Several methods are available: Elek plate, rapid enzyme immunosorbent assay (EIA) or PCR.
- Biotyping – colonial appearance on tellurite, and also biochemical tests (e.g. Hiss serum sugars) subdivide *C. diphtheriae* into the biotypes var. *gravis*, *intermedius* and *mitis*. These biotypes correspond with clinical severity. *Gravis* and *intermedius* (and some *mitis*) biotypes are usually toxigenic. The fourth biotype, var. *belfanti* is rare and cannot produce the lethal exotoxin.

Treatment

- Antibiotics – if high clinical suspicion, treat immediately with IV penicillin for 14 days. Alternatives: erythromycin, azithromycin or clarithromycin. Confirm elimination by nasopharyngeal swab; if cultures are positive give a further 10 days of antibiotics.
- Antitoxin may be given, at different doses depending on site and severity (see guidelines below). Firstly test the patient with a trial dose to exclude hypersensitivity to horse serum.
- Infection control – isolate and barrier nurse the case. Identify close contacts, take nose and throat swabs, and arrange clinical surveillance for 7 days. Provide prophylactic antibiotics (single dose of benzylpenicillin or seven-day course of erythromycin) and booster vaccination for close contacts.
- Notification – diphtheria is a notifiable disease, therefore contact the CDSC.

Prevention

Diphtheria toxoid is part of the triple vaccine DTP (diphtheria, tetanus, polio) given at 2, 3, and 4 months as part of the UK immunization schedule. Immunity can be assessed by the Schick test, which is no longer used in the UK. Note that diphtheria can occur in immunized individuals.

Guidelines

- Laboratory guidelines for the diagnosis of infections caused by *C. diphtheriae* and *C. ulcerans*. *Commun Dis Public Health* 1999;2:250–7. (also available at www.HPA.org.uk)
- Control of diphtheria: guidance for consultants in communicable disease control. *Commun Dis Public Health* 1999; 2: 242–9.

Non-diphtheria corynebacteria

Corynebacteria are also known as coryneforms or diphtheroids. They are environmental organisms found in water and soil, and commensals of the skin and mucous membranes of humans and other animals. In the hospital environment they may be cultured from surfaces and equipment. Thus corynebacteria are frequently considered contaminants but may cause severe disease in hospitalized or immunocompromised patients.

Classification

Corynebacteria are classified according to cell wall composition and biochemical reactions into the following groups:

- non-lipophilic fermentative, e.g. *C. ulcerans*, *C. pseudotuberculosis*, *C. xerosis*, *C. striatum*, *C. minutissimum*, *C. amycolatum*, *C. glucuronolyticum*
- non-lipophilic non-fermentative, e.g. *C. pseudodiptheriticum*
- lipophilic, e.g. *C. jeikeium*, *C. urealyticum*.

Clinical features

Infections may be classified into two groups:

- community-acquired infections, e.g. pharyngitis, native valve endocarditis, genitourinary tract infections, periodontal infections
- nosocomial infections, e.g. intravascular catheter-associated bacteraemia, endocarditis, prosthetic device-related infections, surgical site infections.

Diagnosis

- Microscopy – club-shaped Gram-positive rods. Cells demonstrate variable size and appearance from coccoid to bacillary forms depending on the stage of their life cycle. Corynebacteria typically aggregate to form 'Chinese letter' arrangements.
- Culture – corynebacteria grow readily on blood agar and blood culture media. Thioglycolate broth may be used for wound cultures. Special media used for species identification include tryptic soy agar with or without 1% Tween-80 to assess lipid -enhanced growth.
- Identification – corynebacteria are catalase-positive, nitrate-positive, and urease-positive. They can be identified to species level by the API CORYNE system. The CAMP test (named after Christie, Atkins, and Munch–Petersen) may be used. In this test a streak of β-lysin producing *S. aureus* is plated onto blood agar and the test strain is plated perpendicular to it. A positive reaction is seen if CAMP factor (a haemolysin secreted by some corynebacteria) enhances the haemolysis produced by *S. aureus*.

- Susceptibility testing is problematic but isolates are uniformly sensitive to vancomycin, teicoplanin, and daptomycin.

Infections caused by various corynebacteria

- *C. ulcerans* is primarily a cause of bovine mastitis. However, it has the potential to produce diphtheria toxin and cause an exudative pharyngitis indistinguishable from *C. diphtheriae*. Several reported outbreaks of diphtheria have been found to be due to *C. ulcerans*.
- *C. pseudotuberculosis* is an animal pathogen that causes caseous lymphadenitis in sheep. Human disease is rare but granulomatous lymphadenitis has been seen in farm workers and vets.
- *C. xerosis* is a commensal of the human nasopharynx, conjunctiva, and skin. It may cause severe invasive disease immunocompromised patients.
- *C. striatum* is a commensal of the skin and mucous membranes. It can rarely cause severe invasive disease in hospitalized patients. It may not be correctly identified by the API CORYNE
- *C. minutissimum* is a skin commensal which was previously though to cause eryhthrasma. Bacteraemia and endocarditis may occur in patients with indwelling catheters or immunocompromise.
- *C. amycolatum* is another skin commensal. There are case reports of invasive disease. It may not be correctly identified by the API CORYNE.
- *C. glucuronolyticum* – normal flora of genitourinary tract. May cause urinary tract infection and prostatitis.
- *C. pseudodiphtheriticum* – normal flora of upper respiratory tract. Primarily associated with respiratory tract infections in immunocompromised patients.
- *C. jeikeium* colonizes the skin of hospitalized patients. It may cause severe nosomial infections, e.g. bacteraemia, endocarditis, meningitis, CSF shunt infections, prosthetic joint inections. Risk factors include immunocompromise (malignancy, neutropenia, and AIDS), indwelling catheters and devices, prolonged hospital stay, broad-spectrum antibiotics, and impaired skin integrity. *C. jeikeium* is resistant to many antibiotics, and vancomycin is the treatment of choice.
- *C. urealyticum* colonizes the skin of hospitalized patients. It causes chronic and recurrent urinary tract infections mainly in the elderly or immunosuppressed.

Listeria

Listeria monocytogenes is the main species in this genus, and affects pregnant women, their babies, the immunocompromised (especially those with impaired cell-mediated immunity) and the elderly. *L. ivanovii* occasionally causes human infection. Generally *L. innocua*; *L. welshimeri*, and *L. seeligeri* are non-pathogenic to humans. Up to 5% of healthy adults carry *Listeria* spp. in the gut. While listeria infections are rare in the general population, they can cause life-threatening bacteraemia and meningoencephalitis in susceptible groups. Clinical infections have high mortality rates.

Epidemiology

Disease is mainly sporadic, but may be part of an epidemic associated with contaminated foodstuffs such as pâté, unpasteurised milk, chicken, or soft cheese. Hospital outbreaks have been reported. Vets or farmers may become infected through direct animal contact. Human–human transmission occurs vertically (i.e. mother–baby), and cross-infection in neonatal units has been reported.

While the number of pregnancy-associated cases of listeriosis has been relatively stable, there was a dramatic rise in non-pregnancy-associated listeriosis between 2001 and 2004, especially in people over 60 years old. The reasons for this are unclear.

Pathogenesis

Animal studies have identified listeriolysin O: this is important for bacterial survival after phagocytosis, and its production is related to extracellular iron. In rodents, T lymphocytes are important in protective immunity rather than antibodies. T cells attract monocytes to the infection, activate them and destroy the listeria, resulting in granuloma. The organisms themselves show tropism for the brain itself, particularly the brainstem and meninges. In humans, gastrointestinal disease (e.g. low gastric pH or disrupted normal flora) may help establish listeria infection in the bowel.

Clinical features

- **Pregnancy** – maternal listeriosis is rare before 20 weeks' gestation. After this, infection may be asymptomatic or present with mild symptoms such as fever, back pain, sore throat, and headache. Fever may result in reduced fetal movements, premature labour, stillbirth, abortion, or early-onset neonatal disease.
- **Neonate** – early neonatal disease occurs <5 days post delivery, usually presents with septicaemia, and has a mortality of 30–60%. Late neonatal disease occurs >5 days post delivery, usually presents as meningitis, and may be hospital acquired. Mortality in late disease is lower (approx 10%).
- **Adults –** the main syndromes are CNS infection and meningitis, septicaemia, and endocarditis. Rare manifestations include other CNS disease (such as encephalitis, cerebritis, CNS abscesses), arthritis, hepatitis, endophthalmitis, continuous ambulatory peritoneal dialysis (CAPD) peritonitis, gastroenteritis, pneumonia. Risk factors include immunosuppression due to steroids, cytotoxic therapy, and HIV. Note that in *Listeria* meningitis, CSF biochemistry may be indistinguishable from bacterial meningitis. The Gram stain is often negative for organisms, but *Listeria* may sometimes be cultured from blood.

- **Neonatal disease** – early = 30–60% (20–40% survivors developing long-term sequelae such as lung disease or CNS defects). Late = 10%.
- **Adult disease** – CNS = 20–50%; bacteraemia = 5–20%; endocarditis = 50%. Up to 75% survivors of CNS infection have sequelae such as hemiplegia or CNS defects.

Diagnosis

- Microscopy – *Listeria* are short intracellular Gram-positive rods. However, in clinical specimens they may appear Gram-variable, and look like diphtheroids, cocci, or diplococci.
- Culture – *Listeria* grow on blood agar but selective media are available. Colonies are sometimes β-haemolytic on blood agar, and can be mistaken for streptococci or enterococci.
- Identification – they exhibit tumbling motility at 25°C. They are non-sporulating, and catalase-positive, aesculin-positive, and oxidase-positive. They grow optimally at 30–37°C, but better than most bacteria at 4–10°C (refrigeration temperature). *L. monocytogenes* and *L. seeligeri* show enhanced haemolysis in the presence of *S. aureus* (positive CAMP test). *L. ivanovii* produces a positive CAMP test. Species can also be differentiated by fermentation of D-xylose, L-rhamnose and α-methyl-D-mannoside.
- Typing techniques in current use include phage typing, serotying, PFGE and multi-locus enzyme electrophoresis (MLEE). Serotyping of *L. monocytogenes* with rabbit antisera results in 13 serovars: serovar 4 is commonest in human infections (but serovars 1/2a and 1.2b are also important).

Treatment

Ampicillin ± gentamicin is the usual regimen for meningitis, with co-trimoxazole or meropenem as an alternative for patients who are penicillin allergic. There are no randomized controlled trials to establish the most effective drug or duration of therapy. In meningitis, antibiotics are usually given for at least 14 days (longer in immunocompromised). Most other clinical syndromes should be treated with ampicillin, with consideration given to adding gentamicin for synergy. Vancomycin may be given for bacteraemia, but has been associated with relapse of disease. NB cephalosporins should never be used to treat listeriosis.

Prevention

Health education and dietary advice to pregnant women, immunocompromised and others who are at risk of disease. Co-trimoxazole prophylaxis prevents listeria infections.

Erysipelothrix rhusiopathiae

E. rhusiopathiae is a thin, pleomorphic, non-sporing, Gram-positive rod. It was first isolated in mice by Robert Koch in 1878 and from swine by Louis Pasteur in 1882. It was identified as a human pathogen in 1909.

Epidemiology

E. rhusiopathiae is found in a wide variety of animals and invertebrates – the reservoir is thought to be swine. The organism in transmitted to humans by direct contact. Most human cases are associated with occupational exposure, e.g. fishermen, fish handlers, farmers, vets, butchers, abattoir workers.

Clinical features

There are three clinical presentations:

- **erysipeloid** – a localized skin lesion. The organism enters the skin by trauma and after an incubation period of 2–7 days, pain and swelling of the affected digit occurs. The lesion is well defined, slightly raised, and violaceous. It spreads peripherally with central fading. Regional lyphadenopathy and lymphangitis may occur
- **diffuse cutaneous eruption** – this is rare and caused by progression of the primary lesion. Fever and arthralgia may occur. Recurrence is common
- **bacteraemia** is also rare but frequently associated with endocarditis.

Diagnosis

- **Microscopy** – *E. rhusiopathiae* is a straight to slightly curved Gram-positive rod (1–2.5 micrometre)– it decolorizes readily and may appear Gram-negative. Rods may be arranged singly, in V-shaped pairs, short chains or non-branching filaments.
- **Culture** – colonial and microscopic appearances vary with the medium, pH and incubation temperature. Incubation in 5–10% CO_2 improves culture.
- **Identification** – *E. rhusiopathiae* is catalase-, oxidase-, indole-, Vosges–Proskauer- and Methyl-red-negative.
- **Drug susceptibility** – *E. rhusiopathiae* is usually susceptible to penicillins, cephalosporins, clindamycin, imipenem, and ciprofloxacin. It is resistant to vancomycin, teicoplanin, sulfonamides, co-trimoxazole, and aminoglyocisides.

Treatment

- Penicillin is the treatment of choice.
- Alternatives include ampicillin, cephalosporins, and ciprofloxacin.

Rhodococcus equi

R. equi (previously known as *Corynebacterium equi*) was identified in 1923 as an animal pathogen causing pneumonia in horses. Since then it has been found in a wide variety of animals. The first human case was reported in 1967 – an immunosuppressed patient who presented with a cavitatory pneumonia. Since the 1980s the rise in the numbers of immunosuppressed patients (AIDS and transplantation) has been mirrored by and increase *R. equi* infections.

Clinical features

- Necrotizing pneumonia is the most common clinical presentation (80%) and is characterized by a cavitation on the CXR. Blood cultures are positive in 50% of HIV patients and 25% of solid organ transplant recipients.
- Extra-pulmonary infection may affect the brain or present as subcutaneous or organ abscesses. Bacteraemia may also occur, usually associated with intravenous catheters.

Diagnosis

- *R. equi* is a Gram-positive obligate aerobe that is a non-sporing and non-motile.
- Microscopy – it may appear coccoid or bacillary on Gram stain, depending on growth conditions. It can be acid-fast.
- Culture – *R. equi* grows optimally at 30°C and produces salmon-pink colonies. Selective media include colistin nalidixic acid agar (CNA), phenyl ethanol agar (PEA), or ceftazidime novobiocin agar.
- Identification: *R. equi* is catalase, lipase, urease and phosphatase-positive. It differs from other coryneforms by its lack of ability to ferment carbohydrates or liquefy gelatin. It can be identified using the API CORYNE, ribotyping or PCR RFLP (restriction fragment length polymorphism).

Treatment

Optimal treatment has not been determined by clinical trials. *R. equi* is susceptible to vancomycin, erythromycin, fluoroquinolones, rifampicin, imipenem, and aminoglycosides. Combinations of two or three antimicrobials are usually used until antimicrobial susceptibility results are available.

Arcanobacterium haemolyticum

Arcanobacterium haemolyticum is a β-haemolytic, catalase-negative Gram-positive rod which pits the agar when a colony is removed. Identification can be confirmed by the 'API CORYNE'. It causes acute pharyngitis, and has also been associated with infective endocarditis and skin sepsis. It is sensitive to most antibiotics, except co-trimoxazole. Treatment is usually with penicillin or erythromycin.

Clostridium botulinum

C. botulinum is widespread in the soil and environment. It produces one of the most potent toxins known, which causes botulism.

Pathogenesis

Toxins A–G have identical pharmacological effects, despite possessing different antigens. All can cause human disease, but A, B, and E are most common. Note that the type-specific antibody must be given to a patient with suspected botulism (see below).

Clinical features

- **Food-borne botulism**, the preformed toxin is ingested from food (hams, sausages, tinned fish, meat, and vegetables (particularly home-preserved), honey). The food itself may not appear spoiled. Botulinum toxin is absorbed from the human GI tract and blocks the release of acetylcholine mainly in the peripheral nervous system. Initial symptoms include nausea and vomiting, diplopia, and bilateral ptosis (due to oculomotor muscle involvement), followed by progressive descending motor loss with flaccid paralysis. Speech and swallowing become difficult, but the patient maintains consciousness and has in normal sensation. Botulism is fatal in 5–10% cases. Death is usually due to cardiac or respiratory failure.
- **Wound botulism** causes a similar clinical picture, but is due to growth of the organism. Outbreaks have occurred in intravenous drug users.
- **Intestinal botulism** is also due to organism proliferation in the gut and toxin production *in vivo*.
- **Infant botulism** presents as the 'floppy child syndrome' usually in babies <6 months, because the gut is not yet resistant to colonization.

Diagnosis

If there is a suspected clinical case, involve experts and always alert lab staff as the toxin is dangerous. Human samples (blood, faeces, vomit) and food should be tested for the organism and the toxin. *C. botulinum* is a motile strictly anaerobic rod, with optimal growth at 35°C, but some strains able to grow as low as 1–5°C. The oval subterminal spores are very hardy: some spores persist despite boiling at 100°C for several hours. Moist heat at 120°C for 5 min usually destroys spores.

Treatment

Involve the intensive care unit, as the patient is likely to need organ support. A polyvalent antitoxin is available to neutralize unfixed toxin. In food-borne disease, any unabsorbed toxin should be removed from the stomach and GI tract. In wound botulism, give benzyl penicillin and metronidazole, and surgical debridement. Antibiotics are not recommended for food-borne or intestinal botulism.

Prevention

The canning industry must ensure that food in all parts of the can is adequately heated. Avoid home-canning of food. People who have eaten food suspected of causing botulism should be given one dose of polyvalent antitoxin prophylactically.

Table 4.3 Diseases caused by *Clostridium* spp.

Organism	Clinical syndrome	Toxin production
C. botulinum[a]	Botulism	Neurotoxin
C. tetani[a]	Tetanus	Neurotoxin
C. difficile	Antibiotic-associated diarrhoea/pseudomembranous colitis	Toxin A and B
C. perfringens[a]	Type A causes gas gangrene	Histiotoxic
C. novyi[a]	Type A causes gas gangrene	Histiotoxic
C. sporogenes	Debate re pathogenicity	
C. septicum	Gas gangrene	Histiotoxic
C. histolyticum	Gas gangrene	Histiotoxic
C. sordellii	Gas gangrene	Histiotoxic

[a]Clusters in injecting drug users in Europe in the last 5 years.

Box 4.3 Botulinum toxin as a potential agent of bioterrorism

Several countries have attempted to weaponize the potent toxin for airborne dispersal, which would result in toxin inhalation. Contamination of food would also be possible. Water dispersal is unlikely, as standard water treatment protocols neutralize the toxin. There is a vaccine against botulism, but it is not widely used due to concerns re effectiveness and adverse effects. The HPA Centre for Infections has issued interim guidelines in the event of deliberate release available at www.HPA.org.uk. See 📖 Bioterrorism, p.876.

Clostridum tetani

C. tetani causes ~10 cases of tetanus/year in the UK. However, this vaccine preventable disease still causes considerable morbidity and mortality in the developing world. Tetanus is a notifiable disease.

Pathogenesis

Resilient spores survive in soil and GI tract of horses and other animals for a long time. Transmission usually occurs via introduction of spores into open wounds (particularly in injecting drug users), patients with recent abdominal surgery, patients with ear infections (otogenic tetanus), and neonates after cutting the umbilical cord (tetanus neonatorum). *C. tetani* produces tetanospasmin (powerful neurotoxin which diffuses to the CNS and causes localized or generalized disease), and tetanolysin (oxygen-labile haemolysin).

Clinical features

Localized tetanus involves muscle rigidity and painful spasms near the wound site. Usually a prodrome of generalized tetanus, with symptoms summarised by ROAST (rigidity, opisthotonus, autonomic dysfuction, spasms, and trismus)

Diagnosis

Tetanus is a clinical diagnosis. There are three microbiological tests:

- isolation of *C. tetani* from the infection site. *C. tetani* is a motile, obligate anaerobe which classically produces 'drumstick' terminal spores. It often stains Gram-negative. *C. tetani* produces a thin spreading film on enriched blood agar, due to the motility by peritrichous flagella. If *C. tetani* is suspected, involve the HPA Anaerobe Reference Laboratory
- presence of tetanus toxin in serum (performed at HPA Colindale Food Safety Laboratory)
- low/no antibody levels to tetanus toxin is supportive of the diagnosis.

Treatment

Involve ICU early. Give tetanus immunoglobulin (TIG); wound debridement, and antimicrobials including metronidazole or penicillin. Vaccination with tetanus toxoid following recovery is important to prevent future episodes. See Table 4.4.

Tetanus-prone wound risk factors include the following

- Puncture-type wound,
- contact with soil or manure,
- clinical evidence of sepsis
- significant degree of devitalized tissue
- any wound with delay of >6 h before surgical treatment.

Prevention

Tetanus immunization, introduced into the UK in 1961, now involves the combined tetanus/low-dose diphtheria vaccine (Td) (previously single antigen vaccines (T) were given). Five doses of tetanus toxoid are considered to give lifelong immunity (usually three as DTP as part of childhood immunizations and two doses of Td later). See the Department of Health's 'Green Book': Immunisation against Infectious Disease (2006) for more information.(http://www.dh.gov.uk/en/PublicHealth/Healthprotection/Immunisation/Greenbook/dh_4097254).

Table 4.4 Recommendations for vaccination

Immunization status	**Clean wound**	**Tetanus-prone wound**	
	Vaccine	Vaccine	TIG
Full, i.e. 5 doses	No	No	Only if high risk
Primary immunization complete, boosters incomplete but up to date	No	No	Only if high risk
Primary immunization incomplete/boosters not up to date/never immunized/status unknown or uncertain	Yes – 1 dose and plan to complete schedule	Yes – 1 dose and plan to complete schedule	Yes – 1 dose in a different site

Other clostridia

These anaerobic Gram-positive spore-forming organisms are responsible for a variety of conditions, many of which involve toxin production (Table 4.3). The rods are pleomorphic, but typically large, straight, or slightly curved, with rounded ends.

- *C. perfringens* causes gas gangrene. It is occasionally isolated from blood cultures, and may be associated with food poisoning (enterotoxin production), endocarditis, or a contaminant. In developing countries it may cause enteritis necroticans ('pig bel').
- *C. histolyticum* and *C. sordellii* may be associated with gas gangrene
- *C. novyi* gas gangrene is due to *C. novyi* type A (*C. novyi* types B, C, and D are differentiated by toxin permutation and soluble antigen production, and do not cause human disease). Compared to *C. perfringens*, *C. novyi* bacilli are larger and more pleomorphic. It is a stricter anaerobe, and has peritrichous flagella, but motility is inhibited in the presence of oxygen. The oval spores are central or subterminal. There are at least four toxins which possess haemolytic, necrotizing, lethal, lipase, and phospholipase activities. There was a large outbreak among injecting drug users in Scotland in 1999–2000.
- *C. sporogenes* is probably not pathogenic in its own right. It is usually encountered in a mixed wound culture containing accepted pathogens, and may have a role in enhancing local conditions and accelerating an established anaerobic infection.
- *C. septicum* usually lives in the soil, human, or animal gut and can cause gas gangrene in humans and animals. *C. septicum* bacteraemia is seen with breakdown of gut integrity, e.g. in leukaemia. Gram stain appearance of the organism may be variable, with long, short, and filamentous Gram-positive rods, together with some older Gram-negative cells. Spores start off as swollen Gram-positive 'citron bodies' then tend to be oval, bulging, and either central or subterminal. *C. septicum* grows well on ordinary media at 37°C, and has numerous peritrichous flagella, hence is actively motile. Colonies are often initially transparent and 'droplet-like', with projecting radiations, then become grey and opaque with time. The α exotoxin has lethal, haemolytic, and necrotizing properties, and can be demonstrated in cultures.
- *C. difficile* is an increasing nosocomial infection, which is now subject to mandatory reporting. It can cause *C. difficile*-associated diarrhoea (CDAD) and pseudomembranous colitis. Clinical features vary, and diagnosis is usually by toxin tests rather than culture. Infection control measures are paramount to control the spread of this organism.

Mobiluncus

Mobiluncus is a genus of anaerobic Gram-positive rod shaped bacteria. These organisms are found in the female genital tract in association with Gardnerella vaginalis. There are two named species: *M. curtisii* (smaller, Gram-variable, and slightly bent), and *M. mulieris* (larger, Gram-negative and crescent-shaped).

Pathogenesis

Adherence to vaginal squamous epithelial cells is important, and may be caused by a glycocalyx.

Clinical features

Mobiluncus spp. has been detected in >97% of women with bacterial vaginosis (together with mixed anaerobic flora), and ~5% of healthy controls. It has also been found in other sites, such as breast abscesses and mastectomy wounds.

Diagnosis

These curved Gram-variable rods grow slowly under anaerobic conditions. Electron microscope studies have described a Gram-positive cell wall. Their characteristic corkscrew motility is due to multiple flagella. They need an enriched media for growth, and are oxidase-, catalase- and urease-negative. Gram stain morphology can reliably differentiate the two species.

Treatment

Metronidazole is an appropriate treatment for bacterial vaginosis, so is used whether *Mobiluncus* is isolated or not. *Mobiluncus* is usually susceptible to penicillin, erythromycin, clindamycin, and vancomycin.

Actinomyces

Actinomyces species are mouth commensals that may cause the chronic granulomatous infection actinomycosis. The main species of human importance are *A. israelii* and *A. gerencseriae*. Others include *A. meyeri* (isolated from brain abscesses); *A. viscosus* (found in dental caries); and also *A. naeslundii* and *A. odontolyticus*.

Pathogenesis

Actinomycosis is endogenously acquired, and those with dental caries are at increased risk. It is unclear why males are affected more than females. Historically rural farm workers were affected more than those living in towns, purportedly because of poor dental hygiene. Abscesses, tissue destruction, fibrosis, and sinus formation are typical findings. The masses of mycelia in relatively young lesions may be visible as yellow sulphur granules; later on they form dark brown, hard granules due to calcium phosphate deposition.

Clinical features

Most human cases of actinomycosis are in the cervicofacial area, especially around the jaw. Infection may follow dental procedures. Haematogenous spread to the liver, brain, and other organs is well recognized. In addition to facial disease, clinical presentations include thoracic actinomycosis (due to aspiration of oral actinomyces; characterized by chest wall sinuses and bony erosion of the ribs and spine), appendix or colonic diverticula actinomycosis, pelvic actinomycosis (linked with intrauterine contraceptive devices (IUCDs)), cerebral actinomycosis, and 'punch actinomycosis' (knuckle infection due to human bite).

Diagnosis

Tissue biopsies of suspect lesions are stained with fluorescein-conjugated specific antisera, to demonstrate characteristic sulphur granules and mycelia. Any sulphur granules available should be crushed and stained with Gram stain – branching Gram-positive rods.

Actinomyces often fail to grow aerobically, so plates should be incubated anaerobically and under micro-aerophilic conditions (i.e. 5–10% CO_2). *A. israelii* form large 'molar teeth'-shaped colonies, from 2 to 10 days. Further identification can be confirmed at a reference laboratory by biochemical tests, fluorescent antisera staining, or gas chromatography of metabolic products of carbohydrate fermentation. Note that sputum often contains oral actinomyces.

Treatment

Surgical involvement is vital, and debridement reduces scarring, deformity, and the recurrence rate. Removal of an IUCD is the primary treatment for pelvic disease. Actinomycosis is usually treated with penicillin or ampicillin, for up to six months. Broad spectrum antibiotics e.g. co-amoxiclav, or ceftriaxone and metronidazole may be needed if there are concomitant pathogen. Despite large doses of antibiotics given for long periods, recurrence is common. The issue seems to be one of tissue penetration, rather than drug resistance.

Nocardia

Nocardia species are environmental saprophytes which occasionally cause chronic granulomatous infections in humans and animals. Thirty three species of *Nocardia* have been shown to cause disease in humans with members of the *Nocardia asteroides* complex (*N. asteroides sensu stricto, N. farcinica* and *N. nova*) being the most common worldwide. Other species isolated from clinical samples include *N. abscessus, N. brasiliensis, N. cyriacigeorgica, N. otitidiscaviarum, N. paucivorans, N. pseudobraziliensis,* and *N. transvalensis.*

Pathogenesis

Pulmonary nocardiasis is acquired through inhalation of the bacilli. Cutaneous nocardiosis occurs as a result of inoculation injury. Disseminated or CNS nocardiosis occurs following haematogarneous spread. Pulmonary and disseminated disease is more common in immunosuppressed patients.

Clinical features

Pulmonary nocardiasis is more common in the immunosuppressed and those with pre-existing lung disease, particularly alveolar proteinosis. Presentation and clinical/radiological findings are variable, making the diagnosis difficult. Patients tend to develop multiple lung abscesses, and the course may be acute or chronic. Secondary abscesses, mainly in the brain, occur in approx one-third of patients with pulmonary nocardiasis. Other clinical presentations include cutaneous disease (e.g. post trauma) with lymphatic involvement (sporotrichoid), which may progress to a fungating mycetoma.

Diagnosis

These branching, aerobic Gram-positive rods are weakly acid-fast when decolourised with 1% sulphuric acid (modified Ziehl–Neelsen stain). Other specialist stains that aid the diagnosis of *Nocardia* include the Gomori methenamine silver method. Colonies of *Nocardia* may be coloured (orange/cream/pink) and the surface may be dry or chalky. Nocardia can take up to a month to grow on standard media (e.g. Lowenstein–Jensen media, brain–heart infusion agar, and trypticase–soy agar with blood enrichment). *Nocardia* organisms can be differentiated from *Actinomyces* because they are strict aerobes (whereas *Actinomyces* organisms are facultative anaerobes) and *Nocardia* grow over a wide range of temperatures (whereas *Actinomyces* only grow at 35–37°C).

Treatment

Seek expert advice. Usually a long course (e.g. >3 months in normal host, 6 months if immunocompromised) of a sulfonamide ± trimethoprim, e.g. as co-trimoxazole. Alternatives include minocycline, imipenem and amikacin. In refractory cases, involve the reference laboratory for sensitivity testing.

Actinomadura and *Streptomyces*

Actinomadura and *Streptomyces* species are aerobic filamentous actinomycetes implicated in mycetoma, also known as Madura foot. This is a chronic granulomatous condition that mainly occurs in Africa, Asia, and Central America. Mycetoma can be divided into actinomycetoma (bacterial) or eumycetoma (fungal), which has important treatment implications. The important subspecies of *Actinomadura* and *Streptomyces* are *Actinomadura madurae*, *Actinomadura peletieri*, and *Streptomyces somaliensis*. Other causal organisms include species of *Madurella*, *Exophila*, *Acremonium*, *Pseudallescheria*, and *Nocardia*.

Diagnosis

Clinically, grains seen within host tissues or in the discharge from sinus tracts, are diagnostic of mycetoma. These grains are colonies of the organism and should be crushed in KOH and Gram stained to distinguish between actinomycetoma (which have Gram-positive filaments) and eumycetoma (septate fungi). These grains should be rinsed in 70% alcohol before culture, to try to eliminate any surface contaminants, and appropriate plates set up at 26°C and 37°C. Macroscopically, grains are often red. *Actinomadura* spp. show many similar properties to the *Actinomyces* spp. but strictly *Actinomadura* are not acid-fast when decolourised with 1% sulphuric acid.

Clinical features

Mycetomas usually involve the hand or foot and arise from traumatic inoculation from soil or plants, usually via thorns or splinters. They are chronic granulomatous infections of the skin, subcutaneous tissue, and bone, and may progress to sinus formation.

Treatment

Actinomycetoma tend to respond to therapy better than eumycetoma, which usually requires surgery. Seek expert advice as the regimen depends on the cause, and courses may be up to 9 months. A combination of streptomycin with dapsone, rifampicin, co-trimoxazole, or sulphonamides is used.

Gram-negative cocci – overview

The Gram-negative cocci include a variety of pathogenic and non-pathogenic species:

Table 4.5 Gram-negative cocci

Organism	Microbiology	Syndrome
Neisseria meningitidis	Aerobic Gram-negative diplococci, oxidase positive, grow at 37°C on blood and chocolate agar, glucose and maltose positive	Meningitis Septicaemia
Neisseria gonorrheae	Aerobic Gram-negative diplococci, oxidase positive, grow at 37°C on blood and chocolate agar, glucose positive	Gonorrhoea
Non-pathogenic *Neisseria* spp.	Aerobic Gram-negative diplococci, oxidase positive, grow at 22°C on nutrient agar	Oral commensals – can rarely cause invasive infections
Moraxella catarrhalis	Aerobic Gram-negative cocci oxidase positive, grow at 37°C on blood and chocolate agar	Respiratory pathogen
Anaerobic Gram-negative cocci e.g. *Veillonella* spp.	Anaerobic Gram-negative cocci	

NB *Acinetobacter* spp. are Gram-negative rods that may appear coccoid or bacillary. Unlike *Neisseria* spp. they are oxidase negative. They are discussed further on 📖 p343.

Neisseria meningitidis

Epidemic cerebrospinal fever was first described in 1805 by Vieusseaux. In 1887, Weichselbaum isolated *N. meningitidis* from CSF. In the late 19th century meningococcal carriage was described. In 1909, different serotypes of *N. meningitidis* were recognized

Epidemiology

Humans are the only known reservoir of *N. meningitidis*, and ~ 20% of the population carry the organism in their throat. However, half of these carriage strains are non-capsulate and thus non-pathogenic. During outbreaks, the carrier rate of an epidemic strain may reach 90%. Risk factors for meningococcal disease:

- lack of bactericidal antibody
- age – bimodal distribution: 3 months to 3 years and 18–23 years
- travel to endemic areas, e.g. Africa, Mecca
- complement deficiencies
- splenectomy
- host genetic polymorphisms, e.g. *MBL*, *TNFA*, *FcIRIIa* and *PAI-1*.

Pathogenesis

To cause infection, the organism must cross the nasopharyngeal mucosa and enter the circulation. The Type IV pilus (encoded by *pilC*) is involved in mucosal colonization. The polysaccharide capsule is important in avoiding host immunity (and defines the serogroup of the isolate Table 4.5). Various secretion systems help deliver toxins.

Clinical features

- Meningitis and septicaemia
- Other acute infections – purulent conjunctivitis (which occasionally becomes systemic), purulent mono-arthritis, endocarditis, pericarditis and pneumonia
- Chronic septicaemia with joint and skin involvement is also recognized

Diagnosis

- CSF examination – in meningitis the CSF pressure is elevated and the CSF appears turbid. The CSF white cell count and protein are normally raised and the the CSF glucose level is low (compared to serum glucose). In very early infection, CSF results may be normal as the meningeal reaction has not had time to take place.
- Microscopy – Gram-negative intracellular diplococci. Note that in meningococcal meningitis, CSF usually has a higher yield than blood cultures. If the Gram stain is negative, a methylene blue stain may pick up scanty meningococci.
- Culture – transparent, non-pigmented, non-haemolytic colonies. May be mucoid if capsule production. Oxidase positive. Identified by API NH.
- Serogrouping – capsular polysaccharide antigens are identified by slide agglutination test using polyclonal antibodies. There are at least 13 xerogroups; the most common ones are summarised in Table 4.6.

- Serotyping – identification of (PorB) class 2/3 outer membrane protein by a dot-blot enzyme-linked immunosorbent assay (ELISA) using monoclonal antibodies.
- Serosubtyping – identification of (PorA) class 1 outer membrane protein by a dot-blot ELISA using monoclonal antibodies.
- Multi locus sequence typing (MLST) is being evaluated for routine surveillance.
- Meningococcal PCR (send to meningococcal reference lab).

Microbiology

N. meningitidis is a fastidious Gram-negative diplococcus. It produces a capsule which forms the basis of the serogroup typing system. There are now at least 13 serogroups but the most common ones are summarized in Table 4.6.

Table 4.6 Major serogroups of N. meningitis

Serogroup	Pattern of disease	Vaccines
A	Epidemic meningitis, associated with different clones	Yes
B	Epidemic strains (and outbreaks)	A meningitis B vaccine is currently undergoing clinical trials
C	Local outbreaks	MenC vaccine introduced 1999
W-135	Pilgrims returning from the Haj	Yes
X, Y, Z, 29E, Z′	Rare	

Treatment

- See management of acute bacterial meningitis (see 🕮 Acute meningitis, p.754) and septicaemia. Reduced susceptibility to penicillin in some countries has resulted in empirical therapy for meningitis being a 3rd-generation cephalosporin.
- After treatment, rifampicin (or ciprofloxacin) should be given for nasopharyngeal eradication.

Infection control issues

Inform public health who will arrange chemoprophylaxis of household or kissing contacts of the case. Note that rifampicin is only effective in eradicating carriage in 80–90% of people treated, and rifampicin-resistant strains, which have caused disease in contacts, have been reported. The alternatives are ciprofloxacin or ceftriaxone.

Neisseria gonorrhoeae

N. gonorrhoeae only infects humans and causes the sexually transmitted infection gonorrhoea (see Gonorrhoea, p.748). This is the second-most common bacterial sexually transmitted infection (STI) in the UK. Increasing rates of antimicrobial resistance, together with its persistence and association with poor reproductive health outcomes have made it a major public health concern.

Pathogenicity

Gonococci are divided into four Kellogg types, by colonial appearance, ability to auto-agglutinate, and virulence. Kellogg types T1 and T2 are more virulent and possess many fimbriae, while types T3 and T4 are non-fimbriate and avirulent. In gonococci the fimbriae are associated with attachment to mucosal surfaces and resistance to killing by phagocytes. Epidemiological typing of gonococci uses both auxotyping (nutritional requirements of arginine, proline, hypoxanthine, uracil, etc) and monoclonal antibodies against specific proteins.

Clinical features

Gonorrhoea commonly presents as a purulent disease of the urethral mucous membrane and also the cervix in females. Secondary local complications (e.g. epididymitis, salpingitis, pelvic inflammatory disease, p.744) and metastatic complications (e.g. arthritis) may occur if the primary infection is inadequately treated. Other manifestations of disease include disseminated gonococcal infection (skin lesions, painful joints, and fever), ophthalmia neonatorum (purulent conjuncitivitis of the newborn), perihepatic inflammation (Fitz-Hugh–Curtis); and rarely endocarditis or meningitis. Rectal or pharyngeal infection is often asymptomatic, and identified through contact tracing. If cultured, gonococcus should always be treated, as it is never a commensal.

Diagnosis

- Culture – The only definitive test for legal purposes is culture. Urethral swabs from males and endocervical swabs from females should be Gram stained and then immediately inoculated onto selective media and placed in enriched CO_2 conditions. Typical Gram stain appearance of *N. gonorrhoeae* (Gram-negative diplococci in association with neutrophils) from urethral/endocervical swabs, together with a consistent clinical presentation, is regarded as adequate for treatment in many cases. However, culture is critical for legal cases and for antimicrobial sensitivity testing. After 24–48 h, oxidase-positive colonies appear, and identification can be confirmed by testing for acid production from sugars (APINH). Many laboratories still test for β-lactamase production by the chromogenic cephalosporin (nitrocefin), acidometric, and paper strip methods, although all patients are likely to be treated with a 3rd-generation cephalosporin according to current guidelines
- Non-culture methods – Rapid non-culture tests are increasingly available, mainly based on detection of nucleic acid by hybridization or amplification. These are generally very sensitive and specific, and LCR (ligase chain reaction) has the advantage of being performed on urine.

- If gonococcus is isolated from a prepubertal girl with vulvovaginitis, it may indicate sexual abuse. The case should be dealt with sensitively by a paediatrician and senior laboratory staff should be involved. Thorough documentation is required, since evidence may be needed in court.

Treatment

Current UK guidelines from the British Association of Sexual Health and HIV (http://www.bashh.org/ceguidelines.htm) recommend ceftriaxone or cefixime as first-line therapy. Spectinomycin can also be given (except for pharyngeal infections). The cephalosporins have replaced the fluoroquinolones due to increasing resistance rates. Ciprofloxacin resistance (MIC ≥1 mg/L) was found to be 14% in 2003 according to the Gonococcal Resistance to Antimicrobials Surveillance Project (GRASP). Rates of ciprofloxacin resistance were even higher in men who have sex with men (MSM). Rates of azithromycin resistance and multi-antibiotic resistance are also rising. Alarmingly, a fluoroquinolone was still prescribed as first–line therapy to almost 25% of patients in 2004 (against recommendations). Many of these patients were likely to be infected with resistant organisms, which would not only result in an adverse clinical outcome for the patient, but also result in transmission of this resistant strain to other contacts.

Control

- Prompt and adequate diagnosis and treatment
- Effective contact tracing
- Prevention – condoms and barrier methods
- Prevent ophthalmia neonatorum by putting 1% aqueous silver nitrate into all newborn babies' eyes, in areas of high prevalence
- Screening of high-risk individuals
- Sex education/awareness of STIs

Non-pathogenic *Neisseria*

The non-pathogenic *Neisseria* species are upper respiratory tract commensals and include: *N. lactamica, N. polysaccharea, N. subshara, N. cirerla* (meningitis, endocarditis, bacteraemia, ocular infections, pericarditis, osteomyelitis, empyema). If invasive infection does occur, full susceptibility testing should be performed as penicillin resistance is increasing.

Microbiology

N. lactamica and *N. polysaccharea* are the species most commonly isolated from nasopharyngeal swabs during meningococcal surveys. Colonies appear similar to *N. meningitidis*, and they also grow on selective media, unlike the nasopharyngeal commensals. *N. lactamica* is easy to distinguish as it produces acid from glucose, maltose and lactose, and gives a positive ONPG (orthonitrophenyl-β-D-galactopyranoside) test result for β-galactosidase.

Clinical features

They can occasionally cause invasive diseases such as *N. lactamica, N. polysaccharea, N. subflava, N. sicca, N. mucosa, N. flavescens, N. elongata, N. cinerea,* and *N. weaveri.*

Neisseria spp. are naturally competent for DNA uptake, so the pathogenic neisseria can take up DNA encoding for virulence factors or antibiotic resistance from the non-pathogenic neisseria that are part of the normal flora. For example, one mechanism by which *N. meningitidis* and *N. gonorrhoeae* have acquired penicillin resistance is the interspecies transfer of *penA* from the non-pathogenic neisseria in the throat. Studying exactly what constitutes 'normal flora' – not just in the throat, but also the GI tract, skin, vagina etc – is likely to increase our understanding of the evolution of pathogens.

Moraxella

For decades, *M. catarrhalis* was regarded as an upper respiratory tract commensal. However, since the 1970s it has been recognized as an important and common respiratory tract pathogen.

Microbiology

M. catarrhalis grows well on many media including blood and chocolate agar. It shows the 'hockey-stick' sign, in that it slides across the agar surface when pushed and can be difficult to pick up onto a loop. *M. catarrhalis* is oxidase-positive, catalase-positive, DNase-positive, and produces butyrate esterase.

Clinical features

M. catarrhalis causes otitis media, lower respiratory tract infections in COPD patients, pneumonia particularly in the elderly, nosocomial respiratory tract infections, sinusitis and occasionally bacteraemia. Outer membrane proteins (OMPs), lipo-oligosaccharide (LOS), and pili are probably important in pathogenesis.

Treatment

Almost all strains of *M. catarrhalis* produce β-lactamase, which is inducible. Regardless of the results of ampicillin susceptibility testing, ampicillin should not be used. Suitable agents include co-amoxiclav, cephalosporins, fluoroquinolones, tetracyclines, and macrolides.

Anaerobic Gram-negative cocci

Veillonella spp. organisms are part of the normal flora of the gastrointestinal tract of humans and animals. *Veillonella* may be isolated from a variety of clinical conditions, though their role in causing infection is unclear. The most common species is *Veillonella parvula*, which fluoresces red under ultraviolet (UV) light. *Veillonella* are able to use some of the lactic acid produced by streptococci, lactobacilli, and other bacteria that may induce dental caries. They are associated with supragingival dental plaque and also found as part of the tongue microflora. They are generally regarded as minor components of mixed anaerobic infections.

Acidaminococcus spp. and *Megasphaera* spp. are other anaerobic Gram-negative cocci found in the human gut. They are considered non-pathogenic.

Escherichia coli

E. coli is the type species of the genus *Escherichia* in the family *Enterobacteriaceae*, and contains a variety of strains ranging from commensal organisms to highly pathogenic variants. Infections tend to infect the gut and urinary tract but almost any extra-intestinal site may be involved. *E. coli* is often used as a marker of faecal contamination, e.g. in food and water testing, as it does not otherwise exist outside the animal body.

Pathogenesis

- O and K polysaccharide antigens protect *E. coli* from complement and phagocytic killing, unless antibodies are present. Phagocytosis is usually successful if there are antibodies to K antigens present alone, or to both O and K antigens.
- Haemolysin is more commonly produced by strains causing extra-intestinal infections, and is thought to increase virulence.
- The ColV plasmid, harboured by some *E. coli*, encodes an aerobactin-mediated iron uptake system. This is more common in strains isolated from cases of septicaemia, pyelonephritis, and lower UTIs than in commensal faecal strains.
- Fimbriae – type 1 fimbriae adhere to cells containing mannose residues, possibly contributing to pathogenicity, but their role in UTIs is debated. Other filamentous proteins may cause a mannose-resistant haemagglutination, e.g. CFAs (colonization factor antigens) in human enterotoxigenic *E. coli* (ETEC), K88 in pigs, and K99 in calves and lambs. P. fimbriae bind specifically to receptors on P blood group antigens of human erythrocytes and uroepithelial cells.
- Other – enteric strains demonstrate specific interactions with the intestinal mucosa, release toxins, and may harbour plasmid-encoded virulence factors.

Epidemiology

Serotying of *E. coli* is based on O (somatic), H (flagellar) and K (surface / capsular) antigens, as detected in agglutination reactions.

- There are >160 O antigens, and cross-reactions occur between *E. coli* O antigens and O antigens of other species e.g. *Citrobacter*, *Salmonella*.
- H antigens are usually monophasic, and are determined from cultures in semi-solid agar.
- K antigens traditionally were those that prevented O agglutination (thus agglutination tests are done on boiled samples). K antigens are the acidic polysaccharide capsular antigens, and divided into groups I and II.

Clinical features

UTIs

E. coli is the commonest cause of community-acquired uncomplicated UTIs (see 📖 Urinary tract infections: introduction, p.716), and also causes nosocomial UTIs. Clinical manifestations range from urethritis and cystitis to pyelonephritis and sepsis. Many uropathogenic strains originate in the patient's own gut, and cause infection by the ascending route. Specific P fimbriae or '**p**ili **a**ssociated with **p**yelonephritis (known as the PAP pilus), which attach to uroepithelial cells, are important in pathogenesis. These uropathogenic strains may contain additional virulence factors such as haemolysin, ColV plasmids and resistance to complement-dependent bactericidal effect of serum.

Enteric infections

E. coli is responsible for many cases of diarrhoeal disease ranging from acute gastroenteritis, particularly in the tropics ('traveller's diarrhoea' 📖 Viral gastroenteritis, p.454), to life-threatening haemorrhagic colitis. The strains involved fall into 4–5 groups, with different pathogenic mechanisms (see below and Table 4.7).

Bacteraemia

The usual sources of nosocomial *E. coli* bacteraemia are the urogenital, gastrointestinal, and respiratory tracts, and foreign bodies such as IV lines and endotracheal tubes. The hallmark of cases of Gram-negative bacteraemia is the systemic reaction to lipopolysaccharide or endotoxin, which may be fatal.

Neonatal sepsis

E. coli may cause neonatal meningitis and septicaemia, especially in premature babies. The strains responsible may express the K1 or K5 surface/capsular antigens, which have enhanced virulence.

Other non-enteric infections

E. coli may cause postoperative wound infections and deep abscesses. Respiratory tract infection is usually opportunistic, often in debilitated patients such as diabetics or alcoholics. Nosocomial pneumonia (±empyema) is usually due to aspiration rather than haematogenous spread, and may be associated with high mortality.

Diagnosis

E. coli are usually smooth colourless colonies on non-selective media, and may appear haemolytic on blood agar. Most *E. coli* ferment lactose (and produce acid and gas in 24–48 h), but approximately 5% are non-lactose fermenters (NLFs). *E. coli* are usually motile, and those responsible for extra-intestinal infections often have a polysaccharide capsule. E.coli are usually positive for indole production, ornithine decarboxylase, lysine decarboxylase and methyl red. They are usually negative for urease, citrate utilization, H_2S production and Voges Proskauer test.

Treatment

The management of *E. coli* depends on the site and severity of the infection. Simple *E. coli* UTIs may respond to trimethoprim or ampicillin. Many hospital acquired *E. coli* infections are due to multi-resistant organisms, and may require treatment with a cephalosporin, fluoroquinolone, aminoglycoside, piperacillin-tazobactam or carbapenem. Susceptibility data often varies geographically (due to prior antibiotic usage) so follow your hospital antibiotic policy. Be guided by antibiotic susceptibility results, and use targeted therapy when possible. Antibiotics may be harmful in cases of *E. coli* O157.

Table 4.7 Clinical features and pathogenic mechanisms of different *E Coli*

Abbreviation	Full name	Clinical features	Pathogenesis
EHEC/VTEC/STEC	Enterohaemorrhagic *E. coli* Verotoxin-producing *E. coli* Shigatoxin-producing *E. coli*	Haemorrhagic colitis/haemolytic uraemic syndrome (HUS)	**Verotoxins** (VT1 and 2), also called shiga-like toxins (SLT1 and 2), are phage-encoded toxins thought to target vascular endothelial cells. The A subunit mediates biological activity, while B is responsible for binding and toxin uptake. Risk of developing HUS depends on type of shigatoxin, plus host and environmental factors
ETEC	Enterotoxigenic *E. coli*	Traveller's diarrhoea	**ST (heat-stable enterotoxin)** causes ↑cGMP, thus altering ion transport and ↑fluid secretion by mucosal cells of small intestine
			LT (heat-labile enterotoxin) – B polypeptide binds to mucosal surface of small intestine, allowing the A polypeptide to enter the cell and catalyse adenosine diphosphate ribosylation of the guanine nucleotide component of adenylate cylase, thus ↑cAMP and ↑fluid secretion (as with *V. cholerae*)
			Colonization/adherence factors – see text
EIEC	Enteroinvasive *E. coli*	Disease similar to shigella-like dysentery	
EPEC	Enteropathogenic *E. coli*	Childhood diarrhoea	
EAEC	Enteroaggregative *E. coli*	Traveller's diarrhoea, especially in Mexico and N. Africa	

Klebsiella

Klebsiella species are usually harmless colonizers of the human gut. The classification can be confusing, but the main species defined by DNA hybridization studies are *Klebsiella pneumoniae* subsp. *aerogenes* (formerly *Klebsiella aerogenes*), *Klebsiella pneumoniae* subsp. *pneumoniae* (formerly *Klebsiella pneumoniae*) and *Klebsiella oxytoca*. Other rare respiratory subspecies include *Klebsiella pneumonia* subsp. *ozaenae* and *Klebsiella pneumonia* subsp. *rhinoscleromatis*. Note that some *Klebsiella* have recently been reclassified as *Raoultella*, e.g. *Raoultella ornithinolytica* and *Raoultella terrigena*.

Pathogenesis

Klebsiella organisms that express capsular K antigens are resistant to complement-mediated serum killing. No particular capsular subtype has been linked to a greater risk of infection. Those with O antigens are resistant to phagocytosis. *Klebsiella* spp. have two iron uptake systems: one system uses aerobactin (related to virulence) and the other uses enterochelin (plasmid encoded).

Epidemiology

Common capsular (K) types in the UK are K2, K3 and K21. There are about 80 K antigens recognized overall, some of which cross-react with *H. influenzae* and *S. pneumoniae*. There are also five different somatic O antigen types, but these are rarely used for typing. There is an association between antigenic structure, habitat, and biochemical reactivity, for example capsular types 1–6 are most common in the human respiratory tract. For epidemiological investigations, capsular serotyping, bacteriocin typing, and phage-typing may be useful.

Clinical features

Klebsiella infections are rare in the immunocompetent normal host. They tend to cause nosocomial and opportunistic infections, such as UTIs, pneumonia (lobar), other respiratory infections (bronchitis, bronchopneumonia), surgical wound infections, and bacteraemia, in those with risk factors such as diabetes, COPD, or alcoholism. The likely focus in cases of nosocomial bacteraemia is the urinary tract, intravascular lines, lower respiratory tract, biliary tract, and surgical wound site. Severe pneumonia with 'redcurrant jelly' sputum and multiple lung abscesses is called Friedlander's pneumonia, and has a high mortality.

Diagnosis

Klebsiella spp. are facultatively anaerobic, catalase-test-positive, oxidase-test-negative, and ferment glucose. Organisms are capsular, which may give colonies a mucoid appearance. The capsule is made of glucuronic acid and pyruvic acid, and there are 80 or so capsular 'K' antigens. On Gram stain, organisms may look thicker than other Gram-negative rods, because of the prominent polysaccharide capsule. They are lactose-fermenters and usually fimbriate but non-motile. They are H_2S and indole-negative (except *K. oxytoca* which is indole-positive), Voges–proskauer (VP)-positive, they can grow in KCN, and they can use citrate as a sole carbon source. Different species of *Klebsiella* are usually recognized by different biochemical tests (Table 4.8).

Treatment

Most *Klebsiella* species are inherently resistant to ampicillin and most other penicillins. Many are now multi-resistant, including cephalosporin resistance due to extended spectrum; β-lactameses (ESBLs). Aminoglycoside susceptibility varies between regions. Treat according to local hospital policy and sensitivity data. The carbapenems and fluoroquinolones may be the only options.

Table 4.8 Biochemical reactions useful to distinguish *Proteus, Providencia*, and *Morganella*

	Proteus mirabilis	***Proteus vulgaris***	***Providencia alcalifaciens***	***Providencia rettgeri***	***Providencia stuartii***	***Morganella morganii***
Ornithine decarboxylase	+	–	–	–	–	+
Gas from glucose	+	+	V	–	–	+
H_2S production	+	V	–	–	–	–
Indole formation	–	+	+	+	+	+
Urease formation	+	+	–	+	–	+

+ = most strains positive.

– = most strains negative.

V = variable.

Adapted from Greenwood p281.

Greenwood et al. Medical Microbiology; A guide to microbial infections: pathogenesis, immunity, laboratory diagnosis and control. 15th edition (2000) Churchill Livingston.

Proteus

P. mirabilis is most commonly isolated from community UTIs, while *P. vulgaris* and *P. myxofaciens* tend to cause nosocomial infections. *Proteus* belongs to the tribe Proteae

Epidemiology

Proteus is probably the second most common enterobacteria encountered in many diagnostic laboratories (after *E. coli*). This is because of the huge numbers isolated from urine samples: approximately 10% of uncomplicated UTIs are due to *Proteus* (usually *P. mirabilis*).

Pathogenesis

Factors that contribute to the ability of *Proteus* to colonize and infect the urinary tract include:

- production of the enzyme urease which splits urea into ammonium hydroxide. This increases urinary pH and encourages struvite stone formation. These stones act as a nidus for persistent infection and also obstruct urinary flow.

- fimbriae help uroepithelial colonization
- flagella-dependent motility helps spread in the urinary tract
- uropathogenic *Proteus* synthesizes several haemolysins.

Clinical features

In addition to urine infections, *Proteus* also causes bacteraemia, wound infections and respiratory infections in debilitated hospital patients. The human GI tract is the main reservoir of infection for patients who subsequently become infected.

Diagnosis

Proteus organisms rapidly hydrolyse urea. The presence of hundreds of flagella on each organism makes them extraordinarily motile, which appears as 'swarming' on agar plates, and can produce the Dienes phenomenon (a line of inhibited growth where 2 strains meet). *Proteus* organisms give positive methyl-red reactions, are usually VP-negative (except some strains of *P. mirabilis*), and can grow in the presence of KCN. Most *P. mirabilis* strains are indole-negative, while the other subspecies are indole-positive (see Table 4.8).

Phage-typing, bacteriocin typing and serotyping schemes have been developed. The Dienes phenomenon may be exploited for typing, in that two test organisms are viewed as identical if they show no line of demarcation where the swarming growths meet (after inoculation onto the surface of an agar plate).

Treatment

Antibiotic resistance is increasing, but the indole-negative *P. mirabilis* is generally more sensitive than the indole-positive species. Prescribe according to local policy until sensitivity results are available. Some organisms carry the AmpC β-lactamase which is inducible by cephalosporins. Amikacin, new quinolones and carbapenems may be the only options. Note that *Proteus* is inherently resistant to colistin.

Enterobacter

The genus *Enterobacter* includes *E. aerogenes*, *E. cloacae*, *E. sakazakii*, *E. taylorae*, *E. gergoviae*, *E. asburiae*, *E. hormaechei*, *E. camerogenus*, and *E. agglomerans* (now known as *Pantoea agglomerans*, see page 337). The genus was previously known as *Aerobacter* species, and belongs to the tribe *Klebsiellae*.

Epidemiology

Enterobacter organisms are common human gut commensals, which rarely cause infection in the normal host.

Clinical features

E. aerogenes, *E. cloacae* (and occasionally *E. taylorae*) colonize hospital inpatients and cause nosocomial opportunistic infections, such as wound infections, burn infections, pneumonia, and UTIs. Risk factors for infection include indwelling lines, frequent courses of antibiotics, a recent invasive procedure, diabetes and neutropenia. They can often be isolated from diabetic ulcers. *Enterobacter* infections have been associated with intravenous fluid contamination. *E. sakazakii* has been implicated in severe neonatal meningitis (mortality rate 40–80%), and there have been outbreaks associated with dried-infant formula.

Diagnosis

In common with the other *Enterobacteriaceae*, *Enterobacter* species are facultative anaerobes that give a positive catalase result and a negative oxidase result. They ferment glucose (with the production of acid and gas) and also lactose. They do not produce H_2S on triple sugar iron media, they are indole-negative and methyl-red-negative, they are VP-positive, and they can grow in the presence of KCN. They use citrate as a sole carbon source and are ONPG-positive. Unlike *Klebsiella*, they are usually motile and are less likely to be heavily capsulated. The two most-important clinical species are *E. aerogenes* (which usually decarboxylates lysine but not arginine) and *E. cloacae* (usually decarboxylates arginine but not lysine).

Treatment

Enterobacter organisms (except *E. sakazakii*) are usually resistant to 1st-generation cephalosporins, and readily develop resistance to 2nd- and 3rd-generation cephalosporins due to inducible β-lactamases such as AmpC. Carbapenems are the mainstay of treatment. *E. sakazakii* tends to be more sensitive to antibiotics overall, and ampicillin and gentamicin in combination are the usual treatment of *E. sakazakii* neonatal meningitis.

Citrobacter

C. diversus, *C. freundii*, and occasionally *C. amalonaticus* are associated with nosocomial respiratory and urinary tract infections. Their role as primary pathogens or secondary infections/colonizers is debated. Note that *C. koseri* is a synonym for *C. diversus*. *Citrobacter diversus* has also been associated with outbreaks of neonatal meningitis.

Pathogenesis

Animal studies on neonatal meningitis showed that pathogenic strains of *C. diversus* were more virulent and had an extra outer membrane protein compared to non-pathogenic strains.

Clinical features

The clinical significance of isolation of *Citrobacter* species from the urinary and respiratory tracts of debilitated hospital patients is often unclear. When isolated from blood cultures, it is usually one of a number of species present, and such polymicrobial infections are often associated with a poor clinical outcome (probably due to the patient's general debilitated state rather than organisms' virulence). However, *Citrobacter* is a recognized cause of endocarditis, and in neonates, *Citrobacter* organisms (particularly *C. diversus*) can cause severe meningitis and brain abscesses.

Diagnosis

Citrobacter is so named because the organisms can grow on Simmons citrate media. They are usually motile, methyl red-positive, VP-negative and slowly hydrolyse urea. They are usually non-lactose fermenters, but may appear as late lactose fermenters. *C. freundii* may be mistaken for *Salmonella* as it is produces H_2S. Note there is considerable cross-reactivity with the O antigens of other *Enterobacteriaceae*.

Treatment

Like many of the other *Enterobacteriaceae* that cause nosocomial infections, *Citrobacter* tend to be multi-resistant, so reliance on laboratory antimicrobial susceptibility testing is paramount. *Citrobacter freundii* have the inducible AmpC β-lactamase. Plasmid-mediated ESβLs are becoming more common. Treatment options may include aminoglycosides, antipseudomonal penicillins, carbapenems, and quinolones.

Serratia

There are many named species of *Serratia*, which belong to the tribe *Klebsielleae*. *S. marcescens* is the main one that causes human disease. Infections with *S. liquifaciens*. *S. rubidaea* and *S. odorifera* are very uncommon.

Epidemiology

Unlike the other *Enterobacteriaceae, Serratia* is more likely to colonize the respiratory and urinary tracts of hospital patients (rather than the gut). However, in neonates the GI tract may be the reservoir for cross-contamination.

Clinical features

Serratia species are opportunistic pathogens, particularly in the healthcare setting, and cause respiratory and urinary tract infections, bacteraemias, skin and wound infections. Patients with intravascular catheters and urinary catheters are at increased risk. *Serratia* infections have been associated with contaminated intravenous therapy, and septic arthritis in patients who have had intra-articular injections. *Serratia* also causes endocarditis and osteomyelitis in IVDUs, and cellulitis in patients on haemodialysis.

Diagnosis

Serratia can be recognized by production of a characteristic red/deep pink pigment. They are slow or non-lactose fermenters, and usually motile. They have the characteristics of the *Enterobacteriaeceae*. Like *Enterobacter*, most *Serratia* do not produce H_2S or lactose on triple sugar iron media, are VP-positive, grow in the presence of KCN, and use citrate as a sole carbon source. *Serratia* can be differentiated from the other *Enterobacteriaceae* by production of an extracellular DNase.

Treatment

Serratia are often multi-resistant to antibiotics. Treat according to local epidemiology until sensitivity results are available. Options are often limited to amikacin, piperacillin-tazobactam, and carbapenems. Efforts focused on good infection control practice, especially handwashing, are vital in reducing horizontal transmission between patients. Note that *Serratia* organisms are inherently resistant to colistin.

Salmonella

Salmonellae belong to the family *Enterobacteriaceae*. There are seven subspecies and over 2400 serovars. The correct nomenclature is *Salmonella enterica*, followed by the serotype (e.g. *Salmonella enterica* serotype Typhimurium). This is commonly abbreviated to S Typhimurium (serotype not italicized).

Epidemiology

Salmonellae are commensals and pathogens of a wide range of domesticated and wild animals. Some species, e.g. S. Typhi and S. Paratyphi are well adapted to humans and have no other host. Others are more adapted to animals and rarely affect humans e.g. S. Arizonae and reptiles. In humans, salmonellae can be divided into those that cause enteric fever (S. Typhi and S. Paratyphi) and the non-typhoidal *Salmonella* spp. (NTS). Salmonellae are usually transmitted by the faeco-oral route.

Pathogenesis

- Infection begins with ingestion of organisms in contaminated food and water.
- Salmonellae express an array of distinct fimbriae that help them to adhere to the intestinal wall.
- They also encode a type III secretion system (T3SS) within salmonella pathogenicity island 1 (SPI-1) that is needed for bacteria-mediated endocytosis and intestinal epithelial evasion.
- A number of SPI-1 translocated proteins (SipA, SipC, SopE and SopE2) promote membrane ruffling and *Salmonella* invasion.
- Salmonellae are also adapted to survival and replication in the intracellular environment.

Clinical features

- Gastroenteritis (p.688)
- Enteric fever (p.690)
- Bacteraemia and endovascular infection
- Salmonellosis in HIV – 20 to 100-fold increased risk. More likely to have severe invasive disease (enterocolitis, bacteraemia, meningitis)
- Localized infections
- Chronic carrier state

Diagnosis

- S. Typhi and S. Paratyphi are biohazard group 3 organisms.
- Salmonellae are facultative anaerobic Gram-negative rods, which grow readily on routine media. Their growth on specialized media is summarised in Table 4.9. They are motile, oxidase-negative, urease-negative, non-lactose fermenters (NLF).
- Salmonellae possess lipopolysaccharide somatic (O) heat-stable antigens, and flagellar (H) heat-labile antigens. Usually the H antigens exhibit diphasic variation, so can exist in phases 1 and 2 (Table 4.10).
- S. Typhi, S. Paratyphi C and some strains of S. Dublin and *Citrobacter* produce the Vi polysaccharide capsule, which may mask the O antigens. If only the Vi antiserum is positive, heat the bacterial suspension in boiling water to remove the capsule and test it again using the same antisera. Rough strains, in which the O antigens are absent, tend to cross-agglutinate with different antisera.

- Most diagnostic laboratories identify the organism as *Salmonella* by biochemical tests (e.g. API 20E or shorter panel), and partially determine the antigenic structure with different Poly-O and Poly-H antisera (Table 4.10). This identifies causes of enteric fever or invasive serotypes.
- All Salmonella should be submitted to the HPA reference laboratory for confirmation of serotype and further epidemiological investigations as necessary.

Treatment

- Enteric Fever. First-line treatment for imported cases of typhoid fever in the UK is now ceftriaxone. When susceptibility results are available, options may include ciprofloxacin, azithromycin, ampicillin or co-trimoxazole.
- Non-typhoidal salmonella. Gastroenteritis does not usually require treatment, except in the immunosuppressed, neonates, the elderly and those at risk of bacteraemia. Suitable antibiotics include ampicillin, ciprofloxacin, trimethoprim or chloramphenicol, depending on susceptibility results. Invasive disease due to NTS (e.g. bacteraemia, meningitis) always requires therapy. Cefotaxime and ceftriaxone penetrate the CSF well so are often used for salmonella meningitis.
- Chronic asymptomatic carriers. Management of chronic asymptomatic carriers is debated. Good personal hygiene should prevent spread of disease. In the absence of biliary disease, prolonged antibiotics (e.g. ampicillin, ciprofloxacin) may cure 80% of carriers. Cholecystectomy may be considered for patients with gallstones or chronic cholecystitis, but there is a risk of spreading the organisms during surgery.

Table 4.9 Appearance of *Salmonella* spp. on different media

Agar	*Salmonella* spp.
MacConkey agar	Non-lactose fermenters appear white
CLED cysteine lactose electrolyte deficient	Non-lactose fermenters appear blue
DCA deoxycholate	Yellow or colourless, often with a dark centre
XLD agar Xylose lysine deoxycholate	*Salmonella* appear red, some with black centres
SSA *Salmonella Shigella*	NLF appear colourless, some with black centres
Hektoen agar	*Salmonella* are blue-green. *S.* Typhimurium and others that reduce sulphur produce a black precipitate
Brilliant green agar	Red-pink-colonies surrounded by brilliant red zones
Selenite broth	Growth of *Salmonella* results in a cloudy tube
Tetrathionate broth	Tetrathionate-reducing bacteria (*Salmonella* and *Proteus*) can grow

Table 4.10 Antigenic structure of some *Salmonella* spp.

Serotype	O antigen	H (phase 1)	H (phase 2)
Typhi	9,12 (Vi)	d	–
Paratyphi A			
Paratyphi B	1,4,5,12	b	1,2
Paratyphi C	6,7[Vi]	c	1,5
Typhimurium	1,4,5,12	i	1,2
Enteritidis	1,9,12	g,m	1,7
Virchow	6,7	r	1,2
Hadar	6,8	Z10	e,n,x
Heidelberg	1,4,5,12	r	1,2
Dublin	1,9,12 (Vi)	G,p	–

Shigella

The genus *Shigella* is divided into four species: *S. dysenteriae*, *S. flexneri*, *S. boydii*, and *S. sonnei*, based on serology and biochemical reactions (Table 4.11). The organisms cause bacillary dysentery by an invasive mechanism identical to Enteroinvasive E. coli (EIEC). *Shigella* belongs to the tribe *Escherichiaeae*, and DNA hybridization studies show that *E. coli* and *Shigella* are a single genetic species.

Epidemiology

There are 10 serotypes of *S. dysenteriae* and 15 serotypes of *S. boydii*. *S. flexneri* can be divided into six serotypes by group- and type-specific antigens, and each serotype can be further subdivided. *S. sonnei* must be typed by other means, such as colicine production or plasmids, as they are serologically homogenous. Most cases of shigellosis in the UK occur in young children, although infection occurs in any age after travel to areas where hygiene is poor. *S. sonnei* is endemic in the UK, while *S. boydii* and *S. dysenteriae*, and most *S. flexneri* infections, originate outside the UK.

Pathogenesis

The infecting dose of *Shigella* is only 10–100 organisms, hence the illness can be transferred from person to person (faeco-oral). When one member of a family has acquired the disease, the secondary attack rate is high. Infection can spread rapidly in institutions, especially among young children. It is commonly spread by food and water.

Dysentery results from invasion of the wall of the large bowel, with accompanying inflammation and capillary thrombosis. As the organisms invade and multiply within epithelial cells, cell death results in ulcer formation. Invasiveness is linked to the presence of a 140MDa plasmid. Note that the organisms rarely invade deeper than the mucosa, hence positive blood cultures are uncommon. Some strains also produce an exotoxin, which results in water and electrolyte secretion from the small bowel (and has some similarities with the cholera toxin). This may explain the brief watery diarrhoea that can precede bloody diarrhoea.

Clinical features

See Gastroenteritis, p.454. *S. dysenteriae* usually causes a more-severe illness, possibly with marked prostration and paediatric febrile convulsions. *S. dysenteriae* may also be associated with toxic megacolon and the haemolytic uraemic syndrome. *S. flexneri* and *S. boydii* may also cause severe disease, while *S. sonnei* usually causes mild symptoms. The severity of *S. dysenteriae* infection may be due to an exotoxin (previously thought to be a neurotoxin), but its exact role in pathogenesis is uncertain. *Shigella* rarely invades other tissues, hence septicaemia and metastatic infection is unusual.

Diagnosis

Shigella organisms are non-motile, non-capsulated Gram-negative rods. Most appear as non-lactose fermenters after 18–24 h incubation on MacConkey or DCA (desoxycholate citrate) agar, but *S. sonnei* is the only late lactose fermenter. *Shigella* is urease, citrate and H_2S-negative. *S. dysenteriae* is the only one that cannot ferment mannitol. Suspicious colonies should be confirmed with species-specific antisera, followed by type-specific antisera for all except *S. sonnei*. Direct microscopy of a stained faecal smear (usually methylene blue) will reveal numerous polymorphonuclear leucocytes (Box 4.5). *Shigella dysenteriae* type 1 is a biohazard group 3 organism.

Treatment

Most cases of shigella are mild and self-limiting, so are treated with oral rehydration therapy rather than with antibiotics. Antibiotics may be indicated in severe infections, patients at extremes of age, or the immunocompromised. Options include ciprofloxacin, ampicillin, co-trimoxazole, tetracycline, or cephalosporins, according to *in vitro* susceptibility testing. Antibiotics are unlikely to reduce the period of excretion.

Box 4.5 Positive leucocyte test

Remember this indicates invasive diarrhoea/colitis from a number of causes. Consider the following:

- *Shigella*
- *Salmonella*
- *Campylobacter enteritis*
- Idiopathic ulcerative colitis
- *C. difficile* diarrhoea (but poor sensitivity in addition to poor specificity).

Table 4.11 Biochemical reactions of *Shigella*

Shigella spp.	Gas from glucose	ONPG	Indole	Catalase	Acid from		
					Lactose	Mannitol	Dulcitol
dysenteriae 1	-	+	-	-	-	-	-
dysenteriae 2–10	-	V	V	+	-	-	-
flexneri 1–5	-	-	V	+	-	+	-
flexneri 6	V	-	-	+	-	V	V
boydii	-	-	V	+	-	+	V
sonnei	-	+	-	+	(+)	+	-

V = variable
(+) = positive after incubation for ≥48 hours
ONPG = ortho-Nitrophenyl-β-galactoside

Other *Enterobacteriaceae*

Hafnia alvei

Hafnia alvei (formerly an *Enterobacter*) belongs to the tribe *Klebsielleae*. *Hafnia* is found in human and animal faeces, sewage, soil, water and dairy products. It usually produces greyish colonies on blood agar and ferments fewer sugars than *Enterobacter*. All *H. alvei* are lysed by a single phage, which does not act on any other enterobacteriaceae. *H. alvei* occasionally causes opportunistic/nosocomial infections, and antibiotic sensitivities are usually similar to those of the *Enterobacter* group.

Pantoea agglomerans

Pantoea agglomerans (previously known as *Erwinia herbicola or Enterobacter agglomerans*) is similar to many plant pathogens. It occasionally causes opportunistic infections in humans (UTIs, bacteraemia, and chest infections), and has contaminated intravenous fluids in the past. Colonies are yellow and may be isolated from superficial skin swabs and respiratory specimens, when they are usually regarded as 'normal flora'.

Edwardsiella tarda

Edwardsiella tarda infections in humans probably originate from contact with cold-blooded animals. These organisms are motile and ferment glucose to produce gas. They are H_2S-positive, which together with the fact that they do not ferment lactose means they may be mistaken for *Salmonella* species on enteric media. *Edwardsiella* species rarely cause disease, but are occasionally associated with a *Salmonella*-like gastroenteritis, which usually resolves without antibiotics. There are case reports of bacteraemia, liver abscess, soft tissue infection, and meningitis. Treatment should be guided by disc susceptibility testing.

Morganella morganii

This belongs to the tribe *Proteeae*. Thus they are motile, deaminate phenylalanine rapidly, give positive methyl-red reactions, are usually VP-negative, and can grow in the presence of KCN. Most are indole-positive and hydrolse urea rapidly (Table 4.8). *Morganella* organisms cause hospital-acquired infections, which are often multi-resistant so treatment is with carbapenems.

Providencia alcalifaciens, Providencia stuartii, and Providencia rettgeri

These also belong to the tribe *Proteeae*. Thus they are motile and deaminate phenylalanine rapidly. They give positive methyl-red reactions, are usually VP-negative and indole-positive, and can grow in the presence of KCN. Most *P. rettgeri* hydrolse urea rapidly, while the others are urease-negative. (Table 4.8) *Providencia* causes nosocomial infections in debilitated patients, and treatment is with carbapenems.

Overview of Gram-negative rod non-fermenters

These organisms derive energy from carbohydrates by oxidative (rather than fermentative) metabolism.

Pseudomonads

The pseudomonads are a large and diverse group of aerobic, oxidative Gram-negative rods (GNRs). Most are saprophytes found in soil, water, and moist environments. *Pseudomonas aeruginosa* is the species most commonly associated with human disease, particularly nosocomial infections. Other opportunistic species of *Pseudomonas* include *P. putida*, *P. fluorescens* (which has been associated with blood transfusions), and *P. stutzeri*. Organisms which have been allocated to new genera include *Burkholderia* (*B. cepacia* and *B. pseudomallei*), *Stenotrophomas* (*S. maltophilia*), *Comamonas* (below), and *Brevundimonas* (below).

Delftia acidovorans

Formerly known as *Comamonas acidovorans or Pseudomonas acidovorans*, this rare organism may cause endocarditis in drug users. Confusion arises as it may grow on *B. cepacia*-selective media, and may be resistant to colistin and gentamicin.

Brevundimonas

B. diminuta and *B. vesicularis* are rare and of doubtful clinical significance.

Glucose non-fermenters

This diverse group is taxonomically distinct from the oxidative pseudomonads and the carbohydrate-fermenting *Enterobacteriaceae*. They are mainly opportunistic pathogens, and often multi-resistant to antibiotics. Identification difficulties arise because they tend to be biochemically inert.

Eikenella corrodens

This oral commensal can cause endocarditis ('E' in HACEK, see 📖 HACEK organisms, p.353), meningitis, skin and soft tissue infections (particularly from human bites), and pneumonia. It is a facultative anaerobe, requiring incubation in CO_2. The colonies pit ('corrode') the surface of the agar.

Flavimonas

F. oryzihabitans is found in soil, water and damp environments, and most commonly causes central line-associated bacteraemias in immunocompromised patients.

Flavobacterium

This group of yellow-pigmented organisms is so genetically diverse that many have been re-classified. *F. meningosepticum* is now *Chryseobacterium meningosepticum*, and has caused epidemics of adult and neonatal meningitis with high mortality. Some other flavobacteria now belong to the genus *Sphingobacterium* (see below).

Chryseobacterium

Other than *C. meningosepticum* (see above), isolation of these organisms from clinical samples usually reflects colonization rather than infection. As noted above, *C. meningosepticum* has caused epidemics of adult and neonatal meningitis with high mortality. *In vitro* testing may not correlate with antibiotic clinical efficacy. There is evidence for treatment with vancomycin ± rifampicin, or ciprofloxacin, or levofloxacin.

Sphingobacterium

These contain high amounts of sphingophospholipid compounds in their cell membrane. Most human isolates of this genus are *S. multivorum* and *S. spiritivorum,* which can cause nosocomial infections in various sites.

Shewanella

S. putrefaciens (formerly *Pseudomonas putrefaciens*) is commonly isolated from water and the environment, but rarely causes human disease. It is usually found as part of a polymicrobial infection, typically from cellulitis complicating a leg ulcer or burn.

Roseomonas

These are also known as the 'pink-pigmented coccoid' group. *R. gilardii* is the most common species isolated from humans, and has been reported to cause community-acquired bacteraemia.

Chryseomonas

Infection with the rare *C. luteola* is usually associated with peritoneal dialysis catheters or indwelling lines, and may result in peritonitis, endocarditis, bacteraemia, or meningitis.

Ochrobactrum

Previously called *Achromobacter*. *Ochrobactrum anthropi* causes nosocomial opportunistic infections, particularly catheter-related bacteraemia.

Oligella

O. urethralis (formerly *Moraxella urethralis*) is a genito-urinary tract commensal, while *O. ureolytica* is usually found in patients with long-term indwelling urinary catheters. They are of low pathogenicity.

Alcaligenes

There are three clinically relevant species: *A. xylosoxidans* (sometimes called *Achromobacter xylosoxidans*), *A. faecalis* (formerly *Alcaligenes odorans*), and *A. piechaudii*. They are found in soil and water, and the GI and respiratory tracts of hospital patients. Nosocomial outbreaks have occurred (generally, but not exclusively in immunocompromised patients) with a wide range of clinical manifestations. *A. xylosoxidans* is often multi-resistant to antibiotics, and carbapenems or co-trimoxazole may be required as therapy.

Agrobacterium

These plant pathogens are usually nonpathogenic to humans, with <50 case reports of human disease in the literature. These are mainly due to *A. radiobacter* and *A. tumifaciens.*

Sphingomonas

S. paucimobilis (the only species) has been implicated in nosocomial outbreaks associated with contaminated water. It may be confused with flavobacteria as it produces a non-diffusible yellow pigment.

Pseudomonas aeruginosa

P. aeruginosa is widespread in soil, water, and other moist environments. Humans may be colonized with *P. aeruginosa* at moist sites such as the perineum, ear, and axilla. It is a highly successful opportunistic pathogen, especially in the hospital setting. This success is largely due to its resistance to many antibiotics, ability to adapt to a wide range of physical conditions, and minimal nutritional requirements.

Epidemiology

P. aeruginosa is found almost anywhere in the environment, including surface waters, vegetation, and soil. It usually colonizes hospital and domestic sink traps, taps, and drains. It also colonizes moist areas of human skin, leading to 'toe web rot' in soldiers stationed in swampy areas, and otitis externa in divers in saturation chambers.

Pathogenesis

The broad range of conditions caused by *P. aeruginosa* may be explained by the fact that the pathogen is both invasive and toxigenic. *P. aeruginosa* has low intrinsic virulence in man and animals. Infection occurs when host defences are compromised or the skin/mucous membranes are breached (e.g. neutropenia, burns patients, intensive care patients, indwelling devices), or when a relatively large inoculum is introduced directly into the tissues. The process can be divided into three stages:

- bacterial attachment and colonization
- local invasion
- dissemination and systemic disease.

Different virulence factors are produced depending on the site and nature of the infections, and include:

- exotoxins – exotoxin A and exo-enzyme S
- lipopolysaccharide (endotoxin)
- cytotoxic substances – proteases (elastase and alkaline phosphatase), cytotoxin (previously called leukocidin), haemolysins, phospholipases, rhamnolipids, pyocyanin
- porins
- pili and fimbriae (important in epithelial adherence, e.g respiratory).

Clinical features

P. aeruginosa causes a wide spectrum of conditions:

- Community-acquired infections are rare, and tend to be mild and superficial. Examples include otitis externa, varicose ulcers, and folliculitis associated with jacuzzis.
- Nosocomial infections with *P. aeruginosa* tend to be more severe and more varied than community infections. *P. aeruginosa* may account for ~10% of all hospital-acquired infections. Examples include pneumonia, urinary tract infections, surgical wound infections, bloodstream infections, and respiratory infections.
- Cystic fibrosis patients (Box 4.5), burns patients and mechanically ventilated patients are at particular risk.
- Other conditions associated with *P. aeruginosa* include endocarditis (IDUs and prosthetic valves), eye infections, bone and joint infections, postoperative neurosurgical infections, and eye and ear infections.

Diagnosis

This non-sporing, non-capsulate, motile Gram-negative rod is a strict aerobe (hence often used in testing anaerobic cabinets). However, note that it can grow anaerobically in the presence of nitrate. It grows on many different culture media, and produces a characteristic 'freshly cut grass' odour. The typical green-blue colour is due the diffusible pigments pyocyanin (blue phenazine pigment) and pyoverdin (yellow-green fluorescent pigment; principle siderophore). Other pigments include pyorubrin (red) and pyomelanin (brown). Note that ~10% do not produce detectable pigments even on pigment-enhancing media. *P. aeruginosa* is oxidase-positive (usually within 10 s), and appears relatively inactive in carbohydrate fermentation tests (only glucose is used). It grows best at 37°C and also at 42°C, but not at 4°C. Confusion occasionally arises in differentiating *P. aeruginosa* from other *Pseudomonas* spp. with commercial kits: growth at 42°; flagella stains and differential sugar fermentation tests may prove useful.

For epidemiological studies, serotyping may be useful; however, of the 21 internationally accepted O serotypes, four account for ~50% of clinical and environmental isolates. PFGE may help discriminate between serotypes. Other typing schemes are based on phage susceptibility and bacteriocin production.

Treatment

Antipseudomonal agents include the fluoroquinolones (these are the only oral option), ceftazidime, ticarcillin, piperacillin, carbapenems (imipenem, meropenem), aminoglycosides (gentamicin, tobramycin, amikacin), polymixins (colistin), and aztreonam. Theoretically, the use of dual therapy should reduce the development of antibiotic resistance and may also have the potential for bacterial synergy, but there is little clinical evidence for this.

Box 4.6 *P. aeruginosa* in cystic fibrosis (CF) patients

P. aeruginosa colonizes up to 80% of CF patients and causes chronic lung infection. Once established, it is refractory to treatment. Many *P. aeruginosa* isolates appear mucoid, which is due to production of an alginate-like exopolyscaaharide capsule (glyocalyx). This may form a biofilm, which renders the infection less amenable to antibiotic treatment. Isolates may have atypical growth requirements, such as appearing auxotrophic for specific amino acids, non-motile, and susceptible to semi-synthetic penicillins. Primary culture plates often show mixtures of colonial forms, but these variants are usually genetically identical.

Acinetobacter

Acinetobacter spp. are becoming increasingly important as a cause of nosocomial infections, and are often multi-resistant to antibiotics. Increasing antibiotic-selective pressure and the ability to survive well in the hospital environment (including on curtains and in dust) have contributed to its success as an opportunistic pathogen. There are ~19 genospecies, based on DNA–DNA hybridization studies; seven of these have species names (Table 4.11).

Epidemiology

In the UK there have been outbreaks of two multi-resistant clones (OXA-23 clone 1 and the 'SE clone' which is OXA-51-like). These are now widespread, particularly in London and South-East England. Nosocomial spread in ICUs is common, and may occur via equipment (particularly ventilators), gloves, contaminated solutions, and colonized healthcare workers. There are reports of *Acinetobacter baumanii* infections in casualties returning from Iraq.

Risk factors

Risk factors include the following:

- community-acquired infections – alcoholics, smokers, chronic lung disease, diabetes, and living in a tropical developing country
- hospital-acquired infections – intensive care, ventilation, urinary catheter, intravenous lines, increased length of stay, treatment with broad-spectrum antibiotics, total parenteral nutrition (TPN), surgery, wounds.

Pathogenesis

This organism has very few virulence factors, which explains why it only causes opportunistic infections. It occurs naturally as a saprophyte in soil and water, and occasionally colonizes moist human skin. The ability to survive in the environment is probably related to the capsule, the production of bacteriocin, and prolonged viability under dry conditions.

Clinical features

Acinetobacter spp. are able to infect almost every organ system, though it is vital to distinguish true infection from pseudo-infection (e.g. pseudo-bacteraemia due to skin colonization). The most common site of infection is the respiratory tract, where it causes nosocomial pneumonia, particularly ventilator-associated pneumonia (VAP), adult community-acquired pneumonia, and community-acquired tracheobronchitis and bronchiolitis in children. Other sites include the urinary tract, intracranial (usually post-neurosurgery) tissue, soft tissue (burns, wounds, and line-associated cellulitis), eye infections, endocarditis, and bone. Nosocomial bacteraemia is usually associated with the respiratory tract or intravenous catheters, and has a reported mortality rate of 17–46%. *A. baumannii* bacteraemia tends to be more severe.

Diagnosis

Acinetobacter spp. classically appear as Gram-negative coccobacilli, although they may retain crystal violet so appear Gram-positive. They are generally encapsulated, non-motile organisms, which readily grow on routine media as white, mucoid, oxidase-negative, catalase-positive, colonies. Misidentification may arise using API profiles as they are biochemically relatively unreactive, but acidification of glucose, haemolysis of red blood cells, and ability to grow at 44°C are reliable characteristics.

Treatment

Treatment may require several different approaches:

- localized cellulitis associated with a foreign body or indwelling line may respond to removal of foreign body alone (antibiotics not needed)
- choice of antibiotics for more serious infections is becoming limited – e.g. the OXA-23 clone 1 is resistant to virtually all antibiotics including carbapenems, and many isolates of the SE clone are also carbapenem resistant. Susceptibility testing should include carbapenems, aminoglycosides (including amikacin), sulbactam, polymixins, e.g. colistin, and tigecycline. There is some evidence to support combination therapy with rifampicin and colistin, +/– imipenem
- liason with reference laboratories re typing and antibiotic options
- review of infection control practices and antibiotic prescribing in the case of an outbreak.

Interim Working Party guidelines on the control of multi-resistant acinetobacter (MRAB) outbreaks are available on the HPA website (http://www.hpa.org.uk/infections/topics_az/*acinetobacter*_b/guidance.htm).

- MRAB is defined as an isolate of *Acinetobacter* spp. that is resistant to any aminoglycoside *and* to any third-generation cephalosporin.
- MRAB-C is defined as a MRAB that is also resistant to the carbapenems. These harbour metallocarbapenemases such as VIM and IMP (see 📖 Carbapenems, p.72), and are thought to have originated in Korea.

Table 4.11 Genomic species of *Acinetobacter*

Genospecies	Species name
1	*A. calcoaceticus*
2	*A. baumannii*
4	*A. haemolyticus*
5	*A. junii*
7	*A. johnsonii*
8	*A. lwoffi*
12	*A. radioresistens*
Other	A. species unnamed (>14)

Stenotrophomonas maltophilia

Previously called *Pseudomonas maltophilia* or *Xanthomonas maltophilia*, this organism is becoming increasingly seen as a cause of nosocomial infection. It is an opportunistic pathogen, of relatively low virulence, but an amazing ability to survive in a wide range of environments. It is frequently multi-resistant to antibiotics.

Epidemiology

Ubiquitous in the environment, *S. maltophilia* has been isolated from multiple sources in hospitals, including water (tap and distilled), nebulizers, dialysis machines, solutions, intravenous fluids, and thermometers, etc. Transmission of nosocomial infection has been associated with hospital water or contaminated disinfectant solutions. Studies have shown that most outbreaks result from antibiotic-selective pressure (especially the extensive use of imipenem, to which *S. maltophilia* is resistant) and exposure to multiple environmental strains, rather than cross-infection.

Risk factors

Risk factors for nosocomial infection include: intensive care, increased length of stay, treatment with broad-spectrum antibiotics, malignancy (especially if immunosuppressed), instrumentation (e.g. urinary catheter, intravenous lines, intubation, TPN, CAPD), patients with COPD, IDUs and neutropenia.

Pathogenesis

Potential virulence factors include those involved in adherence to plastics, and production of exoenzymes such as elastase and gelatinase.

Clinical features

S. maltophilia can cause a variety of infections, ranging from superficial to deep-tissue to disseminated disease. It is most commonly isolated from the respiratory tract, and distinguishing true infection from colonization can be difficult. *S. maltophilia* pneumonia has a high mortality, especially when associated with bacteraemia or GI obstruction. Other common sites of *S. maltophilia* infection include skin and soft tissue, intra-abdominal, the urinary tract, the eyes (especially in contact-lens wearers), and implants. Endocarditis is also reported.

Diagnosis

This motile non-lactose-fermenting Gram-negative aerobic rod grows readily on standard media. It is often pale yellow on blood agar, with an ammonia-like smell. Most are oxidase-negative, catalase-positive, DNase positive and can hydrolyse aesculin and ONPG. It is the only pseudomonad that gives a positive lysine decarboxylase reaction. Resistance to imipenem may be a useful marker. Note that *S. maltophilia* is increasingly isolated from sputum from patients with cystic fibrosis. It grows well on colistin-containing media, so may be misidentified as *B. cepacia*.

Treatment

Unfortunately, results of antibiotic susceptibility testing correlate poorly with treatment outcome. The drug of choice is co-trimoxazole (see 📖 Co-trimoxazole, p.97). Other options to consider include ticarcillin-clavulanate, doxycycline, minocycline, newer-generation quinolones, and third-generation cephalosporins. There is clinical evidence that co-trimoxazole and moxifloxacin are synergistic. Most strains are resistant to aminoglycosides.

Burkholderia cepacia

Previously classified as *Pseudomonas cepacia*, this opportunistic pathogen is a particular problem in cystic fibrosis (CF) patients. Other risk factors for infection include chronic granulomatous disease (CGD), and sickle cell haemoglobinopathies. Note that there are ten phylogenetically similar but genomically distinct species, known as the *Burkholderia cepacia* complex (Table 4.12).

Epidemiology

B. cepacia is ubiquitous in the environment and has been isolated from multiple sources in hospitals. Environmental transmission may occur via contact with respiratory equipment, water supplies, or disinfectants. However, transmission by close contact with colonized patients to other CF patients is more significant, and patients should be segregated into separate groups (e.g. for outpatient clinics and summer camps), which can be a highly emotive issue.

Pathogenesis

This is a relatively poorly virulent organism, which can survive in a wide range of environments. Virulence factors include adherence to plastics, and production of elastase, gelatinase, adhesin (a mucin-binding protein), siderophores, haemolysin, and exopolysaccharide. Resistance to non-oxidative neutrophil killing may be important. One successful epidemic strain had giant cable pili to help attach to respiratory mucin.

Clinical features

This is a significant pathogen in CF patients, with mortality rates >50% in the first year post infection. The three main patterns of infection are:

- chronic asymptomatic carriage
- progressive deterioration over months, with frequent hospital admissions, recurrent fevers, and weight loss
- necrotizing pneumonia and bacteraemia, associated with rapid deterioration, which is occasionally fatal. Risk factors for this pattern include females with poor lung function and severe CXR changes.

In other patients, *B. cepacia* can cause a range of other infections, from superficial to deep-tissue to disseminated, but these are rare.

Diagnosis

B. cepacia are motile, non-lactose fermenting, Gram-negative, aerobic rods. Selective media are necessary for culture and the colour and shape of the colonies vary with the particular strain and media.

Treatment

Although agents such as co-trimoxazole, chloramphenicol, minocycline, and the carbapenems show good *in vitro* activity against *B. cepacia,* their activity against strains from CF patients is decreased. Also, clearance of these drugs is increased in the CF population, and monitoring of serum levels should be considered to ensure adequate dosing. Use of combination therapy is debated, and specialist units usually produce strict antibiotic treatment protocols. Note that almost all *B. cepacia* are constitutively resistant to polymyxin. Immunotherapy with specific *B. cepacia* antigens may prove beneficial.

Table 4.12 *Burkholderia cepacia* complex

Genomovar	Name	Notes
I	*B. cepacia*	*Type species*
II	*B. multivorans*	Common in CF, associated with epidemic spread in a number of CF centres world wide
III[a]	*B. cenocepacia*	*Most common in CF* – see footnote
IV	*B. stabilis*	
V	*B. vietnamiensis*	
VI	*B. dolosa*	
VII	*B. ambifaria*	
VIII	*B. anthina*	
IX	*B. pyrrocinia*	
IX	*B. ubonensis*	

[a] Genomovar III has been linked to increased transmissibility between patients, and with a poorer prognosis and higher mortality for some patients.

Burkholderia pseudomallei

Burkholderia pseudomallei (formerly known as *Pseudomonas pseudomallei*) causes melioidosis, which is endemic in parts of SE Asia, Northern Australia, and the Caribbean. It is a major cause of community-onset septicaemia in NE Thailand.

Epidemiology

In endemic areas, *B. pseudomallei* can be cultured from moist soil, surface water (rice paddies), and the surface of many fruit and vegetables. It is also carried by rodents.

Pathogenesis

B. pseudomallei can survive and multiply within phagocytes, hence a long course of antibiotics is recommended, and antibiotics active *in vitro* do not always lead to clinical cure.

Clinical features

B. pseudomallei is usually acquired through inhaling contaminated particles, or cutaneously through skin abrasions. This organism has been called the 'great imitator' as it can cause suppurative infections of almost any organ. Manifestations range from subclinical infection to localized lung infection (cavitating pneumonia with profound weight loss, resembling TB) to overwhelming septicaemia. Bone and joint infections are common, as is parotid gland infection in children. Symptoms may occur years after exposure, due to the intracellular nature of the organism.

Diagnosis

B. Pseudomallei is a Hazard Group 3 organism. Note that laboratory-acquired cases may occur, so all specimens should be handled in Containment Level 3 if meliodosis is suspected clinically.

- Culture – *B. pseudomallei* colonies grow well on blood or Ashdown medium nutrient agar, after 1–2 days. They appear either wrinkled and dry, or mucoid, and after prolonged incubation may turn orange. Gram staining often shows small bipolar Gram-negative rods. *B. pseudomallei* is a strict aerobe and oxidizes glucose and breaks down arginine. Isolates are characterstically resistant to gentamicin and colistin. The API 20NE reliably identifies most isolates, but early involvement of the reference laboratory is recommended for confirmatory tests e.g. Indirect haemagglutination, PCR, IgM- and IgG-specific ELISAs, Serology (but note there are problems with sensitivity and specificity).

Treatment

The antibiotic of choice is ceftazidime IV for 10–14 days, or until clinical improvement occurs (with imipenem or meropenem as alternatives). This should be followed by prolonged oral therapy (e.g. co-trimoxazole plus doxycycline for 20 weeks, plus chloramphenicol for the first 4 weeks) *or* co-amoxiclav to prevent relapse. Note that resistance to these oral agents may develop during treatment: seek expert advice. In case of β-lactam allergy, this oral regimen may be given IV in cases of sepsis, but is less effective than ceftazidime.

Burkholderia mallei

B. mallei causes glanders, which is a rare disease of horses in Asia, Africa, and the Middle East. It is a Hazard Group organism, but has not been isolated in the UK since the 1940s. In humans it causes symptoms similar to melioidosis.

Overview of fastidious Gram-negative rods

These organisms often require specialist supplements or media for culture. They can be divided by appearance on Gram stain as follows:

- **Coccobacilli**:
  - *Haemophilus*
  - HACEK organisms
  - *Gardnerella*
  - *Bordetella*
  - *Brucella*
  - *Yersinia*
  - *Pasteurella*
  - *Francisella*
- **Rods with pointed ends**
  - *Legionella*
  - *Capnocytophaga*
- **Curved rods**
  - *Vibrio*
  - *Aeromonas, Plesiomonas*
  - *Campylobacter*
  - *Helicobacter.*

Streptobacillus moniliformis causes rat bite fever, as does *Spirillium minor.*

Haemophilus influenzae

Haemophilus influenzae is a small, fastidious Gram-negative coccobacillus belonging to the family *Pasteurellaceae*. It is highly adapted to humans and found in the nasopharynx of 75% of healthy children and adults. It was first isolated in 1890 by Pfeiffer and mistakenly thought to be the cause of influenza. It was also the first living organism to have its genome fully sequenced.

Epidemiology

Haemophilus influenzae capsular type 3 (Hib) used to be a common cause of meningitis in childcare. The annual incidence of invasive Hib disease dropped dramatically after introduction of the Hib conjugate vaccine in 1993, but started to rise again in 1999. In 2003 a booster campaign was implemented for children aged 6 months to 4 years.

Pathogenesis

H. influenzae inhabits the upper respiratory tract of humans; 25–80% of healthy people carry non-capsulate organisms, while 5–10% carry capsulate strains (~50% of which are capsular type b). In addition to the polysaccharide capsule that facilitates invasion, virulence factors of capsular type b include fimbriae (involved in attaching to epithelial cells), IgA proteases (help colonization), and outer-membrane proteins (involved in invasion also). There is evidence that simultaneous viral infection may initiate invasion.

Clinical features

- **Invasive infections** – e.g. meningitis, epiglottitis, bacteraemia with no clear focus, septic arthritis, pneumonia, cellulitis. These are mostly caused by capsular type b, but types e and f and non-capsulate strains can also cause serious disease. Infections generally occur between 2 months and 2 years of age, as babies < 2 months are protected by maternal antibody.
- **Non- invasive infections** – e.g. otitis media, sinusitis, purulent exacerbations of COPD. These local infections are usually associated with non-capsulate organisms. There may be an underlying abnormality (anatomical or physiological). Intercurrent virus infection may precipitate infection.

Diagnosis

- Culture – These organisms only grow in the presence of X and V factors. X factor (haemin) is needed to synthesize some respiratory enzymes that contain iron (e.g. cytochrome c, cytochrome oxidase, catalase, peroxidase). V factor is NAD(P): nicotinamide adenine dinucleotide (phosphate), and required for oxidation–reduction processes in metabolism. Blood agar (BA) contains both X and V, but *H. influenzae* grows poorly. NAD supplementation improves growth on BA, as will streaking an organism that excretes NAD, e.g. *S. aureus* – this phenomenon is called satellitism. *H. influenzae* grows well on chocolate agar, which is made by heating BA at 70–80°C for a few minutes to inactivate the NADase which normally limits utilization of V factor. Growth is also better in CO_2-enriched conditions. Antibiotic susceptibility testing with discs may be unreliable: nitrocefin strips are recommended to test for β-lactamases.

- Antigen detection – As well as culture, *H. influenzae* may be diagnosed by antigen detection (e.g. with latex agglutination, but beware of cross-reactions with *S. pneumoniae* and *E. coli*, so culture is needed for confirmation). Molecular tests, e.g. PCR, are available but not yet widely used.
- Capsule detection – Encapsulated strains of *H. influenzae* are responsible for most invasive infections (e.g. meningitis and epiglottitis), while respiratory infections and otitis media are usually associated with non-encapsulated strains. The polysaccharide capsule can be demonstrated by the Quellung reaction with type-specific antisera.
  - Antigenic type – there are six antigenic types (a–f). *H. influenzae* type b (Hib), which is a polymer of ribosyl ribitol phosphate, causes the most severe invasive infections.
  - Biotypes – there are eight biotypes of *H. influenzae* (I–VIII), based on indole, ornithine decarboxylase, and urease reactions. The most common are biotypes I–III, and most invasive (type b) organisms are biotype I.

Treatment

First-choice antibiotics for life-threatening *H. influenzae* infections are 3rd-generation cephalosporins, e.g. ceftriaxone. They are bactericidal, penetrate the CSF and are clinically effective. Alternatives include co-trimoxazole, ampicillin (but ~25% of UK type b strains produce a β-lactamase), or chloramphenicol. Less-serious *H. influenzae* infections can be treated with oral ampicillin (but ~20% of UK non-capsulate strains are β-lactamase-positive), co-amoxiclav or clarithromycin.

Prevention

- Hib conjugate vaccine (capsular polysaccharide and protein) was introduced in the UK in 1992. Given at 2, 3, 4, and 12 months. There is no evidence for serotype replacement in Europe after introduction of this vaccine. For detailed information about use of the Hib vaccine see the Green book guidelines at http://www.dh.gov.uk/prod_consum_dh/groups/dh_digitalassets/@dh/@en/documents/digitalasset/dh_4087384.pdf.
- Household contacts – chemoprophylaxis is no longer recommended to household contacts of an invasive case of *H. influenzae*. If all children <4 years old in the household have been fully vaccinated against *H. influenzae*. If one child <4 years old has been unvaccinated or incompletely vaccinated, then ALL household contacts should receive chemoprophylaxis with rifampicin regardless of age or vaccination status.
- Playgroup/nursery school contacts – chemoprophylaxis should be offered to all room contacts (teachers and children) if two cases of Hib occur within 120 days. Unvaccinated children <4 years old should be vaccinated.
- Cases of Hib <4 years old should receive vaccine and chemoprophylaxis before they are discharged from hospital, to eliminate carriage, as there are reports of infection failing to induce immunity.

Reference

Fleischmann RD, Adams MD, While O, et al. Whole-genome random sequencing and assembly of Haemophilus influenzae Rd. *Science* 1995; 269(5223):496–512.

Other *Haemophilus* spp.

Haemophilus species other than *H. influenzae* have been considered rare causes of human disease in the past. However they may cause infections more commonly than was previously believed. Most are normal flora of the human mouth and upper respiratory tract. They are associated with infections such as endocarditis, respiratory tract infection, septicaemia, brain abscess, meningitis, and soft tissue infection.

H. parainfluenzae

H. parainfluenzae is increasingly recognized as a cause of human infection. Clinical infections are similar to those caused by *H. influenzae*, but *H. parainfluenzae* tends to be less virulent than *H. influenzae*. *H. parainfluenzae* has been reported as a cause of pharyngitis, epiglottitis, otitis media, conjunctivitis, dental abscess, pneumonia, empyema, septicaemia, endocarditis, septic arthritis, osteomyelitis, meningitis, abscesses elsewhere, and urinary and genital tract infections. *H. Parainfluenzae* differs from *H. influenzae* in that it is V factor dependent only and catalase positive.

H. haemolyticus and *H. parahaemolyticus*

It is commonly thought that these species rarely cause human disease. However, recent work suggests that standard methods do not reliably distinguish *H. haemolyticus* from *H. influenzae*, so it may be more common than previously considered.

Aggregatibacter aphrophilus

The species *H. aphrophilus* and *H. paraphrophilus* have been recently reclassified as a single species, *Aggregatibacter aphrophilus*. Both of these species require CO_2 for growth. *H. aphrophilus is* V factor independent *and* has been linked with sinusitis, otitis media, pneumonia, empyema, bacteremia, endocarditis, septic arthritis, osteomyelitis, meningitis, soft tissue abscesses elsewhere, and wound infections. *H. paraphrophilus* is V factor dependent and has been documented as a cause of laryngitis, epiglottitis, endocarditis, brain abscess, hepatobiliary infection, osteomyelitis, and paronychia.

H. ducreyi

This causes chancroid, a sexually transmitted infection, common in Africa and SE Asia. It presents as a painful penile ulcer associated with inguinal lymphadenopathy. Microbiological diagnosis may be made when Gram-negative coccobacilli are isolated from a lymph node aspirate or from ulcer swabs. Treatment options include tetracyclines, erythromycin, and co-armoxiclar mentin.

H. influenzae biogroup *aegyptius*

H. influenzae biogroup *aegyptius* was previously known as *H. aegyptius* or the Koch–Weeks bacillus. It is very similar biochemically to *H. influenzae* biotype III, but can be differentiated by PCR. It causes Brazilian purpuric fever (conjunctivitis leading to fulminant septicaemia, with a high mortality) and epidemic purulent conjunctivitis. Combination therapy with ampicillin and chloramphenicol is recommended.

HACEK organisms

The HACEK organisms (Box 4.7) are rare causes of endocarditis (see 📖 Infective endocarditis, p.670), which tends to be insidious in onset (mean time to diagnosis ~3 months). Most HACEK organisms are part of normal human mouth flora and are occasionally associated with periodontitis and infections elsewhere (e.g. joints). They grow slowly and may need prolonged incubation (14 days) in CO_2 supplementation, so a high index of suspicion and close liaison with the laboratory are crucial.

Box 4.7 The HACEK organisms

- *Hæmophilus influenzae*
- *Hæmophilus parainfluenzae*
- *Hæmophilus parahaemolyticus*
- *Aggregatibacter aphrophilus*[1]
- *Aggregatibacter segnis*[2]
- *Actinobacillus* or *Aggregatibacter actinomycetemcomitans*
- *Cardiobacterium hominis*
- *Eikenella corrodens*
- *Kingella kingae*

Aggregatibacter actinomycetemcomitans

A. actinomycetemcomitans, a mouth commensal, is the major pathogen of the genus *Aggregatibacter*. There are two other Aggregatibacter species: *Aggregatibacter aphrophilus* (includes *H. aphrophilus* and *H. paraphrophilus*) and *Aggregatibacter segnis* (formerly *H. segnis*).

Diagnosis

Aggregatibacter may be difficult to culture, as it is fastidious and grows slowly, so blood cultures should be incubated for at least 14 days. Growth is enhanced by CO_2 supplementation (5–10%). *Aggregatibacter* may form 'granules' in blood cultures/broth (the media remains clear). On Gram stain, they look coccoid/coccobacillary, resembling *Haemophilus*. *A. actinomycetemcomitans* is urease-negative, indole-negative, catalase-positive and reduces nitrate. They do not grow on MacConkey, and are biochemically similar to *Pasteurella* spp.

Pathogenesis

Periodontal disease caused by *A. actinomycetemcomitans* is associated with the ability to invade and multiply within gingival epithelial cells and the production of a leuokotoxin that lyses neutrophils. Other potential virulence factors include a bacteriocin, endotoxin, chemotaxis-inhibiting factor, and fibroblast-inhibiting factor.

Clinical

A. actinomycetemcomitans can cause endocarditis, joint infections, and severe periodontal disease. *Aggregatibacter* has also been found (together with some *Haemophilus* spp. Fusiforms, and anaerobic streptococci) in actinomycotic lesions. Their contribution to the pathogenesis of actinmycoses is unclear.

1. Includes. *H. aphrohilus* and *H. paraphrophilus*
2. Formerly *H. segnis*

Treatment

A. actinomycetemcomitans is usually susceptible to the third-generation cephalosporins, which are considered the drugs of choice for endocarditis (4 weeks for native and 6 weeks for prosthetic valve endocarditis. Periodontitis requires mechanical debridement with antibiotic treatment (e.g. tetracyclines).

Cardiobacterium hominis

C. hominis is the only species in the genus. It is normal flora in the human mouth, nose and throat and occasionally other mucous membranes and the GI tract. Unlike the other HACEK organisms, it rarely causes diseases other than endocarditis.

Diagnosis

This Gram-negative rod has a pleomorphic appearance and may be difficult to decolourise during Gram staining. Culture is enhanced in 5–10% CO_2 and high humidity. It grows well on blood agar and chocolate, with slight β-haemolysis, but poorly on MacConkey agar. It is catalse-negative and oxidase-positive. It produces indole (although positivity is weak with many strains), which differentiates it from other HACEK organisms.

Treatment

Sensitivity testing is difficult because of the slow growth, but it is usually susceptible to β-lactams, tetracycline, and chloramphenicol. However, a β-lactamase-producing isolate that was also resistant to cefotaxime, has been reported. Hence the current first-line recommendation of cefotaxime for HACEK organisms may not be optimal for *C. hominis* endocarditis: an alternative regimen is co-amoxiclav and gentamicin.

Eikenella corrodens

E. corrodens exists as normal mouth and upper respiratory tract flora.

Diagnosis

This facultative anaerobic Gram-negative rod is oxidase-positive, catalase-negative, urease-negative, indole-negative, and reduces nitrate to nitrite. About 50% of strains create a depression in the agar ('corroding bacillus'). As with the other HACEK organisms, culture is slow and enhanced in 5–10% CO_2.

Clinical features

E. corrodens causes subacute endocarditis, but is more commonly found as part of mixed infections (e.g. human bite wounds, head and neck infections, respiratory tract infections). It often co-exists with *Streptococcus* spp. There are reports of *E. corrodens* causing a variety of other infections. Infections are usually indolent, taking >1 week from time of injury to clinical symptoms of disease. Suppuration is common, and may smell like an anaerobic infection.

Treatment

E. corrodens is usually susceptible to the β-lactams, tetracyclines, and fluoroquinolones. It is uniformly resistant to clindamycin, erythromycin, and metronidazole, and often resistant to aminoglycosides.

Kingella kingae

There are four species of *Kingella*, which all colonize the respiratory tract and rarely cause human disease: *K. kingae* (previously known as *Moraxella kingae*), *K. indologenes* (now known as *Suttonella indologenes*), and *K. oralis*. *K. kingae* is the most common, and a recent increase in cases is likely to be due to increased awareness of the organism and improved diagnostics.

Diagnosis

Kingella organisms have been misidentified as *Moraxella or Neisseria* in the past. They are short Gram-negative rods with tapered ends, which sometimes appear coccoid. They tend to resist decolourisation, so may look Gram–positive. *Kingella* is catalase-negative, oxidase-positive, urease-negative, and ferments glucose. *K. kingae* grows on blood and chocolate agar, but not MacConkey. To increase the chance of recovering *K. kingae* from joint fluid, the fluid should be inoculated into blood culture bottles rather than just plated out directly onto agar plates.

Clinical features

Most cases of invasive disease due to *K. kingae* occur in children aged between 6 months and 4 years. *K. kingae* most commonly causes bacteraemia, endocarditis (of native and prosthetic valves), and skeletal infections, e.g. septic arthritis. *K. indologenes* and *K. denitrificans* also cause endocarditis. *K. oralis* is found in dental plaque, but its relationship with periodontal disease is unknown.

Treatment

Most *Kingella* spp. appear susceptible to penicillins and cephalosporins; however, a β-lactamase-producing isolate of *K. kingae* has been reported. Alternatives include aminoglycosides, co-trimoxazole, tetracyclines, erythromycin and quinolones.

Gardnerella

Gardnerella vaginalis is found in the female genital tract and is associated with bacterial vaginosis (BV)/non-specific vaginitis (see 🕮 Bacterial vaginosis, p.734). It is usually classified with GNR, although it is usually susceptible to vancomycin. Of interest, electron microscope studies have noted the cell wall to be either Gram-negative, Gram-positive, or to show an atypical laminated appearance.

Epidemiology

There is debate whether specific biotypes have been associated with BV. Newly acquired strains of *G. vaginalis* may precipitate BV, rather than overgrowth of previously colonizing biopsies.

Pathogenesis

Adherence of *G. vaginalis* to vaginal and urinary epithelial cells may play a role in the pathogenesis is of BV and UTIs. Pili have been seen on *G. vaginalis*, and haemagglutinating activity has been shown. *G. vaginalis* also produces a cytolytic toxin (haemolysin). It is serum resistant, which may aid survival during bloodstream invasion at childbirth.

Clinical fetaures

- BV – *G. vaginalis* is almost universally present in women with BV, along with mixed anaerobic flora.
- UTI – *G. vaginalis* is isolated from <1% of UTIs, and because of its presence in the femal genital tract could represent vaginal contamination. However it has been found from suprapubic aspirates, and also in association with renal disease and interstitial cystitis.
- Bacteraemia – this rare event is associated with female genital tract conditions, such as chorioamnionitis, postpartum endometritis, and septic abortion. Neonatal infection has also been reported.

Diagnosis

Gardnerella is a facultative anaerobe, which appears as a pleomorphic Gram-variable rod. It is oxidase- and catalase-negative, non-encapsualted, and non-motile. It needs enriched media for growth. It is also urease-, indole- and nitrate-negative. Note that *G. vaginalis* is susceptible to SPS (sodium polyanetholesulfonate), which is found in most blood culture bottles, so bacteraemia figures may be underestimated. In clinical practice, BV is diagnosed using the Amsel criteria.

Treatment

G. vaginalis is usually susceptible to penicillin, clindamycin, and vancomycin, and resistant to colistin, cephalexin, the tetracyclines, and nalidixic acid. Metronidazole is usually the preferred treatment for BV.

Bordetella

Bordetella pertussis and *B. parapertussis* cause whooping cough, which is a notifiable disease in England and Wales. The other species only cause human infections under special circumstances: these are *B. bronchiseptica* (causes kennel cough in dogs and snuffles in rabbits), *B. avium* (bird pathogen), *B. hinzii*, *B.holmesii*, and *B. trematum*.

Epidemiology

The organism is spread by droplet infection and is highly infectious. Pertussis has the highest incidence in infants but also occurs in adolescents and adults. Morbidity and mortality are higher in females than males, and in those <6 months old.

Pathogenesis

B. pertussis produces a number of biologically active substances that are thought to play a role in disease:

- Surface components e.g. filamentous haemagglutinin (FHA), pertactin and fimbriae
- Toxins such as pertussis toxin (PT), adenylate cyclase toxin (ACT), tracheal cytotoxin (TCT) and dermonecrotic toxin (DNT).
- Other products e.g. tracheal colonisation factor and BrKA (Bordetella resistance to Killing)

Clinical features

- Classic (severe) pertussis is defined by the World Health Organization (WHO) as ≥21 days cough with paroxysms, associated whoops or post-tussis vomiting, and culture confirmation.
- Mild pertussis is any laboratory-confirmed disease that does not meet the criteria for classic disease.

Diagnosis

These tiny coccobacilli occur singly or in pairs. *B. pertussis* and *B. parapertussis* are non-motile.

- Bordet–Genou agar: pearly colonies on day 3–4
- CCBA agar (charcoal cephalexin) – *B. pertussis* produces glistening greyish white colonies on CCBA. It does not grow on nutrient agar, and grows poorly on BA. *B. parapertussis* colonies are larger and duller, and become visible sooner.
- Culture for *B. pertussis* lacks sensitivity, and enhanced diagnostic methods are available at the HPA Respiratory and Systemic Infection Laboratory
- PCR – one PCR targets the toxin promoter, while another PCR targets the insertion sequence IS481 (occurs in *B. pertussis*; *B. holmesii*, and some *B. bronchiseptica*). Criteria for reference laboratory PCR: pernasal swab or nasopharyngeal aspirate (NPA) from an acutely ill child age ≤12 months on PICU or paediatric ward, with respiratory illness compatible with pertussis
- Serology – anti-pertussis toxin (PT) IgG antibody levels are determined using an EIA, on paired sera or single samples taken >2 weeks after onset for any individuals with prolonged cough.

Treatment

A Cochrane review on antibiotics for pertussis (2007) found that short-term antibiotics (azithromycin × 3–5 days, or clarithromycin/erythromycin × 7 days) were as effective as long-term (erythromycin × 10–14 days) in eradicating *B. pertussis* from the nasopharynx, but had fewer side effects.[1] Co-trimoxazole x 7 days was also effective. There were no differences in clinical outcome or microbiological relapse between short- and long-term antibiotics. In 1994, an erythromycin-resistant strain was reported from the USA, but this has not become a clinical problem.

Antibiotic prophylaxis to contacts >6 months old did not significantly improve their clinical symptoms or the number of cases developing culture-positive *B. pertussis*. The Cochrane authors concluded there is insufficient evidence to determine the benefit of prophylactic treatment of pertussis contacts.

Vaccine

There were major epidemics of whooping cough in 1977/79 and 1981/83, after immunization coverage dropped from >80% to 30%, following a report linking the vaccine to brain damage. Coverage is now back up at ~95%. Acellular pertussis (aP) vaccine is given in the primary immunization course as DTaP/IPV/Hib, aged 2, 3, and 4 months. A further booster as dTaP–IPV, is given with preschool boosters, as vaccine immunity wanes over time.

Reference

Altunaiji S, Kukuruzovic R, Curtis N, Massie J. Antibiotics for whooping cough (pertussis). *Cochrane Database Syst Rev* 2007;Issue 2:CD004404. http://www.cochrane.org/reviews/en/ab004404.html

Brucella

Brucella spp. (table 4.13) cause brucellosis, which is also called undulant fever, Mediterranean fever or Malta fever (or contagious/infectious abortion in cattle). This zoonosis is transmitted via contaminated or untreated milk and milk derivatives, or direct contact with infected animals or their carcasses. *Brucella* species survive well in aerosols and resist drying, so are candidates for agents of bioterrorism (see Bioterrorism, p.876).

Table 4.13 Main species of *Brucella* – note that the host relationship is not absolute, and man and domestic animals may be susceptible to infections by different species

***Brucella* spp.**	**Animal infected**	**Human manifestations**
B. abortus	Cattle; bison and elk in N America	Brucellosis
B. suis	Pigs (swine brucellosis)	Brucellosis
B. melitensis	Goats/sheep	Brucellosis
B. canis	Dogs (mainly beagles in USA)	Mild disease only
B. ovis	Sheep (Australia and New Zealand)	No evidence this species infects man

Epidemiology

Brucellosis is still endemic in Africa, the Middle East, central and south-east Asia, south America, and in some Mediterranean countries. It has been virtually eliminated from most developed countries. Human–human transmission has been documented but is rare – methods include breast-milk, sexual transmission, and congenital disease. Most human cases seen in the UK are due to *B. melitensis* from unpasteurized goats milk and cheeses. Brucellosis also occurs through occupational exposure of laboratory workers, vets, and slaughterhouse workers. A careful epidemiological patient history is crucial, regarding travel, dietary habits, and possible exposure.

Pathogenesis

After ingestion (or entry via skin abrasions, or inhaling infected dust), the bacteria live in the regional lymph nodes during the incubation period (usually 2–8 weeks). They then enter the circulation, and subsequenlty localize in different parts of the reticulo-endothelial system, forming granulomatous lesions which may result in complications in many organs. Brucella organisms surviving within granuloma may cause relapses of actue disease, or result in chronic brucellosis.

Clinical features

Brucellosis has a wide variety of clinical presentations. The 'undulant' or wave-like fever rises and falls over weeks in ~90% of untreated patients. Malodorous perspiration is said to be pathognomonic. Osteoarticular disease occurs in ~20%, with epididimo-orchitis in ~6%. Other symptoms include weakness, headaches, depression, myalgia, and bodily pain.

Sequelae are also variable, and include granulomatous hepatitis, anaemia, leukopenia, thrombocytopenia, meningitis, uveitis, and optic neuritis.

Diagnosis

- Hazard Group 3 organism
- Culture from blood or bone marrow. These coccobacilli or short bacilli may occur singly, in chains, or in groups, and can take up to 8 weeks to grow. They are non-motile, non-sporing, and non-capsulate. They are aerobic, and *B. abortus* requires 5–10% CO_2 to grow. The three main species (*B. melitensis*, *B. abortus*, *B. suis*) can be differentiated biochemically and by antigenic structure. Each of these three species can be further divided into biotypes: there are >9 biotypes of *B. abortus*, >3 of *B. melitensis*, and >5 of *B. suis*
- Molecular techniques/real-time PCR has been developed
- Serology – raised (1:160) or a rising antibody titre in symptomatic patients suggests the diagnosis of active brucella. Demonstration of antibodies with various tests including the SAT (standard agglutination test), mercaptoethanol test, classic Huddleson, Wright, and/or Bengal Rose reactions. At the HPA Reference Laboratory, Liverpool, all sera are screened with a brucella antibody assay and specific IgG/IgM enzyme immunoassays. Positive samples then undergo further testing with in-house micro-agglutination and complement fixation in-house assays
- Histological evidence of granulomatous hepatitis (hepatic biopsy)
- Radiological alterations in infected vertebrae – the Pedro Pons sign (preferential erosion of antero-superior corner of lumbar vertebrae) and marked osteophytosis are suspicious of brucellic spondylitis
- Dye inhibition test (basic fuschin and thiamin dyes) can differentiate individual *Brucella* spp.

Treatment

Drugs must penetrate macrophages and be active in an acidic environment. Doxycycline and one of the following agents (gentamicin, streptomycin, rifampicin) are suitable regimens, and the combination is usually given for at least 6 weeks. Fluoroquinolones or tetracyclines may also be used in combination. Intensive treatment of the acute disease is recommended, to prevent progression to chronic forms which are more difficult to treat. Antibody levels may be measured to monitor response to therapy. A triple combination of doxycycline, together with rifampin and co-trimoxazole has been used succefully to treat neurobrucellosis.

Prevention

Good standards of hygiene in the production of raw milk and its products, or pasteurization of all milk, will prevent brucellosis acquired from ingestion of milk. Also avoid contact with infected animals. Vaccination of young cattle helps to protect animals against *B. abortus*, but is not completely effective. However it helps to limit the spread of disease, and thus aids eradication. Only by testing all animals, and slaughtering those with positive results, can the disease be truly eradicated.

Yersinia pestis

Y. pestis causes plague. There are 3 clinical syndromes: bubonic, pneumonic and septicaemic (Table 4.14).

Epidemiology

Y. pestis occurs worldwide, but most cases of plague are reported from developing countries of Africa and Asia. There are ~10 cases annually from rural areas of the USA. The last outbreak of plague acquired in the UK was in 1918.

Pathogenesis

The somatic (heat-stable) and capsular (heat-labile) antigens are important in virulence and immunogenicity. Somatic antigens V and W help resist phagocytosis, and the capsular antigen containing the immunogenic fraction (F1) is antiphagocytic also. Other virulence factors include a lipopolysaccharide endotoxin, the ability to absorb iron as haemin, and temperature-dependent coagulase and fibrinolysin.

Diagnosis

- Hazard Group 3 organism
- Culture – this short Gram-negative coccobacillus occurs singly or in pairs (or as chains in fluid culture). Old cultures are pleomorphic and may even resemble yeast cells. They are non-sporing and non-motile, and often capsulate at 37°C. Methylene blue shows bipolar staining. *Yersinia* spp. grow between 14°C and 37°C, with optimal growth at 27°C. Small non-haemolytic colonies are seen on blood agar at 24 h, which are catalase-positive and oxidase-negative. Although *Y. pestis* grows on MacConkey, it tends to autolyse after 2–3 days. Organisms are citrate-, indole- and urease-negative. It is usually cultured from a bubo aspirate, but may also grow from blood, CSF, or sputum. Cefsulodin Irgasan Novobiocin (CIN) agar is selective for *Yersinia* (and *Aeromunas* species).
- Direct immunofluorescence is a more-rapid diagnostic method.
- Serological tests for yersiniosis (acute and convalescent) include the complement fixation test and haemagglutination of tanned sheep red cells to which F1 capsular antigen has been adsorbed.

Treatment

Early antibiotic therapy for suspected cases (e.g. streptomycin, gentamicin, or doxycycline) reduces the otherwise high mortality to ~10%. Contacts may also be given antibiotic prophylaxis. Patients with pneumonic plague should be isolated until they are sputum smear-negative (usually ~ 3 days since starting treatment). There is no vaccine currently available.

Control

Flea and rodent control are important. See the HPA website for guidelines for action in the event of a deliberate release of plague.
http://www.hpa.org.uk/infections/topics_az/deliberate_release/Plague/PDFs/plague_guidelines.pdf.

Table 4.14 Features of the main clinical forms of plague

	Bubonic	**Pneumonic**	**Septicaemic**
Transmission	Rat flea bites	Respiratory aerosols from rat fleas	Primary infection
	Xenopsylla cheopis	Person-to-person spread in crowded unhygienic conditions, during epidemics	Complication of bubonic or pneumonic plague
		May arise as a complication of bubonic or septicaemic plague	
Diagnostic specimen	Fluid from buboes	Sputum	Blood culture/blood films
Clinical symptoms	Fever, painful buboes, inguinal lymphadenopathy	Cough or haemoptysis ± bubo	Fever, hypotension, no buboes
Incubation period	2–8 days	1–4 days (maximum 6 days)	2-8 days
Mortality if untreated	~60%	High mortality (approaching 100%)	High mortality (approaching 100%)

Yersinia enterocolitica

Y. enterocolitica resembles *Y. pestis* and *Y. pseudotuberculosis* on culture and morphologically, but differs antigenically and biochemically. The most common serotypes causing human infection in Europe are 3 and 9. Serotypes cultured from healthy individuals are probably non-pathogenic.

Epidemiology

Y. enterocolitica is acquired from eating infected meat or milk. Patients with conditions associated with iron-overload (e.g. haemochromatosis) and the immunosuppressed are at increased risk of *Yersinia* infections.

Pathogenesis

See 📖 *Yersinia*, p.361.

Clinical features

Y. enterocolitica usually presents as a febrile illness associated with bloody diarrhoea, and may mimic salmonellosis, shigellosis, or appendicitis. Other presentations include mesenteric lymphadenitis and septicaemia, which may be fatal in the elderly. Secondary complications include erythema nodosum, polyarthritis, peritonitis, Reiter's syndrome, meningitis, osteomyelitis, and hepatic, renal, and splenic abscesses. *Y. enterocolitica* has been cultured from pseudotuberculous lesions in animals.

Treatment

Gastroenteritis usually resolves without antibiotics. In severe infection, the recommended regimen is doxycycline plus an aminoglycoside. Alternatives include cefotaxime, co-trimoxazole and fluoroquinolones. Note resistance to penicillin. If the patient is on desferrioxamine this should be stopped as it may increase the severity of infection.

Yersinia pseudotuberculosis

Epidemiology

Strains of *Y. pseudotuberculosis* can be differentiated by somatic and flagellar antigens, some of which are shared with *Y. pestis*. Most human infections with *Y. pseudotuberculosis* are due to serotype 1.

Pathogenesis

See 📖 *Yersinia*, p.361.

Clinical features

Y. pseudotuberculosis causes a fatal septicaemia in animals and birds. Humans usually acquire the infection from contact with water polluted by infected animals, or eating contaminated vegetables – infection due to direct contact with animals is rare. In humans, yersiniosis infection ranges from asymptomatic to a fatal typhoid-like illness with fever, purpura, and hepatosplenomegaly. Mesenteric adenitis ± erythema nodosum may mimic appendicitis and tends to infect males aged 5–15 years.

Diagnosis

This small Gram-negative rod is slightly acid fast. It grows poorly on MacConkey (like *Y. pestis*), but can be differentiated from *Y. pestis* because *Y. pseudotuberculosis* can produce urease and is motile at 22°C.

Treatment

Y. pseudotuberculosis shows *in vitro* susceptibility to ciprofloxacin, tetracyclines, aminoglycosides, sulphonamides, and penicillin. Mesenteric adenitis is usually self-limiting.

Pasteurella

The genus Pasteurella includes the species *P.multocida, P.haemolytica* (now known as *Mannheimia haemolytica*), *P. canis, P. stomatis*, and *P. pneumotropica*. *Pasteurella* live in the mouth, and GI and respiratory tracts of many animals (especially dogs and cats) ± humans. *P. multocida*, the most frequent species causing human infections, usually causes skin and soft tissue infections.

Epidemiology

Fifteen serotypes of *P. multocida* have been identified, based on four capsular antigens and 11 somatic antigens. PFGE can be used to compare strains. In addition to acquiring infection through animal bites, humans can also become infected through inhaling air which has become contaminated by infected animals' coughing.

Pathogenesis

In animals, *P. multocida* causes hemorrhagic septicaemia, which is usually fatal. Most virulent *Pasteurella* strains have a polysaccharide capsule, which is antiphagocytic and protects against intracellular killing by neutrophils. Also, some strains produce a leukotoxin and some bind transferrin.

Clinical features

- *P. multocida* causes skin and soft tissue infections after animal bites, most commonly a localized abscess with cellulitis and lymphadenitis. *P. multocida* has also been associated with upper and lower respiratory tract infections. Other sites of infection are uncommon – these include meningitis post head injury, bone and joint infections, septicaemia, endocarditis, and intra-abdominal infections.
- *P. haemolytica* is non-pathogenic for humans. It causes pneumonia in sheep and cattle, septicaemia in lambs, and also infects poultry and domestic animals.
- *P. pneumotropica* may be isolated from the respiratory tract of laboratory animals. There are reports of it causing human infections, e.g. animal bite wound infections, septicaemia, and upper respiratory tract infections.

Diagnosis

P. multocida is a facultative anaerobic Gram-negative coccobacillus, which appears pleomorphic in culture and does not grow on MacConkey agar. At 37°C, organisms are capsulate, non-sporing, and non-motile. They show bipolar staining with methylene blue. Most are fermentative, and oxidase-positive, catalase-positive, and indole-positive. *P. haemolytica* can be differentiated from *P. multocida* as it is haemolytic on blood agar and can grow on MacConkey agar.

Treatment

Penicillin is the mainstay of treatment, and there is a wealth of clinical experience to support this. It is resistant to oral first-generation cephalosporins, flucloxacillin, clindamycin, and erythromycin. It is sensitive *in vitro* to fluoroquinolones, which may be appropriate in penicillin-allergic patients.

Francisella

Francisella tularensis is primarily an animal pathogen (rabbits and hares), which occasionally infects humans as accidental hosts. The resulting infection is called tularaemia, and may be either ulceroglandular or typhoidal/pulmonary. Only two of the five subspecies *F. tularensis* subspecies are clinically important (type A is highly virulent and type B less virulent). It is also a potential agent of bioterrorism (see 📖 Bioterrorism, p.876).

Epidemiology

Tularaemia is endemic in N America and parts of Europe, Asia, northern Australia, and Japan. Most cases in man are sporadic, though outbreaks have been reported. It may survive for days in moist soil and in water polluted by infected animals, and for years in culture at 10°C. Organisms are killed in 10 min after exposure to moist heat at 55°C. For epidemiological investigations, the most discriminatory typing system is based on VNTR (multiple-locus variable-number tandem repeats).

Pathogenesis

There is evidence from animal experiments of intracellular multiplication of *F. tularensis*. Virulence has been associated with the capsule and also citrulline ureidase activity.

Clinical features

Infection ranges from asymptomatic to septic shock, depending on virulence of particular strain, host immune response, route of entry, and degree of systemic involvement. There are two main forms:

- **ulceroglandular** form of tularaemia (due to *direct contact* with infected animal) – acute-onset fever, headache, and rigors, usually followed by glandular lesions and skin ulceration, ± eye involvement.
- **typhoidal/pulmonary** form of tularaemia (results from *indirect contact* through bites from ticks/mosquitoes/biting flies, inhaling infected dust, or eating contaminated food or water) – acute-onset fever, headache, and rigors, usually followed by respiratory or typhoid-like symptoms.

Diagnosis

F. tularensis is a Hazard Group 3 organism. It is a small, non-motile, non-sporulating, capsulate Gram-negative cocco-bacillus, which shows characteristic bipolar staining with carbol fuchsin (10%). It stains poorly with methylene blue. It is a strict aerobe, and culture requires the addition of egg yolk or rabbit spleen to agar. After culture, the bacilli may appear filamentous and larger. Traditional microbiological methods are slowly being replaced by immunological and molecular tests, including ELISA and immunoblots for antibodies (but tests relying on antibody detection are limited in early clinical stages of disease). If a case is suspected, involve the HPA Special Pathogens Reference Unit (SPRU) at Porton Down, who will process tissue biopsies, wound swabs, or specimens from bacterial culture.

Treatment

Seek expert advice. Streptomycin or gentamicin are antibiotics of choice, with the addition of chloramphenicol for meningitis. Relapse is more common if tetracycline or chloramphenicol are used, as these are bacteriostatic for *F. tularensis*. The live vaccine (LVS) is based on an attenuated strain of *F. tularensis*. Post-exposure prophylaxis with doxycycline or ciprofloxacin may be considered after potential inhalation.

Legionella

This organism is named after the outbreak of pneumonia affecting >180 members of the American Legion at a convention in Philadelphia in 1976. The *Legionellaceae* naturally live in water and only accidentally infect humans. This may result in either Legionnaires' disease or Pontiac fever. There are 52 different genetically defined species of *Legionella*, of which ~50% of infect humans. *Legionella pneumophila* serogroup 1 is the most pathogenic, and accounts for ~95% of human cases.

Epidemiology

Legionella is acquired via inhalation of contaminated aerosols (e.g. from spas, showers, air-conditioning systems, water-storage tanks, nebulizers) or drinking water. Water systems are more likely to be contaminated with *Legionella* if the temperature is outside the recommended range (it should be <20°C or >55°C), if the flow is obstructed, or if biofilms have formed. Note that *Legionella* is an intracellular organism and can survive in amoebae, within the environment. The incubation period is 2–10 days, and occasionally symptoms may develop up to 3 weeks post exposure. It is not transmitted person to person. Most cases are isolated, but outbreaks can occur.

Pathogenesis

After the infection is established, pneumonic consolidation develops, characterized by proteinaceous fibrinous exudates pouring into the alveoli. The mechanism of distant toxic changes (e.g. confusion, hallucinations, focal neurology) is poorly understood. *Legionella* organisms are engulfed by monocytes, and may survive intracellularly for prolonged periods of time.

Clinical features

In addition to the two main clinical syndromes below, rare conditions (e.g. prosthetic valve endocarditis, wound infections) have been reported:

- **Legionnaires' disease** – this rapidly progressive pneumonia is characterized by fever, respiratory distress, and confusion, and has a mortality rate of >10% in healthy people. Risk factors include age >50 years, hospital admission, immunosuppression, and smoking. Men are affected more than women
- **Pontiac fever** – this is a brief flu-like illness, which has a high attack rate but low mortality.

Diagnosis

- **Microscopy/culture** – these short rods/coccobacilli may be difficult to see by Gram stain, so fluorescent antibody stains or silver impregnation may help. *Legionella* grows best on media such as BCYE (buffered charcoal yeast extract), which contains iron plus cysteine as an essential growth factor. Some strains prefer 2.5–5% CO_2 at 35–36°C. *L. pneumophila* colonies usually appear by day 5, but other species may require 10 days. Colonies may autofluoresce under ultraviolet (UV) light. Serogroups can be differentiated by slide agglutination or fluorescent antibody tests, which are available at reference laboratories.

- **Antigen detection** – *Legionella* urinary antigen test (ELISA) only detects serogroup 1 of *L. pneumophila*.
- **Antibody detection** – FAT (fluorescent antibody test), RMAT (rapid micro-agglutination test), or ELISA (enzyme-linked immunosorbent assay). A >4-fold rise, or titre 1:256 is usually diagnostic. Remember that antibody may take >8 days to develop after onset of infection, and may persist for months/years post infection. Also note some cross-reactivity with *Campylobacter*.

Treatment

Conventional susceptibility tests in broth and agar are unreliable, and methods have not been standardized. In addition, many antibiotics with excellent *in vitro* activity against *Legionella* (eg, β-lactams and aminoglycosides) are ineffective. Essentially, the macrolides, quinolones, tetracyclines, and rifampicin are effective as they have good intracellular penetration. Preferred treatment is intravenous erythromycin or oral clarithromycin, with ciprofloxacin as an alternative. In severe infections, the dose of erythromycin may be doubled, or rifampicin added.

Control and prevention

This relies on good design and maintenance of water systems to prevent growth of *Legionella* organisms, and subsequent treatment of the source (e.g. contaminated water systems) if a case occurs. The main approaches to control are:

- physical: – heat, UV light, sonication: use of compressed air to drain and flush pipes
- chemical – inhibit scale formation, use of biocides to kill the amoebae (such as sodium hypochlorite, ozone), use of charcoal filters
- plumbing – regular maintenance, no dead legs in the system, pumps should be in series and not in parallel, no dead spaces in the heaters, regular flushing of the system, use of correct components.

Guidelines on legionella investigations and control are available (Box 4.8).

Box 4.8 Legionella guidance

The following guidance is available electronically.

HPA website

- Guidelines for investigating single cases of Legionnaires' disease – http://www.hpa.org.uk/cdph/issues/CDPHvol5/No2/guidelines1.pdf

Health and Safety Executive (HSE)

- Control of legionella: revised approved code of practice http://www.hse.gov.uk/lau/lacs/46-2.htm
- Control of legionella: investigation of outbreaks (and single cases) of legionellosis from water systems incorporating cooling towers and evaporative condensers – http://www.hse.gov.uk/lau/lacs/46-4.htm

Department of Health Estates and Facilities Division

- Health technical memorandum 04-01: *The control of Legionella, Hygiene, 'Safe' Hot Water, Cold Water and Drinking Water Systems.* http://195.92.246.148/knowledge_network/documents/HTM_04_01_exec_sum_20061010171128.pdf

Capnocytophaga

The genus *Capnocytophaga* in the family *Flavobacteriaceae* may be divided into two groups:

- species associated with dog-bite infections (and occasionally bites from other animals such as rabbits or cats), and normally found in dogs' mouths: *C. canimorsus* (formerly known as dysgonic fermenter, 2 DF2) and *C. cynodegmi*
- species found in the human mouth – *C. ochracea* (These were the dysgonic fermenter group[1] (DF1)), *C. gingivalis*, *C. sputigena*, *C. haemolytica*, and *C. granulosa*.

Epidemiology

While *C. canimorsus* and *C. cynodegmi* are most commonly associated with bites, there are reports of infections occurring merely after exposure to dogs, with no bites or scratches.

Pathogenesis

Species found in the human mouth produce a variety of enzymes that help in the invasion of periodontal tissue (e.g. acid and alkaline phosphotases, aminopeptidases, IgA, proteases, and trypsin-like enzymes). These are thought to be important in periodontitis.

Clinical features

All species may cause a wide range of infections in normal and immunocompromised hosts. Among animal bite infections, *C. canimorsus* is more common and more severe than *C. cynodegmi*, with a mortality approaching 30%. Particular risk factors include asplenic patients, alcoholics, and those on steroids. Asplenic patients with *C. canimorsus* infection may present with shock, disseminated purpuric lesions, and disseminated intravascular coagulation (DIC). Fulminant infections with *C. canimorsus* may also occur in healthy people, although infections tend to be milder. Meningitis, endocarditis, pneumonia, corneal ulcer, cellulitis, and septic arthritis due to *C. canimorsus* have also been reported.

Species found in the human mouth may be important in localized juvenile periodontitis. They have also been found in the female genital tract and associated with intrauterine infection, amnionitis, and neonatal infections in premature babies. Rarely, they cause severe infections as opportunistic pathogens (e.g. endocarditis, eye infections, hacteraenis, peritonitis), in both immunocompent and immunosuppressed patients.

Diagnosis

These long thin delicate GNRs are typically fusiform, but older cultures often show pleomorphic sizes and shapes. They are facultative anaerobes, and grow best with CO_2 enrichment. On blood or chocolate agar they may appear yellowish, with a spreading edge with finger-like projections due to the typical gliding motility. Note they do not grow on MacConkey agar and they do not produce indole. Differentiating each individual species is more difficult and generally requires reference laboratory assistance. In general, species from the human mouth are oxidase-negative and catalase-negative, while those from animals' mouths are oxidase-positive and catalase-positive.

C. canimorsus is more fastidious than the others, and may be difficult to grow from blood cultures (even when organisms have been seen on Gram stain). In this situation, culture on enriched agar (e.g. heart infusion agar with rabbit or sheep blood) for 14 days in 10% CO_2 may help.

Treatment

Co-amoxiclav is usually recommended for these infections. Asplenic patients should be given penicillin or co-amoxiclav prophylaxis after a dog bite, as the organism may take a while to grow and be identified, and the mortality rate is high. While the animal-bite-associated species are sensitive to penicillin, resistance to the β-lactams has been reported in the human-mouth species. For instance, *C. hæmolytica* and *C. granulosa* are often resistant to β-lactams and aminoglycosides. All species are usually sensitive to clindamycin, erythromycin, tetracyclines, and the quinolones.

Vibrios

The genus *Vibrio* (family *Vibrionaceae*) includes over 30 species. The most important ones that result in human infections are *V. cholerae*, *V. parahaemolyticus* and *V. vulnificus*. Other species, such as *V. alginolyticus*, *V. damsela* (now known as *Listonella damsela*), *V. fluvialis*, *V. hollisae* (now known as *Grimontia hollisae*) and *V. mimicus*, occasionally cause opportunistic infections.

Vibrio cholerae

There are ~20 cases of cholera (see 📖 Cholera, p.692) imported into the UK every year. These are most commonly O1-El Tor. In the mid-1990s a new serogroup (O139) appeared in the Bay of Bengal – this was the first time a non-O1 serogroup had resulted in epidemic cholera.

Epidemiology

Cholera is prevalent in Central and South America, Africa, and Asia. There are more than 130 different O (somatic antigen) serogroups of *V. cholerae*. Serogroup O1 (the 'cholera vibrio') causes epidemic cholera, and some strains of non-O1 (the 'non-cholera or non-agglutinable vibrios) can also cause diarrhoea. Serogroup O1 is usually acquired by the faecal-oral route, while non-O1 *V. cholerae* may be associated with consumption of seafood or exposure to saline environments. The two biotypes of serogroup O1 (**El Tor**, which is the most common, and Classical) can be distinguished by susceptibility to phage, and the fact that El Tor is haemolytic and resistant to polymixin B. The subtypes of serogroup O1 are Ogawa (most common), Inaba, and Hikojima (which possesses determinants of both other subtypes).

Pathogenesis

The potent cholera enterotoxin, produced by serogroup O1 and some non-O1 strains, comprises five B (binding) subunits and one A (active) subunit. Insertion of the B subunits into the host cell membrane forms a channel for subunit A to enter the cell. By causing the transfer of ADP ribose from NAD to another protein, adenylate cyclase is irreversibly activated and cAMP is overproduced. The resulting hypersecretion of Cl^- and HCO_3^- causes loss of massive water and electrolytes (rice-water stool). Other features important in pathogenesis of serogroup O1 include production of mucinase and other proteolytic enzymes (which help the organism reach the enterocytes), the motility of the organism, and adhesive haemagglutinins (aid close adherence to enterocyte surface). Non-O1 strains may produce other enterotoxins, cytotoxins, haemolysins, and colonizing factors.

Cholera is transmitted by contaminated food or water, and requires a large infective dose. Humans are the only host. Only a handful of those infected are symptomatic (ratios quoted are 40 asymptomatic carriers:1 symptomatic individual for El Tor and 5:1 for classical), which underscores the need for good hygiene.

Clinical features

V. cholerae usually causes the typical profuse watery diarrhoea of cholera, which may rapidly lead to hypovolaemic shock and death from dehydration. Milder cases are similar to other causes of secretory diarrhoea, and asymptomatic infections also occur. Non-O1 *V. cholerae* usually causes mild, sometimes bloody diarrhoea, but may occasionally be severe and resemble cholera. Patients exposed to aquatic environments may suffer from wound infections, and bacteraemia and meningitis have been reported.

Diagnosis

During an epidemic, cholera is a clinical diagnosis. Otherwise, diagnosis is based on high clinical suspicion together with culture or dark-field microscopy of stool (comma-shaped organisms are seen moving around, which ceases when diluted O1 antisera is added). Vibrios are short, curved or 'comma-shaped', aerobic Gram-negative rods, which are motile by a single polar flagellum. They ferment both sucrose and glucose but not lactose, and reduce nitrate to nitrite. Most are oxidase-positive and produce indole. The growth characteristics of vibrios are summarised in Table 4.15. *V. cholerae* is non-halophilic (i.e. can grow on media without added salt), provided the necessary electrolytes are present. *V. cholerae* can grow at 42°C (along with *V. parahaemolyticus* and *V. alginolyticus*). Vibrios are tolerant to alkali but have a low tolerance to acid. *V. cholerae* is usually VP-positive. Vibrios accumulate on the surface of alkaline peptone water. If a loopful is inoculated onto TCBS (thiosulphate citrate bile salts), *V. cholerae* appear as a yellow sucrose-fermenter, which is oxidase-positive. *V. cholerae* is killed by most detergents and by heating at 55°C for 15 min. However, it can survive for up to 2 weeks in salt water at ambient temperatures, and also on chitinous shellfish for 2 weeks, even if refrigerated.

Treatment

Rehydration is key. Antibiotics (e.g. azithromycin, ciprofloxacin) reduce the duration of disease and period of excretion of *V. cholerae* in the stool of infected patients. In the UK, a killed oral vaccine is licensed for relief workers and travellers to remote endemic areas. Results of trials of a live oral vaccine ('Peru-15') are promising. However, the most important preventative strategies are improvement of sanitation and food and water standards.

Table 4.15 Growth characteristics of *Vibrio* spp.

Species	TCBS	Biochemistry	Salt requirement	Growth at 42°C
V. cholerae	Yellow	Oxidase-, nitrate-, lysine-, ONPG-positive	Not halophilic (0–3% NaCl)	Yes
V. parahaemolyticus	Green	Lysine-, indole-positive	Halophilic (3–6% NaCl)	Yes
V. vulnificus	Green (85%), yellow (15%)	Lactose-, lysine-, salicin-positive	Halophilic (3–5% NaCl)	No
V. alginolyticus	Yellow	Lysine-, VP-positive	Halophilic (3–10% NaCl	Yes

Vibrio parahaemolyticus

V. parahaemolyticus is ubiquitous in fish and shellfish, and the waters they inhabit. Outbreaks of diarrhoea occur infrequently in the UK.

Epidemiology

V. parahaemolyticus infection is common in SE Asia, particularly Singapore and Japan. However it also occurs in the UK and USA, particularly during summer months.

Pathogenesis

Kanagawa-positive strains (see below) of *V. parahaemolyticus* adhere to human enterocytes and produce a heat-stable cytotoxin.

Clinical features

V. parahaemolyticus is usually acquired through ingesting seafood, and causes acute explosive diarrhoea. Extra-intestinal infections arise from handling contaminated seafood or exposure to the aquatic environment, the most common being wound infections.

Diagnosis

This organism is halophilic, hence will not grow on CLED agar. Clinical strains of *V. parahaemolyticus* usually appear as green, non-sucrose fermenting colonies on TCBS agar, but isolates from estuary and coastal waters may ferment sucrose. Stool samples should be enriched in alkaline peptone water containing 1% NaCl. The Kanagawa phenomenon refers to haemolysis of human erythrocytes on Wagatsuma's agar, by strains of *V. parahaemolyticus* which cause gastroenteritis.

Treatment

Rehydration is the main intervention for patients with diarrhoea. Severe infections require treatment with fluoroquinolones, doxycycline, or 3rd-generation cephalosporins. Antibiotics do not shorten the duration of symptoms. Prevention strategies involve good food hygiene standards.

Vibrio vulnificus

V. vulnificus has been called the 'terror of the deep' due to the severe fulminant infection it can cause.

Epidemiology

Infections are most common in areas with higher water temperatures, such as the mid-Atlantic and Gulf coast states of the USA. Septicaemia arises from eating contaminated raw shellfish, while wound infections are due to injuries sustained in aquatic environments.

Pathogenesis

The polysaccharide capsule helps resist phagocytosis and bactericidal effects of human serum. The association with liver disease (with increased serum iron levels) may be explained by the ability of virulent strains to use transferrin-bound iron. Toxin production is also important.

Clinical features

There are 3 main infections associated with *V. vulnificus*:

- fulminant septicaemia, followed by cutaneous lesions. This is associated with a high mortality (50%). Immunosuppressed patients are at increased risk, particularly elderly male alcoholics with liver dysfunction
- wound infection, rapidly progressing to cellulitis, oedema, erythema, and necrosis. Patients may develop septicaemia, and it may be fatal
- acute diarrhoea, usually in those with mild underlying conditions. Mortality is rare.

Diagnosis

This organism is also halophilic (see *V. parahaemolyticus* above). See Table 4.15 for further growth characteristics.

Treatment

Early treatment with ceftazidime and doxycycline is key.

Other *Vibrio* Species

- *V. alginolyticus* is the most common vibrio organism found in seafood and seawater in the UK. It is a halophilic organism, which will not grow on CLED but grows in 10% NaCl. Colonies are large and yellow (sucrose-fermenting) on TCBS agar, and there is swarming on non-selective media. It can cause opportunistic wound infections, with mild cellulitis and a seropurulent exudate. Most infections are associated with exposure to seawater and are self-limiting. Little is known about the pathogenic mechanisms.
- *V. fluvialis* is phenotypically similar to *Aeromonas hydrophila* (see 🕮 *Aeromonas*, p.376). It has been implicated in outbreaks of diarrhoea, and is acquired from contaminated seafood.
- *V. damsela* (now known as *Listonella damsela*) is a halophilic, marine organism that can cause severe wound infections. It is acquired in warm coastal areas.
- *V. hollisae* (now known as *Grimontia hollisae*) has been associated with diarrhoea and bacteraemia in areas of warm seawater in the USA. It is acquired from raw seafood.
- *V. mimicus* is associated with gastroenteritis from eating raw oysters. There are also reports of ear infections. It occurs in environments similar to *V. cholerae*.

Aeromonas

Aeromonas spp. are aquatic organisms, which have been implicated in causing diarrhoea. They also cause soft tissue infections and sepsis in the immunocompromised. *A. hydrophila*, *A. sobria*, and *A. caviae* are the main species, and *A. salmonicida* is an economically important fish pathogen. The genus *Aeromonas* has undergone a number of taxonomic and nomenclature revisions recently, and been moved from the family *Vibrionaceae* to the new family *Aeromonadaceae* .

Epidemiology

Aeromonas diarrhoea is more common in the summer months when water concentrations of aeromonads are higher. Outbreaks may occur, and *Aeromonas* infection is being increasingly recognized as a cause of traveller's diarrhoea.

Pathogenesis

Gastroenteritis is the most common disease associated with *Aeromonas*, but its role is debated (Box 4.9). It is unclear whether most faecal isolates recovered from symptomatic patients actually cause diarrhoea. It may be that only specific subsets of *Aeromonas* are pathogenic, and new biotyping schemes are needed to differentiate environmental from clinical strains.

Clinical features

Diarrhoea tends to be watery and self-limiting, but is occasionally more severe. Chronic colitis following diarrhoea has been reported. In addition to gastroenteritis, there are reports of aeromonas septicaemia in the immunocompromised, and wound infections in healthy people and those undergoing leech therapy. The main skin-associated aeromonad is *A. hydrophila*. There are rare reports of nosocomial bacteraemia, peritonitis, meningitis, and eye and bone and joint infections.

Diagnosis

This facultatively anaerobic GNR is usually β-haemolytic on blood agar, and ferments carbohydrates to produce acid and gas. It grows readily on MacConkey agar, and lactose fermentation is variable. Growth on TCBS agar is also variable. It is oxidase-positive, so can be distinguished from the oxidase-negative *Enterobacteriaceae*. Selective techniques are needed to isolate it from a mixed culture. Suitable plates for detection of *Aeromonas* from stool include cefsulodin-irgasan-novobiocin (CIN) agar or blood agar containing ampicillin. Note that not all laboratories routinely culture stools for *Aeromonas,* and some enteric media actually inhibit its growth.

Treatment

There are no controlled trials, but clinical improvement has been seen with antibiotics that are active *in vitro*, such as fluoroquinolones, co-trimoxazole, and aminoglycosides (except streptomycin). Resistance to the carbapenems has been reported.

Box 4.9 Evidence that supports the causative role of *Aeromonas* in diarrhoea

- Symptomatic people have a higher carriage rate of *Aeromonas* than asymptomatic individuals.
- Most symptomatic patients harboring *Aeromonas* do not have other enteric pathogens.
- *Aeromonas* produces an enterotoxin.
- Antibiotics active against *Aeromonas* usually improve patient symptoms.
- Evidence of a specific secretory immune response coincident with diarrhoeal disease.

Plesiomonas

Plesiomonas shigelloides, the only species in the genus, is associated with outbreaks of gastroenteritis in warm climates. In the literature it has been known as *Pseudomonas shigelloides*, C27, *Aeromonas shigelloides*, and *Vibrio shigelloides*. The taxonomic status has varied – it is related to *Proteus*, but is currently placed in the family *Enterobacteriaceae*

Epidemiology

P. shigelloides is found in soil and water (mainly fresh water, but also salt water in warm weather). It is usually transmitted to humans via water or food (e.g. shrimp, chicken and oysters), and also colonizes many animals. Most patients recently travelled abroad.

Pathogenesis

There is no animal model, and no pathogenic mechanism has been identified. Volunteer studies have been largely unsuccessful in causing disease. Hence it has been difficult to prove a causal relationship.

Clinical features

Symptoms vary from mild self-limiting diarrhoea to mucoid bloody diarrhoea with features of entero-invasive disease. It has occasionally resulted in serious extra-intestinal infection such as osteomyelitis, septic arthritis, endophthalmitis, spontaneous bacterial peritonitis, pancreatic abscess, cellulitis, cholecystitis, and neonatal sepsis with meningitis. Bacteraemia is rare and usually in the immunocompromised.

Diagnosis

This motile, facultatively anaerobic GNR does not ferment lactose. It grows readily at 35°C on most enteric agars, such as MacConkey, but does not grow on TCBS. It appears non-haemolytic and is oxidase-positive. Selective techniques are needed to isolate it from a mixed culture.

Treatment

The role of antibiotics is unclear, and results of studies conflicting. *In vitro* it is usually sensitive to the quinolones, cephalosporins, and imipenem.

Campylobacter

Campylobacter organisms are spiral-shaped flagellate bacteria belonging to rRNA superfamily VI. *C. jejuni* is the commonest cause of diarrhoea in most developed countries. *C. coli* also causes diarrhoea. *C. fetus* is the type species of the genus and causes abortion in sheep and cows. It occasionally causes septic abortions in humans and bacteraemia in the immunocompromised. Some species including *C. lari* and *C. upsaliensis* cause diarrhoea in children in developing countries, while species such as *C. concisus* and *C. rectus* are associated with periodontal disease.

Pathogenesis

Campylobacter organisms are ingested (faeco-oral transmission), then colonize (and usually invade) the jejunum, ileum, and terminal ileum, occasionally extending to the colon and rectum. Mesenteric lymph node involvement and transient bacteraemia may occur. Histological findings of acute inflammation ± superficial ulceration are the same as in *Salmonella*, *Shigella* or *Yersinia* infections.

Clinical features

Campylobacter gastroenteritis is variable in terms of symptoms and severity. In severe cases, GI haemorrhage, toxic megacolon, and Haemolytic Uraemic Syndrome (HUS) have been reported. Other complications include meningitis, deep abscesses, cholecystitis, and reactive arthritis. Approx. 25% cases of Guillain-Barré syndrome (GBS) have documented preceding *Campylobacter* gastroenteritis – the LOS cell surface structures act as critical factors in triggering GBS through ganglioside mimicry.

Diagnosis

This small, spiral GNR has a single unsheathed flagella at one or both poles and is extremely motile. It is micro-aerophilic and grows best at 42°C. Like *Helicobacter*, *Campylobacter* organisms undergo coccal transformation under adverse conditions and are biochemically inactive. However they are oxidase-positive. *C. jejuni* is the only species that hydrolyses hippurate. Typing methods include serotyping (Penner scheme for O antigens and Lior scheme for heat-labile surface and flagellar antigens), biotyping, phage typing, and newer molecular methods.

Treatment

Rehydration and symptom relief is usually adequate, as *Campylobacter* infection is usually self-limiting in 5–7 days. However, in severe dysenteric disease, erythromycin or ciprofloxacin may be prescribed. Resistant strains, especially *C. coli* may respod to trimethoprim or co-trimoxazole. Good hygiene standards are important in prevention. Infective organisms may be excreted in the stool for ~3 weeks after resolution of diarrhoea. There is no vaccine.

Helicobacter

The genus *Helicobacter* contains up to 17 species, which colonize the stomachs of different animals. *H. pylori* is a spiral-shaped flagellate bacteria belonging to rRNA superfamily VI, which colonizes humans (it is found in approx 50% of the world population). *H. pylori* was discovered in 1983 in Australia by Warren and Marshall, who went on to receive the Nobel Prize for Medicine in 2005. Its importance in the pathogenesis of peptic ulcer disease soon became clear. *H. cinaedi* and *H. fennelliae* are associated with proctitis in homosexual men.

Pathogenesis

As with other bacteria in rRNA superfamily VI, *H. pylori* is adapted to colonizing mucous membranes (in this case the gastric mucosa only) and penetrating mucus. The cagA protein is important in virulence. After phosphorylation by tyrosine kinase, cagA is injected into epithelial cells by a type IV secretion system. This alters signal transduction and gene expression in host epithelial cells

Clinical features

H. pylori is associated with 95% of duodenal and 70% of gastric ulcers. Epidemiological studies have highlighted the association of *H. pylori* and gastric cancer, and WHO classifies *H. pylori* as a group 1 carcinogen.

Diagnosis

This GNR is shaped like a helix (hence its name) and has a tuft of sheathed unipolar flagella. It is strictly micro-aerophilic and requires CO_2 for growth. It is relatively inactive biochemically, except for strong urease production. Under adverse conditions, it undergoes coccal transformation. Options for testing patients are as follows:

- Serology – if positive, this indicates the patient has been infected
- biopsy of stomach or duodenum for histology ± urease test ± culture
- urea breath tests – the patient drinks ^{14}C- or ^{13}C-labelled urea, which is metabolized by *H. pylori*, producing labelled CO_2 that can be detected in the breath. This test is also used to assess effectiveness of treatment
- rapid urease test (the enzyme urease produced by *H. pylori* catalyses the conversion of urea to ammonia and bicarbonate, which is reflected by a rise in pH.) This is usually performed on a biopsy sample
- faecal antigen tests.

The urea breath test or stool antigen test have greater sensitivity and specificity than serology for diagnosis, and can also be used to confirm eradication. The patient should receive no antibiotics for 4 weeks before the tests, and no proton pump inhibitor (PPI) for 2 weeks before the tests. Molecular typing of *H. pylori* is more useful than serotyping.

Treatment

NICE has issued clinical guidelines on Managing dyspepsia in adults in primary care (August 2004).[1] Triple therapy is given in specific circumstances and consists of a proton pump inhibitor (PPI) e.g. omeprazole and two antibiotics (e.g. amoxicillin, clarithromycin, or metranidende).

Eradication of *H. pylori* is beneficial in duodenal and gastric ulcers and low-grade MALToma (mucosal associated lymphoid tissue), but *not* in gastro-oesophageal reflux disease (GORD). In non-ulcer dyspepsia, 8% of patients benefit. Triple treatment achieves >85% eradication. Essentially, any dyspeptic patient with no alarm symptoms should receive a PPI for one week. If symptoms relapse they should be tested and treated for *H. pylori*, using the breath test or stool antigen test. First-line 'triple therapy' is with a PPI for 1 week, and two antibiotics (amoxicillin or metronidazole, together with clarithromycin). Clarithromycin or metronidazole should not be given if they have been used for any infection in the previous year. Approximately 10% of patients fail treatment, possibly due to antibiotic resistance (Box 4.10).

A Cochrane review (2006) of eradication therapy for peptic ulcer disease in *H. pylori*-positive patients found that treatment had a small benefit in initial healing of duodenal ulcers, and a significant benefit in preventing the recurrence of both gastric and duodenal ulcers, once healing had been achieved.[2] Other treatment includes probiotics (which improved eradication rates and reduced adverse events in a recent meta-analysis), and bismuth compounds.

Box 4.10 Antibiotic resistance in *H. pylori*

Antibiotic resistance varies geographically. Metronidazole resistance varies from ~50% in Europe to 90% in developing countries. Clarithromycin resistance is <10% in Europe, but may be rising. A recent meta-analysis has shown that pretreatment clarithromycin resistance may reduce the effectiveness of therapy by 55%. To date (2006), no strains with resistance to amoxicillin or to tetracycline have been detected in the UK. Levofioxacin is being investigated as an alternative antibiotic in resistant strains.

References

1 NICE. *Managing Dyspepsia in Adults in Primary Care*. London: NICE. Marygedi P et al. Eradication of Helicobacter pylori for non-ulcer dyspepsia. Cochrane Database of Systematic Reviews 2006, Issue 2. Art no. CD0002026. DOI:10:1002/14651858. CD002096.pub4

2 Moaygedi P. *et al*. Eradication of Helicobacter pylon for non-ulcer dyspepsia. Cochrane Database of Systematic Reviews 2006. Issue 2. Art. no CD002026. DOI: 10:1002114651858.CD002096.pub4

Bacteroides

More than 30 genera of anaerobic GNR are recognized but human infections are largely restricted to four of these: *Bacteroides*, *Prevotella*, *Porphyromonas* and *Fusobacterium* (Table 4.16). These organisms are found in the mouth, gastrointestinal tract and vagina and are among the most important constituents of 'normal flora'. They may cause a variety of infections in humans, particularly polymicrobial infections and abscesses. *Bacteroides fragilis* is the most important species; it is found in the gastrointestinal tract and is associated with a wide variety of infections.

Pathogenesis

Virulence factors of *Bacteroides spp.* include:

- capsular polysaccharide – inhibits opsonisation / phagocytosis, promotes abscess formation and promotes adherence to epithelial cells
- pili and fimbriae – promote adherence to epithelial cells and mucus
- succinic acid – inhibits phagocytosis and intracellular killing.
- Enzyme production – contribute to tissue damage and /or promote invasion and spread e.g heparinase, fibrinolysin, hyaluronidsase, neuraminidase
- Synergy between anaerobic and facultative bacteria – see Box 4.11

Clinical features

- Intra-abdominal infections – *B. fragilis* is the most common anaerobic isolate in intra-abdominal abscesses
- Diarrhoea – enterotoxin-producing strains have been implicated in diarrhoea in children.
- Bacteraemia – *B. fragilis* is the most common isolate in anaerobic bacteraemias. The source is usually intra-abdominal and associated with abscesses, malignancy, bowel perforation or surgery. Septic shock is less common in *B. fragilis* bacteraemia than in bacteraemia caused by aerobic Gram-negative bacilli; this is presumably related to the absence of lipid A in the endotoxin of *B. fragilis*.
- Endocarditis – associated with large vegetations and high frequency of thromboembolic complications
- Skin and soft tissue infections – often found as part of mixed flora in diabetic and decubitus ulcers. *B. fragilis* has also been isolated from cutaneous abscesses of the lower limbs
- Bone and joint infections – *B. fragilis* may rarely cause osteomyelitis and septic arthritis.
- CNS infections – anaerobic meningitis is rare and most laboratories do not culture CSF anaerobically. In the cases of anaerobic meningitis that have been described, *B. fragilis* is the most common isolate. In contrast, anaerobes are frequently implicated in brain abscesses.

Diagnosis

- *Bacteroides* are non-spore forming, non-motile anaerobic Gram-negative rods.
- On blood agar, *Bacteroides* appears as glistening, non-hemolytic colonies which are aerotolerant.
- Gram stain may reveal pale pink, pleomorphic coccobacilli, with irregular or bipolar staining.

- They can be differentiated from the other anaerobic GNR by growth in 20% bile.
- The MASTRING™ ID (Mast Diagnostics) may be used to identify *B. fragilis* in the laboratory. This is a ring containing 6 antibiotic discs that is placed on the culture plate and incubated anaerobically at 37°C for up to 3 days. *B. fragilis* is usually sensitive to erythromycin and rifampicin and usually resistant to penicillin G, vancomycin, kanamycin and colistin.

Treatment

- Drainage of abscesses and debridement of necrotic tissue is the mainstay of treatment for anerobic infections. However, some abscesses (e.g. brain, liver and tubo-ovarian) have been managed with antimicrobial therapy alone.
- The choice of antibiotics to treat anaerobic infections is usually empirical, as most of the infections are polymicrobial and require broad spectrum therapy
- *Bacteroides* is usually sensitive to antimicrobials such as metronidazole, clindamycin, chloramphenicol, carbapenems, cefoxitin and β-lactam/β-lactamase inhibitor combinations (e.g. co-amoxiclav, piperacillin-tazobactam).

Table 4.16 Characteristics of anaerobic Gram-negative rods

Organism	Growth in 20% bile	Pigmented	Fluorescence	Resistant to	Sensitive to
Bacteroides fragilis	Yes	No	No	Penicillin Vancomycin Kanamycin Colistin	Erythromycin Rifampicin
Fuso-bacterium	Variable	No	No	Erythro-mycin Vancomycin	Colistin Penicillin Kanamycin
Prevotella	No	Brown/black	Brick red	Erythro-mycin Vancomycin	Erythro-mycin Rifampicin Colistin Penicillin
Porphyro-monas	No	Brown/black	Brick red	Erythro-mycin Vancomycin	Erythro-mycin Rifampicin Penicillin Vancomycin

Prevotella and Porphyromonas

Prevotella and *Porphyromonas* formerly belonged to the genus *Bacteroides* (📖 p.381) but were reclassified in 1990.

- The genus *Prevotella* includes *P. melaninogenica*, *P. bivia*, *P. oralis* and *P. buccalis*.
- The genus *Porphyromonas* includes *P. gingivalis P. endodontalis*, and *P. asaccharolytica*.

Pathogenesis

- Virulence in *P. melaninogenica* is associated with the capsular polysaccharide, which inhibits opsonophagocytosis, promotes abscess formation and also promotes adherence to epithelial cells.
- In *P. gingivalis*, pili and fimbriae aid adherence to epithelial cells and mucus.
- Production of various enzymes may also aid evasion of the host immune response, or promote tissue destruction.

Clinical features

Prevotella and *Porphyromonas* contribute to the formation of abscesses and soft tissue infections in various parts of the body. They also cause infections of the oral cavity (such as periodontal and endodontal disease), female genital tract infections, osteomyelitis of the facial bones and human-bite infections.

Diagnosis

- These are non-spore forming, non-motile anaerobic GNR.
- They are usually isolated (along with other anaerobes) from abscesses and soft tissue infections.
- *Prevotella* and *Porphyromonas* may both appear pigmented – usually brown/black
- Young unpigmented colonies can show brick-red fluorescence under UV light.
- Gram stain reveals small, pale pink cocco-bacilli.
- *Prevotella* and *Porphyromonas* are both inhibited by 20% bile.
- *Prevotella* are *moderately* saccharolytic whereas *Porphyromonas* are asaccharolytic.

Treatment

- The mainstay of treatment for anaerobic infections is surgical drainage of abscesses and debridement of necrotic tissue.
- *Prevotella* and *Porphyromonas* are usually sensitive to agents such as metronidazole, clindamycin, chloramphenicol and cefoxitin. Penicillin resistance is common, but isolates are usually susceptible to co-amoxiclav and other β-lactam . β-lactamase inhibitor combinations

Box 4.11 Synergy in Anaerobic Infections

- Infections involving anaerobes usually contain multiple anaerobic bacteria as well as facultative anaerobic bacteria.
- Evidence suggests true synergy between anaerobic and facultative bacteria, with formation of abscesses occurring more readily infections involving both groups of bacteria than either alone.
- Facultative organisms may lower the oxidation-reduction potential in the microenvironment promoting more favourable conditions for anaerobes.
- Anaerobic bacteria may inhibit phagocytosis of facultative bacteria
- *B. fragilis* produces β-lactamases in abscess fluid that may protect other normally susceptible bacteria from antimicrobials

Fusobacterium

Fusobacterium spp. colonise the mucous membranes of animals and humans, and occasionally cause infections of the oral cavity and head and neck. Clinically, the most important species are:

- *F. nucleatum* (subspecies *nucleatum, polymorphum and fusiforme*)
- *F. necrophorum* (subspecies *necrophorum and fundiliforme*).

Epidemiology

Fusobacteria are commensals of the oral cavity. As with other obligate anaerobes, the significance of these organisms is being increasingly recognized. However, Fusobacterium infections are relatively rare in the UK.

Pathogenesis

Fusobacterium spp. produces lipopolysaccharide (LPS) endotoxin which is biologically active. They also produce metabolites that are important to oral spirochaetes.

Clinical features

- *F. necrophorum* causes severe systemic infections such as Lemierre's disease (see 📖 p.635), post-anginal sepsis and necrobacillosis.
- Lemierre's disease is a severe systemic disease which occurs in previously healthy young adults, and usually presents initially as severe sore throat, followed by fever, cervical lymphadenopathy and unilateral thrombophlebitis of the internal jugular vein. Metastatic infection with spread to the lungs or bones or brain may occur. If untreated the condition leads to death in 7-15 days.
- Other species commonly isolated from oral infections include *F. periodonticum, F. alocis, F. sulci and F. naviforme.*
- Species found in the gastrointestinal or genitourinary tracts (e.g. *F. mortiferum, F. necrogenes F. varium* and *F. gonidiaformans*) may cause intra-abdominal infections, osteomyelitis, ulcers and skin / soft tissue infections.
- *F. ulcerans* was originally isolated from tropical ulcers, but may be found in other sites.

Diagnosis

- *Fusobacterium* are long, thin, GNR with pointed ends ('fusiform') that are often arranged in pairs. They are non-spore forming and non-motile
- They may be haemolytic on blood agar and may grow in the presence of 20% bile.
- They can be identified using commercial tests e.g. MASTRING™ ID (Mast Diagnostics) and the API 20A or Rapid ID 32A (Biomerieux).
- Susceptibility testing is vital as there have been reports of penicillin resistant strains.

Molecular techniques (e.g. PCR) have been developed.

Treatment

- The mainstay of treatment for anaerobic infections is surgical drainage of abscesses and debridement of necrotic tissue.
- Lemierre's syndrome and other severe invasive disease is usually treated with a combination of penicillin and metronidazole, for 2-6 weeks.
- Alternatives include clindamycin monotherapy or chloramphenicol.

Spirochaetes – an overview

The spirochaetes are a group of helical organisms sharing many properties with Gram-negative bacteria. The vast majority are non-pathogenic but a few are important causes of disease in humans (see Table 4.16). There are aerobic and anaerobic species, both free living and parasitic. Axial filaments, fixed at each end of the organism, run along the outside of the protoplasm within the outer sheath and give the characteristic coiled appearance. These are similar to bacterial flagella and are capable of constricting, warping the cell bod, and enabling the bacterium to move by rotating it in space.

Table 4.16 Overview of spirochaetes of clinical significance

Genus	Species	Clinical disease	Morphology	Culture	Diagnosis
Treponema	*T. pallidum* subsp. *pallidum*	Syphilis	Morphologically i dentical. Thin helical cells 10 micrometre by 0.15 micrometre. Visible on dark-field microscopy	Cannot be cultured *in vitro*; remain motile in specific enriched media at 35°C for several days	Direct detection only means in primary syphilis; mainstay is serology; cross-reactivity between species
	T. pallidum subsp. *pertenue*	Yaws			
	T. pallidum subsp. *endemicum*	Endemic syphilis			
	T. carateum	Pinta			
Borrelia	*B. recurrentis*	Louse-borne relapsing fever	Helical. 3–20 micrometre by 0.25 micrometre. Can be stained with aniline dyes	Can be cultured but not practical	Demonstration of spirochaetes in peripheral blood smears; immunological and PCR-based tests available
	B. hermsii and others	Tick-borne relapsing fever			
	B. burgdorferi	Lyme disease		Culture possible from biopsy of rash.	Serology; can remain positive for years
Leptospira	*L. interrogans*	Leptospirosis	Motile, 10 micrometre by 0.1 micrometre. Stain poorly – visible on dark-field or phase contrast	Specialized media. Allow minimum 6 weeks.	Serology; molecular techniques available.

Treponema species

Four members of the genus *Treponema* cause human disease: *Treponema pallidum* subsp. *pallidum* (syphilis) and three 'non-venereal' treponematoses.

Microbiology

Morphologically identical, *Treponema* species appear as motile helical rods on dark-field microscopy. They are thin, helical cells around 10 micrometre long and 0.15 micrometre wide. They cannot be cultured *in vitro* (unlike the non-pathogenic treponemes), but remain motile in specific enriched media for several days at 35°C. Organisms remain viable after freezing. The organisms all share a significant degree of DNA homology and are very similar antigenically, thus all cause positive serological tests for syphilis.

Epidemiology and clinical features

- ***Treponema pallidum* subsp. *pallidum*** – the causative agent of syphilis. An increasing incidence in the UK, beginning in the 1960s, plateaued in the mid-90s but several large outbreaks between 1998 and 2003 saw diagnoses of infectious syphilis in men rise 15-fold. Transmission: sexual contact, direct vascular inoculation (IVDU, transfusions), direct cutaneous contact with infectious lesions or transplacental infection (congenital syphilis, p.824). Interacts with HIV in both acquisition and diagnosis. See p.752 for clinical features.
- ***Treponema pallidum* subsp. *pertenue*** – the causative agent of yaws, a chronic non-venereal disease endemic in the humid tropics (Central Africa, South America, South-East Asia, and parts of the Indian subcontinent). Acquired in childhood through contact with infectious skin lesions. Incubation: 3 weeks. Affects the skin (papular skin lesions which may ulcerate) and bones (periosteitis, dactylitis). Primary stage: lesion at inoculation site; secondary stage: dissemination of treponemes causing multiple skin lesions; latent stage: usually asymptomatic (most patients remain non-infectiously latent for their lifetime); tertiary stage (<10% patients 5–10 years later): bone, joint, soft tissue deformities.
- ***Treponema pallidum* subsp. *endemicum*** – the causative agent of non-venereal endemic syphilis or 'bejel'. Endemic in dry subtropical or temperate areas of the Middle East, India, Asia and parts of Africa. Infection occurs in childhood and is associated with poor standards of hygiene. Transmission: contact with mucosal lesions or contaminated eating utensils/water. Incubation: 10–90 days. Primary lesions (1–6 weeks): patches in mouth followed by skin lesions resembling the chancres of venereal syphilis; secondary stage (6–9 months): macerated patches on lips and tongue, anogenital hypertrophic condyloma lata, painful osteoperiostitis of long bones; tertiary stage: destruction of cartilage and bone, gummata of skin, bones and nasopharynx. CNS/CVS (cardiovascular system) disease is very rare.
- ***Treponema carateum*** – the causative agent of pinta, the most benign of the endemic treponematoses affecting only the skin. Endemic to South/Central America. Spread by contact with infected skin. Incubation 2–3 weeks. Primary lesion: papule or erythematous plaque on exposed surfaces of the legs, foot, forearm, or hands which slowly

enlarges becoming pigmented and hyperkeratotic. May be associated with regional lymphadenopathy. Secondary lesions: disseminated lesiosn of similar appearance appearing 3–9 months later. Late/tertiary pinta: disfiguring pigmentary changes, and atrophic lesions.

Diagnosis

- Direct detection – culture (the gold standard) is expensive and time-consuming and used primarily in research. Direct detection of organisms via dark-field microscopy, or preferably immunofluorescence of material scraped from a lesion, is the only means of diagnosis in primary syphilis.
- Serological diagnosis – the mainstay. Serological tests fall into two groups. Both show cross-reactivity between the four Treponema species:
  - non-treponemal tests (e.g. Venereal Disease Research Laboratory (VDRL)) – detect antibodies to cardiolipin produced as a response to treponemal infection and are not specific but are very sensitive. Samples with very high antibody titres may give false-negative results (the 'prozone' phenomenon), e.g. in early infection or HIV. Poor sensitivity in late-stage infection. Antibody titre tends to decline to negativity, with adequate therapy
  - treponemal tests (e.g. treponema pallidum haemagglutination assay TPHA) – use specific treponemal antigens and are consequently more specific. They are able to detect late-stage infection and tend to remain positive after adequate therapy.
- Traditional WHO recommendations for diagnosis are a sensitive non-treponemal screening test, with positive samples followed up using a more-specific and ideally equally sensitive treponemal assay. In the UK, a combination of sensitive screening tests for treponemal antibody – VDRL and TPHA – is used initially, allowing sensitive and specific screening of all but very early primary disease. Positive samples are followed up using fluorescent treponemal antibody tests. Enzyme-immunosorbent assays (EIAs) for treponemal IgG and IgM are beginning to supersede this technique: more easily automated, more objective, at least as good as the VDRL/TPHA combination, and a useful method for detecting antibody in patients with HIV. Positive EIAs are confirmed using TPHA. Discrepant results are repeated using the fluorescent treponemal antibody-absorption test (FTA-Abs). Seronegative patients at recent risk of acquiring disease should be followed up because of the seronegative window in early primary syphilis (see 📖 Syphilis, p.752).
- Molecular techniques – PCR tests are available but not yet widely used.

Treatment

- Early syphilis and pinta/yaws/bejel – prolonged antibiotic therapy is required due to the slow dividing rate of *T. pallidum* (averages one doubling *in vivo* per day). Highly sensitive to penicillin and a long-acting depot injection of benzathine benzylpenicillin is the standard therapy. A single dose is sufficient for early infection. Alternative: 15-day course of azithromycin (increasing reports of resistance) or doxycycline.

- Late syphilis – weekly doses of benzathine penicillin over three weeks. Alternatives: doxycycline, ceftriaxone. Consider the possibility of re-infection in cases of treatment failure.
- Neurosyphilis – IV penicillin, as benzathine benzylpenicillin achieves no measurable CSF levels and there are a number of reports of patients treated with benzathine benzylpenicillin developing neurosyphilis.

Borrelia species

These are helical bacteria around 0.25 micrometre in diameter and between 3 and 20 micrometre long. Those causing human disease are transmitted by insect vectors.

- Louse-borne relapsing fever – *B. recurrentis*
- Tick-borne relapsing fever – variety of species
- Lyme disease – *B. burgdorferi*

Relapsing fever

Relapsing fever is caused by several *Borrelia* species transmitted by arthropods, characterized by recurring episodes of fever. Two distinct clinical forms were recognized as far back as ancient Greece: epidemic louse-borne and endemic tick-borne relapsing fever. The presentation of abrupt fever, muscle aches, and joint pains with crisis, remission, and then relapse are similar for both but the periodicity tends to be characteristic (e.g. 5.5 days for louse-borne versus 3.1 days for tick-borne). The recurrent nature is thought to be due to antigenic variation of the spirochaetal outer membrane proteins.

Epidemiology

- Tick-borne relapsing fever is worldwide and transmitted by soft-bodied *Ornithodorus* ticks. Most tick species carry a distinctive borreliae. Epidemiology depends on the local vector, e.g. *O. hermsii* is the commnest vector in California and Canada and lives in dead trees and on rodents and transmits *B. hermsii*. Infection is passed down the tick generations, thus disease tends to be endemic.
- Louse-borne relapsing fever has occurred in Africa, Middle East, and Asia. Human body-louse inhabits only humans and *B. recurrentis* is not transmitted vertically within lice thus is maintained by passage from louse to human and then back to another louse, which remains infective for its entire life. Therefore infection is associated with poverty and overcrowding, and disease tends to be epidemic.

Clinical features

- Incubation and symptoms are similar in both conditions – 3–8 days after exposure there is the abrupt-onset fever, headache, myalgia, arthralgia, chills, weakness, anorexia, epistaxis, cough/haemoptysis, and weight loss. Examination findings include hypotension, hepatosplenomegaly, lymphadenopathy, nuchal rigidity, jaundice, photophobia, injected conjunctiva, and iritis.
- Tick-borne disease – the primary episode lasts 3–6 days and is followed by a critical episode than may cause fatal shock. The first relapse occurs

7–10 days later. Subsequent relapses are less severe. The average number of relapses experienced is 3 but can be as many as 10.
- Louse-borne disease – there are fewer relapses than with tick-borne infection, and hepatic or splenic involvement is more common as are neurological manifestations (coma, hemiplegia, meningitis, seizures).

Diagnosis
Organisms can be cultured but isolation is not practical. Serological tests are not diagnostically useful. Approx 5% of patients have positive VDRL. Most useful is demonstration of spirochaetes in peripheral blood smears (and other body fluids – marrow aspirates, CSF). Unlike the other spirochaetes, borreliae stain well with acid aniline dyes such as Giemsa. They are most likely to be found during febrile episodes when the sensitivity of blood smears is around 70% for louse-borne fever (less for tick). Multiple thick and thin smears may need to be examined. Immunological and PCR-based tests are available. Other lab findings iclude deranged clotting tests, elevated liver function tests (LFTs).

Treatment
- Tick-borne relapsing fever – tetracycline is the drug of choice, given for 7–14 days. Other: doxycycline 7 days, erythromycin 10 days.
- Louse-borne relapsing fever – a single dose of doxycycline (preferred), tetracycline, erythromycin, or penicillin G.
- Jarisch–Herxheimer reactions can occur (usually within the first 2 h after antibiotic administration), particularly in louse-borne relapsing fever. Features: sweating, tachycardia, hypertension followed by profound hypotension. It can be fatal and appears to be mediated partly by tumour necrosis factor (TNF)-α. Preadministration of steroids does not appear to limit the reaction significantly. Anti-TNF-α antibodies may help.

Lyme disease
Caused by infection with, and the host immune response to *Borrelia burgdorferi*. Acquired by the bite of ixodes (hard) ticks and co-infection with other tick-borne organisms can occur (e.g. babesiosis).

Epidemiology
Ticks acquire a spread infection through feeding on infected animals (particularly deer). A tick must be attached to a person for 2–3 days to pass on infection, as only small numbers of bacteria are present in the tick until it feeds – the act of feeding sees bacteria multiply and pass to the salivary glands. Until this multiplication occurs ticks are rarely able to pass on infection; 85% of human infections occur while the tick is in the nymph stage (spring to summer), 15% when it is in the adult stage (autumn). Cases are commonest in children aged 5–9 years and adults aged 60–69 years – only 40% give a definite history of tick bite. Cases occur across Europe, China, Japan, Australia, and parts of the USA. It is relatively rare in the UK, most cases occurring in the south (New Forest and Salisbury Plain), East Anglia, Cumbria, and the Scottish highlands.

Clinical manifestations

Like its cousin, syphilis, it is a great imitator. Clinical features may be a result of direct bacterial infection (particularly in the early stages of disease), or a consequence of an immune response leading to symptoms in many organs (e.g. arthritis). They differ with the strain of *Borrelia* involved (see below). Asymptomatic infection occurs – <10% of those in endemic areas with no history of infection are seropositive. Features may be seen in three overlapping stages:

- **early localized** – around 7 days after tick bite <75% of patients develop erythema chronicum migrans; an expanding painless annular skin lesion centred on the bite with or without local lymphadenopathy may occur. It is probably a result of the inflammatory response to the organism in the skin. Multiple lesions can occur and do not necessarily represent multiple bites. Lasts 2–3 weeks untreated
- **early disseminated** – weeks to months after the bite patients develop more severe constitutional symptoms, malaise, generalized lymphadenopathy, hepatitis, arthritis (50% – initially intermittent and migratory it may evolves into a chronic monoarticular arthritis in 10% of those affected), neurological features (15% – include meningitis, meningo-encephalitis, cranial nerve lesions, and neuropathy), cardiac features (10% – atrioventricular (AV) block, pericarditis, congestive cardiac failure (CCF))
- **late persistent** (but can occur within the first year) – arthritis, late neurological manifestations including focal deficits, fatigue, and neuropsychiatric problems. Acrodermatitis chronica atrophicans is a decolouration of the skin, seen in the extremities and similar to the skin changes of peripheral vascular disease.

Microbiology

Three members of the *Borrelia burgdorferi sensu lato* complex cause Lyme disease: *Borrelia garinii* and *afzelii* in Asia, and *Borrelia burgdoferi sensu stricto* in North America. *Borrelia garinii* and *afzelii* are the commonest European clinical isolates. These differences account for the variation in clinical manifestations across the world (*B. garinii* associated with neurological disease, *B. afzelli* cutaneous manifestations, and *B. burgdorferi sensu stricto* joint symptoms).

Diagnosis

- Culture of the organism is possible from skin specimens taken from those patients with erythema migrans. It is difficult to identify spirochaetes in histological section.
- Serology – the diagnostic method of choice – CDC recommend a screening ELISA for anti-Lyme antibody with positive titres confirmed by Western blot. Specific response is sensitive but develops late (30% positive in the acute phase, 70% at 2–4 weeks, 90% at 4–6 weeks). Prompt antibiotic therapy may prevent a good antibody response. Some patients remain positive for years after illness, thus active and inactive infection cannot be distinguished. Tests must be interpreted with caution in those without a positive travel history or a presentation consistent with Lyme disease. Serology is nearly always positive in those with extracutaneous lesions. False positives may be seen in

association with rheumatoid disease, infectious mononucleosis, and syphilis.

- PCR for *Borrelia burgdorferi* DNA is more sensitive than culture or microscopy in the examination of blood or joint fluid, but has not been standardized for routine diagnosis.
- CSF – spinal fluid should be obtained in those patients with neurological symptoms in whom the diagnosis is not obvious. CSF serology may be helpful.

Treatment

- Patients with a good history and classic erythema migrans should be treated regardless of serology results. Patients probably remain at risk of infection even after treatment for one episode.
- Early-stage skin manifestations, arthritis, or Bell's palsy – doxycycline or amoxicillin PO for 28 days. If arthritis persists repeat course or consider IV ceftriaxone for 14–30 days.
- 3rd-degree heart block – IV ceftriaxone for 14–28 days.
- Neurological disease – cranial nerve palsies: 30-day oral regime as above; parasthesia/radiculopathy: 14 days IV ceftriaxone; encephalitis/encephalopathy: 28 days IV ceftriaxone.

Prevention

Tick avoidance and prompt removal of any attached ticks. Some specialists recommend the use of a single dose of prophylactic doxycycline within 3 days of removal of a tick that has been attached for at least 24 h in high-risk areas/hyperendemic regions. Vaccines are available.

Leptospira species

Leptospira are motile, obligately aerobic spirochaetes measuring 0.1 micrometre in diameter and around 10 micrometre long. They stain poorly but can be visualized on dark-field or phase-contrast microscopy. Two species are identified: *L. interrogans* (includes all human pathogens), and *L. biflexa* (a saprophytic species). *L. interrogans* has many serotypes, and antigenically related organisms are grouped into serovars (a synonym for serotype) for classification. Although more-recent DNA analysis does not correlate well with serological classification, serological classification will continue to be used for the foreseeable future. The 'type' strain is *L. interrogans* serovar *ictero-haemorrhagiae*, and the type disease, leptospirosis.

Leptospirosis

A biphasic disease with initial septicaemia and a secondary phase characterized by immune phenomona (vasculitis, aseptic meningitis). Weil's disease is a severe form characterized by jaundice and acute renal failure.

Epidemiology

Leptospira are found worldwide. The primary reservoirs of most leptospiral serovars are wild mammals. These continually re-infect domestic populations and at least 160 mammalian species are affected. The organism has been recovered from rats, pigs, dogs, cats, and cattle among others, but rarely causes disease in these hosts. Rodents are the most-important reservoir and rats the commonest worldwide source. There are associations between particular animals and serovars (e.g. *L. interrogans* serovar *ictero-haemorrhagiae* and rats). Humans are incidental hosts and onwards transmission is rare. Transmission occurs when people come into contact with infected animal urine, e.g. canoeing, swimming in lakes and rivers, farming. It is primarily a disease of tropical and subtropical regions, and infection in temperate regions is uncommon.

Pathogenesis

After gaining entry via the skin or mucous membranes, the organism replicates in blood and tissue. Leptospiraemia particularly affects the liver and kidney causing centrilobular necrosis and jaundice, or intersititial nephritis and tubular necrosis respectively. Renal failure may occur, exacerbated by hypovolaemia. Other organs affected include muscle (oedema and focal necrosis), capillaries (vasculitis), eye (chronic uveitis).

Clinical features

Incubation is 7–12 days. The majority of patients (90%) develop mild disease without jaundice; 5–10% develop the severe form, Weil's disease. Disease is biphasic. The first phase ('septicaemic' – the organism can be cultured from blood, CSF and most tissues) lasts 4–7 days and is characterized by a flu-like illness, fever, chills, weakness, myalgia, cough, haemoptysis, rash, meningism, and headache. A 1–3-day period of improvement follows and patients may become afebrile. The second stage then starts ('immune' or 'leptospiruric phase' in which antibodies may be detected and the organism isolated from urine). Features are due to the immunological response to infection and may last up to a month. Disease may be anicteric or icteric. Aseptic meningitis is the most-important feature of anicteric disease and is seen in 50% of cases. Death is rare in this group. Icteric disease is characterized by jaundice, hepatosplenomegaly, nausea/vomiting, anorexia and diarrhoea/constipation. Organisms can be isolated from the blood <48 h after jaundice onset. Other features: uveitis (<10% – can occur up to a year after initial illness), subconjunctival haemorrhage is commonest ocular complication (92% of patients), renal impairment (uraemia, pyuria, haematuria, oliguria), pulmonary manifestations. Weil's disease is characterized by jaundice, renal failure, hepatic necrosis, lung disease, and bleeding. It starts at the end of stage one. Overall mortality is 10%, up to 40% in those with hepatorenal involvement.

Diagnosis

- Direct examination – dark-field examination of blood, CSF, or urine may demonstrate leptospira but there is a high false-positive rate (misinterpretation of fibrils and red cell fragments). It is not recommended.
- Culture – there has been little change in culture techniques over the years. It is difficult, insensitive and requires several weeks of incubation. Specialized culture media are required (e.g. Ellinghausen-McCullough-Johnson-Harris (EMJH) which contains 1% bovine serum albumin and Tween 80, a fatty acid source). They should be inoculated within 24 h of specimen collection (either blood or CSF in heparin or sodium oxalate). Leptospiral culture can be established by subculture of routine blood culture samples. Organisms can be isolated from blood and CSF in the first week of illness. In the second phase of illness they can be found only in the urine where they may be isolated for up to 1 month. Cultures can be reported as negative after a minimum of 6 weeks – continuing for as long as 4 months is preferable.
- Molecular techniques – quantitative PCR assays to detect leptosiral DNA have been developed. They are sensitive, can distinguish different species, allow early diagnosis, and organisms can be detected after antibiotic therapy has been initiated.
- Serology – the mainstay of diagnosis. Commercial tests using genus-specific antigens are used to screen sera, and positive reactions confirmed in a reference laboratory using the microagglutination test (MAT) with live leptospira (killed have lower sensitivity). The MAT detects agglutinating antibodies in patient serum and is relatively serovar specific so a large number of antigens must be tested. Interlaboratory variation is high. A positive MAT is considered to be a fourfold increase in antibody titre, or a switch from seronegative to a titre of 1/100 or over. Early samples tend to cross-react; convalescent samples are more specific and diagnostic. Enzyme immunoassay for IgM is useful for diagnosing current infection but cross-reactions occur.

Treatment

- Mild disease – doxycycline, ampicillin, amoxicillin
- Severe disease – ampicillin, penicillin G, ceftriaxone
- Prophylaxis – doxycycline reduces morbidity and mortiality in endemic areas, but has no impact on infection rates as measured by seroconversion. It is likely to be useful in cases of accidental lab exposure or military/adventure travel. Vaccines are available against specific serovars.

Overview of *Rickettsia*

Microbiology

The genus *Rickettsia* is a member of the *Rickettsiales* order. All are maintained in a cycle involving mammal reservoirs and arthropod vectors. *Rickettsia* organisms are fastidious, obligate intracellular Gram-negative coccobacilli (0.3 micrometre by 1–2 micrometre). They survive only briefly outside a host. Isolation is usually only performed in reference laboratories.

Epidemiology

Zoonotic reservoirs are varied and include wild rodents, dogs, and livestock. Humans are incidental hosts with the exception of louse-borne typhus where humans are the main reservoir. *R. rickettsii*, *R. typhi*, *R. tsutsugamushi (Orientia tsutsugamushi)*, and *R. akari* can exist as vector commensals. *R. prowazekii* however kills its human body louse vector within 3 weeks. See Table 4.17 for geographical distribution.

Diagnosis

- Culture – usually only performed in reference laboratories. Blood or biopsy tissue from skin lesions should be frozen at –70°C. Organisms may be isolated in small lab animals or in embryonated eggs. They are highly infectious if aerosolized and have been responsible for lab-acquired infections (some fatal).
- Detection of antigen – direct immunofluorescence of skin lesions in cases of Rocky Mountain spotted fever (RMSF) can identify organisms at the time the rash appears (day 3 to 5 of illness). Sensitivity is around 70%, with specificity approaching 100%. Organisms are most likely to be in a blood vessel near the centre of the lesion – the biopsy should include this to increase chances. Availability is limited.
- Serology – the main means of confirming diagnosis. Antibodies first appear around day 7–10 after infection. A fourfold rise on a convalescent sample is required for diagnosis but a single titre over 1:64 is very suggestive of infection. Currently the most sensitive and specific serological assay is the micro-immunofluorescence test – it requires trained personnel and a fluorescent microscope. Latex agglutination tests are available for the diagnosis of RMSF – a single positive test is considered diagnostic – they rarely produce positive reactions in convalescence. Both complement fixation and the classic Weil–Felix test (see Box 4.11) are now considered to be neither sufficiently sensitive nor specific. Cross-reactions among rickettsial species occur and vary from patient to patient. They tend to be strongest between rickettsial subgroups – e.g. difficult to distinguish spotted fevers from each other.

Box 4.11 Weil–Felix test

In 1915 in Poland Weil and Felix found that serum from patients with typhus agglutinated certain strains of *Proteus vulgaris*. It has been the mainstay of diagnosis for many years.

Table 4.17 Overview of Rickettsial Desease

		Species Syndrome	Vector (geography)	Clinical features
Spotted fever group	*R. rickettsii*	Rocky Mountain spotted fever	Ixodid ticks (Western hemisphere)	Incubation ~ 7 days. Fever, headache, myalgia, eschar, rash. Multisystem involvement; 20% untreated mortality
	R. conorii	Mediterranean Spotted Fever	Ixodid ticks (Mediterranean, Africa, and India)	Incubation ~ 5 days. Eschar and local lymphadenopathy. Rash. Mild.
	R. akari	Rickettsial pox	Mite (USA, Africa, Korea and CIS (Commonwealth of Independent States))	Incubation ~ 7 days. As for *R. conorii* plus vesicular rash resembling chickenpox
Typhus group	*R. prowazekii*	Epidemic typhus	Body louse (S America, Africa, Asia)	Incubation ~ 10 days. Fever, headache, neurological and GI symptoms. Rash. 20–50% untreated mortality
		Brill–Zinsser disease	Nil – recurrence years after primary attack	Similar, milder illness than epidemic typhus developing years after recovery – in west was seen in E European immigrants after World War 2. Lasts around 2 weeks
	R. typhi	Murine (endemic) typhus	Flea (worldwide where human/rat co-exist)	Similar to epidemic typhus but much milder
	R. tsutsugamushi (Orientia tsutsugamushi)	Scrub typhus[a]	Mite (S Pacific, Asia, Australia)	Eschar common. Similar to epidemic typhus

[a] So-called 'scrub' typhus as the vector is harboured in scrub vegetation. The chigger mites stay within several metres of where they hatch and are trans-ovarially infected. Infection therefore occurs in very focused rural 'mite islands'.

Rickettsial diseases

Rickettsiae replicate within the cytoplasm of infected endothelial and smooth muscle cells of capillaries, and small arteries. They cause a necrotizing vascultitis with consequent protean manifestations. The classic triad of fever, headache, and rash with the appropriate travel and exposure history should alert to the possible diagnosis. An eschar (black, ulcerated lesion) may develop at the bite site. Severity varies greatly with species – any organ can be involved.

Spotted fevers

Rocky Mountain spotted fever

- Clinical features – the most virulent spotted fever with 20% mortality if untreated. Fever, myalgia, and headache follow a 2–14-day incubation. GI involvement may suggest an acute surgical abdomen. Maculopapular rash (90% cases – more likely to be spotless in elderly or black) appears around day 3–5, often starting at the hands and may become petechial or necrotic. Gangrene is seen in 4%. Severe multisystem involvement is common including lung (pneumonia, effusions, odema), nervous system (meningitis, focal deficits, e.g. deafness), and renal impairment. Thrombocytopenia and DIC can occur. Death at around 10 days (less in fulminant cases which are seen more frequently in black males with glucose-6-phosphate dehydrogenase (G6PD) deficiency).
- Treatment – tetracycline, chloramphenicol (preferred in pregnancy), or doxycycline given for 7 days, continuing for 2 days after the patient becomes afebrile. It is recommended that doxycycline is used even in children with suspected Rocky Mountain spotted fever given the life-threatening nature of the disease. No demonstrated benefit from steroid therapy.

Other spotted fevers

- Clinical features – Mediterranean spotted fever (also known as African tick typhus) is a much milder disease; 5–7 days after inoculation patients develop an eschar with local tender lymphadenopathy and a generalized maculopapular rash. Mortality is very rare, although severe disease can occur in patients with diabetes, cardiac disease, G6PD deficiency, and in the elderly. Rickettsialpox is similar in presentation with the addition of a vesicular rash that resembles chickenpox.
- Treatment – as above. A single dose of 200 mg doxycycline has been proposed and seems effective.

Typhus group

Epidemic typhus

- Clinical features – unusual in that humans are the reservoir and outbreaks are thus commonest in conditions of crowding – especially winter and war. The louse feeds on an infected person, bites and defaecates on the next, and infected faeces are scratched into the bite. After 1 week incubation abrupt onset of headache and fever is followed by maculopapular rash at day 5. This involves the entire body within a few days. Neurological features are common as is

multisystem involvement. Mortality is 20–50% untreated and is low in children, high in the elderly.
- Treatment – as for Rocky Mountain spotted fever. Early therapy nearly eliminates fatal illness.

Murine typhus
- Clinical features – longer incubation (up to 2 weeks) and patients rarely recall flea exposure. Fever, headache, and myalgia are followed by rash in 50%. Some may develop multi-system features but this is less common than with epidemic louse-borne typhus. Mortality is less than 1%.
- Treatment – as for Rocky Mountain spotted fever.

Scrub typhus
- Clinical features – not as severe as epidemic typhus. An individual is inoculated by the bite of the chigger mite and develops abrupt fever and headache 6–18 days later. Usually tender lymph nodes and an eschar at the inoculation point. Severity varies widely – neurological features can occur. Untreated mortality varies – up to 30%. Many serotypes (unlike the other organisms) so people may become infected again.
- Treatment – as for Rocky Mountain spotted fever, but resistance to doxycycline and chloramphenicol has been seen in Northern Thailand. Treatment may need to be prolonged to avoid relapse (2 weeks).

Coxiella burnetii (Q fever)

Q fever is the name coined by the medical officer in Queensland, Australia who first investigated the outbreak of febrile illness that hit 20 employees of a Brisbane meat works.

Microbiology

Coxiella burnetii was previously grouped along with the rickettsiae but is now reclassified as a Gammaproteobacteria, along with Legionella. It is a significantly hardier organism and may be transmitted by aersosol or infected milk. It can form spores and is able to survive outside a host for some time – over 40 months in skimmed milk at room temperature! It grows in the phagosomes of infected cells rather than the cytoplasm (as rickettsia) – appreciating the more acidic environment the phagosome affords.

Epidemiology

Found around the world, and a zoonosis, the organism is usually acquired from occupational exposure to cattle or sheep but can be caught from many different animals – exposure to parturient cats is an important risk factor! Acquisition from unpasteurized dairy products has occurred and person-to-person spread is possible but unusual. It exists in ticks but this is thought to be an insignificant route of human infection – it is likely they maintain the organism and infect those animals from which man may acquire it. Infected ungulates are usually asymptomatic although abortion/stillbirth may result. Organisms from a heavily infected placenta may be found in the soil for 6 months, and the air for 2 weeks after parturition.

Clinical features

Humans are infected by inhalation (occasionally ingestion). Organisms proliferate in the lungs and bacteraemia follows. Presentation is 2–5 weeks after infection and ranges from a self-limited febrile illness (commonest) to pneumonia. The pneumonia may be an incidental finding as part of a fever of unknown origin (PUO) or a severe atypical pneumonia with dry cough, fever, fatigue, pleuritic chest pain pleural effusion, and diarrhoea. It may be rapidly progressive, resembling legionnaire's disease. Hepatomegaly and rashes are common. Acute Q fever can be complicated by behavioural disturbances, Guillain–Barré syndrome, myocarditis, arthritis, glomerulonephritis, among others. Autoantibodies are often found (anti-mitochondria, smooth muscle). Mortality is around 1% and associated with myocarditis.

Chronic Q fever can occur. The commonest manifestation is culture-negative endocarditis which may be accompanied by many extracardiac manifestations. Hepatitis, osteomyelitis, vascular graft infection, and neurological infection are also recognized. Immunocompromised and pregnant hosts (who have increased risk of miscarriage) are at higher risk.

Diagnosis

- Culture – difficult and hazardous for lab staff – a category 3 organism
- Serology – the organism has two biological phases. Antibodies to phase II are produced first. Phase I antibodies appear weeks later. If antibodies to both phases are present simultaneously, chronic infection (specifically endocarditis) should be considered. Cross-reactions occur with *Bartonella* infection. The complement fixation test is most widely used. Fourfold rise between acute and convalescent titres is considered diagnostic of Q fever
- Molecular – PCR tests exist but are not in widespread use

Treatment

- Pneumonia – infection is nearly always self-limited. Even without therapy people begin to recover at around 2 weeks. However, treatment is indicated in all cases to reduce chance of chronic disease: tetracycline, doxycycline, or chloramphenicol for 2–3 weeks.
- Endocarditis – combination antibiotic therapy e.g. tetracycline with rifampicin for a prolonged period (a year, even lifelong has been mooted in some cases). Valve replacement may be required.

Bartonella species

Microbiology

Bartonellae are Gram-negative intracellular organisms belonging to the genus *Bartonella*. There are 14 species, four of which have been found to cause human disease.

Clinical syndromes

- **Oroya fever *(B. bacilliformis)* –** develops 3–12 weeks after inoculation by the fly vector and may be mild or abrupt and severe (high fever, sweats, headache, confusion, anaemia due to erythrocyte invasion). Complications: abdominal pain, thrombocytopenia, seizures, dyspnoea, hepatic and GI dysfunction, and angina can occur. High mortality rate. Survivors have a high incidence of opportunistic infections (e.g. salmonella, toxoplasmosis). Asymptomatic bacteraemia with *B. bacilliformis* occurs in 15% of survivors.
- **Veruga peruana *(B. bacilliformis)*** – a late-stage manifestation characterized by crops of skin lesions weeks to months after untreated acute infection. Initially of a miliary appearance they become nodular, then 'mulaire' – erythematous round lesions, 5 mm diameter. Lesions occur on mucosal surfaces and internally. Histology demonstrates neovascular proliferation with occasional organisms.
- **Cat scratch disease *(B. henselae)*** – the commonest cause of lymphadenopathy in children and young adolescents; 3–10 days after inoculation a papule or pustule may be visible at the site. Most present at 2–3 weeks with the onset of regional lympadenopathy and low-grade fever. Rarer features: headache, sore throat, and skin rash. Lymphadenopathy settles over 2–4 months, even without treatment. Complications (commoner in the immunocompromised): encephalopathy, retinitis, bone and skin involvement, granulomatous hepatitis. Conjunctival exposure may present as the Parinaud's oculoglandular syndrome (ocular granuloma or conjunctivitis, preauricular lymphadenopathy). Other atypical presentations: fever of unknown origin (PUO), osteomyelitis, hepatic and splenic granulomas. Diagnosis is based on history (cat exposure) but biopsy may be necessary to exclude lymphoma. Organisms may be visible on Warthin–Starry silver stain. Culture is possible from blood and tissue and should be attempted in cases of PUO, neuro-retinitis, or encephalitis after cat exposure, especially in the immunocompromised.
- **Bacillary angiomatosis (BA)** – unusual vascular proliferation caused by *B. henselae* or *B. quintana* infection. Usually occurs in the immunocompromised, mostly HIV patients with CD4 $<100/mm^3$, but also transplant patients and those on chemotherapy. Lesions begin as small papules that grow to form round red to purple nodules which can ulcerate. They can also appear as flat hyperpigmented plaques. Occur on skin, liver, spleen, bone, mucosal surfaces, heart, CNS and bone marrow. Pathogenesis: the organism's outer membrane adhesin binds to endothelial cells, and induces endothelial proliferation and new vessel formation. Numerous organisms are visible in lesions stained with Warthin–Starry silver stain. Diagnosis: lesion biopsy.

All patients should be treated – 6–8 weeks erythromycin or doxycycline for cutaneous disease, longer if recurrence occurs. Skin lesions may be excised. Without therapy, systemic infection (fever, abdominal pain and anorexia) can occur.

- **Bacterial peliosis (BP)** – this is characterized by blood-filled cystic lesions scattered throughout a visceral organ. Cases involving the liver (peliosis hepatic) and spleen present with weight loss, diarrhoea, abdominal pain, nausea, fever, hepatosplenomegaly, and elevated liver enzymes. Caused by *B. henselae* or *quintana* (less common, affecting bone) infection. Most patients also have bacillary angiomatosis and previous cat exposure.
- **Fever, bacteraemia, endocarditis *(B. quintana)*** – acquired by scratching infected louse faeces into skin lesions. Epidemics have occurred across the world, usually in conditions of overcrowding and poor sanitation (e.g. soldiers in World War I – 'trench fever'). Bacteraemia has also been described in homeless alcoholics. Incubation is 3–40 days, followed by a relapsing fever with headache, rash, and splenomegaly. It is a recognized cause of culture-negative endocarditis in those with HIV, and immunocompetent alcoholics. *B. henselae* bacteraemia occurs in the immunocompetent as well as those with HIV. Diagnosis is by culture. PCR and serology are available where culture negative. Treatment is by macrolides and tetracyclines – for 4–6 weeks. Endocarditis may need surgery and very prolonged antibiotic courses.

Diagnosis

- Direct examination – Giemsa-stained blood films may be used to detect *B. bacilliformis* in areas of endemic Oroya fever due to the large number of organisms present. This is not feasible in the detection of *B. henselae* or *B. quintana* due to the low level of blood-borne organisms – they may be detected by silver staining of lesions in BA (bacillary angiomatosis or BP and in lymph nodes in the early stages of cat scratch disease.
- Culture – *B. henselae* grows on chocolate agar with characteristic white, dry, cauliflower-like colonies. These become visible 5–14 days after incubation at 37°C in 5% CO_2. On Gram stain they are small, curved bacilli 2 micrometre by 0.5 micrometre, and display twitching motility when mounted in a saline drop. They are non-reactive for many standard biochemical tests.

Colonies with the appropriate morphological characteristics can have their identity confirmed by cellular fatty acid analysis, immunofluorescent antibody, or using commercial enzymatic substrate kits.

- Molecular – PCR or DNA-hybridization techniques can be used to speciate isolates. Direct detection of *Bartonella* organisms in pus or tissue is possible using PCR with wide-ranging sensitivity, depending on technique and sample.
- Serology – enzyme immunoassay or immunofluorescence kits may be used to demonstrate anti-*Bartonella* antibodies in culture-negative endocarditis, those with cat scratch disease, or HIV-associated aseptic meningitis etc. There is substantial cross-reactivity between *B. henselae* and *B. quintana* as well as with certain *Chlamydia* species.

Mycoplasma

Microbiology

Small (0.2 micrometre diameter). They lack a rigid cell wall and are bound only by a trilaminar membrane. They have a small genome with consequent limited biosynthetic capabilities, and require enriched media for growth. They can be distinguished by their differing phenotypic characteristics. Found in a wide range of animals and plants. Most found in humans are in the genus *Mycoplasma* with one (*U. urealyticum*) in the genus *Ureaplasma*. Whereas *M. pneumoniae* has a clear role in disease, most of the other organisms can be isolated in asymptomatic individuals. Several species have specialized structures at the cell ends which act as adhesins.

Mycoplasma pneumoniae

One of the commonest causes of respiratory tract infections and a cause of atypical pneumonia (3–10%). Droplet transmission. Infections occur all year round (peaks in winter) and affect all ages but most significant clinical disease (e.g. pneumonia) is seen from age 5 years to young adulthood. Once exposed, the organism attaches to respiratory tract epithelial cells and multiplies locally. Incubation is for 2–3 weeks and presentation is usually insidious with flu-like symptoms. Pneumonic symptoms follow – purulent sputum, haemoptysis. Multiple lobes may be involved but without consolidation and CXR infection is usually more dramatic than clinical presentation suggests. Pleural effusions in 20%. Disease usually self-limited with resolution over 3–10 days without antibiotics – CXR abnormalities may take 6 weeks to clear. Antibiotic therapy (e.g. erythromycin) speeds resolution but rarely eradicactes the organisms from the respiratory tract – recurrences can occur. Rare complications include: pleuritis, pneumothorax, lung abscess, haemolytic anaemia (secondary to cold agglutinins), thrombocytopenia, arthritis, rashes (e.g. erythema nodosum and multiforme), Guillain–Barré syndrome. Most complications are immune mediated with the exception of neurological complications such as meningoencephalitis which are thought to be due to direct invasion. Humoral and cellular immunodeficiency predisposes to more severe disease.

U. urealyticum and *M. hominis*

May be part of the commensal flora in male and female urogenital tracts. Sexually transimitted – the rate is related to sexual activity and rates are much lower among women using barrier contraception. They are implicated in endometritis and chorioamnionitis, are statistically linked with prematurity, low birth weight, and infertility, and are frequently recovered in culture along with other genital tract pathogens. Both may be isolated from blood cultures in women with post-partum fever (10% of cases). In men they are causes of non-gonococcal urethritis and a rare cause of epididymitis. In neonates, they are a cause of neonatal meningitis and there is an association between neonatal *U. urealyticum* colonization of the respiratory tract and the development of chronic lung disease of the newborn. Both may cause septic arthritis and subcutaneous abscesses in those with immunodeficiency. Sternal wound infections with *M. hominis* have occurred in heart and lung transplant patients.

Other *Mycoplasma* organisms

M. genitalium is a cause of non-gonococcal urethritis and may also have a role in respiratory tract disease. *M. fermentans*, *M. penetrans*, and *M. pirum* have been isolated in those with HIV and are unusual in their ability to actively invade cells. Certain organisms found in animal hosts have caused human disease in cases of sufficient exposure and significant predisposing comorbidity (e.g. *M. arginini* – many animals, *M. canis* – in, you guessed it, dogs).

Diagnosis

- Direct detection – indirect fluorescent antibody tests to detect *M. hominis* in genital samples have been developed (not widely used).
- Culture – *M. pneumoniae* can be recovered from respiratory tract specimens. Genital mycoplasmas can be isolated from many specimens. Fastidious organisms, they should be inoculated to culture media as soon as possible. Several media are used for culture of mycoplasma – most are diphasic (media with agar overlayed by media without agar). All species grow at 35–37°C but differ in their optimal pH and atmospheric conditions as well as their substrate utilization.
  - *M. pneumoniae* isolates should be kept for at least 4 weeks, initially in selective broth. Growth is indicated by change in pH. Positives are subcultured to agar. Colonies should be visible by 1 week and identity confirmed by serological methods, or enzyme substrate tests (e.g. tetrazolium reduction).
  - Genital mycoplasma samples are inoculated into broth and onto agar and should be kept for 8 days (although *M. genitalium* and *M. fermentans* can take longer and are not routinely looked for). Broths that exhibit a change in colour are plated to the appropriate agar. Plates are examined by microscope each day for colonies. Selective plates and colonial morphology are usually sufficient to allow identification. Anti-sera are available to confirm M. hominis.
- Molecular – PCR and DNA probe techniques have been developed but are not in wide use.
- Serology – the time required to culture these organisms means diagnosis is often made by serology. *M. pneumoniae* complement fixation tests detect mostly the early IgM – a fourfold rise between acute and convalescent (7–10 days) samples is diagnostic of recent infection. Enzyme immunoassays are available and detect IgM and IgG.

Treatment

- *M. pneumoniae* is sensitive to a wide range of agents – tetracyclines, quinolones or macrolides.
- *M. hominis* is usually resistant to erythromycin. Some genital mycoplasma isolates have been found to be tetracycline resistant and carry the *tetM* resistance determinant (also found in other genital tract organisms e.g. group B streptococci).
- *U. urealyticum* – usually resistant to clindamycin and less susceptible to quinolones. Erythromycin effective.

Chlamydia

Small, obligately intracellular (they are unable to produce ATP themselves), Gram-negative organisms. Outside a host cell they are tiny (300 nm diameter), inactive 'elementary' bodies. They infect cells primarily by receptor-mediated endocytosis. They are able to inhibit lysosome fusion and reside in a membrane-protected 'inclusion' body where they activate and increase to a diameter of around 800 nm. Three species produce human disease: *Chlamydia trachomatis*, *C. psittaci*, and *C. pneumoniae*. They differ antigenically, in host cell preference and in antibiotic susceptibility.

The family *Chlamydiaceae* was reorganised in 1999 on the basis of genetic similarities. *C. trachomatis* remains in the genus *Chlamydia*, but psittaci and pneumoniae were moved to a new genus, *Chlamydophila*.

Chlamydia trachomatis

Clinical features

There are 15 different serovars causing distinctive clinical syndromes. Natural infection confers only short-lived protection against re-infection.

- **Lymphogranuloma venereum** – serovar L1, L2, L3 – endemic in Africa, India, SE Asia, S America, and the Caribbean. Sexually transmitted – the organism enters through skin abrasions (it cannot infect squamous epithelial cells) and causes a small papule or ulcer (primary lesion) on genital mucosa or nearby skin 3–30 days after infection. It heals rapidly, and perhaps weeks later the patient develops secondary symptoms: lymphadenopathy (usually inguinal or femoral), fever, headache, myalgias, proctitis (can resemble inflammatory bowel disease), and occasionally meningitis. Nodes may coalesce forming abscesses and buboes.
- **Trachoma** – serovar A, B1, B2, C – chronic follicular keratoconjunctivitis which leads to corneal scarring and is the commonest cause of preventable blindness in the developing world (around 9 million blind, and 500 million affected). First infections usually acquired in childhood, resolves spontaneously but multiple re-infections and the consequent host immune response result in conjunctival scarring and corneal damage. The inner surface of the eyelid scars, and inturning eye lashes further abrade the cornea. There is a WHO grading of trachoma (Box 4.12).
- **Inclusion conjunctivitis** – serovars D to K – sexually transmitted eye infection of adults (of whom slightly over half have concurrent genital tract infection) and a cause of neonatal conjunctivitis (probably from mother's genital tract but can occur even if delivered by Caesarean section – 5 days to 6 weeks after delivery). No corneal scarring.
- **Neonatal pneumonia** – serovars D to K – most acquired from the mother's genital tract. Seen in 10–20% of infants born to infected mothers. Usually symptomatic by 8 weeks with nasal congestion, cough, etc. and only moderately ill.
- **Sexually transmitted infections** – serovars D to K – epididymitis (along with *N. gonorrhoeae* the common cause in the under-35s), urethritis, salpingitis, cervicitis with consequent pelvic inflammatory disease, and infertility. Reactive arthritis and Reither's syndrome may follow.

Box 4.12 WHO grading of trachoma

1 Follicular inflammation (follicular) – 5+ follicles in upper tarsal conjunctiva
2 Follicular inflammation (intense) – pronounced tarsal conjunctival inflammation obscuring at least half of deep tarsal vessels
3 Trachomatous conjunctival scarring – scars on tarsal conjunctiva
4 Trachomatous trichiasis – 1+ eyelash rubs on eyeball
5 Corneal opacity obscuring pupil

Diagnosis

Trachoma may be diagnosed on clinical grounds. Other clinical presentations require lab identification for a definitive diagnosis.

- Direct detection – microscopy of certain Giemsa-stained clinical specimens (particularly neonatal conjunctivitis) may allow direct visualization if there are sufficient bacterial inclusion bodies in the cytoplasm. Monoclonal antibodies and immunofluorescence increase sensitivity (in one study comparing Giemsa staining and immunofluorescence in detecting *C. trachomatis* in a series of infants with neonatal conjunctivitis, sensitivity was 42% and 100% respectively). Enzyme immunoassay can demonstrate chlamydial antigen in respiratory or genital secretions.
- Culture – being obligate intracellular organisms, the techniques for culturing are similar to those used in virus cuture. The infective elementary bodies must first be extracted by centrifugation before infecting a cell line. Inclusion bodies containing the organism can be seen at 48–72 h, and species-specific monoclonal antibodies used to confirm identity.
- Serology – most useful for epidemiological studies. Chlamydial complement fixation tests do not distinguish between species; microimmunofluorescence does.
- Molecular tests for chlamydial DNA are the preferred initial investigation for the diagnosis of *C. trachomatis* in all specimen types. PCR- and ligase chain reaction (LCR)-based tests are available, and sensitivity is a high as 99% for some of these but varies with specimen type and quality. The notable advantage is it allows the diagnosis of urethritis on non-invasively acquired specimens.

Treatment

- Trachoma – transmission is by flies or eye-to-hand in endemic areas, thus hygiene is important for control (rates fall quickly with socioeconomic improvement). Systemic therapy (erythromycin or doxycycline) is effective in areas of low transmission (where re-infection is less frequent). Mass treatment at village level has been shown to be effective. Eyelid surgery can prevent further mechanical damage.
- Lymphogranuloma venereum – buboes should be aspirated. Doxycycline for 3 weeks.
- Genital and ocular infections in adults – single-dose azithromycin, or doxycyline for 7 days. Longer courses of amoxicillin may be effective in pregnant women.

- Neonatal infections – topical eye therapy is not recommended as it does not eliminate carriage. Erythromycin orally for 14 days for both conjunctivitis and pneumonia. Prenatal screening of mothers and treating those infected with *Chlamydia* is 90% effective in preventing infants from acquiring infection.

Chlamydophila psittaci

C. psittaci can infect many kinds of birds. The classic term for infection caused by this bacteria, 'psittacosis' (derived from the Greek word for parrot) is therefore not so accurate a description as 'ornithosis'. It is an occupational disease of zoo workers, petshop workers, and poultry farmers. Human-to-human transmission occurs but is very rare. Infection is primarily acquired by inhalation of organisms from aerosolized avian excreta or respiratory secretions from sick birds (mouth-to-beak resuscitation has been implicated in acquisition). Transient exposure is sufficient (e.g. petshop customers). The disease is found worldwide. Incubation is 5–14 days and presentation is with fever, chills, malaise, cough, headache, breathlessness, mild pharyngitis, and epistaxis. Less commonly, nausea, vomiting, and jaundice may be seen. Examination may demonstrate the features of an atypical pneumonia. Other features are bradycardia, peri/myocarditis culture-negative endocarditis, splenomegaly, meningitis, encephalitis, Guillain-Barré syndrome, rashes, acute glomerulonephritis, severe respiratory failure, sepsis, and shock. Relapses can occur. Diagnosis is by serology: the demonstration of a fourfold rise in CF antibody titre is considered diagnostic. Antibodies may cross-react with other chlamydial species. Culture is possible but avoided due to the risks to lab staff. Other tests: ELISA, PCR available. Treatment is with tetracycline or doxycycline for 2–3 weeks (reduces the risk of relapse). Erythromycin may be used in children and the pregnant.

Chlamydophila pneumoniae

The cause of 3–10% of community-acquired pneumonia cases among adults. Adolescents tend to experience a mild pneumonia/bronchitis, whereas older adults can experience more severe disease and repeated infections. Fifty per cent of young adults have serological evidence of previous infection. Unlike *C. psittaci*, human-to-human transmission by respiratory secretions is the norm. In most populations infection is more common in males and this may reflect cigarette use. Incubation is 3–4 weeks and symptoms of a URTI are followed by bronchitis or pneumonia 1–4 weeks later. Most infections are asymptomatic or cause only mild symptoms. Other features: hoarse voice, non-productive cough, headache. Fever is often absent. Symptoms can be very prolonged even with appropriate treatment. Diagnosis is by serology (preferably microimmunofluorescence as CF tests cross-react with other chlamydial species). A definite case requires a four-fold rise in titre – single elevated IgG titres may be seen in the uninfected elderly as a consequence of repeated infections. Antibody tests may be negative in the early weeks after infection – it can take as long as 8 weeks for a significant IgG response to develop after primary infection. Other tests: PCR and cell culture tests are available. Treatment: doxycycline or erythromycin.

Mycobacterium tuberculosis

The *Mycobacterium tuberculosis* complex comprises five species: *M. tuberculosis*, *M. bovis*, *M. africanum*, *M. microti*, and *M. ulcerans*. The term tuberculosis (TB) describes a broad range of clinical diseases caused by *M. tuberculosis* (and less commonly *M. bovis*).

Epidemiology

M. tuberculosis infects one-third of the world's population and is the most-frequent infectious cause of death worldwide, accounting for 3 million deaths per year. Most cases occur in the developing world, with 13 countries accounting for 75% of cases. In the developed world, despite a general downward trend, there has been an increase in incidence in certain groups, e.g. immigrants from high-prevalence countries and HIV-infected patients.

- Transmission – infection is acquired by inhalation of infectious droplet nuclei. Occasionally due to skin inoculation (e.g. pathologists, laboratory personnel) or sexual transmission – determined by closeness of contact with infectious (sputum smear positive) source. Ninety per cent of primary infections are asymptomatic (person clinically well with positive tuberculin skin test).
- HIV infection increases the risk of developing all forms of TB. The risk of disease progression is highest at the extremes of age and in the immunocompromised. Immunocompetent people have a 5–10% lifetime risk of progression; this increases to 7–10% per year in HIV-infected patients.

Immunology

Tuberculosis is the prototype of infection that elicits a cellular immune response. CD4+ T lymphocytes recognize mycobacterial antigens in association with major histocompatibility (MHC) class II molecules, presented by macrophages. The T cells becomes activated and proliferate, resulting in the production of cytokines (interferon-γ, migration inhibition factor) which attract and activate more macrophages at the site of infection. Macrophages also secrete a number of cytokines (TNF-α, platelet-derived growth factor, transforming growth factor-β, fibroblast growth factor). The interplay of these factors determines the nature of the pathological and clinical features. The Langhan's giant cell consists of epithelioid cells (activated macrophages) oriented around mycobacterial antigens, and is a characteristic feature of tuberculous granulomas. When the population of activated lymphocytes reaches a certain size, a cutaneous delayed hypersensitivity reaction to tuberculin occurs.

Pathogenesis

Primary infection results from inhalation of infectious droplet nuclei which lodge in the alveoli and multiply, resulting in the formation of a Ghon focus. Involvement of the regional lymph nodes produces the Ghon complex. Depending on the host's immune response the infection will either become quiescent or progress and/or disseminate. Reactivation of disease may occur in later life, particularly in the immunosuppressed.

Clinical features

Pulmonary tuberculosis is the most common presentation. Tuberculosis may also disseminate (military TB) or affect almost any other organ (extra-pulmonary TB): pleural cavity, pericardium, lymph nodes, GI tract, and peritoneum, GU (genitourinary) tract, skin, bones and joints, and CNS.

Diagnosis

- Diagnosis is based on a combination of compatible clinical syndrome, supportive radiological investigations, and detection of acid-fast bacilli or culture of *M. tuberculosis* from clinical specimens. The gold standard for diagnosis is culture. Patients are often treated on the basis of a presumptive diagnosis.
- *M. tuberculosis* is an aerobic, non-sporing, non-motile, weakly Gram-positive bacillus with a thick cell wall containing mycolic acid, which renders it acid-fast.
- Samples of sputum or tissue are liquefied, decontaminated, neutralized, centrifuged, and the deposit inoculated into solid or liquid media. Normally sterile samples, e.g. CSF, need not be decontaminated, as loss of mycobacterial viability may occur.
- Acid-fast stains:
  - Auramine stain (fluorochrome phenolic auramine or auramine-rhodamine stain, acid-alcohol decolourization, potassium permanganate counterstain).
  - Ziehl–Neelsen (ZN) stain (carbol fuschin stain, decolourize with acid-alcohol, counterstain with methylene blue)
  - Kinyoun stain (ZN stain modified to make heating unnecessary)
- Molecular methods:
  - PCR-based tests may be used to detect M. tuberculosis in clinical specimens, e.g. Amplicor test (Roche), amplified M. tuberculosis Direct Tests (MDT, GenProbe), strand displacement amplification (BD ProbeTec-SDA), ligase chain reaction (Lcx, Abbott systems). Although specificity is high, sensitivity is lower: 90–100% and 60–70% for smear-positive and -negative specimens
  - identification of mycobacterial species e.g. high-pressure liquid chromatography (HPLC) of mycolic acids, DNA sequencing of 16S rRNA, PCR restriction enzyme assay (PRA), DNA probe hybridization (LiPA MYCOBACTERIA, Innogenetics)
  - identify drug resistance mutations e.g. rpoB mutations (InnoLiPA Rif.TB assay, Innogenetics)
  - typing for epidemiological studies e.g. IS6110 RFLP, MIRU, spoligotyping.
- Culture methods:
  - solid media e.g. Lowenstein–Jensen media, Middlebrook agar
  - liquid culture e.g. Kirschner broth, BACTEC 460 system, MB/BacT (Organon Teknika), Mycobacterial Growth Indicator Tube (MGIT, Becton Dickinson), MODS

- lysis-centrifugation method
- solid media, e.g. Lowenstein–Jensen media, Middlebrook agar
- liquid culture, e.g. Kirschner broth, BACTEC 460 system, MB/BacT (Organon Teknika), Mycobacterial Growth Indicator Tube (MGIT, Becton Dickinson), MODS
- Identification and susceptibility testing is usually done at reference lab level. For definitions of drug resistance see Box 4.13.
- Tuberculin skin test – purified protein derivative (PPD) is a standardized protein precipitate of tuberculin. The Mantoux test is a quantitative tuberculin test which is performed by intracutaneous injection of five tuberculin units (TU) of PPD in 0.1 mL of solution. The reaction is usually read after 48–72 h. A positive result is defined as >10 mm of induration. Reactions of 5–10 mm may be due to BCG (bacilli Calmette–Guérin) vaccination, but are also suspicious of TB in low-prevalence areas. False-positive results can occur with non-tuberculos mycobacteris (NTM) infections. False-negative reactions can occur in up to 20% of patients with TB and in HIV-infected patients. Delayed reactivity (>10mm induration after 6 days) may occur in certain populations e.g. Indochinese immigrants.
- Interferon gamma-release assays (IGRA) – this is a relatively new method of detecting T cells specific for *M. tuberculosis* antigens. It has a sensitivity >80% and is more specific than the tuberculin skin test. Sensitivity remains high in children <3 years and in HIV co-infection, and is not confounded by BCG vaccination or infection with NTM. The main limitation of the test is that it cannot distinguish active from latent infection.

Treatment

Treatment is with combination chemotherapy for several months. For most types of tuberculosis, the usual regimen is a 2-month intensive phase with three or four drugs, followed by a 4-month continuation phase. The choice of drugs and duration of treatment depend on the likelihood of drug resistance, the site of infection and the patient's HIV status. For details see 📖 Antituberculous agents—1st line, p.110 and British Thoracic Society (BTS) or CDC guidelines (Box 4.14).

Prevention

BCG vaccine is a live attenuated vaccine derived from *M. bovis*. It is given to infants and children in high-prevalence areas and results in a 60–80% reduction in the incidence of TB. It should only be given to infants <12 weeks or children who are tuberculin skin test (TST)-negative. Although it does not prevent infection, BCG vaccination reduces the risk of disseminated disease in children. BCG is contraindicated in HIV-infected individuals. Vaccination can occasionally cause disseminated BCG infection, usually in immunosuppressed patients. Intravesical BCG (used to treat bladder cancer) can cause liver or lung granulomas, psoas abscess, or osteomyelitis. Newer vaccines include MVA (modified vaccinia Ankara DNA vaccines which use a prime-boost strategy to induce *M. tuberculosis* specific immune responses. These vaccines look promising in phase II clinical trials.

Chemoprophylaxis

Chemoprophylaxis is given to individuals at increased risk of tuberculosis, e.g. contacts of active cases, recent TST conversion, abnormal CXR, HIV infected or certain medical conditions, e.g. renal transplant myeloproliferative or haematological malignancies. For detailed indications see BTS and CDC guidelines (Box 4.14). The usual regimen is isoniazid 300 mg for 6–12 months. The risk of isoniazid hepatotoxicity increases with age and daily alcohol consumption.

Box 4.13 Drug-resistant tuberculosis

- Mono-resistant = resistance to one drug
- Poly-resistant = resistance to >1 drug (but not MDR)
- Multi-drug resistant (MDR) = resistance to at least isoniazid and rifampicin
- Extensively drug resistant (XDR) = resistance to rifampicin, isoniazid, a quinolone, and an injectable agent).

In 2006, an outbreak of XDR tuberculosis was described in Tugela Ferry, South Africa. All patients tested for HIV were found to be positive and 52/53 died with a median survival of 16 days. Genotyping of the outbreak strains showed that 85% were similar, suggesting recent nosocomial transmission.

Box 4.14 Guidelines for tuberculosis treatment

- UK NICE guidelines. Clinical guideline 33. *Tuberculosis: clinical diagnosis and management and measures for its prevention and control.* www.nice.org.uk/CG033
- British Thoracic Society guidelines. Chemotherapy and management of tuberculosis in the United Kingdom. *Thorax* 1998;53:536–48. www.brit-thoracic.org.uk/
- American Thoracic Society Guidelines. ATS/CDC/IDSA: treatment of tuberculosis. *Am J Respir Crit Care Med* 2003;167:603–662. www.thoracic.org
- World Health Organization guidelines. Treatment of tuberculosis: guidelines for national programmes. WHO/CDS/TB 2003.313. Geneva: World Health Organization, 2003. http://whqlibdoc.who.int/hq/2003/WHO_CDS_TB_2003.313_eng.pdf

Mycobacterium leprae

Leprosy or Hansen's disease is caused by infection with *Mycobacterium leprae*, an obligate intracellular parasite whose only natural hosts are humans and armadillos. Experimental infections can be induced in the mouse footpad. The clinical manifestations of leprosy include skin lesions, deformities, and peripheral neuropathy, making it one of the most socially stigmatizing diseases. Leprosy exhibits a spectrum of clinical features ranging from lepromatous (multibacillary) to tuberculoid (paucibacillary) forms.

Epidemiology

Worldwide, there are an estimated 6 million people living with leprosy, 3 million of whom are untreated. Although Africa has the highest prevalence Asia has the greatest number of cases. Leprosy is associated with poverty and rural residence but not HIV infection. Distribution in endemic countries is non-homogeneous, suggesting that genetic factors may play a role in disease expression. The mode of transmission remains uncertain, but is thought to be human-to-human, via nasal droplet infection. Leprosy may also be acquired by direct inoculation into the skin. The incubation period is long, with an average of 5–7 years (range 2–40 years) and the peak onset is in young adults.

Pathogenesis

M. leprae has a dense, mainly lipid, capsule outside the cell wall, which is rich in *M. leprae*-specific phenolic glycolipid 1 (PGL-1). This has been implicated as a scavenger for free radicals, allowing intracellular survival and limiting antibiotic penetration. PGL-1 and lipoarabinomannan have been implicated as causing immunological hyporesponsiveness of both lymphocytes and macrophages in the anergic, highly bacillary lepromatous form of leprosy.

Clinical features

Clinical manifestations of leprosy are largely confined to the skin, upper respiratory tract, and and peripheral nerves. Most of the serious sequelae are a result of peripheral nerve damage resulting in deformities (e.g. ulnar, median, and peroneal nerve palsies), loss of peripheral parts of digits, and plantar ulceration.

- **Lepromatous leprosy** – characterized by symmetric skin nodules, plaques, and a thickened dermis that typically occur in cool areas of the body, e.g. earlobes and feet. This condition is multibacilliary (infectious) and associated with poor cell-mediated immunity. Involvement of the nasal mucosa results in congestion, epistaxis and, rarely, septal collapse ('saddle nose' deformity). May also cause loss of eyebrows and eyelashes, trichiasis, corneal scarring, uveitis, lagophthalmos, testicular dysfunction, and amyloidosis.
- **Tuberculoid leprosy** – characterized by hypopigmented, anaesthetic skin plaques, and asymmetric peripheral nerve involvement. It is typically paucibacillary (non-infectious) and associated with a good cell-mediated immune response.

- **Borderline leprosy** – the majority of patients have manifestations intermediate between the two polar forms, a condition termed borderline leprosy.
- **Reversal reactions** – an abrupt increase in inflammation within previously quiescent skin lesions, as well as new skin lesions, neuritis, and low-grade fever may develop in borderline leprosy patients either before (downgrading reaction) or after (reversal reacion) the initiation of therapy. If the neuritis is not treated promptly, irreversible nerve damage may occur.
- **Erythema nodosum leprosum** – affects >50% of lepromatous and borderline leprosy patients after initiation of therapy. Clinical features include painful nodules (usually on extensor surfaces, may pustulate or ulcerate), neuritis, fever, malaise, anorexia, uveitis, lymphadenitis, oorchitis, anaemia, leucocytosis, glomerulonephritis.

Diagnosis

A firm diagnosis of leprosy requires the presence of a characteristic peripheral nerve abnormality or the demonstration of acid-fast bacilli in skin biopsies or split skin smears. In atypical cases two of the following three criteria are required: a clinically compatible skin lesion, dermal granuloma on skin biopsy, hypoaesthesia within the lesion. Skin biopsies should be taken from skin plaques or nodules in lepromatous patients and from the periphery of lesions in tuberculoid patients. Nerve biopsies may result in loss of function and should only be performed if there is sufficient diagnostic uncertainty to warrant it.

Microbiology

M. leprae is a Gram-variable, acid-fast bacillus that is best visualized by a modified Fite stain (as it may be decolourized by the Ziehl–Neelsen stain). Viable bacilli stain brightly, whereas dead bacilli stain irregularly.
M. leprae grows best at temperatures <37°C in humans and armadillos.
Unique properties of *M. leprae* include loss of acid-fastness by pyridine extraction, presence of dopa oxidase activity, multiplication in the mouse footpad with a doubling time of 12–14 days.

Experimental infection of the mouse footpad can be used to assess antimicrobial susceptibility.

Treatment

Treatment requires combination therapy with two or more agents, e.g. dapsone, clofazimine, and rifampicin. Ethionamide, prothionamide, and certain aminoglycosides have also been used. Newer agents such as minocycline, clarithromycin fluoroquinolones look promising. For treatment regimens see 📖 Antileprotics, p.116.

Prevention

Dapsone prophylaxis of household contacts of leprosy patients is not recommended as although it reduces the subsequent prevalence of tuberculoid leprosy, it only forestalls the development of lepromatous leprosy.

Non-tuberculous mycobacteria

This group of organisms comprises about 50 species of mycobacteria, excluding those in the *M. terrae / M. nonchromogenicum* complex and *M. leprae*. Other names for non-tuberculous mycobacteria (NTM) include atypical mycobacteria, opportunistic mycobacteria or mycobacteria other than tuberculosis (MOTT).

Classification

NTM were previously classified according to growth rate, colonial morphology, and pigmentation (Runyon classification). This has been superseded by molecular methods but nonetheless remains useful to separate NTM into three groups:

- **rapidly growing mycobacteria** (≤7 days incubation) e.g. *M. fortuitum* complex, *M. chelonae/abscessus* group, *M. mucogenicum*, and *M. smegmatis*
- **slow-growing mycobacteria** (>7 days incubation) e.g. *M. avium* complex, *M. kansasii*, *M. xenopi*, *M. simiae*, *M. szulgai*, *M. scrofulaceum*, *M. terrae/nonchromogenicum* complex, *M. malmoense*, *M. haemophilum*, and *M. genavense*
- **intermediately growing mycobacteria** (7–10 days), e.g. *M. marinum*, *M. gordonae*.

Clinical features

The NTM can cause a wide spectrum of diseases (Table 4.18).

Table 4.18 Clinical syndromes caused by NTM.

Syndrome	Most common causes
Chronic bronchopulmonary disease (adults, CF patients)	*M. avium* complex, *M. kansasii*, *M. abscessus*
Cervical lymphadenitis (children)	*M. avium* complex
Skin and soft tissue infections	*M. fortuitum* group, *M. chelonae*, *M. abscessus*, *M. marinum*, *M. ulcerans*
Bone and joint infections	*M. marinum*, *M. avium* complex, *M. kansasii*, *M. fortuitum* group, *M. abscessus*, *M. chelonae*
Disseminated infection (HIV positive)	*M. avium*, *M. kansasii*
Disseminated infection (HIV negative)	*M. abscessus*, *M. chelonae*
Catheter-related infections	*M. fortuitum*, *M. abscessus*, *M. chelonae*

Diagnosis

Because the signs and symptoms of NTM lung disease are often variable and non-specific, diagnosis requires multiple positive respiratory cultures. Diagnosis of NTM infections at other sites requires positive cultures from pus, tissue biopsies or blood cultures.

- Microscopy – the acid-fast stains used for identifying *M. tuberculosis* (p.408) also work well for identifying NTM.
- Culture – appropriate culture media include Middlebrook 7H10 or 7H11 agar or BACTEC broth. Samples from skin and soft tissue infections need to be plated at 28–30°C as well 35–37°C, as some species only grow at low temperatures, e.g. *M. chelonae*, *M. haemophilum*, and *M. marinum*. *M. xenopi* grows best at 42°C. Other species have special growth requirements, e.g. *M. genavense* (BACTEC broth for 6–8 weeks) and *M. haemophilum* (iron supplementation).
- Identification – although traditional biochemical and other standard tests may be performed, identification of NTM increasingly uses rapid molecular methods:
  - HPLC of mycolic acids
  - PCR-RFLP analysis of the heat-shock protein gene
  - genetic probes for mycobacterial RNA
  - 16S ribosomal DNA sequencing
- Drug susceptibility testing – various methods are used:
  - agar disk elution
  - broth microdilution
  - E-test
  - BACTEC radiometric detection
- Strain comparison – for epidemiological studies, standard biochemical identification and susceptibility testing have been superseded by molecular methods.

Treatment

Treatment of NTM infections may be medical, surgical, or a combination of the two. The choice of drugs and duration of treatment depend on the causative organism, site of infection and patient's HIV status:

- UK guidelines – BTS guideline: Management of opportunist mycobacteria. *Thorax* 2000; 55: 210–218 www.brit-thoracic.org.uk
- US guidelines – ATS/IDSA guideline: Treatment of non-tuberculous mycobacteria. *Am J Respir Crit Care Med* 2007;175:367–416. www.thoracic.org

Overview of virology

Table 4.19 Summary of viruses and common clinical syndromes

Group	Virus	Consider in the differential of. . .								
		Respiratory	Rash	Hepatic	Muscular	Diarrhoea	Encephalitis	Meningitis	Shock	Other
DNA viruses										
Adenoviridae	Adenovirus (p.481)	•••				•••	••	••		Cystitis, conjunctivitis
Herpesviridae	HSV-1 and -2 (p.441)	•	•••				•• (1)	•• (2)		Genital, eye, dissemination
	EBV (p.447)	•	•	•	•••		•	•		
	VZV (p.444)	•	•••	•			•	•		
	CMV (p.451)	•	•	••	••	•	•	•		Congenital
	HHV-6 and -7 (p.439–40)		••							Fever alone
	HHV-8 (p.483)		•							Kaposi sarcoma
Poxviridae	Molluscum		•••							
(also smallpox) (p.489)	Orf		•••							
	Monkeypox		•••							
Parvoviridae	Parvovirus B19 (p.436)		•••				-			Arthropathy, aplastic anaemia, congenital

Table 4.19 Summary of viruses and common clinical syndromes

Group	Virus	Consider in the differential of. . .								
		Respiratory	**Rash**	**Hepatic**	**Muscular**	**Diarrhoea**	**Encephalitis**	**Meningitis**	**Shock**	**Other**
Papovaviridae	HPV (p.485)									Warts, epithelial tumours of skin
	Polyomavirus (p.488)	●●								Progressive focal neuro deficits
Hepadnaviridae	Hepatitis B (p.459)			●●●						
RNA viruses										
Orthomyxoviridae	Influenza (p.419)	●●●				●	●			
Paramyxoviridae	Parainfluenza (p.425)	●●●								
	Mumps (p.432)						●●	●		Parotitis, epididymo-orchitis, GBS
	Measles (p.430)	●●	●●●				●	●		SSPE
	RSV/metapneumovirus (p.426)	●●●								
Coronaviridae	Coronavirus (p.428)	●●●				●●				
Picornaviridae	Poliovirus (p.492)						●			Paralysis, myocarditis

	Non-polio enteroviruses (p.474)	••	••			•••	•••		Myopericarditis, conjunctivitis
	Hepatitis A (p.457)			•••					
	Rhinovirus (p.630)	•••							
Reoviridae	Rotavirus (p.454)				•••				
Retroviridae	HTLV-1 and -2 (p.495)								Spastic paraparesis, leukaemia/lymphoma
	HIV-1 and -2 (p.469)		•••			•			Immunodeficiency
Togaviridae (p.476)	Rubella (p.434)	•	•••						Congenital
	Alphaviruses (p.477)		•••			•••			Arthralgia
Flaviviridae	Yellow fever (p.501)		••	•••		•		•	Haemorrhage
	Dengue (p.503)		••	••		•		•••	Haemorrhage
	Hepatitis C (p.462)			•••					
	Japanese encephalitis (p.497)					•••			
Bunyaviridae (also Rift Valley fever)	Hantavirus (p.479)	•						•••	Haemorrhage, renal failure
	CCHF (p.479)							•••	Haemorrhage
	Californian encephalitis (p.479)					•••			

Table 4.19 Summary of viruses and common clinical syndromes

Group	Virus	Consider in the differential of. . .								
		Respiratory	Rash	Hepatic	Muscular	Diarrhoea	Encephalitis	Meningitis	Shock	Other
Arenaviridae	Lassa (p.508)					••	•	•	•••	Haemorrhage
	LCMV (p.476)		•••				••	••		Congenital
Filoviridae	Marburg, Ebola (p.507)		••						•••	Haemorrhage
Rhabdoviridae	Rabies (p.509)						•••			
Other	Astrovirus and calcivirus (p.456)					•••				

••• very frequently seen, •• commonly seen, • occasionally seen.

'Congenital' indicates viruses with the potential to cause significant sequelae if infecting the developing fetus.

CCHF: Congo–Crimean haemorrhagic fever; CMV: cytomegalovirus; EBV: Epstein–Barr virus; GBS: Guillain–Barré syndrome; HHV: human herpesvirus HIV: human immunodeficiency virus; HPV: human papillomavirus; HSV: herpes simplex virus; HTLV: human T-cell lymphotrophic virus; LCMV: lymphocytic choriomeningitis virus; RSV: respiratory syncytial virus; SSPE: subacute sclerosing panencephalitis; VZV: varicella zoster virus.

[a] HIV may of course present in a multitude of ways - however the commoner presentations of primary HIV infection (as opposed to a subsequent opportunistic infection) are rash and encephalitis

Influenza – introduction

One of the commonest infectious diseases of man, primarily causing epidemics of upper respiratory tract infection. Global pandemics may follow dramatic antigenic changes – 21 million died in the 1918–1919 pandemic.

The virus

Members of family *Orthomyxoviridae*. Negative sense single-strand (ss)RNA viruses.

The three distinct influenza viruses:

- influenza A causes the typical influenza syndrome and can precipitate pandemics
- influenza B is similar clinically but does not cause pandemics
- influenza C causes an afebrile common cold-like syndrome and does not occur in epidemics.

All have host cell-derived envelopes embedded with glycoproteins important to viral entry and exit. These have haemagglutinin (HA) or neuraminidase (NA) activities, are key antigenic components and may alter gradually by mutation (antigenic drift – see Box 4.15).

At least 16 HA and 9 NA variants have been identified in influenza A viruses. In the UK in 2008 three subtypes of influenza A were circulating: H1N1, H1N2 and H3N2. Electron microscopy (EM) appearance of the viral particles is variable: spherical to filamentous. Viruses are named by their type, place of initial isolation, strain, year, and antigenic subtype, e.g. A/Victoria/3/75/H3N2.

Epidemiology

- Outbreaks are associated with excess rates of pneumonia and influenza-related illness and mortality, peaking in the winter and varying with the viral type responsible. Not all influenza-related deaths present as pneumonia. Sporadic cases of severe disease have occured in some countries as a result of human acquisition of viral strains adapted to birds - see Box 4.16.
- Attack rates are highest in the young, mortality highest in the elderly, both are increased in those with pre-exisiting medical problems, e.g. cardiovascular, pulmonary, and renal impairment or immunodeficiency.
- Person-to-person transmission occurs by dispersion in small-particle aerosols. Virus is present in large quantities in the secretions of infected people. One individual can infect a large number of others contributing to the explosive nature of outbreaks. Outbreaks can occur in an epidemic or pandemic fashion:
  - **epidemics** – confined to a single location, e.g. a town or country, and occur almost exclusively in winter. They start abruptly with cases seen initially in children and then adults, peak within 3 weeks and last around 6 weeks. Different strains may circulate simultaneously
  - **pandemics** – severe outbreaks that spread to all parts of the world. Caused only by type A viruses. Associated with the emergence of a new virus to which the population has no significant immunity, and characterized by rapid transmission across the world, often out of the usual patterns of seasonality and with high levels of mortality

among healthy young adults. Influenza virus' capacity to continue to cause human disease on such a scale is a function of frequent antigenic changes (see Box 4.15). In the early years after a pandemic, disease is clinically severe, becoming milder as herd immunity improves. At present (winter 2008), late in the inter-pandemic cycle, severe disease is rare.

Pathogenesis

- Virus enters respiratory epithelial cells, replicates and progeny are released – the cell dies. Viral shedding may start within 24 h of infection – illness follows 24 h later.
- There is diffuse inflammation of the trachea and bronchi with an ulcerative, necrotizing tracheobronchitis in severe cases. Primary viral pneumonia is uncommon but is severe when it occurs. Bacterial superinfection is common, facilitated by damage to the mucociliary escalator, and virus-induced defects in lymphocyte and leucocyte function. Viral levels fall rapidly after 48 h of illness, becoming undetectable by 5–10 days.

Box 4.15 Antigenic variation

Alteration of viral antigenic structure allows the production of variants to which there may be little or no herd immunity. This variation involves changes in the HA and NA glycoproteins and takes place by two mechanisms:

- **antigenic drift** – relatively minor changes that occur every year or so through a gradual accumulation of amino acid changes. There is a selective pressure for those changes which are less well recognized by host antibody and these viruses begin to predominate
- **antigenic shift** – major antigenic change that may herald flu pandemics due to lack of population immunity. Type A viruses are able to infect a variety of species: humans, pigs, horse, birds. Luckily viruses adapted for birds are fairly limited in their ability to replicate in humans (divergent evolution?). However, evidence suggests pandemics have been caused by viruses containing both human and avian viral sequences, probably facilitated by recombination within a third party (e.g. pig) easily infected by both human and avian viruses. This is consistent with the observation that pandemics often arise in Asia where humans, pigs, and birds live in close proximity. Other pandemics may have been caused by direct viral adaptation to humans (1918 pandemic probably due to a pig virus).

Influenza – clinical features and diagnosis

Clinical features

- Uncomplicated disease – 1–2 day incubation is followed by an abrupt onset of symptoms. Fever, chills, headache, malaise, myalgia, eye pain, anorexia, dry cough, sore throat, and nasal discharge. After around day 3, respiratory features dominate as fever and other systemic features settle. Convalescence may take two or more weeks. Elderly patients may present with fever and confusion and few respiratory features. Attack rates are highest in children, who may present with croup.
- Respiratory complications – outside of a pandemic these complications are more common in the elderly. Primary viral pneumonia occurs more commonly in those with pre-existing cardiac and lung disorders and presents with worsening cough, breathlessness, and cyanosis shortly after disease onset (resembles acute respiratory distress syndrome). Mortality is high. Secondary bacterial pneumonia develops shortly after an initial period of improvement following the influenza syndrome. Pathogens: *Strep. pneumoniae, H. influenzae* and less commonly *S. aureus*. Other respiratory complications include croup, COPD, and cystic fibrosis exacerbations.
- Immunocompromised patients – more severe disease may occur in some groups of HIV-infected patients, but this has not been widely observed. HIV-infected children with low CD4 counts may shed virus for prolonged periods. Severe disease does occur in immunosuppressed children with cancer, bone marrow transplant recipients, and those with leukaemia. Viral shedding may be prolonged in these groups.
- Non-pulmonary complications – myositis, pericarditis, myocarditis, toxic shock syndrome, encephalitis, GBS.

Diagnosis

In the context of a community outbreak, the diagnosis of influenza can be made with some confidence on clinical criteria alone – 85% accuracy in some studies. Outside of outbreaks or in the institutional setting, laboratory diagnosis is required.

- Viral culture – virus is readily isolated from sputum, throat, or nasal swabs. It is cultured in cell lines and detected within 3–5 days by its cytopathic effect.
- Viral antigen detection – rapid detection within 1–2 days is possible with immunofluoresence or ELISA. Sensitivity approaches that of culture. Reverse transcription (RT)-PCR techniques are in increasing use.
- Serology – acute and convalescent (10–20 days apart) samples showing a fourfold rise in antibody titre can be considered diagnostic but are not useful in clinical decision making.

Box 4.16 Avian influenza H5N1

H5N1 has been around since the 1950s when it killed a number of chickens in Scotland. No human cases occurred. In 1997 18 people were infected and six died during a H5N1 outbreak among poultry in Hong Kong. From 2003, cases of H5N1 human infection have been reported, predominantly in southeast Asia, most related to direct or close contact with H5N1-infected poultry or H5N1-contaminated surfaces. Person-to-person spread has been reported extremely rarely. This is probably a consequence of differences in binding preference demonstrated by the avian and human flu viral HA molecules. Avian HA tends to prefer sialic acid (2–3) galactose which, in humans, is found in the terminal bronchi and alveoli. Human viruses prefer sialic acid (2–6) galactose found on epithelial cells of the upper respiratory tract. The case fatality rate has increased with time (well over 50%), although it is likely that many milder cases have gone unreported. Non-respiratory symptoms (e.g. diarrhoea) have been reported. Death is a result of respiratory failure probably due to a severe 'cytokine storm' effect precipitated by the virus. H5N1 remains much better adapted to birds than other hosts, and the precise nature of the genetic alterations necessary to produce a human flu strain capable of causing a pandemic cannot be known in advance. The virus that has caused human illness and death in Asia is resistant to amantadine and rimantadine. Oseltamavir and zanamavir are probably effective but there is evidence that resistance develops rapidly with their use. There is no human vaccine available as yet. Cases of H5N1 infection in wild birds and poultry have occurred across Europe, the Middle East, Asia, and parts of Africa. Human cases have occurred in parts of Europe and North Africa, as well as Asia.

- Up to date information: www.who.int/csr/disease/avian_influenza/en/
- UK flu pandemic action plans: www.dh.gov.uk/PandemicFlu/fs/en

The 2009 H1N1 swine influenza pandemic

In March 2009 an outbreak of H1N1 influenza A was detected in Mexico. By June WHO had issued its highest level pandemic alert: widespread community transmission on at least two continents. The virus was a reassortment of 2 swine strains, 1 human and 1 avian flu strain. It quickly displaced the seasonal flu variant in Europe and the US, accounting for over 99% of sub-typed influenza A isolates at its peak. The rate of infection was highest in those under 24 years of age with infection proving uncommon in the over 65s, possibly as a result of pre-existing immunity gained in pre-1957 infections. Hospitalisation rates among confirmed cases in the USA were around 2–5% at the peak of the 2009 outbreak - taking into account unreported mild cases the overall rate is thought to be around 0.3%. The highest death rates once hospitalised were seen in adults over 50 years. Those most at risk of severe illness were those with co-morbidities (chronic lung disease, immunosuppression, cardiac disease, diabetes), the obese, and the pregnant. Prophylaxis and treatment guidelines have evolved rapidly as more data becomes available and as the level of transmission varies. The UK 2009/10 pandemic saw negligible oseltamivir resistance. Vaccination is the key prevention strategy and has been recommended for those at risk of severe disease, health care workers and children aged between 6 months and 5 years. For the latest guidance see www.hpa.org.uk.

Influenza – treatment and prevention

Treatment

- The cornerstone of influenza management is an effective vaccination strategy (see below). With this understanding, guidance is produced in the UK by the National Institute for Health and Clinical Excellence (NICE) on the use of antiviral agents in treatment and prevention (Box 4.17). Amantadine (see p.132) is licensed but no longer recommended for the treatment or prophylaxis of influenza. A. Zanamivir (see p.132) is an inhaled neuraminidase inhibitor and used for both prophylaxis and treatment of types A and B. NB - A study[1] of influenza A isolates collected in late 2007 and early 2008 demonstrated that the global prevalance of oseltamivir resistance among H1N1 viruses was 6.4% - a large increase from the previous season. By January of 2009 the overwhelming majority of UK H1 isolates had become resistant. H3 isolates were mostly sensitive and *both* remained susceptible to Zanamivir.
- General measures – adequate hydration, antipyretics (not aspirin in children), and decongestants.
- Severe disease – there is little information available on the effectiveness of antiviral agents in severe disease. The rapidly progressive nature of secondary bacterial infections argues for early presumptive use of antibiotics where suggested by the clinical scenario.

Prevention

Vaccines

- Inactivated vaccines are the main control measure and the Department of Health (DH) recommends annual vaccination for certain risk groups. These are: those with diabetes, immunocompromise, cardiac, lung or renal disease, all people over 65 years of age, and healthcare workers.
- They are prepared each year and are usually trivalent, containing two type A, and one type B strain. Strains are collected continuously across the world, and those to be included in the year's vaccine chosen by educated guess. It takes 6–9 months to produce a vaccine once its components are decided.
- Two doses are required in children under 9 years who have not been previously vaccinated. Otherwise a single dose is sufficient, usually given in October. The main contraindication to vaccination is hypersensitivity to hens' eggs
- Protection is around 70% and lasts for 1 year. Diminished responses are seen in organ transplant recipients receiving immunosuppressive therapy. Protection is reduced in the elderly.

Antiviral agents

- See NICE guidance (Box 4.17).

Box 4.17 NICE guidance on antiviral agents in the treatment and prophylaxis of influenza infection where either influenza A or B are known to be circulating in the community

Treatment

- Amantadine is not recommended for the treatment of influenza.
- Zanamivir and oseltamivir are recommended for the treatment of influenza *only* in children or adults considered to be 'at risk'. 'At-risk' groups are defined as those with chronic respiratory disease (including asthma and COPD), significant cardiovascular disease (not isolated hypertension), chronic renal disease, immunocompromise, diabetes mellitus, and those aged 65 years or older.
- Where indicated, zanamivir and oseltamivir are recommended for the treatment of at-risk adults, and oseltamivir is recommended for the treatment of at-risk children who present with flu-like illness and can start therapy within 48 h of symptom onset.

Prophylaxis

- Amantadine is not recommended for either post-exposure or seasonal prophylaxis of influenza.
- Oseltamivir is recommended for the post-exposure prophylaxis of influenza in at-risk people (see above) aged 13 years or older who are not effectively protected by vaccination and who have been exposed (close contact, e.g. same home environment) to someone with influenza-like illness and are able to begin prophylaxis within 48 h of exposure.
- People who are not effectively protected by vaccination include those who have not been vaccinated since the previous influenza season, those for whom vaccination is contraindicated, or has yet to take effect, and those who have received a vaccine that is not well matched to the strain of influenza virus circulating.
- Regardless of vaccination status, oseltamivir is recommended for the post-exposure prophylaxis of influenza in at-risk people, aged 13 years or older who can begin prophylaxis within 48 h, if they live in a residential care establishment where a resident or staff member has a flu-like illness.
- Oseltamivir is not recommended for post-exposure prophylaxis in healthy people aged under 65 years.
- Oseltamivir is not recommended for the seasonal prophylaxis of influenza.

Further reading

National Institute for Health and Clinical Excellence. www.nice.org.uk (accessed 4 August 2008).

Reference

1. Sheu TG et al. Surveillance for neuraminidase inhibitor resistance among human influenza A and B viruses circulating worldwide in 2004–2008. *Antimicrobial Agents Chemotherapy*, July 14 2008

Parainfluenza

A group of five viruses causing a spectrum of respiratory illnesses, from upper respiratory tract symptoms in healthy children to severe pneumonia in the immunosuppressed.

The viruses

- *Paramyxoviridae* and members of either genus *Paramyxovirus* (parainfluenza types 1 and 3) or *Rubulavirus* (types 2, 4A and 4B).
- Negative sense ssRNA viruses, spherical in appearance with a host cell derived envelope.

Epidemiology

- Sixty per cent of childhood croup cases in which virus is isolated are due to parainfluenza. Second only to RSV as a cause of childhood respiratory hospitalization; 10% of adult acute respiratory illness can be attributed to parainfluenza. Nosocomial and residential care outbreaks have occurred.
- Parainfluenza type 3 is the most frequently isolated member, and like RSV is commonly seen in the first 6 months of life. It is endemic and may be isolated throughout the year, peaking in Spring.
- Types 1 and 2 occur in autumn, often alternating each year.
- Type 4 viruses are rarely isolated.

Pathogenesis

- Transmission is by droplet spread. Replication occurs in cells of the respiratory epithelium. Clinically, illness most frequently involves larger airways of the lower respiratory tract, causing croup.
- Re-infection may occur and tends to cause milder upper airway disease, probably representing waning of immunity – antigenic variation is not progressive (unlike influenza virus). Mucosal immunity is most important for resisting infection. CD8 T cells are important in viral clearance.

Clinical features

- Healthy individuals – children: upper respiratory tract illness (those under 5 years), otitis media, croup, bronchiolitis (infants under 6 months); adults: upper respiratory tract infection.
- Immunocompromised – severe disease in recipients of BMT or lung transplants in all age groups. May cause pneumonia with high rates of mortality. Other immunodeficiencies are associated with prolonged viral shedding.

Diagnosis

- Viral isolation by tissue culture and immunofluorescence is the standard.
- Multiplexed RT-PCR-based tests are faster and can distinguish viral type.
- Paired serology can confirm a diagnosis but is unhelpful.

Treatment

- No specific antiviral therapy
- Some advocate inhaled steroids for the clinical treatment of croup

Respiratory syncytial virus

RSV is a major cause of lower respiratory tract infection (LRTI) in young children.

The virus

- A member of the *Paramyxoviridae* family. An enveloped (bilipid layer derived from the host cell) ssRNA virus which may survives up to 24 h in patient secretions depositing on non-porous surfaces, and around an hour on porous surfaces (tissues, fabric, skin).
- RSV isolates fall into two antigen groups (A and B), which differ in envelope proteins and non-structural protein-1. Several strains from both groups may circulate in the same outbreak.

Epidemiology

- RSV infection has been found worldwide. In temperate parts, outbreaks are annual, occurring in the winter months. Mild outbreaks may be followed by a more-severe one the next year.
- The major cause of childhood pneumonia/bronchiolitis (90% of children admitted with a LRTI in the peak of an epidemic); 95% of children are seropositive by age 2 years. Naturally acquired immunity to RSV is incomplete but subsequent infections rarely produce severe illness.
- Boys and those under 2 years experience the most-severe illness – severity is also affected by socioeconomic factors.
- An important nosocomial infection – virus may spread from an infected infant (secretions, staff) or an infected adult with mild symptoms.

Pathogenesis

- Incubation between 2 and 8 days. Inoculation is by nose and eye, with infection confined to the respiratory tract. Infants often have evidence of pneumonia as well as bronchiolitis.
- Lymphocytic infiltration of the areas around the bronchioles with wall and tissue oedema is followed by proliferation and necrosis of the bronchiolar epithelium: bronchiolitis. Sloughed epithelium and mucus blocks small airway lumens leading to air trapping and hyperinflation. Air absorbed distal to obstructed airways leads to multiple areas of atelectasis.
- Disease is due to the vulnerability of the small airways of the very young to inflammation and obstruction (resistance to air flow being inversely related to the cube of the radius), as well as immunopathology, perhaps immune complex formation. The severest forms are experienced by infants when maternally derived specific antibody is at a high level, and severe disease was seen in children vaccinated with a trial inactivated vaccine in the 1960s.

Clinical features

- Young children – pneumonia and bronchiolitis (see 📖 p.652) are the commonest manifestations of RSV infection in infants. Tracheobronchitis and croup are less common. All may occur in association with fever and otitis media (RSV is present in 75% of middle ear effusions from children with respiratory RSV infection). Rarely asymptomatic. Those with LRTI may have a preceding URTI with nasal congestion and pharyngitis. Cough and fever are common in young children Clinical findings include wheeze and crepitations. It is difficult to differentiate pneumonia from bronchiolitis, and many infants have both. Minimal CXR changes regardless of severity. Hypoxia may be profound. Duration of illness is 7–21 days. Acute complications: apnoea, secondary bacterial infection. RSV has been shown to be a contributing factor to sudden infant death syndrome. Long-term studies suggest those hospitalized with RSV LRTI may have a higher rate of later reactive airway disease.
- Older children and adults – a severe 'common cold' with nasal congestion, cough, fever, earache (in children); <50% of infected older people develop pneumonia (particularly those in residential homes); 2–5% of hospitalized adults with pneumonia have RSV. Secondary infections cause URTI or tracheobronchitis.
- Severe disease – young infants, the premature and those with underlying disease (congenital and cardiopulmonary disease, e.g. cystic fibrosis) are at risk of severe RSV; <66% of deaths occur in those with underlying disease. Prematurity is a risk into the third year of life. Immunodeficient patients (including those with transplants and on chemotherapy) have extensive pulmonary infiltration and prolonged viral shedding.

Diagnosis

- Clinical diagnosis can be made with some confidence in children during an outbreak. Serology is only useful epidemiologically.
- Cell culture – NPA provides the best sample with a high rate of virus isolation. It should be inoculated into cell lines as soon as possible. Infection is characterized by the typical syncytial appearance, and the cytopathic effect is visible at around day 3–7. The major advantage of culture techniques is they can identify other pathogens.
- Rapid tests – immunofluorescence antibody test (IFAT), PCR, and enzyme immunoassays are all available.

Treatment

- Supportive care – oxygen to maintain saturation at 92% or above. Other therapies of potential benefit: heliox, inhaled nitric oxide. No specific benefit shown with bronchdilators, steroid or antibiotics.
- Ribavirin – indicated for RSV LRTI in hospitalized infants considered at high risk of complicated or severe disease (underlying cardiac, pulmonary, or immunosuppressive conditions). Given as an aerosol for 8–20 h each day for 2–5 days. Follow-up over a year shows better pulmonary function and a reduced incidence of reactive airway disease at 1 year in those treated with ribavirin.

Prevention

- Immunotherapy – active immunization is not available. Palivizumab (RSV monoclonal antibody) reduces morbidity in infants at risk of severe RSV. Administered to those at risk once a month during outbreaks, it significantly reduces disease severity and hospital admissions for respiratory illness. It has a role in prophylaxis to the high-risk exposed in hospital.
- Infection control – vital in hospitalized cases. Handwashing, eye–nose goggles, and glove use reduce nosocomial inections. Infected patients should be isolated or cohorted, especially on wards with high-risk patients.

Other respiratory viruses

Coronavirus

- Enveloped viruses of the family *Coronaviridae*. Large positive-sense single-stranded RNA genome. They are found worldwide.
- The name derives from Latin 'corona' (crown) reflecting the EM appearance of the viral spike protein that populates the surface of the virus and determines its host tropism.
- Human respiratory strains cause colds in adults, have been isolated from infants with pneumonia, are associated with bouts of wheeze in children with asthma or recurrent bronchitis, and may be a contributor to exacerbations of COPD. Enteric viruses may cause gastroenteritis in infants and have been associated with outbreaks of necrotizing enterocolitis.

SARS-coronavirus

- Severe acute respiratory syndrome (SARS) was recognized in China in November 2002 and had spread to affect 29 countries across the world by February 2003. The epidemic had died out by July 2003; 8422 cases were reported with a fatality rate of 11% (43% in those over 60 years of age). Between July 2003 and May 2004 there were four small and rapidly contained outbreaks of SARS, three of which were associated with laboratory releases and the fourth thought to be due to an animal source. The possibility of SARS re-emergence remains.
- It is caused by a novel coronavirus. Animals are thought to be the main reservoir. Transmission is by droplets and contact with contaminated surfaces – nosocomial transmission was common in the early stages of the outbreak. Virus is present in stool and may cause diarrhoea. Incubation is 2–10 days.
- Inter-epidemic case definitions proposed by the WHO and adopted by the HPA define a possible case as an individual meeting the clinical criteria, within 10 days of onset of illness with *either* a history of travel to an area classified by WHO as a potential zone of re-emergence of SARS (mainland China, and Hong Kong SAR) *or* a history of exposure to laboratories or institutes which have retained SARS virus isolates and/or diagnostic specimens from SARS patients.

- Clinical criteria: fever of ≥38°C, one or more symptoms of lower respiratory tract illness (cough, difficulty breathing, shortness of breath), radiographic evidence of pneumonia or ARDS, or autopsy findings consistent with the pathology of pneumonia or ARDS without an identifiable cause and no alternative diagnosis to fully explain the illness. Note that this definition is for public health, not diagnostic purposes.
- Confirmation is by PCR, seroconversion, or virus isolation, and details of these and further case definitions are available on the HPA website.[1]

Metapneumovirus

- A newly identified virus (family *Paramyxoviridae*) first reported in June 2001 as a cause of respiratory tract disease in Dutch children.
- Clinical features are indistinguishable from those caused by RSV and it is now known to be a cause of respiratory tract disease in both children and adults worldwide; 98–100% of people are seropositive by age 10 years, it is a significant pathogen in LRTI of children and is implicated in nosocomial spread of infection in hospital wards.
- Symptoms are identical to those of RSV in children (mild URTI to severe cough, bronchiolitis, and pneumonia) and older adults (cough, fever, respiratory distress).
- Diagnosis is by PCR of respiratory secretions or serology. There is as yet no specific treatment (experimental evidence suggests a role for ribavirin) or vaccine.

Reference

1 Health Protection Agency. *Severe acute respiratory syndrome (SARS)*. http://www.hpa.org.uk/ accessed Sept. 2008.

Measles

An acute highly infectious disease of children characterized by cough, coryza, fever, and rash.

The virus

- A member of the family *Paramyxoviridae*, genus *Morbillivirus*; ssRNA, enveloped virus; covered with short surface projections: the haemagglutinin (H) and fusion (F) glycoproteins.
- Humans are the only natural host.

Epidemiology

- Found in every country in the world. Without vaccination, epidemics lasting 3–4 months would occur every 2–5 years.
- Airborne, spread by contact with aerosolized respiratory secretions, and one of the most communicable of the infectious diseases. Sensitive to light and drying but can remain infective in droplet form for some hours.
- Patients are most infectious during the late prodromal phase when coughing is at its peak. Immunity after infection is lifelong.

Pathogenesis

- Virus invades the respiratory epithelium and local multiplication leads to viraemia and leucocyte infection. Reticulo-endothelial cells become infected and their necrosis leads to a secondary viraemia. The major infected blood cell is the monocyte.
- Tissues that become infected include the thymus, spleen, lymph node, liver, skin, and lung. Secondary viraemia leads to infection of the entire respiratory mucosa with consequent cough and coryza. Croup, bronchiolitis, and pneumonia may also occur.
- Koplik's spots and rash appear a few days after respiratory symptoms – may represent host hypersensitivity to the virus.

Clinical features

- Incubation is 2 weeks (longer in adults than children). A prodromal phase (coinciding with secondary viraemia) of malaise, fever, anorexia, conjunctivitis, and cough is followed by Koplik's spots (blue-grey spots with a red base classically found on the buccal mucosa opposite the second molars. Severe cases may involve the entire mucosa) then rash. Patients feel most ill around day 2 of the rash. From late prodrome to resolution of fever and rash is around 7–10 days. Rash begins on the face and proceeds down involving palms and soles last. Erythematous and maculopapular and may become confluent. It lasts around 5 days and may desquamate as it heals.
- Complications – bacterial superinfection (pneumonia and otitis media); acute encephalitis – seen in 1 in 2000 and probably due to host hypersensitivity to virus. Characterized by fever recurrence, headache, seizures, and consciousness changes during convalescence; SSPE – chronic degenerative neurological condition occurring years after measles due to persistent CNS infection with measles virus despite a vigorous host immune response; spontaneous abortion and premature labour – unlike rubella there is no association with fetal malformations but the disease can be more severe in pregnancy and infants can

acquire it. Infants born to mothers with active infection should be given immunoglobulin at birth.

- Prior to vaccination measles was a common cause of viral meningitis and remains so in unvaccinated populations, usually occurring with rash.
- Special conditions:
  - modified measles – a very mild form of the disease seen in people with some degree of passive immunity, e.g. those receiving immunoglobulin or babies under 1 year.
  - atypical measles – seen in those who received early killed measles vaccines and are later infected by wild-type virus. The rash is atypical and may resemble Henoch–Schönlein purpura (HSP), varicella, spotted fever, or a drug eruption. High fever, peripheral oedema, pulmonary infiltrates, and effusions may occur. The disease is more severe, has a longer course, and is thought to be due to hypersensitivity to virus in a partially immune host. It is rare now but those who have received only killed vaccine should be offered live.
  - immunocompromised patients (including the malnourished) may experience severe disease, e.g. primary viral giant cell pneumonia, encephalitis, SSPE-like encephalitis. They may not develop rash, making diagnosis difficult. Immunocompromised people should be passively immunized following exposure even if previously vaccinated.

Diagnosis

Diagnosis can usually be made clinically. Lab confirmation is useful in atypical cases or the immunocompromised.

- Virus isolation – possible in renal cell lines, growth slow. Useful in the immunodeficient where antibody responses may be minimal.
- Serology – a fourfold increase in measles antibody titre between acute and convalescent specimens is diagnostic. ELISA is capable of detecting specific IgM on a single sample.
- Other – immunofluoresent microscopy of cells in secretions, RT-PCR.

Treatment

- Supportive therapy and treatment of bacterial superinfection.
- Vitamin A 200,000 IU given orally to children for 2 days has been shown to reduce severity but should be avoided at or after immunization when it appears to reduce seroconversion.

Prevention

- Measles vaccine is given as part of measles, mumps, rubella (MMR) (12 months and preschool). It can be given earlier in at-risk populations but responses are suppressed and an additional dose should be given later.
- Passive immunization with immunoglobulin is recommended for those exposed susceptible people at risk of severe or fatal measles – it must be given within 6 days of exposure to be effective. Such groups include:
  - children with defects in cell-mediated immunity
  - children with malignant disease – particularly if receiving chemo- or radiotherapy
  - children with HIV should be given immunoglobulin after exposure even if already vaccinated – cases have occurred in vaccinated HIV-infected children.

Mumps

Mumps is an acute generalized viral infection of children and adolescents causing swelling and tenderness of the salivary glands and rarely epididymo-orchitis. More-severe manifestations are commoner in older patients. Mumps was recognized as far back as Hippocrates. The name may derive from the English verb 'to mump' – to be sulky.

The virus

- A member of the *Paramyxoviridae* family; single-stranded RNA virus, irregular spherically shaped virion (average diameter 200 nm), nucleocapsid is enclosed by a three-layer envelope.
- Nucleocapsid contains the S (soluble) antigen, antibodies to which may be detected early in infection.
- Glycoproteins on the surface have haemagglutinin, neuraminidase, and cell-fusion activity, and include the V (viral) antigen detected in late infection by complement fixation.

Epidemiology

- Endemic throughout the world. Prior to vaccination, epidemics took place every 2–5 years with 90% of cases occurring in those under 15 years.
- In the US one-third of cases occur in those over 15 years. UK incidence has been increasing among those born between 1981 and 1989 who missed routine MMR (first introduced in 1988) and are now at university. In 2004 there was a dramatic increase of cases in England and Wales among those born before 1987, many of whom had received just one dose of MMR in the 1998 "catch-up" campaign.
- Passive immunity makes infection uncommon in children under 1 year.

Pathogenesis

- Transmitted by droplet spread or direct contact. Most infectious just before parotitis.
- During incubation the virus proliferates in the upper respiratory tract with consequent viraemia and localization to glandular and neural tissue.
- Parotid glands show interstitial oedema and serofibrinous exudate with mononuclear cell infiltration. Cases of orchitis are similar with the addition of interstitial haemorrhage, polymorphonuclear infiltration, and areas of local infarction due to vascular compromise.

Clinical features

- Incubation is 2–4 weeks. A 24-h non-specific prodrome of fever, headache, and anorexia is followed by earache and ipsilateral parotid tenderness. The gland swells over 2–3 days and is associated with severe pain. Swelling can lift the ear lobe up and outward. The other side follows within a couple of days in most cases – unilateral parotid involvement is seen in 25%. Patients experience difficulty in pronunciation and mastication and may develop fever. Once swelling has peaked recovery is rapid – within a week. Complications of parotitis (e.g. sialectasia) are rare. Other salivary glands may be involved.

- **CNS involvement** – the commonest extra-glandular manifestation in children:
  - meningitis is seen in <10% of those with parotitis although <50% of cases of mumps meningitis show no evidence of glandular disease. Onset is 4 to 7 days after glandular symptoms but can occur 1 week before or 2 weeks later. Men are affected more than women. Symptoms resolve 3–10 days later and recovery is complete with no sequelae. CSF findings – typical of viral meningitis (p. 758) – hypoglycorrachia (CSF glucose <40 mg/100 mL) is seen in up to 30% of cases, more than other viral meningitides
  - encephalitis is seen in 1 in 6000 and takes two forms: early onset which represents direct neuron damage due to viral invasion, and a larger late-onset (7–10 days) group representing a post-infectious demyelinating process. Recovery takes around 2 weeks and sequelae (e.g. psychomotor retardation) and death (around 1.4% of cases) may be seen
  - other neurological manifestations include transient deafness, permanent deafness (1 in 20 000), ataxia, facial palsy, transverse myelitis, GBS.
- **Other extraglandular manifestations**:
  - presternal pitting oedema and tongue swelling (thought to be due to lymphatic obstruction by swollen regional glands) (6%)
  - epididymo-orchitis – the commonest extra-glandular manifestation in adults – 20–30% of post-pubertal males with mumps develop it (1 in 6 of those bilateral). Rare before puberty. It may be the only manifestation of mumps. Onset is abrupt with fever and a warm, swollen (up to four times normal) tender testicle with erythema of the overlying skin. Fever resolves at 5 days with gonadal symptoms following. Some degree of atrophy may be seen in 50% once recovered. Infertility is rare
  - oophoritis – seen in 5% of post-pubertal women with mumps – impaired fertility and premature menopause have been reported but are rare
  - other manifestations – migratory polyarthritis, pancreatitis, myocarditis, nephritis, thyroiditis, mastitis, hepatitis.

Diagnosis

- Lab confirmation is required for epidemiological purposes or when disease is atypical
- General features – exposure history and symptoms; leucocytosis may be seen, particularly with meningitis, orchitis, or pancreatitis; serum amylase is elevated in parotitis or pancreatitis (isoenzyme analysis is required to differentiate the source).

- Serology – most reliably determined using ELISA for IgM. Haemagglutination inhibition (convalescent serum prevents the adsorption of chick red cells to mumps-infected epithelial cells) can be positive with antibodies to parainfluenza 3 (another cause of parotitis).
- Virus isolation – present in saliva from 2 days before symptom onset to 5 days after. May be present in CSF up to 6 days after onset.

Treatment

- Symptom control – antipyretics and fluids if persistant vomiting
- No benefit in steroid use has been demonstrated
- Anecdotal evidence that interferon-alfa speeds resolution of orchitis

Prevention

Vaccination is more than 95% effective and takes place at 12–15 months and preschool as part of MMR.

Rubella

An acute mild exanthematous viral infection of children and adults resembling mild measles but with the potential to cause fetal infection and birth defects.

The virus

- A member of the *Togaviridae* family, in the genus *Rubivirus*.
- Spherical in shape with a diameter of 60 nm. Relatively unstable.

Epidemiology

- Unlike measles, rubella is only moderately contagious. Prior to vaccination incidence was highest in the spring among children aged 5–9 years.
- Once termed 'third disease', measles and scarlet fever being the first and second exanthematous infections in childhood.
- After infection or vaccination most people develop lifelong protection against disease. Re-infection occurs (the majority asymptomatic) as demonstrated by rises in antibody titre in previously vaccinated people.
- There have been rare cases of congenital rubella acquired through the re-infection of a vaccinated mother.

Pathogenesis

- Spread is by droplets; patients are at their most contagious when the rash is erupting and virus may be shed from 10 days before to 2 weeks after its appearance.
- Rash may be immune mediated – it appears as immunity develops and viral titres fall.
- Primary viraemia follows infection of the respiratory epithelium; secondary viraemia occurs a few days later once the first wave of infected leucocytes release virions.
- Infants with congenital rubella shed large quantities of virus for many months.

Clinical features

- Incubation is 12–23 days. Postnatal rubella is a mild infection. Many cases are subclinical. Adults may experience a prodrome of malaise, fever, and anorexia. The main symptoms are lymphadenopathy (cervical and posterior auricular) and a maculopapular rash (starting on the face and moving down) which may be accompanied by coryza and conjunctivits and lasts 3–5 days. Splenomegaly can occur.
- Complications are uncommon – arthritis affecting the wrists, finger and knees and resolving over a month may be seen as the rash appears (women more than men); haemorrhagic manifestations occur in 1 in 3000 (children more than adults) and may be due to thrombocytopenia as well as vascular damage; encephalitis occurs in 1 in 5000 (adults more than children), with a mortality of up to 50%.
- Congenital rubella – can be catastrophic in early pregnancy leading to fetal death, premature delivery, and many congenital defects. Rare since the introduction of vaccination. The younger the fetus when infected, the more severe the illness. In the first 2 months of gestation there is an up to 85% chance of being affected by either multiple defects or spontaneous abortion. In the third month there is around a 30% chance of developing a single defect (e.g. deafness), dropping to 10% in the fourth month and nil after 20 weeks. Temporary defects include low birth weight, low platelets, hepatosplenomegaly, hepatitis, meningitis, and jaundice. Permanent defects include hearing loss, cardiac abnormalities, microcephaly, inguinal hernia, cataract, and glaucoma. Developmental defects may become apparent as the infant grows, e.g. myopia, mental retardation, diabetes, behavioural and language disorders.

Diagnosis

- Its mild nature makes clinical diagnosis difficult.
- Serology – positive IgM on a single sample or a fourfold rise in IgG in paired sera is diagnostic. IgM may be positive in cases of re-infection. Serological diagnosis of congenital rubella in neonates may necessitate the analysis of several samples over time to determine whether antibody titres are falling (maternal antibody) or rising (recent infection). Detection of rubella IgM in a newborn's serum indicates infection.
- Intrauterine diagnosis has been made by placental biopsy and by cordocentesis with detection by PCR.

Treatment

- No treatment is indicated in most cases of postnatal rubella.
- Immunoglobulin used to be given to exposed susceptible pregnant women – however it does not prevent viraemia despite suppressing symptoms.
- All women of child-bearing age should be vaccinated before pregnancy.

Prevention

Vaccination achieves a seroconversion rate of 95%. Women should not become pregnant in the 3 months following vaccination.

Parvovirus B19

Parvovirus B19 has a wide variety of clinical manifestations depending on the state of the host: in immunocompetent children, 'slapped-cheek' disease, in those with underlying haemolytic disorders, an aplastic crisis. Parvovirus B19 was found in 1974 whilst evaluating assays for hepatitis B surface antigen using panels of serum samples – sample 19 in panel B gave a 'false-positive' and EM revealed the guilty virus.

The virus

- A member of the family *Parvoviridae*, genus *Erythrovirus* (so called because replication occurs only in human erythrocyte precursors). B19 remains the only known human pathogenic parvovirus. They are non-enveloped and extremely resistant to physical inactivation.

Epidemiology

- Infection common in childhood – 50% are IgG-positive by 15 years, 90% antibody positive by 90 years. Infected children pass virus on to uninfected members of their family. Patients are infectious from 24–48 h before viral prodrome until rash appearance.
- In temperate climates infection is commonest from late winter to early summer. Rates peak every 3 to 4 years. Prevalence higher in Africa.
- Infection may be passed vertically, by respiratory secretions or from blood and blood products (standard thermal treatments and solvent-detergents are not completely effective) although viraemia is rare.

Pathogenesis

- Parvovirus B19 infects erythroid progenitors and erythroblasts. The receptor it uses to infect cells – P antigen – is found on megakaryocytes, endothelial cells, fetal myocardial cells, and erythroid precursors. Those who lack P antigen on their erythrocytes are resistant to infection.
- Infected individuals may experience an acute, self-limited (4–8-day) halt in red blood cell manufacture as infected cells are destroyed. This may be unnoticed In those with normal erythroid turnover (falls in haemoglobin >1g/dL are uncommon) but in those with a high turnover (e.g. haemoglobinopathies, haemolytic anaemia) falls of 2–6 g/dL are not uncommon and aplastic anaemia may develop.
- The infected fetus may have severe manifestations due to high red cell turnover and the immature immune response: anaemia, myocarditis, and heart failure can occur. These effects are reduced by the third trimester.
- Rash and arthralgia associated with some forms of the disease are probably immune complex related.

Clinical features

- 20% of infections are asymptomatic.
- Erythema infectiosum (EI) – the commonest manifestation. The classic slapped cheek-appearing rash (fiery red eruption with surrounding pallor) follows a 5–7-day prodrome of fever, coryza, and mild nausea/ diarrhoea. A second erythematous maculopapular rash may follow on the trunk and limbs 1–2 days later, fading to produce a lacey

appearance. Adults have milder manifestations. Pruritus (especially soles of the feet) can occur.

- Arthropathy – seen in adults, especially women. Symmetric, mainly small joints of hands and feet lasting for 1–3 weeks. May persist or recur for months. May be confused with acute rheumatoid arthritis.
- Transient aplastic crisis (TAC) – the first clinical illness associated with B19 infection – the abrupt cessation of erythropoiesis with absent erythroid precursors in bone marrow. Described in a wide range of haemolytic conditions: sickle cell (nearly 90% of TAC episodes), thalassaemia, pyruvate kinase (PK) deficiency, autoimmune haemolytic anaemia. It has also been seen after haemorrhage, in iron-deficiency anaemia, and those who are otherwise well (who are likely to see deficiencies in other blood lineages: neutropenia, thrombocytopenia). Patients can be severely ill: dyspnoea, confusion, cardiac failure. It does not appear to cause true, permanent aplastic anaemia.
- Pure red cell aplasia (PRCA) – anaemia in the immunosuppressed (HIV, congenital immunodeficiency, patients undergoing transplantation). Administration of immunoglobulin may be beneficial.
- Virus-associated haemophagocytic syndrome – usually healthy patients with cytopenia – characterized by histiocytic hyperplasia and haemophagocytosis. Self-limiting.
- Fetal infection – 10–15% of non-immune hydrops foetalis. Where maternal infection occurs, fetal loss averages 9% and occurs within the first 20 weeks of pregnancy. There is no evidence of long-term abnormality in those who survive.
- Other manifestations – encephalitis, myocarditis, hepatitis, vasculitis, erythema multiforme, glomerulonephritis, idiopathic thrombocytopenia purpura, Henoch-Schönlein purpura.

Diagnosis

- IgM detection – 90% of cases are positive by the time of the rash in EI or by day 3 of TAC. IgM remains detectable for up to 3 months. IgG is detectable by day 7 of illness and remains detectable for life (50% of the population are IgG positive). It is not useful in diagnosing acute infection or in attributing manifestations such as chronic arthropathy to B19.
- Virus detection is possible by DNA hybridization – however, in immunocompetent people viral DNA can only be detected for 2–4 days.
- Immunocompromised people with chronic infection do not mount an immune response and diagnosis relies on detecting DNA by PCR.
- Fetal infection can be confirmed by amniotic fluid sampling, and investigations should include maternal B19 serology.
- Those with aplastic anaemia will show a fall of at least 2 g/dL from baseline Hb.

Treatment

- General measures – e.g. non-steroidal anti-inflammatory drugs (NSAIDS) for arthritis, transfusions for TAC.
- Immunosuppressed patients with persistent infection may benefit from temporary cessation of immunosuppression to allow them to mount an immune response. If this is not feasible, IV IgG over 5 days may help. If disease recurs they may require repeated infusions. Some HIV-infected patients will resolve chronic B19 infection with the initiation of highly active antiretroviral therapy (HAART).
- Intrauterine blood transfusions may help some cases of hydrops.

Prevention

Unlike 'slapped-cheek' patients, those with TAC and PRCA are infectious at presentation and should be separated from high-risk contacts (glove and gown, own room, mask, etc) for 7 days or for the duration of the illness. Pregnant health workers should not care for such patients.

Human herpesvirus type 6

A lymphotropic virus and the single most common cause of hospital visits in infants with fever. The cause of roseola (also known as sixth disease and exanthem subitum).

The virus

- A herpesvirus, originally called B-lymphotropic virus, now shown to grow in many different cell types (T cells, macrophages etc).
- Two subtypes (A and B), which differ in their epidemiology and growth.

Epidemiology

- Nearly all humans infected by age 2 years, probably by saliva exchange.
- Most isolates from healthy people are HHV-6B, the only variant to have been linked to specific clinical syndromes.

Pathogenesis

- Primary infection is via the oropharynx. Regional lymph nodes and mononuclear cells are subsequently infected and virus spreads throughout the body. Incubation is 5–15 days. Like other herpesviruses it causes an initial infection, a lifelong latency, and has the potential for clinical reactivation, especially in hosts who are immunocompromised (see box 4.18). Asymptomatic carriers may continue to excrete the virus for months.
- Specific mechanisms for host immune evasion facilitate persistent infection.

Clinical features

- Infantile fever – fever without rash is the commonest manifestation and may be accompanied by periorbital oedema. 10% of cases in one series of acute febrile illnesses were attributed to HHV-6. Benign febrile convulsions may occur and are more frequent than fever alone explains – perhaps due to viral replication within the CNS.
- Exanthem subitum (6th disease) – an illness of infants and young children. 3–5 days of fever and URT symptoms are followed by the development of rose-pink papules which are mildly elevated, non-pruritic, and blanche on pressure. Rash lasts around 2 days and may be associated with malaise, vomiting, diarrhoea, cough, pharyngitis, and lymphadenopathy. Most infants are asymptomatic.
- Encephalitis – can occur alone or as a complication of exanthem subitum. Virus is frequently detected in the CNS even in the absence of symptoms.
- Immunocompromised hosts – as with other herpesviruses, immune suppression permits replication with the potential for clinical illness. Viral DNA has been isolated in bone marrow transplant recipients with pneumonitis, but often in the presence of other pathogens (such as CMV) with better established pathological pedigrees. There are other anecdotal reports, including associations with certain leukaemias and lymphomas, but no definitive links have been established.

- Other – infectious mononucleosis, hepatitis, chronic fatigue syndrome (causality unproven).

Diagnosis

- Exanthem subitum does not need specific investigation or treatment due to its self-limiting nature. Diagnosis should be confirmed in those patients who are recipients of organ transplants or patients with immunodeficiency, encephalitis, or hepatitis.
- Serology – rarely helpful as nearly everyone is positive by age 2 years. IgM assays are in any case not good indicators of acute infection – paired sera are required. CMV antibodies may cross-react with HHV-6.
- Viral culture – it is very unusual to detect it in the healthy, and high rates of recovery in the immunocompromised regardless of presentation negate its usefulness as a diagnostic tool.
- Quantitative PCR – may prove useful in the future.

Treatment

In vitro data indicates that aciclovir is inactive, ganciclovir responsiveness is variable, and foscarnet is inhibitory. There are no controlled trials.

Box 4.18 HHV-6 and HIV

These viruses have an interesting relationship. Both infect and replicate in CD4+ cells and HHV-6 induces CD4 expression in otherwise CD4-negative lymphocyte populations, rendering them susceptible to HIV infection. It also seems to accelerate HIV 1 transcription and replication. In turn HIV-1 upregulates *HHV-6* gene expression and immune suppression permits HHV-6 spread. However, although there is evidence that HHV-6 disseminates widely in AIDS there is no conclusive evidence that it either causes any opportunistic illness or affects AIDS progression.

Human herpesvirus type 7

Epidemiology

- First demonstrated in 1990 from the peripheral blood mononuclear cells of a healthy adult and researchers are still looking for role in human disease.
- Infects nearly all humans by age 5 years. A commensal inhabitant of saliva.
- Homology to HHV-6 is limited but confused earlier serological studies.

Pathogenesis

- HHV-7 infects activated cord blood and CD4+ cells. CD4 is a receptor for the virus thus HHV-7 interferes with HIV-1 infection.
- Like HHV-6 it encodes genes allowing it to evade the immune system.

Clinical features

- Causation difficult to establish given its ubiquitous nature – probably causes similar fever and rash syndromes as HHV-6.

Herpes simplex virus

The virus

- Herpes viruses are enveloped, containing dsDNA.
- Nine known viruses are similar morphologically but differ clinically and biologically. Classified on this basis into one of three groups:
  - alpha-herpesviruses (HSV-1, HSV-2 and VZV)
  - beta-herpesviruses (CMV, HHV-6 and HHV-7, Simian herpes B)
  - gamma-herpesviruses (EBV and HHV-8).
- HSV genome encodes for around 80 gene products with around 50% sequence homology between HSV-1 and -2.
- HSV is able to infect a wide variety of cells and is cytopathic to those cells in which it completes a full replication cycle, with the exception of certain neuronal cell types. In these cells the virus is capable of establishing latent infection. Only certain viral proteins are produced but reactivation of the viral genome can occur resulting in the production and release of viral particles with consequent infection of adjacent cells.

Epidemiology

- HSV exists worldwide, humans are the only known natural reservoir.
- Nearly all adults have antibodies to HSV-1 by their 40s – prevalence is highest in lower socioeconomic groups.
- HSV-2 antibodies correlate with sexual activity – either that of the individual or their partners – appearing in puberty and being closely related to the number of partners and the presence of a history of STIs.
- HSV-2 seroprevalence is around 22% in the USA (slightly higher in women than men), a little lower than this in the UK and in general higher in developing countries.
- Half of HSV-2 seroconversions are subclinical – more in those who have previously had HSV-1 infection. Many of these will experience symptomatic reactivation.
- Infection with one viral type confers partial immunity to the other.

Pathogenesis

- Infection occurs via close contact with an individual who is shedding virus peripherally. Virus enters through mucosal surfaces or skin breaks and replicates locally, often with no clinical manifestations.
- Viral progeny infect neurons and travel via the axon to the nerve ganglion (e.g. HSV-1 and trigeminal, HSV-2 and sacral nerve root ganglion) where a second phase of replication takes place, virus then spreading peripherally along sensory nerves. This accounts for the large areas that may be involved in clinical disease.

- After primary disease resolution, infectious HSV cannot be detected but viral DNA may be found in up to half of ganglion cells.
- Reactivation mechanisms are not clear – viral subtype, route of infection, and host factors all contribute. Individuals infected with HSV-1 both orally and genitally experience reactivation more frequently orally. On the other hand, HSV-2 local reactivation is 8–10 times more frequent if acquired genitally.
- Individuals with impaired cellular immunity (transplant recipients, patients with AIDS) may develop severe, possibly disseminated life-threatening disease.

Clinical features

- Clinical manifestations are various, influenced by the site of infection, viral type, host age, and immune status. First infections tend to be more severe than reactivation, with more systemic features, longer symptom duration, and a higher rate of complications. Both viral subtypes can cause genital and orofacial infection.
- Cutaneous manifestations (see 📖 Cutaneous manifestations of systemic fungal infection, p.797).
- Visceral infection – viraemia usually results in multiple organ involvement but single organs may be affected. Severe disseminated disease is rare but occurs at increased frequency in women in the third trimester of pregnancy (see 📖 Maternal infections associated with neonatal morbidity, p.826).
  - HSV oesophagitis – may follow disease extension from the pharynx or by viral reactivation via the vagal nerve. Ulceration usually involves the distal oesophagus with retrosternal chest pain and dysphagia. Disease may be extensive and resemble invasive Candida infection
  - pneumonitis – rare. Tends to occur in the severely immunosuppressed; mortality exceeds 80%.
- Encephalitis – >95% of cases caused by HSV-1, incidence peaking in those aged between 5 and 30 years, and those of 50 years old. Higher rates in the immunocompromised. Primary infection may cause encephalitis in children and young adults, virus entering the CNS by neurotropic spread from the periphery. Most adults have evidence of previous infection, and disease can result from reactivation or re-infection with exogenous virus. Onset is usually acute (may be insidious) with a prodrome of headache, behavioural change, and fever. Other symptoms: focal neurological signs (classically temporal lobe), seizures and coma. Diagnosis is by CSF PCR. Electroencephalogram (EEG) and computerized tomography (CT) may demonstrate characteristic focal features but can be normal early in illness. MRI is more sensitive. Brain biopsy for culture and histology are rarely indicated.
- Meningitis – more commonly caused by HSV-2. Cases may complicate genital herpes (women>men) with symptoms following genital lesions by 3–10 days. It is usually benign in the immunocompetent lasting 2–4 days, resolving over 2–3 days. CSF findings: raised white cells (majority lymphocytes, PMN cells in early infection), glucose usually over 50% blood levels but hypoglycorrhachia occurs, raised protein, PCR for HSV DNA is the most sensitive means of diagnosis.

- Other neurological features – sacral radiculopathy, autonomic nerve dysfunction (hyperaesthesia/anesthesia of lower back and sacral area, constipation, urinary retention, transient impotence resolving over 4–8 weeks), transverse myelitis – decreased tendon reflexes and reduced muscle strength in lower extremities, autonomic features.

Diagnosis

- Herpetic ulcerations resemble those of other causes. Lab diagnosis is important to guide therapy where there is doubt.
- Histology – may demonstrate giant cells or intranuclear inclusions.
- Viral isolation – culture allows identification within 48 h.
- Antigen detection – fast. Not as sensitive as viral isolation in asymptomatic patients.
- PCR to detect HSV DNA – very sensitive, particularly when samples have been taken from late-stage lesions and using CSF in CNS infections where culture is less reliable.
- Serology – may be used to identify asymptomatic carriers but changes occur too late to be useful in the diagnosis of acute disease.

Treatment

- Oral aciclovir shortens the duration of primary attacks of cutaneous, oral, and genital herpes. It is less effective against recurrent disease but may be given prophylactically where recurrences are very frequent and in the immunocompromised. Oral valacicovir has better oral bioavailability and is indicated for treatment of primary HSV infections. It is also used in treatment/suppression of recurrent infections in immunocompromised or HIV positive patients.
- Severe and disseminated HSV infection – intravenous therapy.
- Encephalitis – antiviral therapy (IV aciclovir for 10 days) should be given to suspected cases of HSV encephalitis until the diagnosis can be confirmed – it reduces the mortality from nearly 80% to around 25%.

Varicella zoster virus

Varicella zoster virus causes two distinct diseases: a primary infection – chickenpox (varicella), and shingles (herpes zoster) – and the localized recurrence.

The virus

- A dsDNA virus around 200 nm in diameter and a member of the *Herpesviridae* family. The lipid-containing envelope is studded with glycoprotein spikes which are the primary markers for humoral and cell-mediated immunity.
- Spreads between cells by direct contact and may be isolated in many human cell lines – virus can be detected by immunofluorescence 8–10 h after infection.

Epidemiology

- Humans are the only known reservoir.
- Infection is acquired via the respiratory tract and over 90% of people are seropositive by 20 years of age.
- Exposure of a susceptible person to VZV results in chickenpox.
- Infection is common in childhood (90% of cases in those under 13 years) peaking in late winter and early spring. Secondary attack rates in susceptible siblings within a household are around 80%.
- Patients are infectious for 48 h before rash appearance, and 4–5 days after vesicles crust over.
- After infection the virus becomes latent within the dorsal root ganglia.
- Reactivation, causing shingles, occurs in 20% of the population. All ages are affected – highest incidence in the elderly and immunocompromised.
- Mortality – under 2 per 100,000 for children, increasing 15-fold for adults.

Pathogenesis

- After infection and local replication, patients become viraemic.
- Viral replication in the skin precipitates degenerative change of epithelial cells with ballooning and the appearance of multinucleated giant cells and eosinophilic intranuclear inclusions.
- Vesicles contain a cloudy fluid (leucocytes, fibrin, and degenerate cells) and either rupture, releasing infectious virus, or slowly resolve.
- Necrosis and haemorrhage may occur in the upper part of the dermis.
- Histological findings of the skin in chickenpox and herpes zoster are very similar.

Clinical features

Chickenpox

- Incubation is 10–14 days and may be followed by a 1–2-day febrile prodrome before the onset of constitutional symptoms (malaise, itch, anorexia) and rash.
- Skin lesions start as maculopapules (up to 5 mm across), progressing to vesicles which quickly pustulate and form scabs which fall off 1–2 weeks after infection.

- Lesions appear in successive crops over 2–4 days, starting on the trunk and face and spreading centripetally. May rarely involve the mucosa of the oropharynx and vagina.
- Complications – secondary bacterial infection of lesions (Gram-positive organisms) including streptococcal toxic shock, acute cerebellar ataxia (1 in 4000 children under 15 years, onset 7–21 days after rash and resolves over 2–4 weeks), encephalitis (0.2% cases, characterized by depressed consciousness, fever, vomiting, seizures, and may be life-threatening in adults; recovery over 2 weeks; 5–20% experience progressive deterioration and die), cerebral angiitis (after herpes zoster opthalmicus), meningitis, transverse myelitis, pneumonitis (1 in 400 adult VZV cases, often without symptoms; life-threatening in 2nd and 3rd trimester of pregnancy), myocarditis, bleeding, hepatitis. Pregnant women are at increased risk of complicated disease (see Box 4.19)
- Immunocompromised – children (especially those with leukaemia) have many more lesions, often haemorrhagic, with healing extended threefold. There is a greater risk of visceral involvement (lung, liver, CNS). Those undergoing bone-marrow transplantation are at increased risk of infection and nearly half of cases have cutaneous or visceral dissemination. Risks are greatest in those requiring anti-thymocyte globulin or experiencing graft-versus-host disease.

Herpes zoster

- Unilateral vesicular eruption in a dermatomal distribution (most commonly thoracic and lumbar), often preceded by 2–3 days of pain in the affected area. Maculopapular lesions evolve into vesicles with new crops forming over 3–5 days. Resolution may take 2–4 weeks.
- Other manifestations – eyelids (1st or 2nd branch of the trigeminal), keratitis (herpes zoster opthalmicus – sight-threatening and requires ophthalmic referral), intraoral – palate, tonsillar fossa, tongue (maxillary or mandibular branch of trigeminal nerve), Ramsay–Hunt syndrome – pain and vesicles in the external auditory meatus, ipsilateral facial palsy, loss of taste to the anterior two-thirds of the tongue (geniculate ganglion), encephalitis, granulomatous cerebral angiitis (after zoster opthalmicus), paralysis (anterior horn cell involvement).
- Immunocompromised patients – disease is more severe with prolonged lesion formation and recovery with higher risk of cutaneous dissemination and visceral involvement. Disease is rarely fatal. Those with HIV have an increased incidence of complications such as retinitis, acute retinal necrosis, and chronic progressive encephalitis, and may develop chronic herpes zoster.
- Post-herpetic neuralgia – uncommon in young people but occurs in up to 50% of those over 50 years old. Pain can be debilitating and may be constant or stabbing.

Diagnosis

- Vesicular fluid analysis – allows differentiation from certain forms of impetigo (Gram-stain of lesion fluid) and other viral infections (e.g. Coxsackie and HSV) that might present with either widespread or unilateral dermatomal vesicular lesions (PCR, culture or electron microscopy of vesicle contents).

- Demonstrating seroconversion or titre rises between acute and convalescent samples may confirm diagnosis.
- PCR of CSF for VZV DNA can allow diagnosis of CNS infection.

Treatment

Chickenpox

- General measures – hygiene to prevent secondary infection of lesions, management of pruritus to reduce scratching, paracetamol for fever.
- Aciclovir – given within 24 h of onset it significantly reduces the duration and severity of illness but is not recommended for routine use in immunocompetent children:
  - adults presenting within 24 hours may benefit from oral aciclovir; IV therapy should be considered those who present late and show signs of complications or are failing to improve
  - Valaciclovir is not licensed for the treatment of chicken pox but is often used in preference to aciclovir
  - the immunocompromised, neonates, and those with disseminated disease (ophthalmic, encephalitis, pneumonitis) should receive early IV therapy and may also require antibiotics to treat or prevent secondary bacterial infections.

Shingles

- General measures as above
- Opthalmic referral in cases involving the eye
- Oral aciclovir is recommended for:
  - cases occurring outside the thoracic dermatomes
  - where pain is very severe
  - those over 60 years of age
- Oral valaciclovir has better oral bioavailability and is indicated for treatment of herpes zoster
- IV aciclovir may be necessary in the immunocompromised
- Post-herpetic neuralgia is unusual beyond 4 weeks but such cases can be difficult to treat. They usually resolve over 6–24 months. Therapies include tricyclics, counter-irritants, and gabapentin

Prevention

- Live-attenuated vaccines are available. They are recommended for use in non-immune healthcare workers including laboratory staff, and the healthy susceptible contacts of immunocompromised patients where continuing close contact is unavoidable (e.g. siblings of a leukaemic child, or a child whose parent is undergoing chemotherapy). A vaccine has recently been produced for the prevention of shingles in the over 60s (Zostavax® – reduced incidence of shingles by 64% in those aged 60–69 years and reduced post-herpetic neuralgia by 39% in those vaccine recipients who developed shingles).
- Varicella zoster immune globulin (VZIG) does not prevent infection but reduces disease severity. Most effective if given within 72 h. Indicated for seronegative patients with a contact history who are at increased risk of complications: those with defects in cell-mediated immunity, those on significant doses of steroids, those who have received a bone marrow transplant, radiotherapy, or chemotherapy within the last

6 months, organ transplant recipients on immunosuppression, pregnant women, infants under 1 month old whose mothers develop chickenpox between 1 week before and 4 weeks after delivery, or who are exposed to chickenpox and whose mothers have no history of prior infection or are seronegative.

Box 4.19 Infection in pregnancy (📖 p.825)

Pregnant women with varicella are at increased risk of complications. Pneumonitis can be particularly severe and is most likely in the last trimester and in those with COPD, who smoke, are on steroids, or have extensive cutaneous disease.

- If there are signs of pneumonitis together with CXR changes or hypoxia, IV aciclovir should be given. Where the CXR is normal and there are no symptoms patients should be observed and oral aciclovir started if over 20 weeks' gestation.
- If there are no signs of pneumonitis and the rash is less than 24 h old, oral aciclovir should be considered (but with expert advice in those under 20 weeks' gestation).
- If there are no signs of pneumonitis and the rash is over 24 h old, aciclovir should be withheld but the patient reviewed daily.

Infectious mononucleosis

Infectious mononucleosis is a syndrome of sore throat, fever, and lymphadenopathy with atypical lymphocytosis; 80–90% of cases are due to Epstein–Barr virus, which is generally associated with a positive heterophile antibody test (see later). Most of the remainder are caused by cytomegalovirus. Young adults and adolescents are most frequently affected. Other conditions which should be considered are viral hepatitis, acute toxoplasmosis, rubella, streptococcal sore throat, primary HIV, diphtheria.

Epstein–Barr virus

A human herpesvirus, the cause of heterophile-positive infectious mononucleosis and associated with African Burkitt's lymphoma.

The virus

- Enveloped hexagonal nucleocapsids containing dsDNA which encodes around 80 proteins.
- The virus is easily cultivated in human B cells and nasopharyngeal epithelial cells. The cell surface receptor for the virus is the receptor for C3d complement protein.
- Early after infection, Epstein–Barr nuclear antigens (EBNA) are detectable in cell nuclei. Viral DNA may become incorporated into host DNA in transformed cells but most remains in a circular non-integrated form.
- The virus remains latent in most infected cells.

Epidemiology

- Worldwide. Infection acquired earlier in tropical countries than industrialized. By adulthood 90–95% of most populations are antibody positive.
- In the UK 50% seroconvert before the age of 5 years, with a second wave in the teenage years – the group in whom clinical manifestations are most common. Infection does not occur in epidemics and the virus is of relatively low transmissibility. Spread is by intimate contact.
- Incidence is the same for men and women (but female peak age-specific incidence occurs two years earlier) but is 30 times higher in whites than blacks, reflecting the higher rate of early primary infection in blacks.

Pathogenesis

- Virus infects epithelial cells and susceptible B lymphocytes. Incubation is 30–50 days, less in young children. It can persist in the oropharynx of those who have recovered from infectious mononucleosis <18 months.
- EBV-related infectious mononucleosis prompts the synthesis of antibodies against viral antigen and unrelated antigens such as those found on sheep, horse, and beef red cells (heterophile antibodies, see p.449) and, less frequently, platelets, neutrophils, and ampicillin.
- In early illness there is a mononuclear lymphocytosis – most cells bear T-cell markers. The immune response includes both T cells and natural killer (NK) cells. Atypical lymphocytosis resolves by recovery – the virus is, however, not eliminated from the host.

Clinical features

- Infection usually asymptomatic in young children (perhaps mild changes in LFTs). When there are symptoms they are more likely to experience rash, neutropenia, or pneumonia than adults; 50% of cases in adolescence are asymptomatic. Typical case: triad of fever, sore throat, and lymphadenopathy (symmetric involvement of cervical, axillary, and sometimes inguinal nodes) which may be abrupt or follow a 1–2 week prodrome of anorexia and malaise; 5% develop rash (macular, urticarial, petechial, or erythema multiforme-like); 90–100% develop a pruritic maculopapular rash if given ampicillin (may appear after the drug has stopped).
- Examination – pharyngitis (exudative in 33%, with palatal petechiae in 25–60%), hepatomegaly, splenomegaly (50% cases – maximal at day 8 resolving over 10 days), periorbital oedema, and lymphadenopathy. Tachycardia is unusual even in the presence of fever. Tonsillar enlargement can be so great as to threaten the airway. Abdominal pain, particularly in the left upper quadrant, may be related to hepatomegaly or splenic enlargement, which may be rapid. There is a danger of splenic rupture with minor trauma or even spontaneously.
- Fever resolves over 10–14 days. Recovery from fatigue can take longer with good days interspersed with periods of symptom recrudescence.

- Complications – death is unusual in the immunocompetent but may follow neurological complications, airway obstruction (due to tonsillar hypertrophy, <1%), or splenic rupture
  - haematological – thrombocytopenia (50%), autoimmune haemolytic anaemia (0.5–3% cases with 70% having cold agglutinins, usually IgM. Haemolysis apparent day 7–14 and recovers over 6 weeks. Corticosteroids may help); splenic rupture (0.2%, usually in the 2nd or 3rd week of illness and may be abrupt or insidious following rapid increase to 2–3 times normal size. Usually associated with left upper quadrant and shoulder pain; 50% of cases have associated trauma, therefore recommend patients avoid contact sports etc)
  - neurological – encephalitis, cerebellitis, meningitis (may find atypical lymphocytes in CSF), Guillain-Barré, optic neuritis, Bell's palsy; 85% of cases with neurological features recover completely
  - other – hepatitis (90%, jaundice in only 5%), renal (interstitial nephritis), cardiac (pericarditis, myocarditis), lung (rare pneumonia).

EBV associations with other diseases

- **Neoplastic disorders** – most patients with nasopharyngeal carcinoma and nearly all African cases of Burkitt's lymphoma have high EBV antibody titres and EBV DNA has been detected in biopsy specimens from such patients. Associations have also been found between EBV and Hodgkin's lymphoma and both polyclonal B- and T-cell lymphomas (particularly in the setting of immunodeficiency, including renal and bone marrow transplants).
- **Infection in immunocompromised children** – several congenital immunodeficiencies are associated with the development of EBV-associated lymphoproliferative disorders. X-linked immunoproliferative syndrome has a particular association with acute fatal infectious mononucleosis with hepatic necrosis and pancytopenia.
- **EBV and HIV** – EBV is associated with polyclonal B-cell lymphomas seen in AIDS patients, and is the cause of oral hairy leukoplakia. Children with AIDS may develop a lymphoid interstitial pneumonitis.
- **Chronic fatigue syndrome** (CFS) – there is no serological or epidemiological basis to suggest that EBV infection is the cause of CFS. Rare patients have chronic ongoing organ dysfunction due to EBV (fever, pulmonary involvement etc) but these are clearly identifiable.

Diagnosis

General findings

- **Haematological** – peripheral lymphocytosis representing activated T cells responding to virus-induced B-cell proliferation (peaks day 7–21 with lymphocytes/monocytes accounting for 70% of the total white cell count of 20–50,000 leucocytes/mm^3), atypical lymphocytes (about 3% of cells – large, vacuolated, basophilic, eccentric lobulated nucleus – also seen in CMV, rubella, mumps, drug reactions among others), neutropenia (60–70% cases), thrombocytopenia (50% but bleeding is rare).
- **Biochemical** – LFT abnormalities (90% cases), mild elevation of bilirubin (45%), low-level cryoglobulins (IgG and IgM) in 90% of cases.

Specific tests

- **Heterophile antibodies** – EBV-infected B cells produce polyclonal IgM antibodies. Some of these (heterophile antibodies) agglutinate the red blood cells (RBCs) of other species. Such antibodies may be present in the sera of some healthy patients and those with lymphoma. Pre-incubation with guinea-pig cells removes these. Antibody titre is reported as the highest serum dilution at which sheep or horse (more sensitive) red cells are agglutinated; 40% of patients are positive by week 1, and 80% by week 3 of illness. Delayed appearance may be associated with prolonged convalescence. A test remains positive for up to 1 year. The Paul–Bunnell test measures agglutination of sheep RBCs by patient sera, the Paul–Bunnell–Davidsohn test is similar but is performed after pre-absorption of sera with guinea-pig cells, and the monospot test measures the agglutination of formalinized horse RBCs after pre-absorption of sera with guinea-pig cells. Commercial monospot tests have a slightly greater sensitivity than the classic tube test.
- **EBV-specific antibodies** – positive antiviral capsid antigen (VCA) IgM and negative EBNA is a sensitive and specific indicator of acute infection. Positive VCA-IgG, positive EBNA (appears late in illness), and negative VCA-IgM (remains positive for 4–8 weeks) suggest infection between 3–12 months ago. VCA-IgG and EBNA remain positive for life.
- **Virus detection** – virus is easily cultured from patient samples but there is little clinical use in doing so.

Treatment

- The majority of cases are self-limited and do not require specific therapy. The virus is poorly transmitted and isolation is unnecessary. The level of activity a patient undertakes depends on symptom severity. Some may require bed rest. Patients should not participate in contact sports or heavy lifting for between 3 weeks and 2 months.
- Admit to hospital those patients with evidence of splenic rupture, airway compromise, dehydration, significant thrombocytopenia, or hemolytic anaemia, and other major complications.
- Corticosteroids may be indicated in cases of severe thrombocytopenia, haemolytic anaemia, impending airway obstruction, and for CNS or cardiac involvement. Response is usually rapid and doses can be tailed off over 1–2 weeks.
- Immunoglobulin has been used to treat severe immune thrombocytopenia.
- Antiviral drugs – no evidence of benefit in uncomplicated infectious mononucleosis. Aciclovir has been used in the treatment of complicated disease associated with immunodeficiency. It has been shown to reduce the oral hairy leukoplakia in HIV-positive patients and been used to induce a temporary remission of a polyclonal B-cell lymphoma in renal transplant recipient.
- Splenic rupture – urgent surgical intervention – usually splenectomy.
- Tracheotomy may be indicated in some cases of airway obstruction.
- General malaise can persist for up to 3 months. Haematological and hepatic complications settle over 2–3 months. Adults with neurological complications may be left with residual deficits.

Cytomegalovirus

A herpesvirus – the largest virus to infect humans – found across the world and the cause of a wide spectrum of clinical syndromes from congenital disease to pneumonia.

The virus

- dsDNA virus encoding around 230 proteins, many of which are directly involved in downregulating the host immune response (e.g. preventing class I HLA molecule transport to the surface).
- Genome is surrounded by a nucleoprotein core, which in turn is covered by matrix proteins and a lipid envelope.
- Infects by endocytosis. CMV replication takes place in the nucleus of the cell, resulting in large nuclear inclusions useful in diagnosing infection.
- After acute infection CMV persists in a non-replicating form, reactivating with immunosuppression or illness. Some cases of secondary infection can be re-infection as well as simply reactivation.

Clinical features

- **Mononucleosis** – primary infection in a young adult produces an infectious mononucleosis picture (fever, sore throat, lymphadenopathy). The heterophile agglutinin test is negative. Lymphadenopathy and sore throat are milder than with EBV. CMV disease tends to have more systemic features ('typhoidal'). Virus is acquired by intimate contact and typically by transfusion. Complications: interstitial pneumonia (seen in bone marrow transplant patients in whom there is a high mortality despite antiviral therapy. Rare occurrence in healthy people in whom therapy is not required), hepatitis (common and mild in the immunocompetent), Guillain–Barré syndrome (CMV is the precipitant of around 10% of GBS cases), meningoencephalitis, myocarditis, thrombocytopenia, and haemolytic anaemia (common in children with congenital CMV), skin eruptions (usually mild – can be associated with ampicillin).
- **Patients with AIDS** – co-infection with CMV is seen in over 90% of homosexual men with HIV-1 and there is an increased risk of serious CMV disease once CD4 cells fall below 100/mm^3. The incidence of end-organ CMV disease has now fallen by over 80% with the introduction of HAART. CMV retinitis is the commonest manifestation (p.782), and was once seen in nearly one-third of HIV-infected patients. Polyradiculopathy is the commonest CNS manifestation of CMV, characterized by ascending weakness in the legs, and loss of bowel and bladder control. Gastrointestinal manifestations include oesophageal erosions, colitis (fever and diarrhoea which can be complicated by perforation or partial obstruction due to lesions resembling Kaposi's sarcoma), pancreatitis, and cholecystitis. Characteristic inclusion bodies may be seen on biopsy.
- **Immunosuppressive therapy –** agents such as cyclophosphamide and azathioprine are sufficient in themselves to reactivate CMV – corticosteroids, insufficient alone, act synergistically with these agents. Ciclosporin increases CMV disease only in combination with steroids.

- **Transplant recipients** – immunosuppressive regimes render such people prone to severe CMV. It is the commonest pathogen isolated after solid organ transplantation. Some cases represent reactivation of latent infection; many are acquired from the transplanted organ or transfused blood. The severity of the end-organ disease caused by CMV reflects the degree of immunosuppression. The most important source of infection is the transplanted organ, or transfused blood:
  - bone marrow – the commonest life-threatening complication following allogenic BMT is CMV interstitial pneumonia which usually occurs in the first 4 months following the procedure. Onset is rapid – over the course of a few days with fever, non-productive cough, and dyspnoea. Severe cases will require ventilation and that occurring in the context of BMT has a mortality of around 80% even with antiviral therapy (ganciclovir and CMV immune-globulin) – this may reflect an element of graft-versus-host disease in the lung
  - liver – CMV infection, particularly CMV hepatitis, is a leading cause of morbidity in the first 3 months following transplantation. CMV hepatitis has led to liver failure and repeat transplantation. Liver biopsy is required to distinguish it from graft rejection, a complication with the opposite management strategy. A prolonged course of IV ganciclovir at the time of transplantation may reduce the incidence of CMV disease in seronegative recipients
  - kidney – the rate of CMV infection of seronegative recipients following transplantation of a kidney from a seropositive donor is over 80% – such primary infections tend to be more symptomatic than reactivation secondary to immunosuppression. Primary infections may present with 'CMV syndrome': fever, leucopenia, atypical lymphocytes, lymphocytosis, hepatosplenomegaly, myalgia, arthralgia. CMV pneumonia is less severe than with BMT – ganciclovir therapy can be life saving. Significant CMV hepatitis is rare.
- **Congenital infection** – intrauterine CMV infection (seen in around 1% of live births) is less common than perinatal infection but is clinically more severe (fulminant cytomegalic inclusion disease: jaundice, hepatosplenomegaly, petechial rash, multiple organ involvement with the possibility of CNS findings such as microcephaly and cerebral calcification). Diagnosis is best made by the confirmation of infant viruria within the first week of life. Such infections tend to be seen in infants born to primiparous mothers experiencing primary CMV infection during pregnancy. Asymptomatic congenital infection has been reported in infants born to CMV-immune mothers. Perinatal infection follows acquisition from virus carried in the cervix or breast milk. Diffuse visceral and CNS disease does not occur and resembles mild CMV mononucleosis. Perinatal CMV infection may be associated with hearing loss.
- **Infection in pregnancy** – during pregnancy, primary infection of the mother may present with a mild mononucleosis syndrome but is usually asymptomatic. CMV may colonize the cervix – rate of cervical infection increases in the later stages of pregnancy.

Diagnosis

- Culture – slow (up to 4 weeks) and is combined with immunofluorescence to enable the early detection of viral antigens. Culture of virus from saliva or urine does not necessarily indicate active infection (healthy seropositive people may continue to produce virus). Culture from blood is highly suggestive of pathogenic infection.
- Seroconversion (or rise in antibody titre) suggests current infection.
- CMV viraemia detected by isolation of virus from buffy coats corresponds with active infection. PCR to detect CMV DNA correlates well with culture results and can detect virus in CSF, blood, and cells.
- Biopsy of infected tissue may reveal the distinctive appearance of CMV-infected cells (e.g. inclusion bodies).

Treatment and prevention

- Infection in immunocompetent individuals does not require treatment.
- Ganciclovir, foscarnet, and cidofovir (see 📖 Antivirals for cytomegalovirus, p.130) inhibit the CMV DNA polymerase and are effective at treating CMV end-organ disease in the immunocompromised. AIDS related retinitis responds to oral valganciclovir. Those with sight threatening lesions should be considered for intravitreal (injection or implant) or IV ganciclovir. Intravitreal treatment should be accompanied by PO valganciclovir. Induction therapy should last at least 3 weeks. Those not on HAART should be initiated and maintenance valganciclovir continued until CD4 >100 and expert review confirms that retinitis is quiescent. Relapse may occasionally occur in those with CD4 counts above 100.
- Serious CMV disease may be prevented after BMT and solid-organ transplantation with antiviral therapy. Both pre-emptive (initiation of therapy to those with evidence of viral replication in the absence of symptoms) and prophylactic (therapy to all those with positive CMV serology) strategies have shown benefit. Because of the risk of drug toxicity, many centres reserve prophylactic therapy for those most at risk (e.g. CMV-positive donors and -negative recipients).

Drug resistance

For antiviral activity ganciclovir requires phosphorylation by both viral and cellular enzymes. Resistance may be conferred by mutations either or both the viral enzyme responsible for phosphorylating the pro-drug, and the CMV DNA polymerase. Mutations in both these regions may render a virus resistant to both ganciclovir and cidofovir. Cross-resistance between ganciclovir and foscarnet has not been observed. Foscarnet resistance has been reported.

Viral gastroenteritis

Viruses account for over half the diarrhoeal episodes in infants and young children, particularly in poorer, overcrowded parts of the world where viral diarrhoea and dehydration account for millions of deaths each year. Acute viral gastroenteritis is seen in three settings:

- sporadic gastroenteritis of infants (usually rotavirus – see below, sometimes adenovirus – see 📖 p.481)
- epidemic gastroenteritis occuring in semiclosed communities (families, institutions, ships) or as a result of classic water/food-borne infection (mostly caliciviruses, see 📖 p.456)
- sporadic acute gastroenteritis of adults (caliciviruses, rotaviruses, astroviruses, adenoviruses).

All are transmitted feco-orally, but droplet spread and food contamination can also occur. Asymptomatic infection is common. EM is the only way to detect all known viruses – labour intensive and insensitive unless the titre is extremely high. Other methods: antibody-based (ELISA, immunofluorescence), PCR. Useful in identifying the cause of an outbreak rather than individual cases – patients have usually recovered by diagnosis.

Rotavirus

Outnumbers other viral causes of diarrhoea by 4:1, causing around 50% of those cases requiring hospitalization in the developed world.

The virus

- From the family *Reoviridae*, genus *Rotavirus*. Three antigenically distinguishable groups (A to C) cause human disease. Virtually all outbreaks worldwide are caused by group A organisms.
- On electron microscopy they have a wheel-like appearance (Latin '*rota*'). They have no envelope – just an outer capsid and core.

Epidemiology

- Found worldwide. Infectious dose very small – 100 organisms or even less. Transmission is faeco-oral (contaminated food, water, direct contact, or inhalation of aerosol from vomit or faeces) and virus is shed for up to a week after symptom onset. Many strains are avirulent.
- Almost all children are infected within the first 3 years of life. There appears to be no difference in exposure risk in low-income countries (unlike bacterial causes of diarrhoea). Antibody to virus is acquired by 80 to 100% of the population by age 3 years.
- Highest incidence is between 6 months (maternal antibody wanes) and 24 months old. Severe disease may still occur in younger and older children; however those under 2 months are more resistant to disease.
- Infection is seasonal in temperate countries (peaking in winter) with attack rates around 20–30% for a child's first two seasons. Susceptibility continues throughout life – 50% of parents experience infection if they have an infected infant.

Pathogenesis

- Viral capsid protein vp4 binds glycolipids on the surface of villous epithelial cells lining the small intestine leading to virus entry.
- Diarrhoea is induced through several mechanisms – certain strains have non-structural proteins that act as enterotoxins causing diarrhoea directly, loss of villus tip cells decreases absorptive capacity and a decrease in intestinal disaccharidases can cause an osmotic diarrhoea. Rotavirus-induced lactase deficiency lasts up to 2 weeks.

Clinical features

- Incubation 1–2 days. Infants/young children experience fever, vomiting (duration 2–3 days), and watery (bloodless) diarrhoea (duration 4–5 days).
- More-severe cases may have prolonged symptoms and develop an isotonic dehydration requiring hospital treatment. No clear evidence that rotavirus causes any other syndrome. Children with immunodeficiency can develop a gastroenteritis lasting many weeks.
- Asymptomatic viral shedding occurs.

Diagnosis

- Electron microscopy is now rarely used.
- Rapid diagnostic tests with high levels of specificity (albeit less sensitive than EM) include ELISA for rotavirus-specific antigen and latex agglutination kits.
- Other techniques – electrophoresis (no false-positives and allows the identification of strain-identical infections), PCR (allows serotyping).

Treatment

- See 📖 Infectious diarrhoea, p.688 for more information on the management of diarrhoea.
- Fluid replacement is fundamental. Should be administered orally in mild and moderate cases. Feeding early in illness (within 24 h) promotes enterocyte regeneration and reduces gut permeability. Fruit juices and soft drinks should be avoided. Milk can be given to infants, lactose-free, carbohydrate-rich foods to older children.
- There is no role for antibiotics, antimotility drugs etc.

Prevention

- Prevention by hygiene alone is difficult. Asymptomatic shedding is common. They are relatively resistant to common handwashing agents and can survive for some time on hard surfaces and in water.
- For management of hospital outbreaks see 📖 Infection control in the community, p.262.
- There are currently 2 oral vaccines compatible with the traditional EPI schedule available. Several other oral vaccines are in development for use in developing countries. The WHO recommends rotavirus vaccine introduction into the immunisation schedules of those regions in which its efficiency has been demonstrated.

Caliciviruses

- ssRNA viruses detected in a variety of animal species. Norwalk virus – first isolated in 1972 – is the prototype of this group. The name calicivirus' derives from cup-like indentations in the surface of the virus (from the Latin '*calyx*').
- Associated with point-source outbreaks among adults (particularly in closed settings such as ships, hospitals, and the military) and common causes of diarrhoea in children worldwide.
- Incubation is 24–48 h; symptoms last 2–3 days. Vomiting is a dominant feature in most affected people. Secondary attack rates are high. Pathological changes are similar to those seen in rotavirus infection.
- Vehicles for infection include water, food that has come into contact with contaminated water, and even contact with lakes or pools with which an infected individual has been in contact. They are relatively resistant to inactivation by chlorine. Outbreaks last 1–2 weeks but recurrent episodes on ships are common despite efforts to disinfect the ship between cruises. Disease is self-limited – IV fluid replacement may be indicated in rare instances of severe dehydration.
- Antibody may be detected in up to 90% of older children and adults. Resistance to a specific virus lasts a maximum of 2–3 years and there is little if any cross-protection against infection by other caliciviruses.

Astroviruses

- RNA viruses, members of the family *Astroviridae* and important causes of gastroenteritis in children and adults. Name is derived from its 5- or 6-pointed star appearance when examined by EM.
- Cause diarrhoea in schools, daycare institutions, and hospitals, usually in children under 3 years of age. Asymptomatic infection is common and they appear to be less pathogenic in adults than Norwalk viruses.
- Incubation is 3–4 days with symptoms lasting up to 5 days. Diarrhoea, malaise and nausea are dominant – vomiting less so. Disease is self-limited and treatment directed at maintaining hydration.

Hepatitis A virus

Hepatitis A is usually a self-limiting illness caused by hepatitis A virus (HAV), an enterically transmitted picornavirus. Although outbreaks of infectious hepatitis have been recognized for centuries, the virus was only demonstrated in the stool of volunteers infected with HAV in 1973. Since then the virus has been identified, transmitted to primates, propagated in culture, and sequenced.

The virus

HAV is a 27–28 nm spherical non-enveloped virus with a surface structure suggesting icosahedral symmetry.

Purification of the virus yields three distinct types of particle:
- mature virions
- empty capsids or particles with incomplete genome
- less-stable particles with more open structure.

The HAV genome is a single-stranded, positive sense, linear RNA of 7478 nucleotides with a molecular weight of 2.25×10^6. As with other picornaviruses, the coding region is divided into three parts:
- P1, which encodes the four capsid proteins (VP1, VP2, VP3, and VP4)
- P2 and P3, which encode the non-structural proteins.

Although a variety of genotypes (I to VII) have been identified, there appears to be only one major serotype.

Epidemiology

HAV has a worldwide distribution. It is associated with overcrowding and poor sanitation, and endemic in the developing world where it is an infection of childhood. In developed countries, incidence increases with age. Epidemics are associated with food and waterborne transmission. Certain groups are at increased risk:
- children and staff in childcare facilities
- patients and staff in mental health institutions
- men who have sex with men (MSM)
- intravenous drug users
- travellers to endemic areas.

The virus is usually transmitted via the faeco-oral route, although outbreaks have been associated with contaminated blood products.

Clinical features

- **Subclinical infection** – this is common in children (>90% in children <5 years of age).
- **Acute hepatitis** – this occurs more frequently with increasing age. Symptoms include fever, headache, malaise, anorexia, vomiting, weight loss, dark urine, and pale stools. Occasionally diarrhoea, cough, coryzal symptoms, or arthralgia may occur (more common in children). Physical findings include jaundice, hepatomegaly, splenomegaly (5–15%), and spider naevi.
- **Complications** include prolonged cholestasis, relapsing disease, fulminant hepatitis (rare, more common in older patients), extrahepatic disease, and triggering of autoimmune chronic active hepatitis.

Diagnosis

- Liver function tests are elevated with high aspartate transaminase (AST) and alanine aminotransferase (ALT) levels. Bilirubin and alkaline phosphatase are usually only mildly elevated.
- Serology – detection of anti-HAV IgM confirms the diagnosis and remains positive for 3–6 months. Anti-HAV IgG becomes positive at 2–3 months and persists for life.

Treatment

- Acute hepatitis – this is treated symptomatically (avoid paracetamol and alcohol).
- Fulminant hepatitis – patients should be treated with supportive therapy and referred for consideration of liver transplantation.

Prevention

- Passive immunization – pooled immunoglobulin (gammaglobulin) was previously the mainstay of prophylaxis in travellers to endemic regions. However, there are concerns that waning antibody levels may not be adequate to confer protection and the duration of protection is short-lived. Immunoglobulin is now only used for postexposure prophylaxis and for unvaccinated individuals who will be in a high-risk situation in < 2 weeks.
- Active immunization – several inactivated HAV vaccines exist and these have largely superseded the use of immunoglobulin. Two doses of HAV vaccine are given, at 0 and 6–12 months, and provide protection for at least 10 years. Indications for immunization include: travellers to endemic areas, male homosexuals, intravenous drug users, regular recipients of blood products, high-risk employment, military personnel. Hepatitis A vaccine may also be given as postexposure prophylaxis.
- For immunization schedules see Immunisation against infectious disease – 'The Green Book' (2006). http://www.dh.gov.uk/en/Publichealth/Healthprotection/Immunisation/Greenbook/dh_4097254

Hepatitis B virus

Hepatitis B virus (HBV) is a DNA virus that causes acute and chronic viral hepatitis in humans. 'Serum hepatitis' has been recognized since 1833 when shipyard workers in Bremen developed jaundice following smallpox inoculation. In 1965 Blumberg found a serum antigen in the blood of an Aboriginal, initially called the Australia antigen, which was associated with post-transfusion hepatitis. This was eventually identified as the hepatitis B surface antigen (HBsAg).

The virus

Virus structure

The HBV virion (Dane particle) has a diameter of 42 nm. It has an outer envelope that contains HBsAg proteins, glycoproteins, and cellular lipid. HBsAg proteins may also be released from HBV-infected cells as small spherical particles or filamentous forms. Beneath the envelope is the internal core or nucleocapsid, which contains hepatitis B core antigen (HBcAg). The third antigen, hepatitis B e antigen (HBeAg) is a truncated form of the major core polypeptide. It is released from infected liver cells in which HBV is replicating.

Viral genome

HBV has one of the smallest viral genomes, consisting of a 3200 base pair circular DNA molecule. The genome has four long open reading frames:

- C (core or nucleocapsid) gene, encodes HbcAg and HbeAg. Mutations in the pre-core region results in HBV mutants that lack HBeAg
- S (surface or envelope) gene, which includes the pre-S1, pre-S2 and s regions, encodes HBsAg
- P (pol or polymerase) gene, encompasses 3/4 of the viral genome and overlaps part of the C gene, the S gene and part of the X gene. It encodes a polypeptide with DNA polymerase and ribonuclease H activity
- X gene, encodes a polypetide, with several functions.

Epidemiology

HBV infection is a global public health problem with an estimated 400 million people chronically infected and >500,000 deaths per year. Hepatitis B virus may be transmitted by the following routes:

- vertical/perinatal
- sexual
- transfusion of contaminated blood or blood products
- intravenous drug use
- needlestick injury
- horizontally between children in Africa.

Clinical features

- Acute hepatitis
- Fulminant hepatic failure (0.1–0.5% of acute infections)
- Chronic hepatitis (5–10% of adult acute infections)
- End-stage liver disease (15–40% of chronic infections)
- Hepatocellular carcinoma
- Extrahepatic manifestations, e.g. arthropathy

Diagnosis

Initial evaluation should include:

- history and physical examination (risk factors for HBV, clinical evidence of cirrhosis, portal hypertension, liver failure)
- liver function tests: bilirubin, AST, ALT, alkaline phosphatase (ALP), gamma-glutamyl transferase (GGT), prothrombin time (PT), albumin
- HBV markers: HBsAg, HBeAg, HbcAg, anti-HBs, anti-HBe, HBV DNA
- screening for other blood-borne viruses, e.g. HCV, HIV
- screening for hepatocellular carcinoma (HCC): alphafetoprotein (AFP), and liver ultrasound
- screening for oesophageal varies by upper GI endoscopy
- liver biopsy to determine disease activity (necroinflammatory and fibrosis scores).

Treatment

General advice

- Avoid alcohol
- Safe sexual practices
- Hepatitis A vaccination (low prevalence areas)
- Avoid occupations that may spread HBV (e.g. surgery, dentistry)
- HBV immunization of household contacts

Antiviral therapy

Treatment depends on the stage of infection:

- acute HBV – no treatment, refer for transplantation if fulminant hepatic failure
- inactive HbsAg carrier – no treatment, monitor ALT every 6–12 months, screen for HCC every 2 years
- chronic HBV with normal ALT – no treatment, monitor ALT every 3–6 months, screen for HCC every 2 years
- chronic HBV with raised ALT – treat with peginterferon alfa-2a or lamivudine. Adefovir dipivoxil is recommended in patients who are unable to tolerate or have failed therapy with peginterferon alfa-2a or who have developed lamivudine resistance. Entecavir has also recently been approved by NICE for the treatment of chronic HBV.
- compensated cirrhosis – treat with interferon or lamivudine (but poor response)
- decompensated cirrhosis – treat with lamivudine, refer for transplantation.

Endpoints for treatment of chronic HBV

- Biochemical – normalization of ALT
- Virological – HBV DNA level to $<10^5$ copies/mL
- Serological – loss of HbeAg and appearance of anti-HBe
- Histological – reduction in necroinflammatory score ≥ 2 points

Difficult-to-treat patients

- Patients co-infected with other viruses, e.g. HCV, HDV, HIV
- Decompensated cirrhosis
- Renal failure
- Immunodeficiency, e.g. chemotherapy

Prevention

- Education re modes of transmission
- Screening of blood products
- Pre-exposure immunization, e.g. neonates, healthcare workers
- Post-exposure vaccination, e.g. sexual contacts, needlestick recipients
- Hepatitis B immunoglobulin for neonates born to HBeAg-positive mothers and unvaccinated needlestick recipients from HBeAg-positive donor
- For immunization schedules see Immunisation against infectious disease –'The Green Book' (2006). http://www.dh.gov.uk/en/Publichealth/Healthprotection/Immunisation/Greenbook/dh_4097254

Hepatitis D virus

- Hepatitis delta virus (HDV) is a defective virus whose replication requires co-infection with HBV.
- Virions of HDV consist of a core of delta antigen and RNA enclosed in an HBsAg-containing envelope.
- The delta antigen was originally discovered by immunofluorescent staining, as an antigen distinct from HBsAg, HbcAg and HBeAg in the hepatocytes of chronic HBV carriers in Italy. Most patients with delta antigen in the liver have anti-delta antibodies in their serum.
- Serological surveys have shown high anti-delta seroprevalence in southern Italy, Mediterranean countries, and north Africa.
- There is a higher incidence of HDV infection in HBsAg-positive patients with acute and chronic hepatitis, compared with asymptomatic carriers.
- Simultaneous infection with HBV and HDV is more likely to result in severe or fulminant hepatitis compared with HBV infection alone.
- Exacerbations of hepatitis may also occur in HBsAg carriers who subsequently acquire HDV. There is no specific treatment for HDV.

Hepatitis C virus

Introduction

Hepatitis C virus (HCV) is an RNA flavivirus, which was discovered and sequenced in the late 1980s. Prior to this it was recognized as a major cause of post-transfusion hepatitis and called non-A, non-B hepatitis virus.

The virus

Virus structure

HCV is a spherical, enveloped, RNA virus, similar in structure to other flaviviruses. Its genome is a positive single-stranded RNA molecule, approximately 9.7 kB in length which contains a large open reading frame flanked by untranslated regions (UTRs). The open reading frame encodes a polyprotein that is processed into at least 10 proteins: core protein (C), envelope proteins (E1 and E2), NS2a, and six non-structural proteins (NS2, NS3, NS4a, NS4b, NS5a, NS5b). Little is known about the life cycle of HCV; it has only recently been propagated in cell culture.

Genetic diversity

The high level viral turnover, and the absence of proofreading by the NS5b polymerase, results in a rapid accumulation of mutations within the genome. Multiple closely related but distinct HCV variants (quasispecies) may be recovered from the plasma and liver of an infected patient at one time. There is also remarkable genetic heterogeneity among HCV sequences recovered from different patients. At least seven major genotypes (genotypes 1–7) exist and these may be further grouped into subtypes (e.g. 1a, 1b, 1c). Different subtypes predominate in different geographical locations.

Epidemiology

HCV infects an estimated 170 million people worldwide. In developed countries, HCV prevalence is generally low (0.5–2%) apart from in intravenous drug users. In certain geographic areas HCV is more common, e.g. Egypt, Japan, Taiwan, and Italy, and may be related to re-use of needles for injection, acupuncture, or folk remedies.

Routes of transmission

- Transfusion of contaminated blood or blood products
- Intravenous drug use
- Nosocomial transmission, e.g. needlestick, haemodialysis, colonocopy (inadequate sterilization of colonoscopes)
- Sexual transmission or mother-to-child transmission is rare

Natural history and pathogenesis

Viral persistence

Fifteen per cent of acutely infected people clear the virus in the 3–24 months after infection; 85% develop chronic infection. Because of limitations in experimental models, and the infrequent recognition of acute infection, the mechanisms of viral clearance remain poorly understood. An HCV-specific cytotoxic T lymphocyte (CTL) response plays an important role in suppressing HCV RNA levels, but this does not appear to be sufficient to clear the virus. Similarly, viraemia persists despite a broad humoral response to HCV epitopes. This suggests that HCV has developed mechanisms to evade immune responses, e.g. changes in the viral envelope.

Disease progression

HCV leads to hepatic inflammation, steatosis and, eventually to hepatic fibrosis and hepatocellular carcinoma (estimated risk 5–25% after 10–20 years). A number of factors are associated with cirrhosis, including excessive alcohol ingestion, coinfection with HBV or HIV, human leucocyte antigen (HLA) B54, HCV genoytpe 1b, HCV quasispecies complexity, and higher levels of HCV viraemia.

Clinical features

- **Acute hepatitis C** – 75% of infections are anicteric. Symptomatic infection is similar to acute HAV and HBV, but with lower transaminases.
- **Fulminant hepatitis C** is unusual in western countries but occurs in 40–60% of cases in Japan.
- **Chronic hepatitis C** occurs in 85% of patients and is associated with fatigue, malaise, reduced quality of life indices. ALT levels fluctuate independently of symptoms, whereas HCV RNA levels remain fairly constant. Eventually progresses to cirrhosis, decompensated liver disease, and HCC.
- **Extrahepatic manifestations** – essential mixed cryoglobulinaemia, membranoproliferative glomerulonephritis, sporadic porphyria cutanea tarda, Mooren's corneal ulcers, Sjögren's syndrome, lichen planus, idiopathic pulmonary fibrosis, thyroid hormone abnormalities.

Diagnosis

- Serology – laboratory diagnosis is based on detection of antibody to recombinant HCV peptides using an EIA. Third-generation EIAs have an estimated sensitivity of 97% and can detect HCV antibody within 6–8 weeks. Infection is confirmed using recombinant immunblot assay (RIBA), which identifies the specific antigens to which the EIA is reacting, and is positive if ≥2 antigens are detected.
- Detection of viral RNA indicates ongoing infection. HCV RNA can be detected by RT-PCR (Roche AMPLICOR HCV detection kit) or bDNA assays (Chiron Quantiplex HCV-RNA assay). RT-PCR assays are generally more sensitive. Various methods may been used to genotype HCV but phylogenetic analysis of cDNA remains the gold standard.
- Liver biopsy remains the best method for assessing the severity of hepatitis C. Various grading systems exist that quantify the extent of inflammation and fibrosis, e.g. Knodell score. These are used to influence the decision to treat the disease.[1]
- Elastography/fibroscan – this is a new ultrasound method that is used to assess hepatic fibrosis in a non-invasive manner.

Treatment

The treatment of chronic HCV depends on histological stage and HCV genotype. Histologically mild disease does not require treatment. Moderate disease is treated with combination therapy. Cirrhosis may also be treated with combination therapy but complication rates are higher and response rates lower. Decompensated liver disease is an indication for hepatic transplantation.

- Interferon and pegylated interferon (📖 p.136) – interferon-α (IFN-α) has combined antiviral and immunomodulatory activity. It was the first drug to be licensed for the treatment of chronic HCV but sustained virological response rates (SVR, defined as HCV RNA negative 6 months after end of treatment) were disappointing. Problems with IFN-α include the need to administer it by s/c injection three times a week and its side-effects, e.g. flu-like symptoms, bone marrow suppression, mood disturbance, thyroid hormone abnormalities. Pegylated interferon (PEG-IFN) may be given once a week and is now the standard of care (in combination with ribavirin).
- Ribavirin – the mechanism of action of ribavirin, a guanosine analogue, is not fully understood but it has been shown to be of benefit in combination with IFN- α and PEG-IFN. Its main side-effect is a haemolytic anaemia which is reversible.
- Combination therapy – the combination of IFN-α with ribavirin has resulted in improved SVR. The use of PEG-IFN has further improved SVR (genotypes 1 and 4 ~40%, genotypes 2 and 3 70–80%).
- Other therapies that have been tried are iron-reduction therapy, parenteral thymosin-α1, oral amantadine, *N*-acetylcysteine, and ursodeoxycholate, but none have become generally accepted.

Prevention

The main methods to reduce the transmission of HCV infection are:

- screening of blood and blood products for HCV
- improving adherence to universal infection control precautions
- needle exchange programmes for intravenous drug users.

After a documented exposure, e.g. needlestick injury, the recipient should be screened for HCV antibodies immediately and at 6 months. No vaccine currently exists and treatment is not indicated.

Reference

1 NICE technology appraisal guidance 75. Interferim alfa (pegylated and non-pegylated) and ribavirin for the treatment of chronic hepatitis C. http://www.nice.org.uk/TA075guidance

Hepatitis E virus

Hepatitis E virus (HEV) is an unclassified virus that is enterically transmitted and causes acute hepatitis. It is the most common cause of acute hepatitis in certain parts of Asia and is becoming more common in the UK. Infection with hepatitis E may be asymptomatic, or range in severity from mild to fulminant hepatitis; the latter is more common in pregnant women and older men. There is no specific treatment.[1]

The virus

- Structure – HEV is a spherical, non-enveloped particle (30–32 nm diameter) with a surface structure intermediate between the Norwalk agent (a calicivirus) and hepatitis A virus (📖 p.457).
- Genome – the genome is a single-strand positive-sense RNA molecule approximately 7.2 kB in length. It encodes three open reading frames (ORFs): ORF1 encodes non-structural proteins, ORF2 encodes the capsid, and ORF3 encodes an immunogenic protein of unknown function.
- Classification – HEV was originally classified as a calicivirus, because of its morphology and genomic structure. However, its genomic sequence most closely resembles the rubella virus, a togavirus. HEV is the sole member of the genus Herpevirus. Four genotypes have been reported.

Epidemiology

- Geographical distribution – HEV is important in the developing world and is endemic in Southeast and Central Asia, the Middle East, and North and West Africa. Epidemic hepatitis E has been reported in Mexico. It is becoming more common in the UK.
- Host range – HEV has a wide host range and has been shown to experimentally infect New and Old World monkeys, swine, rodents, and sheep. In the UK the reservoir is thought to be pigs.
- Routes of transmission – most epidemics of HEV have been waterborne, but a few foodborne outbreaks have been reported. The epidemiological risk factors associated with sporadic transmission have not been identified. Zoonotic transmission, e.g. from swine, may occur. Recent evidence suggests HEV may be transmitted by blood transfusions.
- Seroprevalence – the presence of anti-HEV antibodies is lower than might be expected in endemic countries (15–60%) and higher than expected in non-endemic regions (1–28%). HEV seroprevalence is also low in infants and children, which is surprising for a virus transmitted by the faeco-oral route. The greatest incremental increase in anti-HEV occurs in young adults, who are also most at risk of clinical disease. In older adults the prevalence is relatively constant at 10–40%.
- Peak incidence in 15–35 year olds – men more commonly affected.

Pathogenesis

The incubation period is 2–9 weeks. The duration of infectivity is not known but the virus has been detected in faeces 1–5 weeks after experimental infection in non-human primates, and protracted viraemia of 45–112 days has been detected by PCR in naturally infected patients. The replicative pathway is not fully understood as HEV does not replicate well in cell culture. The histological changes are characteristic: hepatoctye 'ballooning', cytoplasmic cholestasis, focal cytolytic necrosis, and 'pseudo-glandular' alteration.

Clinical features

Infection with hepatitis E may be asymptomatic, or range in severity from mild to fulminant hepatitis. Acute HEV is clinically indistinguishable from other causes of viral hepatitis, with fever, nausea, vomiting, jaundice, and abdominal tenderness. The only complication is fulminant hepatitis which is more common in pregnancy and older men and carries a high mortality (up to 20%). Outcome can be poor in those with chronic liver disease (up to 70%). Recently chronic HEV infection has been reported in organ transplant recipients.

Diagnosis

Serology-Specific IgM and IgG responses occur early in the infection, usually by the onset of clinical illness. These can be detected using commercial tests based on ELISA or Western blot assays. Anti-HEV IgM can be detected in up to 96% of cases 1–4 weeks after acute infection. This response wanes, so that by 3 months anti-HEV is only detectable in 50% of patient. Anti-HEV IgG is also detectable 2–4 weeks after onset of illness; a single high titre or rising titres suggest recent infection.

Molecular tests-PCR and RT-PCR have been used to study the epidemiology of acute HEV infection, and seem to be more sensitive than serology.

Treatment

There is no specific therapy and treatment is supportive. Patients who develop liver failure should be referred for transplantation.

Prevention

Improved sanitation-Despite the fact the HEV appears to be less prevalent and less easily spread than HAV in developing countries, improved sanitation is likely to be important in the control of an infection that is predominantly spread by the faeco-oral route. Pork should be cooked properly to prevent possible zoonotic transmission.

Immunization-Attempts at passive immunoprophylaxis with pooled immune globulin derived from endemic areas have been generally unsuccessful. Several HEV vaccines are under development.

Reference

Dalton HR, Bendall R, Ijaz S, Banks M. Hepatitis E: an emergency infection in developed countries. *Lancet Inf Dis* 2008; **8**:698–709.

Hepatitis viruses

Acute viral hepatitis is characterized by a necroinflammatory response in the liver. It can be caused by a number of viruses and may be mimicked by many other infectious diseases and non-infectious causes. It is usually a self-limiting disease but can progress to fulminant hepatitis (rare but 1–20% mortality) or chronic liver disease (depending on the cause). A flowchart for the investigation of acute hepatitis given in Fig. 4.1.

Hepatitis viruses

Over the past 30 years at least five primarily hepatotropic viruses have been identified:

- hepatitis A virus (📖 p.457)
- hepatitis B virus (📖 p.459)
- hepatitis C virus (📖 p.462)
- hepatitis D virus (📖 p.461)
- hepatitis E virus (📖 p.465).

Novel hepatitis viruses

Further efforts to identify novel viruses have led to the discovery of other candidate viruses, the significance of which are debated.

Hepatitis G virus (HGBV-C or HGV)

This virus was independently identified by two groups. It is an RNA flavivirus which is spread by the parenteral route, by sexual intercourse and, to a lesser extent, from mother to child. It does not appear to be pathogenic in humans.

Other viruses

Many other viruses can induce hepatitis as one feature of a wider clinical syndrome. These include:

- Epstein–Barr virus (📖 p.447)
- Cytomegalovirus (📖 p.451)
- Herpes simplex virus (📖 p.44˜)
- Measles virus (📖 p.430)
- Rubella virus (📖 p.434)
- Coxsackie B virus
- Adenovirus (📖 p.481)
- Yellow fever virus (📖 p.501).

Differential diagnosis

Other non-viral infectious diseases may cause an acute hepatitis:

- leptospirosis (📖 p.393)
- syphilis (📖 p.387)
- toxoplasmosis (📖 p.567)
- Q fever (📖 p.398).

Finally, non-infectious causes may also mimic viral hepatitis:

- Drug-induced hepatitis
- anoxic liver injury
- alcoholic hepatitis
- cholestatic liver disease
- Wilson's disease
- Budd–Chiari syndrome
- liver tumours.

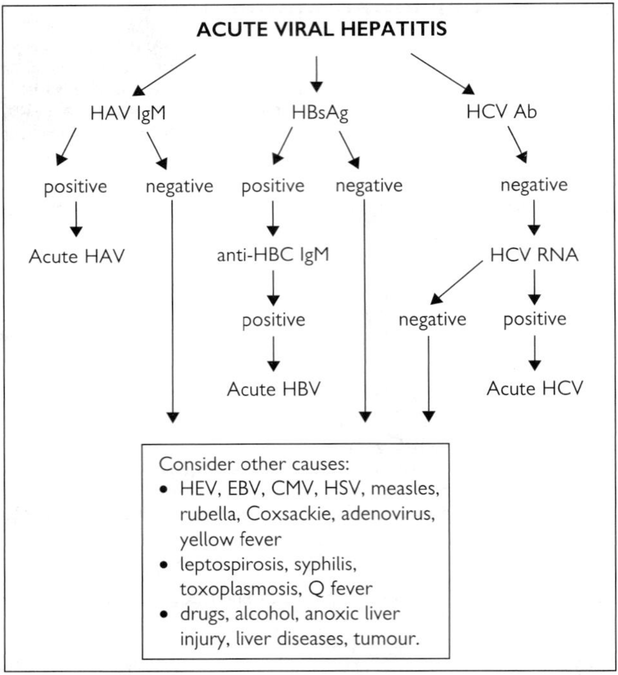

Fig. 4.1. Investigation of acute hepatitis

HIV virology and immunology

HIV is an enveloped RNA virus belonging to the retrovirus family.

Types and subtypes

There are two HIV types:

- HIV-1 which accounts for the majority of infections worldwide. HIV-1 is divided into seven subtypes or clades, designated A–K (referred to as the 'M' subtypes) and O. There are now also circulating recombinant forms (CRFs); the major ones are CRF01_AE and CRF02_AG. A new group of viruses labelled N (for new) was reported in 1998. The majority of infections in Europe and North America are caused by subtype B, whereas subtypes A, C, and D dominate in Africa
- HIV-2 occurs primarily in West Africa. It is less readily transmissible than HIV-1 and associated with lower viral loads, higher CD4 counts, and slower disease progression than HIV-1. Laboratory confirmation is difficult as 20–30% have negative HIV antibody tests and there are no commercially available viral load assays. Treatment is complicated as there are no specific treatment guidelines for HIV-2 infection, HIV-2 is not susceptible to non-nucleoside reverse transcriptase inhibitors (📖 p.142), and may have multiple protease inhibitor (📖 p.144) mutations.

Virus structure

- The HIV-1 virion is composed of (Fig. 4.2):
  - viral envelope, constructed from host cell membrane, into which are inserted HIV-1 envelope proteins (e.g. gp41 and gp120) and host proteins (e.g. MHC class II molecules)
  - matrix, predominantly protein p17
  - core, containing viral RNA associated with protein p7, the enzymes reverse transcriptase, protease, and integrase, and the major structural proteins p6 and p24.
- The genomic structure of HIV consists of *gag-pol-env* genes flanked by two complete viral long tandem repeats (LTRs):
  - gag gene products – p24, p17, p7, p6, p2, p1
  - pol gene products – protease, reverse transcriptase, and integrase
  - env gene products – gp120, gp 41.
- The virus also encodes several other genes of diverse function, e.g. *vif, vpr, tat, rev, vpu, nef* (HIV-1 only) and *vpx* (HIV-2 only).

Virus lifecyle

- Infection is initiated by binding of the envelope protein gp120 to CD4 molecules found on some T cells, macrophages, and microglial cells. Binding is also mediated by a second receptor, usually CCR5 or CXCR4.
- Fusion of the viral envelope with the host cell membrane results in release of the viral core into the cytoplasm.
- Viral cDNA is produced by reverse transcription and integrated into the host genome to form proviral DNA.

- In the presence of appropriate host cell stimulation the viral 5′ LTR produces viral mRNA transcripts.
- New virions are released from the host cell membrane by budding.

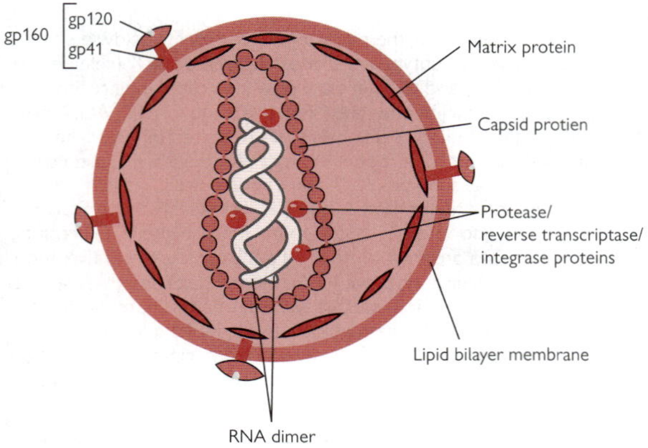

Fig. 4.2. HIV-1 virus structure.

Immunology

HIV infection is associated with a wide array of immune dysfunction:

- CD4+ T-cell depletion is the hallmark of disease and results in the development opportunistic infections.
- After primary infection, CD8+ T cells rebound and may remain elevated until late in disease.
- Cells of the monocyte-macrophage lineage serve as a reservoir of infection, are central to the pathogenesis of HIV-associated CNS disease, and contribute to impaired host defences against intracellular pathogens, e.g. *M. tuberculosis*.
- HIV infection is usually associated with B-cell hyperactivation and hypergammaglobulinaemia.
- Abnormalities of neutrophil function occur at all stages of HIV infection and predispose to *Candida* infections.
- Abnormalities of NK cells also occur increase with disease progression.

See 📖 p.850–866 for clinical manifestations of HIV and 📖 p.140–148 for drug treatment of HIV.

HIV laboratory tests

HIV infection may be established by a number of methods:

- serology (detecting antibodies to the virus)
- detection of viral antigens
- detection of viral RNA/DNA
- virus culture.

HIV serology – the 'HIV test'

- This is the most commonly used diagnostic test.
- Indications – adults in populations with an estimated prevalence >1%, pregnant women, sexual assault victims, occupational exposure, anyone who requests an HIV test.
- The standard test consists of a screening EIA followed by a confirmatory Western blot; this has a sensitivity of 99.5% and a specificity of 99.9% in patients with established disease (>3 months after transmission).
- False-negative results occur in <0.001% (low-prevalence populations) to 0.3% (high-prevalence populations). Causes include testing in the 'window period' before seronconversion (up to 6 months), seroreversion (occurs rarely in advanced disease), 'atypical host response', agammaglobulinaemia, types N, O, or HIV-2 infection, technical or clerical error.
- False-positive results occur in 0.0004% to 0.0007%. Causes include autoantibodies, HIV vaccines, factitious HIV infection, technical or clerical error.
- Alternative serological tests include immunofluorescence assay (IFA), rapid tests (e.g. OraQuick, Uni Gold Recombigen, Reveal G2, Multispot HIV-1/HIV-2), and home kits (e.g. Home Access Express Test).

Viral detection

- This may also be used to establish HIV infection when serological tests are likely to be misleading, e.g. early HIV infection (prior to seroconversion), agammaglobulinaemia, and neonatal HIV infection.
- HIV-1 DNA PCR – qualitative DNA PCR is used to detect cell-associated proviral DNA, including HIV reservoirs in peripheral CD4 cells in patients on HAART. Highly sensitive (>99%) and can detect 1–10 copies of HIV proviral DNA. Not Food and Drug Administration (FDA) approved.
- HIV-1 RNA PCR – these assays are 95–98% sensitive – depends on viral load, threshold of assay, and status of antiretroviral therapy. False-positive tests occur in 2–9%, usually at lower titres, e.g. <10,000 copies/mL.
- p24 antigen – this is sometimes used as an alternative to HIV RNA PCR to diagnose acute HIV infection. It is cheaper but sensitivity is poor (30–90%) compared with quantitative HIV RNA tests. It is also sometimes used as a low-cost viral load test in resource-limited settings.

Quantitative plasma HIV-1 RNA (viral load)

- Quantitative HIV RNA is useful for:
  - diagnosing acute HIV infection (in neonates and prior to seroconversion)
  - predicting rate of disease progression
  - monitoring treatment
  - predicting probability of transmission.
- There are three commercially available assays:
  - RT-PCR assay (Amplicor HIV-Monitor test version 1.5, Roche). Dynamic ranges are 400–750,000 copies/mL (standard assay), or 50–100,000 copies/mL (ultrasensitive assay).
  - branched chain DNA (bDNA) assay (Versant RNA 3.0 assay, Bayer). Dynamic range 75–500,000 copies/mL.
  - nucleic acid sequence-based amplification (NASBA) assay or Nuclisens HIV-QT, bioMerieux). Greatest dynamic range 176–3,500,000 copies/mL and can be used on various body fluids or tissue.
- Testing should be done at baseline and then usually every 3–4 months. The target of therapy is an undetectable viral load. The time to viral load nadir depends on pretreatment viral load, potency of the regimen, adherence, pharmacology, and resistance. An expected response is a reduction in viral load by 1 $\log_{10}$ copies/mL at 1 week. Failure to achieve a reduction in viral load of 1 $\log_{10}$ copies/mL at 4 weeks suggests non-adherence, inadequate drug exposure, or pre-existing resistance. Failure to achieve a reduction in viral load of 1 $\log_{10}$ copies/mL at 8 weeks constitutes virological failure (US guidelines).

CD4 T-lymphocyte count

- The CD4 count is used to assess prognosis for progression to AIDS, formulate the differential diagnosis in a symptomatic patient (see Table 5.18 p.848), and guide therapeutic decisions regarding antiretroviral treatment and prophylaxis for opportunistic infections.
- The standard method involves flow cytometry which relies on fresh cells and is expensive. An alternative system (TRAX CD4 Test Kit) uses EIA technology and may be more suitable for resource-limited settings.
- The normal mean value is 800–1050 cells/mm^3 (range 500–1400 cells/mm^3). There is considerable analytical variability in CD4 test results, e.g. the 95% confidence interval for a true count of 200 cells/mm^3 is 118–337 cells/mm^3. CD4 counts may also be influenced by seasonal and diurnal variation, intercurrent illnesses, and corticosteroids.
- The CD4 count is usually tested every 3–6 months in untreated patients and every 2–4 months in patients on treatment. The CD4 count typically increases by ≥50 cells/mm^3 at 4–8 weeks after viral suppression with HAART, and then 50–100 cells/mm^3 per year thereafter. Factors that correlate with a good response include a high baseline viral load and low baseline CD4 count.

Resistance testing

- The prevalence of ≥1 major resistance mutation in treatment-naive patients is 10–25%, and in patients receiving HAART is about 50%.
- Limitations of resistance testing:
  - resistance assays measure only the dominant species at the time the test is performed, resistant variants that account for <20% in sequestered havens, e.g. CNS, latent CD4 cells, genital tract will not be detected
  - there must be a sufficient viral load to perform the tests, usually ≥500–1000 copies/mL
  - genotypic assays are often difficult to interpret because multiple mutations are required for agents other than lamivudine and NNRTIs
  - phenotypic assays are difficult to interpret because of arbitrary thresholds used to define susceptibility, e.g. for ritonavir-boosted regimens. They are also less sensitive at detecting emerging resistance
  - clinical trials with resistance tests compared with standard care for selection of salvage regimens have shown variable results.
- Indications for resistance testing:
  - acute HIV infection
  - chronic infection, treatment naive
  - virological failure
  - pregnancy.
- Genotypic assays:
  - identify mutations associated with phenotypic resistance
  - may be performed by commercial kits or 'in-house' assays
  - methodology involves amplification of the reverse transcriptase and protease genes by RT-PCR, DNA sequencing of amplicons generated for the dominant species, and reporting of mutations for each gene using a letter-number-letter standard e.g. K103N.
  - interpretation requires expertise – see http://www.iasusa.org or http://hivdb.stanford.edu for more information.
- Phenotypic assays:
  - measure the ability of HIV to replicate at different concentrations of tested drugs
  - methodology involves insertion of reverse transcriptase and protease genes from patient's strain into a laboratory clone, and monitoring of replication at various drug concentrations compared with a reference wild-type virus
  - interpretation is more straightforward as results are reported as IC_{50} for the test strain relative to the reference strain
  - phenotypic resistance testing is more expensive and time consuming than genotypic testing but useful in treatment-experienced patents.

Enterovirus

The genus *Enterovirus* includes polio, Coxsackie A and B, and the echoviruses. Here we consider the non-polio enteroviruses, causes of a large number of different clinical syndromes and accounting for the majority of childhood fever-rash syndromes as well as being important causes of meningitis, myocarditis, and even neonatal sepsis. For polio see Poliovirus, p.492.

The viruses

- Small RNA viruses, family *Picornaviridae* ('pico', very small), identical in appearance to another genus of the same family, the rhinoviruses.
- Most identified in stool during polio research. Historically divided into subgroups: polioviruses, group A and B Coxsackie viruses and echoviruses (**e**nteric **c**ytopathic **h**uman **o**rphan viruses). New enteroviruses are now simply numbered sequentially.
- Some clinical syndromes are caused by many enteroviral types (e.g. rashes and meningitis), others are associated with specific enteroviruses (e.g. Coxsackie B and pericarditis).

Epidemiology

- Found worldwide throughout the year but in temperate climates infections peak in the summer and autumn months.
- 75% of cases occur in those under 15 years old, with attack rates highest in those under 1 year. Over 90% of non-polio enteroviral infection is asymptomatic, or causes only a mild febrile illness.

Pathogenesis

- Infection may take place via the respiratory tract but is mainly faeco-oral.
- Viral replication occurs in respiratory and GI epithelium, passing to regional lymph nodes. Patients become viraemic (usually undetectable).
- Most clear the infection and experience no symptoms. A minority experience further viral replication at distant reticulendothelial sites and a secondary viraemia that coincides with their non-specific illness. This major viraemia sees dissemination to other organs, e.g. CNS.
- Humoral immunity is the main host defence – people with an isolated cell-mediated immune deficiency are not predisposed to severe enteroviral illness. Virus is shed in faeces for weeks after symptoms resolution.

Clinical features

- **Aseptic meningitis** – manifestations of enteroviral meningitis vary with host age and (particularly humoral) immune state:
  - **neonates** – fever, vomiting, rash, anorexia, upper respiratory tract symptoms, and altered mental state. Meningeal signs are uncommon and presentation is usually non-specific. Severe meningoencephalitis is a rare manifestation (death rate 10%) and may be associated with hepatic necrosis, myocarditis, and necrotizing enterocolitis
  - **older children and adults** – severe disease is rare. Commonly manifests as sudden fever (may be biphasic, with a gap as long as 2–10 days), nuchal rigidity, headache, photophobia, and non-specific features (e.g. vomiting, anorexia, diarrhoea, upper respiratory tract

symptoms). Those with humoral immune deficiency can develop a chronic meningoencephalitis which may last for years and is often ultimately fatal. There may be general features of enteroviral infection (e.g. pharyngitis) as well as those specific to the infecting enterovirus (characteristic rashes, pericarditis.
In uncomplicated disease the illness usually lasts a week. Sequelae are rare in those beyond the neonatal period.

- Other neurological syndromes:
  - **encephalitis** – lethargy, drowsiness, seizures, paresis, coma. Features can be focal (particularly with Coxsackie A infection): focal motor seizures, cerebellar ataxia, hemichorea
  - **paralysis** – sporadic flaccid paralysis has been seen with Coxsackie A7 and enterovirus 71, among others. It is less severe than that seen with polio and rarely permanent. Guillain-Barré syndrome is associated with certain Coxsackie A and echovirus types. Other neurological manifestations include transverse myelitis and opsoclonus-myoclonus.
- **Other manifestations of enteroviral disease** – rashes: herpangina, hand-foot-and-mouth disease; respiratory: URTI, epidemic pleurodynia (spasmodic rib pain and fever); cardiac: myopericarditis (📖 p.678); ophthalmological: acute haemorrhagic conjunctivitis (📖 p.780).

Diagnosis

- See above referenced sections for diagnosis of non-neurological disease.
- CSF – clear and pressure is usually normal/slightly raised. White cell count raised: initially neutrophils, but lymphocytes dominate within 6–48 h of symptoms. Higher counts are associated with a greater likelihood of viral identification. A slight rise in CSF protein and depression in CSF glucose may be seen. Viral identification by culture is not sensitive (around 30%) due to the low viral titre in CSF. RT-PCR is more sensitive (60–90%) and over 94% specific for diagnosing enteroviral meningitis.
- Concomitant culture from sites other than CSF may aid diagnosis; however, virus is shed from some parts of the body for several weeks after infection has resolved, and during viral seasons enterovirus is produced by 7.5% of healthy controls.

Treatment

- Exclude bacterial aetiology of suspected meningitis. Otherwise treatment rests on symptom relief.
- Immunocompromised people with persistent enteroviral infection may benefit from immunoglobulin.
- Antiviral agents are in development – one of those, pleconaril, interferes with the viral capsid protein altering attachment and uncoating. It has been shown to reduce illness duration in adults with enteroviral meningitis.

Lymphocytic choriomeningitis virus

- Lymphocytic choriomeningitis virus (LCMV) is found worldwide. It is a ssRNA virus (family *Arenaviridae*) and causes an asymptomatic infection in mice (its reservoir). Human cases are rare and one-third of such infections are subclinical; 50% of clinical infections have neurological involvement.
- Infection is by inhalation or consumption of infected excreta and it typically causes a febrile, self-limited, biphasic disease. Individuals exposed to rodents (living conditions, pet handlers, lab personnel) are at risk for the infection. Transmission of infection during organ transplantation has occurred.
- Primary viraemia leads to CNS infection and a non-specific febrile illness with rash and lymphadenopathy. Secondary viraemia follows a few days later and is associated with meningitis/meningoencephalitis. Clinical course is benign with <1% of cases fatal.
- It is a significant teratogen. Acquired congenitally it produces similar abnormalities to congenital CMV; 35% of congenital infections are thought to lead to fetal loss; 84% of surviving infants have neurodevelopmental sequelae (e.g. cerebral palsy, seizures, mental retardation, visual problems, hydrocephalus).

Togaviridae

A family of viruses containing two key genera: *Rubivirus* (containing just rubella, 📖 p.434) and *Alphavirus*. The alphaviruses are considered here.

The virus

Lipid-enveloped virions containing 11–12 kB positive-stranded RNA.

Epidemiology

- All medically significant alphaviruses are vector borne (they were once called the group A arboviruses). 'New World' viruses predominantly cause encephalitis and circulate in North and South America. 'Old World' viruses cause predominantly fever, rash, and arthropathy. Geographic spread is determined by the distribution of their vectors.
- Only some of the many viruses identified are known to cause human disease, or cause only a rare and mild 'fever-arthropathy'.
- Some are maintained in an animal-vector cycle (e.g. eastern equine encephalitis (EEE): birds-mosquito), with a third party vector transmitting to other animals (e.g. horses) or man. Animal infections may precede human – animal health surveillance can warn of human outbreaks.
- Humans can develop a viraemia significant enough to infect mosquitos with some agents (Venezuelan equine encephalitis (VEE)) but not others (EEE).

Pathogenesis

- Incubation 1–12 days. Bite of an infected mosquito deposits virus in the subcutaneous tissues. VEE has been acquired by aerosol in the lab.
- Non-neurotropic agents cause viraemia, skin lesions (lymphocytic perivascular cuffing and red cell leak from superficial capillaries), arthralgia, and an inflammatory frank arthritis (aspirates are inflammatory – no virus present).
- Neurotropic viruses cause viraemia and fever in early infection, indicating replication in non-neural tissues. Infection of capillary endothelial cells is thought to allow subsequent CNS invasion. Acute encephalitis follows and lesions may be seen throughout the brain and spinal cord. Transplacental spread can occur with VEE and western equine encephalitis (WEE).

Clinical features

- Encephalitis – headache, high fever, chills, nausea and vomiting. Respiratory symptoms may be seen with WEE. In cases with CNS involvement (<1% adults, 4% children) confusion and somnolence follow within a few days. Seizures are more common in the young. Infants may develop bulging fontanelles. CSF protein and lymphocyte counts are high. Sequelae (70% infants after EEE, 30% WEE): mental retardation, behaviour change, paralysis. Case fatality rate: <1% overall, 20% encephalitis cases.
- Fever, rash, arthritis – rapid-onset fever (up to 40°C) and chills – may last several days, remit then recur (saddle-back fever chart). Rash usually appears on day one (can be late) – a face and neck flush evolves to maculopapular lesions on the trunk, limbs, face, palms/soles, and may

be pruritic. Arthralgia lasts a week to a few months, is polyarticular and affects small joints which may be swollen. Other features: headache, photophobia, sore throat. Long-term joint problems may be associated with HLA B27.

Diagnosis

- History – the epidemiology of each disease is fairly specific (Table 4.20).
- Acute and convalescent serology.

Treatment and prevention

- No specific therapy.
- Prevention depends on vector control and mosquito avoidance.
- Vaccines exist for EEE, WEE, VEE, and chikungunya – in limited human use.

Table 4.20 Clinical syndromes caused by Alphaviruses.

Name	Location	Vector	Features
'New World': cause predominantly encephalitis			
Eastern equine encephalitis	Eastern and gulf coast USA, southern Canada and northern South America	Enzoonotic vector: *Culiseta melanura*, breeds in freshwater swamps and feeds on birds (which become infected but may be asymptomatic). Vectors to human: *Aedes* and others	Summer disease of horses, children and elderly. Rare but case fatality over 50%
Western equine encephalitis	North and South America	*Culex tarsalis*	Summertime disease of horses and humans – highest incidence in infants
Venezuelan equine encephalitis	South and Central America	Many different mosquito species	Rainy season; human disease follows horse by 1–2 weeks. Severest in children
'Old World': cause predominantly fever, rash and polyarthritis			
Chikungunya	Africa and Asia	*Aedes stegomyia* and *A. aegypti*.	Sporadic outbreaks (90% population seropositive in sub-saharan Africa)
O'nyong-nyong	Africa	*Anopheles*	
Mayaro	South America, Caribbean	*Haemagogus* and others	Fever and rash
Sindbis	Africa, Scandinavia, CIS, Asia	*Culex* maintains a bird cycle, several species infect man.	Outbreaks occur at times of rainfall and flooding
Ross river	Australia, Oceania		Epidemic polyarthritis. Joint symptoms can last up to 3 years after infection.

Bunyaviridae

The virus

Four genera: *Bunyavirus*, *Phlebovirus*, *Nairovirus*, and *Hantavirus*. They are spherical, lipid membrane-contained viruses containing three negative-sense RNA segments coding for around six proteins.

Epidemiology

- **California encephalitis (CE) viruses** – CE is the commonest childhood CNS infection in the USA. La Crosse virus is the main cause. The virus is maintained in the mosquito vector population by transovarial transmission and each summer the infected mosquito mass increases as they feed on viraemic squirrels, foxes etc.
- **Rift Valley fever (RVF)** – maintained in sub-Saharan Africa by transovarial transmission in some *Aedes* species. Infected eggs in the soil remain viable for years, hatching in heavy rains. Sheep and cattle amplify infection.
- **Congo–Crimean haemorrhagic fever** – tick transmission and may also be acquired through infection with infected animals which may be asymptomatic (cattle and sheep).
- **Hantaviruses** – each viral species has a principal rodent host. They become chronically infected and excrete virus in urine and saliva for months. Infection is acquired by animal bite or aerosols of virus-contaminated urine or faeces.

Clinical features

- **California encephalitis** – most human infections are asymptomatic. Incubation 3–7 days, then fever, encephalitis, or meningoencephalitis. Severity ranges from mild viral meningitis to a severe disease similar to herpes encephalitis. Aphasia, ataxia, paralysis and convulsions (50% cases) can occur. CT usually normal, MRI may be positive, EEG usually abnormal; 90% of La Crosse virus cases occur in those under 15 years old. Mortality in acute disease around 1%. Sequelae include EEG abnormalities (75% at 1–5 years), epilepsy (10%), and emotional lability (10%).
- **Rift Valley fever** – a febrile illness; 10% experience retinitis and vasculitis that can cause a permanent loss of vision; 1% develop fulminant disease after 3–6 days of fever, with haemorrhage and hepatitis – half these patients die. There are rare cases of severe encephalitis.
- **Congo–Crimean haemorrhagic fever (CCHF)** – shock, DIC, bleeding, and thrombocytopenia. Mortality 20–50%.
- **Hantavirus**:
  - haemorrhagic fever with renal syndrome – incubation 5–40 days. Patients develop fever, thrombocytopenia and acute renal failure (interstitial nephritis). In the severe form (Hantaan virus), a toxic phase (headache, back pain, fever, blurred vision, erythematous rash, and petechiae) may be followed by severe shock. Those surviving can have prolonged renal insufficiency (oliguria, electrolyte, and acid–base abnormalities, then polyuria), bleeding, and pneumonitis. Mortality is 5% with Hantaan infection – one-third in the shock

phase, two-thirds in the renal phase. The milder form (Puumala virus) is fatal in less than 1% of clinical cases
- hantavirus pulmonary syndrome – a 4–5-day febrile prodrome is followed by pulmonary oedema and shock. If hypoxia and shock are well managed the vascular leak resolves in a few days. Thrombocytopenia may be seen.

Diagnosis

- Diagnosis of CE and Hantavirus infection is serological.
- CCHF and RVF viruses are readily cultured from the blood of infected patients and antigen detection by ELISA may be useful in severe cases but the aerosol infection risk limits its usefulness. Antibodies can be detected at 5–14 days, coinciding with clinical improvement.

Treatment and prevention

- There are no vaccines in general use, and prevention is by public health measures to reduce vector numbers, and personal avoidance.
- Ribavirin has been used effectively in treating Hantavirus infection, and evidence suggests it has a role in the therapy of RVF and CCHF.
- Supportive measures – anticonvulsants, fluids, circulatory/renal support.

Adenovirus

A cause of acute infections of the respiratory tract and conjunctivae.

The virus

- DNA virus of around 70nm diameter, with a complex outer capsid consisting of 252 subunits forming a 20-sided icosahedron.
- The three different types of subunits (hexons, pentons and fibres) differ immunologically – some antigenic sites common to all adenoviruses, others type specific.
- There are around 50 human serotypes some of which are associated with specific clinical syndromes. Similar viruses found in some animals.

Epidemiology

- Worldwide – different geographical areas see different syndromes associated with different serotypes. Transmission: respiratory and faeco-oral.
- Most people have experienced at least one infection by age 10 years; ~5% of all infectious illness in US infants may be attributable to adenovirus.
- Infecting serotype and consequent disease are related to age. Types 1, 2, 5 and 6 are commoner in young children (cause URTI); types 3, 4, and 7 in young adults (URTI and LRTI); types 8 and 19 cause adult eye infections.

Pathogenesis

- Generally causes lytic infection of epithelial cells resulting in cell death with the release of up to 1 million progeny (up to 5% are infective).
- Latent/chronic infection can be demonstrated in lymphoid cells (e.g. tonsils) where only small numbers of virus are released.
- Oncogenic transformation is demonstrated experimentally in animals and tissue culture. Viral DNA integrates with host DNA. No infectious virus produced.

Clinical features

- Respiratory infection – serological surveys suggest adenovirus causes 10% of all childhood respiratory infections in the developed world. Incubation is around 4–5 days and illness takes the form of mild pharyngitis/tracheitis (cough, fever, sore throat and rhinorrhoea) or, less commonly in infants, bronchiolitis and atypical pneumonia (serotype 7). Most cases improve over 3–5 days.
- Pharyngoconjunctival fever (serotype 3) – a syndrome occurring in outbreaks among children and characterized by the acute onset of conjunctivitis, pharyngitis, fever, and adenitis. Initially affecting one eye, the other usually becomes involved. Symptoms last 3–5 days, bacterial superinfection is uncommon and there is no permanent eye damage. Respiratory involvement rarely progresses to the lungs. Contaminated swimming areas have been implicated in some outbreaks.
- Epidemic keratoconjunctivitis (serotypes 8, 19, 37) – a slow-onset, usually bilateral conjunctivitis, seen in adults and acquired through such routes as contaminated hand towels and ophthalmic solutions.

Incubation is 4–24 days; conjunctivitis lasts up to 4 weeks with the subsequent development of keratitis. The cornea may be involved for several months. Secondary spread to household contacts occurs in 10% of cases.
- Haemorrhagic cystitis (type 7, 11, 21) – seen in children (male more than female) and causes around 3 days of macroscopic haematuria – the means of infection is unclear. It is also seen in children and adults undergoing BMT.
- Infantile diarrhoea (serotype 40, 41) – a common cause of watery diarrhoea and fever lasting up to 2 weeks. Distinct causative serotypes. Certain of the common 'respiratory adenoviruses' may be associated with intussusception.
- Intussusception (serotype 1, 2, 3, 5) – adenovirus was isolated in over 40% of intussusception cases in one series.
- Encephalitis/meningoencephalitis (serotypes 7, 1, 6, 12) – pneumonia is often an associated finding. Chronic meningoencephalitis occurs in patients with hypogammglobulinaemia.
- The immunosuppressed – those undergoing BMT or solid organ transplant are at increased risk of adenoviral infection, both of the organ system transplanted and disseminated disease (e.g. lung, gut, CNS). Dissemination is commonest in children but occurs in adults and has a high mortality. Adenovirus may be detected in those with AIDS (particularly in the urine or GI tract) and diseases such as colitis, parotitis, and encephalitis have been attributed to it. The significance is not clear as most patients are asymptomatic.
- Other – fatal dissemination in neonates, pericarditis, congenital abnormalities.

Diagnosis
- Culture – it is easy to culture virus from respiratory specimens, stool, urine, and conjunctival scrapings. Typical cytopathic changes are seen in human epithelial monolayers at 2–7 days. Isolated virus can be grouped by haemagglutination and serotyped.
- Antigen detection – indirect immunofluoresence allows detection of virus quickly and cheaply and correlates well with culture.
- Serology – infection is demonstrated by a fourfold rise in antibodies between acute and convalescent samples as demonstrated by complement fixation (group specific), viral neutralization (serotype specific), or ELISA.

Treatment and prevention
- Most infections are self-limited.
- There is no proven benefit of antiviral treatments – there are anecdotal reports suggesting a beneficial effect in severe cases from agents such as ribavirin, ganciclovir, cidofovir, or immunoglobulin.
- Effective vaccines have been developed – they have been used in the US military – but they are not generally available.

Human herpesvirus type 8

A herpes virus associated with several human neoplasms, most importantly Kaposi's sarcoma.

The virus

A gammaherpesvirus (like EBV), first identified in Kaposi's sarcoma (KS) tissue from patients infected with HIV.

Epidemiology

- KS was first recognized in the 1870s among those of Mediterranean and eastern European descent. Later, a more-aggressive form was recognized across black populations of eastern Africa. In the 1980s clusters of KS among homosexual men contributed to the recognition of HIV. It has also been seen in renal transplant recipients.
- HHV-8 has since been identified in lesions from all forms of KS. Epidemiological studies suggest a sexual route of HHV-8 transmission.
- >95% of HIV-associated KS cases are in homosexual men in whom it is up to 15 times more common than in those acquiring HIV non-sexually.
- Serological studies are conflicting: two studies in the US gave seropositive rates of 2% or 25% in healthy adults, 25% or 90% in homosexual HIV-positive men without KS, and 80% or 100% of KS patients.
- HHV-8 DNA has been detected in the semen and saliva of HIV-infected patients with KS, and not healthy controls.

Pathogenesis

- The sites of replication are not known but viral DNA has been detected in leucocytes suggesting they may be involved in viral dissemination.
- Encodes cell cycle regulatory proteins and cytokine homologues which probably contribute to pathogenesis of KS and other malignancies.
- KS lesions consist of inflammatory, endothelial, and red blood cells.

Clinical features of Kaposi's sarcoma

- Lesions are vascular, often nodular (0.5–2 cm diameter) and appear on skin, mucous membrane, or viscera (lung and biliary tract particularly). Violaceous or brown/black in pigmented skin. Visceral disease may involve any organ (e.g. gastrointestinal – can bleed, pulmonary – effusions). Lymphoedema may follow lymph node infiltration.
- Endemic KS is slow growing and has little prognostic significance.
- In AIDS patients with KS, the health impact of the lesions is generally of less importance than opportunistic infections.

Diagnosis

Diagnosis is clinical and histological.

Treatment

- Isolated lesions can be observed. Individual lesions of cosmetic impact can be irradiated or intralesional vinblastine administered. Other indications for treatment include lymphoedema, bulky lesions in the oropharynx and pulmonary disease.

- Extensive disease may be treated with recombinant interferon-alfa or single cytotoxic agents (doxorubicin or vinblastine). Life-threatening disease may require combination chemotherapy.
- Other – antiviral agents have not yet been demonstrated to be of benefit. HIV patients with KS require HAART. Taxanes (e.g. paclitaxel) have antiangiogenic properties and may be beneficial.

Other malignancies associated with HHV-8

- Primary effusion lymphoma – an uncommon, aggressive B-cell lymphoma seen in AIDS patients and presenting as lymphomatous effusions arising predominantly in the pleural, pericardial, or peritoneal cavities.
- Castleman's disease – an angiofollicular lymph node hyperplasia. Localized forms are benign and may be cured by surgical excision. Multicentric disease is associated with HIV and is aggressive and usually fatal.
- Other – inconclusive claims have been made of HHV-8 isolation in some skin cancers and certain cases of multiple myeloma.

Human papillomavirus

A group of viruses producing epithelial tumours of the skin and mucous membranes and associated with genital tract malignancies. 70% of cervical cancers in women worldwide are associated with infection by HPV 16 and 18.

The virus

- Non-enveloped dsDNA viruses of the *Papillomavirus* genus of the family *Papovaviridae*. Identified in many vertebrates and highly species specific.
- Genome comprises a non-coding regulatory region, an early region involved in viral transcription and regulation, and a late region that codes capsid proteins. Certain viral early proteins have transforming properties and may contribute to the development of malignancy.
- >100 HPV types have been identified by DNA sequence homology. They vary in the pathological process with which they are associated (Table 4.21).

Pathogenesis

- All types of squamous epithelium may be infected by HPV. Warts develop around 3 months (range 6 weeks to 2 year) after inoculation.
- Infects the basal cells of the stratum germinativum with replication and viral assembly taking place as basal cells mature and move to surface. Virions are shed along with dead keratinocytes.
- Warts and condylomata are associated with proliferation of epidermal layers resulting in acanthosis and hyperkeratosis. Some infected cells may develop characterisitic perinuclear vacuolation – koilocytosis.
- HPV DNA may be found in normal-looking cells, accounting for recurrence after treatment for warts. Excessive proliferation of the basal layer is a premalignant feature. DNA is extrachromosomal in benign disease but usually integrated in malignancy.
- Disease occurs at increased frequency and severity in those with immunodeficiencies (primary, lymphoproliferative disease, HIV) and receiving immunosuppressive therapy, and may be more severe in pregnancy.
- Antibodies are often raised against HPV – their significance is uncertain.

Clinical features

- **Cutaneous warts** – groups at risk of developing warts include butchers and fish handlers. Close personal contact may be important in transmission, with minor trauma at the infection site facilitating infection. Asymptomatic but may become painful. Resolve in 90% by 5 years. Very rarely progress to verrucous carcinoma.
  - Common warts (71%) occur frequently among school-aged children with a well-defined, exophytic appearance. Commonly found on back of hand, between fingers, on palms and soles. May coalesce.
  - Plantar (and palmar) warts (34%) are commonest among adolescents/young adults. Appear as raised bundles of fibres – often painful.
  - Planar warts (4%) are irregular, slightly elevated papules. Seen in childhood.

- **Epidermodysplasia verruciformis –** a rare genetic condition characterized by the appearance of disseminated cutaneous warts in early life (under 10 years old), with a high incidence of malignant transformation (one-third of patients in young adulthood). Associated with a relatively specific group of HPV types. Vary in appearance.
- **Anogenital warts** – the commonest viral sexually transmitted disease (STD) in the UK (over 80,000 new diagnoses in 2006 and increasing every year) with highest rates in women aged 16–19 years, and men aged 20–24 years. Increased in those with many partners, or not using barrier contraception. Around 66% of those having sex with an individual with warts will develop them within 3 months. Young children may develop genital warts from hand contact with non-genital lesions – their presence should, however, prompt the consideration of abuse. Seventy-five per cent of patients with anogenital warts are asymptomatic, with the remainder experiencing itching, burning, and tenderness. Appearance: exophytic papules, which may be sessile or pedunculated, small (<1 mm) or coalesce into large plaques.
  - Men – commonest affected area in uncircumcised is preputial cavity (85%); in the circumcised the penile shaft is more commonly involved. Other sites: urethral meatus, distal urethra, perianal involvement (common in the homosexual men, low in heterosexual men). Increased risk of anal cancer with a history of anal warts.
  - Women – mostly affected over the posterior introitus, the labia, and the clitoris. The application of 3% acetic acid may whiten vulval lesions. The presence of external genital warts should prompt the consideration of cervical HPV, and cervical intraepithelial neoplasia (CIN). Genital warts may spontaneously remit (approximately 10% of cases over 4 months). Lesions can become very large, particularly during pregnancy or when immunosuppressed. They may cause local destruction, enter the spectrum of CIN, or rarely transform into invasive squamous cell carcinoma. CIN has a highly variable outcome depending on the HPV type and grade of tumour (grade 1, 60% regress, 1% become invasive; grade 3, 33% regress, 12% progress).
- **Recurrent respiratory papillomatosis –** disease of larynx and airways existing in two forms: juvenile- and adult-onset. Juvenile infection is probably acquired intrapartum. Median age at onset is 3 years, patients presenting with hoarseness, or an altered cry. Disease may spread to the trachea and lungs resulting in obstruction, stridor, infection, and respiratory compromise. May require surgical excision. Adult disease is associated with a high number of sexual partners and oral-genital contact. Presentation is less aggressive – malignant transformation occurs rarely.
- **Other** – conjunctival papillomas, epidermoid cysts, coinfection with EBV in oral hairy leukoplakia in HIV-infected patients.

Diagnosis

- Diagnosis is usually made from clinical examination assisted by use of the colposcope and 3% acetic acid.
- Biopsy may be indicated to confirm diagnosis (📖 p.738). HPV antigen detection and PCR-based tests are available.

Prevention

- Vaccination – a vaccine against the four types of HPV that cause cervical cancer and warts (16, 18, 6, and 11) was licensed for use in the UK in 2006 (Gardasil®) and another bivalent vaccine (types 16–18, Cevarix®) has chosen by the UK Department of Health for its national vaccination programme in 2008. Both vaccines have high efficacy against infection, CIN, and consequent HPV-associated cervical cancer. Guidelines for its use are awaited. For greatest efficacy young women would need to be vaccinated prior to becoming sexually active.
- Barrier contraception may reduce HPV transmission during intercourse.
- Cervical smears are essential in detecting premalignant change (📖 p.738).

Treatment

- **Cutaneous warts** – most cutaneous warts undergo spontaneous resolution. The two main treatment modalities for hand warts – daily salicylic acid-based preparations or cryotherapy 3 weekly – both achieve cure in up to 70% of cases. Salicyclic acid cures around 80% of deep plantar warts, but only 50% of mosaic plantar warts. Other modalities exist: curettage, cryotherapy, and electrosurgery.
- **Anogenital warts** – treatments include the use of local caustic agents (e.g. podophylum or trichloroacetic acid), crytotherapy, electrosurgery, and surgical excision.
- **Epidermodysplasia verruciformis** – lesions should be carefully observed and malignant lesions treated rapidly.
- **Recurrent respiratory papillomatosis** – endoscopic crytotherapy or laser surgery. There may be roles for interferon-alfa and cidofovir.

Table 4.21 The Clinical Spectra of HPV disease

Clinical manifestation	Most common HPV types
Cutaneous warts	1, 2, 3, 10
Warts in meat and fish handlers	7, 2
Epidermolysis verruciformis	2, 3, 10, 5, 8, 9, 12, 14, 15, 17
Condylomata acuminata	6, 11
Low-grade intraepithelial neoplasia	6, 11
High-grade intraepithelial neoplasia and cervical carcinoma	16, 18
Recurrent respiratory papillomatosis	6, 11

Polyomaviruses

Viruses causing widespread infection which is asymptomatic in the majority, but important causes of disease in the immunosuppressed.

The virus

- Members of the *Papovaviridae* family (small non-enveloped viruses with dsDNA genomes). Two genera: *Polyomavirus* and *Papillomavirus*.
- Polyomaviruses are found in humans, monkeys, and mice, and are relatively species specific. There are two human polyomaviruses, BK and JC (named with the initials of the patient in whom each was first identified). JCV is associated with progressive multifocal leucoencephalopathy (PML) and BKV with post-renal transplant ureteral stenosis. They have around 75% nucleotide homology.

Epidemiology

- 60–80% of European adults have antibodies to JCV, BKV or both – BKV infection probably occurs at around age 4 years and JCV infection at age 10 years. No evidence of perinatal infection.
- Immunosuppressed patients and pregnant women can develop asymptomatic viruria. PML used to be seen only in older patients with haematological malignancy or receiving steroid therapy – now over half the deaths associated with PML occur in those with HIV infection.

Pathogenesis

- Primary viraemia probably leads to the establishment of latent infection in the kidney. Immunosuppression allows viral reactivation and replication leading to viuria (and the renal clinical features described below).
- JCV and BKV sequences have been identified in peripheral blood mononuclear cells in patients with leukaemia, HIV infection, and PML – these cells may provide the means for JCV to reach the CNS.
- Not known whether PML follows reactivation of latent JCV in the CNS or new CNS infection following reactivation of renal JCV. Virus probably directly infects oligodendrocytes leading to demyelination.
- 3 cases of PML occuring in patients treated with natalizumab have been described.[1]

Clinical features

Primary infection is usually asymptomatic but children may experience mild upper respiratory tract symptoms.

- **BKV and JCV viruria** – rare in those without immune impairment:
  - pregnant women – JCV/BKV is found in <3% of pregnant women in the last trimester, ceasing rapidly post-partum. Probably represents reactivation (due to immunosuppression or hormonal change)
  - renal transplant recipients – seen in 10–45% of patients after transplantation representing both reactivation and primary infection of the recipient from a previously infected kidney. Most cases occur in the first 3 months after surgery and are asymptomatic. Some cases of BKV infection have been associated with ureteric stenosis

- bone marrow transplant recipients – BKV viruria occurs in <50% of patients, most <2 months of the procedure and probably due to reactivation. Associated with the development of haemorrhagic cystitis
- other immunodeficiencies – BKV viruria with renal complications has been reported in cases of other causes of immunodeficiency.

- **Progressive multifocal lencoencephalopathy (PML)** – presentation is similar in both those with HIV and those with other immunodeficiencies: rapidly progressive focal neurological deficits including hemiparesis, visual field defects, aphasia, ataxia, and cognitive impairment. Late features include cortical blindness, quadriparesis, dementia, and coma. Abnormalities occur predominantly in cerebral white matter, and less commonly in the cerebellum and brainstem. Spinal cord involvement is rare. Death usually occurs within 6 months of diagnosis, although some experience a 2–3-year fluctuating course.

Diagnosis

Serology is unhelpful given the high seroprevalence, so diagnosis relies on virus detection and pathological findings. Viral culture is difficult as both JCV and BKV are very slow growing.

- Viruria – cytological examination is useful for the detection of viruria although normal appearance does not exclude infection – infected cells have large nuclei with a large basophilic intranuclear inclusion. Although relatively distinctive, these changes can be confused with those caused by other viral infections, e.g. CMV, adenovirus. PCR detects virus; however, it is also positive in a proportion of normal controls and apparently healthy elderly. ELISA-based serological assays are available.
- PML and JCV – brain biopsy allows definitive diagnosis demonstrating multiple asymmetric foci of demyelination, cytopathic change apparent in oligodendrocytes, and EM revealing viral particles within their nuclei. Fluorescent antibody staining allows identification of JCV. CT scan appearance may be less dramatic than the severity of the clinical findings suggests: hypodense non-enhancing white matter lesions. MRI is more sensitive. PCR of CSF to identify JCV DNA should only be used in combination with imaging and clinical findings – sensitivity is variable depending on technique and may be positive in immunosuppressed patients without PML.

Treatment

- Majority of patients with BKV and JCV are asymptomatic and do not require treatment.
- PML patients with HIV may demonstrate marked improvement with the introduction of HAART.
- Treatment of other immunosuppressed patients is supportive and symptomatic.
- Cytarabine has been shown to have no clinical benefit. Evidence is awaited to demonstrate a role for interferon or cidofovir.

Reference

1. Yousy TA, et al. Evaluation of patients treated with natalizumab for progressive multifocal lencoencephalopathy. *New Engl J Med* 2006;354:924–33.

Poxviruses

The largest of all virus groups and unlike most other DNA viruses replication occurs in the infected cell's cytoplasm rather than within the nucleus.

The viruses

- Large, asymmetric virions containing dsDNA and enzymes enabling cytoplasmic replication.
- Extremely resistant to chemical and physical inactivation and remain infective for months at room temperature, or years if frozen.
- See Table 4.22 for viruses and respective hosts.

Vaccinia

- Derived from cowpox in the early 19th century by person-to-person transmission. Now has no natural host. Jenner observed in 1798 that inoculating people with pustular material from cowpox gave protection from smallpox, inventing 'vaccination'. Routine vaccination for smallpox has been discontinued.
- Vaccinia continues to have a role among military personnel, as a vaccine vector for other infections and in immunotherapy. Protection is almost 100% for 1–3 years with disease-attenuating protection for <20 years.
- Complications of vaccination include fever, regional lymphadenopathy, postinfectious encephalitis (1–2 weeks later), and skin eruptions.

Variola (smallpox)

- Unlike vaccinia, variola infects only humans, and occasionally monkeys.
- Two main viral strains: the virulent variola major (mortality 20–50%), and the milder variola minor (mortality <1%).
- Rapid diagnosis can be made using PCR, EM, or gel diffusion techniques on vesicular fluid. It no longer exists in nature (last case: Somalia 1977).
- Incubation is <12 days then a 2-day prodrome is followed by rash (maculopapules, vesicles to pustules and scabs). In fulminant disease death can occur before the rash. Today only two laboratories in the world are known to have isolates of variola.

Monkeypox

- Causes vesicular illness in monkeys, similar to smallpox. Sporadic cases of human infection occur in endemic areas (rainforests of Western and Central Africa and some imported cases have been seen in the USA).
- Rash resembles smallpox and is contagious to other humans. A 1996/97 outbreak in the Congo had a fatality rate of 1.5%.

Parapoxviruses

- Found worldwide. Native to a variety of animals and some members are capable of infecting humans. Can persist within herds for long periods.
- Orf (sheep and goats) and pseudocowpox (cattle) cause lesions in the mouth and skin, and passage to humans is acquired by direct contact with the animal or contaminated objects. Lesions in humans are milder (vesicle to pustules) with prolonged incubation and can last for weeks.
- Diagnosis is from vesicular material, examined by PCR, EM or culture.

Molluscum contagiosum

- The only poxvirus specific for humans in the post-smallpox era. Found worldwide and spread by close human contact.
- Causes small, firm, umbilicated papules on exposed epithelial (children) or genital areas (adults). Usually resolve spontaneously but can persist for months in the context of immunosuppression, or become generalized in atopic patients.
- An opportunistic pathogen in AIDS patients – can cause generalized infection with large atypical lesions resembling basal cell carcinomas.
- Diagnosis is by histology and EM.
- Management is by local therapy (cryotherapy, incision etc) and improving immunological function.

Table 4.22 Genera of the *Poxviridae* with example species

Genera	Viruses	Normal host
Orthopoxvirus	Vaccinia	Man, derived from cowpox
	Variola	Man, (monkeys)
	Monkeypox	Monkeys, (man)
Avipoxvirus	Fowlpox	Chickens
Capripoxvirus	Sheep-pox	Sheep
Leporipoxvirus	Myxoma	Rabbit
Parapoxvirus	Bovine papular stomatitis virus	Cow, (man 'milker's nodule')
	Pseudocowpox virus	Cattle, (man)
	Orf	Sheep and goats, (man)
Suipoxvirus	Swinepox	Pigs
Molluscipoxvirus	Molluscum contagiosum	Man
Yatapoxvirus	Tanapox	Monkeys, (man - resembles monkeypox)

Poliovirus

The virus

- Members of the *Picornaviridae* family, genus *Enterovirus* (📖 p.474).
- There are three serotypes – infection by one confers protective immunity only to that type, with little heterologous protection. Prior to widespread vaccination paralytic disease was caused largely by type 1.
- Wild-type and/or vaccine-strain virus may circulate in different populations depending on regional vaccine use and the level of endemic naturally occurring viral transmission.
- Humans are the only natural host (infections can be achieved experimentally in primates). Wild-type strains vary widely in neurovirulence.

Epidemiology

- Transmission is similar to other enteroviruses.
- Polio was largely sporadic in the 19th century, affecting mostly children <5 years. By 1950 developing world infections were epidemic in nature, most cases in children aged 5–9 years (one-third in those over 15 years). This was attributed to rising standards of hygiene delaying inapparent infections that previously took place in childhood and conferred widespread immunity. The resulting pool of older, susceptible individuals facilitated epidemics – the higher rate of paralytic disease due to the loss protective maternal antibody.
- Rates fell dramatically after vaccine introduction. The last naturally occurring UK case was in 1984. Of the 25 cases of paralytic polio notified since 1983 most were vaccine-associated (see below), some were imported, and wild-type virus was not detected in the remainder.
- The WHO has certified polio transmission to be interrupted in the Americas, Western Pacific, and Europe. India, Pakistan, Afghanistan, and Nigeria are the main remaining reservoirs.

Pathogenesis

- Virus enters through the gastrointestinal tract and replicates in the gut and adjacent lymphoid tissue. Reaches susceptible reticuloendothelial tissue via the bloodstream.
- Asymptomatic cases stop at this point. Type-specific antibodies are formed. Otherwise replication and viraemia occur (the 'minor illness').
- CNS is probably infected by retrograde axonal transport from muscle to nerve to cord. Neurons throughout the grey matter are affected, especially those within the anterior horn of the spinal cord and motor nuclei of medulla and pons. Distribution of lesions is similar in all cases – it is their severity that determines clinical disease.

Clinical features

- Incubation – 9–12 days from acquisition to prodrome ('minor illness' – 2–3 days of fever, headache, sore throat, anorexia, vomiting, abdominal pain), and 11–17 days until the onset of paralysis ('major illness').

- Manifestations:
  - inapparent – 95% of cases are asymptomatic
  - abortive – 4–8% of infections experience just the prodrome and appear as any non-specific viral infection
  - non-paralytic polio – similar to severe cases of abortive but with signs of viral meningitis
  - spinal paralytic polio – frank paralysis occurs in 0.1% of infections. Children experience the classic biphasic illness: the 2–3-day minor illness coincides with viraemia and is followed by 2–5 asymptomatic days before the abrupt onset of the major illness. Headache, fever, malaise, vomiting, neck stiffness, and muscle pains are followed after 1–2 days by flaccid weakness and paralysis of anything from a single muscle to quadriplegia (very rare in infants). Initially hyperactive reflexes become absent. Paralysis is asymmetric with proximal involvement more severe than distal, and the legs more commonly involved than the arms. Bladder paralysis usually accompanies the legs. Occasional cases progress from onset of weakness to quadriplegia and bulbar involvement within hours – more commonly progression is over 2–3 days, halting when the patient becomes afebrile. Sensory loss is rare (consider Guillain–Barré syndrome)
  - bulbar paralytic polio – paralysis of those muscle groups innervated by the cranial nerves resulting in dysphagia, nasal speech, and dyspnoea. Seen in 5–35% of paralytic cases. Medullary circulatory and respiratory centres may become involved. Mixed bulbar and spinal involvement is common
  - polioencephalitis – an uncommon form occurring mainly in infants. Confusion, disturbed consciousness and seizures with spastic paralysis (upper motor neuron involvement).
- Prognosis – prior to vaccination mortality of paralytic disease was 5–10%, rising to 20–60% with bulbar involvement. Two-thirds with paralytic disease have a degree of permanent weakness on recovery, and complete recovery is rare when acute paralysis is severe, particularly if requiring ventilation. Those surviving bulbar disease show the best recovery with significant improvement by 10 days, and ultimately usually attaining normal function. Most reversible muscle paralysis in spinal disease will have resolved by 1 month with some improvement up to 9 months.
- Postpoliomyelitis syndrome – 20–30% of previously paralysed patients experience a new onset of weakness, pain, and atrophy in previously affected muscle groups 25–35 years after acute illness. This is thought to be due to attrition of motor units in innervated muscle that is already less innervated due to the initial disease.
- Risk factors for paralysis – prepubertal male, pregnacy, B-cell immunodeficiency (increases the risk of oral polio vaccine (OPV)-associated disease), strenuous exercise within first 3 days of major illness, intramuscular injections (paralysis localizes to the limb injected or injured within 2–4 weeks before infection), the tonsillectomized (8 times the risk of those with tonsils).
- Complications – respiratory compromise (diaphragmatic and intercostals muscle paralysis, upper airway obstruction, respiratory centre impairment), myocarditis, gastrointestinal haemorrhage, ileus, complications of paralysis.

Diagnosis

- CSF – abnormalities do not distinguish from other viral causes of aseptic meningitis. Unlike other enteroviruses, polio is rarely isolated from CSF.
- Virus isolation – poliovirus can be isolated from the throat in the first week of illness and from faeces for several weeks. In areas of low incidence, it is important to identify the viral strain as either wild-type or vaccine-related. If achievable, CSF viral isolation is valuable as recovery of vaccine-virus in faeces is common and does not demonstrate conclusive aetiology.
- Serology – acute and convalescent sera – but cannot distinguish wild-type from vaccine.

Treatment

- No specific antiviral therapy. Management is supportive.
- Bed rest is essential in the acute phase to reduce extension of paralysis.
- Avoid intramuscular injections.
- Ventilatory assistance is required once vital capacity falls to <50%. Tracheal intubation is indicated in severe cases of bulbar paralysis.
- Physiotherapy can start once the progression of paralysis has ceased.
- Multidisciplinary management of long-term physical sequelae.

Vaccination

- IPV (inactivated polio vaccine) – developed by Salk and introduced in the UK in 1956. The preparations now used contain all three serotypes and have seroconversion rates equal to OPV. Neutralizing antibodies are found in 100% of people after the third dose. Recipients develop little or no secretory antibody and are thus capable of asymptomatic infection and shedding to unimmunized contacts.
- OPV (oral polio vaccine) – developed by Sabin in 1963. It is trivalent – four doses are required to achieve seroconversion to all three serotypes. Non-immune recipients shed virus in faeces for up to 6 weeks. OPV promotes antibody formation in the gut, (providing local protection against viral entry) and boosts community immunity (recently vaccinated children excrete virus which may be acquired by their contacts). However, around 1 dose in every 2.6 million results in vaccine-virus related disease – both in receivers (usually those under 4 months of age, 7–21 days after administration) and their contacts (usually young adults, 20–29 days after administration); 22% of reported cases occur in those humoral immunodeficiency. The syndrome is similar to naturally occurring disease but is more protracted. In the UK there is less than one case a year (last reported in 2000). In 2004 the UK switched to the routine use of inactivated vaccine.
- Developing world – OPV is in widespread use in developing nations due to its lower cost and easy administration. Even those in tropical countries who receive all the recommended doses may fail to seroconvert for all three serotypes – this may be related to the prevalence of diarrhoeal disease and vaccine formulation.

Reference

1 Health Protection Agency. www.hpa.org.uk

Human T-cell lymphotropic virus

Human T-cell lymphotropic viruses 1 and 2 (HTLV) are retroviruses. HTLV-1 is associated with certain forms of adult T-cell leukaemia/lymphoma and neurological disease. HTLV-2 has not yet been definitively linked to a specific disorder.

The virus

- Members of the primate T-cell leukaemia/lymphoma viruses. Non-human primates are thought to constitute the natural reservoir of HTLVs.
- Differ from lentiviruses (retroviruses such as HIV) in that although both are capable of prolonged asymptomatic infection the lentiviruses have a cytopathic effect, whereas HTLV transforms T cells, immortalizing them.
- Spherical virions; 65% nucleotide homology between HTLV-1 and -2.
- HTLV-1 isolates show a high degree of similarity (92–97%) across the world (unlike HIV-1). Small variations seem to reflect geographical origin – it has been classified into five clades.
- HTLV-2 has three subtypes: 2a (among IVDUs in North America), 2b (indigenous groups in Panama, Colombia, and Argentina), and 2c (urban Brazil).

Epidemiology

- HTLV-1 is widely scattered. Endemic in some regions (e.g. 20% of adults are seropositive in some islands of SW Japan), and widespread in immigrants from these areas (parts of the Caribbean, Central and West Africa, Melanesia, Middle East, India, and parts of S America).
- HTLV-2 is found largely among IVDUs and their sexual contacts as well as some Native American populations.
- Transmission occurs sexually (mostly male to female, risk increased in the presence of genital ulceration), via blood products (those containing cellular components – not plasma derivatives – the infectious titre in plasma is extremely low) or from mother-to-child (breast feeding being the predominant route – 15–20% of breastfed children of HTLV-positive mothers acquire it). Most infections are lifelong and asymptomatic.

Clinical features

- Adult T-cell leukaemia/lymphoma (ATL) – proliferative disorder of mature CD4+CD25+ T cells. Integration of the provirus into the cellular genome is monoclonal (T cells originate from a single transformed cell) and the virus appears to be latent in neoplastic cells. 1–4% lifetime risk of an HTLV-1 carrier developing it. Clinical features: lymphadenopathy, hypercalcaemia, lytic bone lesions, skin lesions (nodules through to erythroderma), and hepatosplenomegaly. Patients can be immunocompromised and opportunistic infections are common (Strongyloides stercoralis in Japan). Classified in to four types:
    - smouldering ATL (normal cell count but 5% or more abnormal T-cell morphology with skin lesions and can last for years)

  - chronic ATL (raised cell count with organomegaly, skin or pulmonary involvement but no effusions, bone, or CNS involvement – median survival: 24 months)
  - lymphomatous ATL (lymphadenopathy with histological confirmation but normal cell counts – median survival: 10 months)
  - acute ATL (presents as leukaemia or high-grade non-Hodgkin's lymphoma – median survival: 6.2 months).
- HTLV-1-associated myelopathy (tropical spastic paraparesis) – chronic progressive demyelinating disease affecting the spinal cord and white matter of the CNS; 5% lifetime incidence in HTLV-1 carriers, onset typically at >30 years of age. Clinical features: gait disturbance, leg weakness/stiffness, with moderate to severe spasticity, back pain, bladder/bowel dysfunction, and variable degrees of sensory loss. Progression varies: some have long periods of only mild difficulty walking, others become bed-bound. Unlike ATL, infected lymphocytes are polyclonal and may cause disease indirectly by activating autoimmune T cells, or by infecting CNS glial cells, precipitating a cytotoxic response against them.
- Other disease associations:
  - HTLV-1 – arthropathies, uveitis, polymyositis, infectious dermatitis, Sjögren's syndrome, and possibily mycosis fungoides
  - HTLV-2 – unconfirmed associations with rare haematological malignancies and neurodegenerative disorders.

Diagnosis

- ELISA-based tests are used to screen for HTLV. Positives are confirmed using more-sensitive techniques to distinguish HTLV-1 and -2 (e.g. PCR).
- ATL – diagnosis is by histology (blood cell appearance, skin lesion biopsy), cytogenetics, immunophenotyping, and confirming the monoclonal integration of proviral DNA into malignant cells.
- HTLV-1-associated myelopathy – MRI may show lesions, CSF may reveal atypical lymphocytes.

Treatment

ATL

- Combination chemotherapy. Side-effects may outweigh benefits in indolent disease. Durable remissions are rare with most relapsing within 12 months.
- The role of BMT or stem-cell transplantation has not been established.
- Zidovudine in combination with IFN-α was reported to induce remission in 26% of patients in one series.
- Monoclonal antibodies (coupled to yttrium 90) against the IL-2R chain (expressed by ATL cells) have been reported as effective in some patients.

HTLV-1 associated myelopathy

- No effective treatment.
- Danazol has been reported to improve gait and bladder function.
- Corticosteroids, plasmapheresis, cyclophosphamide and IFN-α may produce transient responses.

Flaviviruses

The Family *Flaviviridae* includes the genus *Flavivirus*, along with the *Pestivirus* genus and the Hepatitis-C-like viruses. Although genetically similar there is no known antigenic relationship between these genera.

This group of viruses derives its name from yellow fever (the type species – *flavus* being Latin for 'yellow'). The *Flavivirus* genus has around 70 members, 30 of which are known to cause human disease. Most are arthropod born or zoonotic viruses. Here we consider those members for which encephalitis is the defining clinical feature. Dengue and yellow fever itself are covered on pages 501 and 503.

Viral structure

- Virions are spheres 50 nm in diameter, with an outer lipid envelope packed with the membrane (M) and envelope (E – involved in cell attachment and containing several epitopes involved in viral neutralization) glycoproteins. Positive sense ssRNA is contained within the nucelocapsid.
- Cross-neutralization assays allow flaviviruses to be classified into one of eight antigenic groups, the most important of which are the JE complex (Japanese encephalitis, St Louis encephalitis, West Nile virus, Murray Valley encephalitis virus), dengue complex, tick-borne virus complex (Central European encephalitis, Russian Spring–Summer encephalitis, Kyasanur Forest disease).

Pathogenesis

- Incubation 4–28 days (usual 1 week). 1 in 250 infections are symptomatic.
- Old age is the most significant risk factor for severe disease.
- Virus replicates locally. Brief viraemia (virus is rarely recovered from blood) followed by CNS invasion. CNS spread occurs cell-to-cell with meningeal inflammation, cerebral oedema, encephalitis (particularly temporal, thalamus, brainstem, and anterior spinal cord) and in particular may resemble polio damage far eastern tick-borne encephalitis (TBE) (damage to motor neurons in the brainstem, cervical and upper lumbar cord).

Viruses causing encephalitis

Japanese encephalitis (JE)

- **Epidemiology** – found throughout Asia (Pakistan to Eastern Russia) and responsible for up to 65% of hospitalized encephalitis cases in areas of high endemicity. Outbreaks have occurred in Northern Australia and Pacific islands. Widespread childhood immunization in China, Japan, and Korea has resulted in a large fall in incidence. Transmitted by *Culex* mosquitos with viral amplification occurring in pigs and aquatic birds. Humans are incidental hosts. Conditions for mosquito breeding are most favourable in rural areas (e.g. rice paddies) where the risk of infection is highest. Most infections are subclinical and occur in childhood (2–10 years of age), thus 80% of young adults are immune in endemic areas. Regions with high vaccine uptake see most infections in the elderly.

- **Clinical features** – although infection is symptomatic in <1% of cases, these experience a severe encephalitis (25% fatality even with intensive care facilities). Main findings at presentation: high fever and altered consciousness (personality change to coma). Early symptoms: lethargy, fever, headache, abdominal pain, vomiting. Over a few days progress to: agitation, delirium, motor abnormalities (facial paralysis, dysconjugate gaze, convulsions, hemiparesis, focal weakness, flaccid and spastic paralysis, ataxia, tremor, choreoathetosis, and other extrapyramidal signs), neck stiffness, coma (some needing ventilatory support). Severe cases may be quickly fatal. Milder cases see improvements after 1 week. Neurological recovery can take weeks to years with one-third having residual problems at 5 years, and psychological sequelae in over 50% children. Secondary complications include infections and pressure sores. Abortion may be precipitated where infection is acquired in the 1st or 2nd trimester.
- **Laboratory findings** – leucocytosis, hyponatraemia, raised CSF lymphocytes, CSF protein normal or slightly raised, EEG shows diffuse delta waves (occasionally seizure activity), CT/MRI may show cerebral oedema and evidence of abnormalities in the brainstem, cerebellum, and spinal cord.
- **Treatment and prevention** – no specific therapy. Anecdotal reports of benefit from interferon-alfa, and immunotherapy need further validation. JE vaccines are available – three doses of an inactivated vaccine over 4 weeks is around 90% effective. Side-effects include angioedema, urticaria, and there have been reports of encephalomyelitis. Because of the low risk of acquiring JE while traveling, it is not recommended routinely. A live vaccine is available only in China.

St Louis encephalitis

- **Epidemiology** – outbreaks have occurred throughout the USA and in parts of Canada and Mexico; sporadic cases occur further afield including S America and parts of the Caribbean. Transmitted in an enzootic cycle (birds) by differing *Culex* species in different geographical areas – human infection is incidental. The Eastern USA sees periodic regional outbreaks in late summer, usually in urban areas where polluted water provides mosquito-breeding areas. Vectors are most active in the evening. Incidence of infection is highest in men and the homeless. In the Western USA, infection occurs at low levels throughout the year, often associated with irrigated areas. The risk of illness is greatest in the elderly.
- **Clinical features** – three broad syndromes: febrile illness, aseptic meningitis, and fatal encephalitis. Most severe and fatal cases occur in adults and the elderly. Early symptoms: malaise, fever, headache, myalgia, upper respiratory, and abdominal symptoms. After several days: lethargy, confusion, tremor, ataxia, and other cerebellar signs, vomiting and diarrhoea, generalized weakness, meningism (children), tremor (eyelids, lips, extremities), cranial nerve palsies. Most patients do not progress to coma, and convulsions are rare; 8% overall fatality (20% of cases in those over 60 years). Children often have residual deficits

but show late recovery. Adults may have residual neurological and psychological disturbance for some months.

- **Laboratory findings** – leucocytosis, hyponatraemia, proteinuria, raised CSF pressure in one-third of cases, raised CSF protein in two-thirds, EEG shows generalized slowing with delta waves.
- **Treatment and prevention** – there is no specific treatment. Bird and mosquito surveillance may allow outbreak prediction and the early use of insecticides in at-risk areas with public health education to reduce mosquito exposure.

Tick-borne encephalitis

- **Epidemiology** – *Ixodes* tick-borne viruses causing encephalitis include central European encephalitis (CEE – principally in Austria and surrounding countries but also Scandinavia, and other parts of Europe), Russian spring–summer encephalitis (RSSE – eastern Russia, Korea, China, Japan), louping ill (Britain – an occupational disease of vets and butchers), and Powassan virus (North America). Incidence is highly variable within countries at risk (e.g. RSSE occur primarily in sylvatic locations). Viruses are transmitted between ticks and vertebrates, and passed vertically to tick offspring. Human infections are incidental and in central Europe tend to occur from April to November. Most cases occur in adults 20–50 years of age. Infection has been acquired by the consumption of unpasteurised milk from infected animals (CEE) and through handling infected meat (louping ill, CEE).
- **Clinical features** – incubation may be as long as a month; 1 in 250 of CEE infections are symptomatic and of those, 5–30% develop neurological features. Illness is biphasic (initial phase may not be reported). Early features: fever, headache, malaise, vomiting. Symptoms resolve at 1 week. Those who develop the second, neurological phase, do so after a 2–10-day remission, after which fever, headache, and vomiting return. Neurological features: aseptic meningitis, encephalitis, myelitis (particularly shoulder girdle and upper limb paralysis which may be permanent), autonomic and bladder disturbances, bulbar involvement; 1% of cases are fatal (mostly elderly), 40% experience sequelae (ataxia, psychological disturbance). RSSE tends to be monophasic and more severe with seizures, coma, brainstem involvement, and serious neurological sequelae much more common. Powassan virus produces a severe disease which may resemble herpes encephalitis. NB – a history of tick bite is only given in half of cases. Neuroborelliosis should be considered – both as a differential, and because ticks and patients may be dually infected.
- **Laboratory findings** – early-phase leucopenia, late-phase leucocytosis, moderate CSF lymphocytosis.
- **Treatment and prevention** – passive immunization with TBE immunoglobulin has been tried but there is a significant risk of exacerbating the disease. Corticosteroids reduce fever duration but prolong hospital stay. Inactivated vaccines are available and are widely used in areas at risk. Local control through the use of acaricides to reduce tick numbers is effective but not practical over large areas.

Other

- West Nile fever – related to JE but produces a febrile illness and arthropathy. Elderly people are at increased risk of more-severe manifestations including hepatitis, pancreatitis, myocarditis. Severe neurological disease leads to death in 5% of cases. It is transmitted between *Culex* mosquitoes and birds, and is the most widespread of the arboviruses (Africa, Europe, Asia, America, Middle East).
- Murray Valley encephalitis – related to West Nile and found in Australia and Papua New Guinea. Causes encephalitis with coma, limb paralysis and respiratory depression in severe cases.
- Kyasanur Forest disease and Omsk haemorrhagic fever – two related viruses transmitted by ticks and found only in the Kyasanur Forest of India and parts of Siberia respectively. A 3–8-day incubation is followed by the abrupt onset of fever, headache, and photophobia with hepatosplenomegaly and petechiae. Gum, GI tract and pulmonary haemorrhage can occur with renal failure in severe cases. Symptoms then remit for up to 3 weeks before the onset of a neurological syndrome. Case fatality: Kyasanur 5–10%, Omsk 3%.

Diagnosis

- Viral isolation – rarely useful with neurotropic flaviviruses. Isolation from blood is only likely in the first week (before the onset of neurological features). Isolation from CSF is possible in early fulminant disease and it may be isolated from tissue samples (brain, spleen, etc) in some cases.
- PCR of acute-phase serum may allow early diagnosis.
- Serology of both blood and CSF is the main means of diagnosis. ELISA for IgM is positive in one or the other by day 10 in nearly all cases of flaviviral encephalitis. Tests can remain positive for months, and in areas where several flaviviruses circulate it can be difficult to distinguish acute from previous infection and between different flaviviruses.

Yellow fever

The type species of the genus *Flavivirus* (📖 p.497).

Epidemiology

- Originated in Africa. Probably introduced into the Americas by mosquito-infested slave-trading ships. Now found in areas of sub-Saharan Africa and South America but has not been documented in Asia. Two patterns of transmission occur:
  - urban (epidemic) – occurs in Africa and represents human-to-human transmission by *Aedes aegypti* mosquitos
  - jungle – infection maintained in monkeys, transmission occurring via *Haemogogus* and *Aedes* mosquitos. Human cases occur when susceptible individuals come into contact with infected mosquitos (e.g. forestry workers). Viraemic individuals may precipitate an urban transmission cycle on returning home.
- African epidemic attack rates can be high (30 people in 1000) with death rates of 20–50%. Declining epidemics reflect changes in viral activity and human immunity (natural and vaccination related). In South America, ~100 cases are reported each year – urban outbreaks are unusual.
- Person-to-person transmission is theoretically possible in any area where the vector exists (including Southern USA).

Pathogenesis

- Virus is inoculated by mosquito and replicates in local lymph nodes, spreading via the bloodstream to other lymphoid sites and tissues.
- Viraemia peaks around day 5–6, corresponding with an increase in inflammatory cytokine production and the onset of symptoms.
- Widespread haemorrhages develop on mucosal surfaces, skin, and other organs. Gastric erosions may precipitate haematemesis and there can be extensive hepatocellular damage with lobular necrosis. Renal impairment may be secondary (prerenal) or due directly to viral infection. Neurological impairment is usually due to oedema and haemorrhage rather than encephalitis. Other features may include coagulation deficiency, myocarditis, and systemic inflammatory response syndrome.

Clinical features

- Incubation – 3–6 days. Mosquitos may become infected if biting within the first 3–5 days of illness.
- Symptoms – range from asymptomatic to a haemorrhagic fever. Commonly biphasic starting with abrupt-onset headache, fever, and myalgia lasting 3–4 days. Most people recover at this point. After a few hours'/days' respite severe cases go on to a second phase of high fever, back pain, nausea, vomiting, abdominal pain, and drowsiness before experiencing jaundice, hepatitis, and bleeding. Haematemesis, melaena, epistaxis, petechial, and purpuric rashes may occur. Patients may become oliguric and uraemic. Other: myocarditis, arrhythmias, shock, metabolic acidiosis, acute tubular necrosis, confusion, seizures, and unconsciousness.

- Laboratory findings – leucopenia in the early stages, thrombocytopenia, coagulation abnormalities, very high transaminases (AST may be more than ALT if myocarditis), normal or slightly raised ALP, uraemia, metabolic acidosis, albuminuria (a characteristic feature of yellow fever hepatitis), raised CSF protein.
- Prognosis – those that survive the critical period commonly get bacterial pneumonia or sepsis. Hepatic recovery is good – chronic hepatitis is not a feature. Death rate in severe cases of haemorrhagic fever can be as high as 50%.

Diagnosis

- Severe yellow fever resembles other viral haemorrhagic fevers circulating in Africa and South America, so lab confirmation is required for conclusive diagnosis.
- Viral detection – frequently possible as patients present while still viraemic allowing the detection of viral antigens, or virus culture.
- Serology – in primary infections IgM detection by ELISA is over 95% sensitive on serum samples taken 7–10 days after illness onset. In secondary infections, positive assays for IgG and IgM can be 100% sensitive at 5 days. Paired serum samples showing a fourfold rise in titre confirm the diagnosis. Cross-reactivity is problematic in those who have been exposed to several flaviviruses. Neutralization assays are most specific but are only offered in specialized laboratories. Complement fixation assays can distinguish between the flavivirus complexes but rise at only 4–6 weeks after onset.

Treatment

- Treatment is supportive – fluid balance, management of coagulopathy, and renal insufficiency, reducing risk of GI bleeding etc.
- In some areas of the world it may be necessary to exclude the patient from mosquitos to prevent onward transmission.
- The 17D vaccine is highly effective and a single dose produces long-term protection in 95% of people. There are rare cases of encephalitis – infants at are higher risk and it is contraindicated in those under 4 months of age. Travellers to at-risk countries should receive the vaccine every 10 years. It is recommended for routine use in 35 African countries but uptake is low.
- Vector reduction is difficult and expensive but has been achieved in some areas.
- In South America surveys to identify dead monkeys can warn of an increased risk of to humans.

Dengue

A flavivirus (📖 p.497) and arbovirus transmitted by *Aedes* mosquitos causing fever (dengue fever, DF) which may be complicated by fluid leak, shock and haemorrhage (dengue haemorrhagic fever, DHF).

Epidemiology

- The virus exists in four distinct serotypes. Infection with one serotype provides brief (around 6 months) cross-protection to all four upon recovery, after which immunity remains to the infecting serotype only. Later secondary infection with one of the other three is then associated with an increased risk of severe disease, DHF.
- Worldwide outbreaks of dengue began at the start of the 20th century. 'Breakbone fever' was recorded in Greece (1928), Australia (1897), Florida (1934), and the tropics. After the second World War, transmission of multiple serotypes greatly increased in Southeast Asia – with a consequent increase in cases of DHF.
- Today dengue occurs in regions of the tropics in which its vector is found (*Aedes aegypti* and to a lesser extent *A. albopictus* and *A. polynesiensis*) – particularly Southeast Asia, parts of South and Central America, and certain parts of sub-Saharan Africa. DHF can be endemic where more than one viral type circulates.
- It takes around 2 weeks for a mosquito to become infective after feeding on a viraemic individual. *Aedes* are day biting and are easily disturbed while feeding. A single mosquito can infect an entire household. They breed in any source of open water (domestic containers, puddles, tyres etc) thus transmission can be very intense in tropical cities.
- Tropical transmission occurs throughout the year, increasing in the rainy season. Epidemic attack rates can be as high as 70%. Around 3 billion people live in areas at risk of dengue, with 100 million cases per year of dengue fever, and 500,000 cases of dengue shock.

Pathogenesis

- Virus disseminates in the blood within 2–3 days of an infected bite. Patients are viraemic for 4–5 days. Malaise reflects the cytokine response.
- Most people infected with virus experience a significant but self-limited febrile illness. Some patients, usually those who have experienced previous infection, develop a severe immunopathological response (DHF), thought to be due to antibody-dependent enhancement: heterologous antibody is non-neutralizing, and enhances viral uptake and increasing the infected cell mass. The consequently exaggerated inflammatory response includes vasoactive cytokines (e.g. TNF-α) which contribute to fluid leak – unlike viral haemorrhagic fevers such as Ebola, structural damage to blood vessels is not a feature.
- It is rare for individuals to have more than two episodes of dengue.

Clinical features

- Dengue fever – asymptomatic in up to 80% of infants and children (often difficult to distinguish from other causes of fever). Tends to be more severe in adults; 4–7 days' incubation followed by abrupt fever, headache, muscle pain, and rash (macular erythema with petechiae on extensor surfaces) with rapid progression to prostration, back and abdominal pain. Defervescence occurs after 2–7 days – it may settle and recur (saddleback fever pattern). Recovery may be followed by prolonged fatigue and occasionally depression. Other features: minor mucosal bleeding (severe in some cases – e.g. pre-existing peptic ulcer), subcapsular splenic bleeds, hepatitis, neurological features (may represent the effects of cerebral oedema or viral encephalitis).

- DHF (dengue shock syndrome) – the early features are identical to mild disease. Severe symptoms tend to occur at defervescence (when, notably, viral load is falling rapidly) with reduced perfusion, central cyanosis, sweating, and other signs of shock. Platelets fall, petechiae develop, with spontaneous bruising and bleeding from mucosal surfaces. Fluid leak occurs (increase in haematocrit, pleural effusions and ascites). The duration of illness is 7–10 days, and with support in the critical period (fluid therapy etc) mortality is under 1%. Without support, death rates can reach 50%. Complications: encephalopathy, hepatic failure, renal failure, dual infections (Gram-negative sepsis, parasitic disease).

Diagnosis

- Clinical – DF is not easily distinguished from other causes of childhood febrile disease and even in cases of shock DHF may resemble yellow fever. Leucopenia, low platelets and abnormal LFTs are common.
- Viral detection – culture is definitive but impractical and rarely achieved if samples are taken more than 1–2 days after defervescence. Detection of viral RNA by RT-PCR is used in some centres.
- Serology – serological techniques are less specific than culture due to cross-reactivity between different flaviviruses. Neutralization assays allow different dengue serotypes to be distinguished in primary infection. Acute infection (as opposed to recent infection) can only be confirmed by demonstrating a rise in anti-dengue immunoglobulin in paired sera, e.g. IgM antibody capture (MAC)-ELISA for IgM (which tends to be specific to dengue complex).

Treatment and prevention

- Supportive treatment – antipyretics, oral rehydration, close observation of fluid status with appropriate interventions (IV fluids, circulatory support), avoid invasive procedures.
- The WHO algorithm for classifying, monitoring, and identifying severe disease (e.g. 20% increase in haematocrit, narrowed pulse pressure etc), and guiding fluid replacement has been responsible for large falls in mortality.[1]
- Prevention – there are several experimental vaccines but none in widespread use. Vector control is effective but expensive and rarely practical in the regions that most need it.

Reference

1 World Health Organization. *Dengue haemorrhagic fever: diagnosis, treatment, prevention and control*, 2nd edition. Geneva: World Health Organization, 1997.

Viral haemorrhagic fevers

Viral haemorrhagic fevers (VHF) are severe, potentially life-threatening illnesses caused by members of several viral families (Table 4.23). Most patients are not severely unwell when they present, and VHF should be considered as a cause of fever/rash/sore throat in patients who have visited at-risk areas (e.g. rural West or Central Africa) within the last 3 weeks – *but* remember malaria is a much more likely diagnosis. Here we consider the filoviruses and arenaviruses. Certain bunyaviruses (📖 p.479) and flaviviruses (📖 p.497) may cause similar presentations.

Management of suspected viral haemorrhagic fever

- Presentation – consider VHF in a patient with a febrile illness who has returned from a tropical African or VHF endemic country within 3 weeks and discuss with a local expert. Patients are infectious only after they develop symptoms. Early clinical features are mild and are not particularly characteristic: fever, cough, headache, sore throat, nausea, vomiting, weakness, abdominal and chest pains. Severe features present later: haemorrhage, encephalopathy, hepatitis, shock.
- Differential diagnosis – *malaria should be excluded as soon as possible*. Also consider typhoid fever, dengue, and rickettsial infection.
- Infection control – strict infection control precautions aim to prevent secondary infection of other patients, hospital and laboratory staff. Onward transmission to other people requires contact with the patient or infected secretions – aerosol transmission is not thought to be a significant means of infection. Medical and lab staff should be meticulous in taking blood, handling fluids, disposing of excreta, and performing invasive procedures. Gloves, water-repellant aprons, face visors, and masks should be used. Sharps and other contaminated equipment should be disposed of extremely carefully.
- Risk assessment:
  - minimum – the patient has not been in an endemic area or left more than 3 weeks before symptom onset
  - moderate – the patient was in a known endemic area within 3 weeks of onset in the absence of any other features listed below or was in an adjacent area within 3 weeks and has severe illness that could be due to VHF
  - high – if (a) the patient was in a known endemic area with 3 weeks of onset and stayed in a house with a VHF case/suspected case, nursed a case/suspected case, had contact with fluids or the body of a case/suspected case (e.g. lab worker), or was moderate risk but has gone on to develop severe illness (organ failure, haemorrhage) or (b) was not in an endemic area within the 3 week window but has cared for a case/suspected case, or handled fluids or the body of a case/suspected case.
- Sample processing – if samples have already been sent to the lab then it should be informed immediately so that they may be stored. Defer further samples, with the exception of malaria screening, until a detailed risk assessment has been performed. If patients are considered moderate to high risk, call the reference labs in London or Salisbury for advice before sending samples.

- Subsequent management – depends on the risk, clinical progress, and local protocols. Minimum-risk patients are usually managed in a standard side room with specimens transported and processed as for other blood-borne viruses. Moderate-risk patients should be isolated and full special precautions taken in acquiring and transporting samples. Only malaria films should be performed (at a minimum of category 3 containment). If positive with a good response to malaria therapy, most moderate-risk patients are recategorized as minimum risk. High-risk patients (especially those with a negative malaria film) should be considered for urgent transfer to a high-security isolation facility (Newcastle, London). Recovering patients may excrete virus in the urine for weeks.
- See http://www.hpa.org.uk/ for the UHF guidelines produced by the Advisory Committee on Dangerous Pathogens

Table 4.23

Virus	Source or vector	Distribution
Arenaviridae		
Junin (Argentine haemorrhagic fever)	Rodent	Agricultural areas of northern Buenos Aries province.
Machupo (Bolivian haemorrhagic fever)	Rodent	Northeastern Bolivian savannah.
Guanarito (Venezuelan haemorrhagic fever)	Rodent	Cleared forest areas of Venezuela.
Lassa	Rodent	West Africa
Bunyaviridae		
Congo–Crimean haemorrhagic fever	Tick, or contact with infected animals.	CIS, Middle East, Africa
Hantavirus (haemorrhagic fever with renal syndrome)	Rodent	Parts of China, Asia, Russia and Europe.
Rift valley fever	Mosquito	Sub-Saharan Africa
Filoviridae		
Ebola	Unknown	DRC, Sudan and Côte d'Ivoire.
Marburg	Unknown	Uganda, Western Kenya
Flaviviridae		
Yellow fever	Human or monkey, via mosquito	South America, sub-Saharan Africa.
Dengue	Human via mosquito	Asia, sub-Saharan Africa, South America
Omsk haemorrhagic fever	Tick	Siberia
Kyansanur forest disease	Tick	Kyansanur forest, India
Togaviridae		
Chikungunya	Mosquito	Africa and Asia

Filoviruses

Named after their characteristic filament-like morphoplogy, filoviruses are elongated structures 80 nm across and 800–1000 nm long. Genetic material is negative sense ssRNA. The two agents identified to date (Ebola and Marburg) show no serological cross-reactivity. Details of their natural history remain elusive. Viral particles are stable and highly infective.

Epidemiology

- Outbreaks emerge abruptly. Source may be traced to a single human or primate index case but no further. No conclusive animal reservoir identified – however, prolonged infection has been demonstrated in bats.
- Exact routes of transmission unknown. Infection acquired parenterally has a high mortality, but most infections probably occur through skin or mucous membrane contact. Aerosol transmission may occur in monkeys but is not thought to be important in human-to-human spread.
- Marburg virus – identified in 1967 when African green monkeys brought from Uganda to Germany developed a haemorrhagic illness subsequently transmitted to humans (7 deaths among 31 cases). Primary cases were associated with close contact with monkey blood or cell culture, and secondary cases with human blood exposure. Cases have occurred in western Kenya and once in Zimbabwe. Mortality is around 25%.
- Ebola – identified in 1976. Four subtypes identified: Zaire, Sudan, and Côte d'Ivoire in Africa, and Reston – first identified in the USA in monkeys imported from the Phillipines. Early Ebola outbreaks were exacerbated through the use of infected needles. Without precautions, rates of infection of household contacts can reach 17%. Case fatality rates range from 50% (Sudan) to over 80% (Zaire subtype).

Clinical features

- Incubation 5–10 days. Abrupt onset of fever, myalgia, and headache. Nausea, vomiting, abdominal pain, diarrhoea, and chest pain follow. Petechiae, haemorrhages, and spontaneous bruising occur as disease progresses. A maculopapular rash can develop around the fifth day.
- In week 2, patients either become afebrile and improve, or develop shock and multi-organ dysfunction with DIC, renal and liver failure.
- Convalescence is prolonged. Virus is detectable in semen/urine for weeks.

Diagnosis

- Clinical clues – travel to at-risk areas or contact with monkeys, maculopapular rash (not seen with other VHF).
- Viral detection – culture is usually positive acutely and viral particles may be seen on EM of histological speciments. Viral antigen detection and PCR-based tests are also available.
- Serology – seroconversion occurs at around day 10.

Treatment and prevention

- There are no specific treatments – management is supportive.
- Prevention – early recognition of outbreaks and barrier nursing.
- Experimental vaccines show promise.

Arenaviruses (Lassa fever)

A family of ssRNA viruses. All are parasites of rodents, each virus showing specificity for a single rodent species. Here we consider Lassa fever. Other members include lymphocytic choriomeningitis virus (a cause of fever and meninigitis) and the *Tacaribe* complex of viruses (causes of VHF in South America).

Epidemiology

- Found in West Africa. The *Mastomys* rodent hosts are chronically infected and show no evidence of disease. Humans are infected through contact with excreta.
- Unlike other arenaviruses, person to person transmission of Lassa can occur. Endemic transmission occurs throughout the year with nosocomial outbreaks (where aerosol and parenteral transmission have been implicated) in the dry season.

Clinical features

- Incubation 7–12 days. Most infections in Africa are mild with severe disease in <10% of cases – mortality in this group can be as high as 25%.
- Symptoms – fever, chest pain, back and abdominal pain, cough, vomiting, diarrhoea, sore throat, conjunctivitis, facial oedema, and CNS features (encephalitis, meningism).
- Severe cases in the second week – shock, fluid leak (facial and pulmonary oedema, ascites, pleural effusions), mild haemorrhages from mucosal surfaces. Pregnancies in maternal infections frequently abort – high maternal mortality.
- Late complications in less-severe cases – cranial nerve deafness (up to one-third of hospitalized cases), pericarditis, uveitis, and orchitis.

Diagnosis

- Viral detection – culture from blood, throat swabs and urine is possible within first 7–10 days.
- Serology – IgM detection is rapid and sensitive – around 75% of patients being positive on admission in Sierra Leone in one study.

Treatment and prevention

- No vaccine available.
- Ribavirin therapy initiated before day 7 reduces mortality significantly (55% to 5% in one study), but improves survival at all stages of illness.
- Contacts should be monitored for development of illness, with early presumptive use of ribavirin if fever develops.
- Nosocomial transmission can be reduced through barrier nursing and isolation wherever possible (person-to-person spread by aerosol appears to occur).
- Rodent control is rarely practical in the countries affected.

Rabies virus

Present throughout history, rabies (Latin 'madness') virus produces near-uniformly fatal encephalitis.

The virus

- Negative-sense ssRNA virus, member of the family *Rhabdoviridae*, genus *Lyssavirus*. Virions are bullet shaped (180 nm long by 75 nm wide).
- Genus has six members – classic rabies is serotype 1. The other five rarely cause human disease (e.g. European bat lyssavirus 2 caused the death of a bat handler in Scotland in 2003).

Epidemiology

- Worldwide with the exception of Antarctica and certain islands. 40,000–70,000 deaths each year, most in Asia. In 1999 100 countries reported cases of rabies; 45 had none.[1]
- Many mammals maintain and transmit rabies – dogs account for 54% of animal cases; also foxes, wolves, raccoons, and bats (4%). Other animal species are susceptible but like man, develop disease and do not participate in onward transmission (camels, horses).
- Epidemiology of human disease reflects the pattern of animal infection. In developing regions most human cases are acquired through dog bites. Developed nations have largely eliminated disease from domestic animals – most human infections follow exposure to rabid wild animals. Cases have followed corneal and solid organ transplantation.
- Animal rabies has been increasing in the US, generally arising in raccoon populations. In 2003 there were two cases of human rabies reported in the US. In the UK the last locally acquired human case of *classical* rabies occurred in 1902. There have been 22 rabies deaths in the UK since 1946 – all were imported, none received post-exposure prophylaxis.

Pathogenesis

- Virus enters through a break in the skin, or across mucosal surfaces, attaching to muscle and nerve cells. Internalization is followed by local replication within muscle cells.
- Rapid administration of antirabies immunoglobulin and immunization are able to prevent spread of the virus at this stage – once virus has entered peripheral nerve it is not possible to prevent replication and spread.
- Nerve innervating the muscle spindle is infected first, replication continuing in peripheral neurons as viral particles migrate along the axon. Virions travel by retrograde axoplasmic flow, unlike herpes simplex which uses microtubular transport systems and therefore infects faster.
- On reaching the cord the rabies virus spreads throughout the CNS, reaching the rest of the body (e.g. saliva) via peripheral nerves.

- The mechanism of CNS damage is uncertain. It may interfere with neurotransmission, with endogenous opioid systems, or act in an excitotoxic manner. Pathological examination of the brain in furious rabies reveals findings characteristic of encephalitis with Negri bodies (round eosinophilic cytoplasmic inclusions up to 7 micrometre across and containing viral nucleocapsids). These are concentrated in hippocampal pyramidal cells, but may be found in cortical neurons. Paralytic rabies affects the spinal cord (inflammation and necrosis) and may cause segmental demyelination. Myocarditis can occur, and Negri bodies may be identified in the myocardium suggesting a direct viral pathology.

Clinical features

- Risk of acquiring rabies is related to the size of the inoculum (e.g. multiple bites, bite directly on skin versus through clothing), and the location of the bite (greater risk with face compared to extremities).
- Incubation – days to years, most develop symptoms within 3 months.
- Initial symptoms – fever, headache, malaise, vomiting with altered sensation at the bite site, subtle personality changes. May see myo-oedema (localized contraction of the muscle when struck with a tendon hammer, disappearing over a few seconds).
- Acute neurological disease develops 4–10 days later and lasts 2–14 days before coma intervenes. Patients die an average of 18 days after symptom onset. There are two main clinical presentations:
  - furious (encephalitic) rabies (80% of cases) – anxiety, biting, hydrophobia (an exaggerated respiratory tract irritant reflex), delirium, agitation, seizures, hyperventilation, pituitary dysfunction (e.g. diabetes insipidus), cardiac arrhythmias, autonomic dysfunction (papillary dilatation, salivation, priapism). Patients eventually enter coma
  - paralytic rabies (20% of cases) – the spinal cord and brainstem are predominantly affected and patients develop an ascending paralysis that may resemble Guillain–Barré syndrome or a symmetrical quadriparesis. Meningeal signs may develop. Disease progression is marked by confusion and a decline into coma.

Diagnosis

- Incubation – no tests are useful during incubation, and history is paramount in determining exposure to a potentially rabid animal and prompting the initiation of prophylactic treatment.
- Symptomatic – the diagnostic standard is direct fluorescent antibody staining of a skin biopsy taken from the nape of the neck – virus localizes in hair follicles; 50% of samples are positive in the first week of illness, increasing thereafter. RT-PCR of tissue, CSF, or saliva is in increasing use. Other standard tests do not distinguish other causes of encephalitis.

Treatment and prevention

- Prevention – reducing disease in animal populations is central to the control of human disease. Vaccination of cats and dogs is a legal requirement in many countries, and vaccination of wild animals is effective in regions that can maintain it.
- Pre-exposure vaccination – generally offered to those in high-risk occupations or travelling to at-risk countries with limited access to medical facilities. Given as three IM (intramuscular) or intradermal injections on day 0, 7 and 21. Booster doses recommended every 2–3 years.
- Post-exposure treatment (PET) – wound care by thorough washing with 20% soap solution and irrigation with a virucidal agent (e.g. iodine) may reduce the risk of rabies by as much as 90%. Local advice should be sought regarding the risk of rabies from the species involved. Observe the responsible animal for 10 days if apparently healthy – it should undergo pathological examination if its behaviour changes. If PET is considered necessary it should be started immediately. It has a good record and appears safe in pregnancy. Immunocompromised patients may not respond sufficiently to vaccination and should have antibody titres checked at 2–4 weeks.
    - **If not previously vaccinated** rabies immunoglobulin (20 IU/kg body weight) should be infiltrated around the wound with any remaining dose given IM at an anatomical site distant from that of vaccine administration. Vaccine should then be given IM deltoid on day 0, 3, 7, 14, and 28.
    - **If previously vaccinated** immunoglobulin should not be given, and vaccine should be given IM deltoid on day 0 and 3.
- Treatment after symptom onset – generally treatment initiated after symptom onset is of no benefit. In October 2004 a 15-year-old girl developing symptoms after a bat bite survived despite no history of vaccination or post-exposure treatment. Her treatment involved early induction of coma, ventilatory support, ribavirin and amantadine[2]. However, ribavirin had been used before without success and it is not clear what elements of this case or treatment regime contributed to recovery.

Reference

1. World Health Organization. Rabies: epidemiology. http://www.who.int/rabies/epidemiology/en/ (accessed 6 August 2008).

2. Willoughby RE et al. Survival after treatment of rabies with induction of coma. *New Engl J Med* 2005;352:2508–14.

Creutzfeldt–Jakob disease (CJD)

One of the transmissible neurodegenerative, or prion (proteinaceous infectious particle) diseases. A prion has been defined as a small infectious pathogen containing protein, and is resistant to procedures that modify or hydrolyse nucleic acid, including UV radiation-this raises issues regarding infection control – see Box 4.20. Prions are, however, fairly sensitive to procedures that digest or denature protein. Human prion diseases have certain properties in common:

- pathologic manifestations confined largely to the CNS
- long incubation times (kuru up to 30 years)
- progressive and fatal
- similar neuropathological features (astrocytosis with little inflammatory response and usually small vacuoles – the spongiform change)
- all result in the accumulation of an abnormal prion protein (PrP), a protease-resistant protein.
- Other prion diseases include:
  - scrapie, a spongiform – encephalopathy of sheep and goats
  - bovine spongiform encephalopathy (BSE)
  - kuru (p.514)
  - Gerstmann-Sträussler-Scheinker syndrome (p.514)
  - familial fatal insomnia (p.515)

Epidemiology

- Familial cases are inherited in an autosomal dominant fashion – penetrance varies. Mean age of onset is around 65 years and they have the longest clinical course.
- Sporadic cases (sCJD, around 90% instances) are rare, with an incidence of around 1 per million worldwide. Age of onset 57–62 years. Cases have occurred in older teenagers. Early onset should prompt the consideration of iatrogenic sources of infection, as cases have occurred in association with cadaveric growth hormone and gonadotropin, dural grafts, corneal transplants, liver transplant, and contaminated neurosurgical instruments. Blood can be demonstrated to have low-level infectivity but there have been no recorded cases of CJD transmission by blood or blood products, nor do epidemiological studies identify previous transfusions as a risk factor for the development of CJD. The illness duration is around 4–8 months.
- New variant CJD (vCJD) is an entity distinct from sCJD, affecting younger patients (mean age 29 years, compared with 65 years for sCJD), with a longer illness duration (14 months), and a distinct clinical presentation. vCJD is attributed to consumption of cattle infected with bovine-spongiform encephalopathy (BSE). A large outbreak of BSE in the UK, thought to be due to cattle feed made from scrapie-infected sheep carcasses, peaked in 1992, 4 years after the introduction of a ban on ruminant feed. Five million animals were slaughtered in the effort to halt the epidemic. By October 2008, 167 human cases of definite or probable vCJD had been reported in the UK of which 164 had died. The number of reported cases has remained relatively stable. Only five cases were confirmed each year between 2005 and 2007.

Pathogenesis

Much of our current understanding of prion disease comes from studies of scrapie. A protein similar to PrP^{sc}, PrP^{C} is present in the brains of uninfected animals and is expressed constitutively – it is thought to have a role in neuronal development. The proteins have very different biological properties (PrP^{C} is sensitive to protease degradation and exists predominantly on the cell surface in an alpha-helix form, PrP^{sc} is resistant to protease degradation and is found predominantly intracellularly with a beta-sheet secondary structure). It is thought that PrP^{sc} is formed by post-translational modification of PrP^{C}. This conformational change is thought to be induced by the interaction of PrP^{sc} with normal PrP protein, resulting in an exponential increase in abnormal forms. Several mutations in the gene encoding prion protein have been identified in familial CJD, and some have associations with particular disease phenotypes (e.g. age of onset, rate of progression, associated symptoms). Prion gene mutations are very rarely found in cases of sporadic CJD.

Clinical features

- sCJD presents as a rapidly progressive dementia with myoclonus. The early clinical picture of around one-third of patients may be dominated by visual or cerebellar features. Myoclonus can be aggravated if the patient is startled. Two-thirds of patients eventually develop extrapyramidal signs; 40–80% develop corticospinal tract signs such as hyperreflexia and spasticity; visual features include cortical blindness and agnosia. Seizures, sensory signs, cranial nerve lesions, and autonomic dysfunction may occur but are uncommon. Average duration from symptom onset to death is 7–9 months. There are rare cases of long duration.
- vCJD cases are dominated by sensory (pain and parasthesia of face, hands, feet, and legs) and early psychiatric features such as depression and delusions which persist until obscured by dementia. They show the spongiform changes and neuronal loss of sCJD but differ in having prominent cerebellar involvement and widespread PrP^{sc}-positive amyloid plaques throughout the cerebellum and cerebrum.

Diagnosis

- Routine tests, including standard CSF examination, are rarely helpful. Other causes of dementia and encephalopathy should be excluded (e.g. syphilis, HIV, nutritional, metabolic).
- Imaging – MRI is more sensitive than CT imaging in picking up abnormalities associated with CJD, and certain findings specific to vCJD, but a normal brain scan does not significantly decrease the diagnostic probability of CJD.
- EEG – between 70% and 95% of patients ultimately develop a typical EEG pattern of slow background waves interrupted by generalized bilaterally synchronous biphasic or triphasic periodic sharp wave complexes. They are said to be 86% specific for CJD. Lack of this pattern after 4 months of symptoms should challenge the diagnosis. These abnormalities may not be seen in familial or variant CJD.
- Certain specific CSF proteins have been noted to have an association with CJD: protein 14-3-3 (sensitivity and specificity around 96% in sporadic cases, but only 50% and 91% respectively in vCJD), tau

protein (the best sensitivity and specificity for vCJD at 80% and 94% respectively). Elevations may be seen in herpes encephalitis, metabolic encephalopathies, and cerebral metastases.
- Tonsil biopsy – patients with vCJD (but not sporadic) may have detectable PrP^{sc} in follicular dendritic cells within lymphoid germinal centres. This may be detected by immunocytochemistry and this is highly sensitive and specific for vCJD.
- The gold standard remains examination of brain material: spongiform change, neuronal loss, reactive gliosis and little inflammatory response. PrP^{sc} can be identified by Western blot of material obtained at autopsy or biopsy.

Treatment

- All prion diseases seem to be invariably fatal and there is no effective therapy. Symptom progression has apparently slowed in individual patients treated with pentosan polysulphate. Trials are in progress assessing the efficacy of pentosan polysulphate and quinacrine.

Kuru

Originally endemic within a specific tribal group of Papua New Guinea, epidemiological studies suggested it was transmitted through ritual cannibalism and there have been no further cases since the practice was abandoned. A prodromal phase of headache and arthralgia is followed by progressive neurological decline (ataxia, tremor, choreoathetosis, myoclonus) and dementia. Cranial nerve abnormalities, weakness, and sensory loss occur late in the disease if at all. Laboratory tests are unhelpful and EEG does not share the characteristic features of that seen in some cases of CJD. The pathologically distinct features of kuru is the presence of PrP^{sc} plaques, predominantly in the cerebellum (similar but not identical to those seen in vCJD).

Gerstmann–Sträussler–Scheinker syndrome

A rare prion disease (incidence <10 cases per 100 million people per year) predominantly familial in nature. It is inherited in an autosomal dominant fashion with nearly complete penetrance. The key feature is spinocerebellar degeneration with dementia developing at a mean age of around 43–48 years. The average illness duration is 5 years. It is associated with several prion protein gene mutations, which may contribute to the clinical heterogeneity – some families have prominent dementia, others have extrapyramidal signs for example. Even within a family the same mutation can produce a varied phenotype. Laboratory tests including EEG are helpful only insofar as they may allow the exclusion of other diagnoses. Definitive diagnosis requires the examination of brain, which shows the finding typical of prion disease together with plaques similar to those of kuru and vCJD. These plaques may have an atypical distribution resembling that of Alzheimer's disease, and in the past some cases may have been diagnosed as Alzheimer's disease.

Fatal familial insomnia

Fatal familial insomnia (FFI) is a prion disease showing autosomal dominant inheritance and presenting in middle age or above (range 35–61 years) with insomnia, autonomic dysfunction (hyperthermia, hypertension etc), and motor abnormalities (ataxia, myoclonus etc). Patients can experience hallucinations, confusion and memory problems, but dementia is rare. Some patients develop endocrine disturbance (increased cortisol, loss of circadian pattern of growth hormone secretion). Average disease duration is 13 months. Neuropathology (neuronal loss, gliosis) is focused in thalamic nuclei, with changes seen occasionally in the cerebellar and cerebral cortices. Spongifrom change is rare. PrP^{sc} can be identified on immunostaining, but concentrations are the lowest of the prion diseases. Changes may be seen in EEG and sleep studies.

Box 4.20 Transmission of prion diseases

Kuru, CJD and BSE have been transmitted to primates via the oral route. Repeated oral inoculations are more effective than single doses. There is not evidence that ingestion is a means of transmission for sCJD, Gerstmann–Sträussler–Scheinker syndrome (GSS) or FFI. The highest concentrations of infectious material are to be found in brain, spinal cord, and eye, with material also present in CSF, lymphoreticular organs, lung, and kidney. There have been four cases of probable blood transfusion-transmitted CJD. Iatrogenic transmission appears to require direct inoculation, implantation, or transplantation of infectious material. In terms of healthcare workers' protection, universal precautions are adequate and care should be taken to identify and safely dispose of material that may carry an infection risk. There are specific guidelines for the performing of autopsies in suspected cases of CJD. Controversy continues on the best means of sterilizing instruments or other materials known to contain prions. For guidance use "Transmissible spongiform encephalopathy agents: Safe working and prevention of infection." Guidance from the Advisory Committee on Dangerous Pathogens TSE Working Group. http://www.advisorybodies.doh.gov.uk/acdp/tseguidance/index.htm.

Overview of fungi

Fungi are aerobic eukaryotes with limited anaerobic capabilities. They have chitinous cell walls and ergosterol-containing plasma membranes (human cell walls contain cholesterol). They may grow as yeasts (single-celled, reproduce by budding), moulds (form multicellular hyphae which grow by branching and extension) or both – dimorphic, growing as yeasts *in vivo*, and at 37°C *in vitro* but as moulds at 25°C. Fungal infection is the seventh commonest cause of infection-related death in the USA, with HIV the commonest predisposing factor.

Reproduction

Fungal reproduction may be sexual or asexual. Virtually all fungi can produce asexual spores by mitosis. Sexual spores are formed by the fusion of two haploid nuclei followed by meiotic division of the diploid nucleus (thus they carry half the chromosome number). Certain fungi can only sexually reproduce with other colonies of a different, compatible mating type (e.g. *Histoplasma* spp.).

Pathogenesis

Fungal infections may be cutaneous (e.g. dermatophyte infection, pityriasis versicolor), subcutaneous (e.g. following traumatic inoculation: sporotrichosis), or systemic (see Table 4.24). Systemic mycoses usually follow inhalation-acquired primary lung infection but may be caused by normal flora in an immunocompromised host (e.g. *Candida albicans* infection may be part of the normal flora, yet cause systemic disease in the immunosuppressed). Disease may be a consequence of toxin production (e.g. aflatoxin) or the host immune response to an infecting agent. Organism characteristics facilitating infection include: good growth at 37°C, the production of substances such as keratinases by dermatophytes (digests keratin in skin, hair, and nails), the ability to change form (exist in nature as moulds but take on yeast forms in a host allowing them to spread and become pathogenic), the ability to adhere to surfaces (e.g. *Candida albicans*), antiphagocytic capsules (*Cryptococcus neoformans*), persistence following phagocytosis allowing dissemination with macrophages. Generally, hosts have a high level of innate immunity to fungi – most infections are mild and self-limiting. This resistance derives from the fatty acid content of the skin, pH of the skin, mucosal surfaces and body fluids, epithelial turnover, competition with the normal bacterial flora, transferrin, cilia of respiratory tract. Cell-mediated immunity (CMI) is important in controlling fungal infection. Humoral responses play a part but patients with defects in CMI experience more-severe fungal infections than those with humoral defects.

Table 4.24 Overview of common fungi causing human disease

Phylum	Organism	Syndrome				Comment
		D	O	SC	SU	
	Yeasts					
Basidiomycota	*Cryptococcus neoformans*	•	•			Mild lung granuloma in healthy
	Trichosporon beigelii		•		•	May disseminate
	Rhodotorula spp.		•			
	Malassezia furfur		•		•	Pityriasis versicolor
Ascomycota	*Candida* spp.		•		•	
	Pneumocystis jiroveci (previously *P. carinii*)		•			
	Dimorphic fungi					
	Histoplasma capsulatum (*Ajellomyces capsulata*)	•				Histoplasmosis
	Blastomyces dermatidis (*Ajellomyces dermatitis*)	•				Blastomycosis
	Sporothrix schenkii	•	•	•		Acquired by local trauma. Rare dissemination
	Coccidioides immitis, *Paracoccidioides brasilensis*	•				In the Americas
	Penicillium spp.		•			Disseminate in immunosuppressed
	Moulds					
	Aspergillus spp.	•	•		•	Allergic, localized and invasive disease
	Epidermophyton spp.; *Trichophyton* spp.; *Microsporum* spp.				•	Dermatophytes – infect skin, hair and nails.
	Fusarium spp.		•		•	Disseminate in immunosuppressed
	Pseudallescheria boydii		•	•		Mycetoma. May also infect any organ or disseminate
	Madurella spp.; *Acremonium* spp.; *Exophiala* spp., etc			•		Mycetoma
Zygomycota	*Mucor* spp.; *Rhizopus* spp.	•	•			Rare invasion, e.g. mucormycosis

D: disseminated infection; O: opportunistic infection; SC: subcutaneous infection; SU: superficial infections.

Epidemiology

- Host factors play an important part in the epidemiology of fungal infection. Immunocompromise leads to a general increase in opportunistic fungal infection, but certain conditions predispose to specific organisms, e.g. rhinocerebral mucormycosis in patients with diabetic ketoacidosis, histoplasmosis in AIDS patients. Less dramatically, the use of antibiotics may increase the rate of candidal vaginitis.
- Environment affects the pattern of fungal disease. Some are found worldwide but are seen mostly in individuals whose lifestyles place them at risk of exposure and inoculation (e.g. a gardener experiencing subcutaneous inoculation of *Sporothrix schenckii* through minor trauma). Others are more likely to be seen in people living in or visiting specific regions (e.g. *Coccidioides immitis* in the desert of southwestern United States).

Candida species

A yeast, and the most common cause of fungal infection: candidaemia is the 9th-most commonly reported bloodstream infection in the UK, and the 4th in the US. *C. albicans* is responsible for 90% of cases of infection, and for 40–50% of cases of fungaemia. Non-albicans species are increasingly associated with invasive candidiasis and tend to be more resistant to certain antifungal drugs.

Mycology and epidemiology

- Small ovoid cells that reproduce by budding. Both sexual and asexual forms exist. Of the over 150 *Candida* species, only nine are frequent human pathogens: *C. glabrata* and *C albicans* account for approximately 70–80% of cases of invasive candidiasis. The others are: *C. guilliermondii (Pichia guilliermondii)*, *C. krusei (Issatchenkia orientalis)*, *C. parapsilosis*, *C. tropicalis*, *C. pseudotropicalis (Kluyveromyces marxianus)*, *C. lusitaniae (Clavispora lusitaniae)*, and *C. dubliniensis*.
- *C. albicans* is ubiquitous and may be found in soil, food, and hospital environments. They are normal commensals of humans (skin, sputum, GI tract, female genital tract etc). The vast majority of human infections are of endogenous origin.
- *C. krusei* is found in many environmental sites. It is fluconazole resistant and often found colonizing patients receiving fluconazole prophylaxis.
- *C. parapsilosis* (adheres well to synthetic materials) and *C. tropicalis* are now more common causes of IV catheter infections and endocarditis than *C. albicans*.
- *C. glabrata* infections of ICU patients are associated with a low survival rate. *Candida* spp. are uncommon laboratory contaminants.

Pathogenesis

The rise of *Candida* spp. infection relates to the increase in medical interventions: the use of antibiotics (suppressing normal bacterial flora and permitting the proliferation of *Candida* organisms), intravenous catheters (providing a route of entry) or prosthetic implants, and GI tract surgery. Immune suppression mediated by disease (e.g. HIV) or therapies such as steroids are also associated with increased rates of infection. The immune response to *Candida* infection is mediated by humoral and cellular mechanisms (cf patients with AIDS who demonstrate high susceptibility to cutaneous infection). *Candida* spp. virulence factors include surface molecules that permit organism adherence to other structures (human cells, extracellular matrix, prosthetic devices), acid proteases, and the ability to convert to a hyphal form.

General points on diagnosis

- Many patients will require early treatment in the absence of a conclusive microbiological diagnosis. Those at risk (e.g. the neutropenic) who remain febrile despite broad-spectrum antibiotic therapy should be suspected of having systemic candidiasis. Therapy should be started early and empirically in such patients. Always consider positive culture results from sterile sites to be significant. *Candida* may contaminate blood cultures but treatment is usually indicated, as distinguishing contamination from infection is very difficult. Blood cultures are positive in 50–60% of cases of disseminated disease. Serology is rarely useful.
- **Culture –** strict aerobes that grow well in routine blood cultures (usually positive in 48–96 h in cases of candidaemia) and on agar plates (smooth white colonies). Yeast forms (Gram-positive) and hyphae may be found on microscopy of clinical specimens (facilitated by 10% KOH). Presumptive identification of *C. albicans* is possible by inoculating organisms from a colony into a small tube of serum – germ tubes should form within 90 min, which tend not to be seen with the other species. There are relatively high rates of false-positive and false-negative germ tube tests. Accurate speciation relies on physiological characterisitics (e.g. fermentation, nitrate utilization), and can be demonstrated on commercial indicator agar preparations (ChromAgar) and with multiparameter kits.
- **Fungal antigen detection assays** are useful adjuncts in the diagnosis and monitoring of invasive fungal infection:
  - 1-3 beta-d-glucan – a cell wall component in a wide variety of fungi excepting cryptococcus and the zygomycetes. It is a broad-spectrum assay that detects Aspergillus, Candida, Fusarium, Acremonium, and Saccharomyces species The commercial blood assay (Glucatell®) has a sensitivity of 75–100% and a specificity of 88–100%. Its negative predictive value is useful but the positive predictive value is limited by the ubiquitous nature of glucan – for example positive tests may follow exposure to the cellulose in certain haemodialysis membranes, or surgical gauze
  - Candida mannan assay – sensitivity ranges from 31–90% (less for non-albicans species).
- **Sensitivity testing –** *in vitro* susceptibility testing can help guide the treatment of candidiasis to a greater extent than for the other fungi. Testing is not always available locally. Knowledge of the infecting species is highly predictive of likely susceptibility and can be used to guide therapy. Susceptibility testing is important in managing deep infections of non-albicans species, particularly if the patient has been previously treated with an azole. *C. albicans* is rarely resistant to azoles, whereas certain non-albicans species may show intrinsic resistance to these and certain other antifungal agents (see Table 4.25).

Superficial *Candida* infections

See 📖 p.798 for details of the cutaneous manifestations of candidal infection.

Mucous membrane infection

Thrush

A form of oral candidiasis characterized by white, creamy patches on the tongue and oral mucosa. Scraping removes the lesions, leaving a sore bleeding surface. Diagnosis can be confirmed using a KOH smear or Gram stain to demonstrate hyphae and yeast forms. Other manifestations include acute atrophic candidiasis (affecting the tongue), chronic atrophic candidiasis (associated with denture use), angular cheilitis (not caused solely by *Candida*) and *Candida* leukoplakia (white plaques affecting cheek, lips and tongue. May be precancerous). Oral thrush is associated with the use of inhaled steroids (often resolves spontaneously even without reduction in steroid use, or with topical therapy), malignancy, and AIDS. Treatment is usually topical (systemic where this fails). Prophylaxis with fluconazole has been effective in the prevention of oral *Candida* infections in cancer and AIDS patients, but remains controversial. It is no longer routinely recommended in AIDS patients due to the development of resistance. It does reduce the rate of clinical candidiasis in patients undergoing immunosuppressive therapy.

Candida oesophagitis

The majority of cases are associated with HIV or the treatment of malignant disease of the haematopoietic or lymphatic systems. May occur in the absence of oral disease. Symptoms include dysphagia, retrosternal chest pain, nausea, and vomiting. Symptoms may be mild even in extensive disease. Diagnosis is made by endoscopy and biopsy. In practice, diagnosis is often made presumptively in those with AIDS or malignancy on the basis of oral thrush and symptoms of oesophagitis. Extensive disease may result in intraluminal protrusions and partial obstruction. Perforation is rare. Severely immunocompromised patients may be co-infected with CMV or HSV. Fluconazole has been demonstrated to have greater efficacy than ketoconazole in AIDS patients with *Candida* oesophagitis.

Gastrointestinal candidiasis

Usually associated with malignant disease, the commonest manifestation being focal invasion of benign stomach ulcers. Diffuse gastric mucosal involvement is rare. Small and large bowel infection also occurs with white plaques, erosions, pseudomembrane, and ulceration visible on endoscopy.

Vulvovaginitis

Candida is the commonest cause of vaginitis, and 75% of women have at least one episode in their lives. Predisposing factors include diabetes, antibiotic therapy, and pregnancy. Oedema and vulval pruritis may be accompanied by discharge which may be scanty or thick. Secondary infection of perineal skin and the urethra can occur.

Invasive *Candida* infections

Candidaemia

Candida bloodstream infection is becoming more common with a rise in the number of susceptible patients and the increased use of indwelling catheters. Venous and arterial vessels may be affected, as well as prosthetic vascular materials. Complications: obstruction (e.g. of the superior vena cava (SVC)), endocarditis, pulmonary venous thrombosis. Disease may be more extensive than symptoms suggest. Patients should always be checked for endocarditis and eye involvement. Management always involves removal of any lines or other prosthetic material.

CNS candidiasis

May infect both brain substance and meninges. Around 50% of *Candida* meningitis cases occur in the context of disseminated disease. Meningitis may present non-specifically or with features typical of meningism. Parenchymal infection takes the form of scattered multiple microabscesses and can have extremely variable clinical presentations. Infection may follow trauma, neurosurgery, or colonization of a ventricular shunt. CSF may show lymphocytosis with low glucose, but these findings are not consistent. Organisms are visible on Gram stain in 40% of cases. The mortality rate is very high without therapy. Hydrocephalus is a common complication.

Cardiac candidiasis

Candida endocarditis usually affects the aortic and mitral valves. Associations include valve disease, chemotherapy, implantation of prosthetic heart valves, prolonged use of IV catheters, heroin addiction (*C. parapsilosis* the commonest cause), and pre-existing bacterial endocarditis. Around 50% of cases follow cardiac surgery (associated with the length of the postoperative course reflecting use of IV lines and antibiotics). Most cases present within the first 2 months post-operatively; <40% of cases are caused by non-albicans species; >70% of patients have positive blood cultures. Prior to antifungal therapy, mortality was 90%. Combined prolonged (6–10 weeks) medical and early surgical treatment has brought this to around 45%. There is a high risk of relapse. Patients should be followed up for at least 2 years postoperatively. It may be appropriate to use long-term suppressive therapy, e.g. fluconazole. Other cardiac manifestations – myocardial microabscesses may be a relatively frequent occurrence in cases of disseminated candidiasis. *Candida* is emerging as a cause of pericarditis.

Urinary tract infection

Probably follows extension of vaginitis in women or acquired by sexual contact with a woman with candidal vaginitis in men. There is frequently a history of recent antibiotic use. Candiduria is a common finding (particularly in those with urinary catheters) and does not equal renal tract infection. *Candida* cystitis is associated with prolonged catheterization. Bladder perforation occurs in severe cases. Renal infection may occur haematogenously or less commonly by retrograde spread (particularly in association with renal tract obstruction or diabetes) causing papillary necrosis, fungal balls, and perinephric abscesses. Surgery may be required to remove fungal balls. Prosthetic material within the tract should be removed. Asymptomatic candiduria should be treated in renal transplant patients, the neutropenic, low birth weight infants, and those undergoing urinary tract interventions (e.g. nephrostomy).

Bone and joint infection

Candida osteomyelitis may affect the vertebrae and discs, wrist, femur, scapula, humerus, and the costochondral junctions. Blood cultures are usually negative. Diagnosis is by aspiration of the affected area. Most cases follow haematogenous spread but may occur secondary to spread from the skin. Surgery may be required. Septic arthritis due to *Candida* occurs in the context of dissemination, trauma, surgery, and intra-articular injection of steroids. It may be a complication of AIDS and rheumatoid arthritis. *C. albicans* is the commonest cause in the context of disseminated infection. Non-albicans species are commoner when infection is local.

Intra-abdominal infection

Peritonitis may complicate peritonial dialysis (PD), GI perforation, and surgery. Infection tends to remain localized – dissemination is extremely rare in cases associated with PD, and around 25% in cases secondary to GI perforation. Other GI organs that can be affected include the gall bladder, spleen, liver, and pancreas. Hepatosplenic infection occurs in the severely immunocompromised. *Candida* peritonitis due to PD can be treated by local instillation of amphotericin B (can be painful) or fluconazole. Catheter removal may be indicated. *Candida* from post-op drains need not always prompt antifungal treatment. Therapy is required for treatment of *Candida* identified in ascites from an undrained abdomen. The failure rate for treatment of liver and spleen infection is high, even with combination therapy. Results may be better with liposomal amphotericin. Candidiasis of the biliary tract may require drainage.

Disseminated candidiasis

Patients at risk of dissemination include those with malignancy (particularly acute leukaemia), burns patients, and those with complicated post-operative courses (e.g. organ transplants, GI tract surgery). Multiple organs tend to be affected, most frequently the kidney, brain, myocardium, and eye. The pathological features are small abscesses and diffuse microabscesses with a granulomatous reaction. Many patients have negative blood cultures, and diagnosis is often made very late or at post-mortem. Tests for *Candida* antigen and serum antibodies have a high rate of false-negative results – diagnosis is largely clinical. A positive blood culture should be considered abnormal. The consensus is that all patients should be treated with antifungals and examined carefully for evidence of ocular involvement or other manifestations of disseminated disease. Cultures should be repeated several times and catheters removed or replaced (but not by passing the new one over a wire at the site of the old). Speciation and isolate sensitivity should be confirmed. Therapy should be continued for 2 weeks after the last positive blood culture.

Other

- Respiratory tract candidiasis – may cause a diffuse infiltrate following haematogenous spread (resembling heart failure or pneumocystis pneumonia (PCP) in the early stages) or a bronchopneumonia due to local inoculation from the bronchial tree. Definitive diagnosis depends on biopsy: recovery of *Candida* from sputum or bronchoalveolar lavage (BAL) alone is not sufficient (high levels of colonization).
- Ocular infection – see 🕮 Uveitis, p.784.

Treatment

The treatment of *Candida* infections is summarised in Table 4.25.

Table 4.25 Treatment of *Candida* infections

Presentation	First-line treatment	Duration	Comment
Candidaemia in non-neutropenic adults	Flucon 800mg loading, then 400mg OD. Alternative: echinocandin, LPAmB of voricon	14 days after last positive blood culture and resolution of symptoms	Remove intravascular catheters. Examine eyes. Initial treatment with echinocandin preferred for moderate-severe illness and those with recent azole exposure
Candidaemia in neutropenic adults	LPAmB IV or echinocandin. Alternative: flucon 800mg load, then 400mg daily	14–21 days after last positive culture, symptom resolution & resolved neutropenia	Gastrointestinal sources of infection are common. Flucon acceptable in those without azole exposure who are not critically ill. Voricon recommended if mold coverage required
Chronic disseminated candidiasis	Flucon in stable patients. AmB for severely unwell	3–6 months and resolution or calcification of radiological lesions	Flucon may be given after 1–2 weeks of AmB therapy if stable or improved
Neonatal candidiasis	AmB IV (alternative Flucon)	At least 3 weeks. Longer if delayed improvement	Perform LP and retinal exam on all neonates with suspected candidiasis. Remove IV catheters

Table 4.25 Treatment of *Candida* infections[1] (*Continued*)

Presentation	First-line treatment	Duration	Comment
Pyelonephritis	Flucon 200–400mg daily. Alternative: AmB with or without 5-FC (esp, cases due to non-*albicans* species)	7–14 days	Remove/replace stents/catheters. If disseminated treat as for candidaemia. Urinary fungal balis may require surgical removal and local AmB irrigation
Endocarditis	LPAmB IV with or without 5-FC. Alternative: echinocandin	At least 6 weeks after valve replacement	Stable patients with susceptible organism and negative blood cultures may complete on flucon. Valve replacement is nearly always necessary. Long-term suppression with Flucon may help if this is not possible
Meningitis	LPAmB IV with or without 5-FC IV, followed by flucon. Alternative: flucon 400–800mg OD	At least 4 weeks after resolution of symptoms and normalized CSF findings	Remove shunts if possible. Flucon has been used as a long-term suppressive therapy
Endophthalmitis	AmB IV or Flucon IV/PO	6–12 weeks after surgery	Vitrectomy is usually necessary if vitreitis present
Oropharyngeal candidiasis	Clo or nystatin or Flucon PO (Refractory disease, AmB IV/PO, Itracon, voricon)	7–14 days after clinical improvement	Long-term suppression with Flucon in patients with AIDS does not appear to lead to resistance
Oesophageal candidiasis	Flucon PO/IV or echinocandin. Alternative: Itracon, voricon		IV therapy may be required in severe cases
Genital candidiasis	Topical nystatin or Clo or short course Flucon PO	1–7 days	10% with complicated vaginal candidiasis (severe or recurrent or caused by non-*albicans* species or in an abnormal host) may need 7–10 days of therapy with a non-azole

Key: AmB (conventional amphotericin B), LPAmB (liposomal amphotericin B), Flucon (fluconazole), Clo (clotrimazole), 5-FC (5-flucytosine), Itracon (itraconazole), Voricon (voriconazole).

Reference

1 Pappas et al. Guidelines for the treatment of candidasis. *Clinical Infectious Disease* 2009:48, p.503–535.

Points to remember in treatment

- Remove infected IV lines, replace infected valves if possible.
- Amphotericin B is the first-line agent for the treatment of disseminated and deep-organ infection – most strains are sensitive. It may be combined with flucytosine in systemic neonatal infection.
- Knowledge of the infecting species is highly predictive of likely susceptibility and can be used to guide therapy. The new azoles (e.g. voriconazole) and candins (caspofungin) are useful against non-albicans species which are commonly fluconazole resistant:
    - *C. glabrata* is less susceptible to azoles and amphotericin B
    - *C. krusei* is intrinsically resistant to ketoconazole and fluconazole and less susceptible to other antifungals, including itraconazole, voriconazole, and amphotericin B. It remains sensitive to the candins
    - *C. lusitaniae* is susceptible to azoles but frequently resistant to amphotericin B.
- Susceptibility testing is important in managing deep infections of non-albicans species, particularly if the patient has been previously treated with an azole when resistance must be considered.
- Azoles are a useful continuation therapy for *C. albicans* infections initially controlled with IV amphotericin B.

Malessezia infections

Lipophilic yeasts (grow in the presence of certain fatty acids), oval or round in shape and normal commensals of the skin which they colonize in late childhood. They are a cause of skin infections and have been known to cause catheter-acquired sepsis.

Pityriasis versicolor

Superficial skin infection characterized by hypopigmented lesions usually confined to the trunk and proximal limbs. Usually caused by *M. globosa* or *M. furfur (Pityrosporum orbiculare)*.

- **Pathogenesis** – clinical infection is usually associated with yeasts transforming to hyphal forms from their round/oval appearance. Although seen at greater frequency in those with Cushing's syndrome, there is no clear association with T-cell suppression. Commoner in the tropics and may be precipitated by sun exposure. A carboxylic acid produced by the yeast may lead to the depigmentation.
- **Clinical features** – non-itchy macules develop on the trunk and proximal limbs and may be hypo- or hyperpigmented. They can coalesce forming scaly plaques.
- **Diagnosis** – direct microscopy of the lesions will reveal yeasts and hyphae. They may fluoresce under UV light. Organisms are best seen in skin scrapings after ink and potassium hydroxide staining.
- **Treatment** – some lesions resolve spontaneously. Otherwise topical treatment for 2 weeks with an azole, terbinafine cream, selenium lotion, or 20% sodium thiosulphate is usually effective. Severe cases may require a course of oral azole.

Malassezia folliculitis

Topical therapy may be effective. Systemic treatment is often necessary. Three clinical presentations:

- itchy papules/pustules on the back and upper chest, sometimes appearing after sun exposure
- multiple small papules across the back and chest in patients with seborrheic dermatitis. Lesions may display erythema and scaling
- multiple pustules across trunk and face in patients with HIV.

Seborrheic dermatitis

Once thought simply to colonize a pre-exisiting area of hyperproliferative skin, *Malassezia* is now implicated in pathogenesis.

- **Pathogenesis** – although it is unlikely that direct invasion precipitates the appearance of seborrhoeic dermatitis, an indirect mechanism such as sensitization may exist. Most cases resolve with a course of azole, and improvement in clinical appearance follows eradication of the organism.
- **Clinical features** – erythema, itch and scaling of chest and upper back. Facial lesions have a greasy appearance and appear around the eyebrows, ears, and nose. Scalp lesions are scaly, and pustules may develop. Patients with AIDS may develop abrupt and widespread lesions.
- **Diagnosis** is clinical and topical treatment is usually sufficient.

Other yeasts

Trichosporon species

Trichosporon beigelii can be part of the commensal flora of humans. It can cause invasive infections in the immunocompromised and has also been identified in prosthetic valve infections. Trichosporonosis is an acute, febrile infection with dissemination to multiple organs. The means of acquisition is not clear. Diagnosis is by biopsy and culture. Blood cultures tend to be positive late in the course of illness. Amphotericin B has been used in treatment.

Rhodotorula species

A cause of disseminated infections in immunocompromised patients, usually acquired through IV catheter infection following bone marrow transplants.

Cryptococcus neoformans

Cryptococcosis is a systemic fungal disease caused by the yeast-like organism *Cryptococcus neoformans*.

The organism

C. neoformans is an encapsulated yeast-like organism that reproduces by budding. The cell is round or ovoid (4–6 micrometre in diameter), and surrounded by a capsule of variable size. There are four capsular serotypes (A to D). Types A and D were previously classified as *C. neoformans* var. *neoformans* and B and C as *C.neoformans* var. *gattii*. Genotypic evidence has led to a recent reclassification. Serotype A are considered var *grubii*, serotype D var *neoformans* and B and C are now considered a separate species: *Cryptococcus gattii (Cryptococcus bacillisporus)*. *C. gattii* tends to cause infections in the immunocompetent whereas *C. grubii* and *neoformans* (the majority of clinical isolates) infect the immunocompromised.

Epidemiology

C. neoformans is a ubiquitous environmental saprophyte. Serotype A (most common) and serotype D are found worldwide and have been isolated from pigeon droppings and nesting places, from contaminated soil and decaying woodchips or, occasionally, from fruit. Infections caused by serotypes B and C are largely restricted to tropical and subtropical areas. The organism has been cultured from eucalyptus trees.

Transmission

Circumstantial evidence suggests that infection occurs by inhalation of aerosolized organisms but there is no evidence of person-to-person transmission or laboratory-acquired infection. Rare routes of transmission include organ transplantation from infected donors or cutaneous inoculation. Cryptococcosis also occurs in animals but there is no evidence of zoonotic transmission.

Risk factors

Cryptococcosis occurs more commonly in patients with defects in T-cell-mediated immunity. Predisposing factors include:

- AIDS (80–90% of cryptococcal infections, AIDS-defining illness in 5–88% of patients, associated with CD4 count <100 cells/mm^3)
- post-transplantation (peak period 4–6 weeks)
- others – corticosteroid therapy, lymphoreticular malignancies (especially Hodgkin's disease), sarcoidosis.

Pathogenesis

A number of potential virulence factors have been identified:
- capsular polysaccharide
- melanin production
- mannitol production
- lack of soluble anticryptococcal factors in CSF.

The inflammatory response to infection is variable. The characteristic lesion consists of cystic clusters of fungi with no inflammatory response that are spread throughout the brain. Less commonly, focal inflammatory lesions (cryptococcomas) are found. In severe infections, the leptomeninges are thickened with distension of the subarachnoid by a white gelatinous material (attributed to capsular polysaccharide).

Clinical features

- **Cryptococcal meningitis** – onset may be acute or chronic, and symptoms may be mild and non-specific. On examination, fever may be absent, there may be minimal or no nuchal rigidity, papilloedema (30%), cranial nerve palsies (20%), blindness. Seizures occur late. Differential diagnosis: other mycoses, TB meningitis, viral meningoencephalitides, meningeal metastases.
- **Pulmonary cryptococcosis** – may be asymptomatic or present with dyspnoea, cough, and chest pain. Physical signs are unusual. May be rapidly progressive in AIDS. Differential diagnosis: tumour (HIV-negatives), *Pneumocystis* pneumonia, pulmonary TB, histoplasmosis.
- **Other sites** – *C. neoformans* may cause skin lesions, bone lesions, oral lesions, vulvar lesions, post-transplant pyelonephritis, prostatic cryptococcosis.

Laboratory diagnosis

- CSF examination – CSF findings include elevated opening pressure (may be >500 mm CSF), low glucose, high protein, high white cell count (>20/mm^3, lymphocyte predominance). CSF abnormalities may be minimal or absent in AIDS. A cryotococcal antigen test should be performed.
- India ink smear – India ink or nigrosin staining of the CSF deposit shows a capsule, double cell wall, and refractile inclusions in the cytoplasm in 20–50% of HIV negative cases and around 75% of AIDS patients.
- Fungal culture – appropriate samples include CSF deposit, sputum, and urine, which are cultured on Sabouraud agar. Positive blood cultures may occur in extensive infections. *C. neoformans* colonies are smooth, convex, and yellow or tan on solid media. It produces brown colonies on birdseed agar (melanin production). Unlike other yeasts, it does not produce pseudomycelia on cornmeal or Tween agar. It can be identified by biochemical tests, e.g. API or biotyping agars. There are commercially available DNA hybridization probes.

- Cryptococcal antigen test – a variety of latex agglutination tests are available with reported sensitivities of ≥90%. They can be performed on both CSF and serum. False-positive tests may occur but titres are usually ≤1:8 and can result from infection with *Trichosporan beigelii* and certain bacterial genera (e.g. capnocytophaga).
- Histopathology – methenamine silver or periodic-acid–Schiff staining show a yeast-like organism with narrow-based buds. Mayer's mucicarmine stain stains the capsule rose red.

Treatment

Meningitis

Multiple studies have compared various treatment regimes. All consist of an induction phase (usually 2 weeks), a consolidation phase (around 6 weeks), and for those patients with long-term immunosuppression (e.g. HIV with low CD4 counts) a long term maintenance phase as relapse is common. Aggressive management of raised ICP by repeated LPs has been associated with significantly better outcomes. Those with raised ICPs at diagnosis, or who develop symptoms suggestive of it, should have daily LPs to reduce the pressure to <20cm CSF or 50% of the opening pressure. Those requiring LPs after 4 weeks will probably need a permanent ventricular shunt.

- Induction should be with amphotericin and flucytosine. This has been shown to be the most rapidly fungicidal and should be continued for 2 weeks, or until the patient is asymptomatic.
- Consolidation follows with high dose fluconazole (400mg daily) for 6-10 weeks.
- Maintenance therapy with lower dose fluconazole (200mg daily) is required for at least a year in patients with HIV. Cessation may then be considered in those who have responded to HAART.
- Selected AIDS patients with mild, asymptomatic CNS cryptococcosis have been treated successfully with a long course of very high dose oral fluconazole (<1g/day) and flucytosine – however many have significant problems with drug side effects (GI and bone marrow toxicity).

Pulmonary disease

Mild to moderate infections in the immunocompetent can be treated with oral fluconazole alone. Those with severe disease, or multi-organ involvement should be treated as for meningitis. Surgical excision may be curative.

Prognosis

Prognosis depends on the severity of the illness at presentation and the nature of any underlying disease. Relapse is rare in those HIV patients who respond to treatment, continue suppressive therapy and experience improved immune function on HAART.

Reference

1 Perfect JR. et al. Clinical practice guidelines for the management of cryptococcal disease. *Clin Infect Dis* 2010;50:291–322. http://www.idsociety.org/content.aspx?id=9200#cryp.

Pneumocystis jiroveci

For many years *Pneumocystis carinii* was thought to be a protozoan but rRNA analysis suggests that it is more closely related to fungi. The human-derived organism was recently renamed *Pneumocystis jiroveci*, whereas the rat-derived organism remains *Pneumocystis carinii.*

Microbiology

P. jiroveci is an unusual fungus that lacks ergosterol in its cell wall and is therefore not susceptible to certain antifungals. It has proved extremely difficult to culture *in vitro*, so biochemical and metabolic studies of the organism have been limited. Three developmental stages exist: the trophic form, the sporocyte, and the spores.

Epidemiology

P. jiroveci is ubiquitous in the environment and has a worldwide distribution. Primary infection occurs in childhood and is probably asymptomatic. The principal mode of transmission is by the airborne route. Once infection is acquired, there is some debate about how long the organism resides in the host. One view is that *P. jiroveci* remains quiescent and reactivates when immune deficiency develops. The other school of thought is that infection is transient but people are frequently exposed to the organism. Risk factors include:

- premature, debilitated infants
- severe protein malnutrition
- primary immunodeficiency, e.g. severe combined immunodeficiency disease (SCID)
- immunosuppressive drugs, e.g. corticosteroids, cytotoxics
- organ transplantation.

Pathogenesis

Once inhaled, the trophic form attaches to the alveolar type I cell and undergoes proliferation. The host immune defects that contribute to the uncontrolled proliferation of *P. jiroveci* and the development of disease are incompletely understood, but impaired humoral and T-cell-mediated immunity both appear to be important. The host immune response results in the production of inflammatory cytokines (e.g. TNF-α and interleukin-1 (IL-1)), which may contribute to lung damage. The principal histological finding is a foamy eosinophilic alveolar exudate. There may be hyaline membrane formation, interstitial fibrosis, and oedema.

Clinical features

- **Pneumonia** – insidious onset of fever, dyspnoea, a non-productive cough, and reduced exercise tolerance. Occasionally there may be sputum production, haemoptysis, or chest pain. Examination reveals tachypnoea, tachycardia, exercise-induced hypoxia, and crackles (in <30% adults). Infants may be cyanosed with respiratory distress.
- **Extrapulmonary disease** – this occurs mainly in the context of advanced HIV infection and is rare. The most commonly affected sites are the lymph nodes, spleen, liver, bone marrow, GI tract, eyes, thyroid, adrenal glands, and kidneys. Clinical findings may vary from incidental

findings at autopsy to severe progressive disease, with or without pulmonary involvement.

Laboratory diagnosis

- Specimens – *P. jiroveci* is rarely found in expectorated sputum but can be found in induced sputum (50–90% sensitivity). Bronchoalveolar lavage (BAL) increases the diagnostic rate to >90%, especially if multiple lobes are sampled or the procedure is directed to the site(s) of radiographic involvement. Transbronchial biopsy may provide further information but is associated with risk of complications, e.g. bleeding, pneumothorax. Open lung biopsy may be helpful if BAL is non-diagnostic.
- Histopathology – a variety of stains have been used to identify *P. jiroveci*, e.g. methenamine silver, Wright Giemsa, calcofluor white. Commercial immunofluorescence tests are more sensitive than histological stains but more expensive. Immunohistochemistry has also been used.
- Molecular diagnostics – PCR amplification and detection of *P. jiroveci* DNA has proved highly sensitive and reasonably specific. However, the detection of a PCR product in a clinical specimen may represent subclinical infection or result from antibiotic pretreatment.

Treatment

The treatment of choice is high-dose co-trimoxazole (15–20 mg/kg/day trimethoprim, 75–100 mg/kg/day sulfamethoxazole) for 14–21 days. Side-effects are common in HIV patients and include: skin rash, fever, cytopaenias, nausea, vomiting, hepatitis, pancreatitis, nephritis, hyperkalaemia, metabolic acidosis, CNS symptoms.

Alternative regimens include:

- trimethoprim and dapsone
- clindamycin and primaquine
- atovaquone
- pentamidine isetionate
- trimetrexate.

Adjunctive corticosteroids are recommended in patients with hypoxia.

Prognosis

Untreated PCP in immunocompromsed patients is fatal. Poor prognostic factors include hypoxia (partial pressure of arterial oxygen (PaO_2) <7kPa), high alveolar–arterial oxygen gradient (>45 mmHg), extensive pulmonary infiltrates. The short-term mortality (1–3 months) in HIV-infected patients has fallen to 10–20%, whereas the mortality in HIV-negative patients remains 30–50%. Patients who recover are at risk of developing recurrent episodes or pneumothoraces.

Prevention

- HIV-infected patients – primary prophylaxis with co-trimoxazole is recommended in HIV-infected patients with a CD4 count <200 cells/mm^3. Secondary prophylaxis is recommended for life but may be discontinued in patients with CD4 count consistently >200 cells/mm^3. Alternative regimens: dapsone ± pyrimethamine + folinic acid or IV pentamidine isethionate.
- HIV-negative patients – although no formal guidelines exist, chemoprophylaxis should probably be considered in all patients with predisposing conditions (see above). Co-trimoxazole is the agent of choice as there is limited clinical experience in HIV-negative individuals with other regimens.

Reference

Briel M, Bucher HE et al. Adjunctive corticosteroids for pneumocystis pneumonia in patients with HIV infection. *Cochrane database* 2006, article No.CD006150

Aspergillus

A mould capable of causing a wide range of disease in both healthy and immunocompromised individuals.

Mycology

- Many species cause invasive disease in humans. The most frequently identified include: *A. fumigatus* (around 90%), *A. flavus*, and *A. niger*.
- Pathogenic species grow better on routine mycological media at 37°C than non-pathogenic, most of which cannot grow at this temperature. Colonies become apparent at 36–90 h. Sporulation occurs up to 2 days later, but may take longer with less-common species.
- Identification of common species is possible by microscopic and colonial appearance, although more-detailed identification requires specialized media and molecular methods.
- Microscopy of pathological specimens may reveal hyphae (best seen on silver stains), but sporulation is not often seen (save in specimens taken from air-containing areas such as the lung), thus the organism cannot be distinguished from other pathogenic moulds by this means.

Epidemiology

The organism is found worldwide, favouring decomposing vegetable material (e.g. potted plants, spices, and particularly around farms). Molecular techniques have demonstrated that colonized individuals and those with aspergilloma tend to pick up several different genotypes over time. Most invasive infections, however, are caused by a single genotype.

Pathogenesis

- Disease spectrum is wide, from superficial infection, to allergic to disseminated. Time from exposure to disease in invasive aspergillosis ranges from 36 h to 3 months. Disease may follow infection by organisms that have already colonized an individual (e.g. in neutropenia).
- Many factors influence disease form and severity – organism growth rate (*A. fumigatus* being the fastest), spore size (*A. fumigatus*' small spores allow them to pass deep into the lung), the hydrophobic coat of conidia (protection from host defence), the ability to adhere to epithelial surfaces (achieved by *A. fumigatus* much more effectively than other species), enzyme/toxin production (e.g. aflatoxin produced by *A. flavus*).
- Host defences include lung macrophages (capable of ingesting and killing conidia), T lymphocytes appear to be important in chronic and allergic disease), complement proteins, and neutrophils (damage hyphae). Corticosteroids impair macrophage and neutrophil killing. Pathological appearances: vascular invasion (with consequent infarction of distal tissue) with acute invasive disease in the immunosuppressed; alveolar consolidation, necrotizing granulomatous pneumonia, and bronchiectatic cavities in chronic invasive aspergillosis; exudative bronchiolitis and eosinophilic pneumonia with allergic bronchopulmonary aspergillosis.

Clinical features of non-invasive *Aspergillus* disease

- **Superficial** – cutaneous infections are rare, usually occurring in neutropenic patients at the site of IV catheter insertion, or in burn wounds. More common is otomycosis – a growth of *A. niger* in those with chronic otitis externa which may cause itch and discomfort. Appearance may be of non-specific inflammation. Cleaning and topical therapy with an agent such as amphotericin B 3% or clotrimazole is curative.
- **Allergic disease** – **allergic bronchopulmonary aspergillosis** (ABPA) occurs in those with asthma or cystic fibrosis and hypersensitivity to airway colonization by aspergillus. Presents with worsening asthma or lung function. Patients may have an eosinophilic pneumonia or airway sputum impaction. Blood tests may reveal eosinophilia early in disease. Oral corticosteroids can help exacerbations, and inhaled steroids may prevent episodes occurring. Oral itraconazole may help those requiring long-term steroids. *Aspergillus* may also cause allergic sinusitis – this is best managed by aeration of the affected sinus.
- **Aspergilloma** – **pulmonary aspergilloma** follows *Aspergillus* colonization of pre-exisiting cavities or cysts left from TB, sarcoidosis, or *Pneumocystis* pneumonia. TB cavities 2 cm or larger have a 15–25% risk of developing an aspergilloma. Some patients are asymptomatic, most have productive cough, haemoptysis, weight loss, wheeze, and clubbing. Culture of sputum may reveal the organism, and precipitating IgG may be detected in serum. Radiological imaging demonstrates the hyphal mass as a cavity surrounded by a rim of air. Complications: massive haemoptysis (may be fatal – embolization may help those in whom surgery is not possible), contiguous spread of infection to

pleura or vertebrae, dissemination. Aspergilloma must be distinguished from chronic invasive disease requiring systemic therapy. Treatment: 10% of cases resolve spontaneously; surgical resection has a role in the treatment of isolated lesions in those with good lung function; amphotericin B has been injected into cavities with some effect; oral itraconazole provides symptomatic relief. Sinus aspergilloma may develop in the ethmoid or maxillary cavities. Surgical drainage is usually sufficient in those with no evidence of mucosal involvement. Medical therapy should be used in combination with surgery in those with invasive disease or involvement of the frontal or sphenoid sinus.

- **Eye** – *Aspergillus* keratitis (and less commonly endophthalmitis) may occur following ocular trauma or haematogenously (e.g. IVDU, endocarditis). Early recognition and treatment is essential for a good outcome. Corneal smears may reveal hyphae, and cultures are usually positive. Keratitis may be treated with topical amphotericin B (27% response when given hourly). Superficial infections may be treated with topical clotrimazole. Oral itraconazole is effective in up to 75% of cases. Surgery is required where medical therapy fails or where there is the threat of ocular perforation or the formation of a descemetocele (a herniation of the posterior limiting layer of the cornea). Vitrectomy may be required to establish the diagnosis and, where possible, intraoperative examination of the specimen for hyphae is useful as it allows immediate administration of intravitreal amphotericin B. Systemic therapy is also recommended.

Clinical features of invasive *Aspergillus* disease

- **Invasive pulmonary disease** – over 80% of patients with invasive disease have pulmonary infection. The immunocompromised tend to experience few symptoms but progress rapidly, whereas the less immune impaired have more symptoms with a slowly progressive, chronic course.
- **Acute invasive pulmonary aspergillosis** – around 25% have no symptoms in the early stages, with the remainder experiencing dry cough and a mild fever. Pleuritic chest pain and haemoptysis may occur, as may breathlessness in those with bilateral disease who can become hypoxic. Signs resemble pulmonary emblism (PE) or mucormycosis. CXR changes can be non-specific (consolidation, cavities, wedge-shaped lesions, lower lobe shadowing), and may appear normal in 10% of cases. High-resolution CT images are useful in aiding early diagnosis. Early in disease, radiological imaging may demonstrate nodules with the 'halo' sign (a zone of ground-glass attenuation surrounding a nodule or mass). In later disease, with neutrophil recovery, nodules may cavitate producing the 'air-crescent' sign. Focal or nodular disease has a better prognosis than diffuse or bilateral infection. In focal disease, the danger is of massive haemoptysis which may occur with no warning.
- **Chronic invasive pulmonary aspergillosis** – occurs less frequently than acute. Predisposing conditions include AIDS, alcoholism, diabetes mellitus, chronic granulomatous diseases, and corticosteroid therapy for chronic pulmonary diseases. Some patients have no identifiable predisposing factors. Presentation is with weeks of chronic productive cough. Other symptoms: haemoptysis, fever, weight loss. Infection

may extend to the chest wall, spine or brachial plexus. CXR shows cavitation and consolidation, and in the absence of previous images to demonstrate the presence or absence of a pre-exisiting cavity can be difficult to distinguish from aspergilloma. Definitive diagnosis can be made through positive culture of a biopsy specimen. Patients usually have strongly positive serum *Aspergillus* antibodies (with the possible exception of AIDS patients) which may help in biopsy-negative cases.

- **Airways** – more common in lung transplant recipients and AIDS patients, *Aspergillus* tracheobronchitis varies from mild inflammation to severe ulcerative disease; 80% experience symptoms (cough, fever, breathlessness, chest pain, haemoptysis) which become more severe with progression. Complications: wheeze or stridor, tracheal perforation, dissemination, airway occlusion, and death. Bronchoscopy allows definitive diagnosis. CXR is usually normal in early disease.
- **Sinus** – invasive *aspergillus* sinusitis can be acute or chronic:
  - acute – early symptoms include fever, cough, nose-bleeding, headaches, discharge, and sinus discomfort. Ulceration is preceded by decreased blood flow to the affected nasal areas – these may be identified by careful examination for loss of sensitivity. Infection may extend to the palate, orbit, or brain. CT or MRI allows identification of the extent of disease and culture or hyphal identification tissue confirms diagnosis
  - chronic – most have no identifiable immunocompromise. Early symptoms include nasal congestion and discharge, loss of smell, headache. It is clinically indistinguishable from other causes of sinusitis. As disease extends, proptosis, loss of vision, ocular pain, and even features of stroke may develop. Radiological features are similar to acute disease. Obtaining a positive culture may require multiple samples.
- **Brain** – 10–20% of cases of disseminated disease develop cerebral aspergillosis. It is rare in the immunocompetent when it is usually secondary to neurosurgery. Severely immunocompromised individuals have a non-specific presentation with confusion and seizures, and death following a few days later. Less severely immune impaired patients tend to have headache and focal neurological features. Fever may occur. Meningitis is rare. Contrast CT appearances are of infarction, or of a ring-enhancing abscess with oedema. Lesions may be deep and surgically inaccessible. Diagnosis rests on culture and microscopy of a biopsy or aspiration sample. This may not be possible and diagnosis can be made presumptively in those with invasive disease elsewhere and typical radiological appearances.
- **Other** – *Aspergillus* endocarditis can occur in isolation or as part of disseminated disease. Blood cultures are usually negative, and valve replacement is necessary to achieve cure. Other sites of infection include pericardial, intestinal, oesophageal, renal, vascular graft, and bone.

Diagnosis

- Radiology – once clinical suspicion of invasive disease has been raised, CT or MRI assessment of the lungs, sinuses, and brain should be performed within 24 h.
- Airway disease – bronchoscopy is useful but biopsy is rarely positive in focal lung disease. A needle biopsy or surgical resection (superior to open biopsy) is appropriate for peripheral pulmonary lesions. Focal lesions near the great vessels should prompt urgent resection given the high risk of massive haemoptysis. Respiratory samples should be examined by microscopy (rapid but does not distinguish species) and cultured. Antigen tests of such samples are available.
- Culture in invasive disease – positive cultures from any site should prompt a thorough evaluation, particularly in the immunocompromised. Definitive diagnosis of invasion requires culture of the organism from a sterile site. Combinations of other positive findings have less certainty as other moulds can give the same appearance. The significant *Aspergillus* species are *A. fumigatus*, *A. flavus*, *A. terreus*, or *A. niger*.
- Antibody testing – *Aspergillus* antibody tests are useful in the diagnosis of ABPA, aspergilloma, and chronic invasive disease. They are not sensitive in the severely immunocompromised.
- Serum antigen tests – the galactomannan (a fungal exoantigen released by all pathogenic *Aspergillus* species during growth) assay has moderate accuracy for diagnosis of invasive aspergillosis in immunocompromised patients. The test is most useful in patients who have haematological malignancy or who have undergone haematopoietic cell transplantation. It can become positive a week before clinical disease is manifest, and the titre corresponds with the tissue burden, a high level having prognostic significance. It is of low sensitivity in paediatric patients with primary immunodeficiencies. False positives may follow absorption of galactomannan from the diet in those with mucositis, or the administration of Tazocin®. The 1–3 beta-d-glucan test has a high negative predictive value which may help exclude invasive disease in adults (see 📖 *Candida* species, p.518).

Treatment[1]

- Invasive aspergillosis is fatal in virtually all cases if untreated. Good outcomes require aggressive diagnosis in those groups at risk, early presumptive treatment, early changes in treatment where the response is poor, and early surgical resection of focal lung lesions located near the hilum and great vessels. Evaluating the response takes longer in those with less immunocompromise.
- Invasive aspergillosis – Intravenous liposomal amphotericin B, voriconazole, or caspofungin are the treatments of choice. Treatment should be continued until the disease has stopped progressing (at least 2 weeks). It may be appropriate to continue therapy with voricoazole or posaconazole while the patient remains immunocompromised. Itraconazole was previously used but is limited by its variable bioavailability (e.g. in those with intestinal problems such as that resulting from graft-versus-host disease (GVHD)) and drug interactions (e.g. agents activating P-450, ciclosporin, protease inhibitors).

- Acute invasive sinusitis – first-line treatment is amphotericin B. Itraconazole is not as effective for this form of disease.
- Chronic invasive sinusitis – surgical debridement and prolonged medical therapy. Relapse is common.
- Cerebral disease – surgery is useful only for diagnosis unless the lesion is superficial and isolated. Otherwise treatment rests on antifungal therapy.
- Surgery – indicated for focal invasive pulmonary disease, persisting lung shadows prior to BMT or aggressive chemotherapy, significant haemoptysis, or lesions near to great vessels and airways.
- Other medical interventions: no significant clinical benefit yet seen with G-CSF, GM-CSF or IFN-gamma.
- See IDSA guidelines for treatment of aspergillosis.[2]

References

1 The Aspergillus Website. www.aspergillus.man.ac.uk (accessed 6 August 2008).

2 Walsh TJ, Anaissie EJ, Denning DW et al. Treatment of aspergillosis: clinical practice guidelines of the Infectious Diseases Society of America. *Clin Infect Dis* 2008;46:327–60.

Mucormycosis

Mucormycosis is a clinical syndrome caused by a number of fungal species belonging to the order *Mucorales* (class *Zygomycetes*).

Mycology

Spore forming and grow rapidly in the mould form in both tissues and the environment. Common species causing mucormycosis include *Rhizopus*, *Rhizomucor*, and *Mucor*. Microscopy allows a degree of speciation (e.g. appearance of the columellae and rhizoids if present).

Epidemiology

All members of the order are widespread in decaying matter. As testament to their ubiquitous nature, they have been isolated from wooden tongue depressors and infection has resulted from their use as splints in neonates. Clinical disease is limited largely to the immunocompromised, transplant patients, those with diabetes mellitus, and trauma patients. Infection may be rhinocerebral, respiratory, cutaneous, disseminated, or localized to specific organs.

Pathogenesis

Infection is acquired via the respiratory tract, or in primary cutaneous infection, via the inoculation of spores into skin abrasions. Spore germination follows in hosts whose immune response is deficient. Macrophages and neutrophils are important in preventing growth, and normal human serum is fungistatic. Invasive disease is favoured by hyperglycaemia, acidosis, and in patients receiving desferrioxamine (an agent which enhances fungal growth experimentally). Cases are not as significantly raised in those with AIDS as might otherwise be expected. Hyphae invade tissues, penetrate blood vessel walls, and may grow along the vessel contributing to thrombosis and necrosis.

Clinical features

- **Rhinocerebral disease** – a disease of the immunosuppressed, seen in diabetic patients (particularly if acidotic), and neutropenic leukaemia patients on antibiotics and almost invariably fatal. Fungal infection causes septic necrosis and infarction of the tissues of the nasopharynx and orbit. Patients develop facial pain or headache with fever and may have orbital cellulitus, with proptosis and conjunctival swelling, evolving cranial nerve defects, and black crusty material apparent in the nasopharynx. Fungal invasion of vessels may lead to retinal artery thrombosis and visual impairment. Other complications: ptosis/pupil dilatation (secondary to cranial nerve lesions), cerebral abscess, cavernous sinus/internal carotid artery thrombosis. X-ray of the sinuses may show mucosal thickening and fluid and bone destruction may be apparent on CT. Features may recur after apparently successful therapy – patients should be monitored.
- **Pulmonary disease** – usually secondary to neutropenia and seen in BMT or leukaemia patients receiving chemotherapy. Symptoms are initially mild: fever, mild shortness of breath, and cough. With progression, haemoptysis may develop and erosion of a blood vessel can cause severe pulmonary haemorrhage. CXR may show infiltration, consolidation, and cavities. Infection may start in one lung segment but often disseminates in the late stages (e.g. multiple lung area, spleen, kidney). Diabetics may develop a milder chronic form of pulmonary infection.
- **Cutaneous disease** – outbreaks have been associated with colonized bandages. The appearance is of cellulitus but if unrecognized the organism penetrates deeper into the skin, and necrosis may follow vascular invasion. Dissemination may follow. May take the appearance of a chronic ulcer. Cases have occurred with minor trauma (e.g. gardeners), major trauma, burns, and insect bites. Skin lesions may also develop following dissemination from a distant site.
- **Gastrointestinal disease** – seen in those suffering from malnutrition although cases have occurred in renal transplant recipients. Any part of the tract may be infected and it is rapidly fatal. Symptoms include abdominal pain, fever, nausea, and vomiting.
- **CNS disease** – rare and usually due to direct invasion from infected sinuses. Cases have occurred in leukaemia patients with no obvious route of acquisition and as a result of open head trauma. Presentation is with decreasing levels of consciousness and multiple focal neurological deficits.
- Other – endocarditis, osteomyelitis, renal infection, allergic sinusitis.

Diagnosis

- Clinical suspicion should be raised by the presence of vascular invasion and tissue necrosis which may manifest as black eschars and discharges. These are markers of advanced disease, and the earlier the diagnosis the better the outcome. Lesions may be apparent only on the nasal mucosa and palate. Diagnosis rests on identifying the organism in tissue biopsy. Swabs are insufficient. Non-septate branching fungal hyphae can be seen on routine haematoxylin and eosin (H&E) stains and help distinguish it from *Aspergillus*. There is usually an associated neutrophilic infiltrate, and tissue necrosis may follow blood vessel invasion with an inflammatory vasculitis. Organisms rarely appear in blood cultures. The differential includes *Aspergillus* infection, rapidly progressive orbital tumour, cavernous sinus thrombosis, pulmonary embolism, and acute leukaemia.

Treatment

- Good outcomes rest on early diagnosis and correction of any predisposing factors, e.g. acidiosis, hyperglycaemia, immunosuppresion. Overall mortality is around 50%.
- Invasive disease – high-dose amphotericin B (1–1.5 mg/kg/day initially moving to alternate day dosing once treatment established. Even higher doses have been used with liposomal preparations) in combination with aggressive surgical debricement of necrotic tissue. Reconstructive surgery may be necessary once recovered. Currently available azoles are not effective and other therapies suggested but unproven include colony-stimulating factors, oxygen therapy, the use of adjunctive rifampicin.
- Primary cutaneous disease – local debridement and topical amphotericin B. Treatment duration should be guided by response.

Eumycetoma

Mycetoma is a chronic, slow-growing destructive infection usually involving hands or feet and characterized by the spread of the infecting organism from its subcutaneous site of implantation to adjacent structures. Serous discharges contain small grains of organism colonies.

- Actinomycetoma – caused by filamentous branching bacteria
- Eumycetoma – caused by fungi

Epidemiology

Found worldwide in tropical regions – most commonly in India, Mexico, parts of sub-Saharan Africa, Yemen among others. Rare in temperate areas but may be seen in the southwestern USA. Causative organism varies with geography: *Madurella mycetomatis* accounts for most cases worldwide; *Madurella grisea* is a common cause in South America; *Pseudallescheria boydii* in the USA; *Leptosphaeria senegalensis* and *Leptosphaeria tompkinsii* are common causes in West Africa. Geographic distribution of the causative agents is related to local climate (e.g. rainfall).

Pathogenesis

Soil fungi enter tissues of the foot or hand after local trauma. Infection spreads along tissue planes destroying connective tissue and bone. Multiple sinuses and tracts form between the surface, each other, and deep abscesses. Inflammation and scarring leads to enlargement and disfiguring of the infected area. Histologically the appearance is of a suppurative granuloma with grains embedded in abscesses. Grain appearance is often characteristic of the specific organism.

Clinical features

Seen most frequently in men aged 20–40 years, often farmers and rural labourers. The foot is the most commonly affected area, other regions including the hand, leg, arm, head, thigh, and even back (carrying contaminated sacks). Early manifestation is a small painless nodule. This quickly increases in size (faster in actinomycetoma) and ruptures forming a sinus. Additional nodules appear in adjacent areas as some areas heal. The cycle of swelling, discharge, and scarring leads to a swollen mass of deformed tissue with multiple discharging fistulas. Lymphatic spread to regional nodes may occur. Cortex of bone may be invaded. Osteolytic lesions can be seen on x-ray. Pathological fractures are less common than would be expected. Constitutional symptoms are rare. Fever implies secondary bacterial infection.

Diagnosis

The appearance is fairly typical: indurated swelling, deformity with multiple sinus tracts draining grainy pus. Grains may be black, white, yellow, red, or pink, depending on the causative organism, e.g. white to yellow grains with *Acremonium* species, *Aspergillus nidulans*, *Aspergillus flavus*, *Cylindrocarpon cyanescens*, *P. boydii*, *Fusarium* species; black grains with *Corynespora cassicola*, *Curvularia* species, *M. mycetomatis*. They can be difficult to spot in tissue sections. Tissue Gram stain can detect the fine branching hyphae of actinomycetoma, but other stains are better for the detection of eumycetoma grains (e.g. periodic-acid–Schiff stain). A good idea of the causative organism can be drawn from grain characteristics. They can be cultured for more-exact diagnosis – biopsy specimens are best to avoid contamination. Serological diagnosis is possible in some centres.

Treatment

Mycetoma at all stages is usually amenable to medical therapy. Surgery usually leads to recurrence or mutilation that is more severe than that pre-existing. Treatment success depends on identification of the causative organism. All cases of actinomycetoma are treated with a combination of streptomycin sulphate and a second agent determined by the causal species (e.g. dapsone for *Actinomadura madurae*). Eumycetoma caused by *Madurella mycetomatis* often responds to ketoconazole. Itraconazole and even liposomal amphotericin B have also been used. Treatment continues in all cases for at least 10 months. There is a role for surgery in combination with effective medical treatment, e.g. for bulk reduction.

Dermatophytes

A group of fungi capable of invading the dead keratin of skin, hair, and nails, causing dermatophytosis (tinea). Also known as 'ringworm', several species infect humans and belong to the *Epidermophyton*, *Microsporum*, and *Trichophyton* genera. Clinical classification is by the body area involved: tinea capitis (scalp hair and the commonest in children), corporis (trunk and limbs), manuum and pedis (palms, soles and the commonest overall worldwide), cruris (groin), barbae (beard area and neck), faciale (face), unguium (nail – also known as onychomycosis and affecting 2.7–4.7% adults in the UK).

Mycology

- May be anthrophilic, zoophilic (causing incidental human infection), or geophilic (found primarily in soil and infrequent causes of human infection outside certain specific tropical regions). All favour humid or moist skin. Cases occur worldwide but incidence is highest in hot, humid regions where they may be the commonest skin infections.
- The commonest anthrophilic species is *Tricophyton rubrum*, a common cause of tinea pedis or tinea cruris in temperate regions, and tinea corporis in the tropics. Spread is through contact with infected descuamated skin scales – primarily through sharing common washing facilities.
- *Epidermophyton floccosum* may cause tinea cruris and foot infections either sporadically or in outbreaks in institutions.
- *T. concentricum* is a cause of tinea corporis in remote parts of the humid tropics, often affecting infants shortly after birth.
- Some other species have fairly specific geographical distributions. *T. tonsurans* is the predominant cause of tinea capitis in the UK, USA, and Mexico, whereas it is *T. violaceum* in India, East Africa, and the Middle East.

Pathogenesis

Infection is transmitted by hardy arthrospores, formed by dermatophyte hyphae. Direct contact between individual people is not necessary. Fungal cells adhere to keratinocytes where they germinate and invade. Host susceptibility to infection appears to be influenced by genetic factors, local moisture, and cell-mediated immunity. Risk factors: moist conditions, communal baths, athletic activities leading to abrasions (wrestling, judo etc), atopy, genetic predisposition, impaired cell-mediated immunity (e.g. Cushing's disease, AIDS – may lead to severe infection, e.g. extensive skin involvement, abscess, and dissemination)

Clinical features

The key feature is an annular scaling patch with a raised margin showing a degree of inflammation, the centre usually less inflamed than the edge. The precise appearance varies with the affected site, the fungal species involved, and the host immune response. Inappropriate application of topical steroids may lead to an infection showing none of the classical signs. Interdigital fungal infection may cause cracks in the skin and lead to bacterial cellulitus. Differential diagnosis includes seborrhoeic dermatitis, psoriasis, eczema, erysipelas, impetigo.

Tinea capitis

A disease of childhood (perhaps due to the presence of medium-chain-length fatty acids in sebum that inhibit growth in post-pubertal adults). Found worldwide. Endemic infections affecting a large number of children tend to be caused by anthrophilic organisms and sporadic cases by zoophilic fungi. Those infections in which the arthrospores are found on the hair surface are termed ectothrix infections, and those in which the spores develop within the hair, endothrix infections. The main clinical findings are scalp scaling and hair loss. It may resemble dandruff. In ectothrix infections hair tends to break a few millimeters above the skin, in contrast to endothrix infections where the hair breaks at the skin surface. Inflammation is variable and may be severe with pustules and an exudative crust. Untreated scalp ringworm usually remits spontaneously after puberty.

Tinea corporis

Commonly known as 'ringworm' and due to infection with one of a number of species. Anthrophilic species produce only mild inflammation and a consequently less well-defined skin lesion (e.g. *T. rubrum*), whereas zoophilic species produce more-inflamed lesions which may contain pustules (e.g. *M. canis*). Tinea barbae affects only the beard area with scaly plaques, pustules, and vesicles. Tinea imbricata is a variant caused by *T. concentricum*, characterized by a rash composed of concentric rings of scales. It is endemic in parts of Southeast Asia, the South Pacific, Central America, and South America.

Tinea pedis

Seen in children and young adults and usually due to infection with *T. rubrum* or *T. mentagrophytes*. Causes toe-web fissures, maceration, scaling of soles, erythema, vesicles/pustules, bullae. Scaling between the toes often called 'athelete's foot' but infection with other organisms (including bacteria) can produce an identical appearance.

Tinea cruris

Usually due to infection with *T. rubrum* or *E. floccosum*. Erythematous lesions with central clearing and raised borders in the groin and less commonly the scrotum. Usually seen in young men.

Tinea unguium (onychomycosis)

Usually occurs in association with infection of the adjacent skin. In the commonest form the nail is invaded from the distal and lateral aspects with onycholysis (separation of nail from nail bed), thick, discolored (white, yellow, brown, black), dystrophic nails. Can occur at any age but is commoner with increasing age. A number of other fungi may cause onychomycosis including *Scopulariopsis brevicaulis*, *Acremonium* spp., *Fusarium* spp.

Diagnosis

Skin scrapings, nail specimens or plucked hairs are treated with KOH and examined by direct microscopy. Samples should be taken from the edge of the lesion. Look for hyphae and arthrospores around the hair shaft. Fungal cultures may be performed (culture on Sabaroud's agar containing antibiotics and antifungal agents to selectively suppress the growth of environmental fungi – growth may take at least 2 weeks). Examination of the lesions under UV light may help in the diagnosis of tinea capitis – hairs infected with *Microsporum audouinii* and *M canis* fluoresce yellow-green, those infected with *Trichophyton schoenleinii* dull green. It is important to identify the organism causing scalp infection as the presence of an anthrophilic species should prompt screening of classmates and the family of affected children. Zoophilic infections rarely spread from child to child.

Treatment

- Generally topical although oral therapy may be considered in extensive or unresponsive disease. Hair and nail disease usually requires oral therapy.
- Infections confined to heavily keratinized areas (palms and soles) – keratolytic agents such as Whitfield's ointment (salicyclic and benzoic acid compound) may be effective.
- Tinea corporis, tinea cruris, and tinea pedis – topical ketoconazole 2%, miconazole 2%, or clotrimazole 1% rubbed into the affected area daily for 2–6 weeks. Extensive or unresponsive disease may need systemic therapy with one of the agents below.
- Nail infections rarely respond to topical therapy – terbinafine and itraconazole are suitable. Treatment may need to last 3–4 months. Drugs are well tolerated – there is a low risk of hepatic injury (1 in 70,000). Terbinafine tends to show the best response rates (70–80% cure for fingernails after 6 weeks and toenails after 12 weeks' treatment).

Other moulds

Pseudallescheria boydii

A fungus found in soil and fresh water throughout the world. The asexual form is called *Scedosporium apiospermum*. It causes two distinct diseases: mycetoma (see 📖 Eumycetoma, p.539) and pseudallescheriasis (all other infections). Pseudallescheriasis may affect lung, bone, joints, and the CNS, as well as causing skin and soft tissue infections. Infection may be acquired by inhalation or through skin trauma. Invasive pulmonary disease reminiscent of pulmonary aspergillosis may be seen in the immunocompromised (e.g. those undergoing BMT). Infections in the immunocompetent have subacute or chronic courses. Those in the immunocompromised tend to be acute and severe. Osteoarticular infection manifests as a painful swollen joint with overlying erythema. Weeks or months may pass between local fungal inoculation and the development of symptoms. Cerebral abscesses may develop in association with lung infection in the immunocompromised. Abscesses are usually multiple. In the immunocompetent they may be associated with instances of near-drowning in polluted water. CNS infection can also occur as a result of contiguous spread from infected paranasal sinuses. There have been cases of indolent meningitis caused by *P. boydii*. Isolation of the organism from sterile sites is diagnostic. It is rarely cultured from blood. Effective antifungal therapy has not been established. Resistance to amphotericin B has been reported. Successful treatment regimes have utilized both surgical debridement and voriconazole.

Scedosporium prolificans

An uncommon cause of human infection with several dozen cases reported worldwide. Immunocompetent patients experience focal, usually osteoarticular, disease. Immunocompromised patients frequently develop disseminated infection, e.g. those undergoing BMT. Fungaemia, skin lesions, myalgia, pulmonary infiltrates, cerebral lesions have all been reported. Diagnosis is made by culture. The organism is intrinsically resistant to most antifungals. Most therapy successes have involved debridement. Disseminated disease carries a high mortality.

Fusarium species

Found commonly in soil and in the healthy causes disease only rarely, usually through traumatic inoculation. *Fusarium* species may cause endophthalmitis, skin infection, musculoskeletal infections, mycetoma, and, particularly in the immunocompromised, disseminated infections. Systemic fusariosis occurs most commonly in patients with acute leukaemia and prolonged neutropenia, and those undergoing BMT. Presentation is with fever and myalgia unresponsive to antibiotics. Skin lesions are seen in up to 80% of cases, often starting as macules and progressing to necrotic papules. In the severely neutropenic, infection progresses rapidly to death. In those whose neutrophils recover, infection is more subacute and progresses slowly or is controlled and cured. Diagnosis is by culture. Unlike aspergillosis, blood cultures are often positive (50% cases). High-dose amphotericin B is the drug of choice. Overall mortality ranges from 50% to 80%, and survival is nearly always associated with recovery from neutropenia.

Dark-walled fungi

Phaeohyphomycosis is a loose term designating infection with moulds with dark walls in culture, but not always in tissue. These organisms are a cause of brain abscess (e.g. *Cladophialophora bantiana*), allergic fungal sinusitis (e.g. *Bipolaris, Exserohilum*), and cutaneous disease.

Sporothrix schenckii

A dimorphic fungus and the cause of sporotrichosis which may take the form of cutaneous infection, granulomatous pneumonitis, or disseminated disease.

Mycology

- Dimorphic. Demonstrating the temperature-dependent conversion is a useful means of identification. Hyphal colonies are initially white and later turn brown/black as they produce pigment.

Epidemiology

- Found across the world – most human infections occur in tropical and subtropical parts of the Americas.
- Animal-to-human transmission has been described. Human-to-human is rare.
- Identified in soil, plants, straw, and wood, and outbreaks have occurred in association with exposure to mine timbers, hay, thorned plants, etc.

Clinical features

- Cutaneous sporotrichosis – fungus is inoculated into skin at sites of minor trauma, with disease usually arising in cooler body areas such as distal extremities. There are no systemic symptoms. Lesions take one of two forms, both of which wax and wane over years:
  - painless smooth or verrucous erythematous papulonodular lesions (0.5–4 cm diameter) which often ulcerate and can be followed by secondary lesions along the line of proximal lymphatics
  - a fixed (plaque) form which does not spread locally. Spontaneous resolution has been described.
- Extracutaneous sporotrichosis:
  - osteoarticular (the most common extracutaneous form) infections involve the extremities, e.g. elbow, knee, hand, foot, not the hip, shoulder, or spine. Most present with primary involvement of a single joint which is swollen and painful, with an effusion and possibly a sinus tract. Other joints may become involved without therapy. There are few systemic features. Repeated joint aspiration and culture or synovial biopsy may be necessary to make a diagnosis
  - pulmonary sporotrichosis – one-third of patients are alchoholic, one-third have pre-existing illness (e.g. diabetes, sarcoid), and one-third are healthy. They may be asymptomatic but productive cough, fever, and weight loss are common, as are raised inflammatory markers. CXR reveals cavitation with or without hilar lymphadenopathy and effusions. Sputum Gram stain and

culture is usually diagnostic, but long-term follow-up with repeated cultures may be necessary for diagnosis. Untreated disease leads to progressive respiratory decline
  - meningitis – rare. CSF: high lymphocytes, high protein, low glucose.
  - also rare – endopthalmitis, sinuses, kidney, testes.
- Multifocal extracutaneous sporotrichosis – in healthy people lesions tend to be single-site. Multifocal disease with systemic features is usually seen in those with immunosuppression. Untreated infection is fatal.
- Patients with HIV – those with low CD4 counts are at greater risk of widespread ulcerative skin lesions and systemic dissemination. Presentation may be with arthritis and resemble seronegative arthropathies such as Reiter's syndrome. Visceral involvement occurs: meningitis, lung abscess, liver and spleen, endopthalmitis, bone marrow, sinus invasion etc.

Diagnosis

- Culture – success may require multiple samples taken from affected sites at different times. A positive blood culture indicates multifocal disease. Although culture of skin lesion fluid may be positive, biopsy is best.
- Histology – a pyogranulomatous response is usually apparent and in the presence of yeast forms can be diagnostic. Again, multiple samples may be necessary. Yeast may be cigar or oval in shape.

Differential diagnosis

- Cutaneous disease – fixed lesions: bacterial pyoderma, foreign body granuloma, dermatophyte infections, cutaneous tuberculosis, and other granulomatous conditions; lymphocutaneous lesions: nocardiosis, leishmaniasis, mycobacterial infections (e.g. *M. chelonae, M. marinum*).
- Osteoarticular disease – TB, gout, rheumatoid arthritis.
- Pulmonary disease – TB and other mycobacteria, histoplasmosis, coccidioidomycosis.

Treatment

- Cutaneous disease – itraconazole or saturated postassium iodide (5–10 drops PO three times a day, increased slowly to 40 drops per dose. Side-effects include anorexia, diarrhoea, and parotid gland enlargement). Treatment course is usually 6–12 weeks. Prognosis is good. There have been cases described in which simply warming the lesion has been curative, reflecting the organism's temperature sensitivity.
- Osteoarticular disease – itraconazole is used as initial therapy. Amphotericin B is curative in two-thirds. Courses are long, and relapse is common. Role of surgery is not known.
- Pulmonary disease – prior to cavitation amphotericin B may be effective. Advanced disease requires surgical resection of cavities and a course of amphotericin B.
- Meningitis – varying response to amphotericin and some advocate combination therapy with 5-flucytosine.
- Immunocompromised patients may require lifelong itraconazole therapy for multifocal extracutaneous disease.

Chromomycosis

A localized chronic fungal infection of cutaneous and subsutaneous tissue caused by several species and producing verrucous lesions.

Mycology

- Several different species cause chromomycosis – all take the appearance of dark brown cells occurring singly or in small clusters. Culture colonies are dark with a grey/green or brown/black surface.
- Agents grow slowly and culture may take 6 weeks.
- Organisms include *Fonsecaea pedrosoi* (the most commonly isolated agent), *Fonsecaea compacta, Phialophora verrucosa, Cladosporium carrionii* (common in Australia, South Africa and Venezuela)

Epidemiology

- Occurs worldwide but commonest in tropical/subtropical areas among barefoot workers.
- Found in soil, decaying vegetation etc, and inoculated into skin by minor trauma, thus feet and legs are the most commonly affected areas.

Clinical features

- Lesions can appear a long time after inoculation.
- Primary lesion usually a small pink papule that may itch and is followed (possibly many months later) by crops of either warty, violaceous nodules, or firm tumours. These tend to enlarge and form groups with ulceration and dark haemopurulent material on the surface. Satellite lesions may occur.
- Some people develop annular papular lesions with active edges and healing in the centre which can become scarred, or form keloid. Fibrosis and oedema of the affected limb may occur in severe cases.
- Complications: secondary infection, lymphoedema (elephantiasis), fistula formation, haematogenous spread (rare), squamous carcinoma in longstanding lesions, late recurrence (years later sometimes).
- Differential diagnosis – blastomycosis, yaws, tertiary syphilis, leishmaniasis, mycetoma, sporotrichosis, *M. marinum*, leprosy.

Diagnosis

- All forms of disease produce characteristic sclerotic bodies which may be identified on biopsy along with pyogranulomata and microabscesses. Microscopy of exudates may reveal hyphal strands.
- Culture is necessary to confirm identity and may take 6 weeks.

Treatment

- Early small lesions – surgical excision or cryotherapy is effective.
- Late disease – most cases present late with large lesions. Local heat and several antifungals have been reported as effective, e.g. itraconazole. Treatment duration may last well over a year. It has been used in combination with flucytosine in cases of relapse.

Histoplasma capsulatum

A dimorphic fungus and an emerging (particularly opportunistic) infection in parts of the USA and elsewhere.

The organism

- A member of the class *Ascomycetes*. First identified in 1905 and mistaken for a protozoan. Its characterization as a fungus in the 1930s led to the re-evaluation of many tuberculosis cases in the USA whose diagnosis had been made on chest x-ray alone.
- *H. capsulatum* contains between four and seven chromosomes. Restriction fragment length polymorphism (RFLP) of certain genes permits strains to be placed in one of six groups which correlate with virulence and geographical location.
- Mating types exist: (+) and (–). These are found in equal ratio in soil, but the (–) predominates in clinical isolates.
- It is dimorphic: the mycelial phase grows at ambient temperatures, the yeast phase at 37°C. These differ in their growth requirements (e.g. mycelial needs calcium, yeast does not). The mycelial phase exists in two forms: macroconidia (<15 micrometre) and microconidia (<5 micrometre, and thought to be the infective form being small enough to reach terminal bronchioles).

Epidemiology

- Found throughout the world but most common in warm, humid environments, e.g. southern United States. It is associated with the presence of bat and bird guano. Birds do not carry the organism – bats can and shed it in their droppings.
- Infection tends to occur when soil disruption (e.g. excavation) releases fungal elements which are then inhaled.
- Disease develops in men more often than women (4:1), which may reflect the association of chronic pulmonary disease with smoking (previously more common in men).

Pathogenesis

- The key pathological determinant is the transition from mycelial to yeast form. Artificially blocking the transition blocks infection in experimental systems. Mycelial forms are rarely seen in established infection.
- Microconidia settle in the terminal airways where they are phagocytosed by neutrophils and macrophages. Transition to the yeast phase takes hours to days, after which they migrate (probably intracellularly) from the pulmonary parenchyma to the local lymph nodes and beyond.
- The resulting inflammatory response produces caseating or non-caseating granulomas. These consist of fungal elements, mononuclear cells, T lymphocytes, and calcium deposits. Excessive granuloma formation may be followed by fibrosis.
- Macrophages are the key mediator of resistance to *H. capsulatum*. Macrophages from HIV-infected individuals are impaired in their ability to kill and inhibit fungal growth. T cells (CD4+ in particular)

are important in acquired immune defence – B cells and antibodies have little role in host resistance. CD4 cells seem to be vital for the activation of mononuclear phagocytes through cytokine release. Cell-mediated immunity limits but does not eliminate infection – infected people contain dormant organisms for years. These pose risk only if the individual subsequently becomes immunosuppressed.

Clinical features

Pulmonary histoplasmosis

- Acute primary infection – most patients are asymptomatic or experience mild flu-like symptoms. Around 10% become very unwell. Severity is affected by: inoculum size, age (the young and the old), underlying disease, and the presence of immunodeficiency. Incubation is 1–3 weeks. Symptoms: high fever, headache, dry cough, substernal chest pain, malaise, weakness and fatigue, arthralgias, erythema nodosum, and erythema multiforme. Examination: some added respiratory sounds, hepatosplenomegaly rarely. CXR: hilar lymphadenopathy and patchy pneumonitis which may calcify with time. Consider sarcoidosis and haemtological malignancy in those presenting with isolated hilar lympadenopathy. Most symptoms settle by day 10 but may persist in those with a large initial inoculum. Lab tests are non-specific – transient rises in ALP and rises or falls in white cell count (WCC). Around 6% of patients infected with histoplasmosis develop pericaridits, probably representing the granulomatous inflammatory response of nearby lymph nodes.
- Acute re-infection – those in endemic areas who experience repeated infection develop mild flu-like symptoms within 3 days of re-exposure. The illness is shorter than primary infection.
- Cavitary pulmonary histoplasmosis – a distinct presentation of acute infection generally seen men over 50 years with existing lung disease (e.g. COPD). Symptoms: low-grade fever, cough, weight loss, night sweats, and chest pain. Any existing pulmonary impairment may be exacerbated. CXR: cavitating lesions, mostly in the upper lobes (90%) near a bulla. Lymphadenopathy is not a feature. Patients may be anaemic, ALP may be raised. Spontaneous resolution is seen in up to 60%. Healing leads to fibrosis of the affected area with consequent respiratory impairment. Recurrence is seen in 20%. Death is rare.
- Histoplasmoma – a mass lesion resembling a fibroma and a rare complication of primary infection. Usually located in the lung. Enlarges slowly over years forming a calcified mass. On CXR it may appear to have a core of calcium or as a collection of calcified clusters.
- Mediastinal granuloma and fibrosis – granulomatous inflammation in response to infection leads to massive enlargement of the mediastinal lymph nodes (up to 10 cm). They may be asymptomatic or cause airway impingement. Fibrotic tissue formed during healing may distort airways (leading to pneuominas and bronchiectasis), the oesophagus, or the superior vena cava. Large nodes can penetrate the airways creating sinuses or fistulas to the pericardium or oesophagus. Rarely, mediastinal fibrosis can develop affecting all structures within the mediastinum.

African histoplasmosis

H. capsulatum var. *duboisii* is found in Africa along with classical var. *capsulatum*. Infection is associated with a distinct clinical presentation, skin and bone being the most frequently affected organs perhaps reflecting an increased incidence of cutaenous inoculation. Patients develop ulcers, nodules, and rashes that may resemble psoriasis. Osteolytic lesions affecting the skull, ribs, and verterbrae may become cystic with time. Infection can become disseminated and affect multiple organs with a pathology that resembles infection with *Coccidioides immitis* more than it does classic disseminated histoplasmosis – this may reflect the larger size of the var. *duboisii* yeast form compared to var. *capsulatum*.

Ocular histoplasmosis

- Uveitis or panophthalmitis occur rarely as part of histoplasmosis.
- Presumed ocular histoplasmosis syndrome (POHS) – visual impairment secondary to macular choroidal neovascularization. Although incidence is epidemiologically associated with areas where histoplasmosis occurs in the USA there is no proven pathological link.

Progressive disseminated histoplasmosis (PDH)

Continued growth of histoplasma in multiple organs. Occurred in ~8% of infections in some outbreaks. Risk factors: age, immunosuppression. Most cases probably represent reactivation of quiescent fungi, although PDH may result from primary infection or re-infection with a large inoculum.

- Acute PDH – associated with a fulminant course and seen in children and the immunosuppressed (HIV and haematological malignancies). Symptoms: abrupt onset of fever, cough, weight loss and diarrhoea.
  - **Children** – chest features dominate. Findings: hepatosplenomegaly, cervical lymphadenopathy, mouth ulcers, jaundice (a few), anaemia (90%), low platelets and WCC (80%), LFTs may be raised, CXR may show enlarged hilar lymph nodes and patchy pneumonitis. Untreated mortality is 100%, and prior to antifungal therapy children died around 6 weeks after symptom onset as a result of DIC, haemorrhage, or secondary bacterial infections.
  - **Adults with HIV infection** – CD4 usually <200 cells/mm^3 with positive histoplasma serology prior to illness. Before HAART up to 25% of AIDS patients in an endemic area could develop infection. Findings: hepatosplenomegaly, lymphadenopathy, cutaneous signs (maculopapular rash, bruising, petechiae). Anaemia, low platelets, and WCC are common (may predate PDH). Rarer manifestations: colonic masses, perianal ulcers, meningitis, encephalitis. CXR may be normal or show diffuse reticulo-nodular shadowing. Untreated fatality is 100%; with therapy, acute survival is 80%. The rare severe form, reactive haemophagocytic syndrome, is nearly always fatal despite therapy.
- Subacute PDH – symptoms are prolonged. Hepatosplenomegaly is common but fever, weight loss, and lab abnormalities are less pronounced. Infective lesions may occur in the GI tract (ulceration), CNS (chronic meningitis, cerebritis and mass lesions), adrenal glands (affected in 80% but overt Addison's seen in only 10%), and vascular structures (endocarditis and infection of aortic aneurysms). CSF in chronic meningitis shows lymphocytosis, elevated protein, and low glucose. Basilar meninges are badly affeceted – hydrocephalus may develop.

- Chronic PDH – seen exclusively in adults. It is distinguished from subacute disease by the mild chronic nature of the symptoms. Malaise and lethargy are the key complaints. Fever is less common. The commonest physical finding is painless mouth ulceration (which may have the appearance of malignancy). Yeasts and macrophages can be identified on biopsy from the centre of the lesion. Disease usually remains undiagnosed for some years until features associated with a single organ become apparent.

Diagnosis

- Culture – isolation of *H. capsulatum* is the only sure way to confirm a diagnosis of histoplasmosis. Specimens are cultured for up to 6 weeks at 30°C in brain-heart infusion agar with blood, antibiotics, and cycloheximide; 90% of positives will exhibit fungus by day 7. Isolates identification is confirmed using a probe for ribosomal DNA. Recovery rates depend on the source of the specimen and burden of infection: 10–15% for sputa from patients with acute pulmonary histoplasmosis, up to 60% from patients with cavitary disease, up to 90% for bronchoscopic specimens taken from AIDS patients with pulmonary disease, and 25–65% for CSF from patients with *H. capsulatum* meningitis (best if large volumes are taken, >20 mL). In endocarditis, blood cultures are often negative, heart valves are frequently positive. The organism is rarely found in pleural or pericardial fluid and is better grown from pleura or pericardium.
- Organism detection – tests are available that can detect polysaccharide antigen in urine or serum. They are positive in 90% of patients with progressive disseminated disease, and 20% of those with acute pulmonary histoplasmosis. It can be performed on BAL from patients with pulmonary disease. It is useful for detecting relapses in those with PDH – particularly in the immunosuppressed – and is more sensitive than serology. It is sensitive and fairly specific in the diagnosis of meningitis from CSF samples. Cross-reactivity with *Blastomyces* and *Coccidioides* species causes false-positive results. PCR tests are available and are rapid and specific.
- Serology – tests are negative in up to 50% of immunosuppressed patients. The detection of complement-fixing antibodies has been widely used in diagnosis. Around 10% of healthy people living in endemic regions have low-level responses. A titre of 1:8 is considered positive. A fourfold rise, or a titre of 1:32 is indicative of active infection and is seen in 75% of patients by 6 weeks of infection. Rising titres in those who have been treated hints at relapse. There is a false-positive rate of around 15%, seen often in those with coccidiomycosis or blastomycosis due to antibody cross-reactivity. The precipitin bands test detects antibodies to the H and M glycoproteins in patient sera. These are released by yeast and mycelial forms. The presence of anti-H antigen indicates active infection but is seen in <10% of patients. Anti-M antigen is detected in 80% of cases but does not distinguish active disease from previous infection.
- Histochemical stains can allow rapid identification of fungus from tissues and body fluids – it may be detected in blood smears from up to 40% of patients with acute PDH.

- Skin test – the histoplasmin skin test is useful only for epidemiological studies. It is positive in up to 90% of those in endemic areas and can be negative in cases of PDH.

Treatment

- Acute pulmonary histoplasmosis – most cases do not require treatment – bed rest and antipyretics are sufficient. Those who have had a high level of exposure, experience respiratory compromise, malaise and continue to have fevers after 1 week should receive therapy. Itraconazole 200 mg daily for 4–6 weeks is usually sufficient. Alternative if not tolerated: amphotericin B IV.
- Mediastinal granuloma – enlarged mediastinal lymph nodes can impinge on major airways and may require treatment as above. Amphotericin B may be preferable if rapid resolution of symptoms is required. Surgical excision may rarely be necessary.
- Mediastinal fibrosis – surgery, steroid, and antifungal therapy have all been tried with little success. Surgery is often difficult and fibrosis may recur.
- Histoplasmoma – require surgical excision only if enlarging. 2–3 months' therapy with an azole may be beneficial afterwards. NB: fluconazole is the only azole that penetrates the CNS.
- Cavitary pulmonary histoplasmosis – cavities with thin walls usually resolve spontaneously. Treatment should be given to those with progressive infiltrates, thick-walled cavities, or persistent cavities impairing respiratory function. Itraconazole 400 mg daily for 6 months leads to improvement in up to 85% of patients. Amphotericin B should be used in the immunosuppressed or if disease progresses on therapy. Relapse is seen in up to 20%. Surgical resection may be necessary in such cases.
- Acute PDH – acute life-threatening PDH should be treated with early amphotericin B. Most patients show significant improvement in the first week. Mild acute PDH can be treated with a 6-month course of itraconazole. Patients with AIDS and those on long-term immunosuppressive therapy should receive lifelong itraconazole.
- Subacte and chronic PDH – itraconazole 400 mg daily gives a 90% success rate. Alternative: amphotericin B.
- Meningitis – treatment is with amphotericin B with weekly lumbar punctures to assess efficacy. Relapses are common with overall cure seen in around 50% (more in the immunocompetent, less in the immunocompromised).
- Endocarditis – amphotericin B and surgical removal of the infected valve.
- Pericarditis – may be seen after acute PDH. The pericardium is rarely infected, and pericarditis does not require antifungal therapy in its own right. Bed rest, NSAIDS and possibly steroids (beware of exacerbating active histoplasmosis lesions) usually suffice. Cardiac tamponade is uncommon. In the rare cases where the pericardium is itself infected, antifungal therapy is indicated.
- Presumed ocular histoplasmosis – antifungal therapy is not required. Laser therapy can prevent neovascularization.
- See IDSA guidelines for further details on the treatment of histoplasmosis.[1]

Prevention

There is no vaccine available at present. Prevention rests on educating those who work in areas where there is a risk of acquiring infection, e.g. workers in buildings and other environments that have served as bat habitation. In HIV-infected patients who have had histoplamosis, secondary prophylaxis is given until the CD4 count is consistently above 200 cells/mm^3.

Reference

1 Wheat LJ, Freifeld AG, Kleiman MB et al. Clinical practice guidelines for the management of patients with histoplasmosis: 2007 update by the Infectious Diseases society of America. *Clin Infect Dis* 2007;45:807–25.

Blastomyces dermatitidis

A dimorphic fungus and cause of blastomycosis, a systemic pyogranulomatous disease.

Mycology

Grows in the mycelial form at room temperature and as a yeast at 37°C. The mycelial form grows on plates by 3 weeks and produces the infectious conidia (2–10 micrometre in diameter). Yeast cells are multinucleate with thick cell walls and have a similar appearance *in vitro* as in clinical specimens. *B. dermatititdis* is the asexual stage of *Ajellomyces dermatitidis*, the sexual form which requires opposite mating types for reproduction.

Epidemiology

Information is limited due to the lack of a sensitive skin test. Endemic areas include southeastern USA, parts of South America, the Middle East, and India. African strains are serologically distinct from American isolates. It is a probably saprobe of soil, favouring decaying wood material in moist areas. Symptomatic disease occurs in less than 50% of infected people.

Pathogenesis

Infection is acquired via the lungs. Conidia are inhaled and convert to the yeast phase. A non-caseating granulomatous response usually follows although respiratory disease may not be apparent. The organism may disseminate to other sites. The histology of cutaneous disease is distinct, producing pseudoepitheliomatous hyperplasia with microabscesses. In appearance it may resemble other skin lesions (e.g. squamous cell carcinoma). Mucosal involvement of the mouth and larynx may take a similar appearance. Protection from infection is mediated primarily by natural resistance (e.g. alveolar macrophages – rates are not greatly increased in those with immunocompromise) and cellular immunity (antibodies confer no protection).

Clinical features

Almost any organ can become infected. Some patients present with an acute pneumonic-type illness. Most experience a more-chronic course – symptoms may be systemic or purely pulmonary.

- Acute infection – following exposure there is an incubation period of 4–6 weeks before the development of non-specific flu-like symptoms: fever, arthralgia and cough (initially non-productive but may later produce purulent sputum). CXR may demonstrate an area of consolidation. There have been reports of spontaneous resolution of such acute pneumonias.
- Chronic infection:
  - pulmonary – chronic pneumonia with productive cough, pleuritic chest pain, haemoptysis, weight loss and low-grade fever. CXR may show infiltrates, mass lesions (which may resemble malignancy), and cavities but effusions are rare. Miliary disease or diffuse pneumonitis is unusual – both have a high mortality
  - skin – the commonest extrapulmonary manifestation, seen in up to 80% of cases and may occur in the absence of respiratory

features. Lesions may be verrucous or ulcerative – both may occur in the same patient. Verrucous lesions resemble squamous cell carcinoma with a crusted grey to violet appearance. Where there is discharge microscopy may reveal yeast forms. Subcutaneous nodules represent cold abscesses and are usually seen in acutely ill patients with severe pulmonary or extrapulmonary disease at another site
- bone and joint – long bones, vertebrae and ribs are common sites and extension may lead to arthritis. Organisms are seen easily in synovial aspirates
- genitourinary tract – around 25% of men have GU involvement, usually of the prostate
- CNS – uncommon in the normal host but seen at increased frequency (e.g. as abscess or meningitis) in patients with AIDS who develop blastomycosis
- other – liver, spleen, gut, thyroid, pericardium.

- The immunocompromised – an unusual opportunistic pathogen in HIV patients. Disease is more severe and more likely to be fatal in those with late-stage AIDS. CNS infection is seen in 40% of cases. Similar severity is seen in patients with immunocompromise due to steroid therapy and chemotherapy. Up to 40% of cases are fatal. Treat suspected cases early – most deaths occur within a few weeks. Relapse is common if immunodeficiency remains, and long-term suppression should be considered.

Diagnosis

- Microscopy of secretions – the characteristic yeast cell may be seen in a wet preparation of sputum or pus. Body fluids such as urine or pleural fluid should be centrifuged and the sediment examined. BAL may be useful in patients who are not producing sputum. When organisms are sparse they may be more easily identified on Papanicolaou preparations.
- Histology – Gomori's methenamine silver and PAS stains readily visualize the fungus.
- Culture – material should be inoculated on to Sabouraud's or more-enriched agar and incubated at 30°C. The mycelial form is not diagnostic, and ideally conversion to yeast at 37°C should be demonstrated. This is not always possible and mycelial identification can be achieved by nucleic acid probes for *Blastomyces* RNA.
- Serology – complement fixation, immunodiffusion and ELISA-based tests are available but sensitivity and specificity vary widely. A negative test does not rule out disease and a positive one does not on its own warrant therapy – rather it should fuel the hunt for the organism.

Treatment

All patients should receive therapy. A small number of cases do resolve spontaneously but these cannot be predicted at presentation. Surgery has a role in the drainage of large abscesses and the resection of necrotic bone tissue in combination with medical therapy.

- Mild-to-moderate disease in the immunocompetent – ketoconazole or itraconazole. Therapy should be continued for at least 6 months.

Relapses do occur (around 10% with ketoconazole), and patients should be followed up for 2 years.
- Severe, including CNS disease – amphotericin B. Relapse is more common in those with immunocompromise, and long-term azole therapy should be considered.
- The role of the newer azole antifungal agents has not been established.
- See IDSA guidelines for further details on the management of blastomycosis.[1]

Reference

1 Chapman SW, Dismukes WE, Proia LA et al. Clinical practice guidelines for the management of blastomycosis: 2008 update by the Infectious Diseases Society of America. *Clin Infect Dis* 2008;46:1801–12.

Coccidioides immitis

A dimorphic fungus found in certain regions of the western hemisphere and a cause of respiratory illness.

Mycology

May exist as a mycelium or a spherule (a structure unique to this organism). The mycelial form is seen on routine laboratory agar and in soil. Once mature, some cells develop a hydrophobic outer layer which renders them capable of prolonged survival (arthroconidia). Once airborne these may be inhaled and deposited in the lungs where they begin to multiply. The resultant spherule consists of a thin wall containing many endospores. This wall eventually ruptures allowing the spread of endospores.

Epidemiology

Endemic to certain regions of the western hemisphere with an arid climate, hot summers, and alkaline soil (e.g. regions of the southern USA, Mexico, Central America, parts of South America). Arthroconidia transport (e.g. in dust storms) has resulted in infections in non-endemic areas. Infections tend to occur when the soil is dry towards the end of the summer. The organism is most easily isolated after the winter rains. The number of new infections varies greatly year to year; 30% of people living in the endemic regions of the USA show evidence of prior exposure.

Pathogenesis

Most infections follow inhalation of arthroconidia. There are rare cases of cutaneous infection (which tend to resolve without treatment). Inflammation follows its conversion to a spherule. The resulting pulmonary lesion consists of neutrophils and eosinophils, and, if infection becomes chronic, granulomas with lymphocytes and multinucleated giant cells. Both acute and chronic lesions may be found at different sites in the same individual. Control of infection relies on the T-cell response, and those with deficient T-cell immunity are at risk of severe disease. The innate immune response appears to be important against arthroconidia and endospores.

Clinical features

Up to two-thirds of infections produce only mild or subclinical disease, and of those producing respiratory symptoms most follow a self-limited course. Complications may occur up to 2 years later and do not correlate with the severity of the original infection.

- Early respiratory infection – symptoms develop 1–3 weeks after exposure: cough, pleuritic chest pain, breathlessness, and fever. Onset is usually slow but can be abrupt. Inhalation of a large number of arthroconidia may result in early symptoms. Weight loss and migratory arthritis can occur, and some develop skin rashes ranging from a fine papular rash early in illness to erythema multiforme and nodosum (particularly in women). Laboratory tests may reveal peripheral blood eosinophilia and raised inflammatory markers. Around 50% of patients have CXR changes: effusions, infiltrates, hilar lymphadenopathy, and cavities. Most infections resolve without complications over several weeks. There are rare cases of severe diffuse coccidioidal pneumonia (due either to massive exposure or haematogenous seeding) leading to respiratory failure and septic shock with a high mortality. One-third of coccidioidal infection in HIV-positive patients (typically those with CD4 <100 cells/mm^3) presents in this manner.
- Pulmonary nodules and cavities – 4% of lung infections result in a nodule which may reach up to 5 cm in diameter. Although usually asymptomatic, a biopsy may be necessary to distinguish it from a neoplastic lesion. Nodules may liquefy and drain via a bronchus to form a cavity. Cavities are usually peripheral and although half close within 2 years, some may cause pain, cough, and haemoptysis as well as providing a focus for the development of mycetoma. Peripheral cavities can rupture causing a pyopneumothorax.
- Chronic fibrocavity pneumonia – associated with diabetes or pre-existing lung fibrosis; some people develop a chronic fibrotic pneumonia with widespread pulmonary infiltrates and cavities involving more than one lobe. As well as local symptoms, people may experience night sweats and weight loss.
- Dissemination – 0.5% of the general population develop disseminated infection. Those with immunodeficiency (solid organ transplants, late-stage HIV infection, those on high-dose steroids, Hodgkin's disease) are at greater risk. Many with disseminated disease do not develop respiratory features and have normal CXRs. Sites of dissemination: skin (causing maculopapular lesions, verrucous ulcers, and abscesses with a predilection for the nasolabial fold), joints (knee, hand, wrist, feet), bone (particularly the vertebrae – which may progress to develop a paraspinous abscess), and meningitis – the most serious manifestation. Meningitis develops a few weeks to a few months after initial infection and is usually fatal within 2 years of diagnosis. CSF findings: elevated pressure, raised protein, low glucose, raised eosinophils. The basilar meninges are usually involved, and hydrocephalus is a common complication in children.

Diagnosis

- Clinical features of coccidioidal infection are not specific and lab tests are required to establish diagnosis. Travel history is vital. Remember that exposure can be subtle (e.g. changing planes within an endemic area). Complications are usually apparent within 2 years of exposure, but infection may have occurred many years previously in the immunodeficient.
- Isolation of the organism – definitive diagnosis is by culture of, or identifying fungal elements within clinical specimens (e.g. biopsy, sputum). These can be collected without risk to personnel – infection is not transmitted from primary specimens. Stains such as silver, perodic-acid–Schiff, or H&E will reveal spherules. *C. immitis* grows on standard microbiological media in aerobic conditions and typically takes the form of a white mould at around 5–7 days. At this point it is highly infectious. Unlike the spherule, the mycelial form is not unique, and identification is made by reference laboratories either antigenically or by the detection of specific ribosomal RNA.
- Serology – most patients are not very symptomatic, and diagnosis is by serology. Serology is important in the diagnosis of coccidioidal meningitis as CSF is usually culture-negative. Tests are highly specific, and even borderline positive results should be treated seriously. Negative tests do not exclude infection. The tube precipitin (TP) antibody (IgM) test detects a fungal cell wall polysaccharide, and 90% of patients will have a positive test in the first 3 weeks of illness; complement fixing antibodies (predominantly IgG) are detected later and for longer than TP antibodies, and their presence in CSF is important in the diagnosis of coccidioidal meningitis; immunodiffusion tests; ELISA for IgG or IgM is sensitive and in increasing use.
- Skin testing – for delayed-type hypersensitivity to coccidioidal antigens is useful epidemiologically but limited as a diagnostic tool.

Treatment

- Newly diagnosed patients should be assessed for the extent of disease and factors that increase the risk of future complications. Amphotericin B is the preferred agent in cases of severe pulmonary disease or those who are deteriorating. Azoles are used in cases of chronic infection (ketoconazole, itraconazole or fluconazole). Surgery may be required in some patients. Debridement and drainage of infected sites is essential in the management of extensive bone infection, and in cases of vertebral infection stabilization may be required.
- Uncomplicated pulmonary disease – may not require therapy in the healthy. Those at risk of dissemination (e.g. the immunosuppressed, pregnant women) should usually receive antifungal therapy. Those at risk of severe pulmonary disease (e.g. diabetics and those with pre-exisiting lung fibrosis) should also be considered for treatment.

- More-severe pulmonary infection – considered to be that associated with weight loss of over 10%, night sweats for three or more weeks, infiltrates that are either bilateral or involve more than half of a lung, persistent hilar lympadenopathy, symptoms persisting for over 2 months. Treatment is with azole antifungals for 3–6 months.
- Diffuse pneumonia – amphotericin B for a few weeks with azole therapy following for at least 1 year.
- Pulmonary cavity – cavities that do not resolve spontaneously should be resected if it is safe to do so to avoid future complications.
- Persistant fibrocavitary pneumonia – treatment is usually started with oral azoles. Up to 60% respond with improved symptoms and CXR. Those who do not may respond to amphotericin B.
- Dissemination – coccidioidal meningitis is treated initially with fluconazole, which achieves a response rate of 70%. Those who do not respond may benefit from intrathecal amphotericin B. Hydrocephalus may require shunting. Occasionally cerebral abscesses develop which may need draining. Other forms of disseminated disease can usually be treated with oral azoles except in those who are showing rapid deterioration or have infection in critical places (e.g. vertebrae) when amphotericin B tends to be preferred. Treatment is continued for at least a year, and for 6 months beyond the end of recovery. Relapses occur in one-third, and lifelong suppressive therapy may be required.
- See IDSA guidelines for further details on the management of cocciodioidomycosis.[1]

Reference

1 Galgiani JN, Ampel NM, Blair JE et al. Coccidioidomycosis. *Clin Infect Dis* 2005;41:1217–23

Paracoccidioides brasiliensis

A cause of chronic progressive systemic mycosis in South America.

Mycology

Paracoccidioides brasiliensis is a dimorphic fungus (at 37°C an oval/round mould of variable size reproducing by budding, below 28°C a slow-growing mould). At 37°C, colonies take around 10 days to appear and have a creamy soft appearance. Mould-form colonies take up to a month to develop.

Epidemiology

Geographical distribution is very limited: it has been found in South America from Argentina as far north as Mexico. Within these regions it tends to cause infections in forest regions with high year-round humidity and mild temperatures. Brazil has the highest number of reported cases. It has been isolated from soil but is not widespread and the organism's ecological niche is not clear. Human infection is probably acquired by inhalation. Most cases occur in men over 30 years of age, yet skin testing demonstrates that men and women are equally exposed. The rare prepubescent cases have an equal sex distribution. Agricultural workers, smokers, and alcoholics are at greater risk.

Clinical features

Most primary cases are subclinical. The organism can remain dormant for prolonged periods, disease becoming apparent only in states of debilitation or immunosuppression. Paracoccidiomycosis tends to cause subacute severe disease in the young and chronic disease in adults, in whom it has a better prognosis with therapy. Infection is acquired via the lungs.

- Lung – patients complain of breathlessness and CXR may reveal nodular infiltrates which are often bilateral. Lesions are concentrated in the mid and lower zones, apices are usually clear. Cavities, fibrosis, areas of emphysema, and right ventricular hypertrophy become more likely as disease becomes chronic.
- Mouth and upper respiratory mucosa – ulcerated lesions of the mouth, lip, gums, tongue, and palate are common. Other features: tooth loss, dysphonia, nasal lesions.
- Cutaneous lesions – warty ulcerated lesions infiltrate the subcutaneous tissue. Appear over legs and orifices.
- Other – lymphadenopathy (sometimes with fistulas), diminished adrenal function, spleen, liver, gut, vascular system, bone, CNS.

Diagnosis

- Microscopy with KOH – reveals the organism (easily identified by its distinctive multiple budding) in over 90% of cases where sputum or exudates is available.
- Biopsy – often diagnostic. Histology shows granuloma with multinucleated giant cells which may contain fungi. Ulcerated lesions may show a pyogenic reaction, and skin lesions may have intraepithelial microabscesses.
- Culture – on Sabouraud-dextrose agar confirms diagnosis. Cultures should be kept for 6 weeks.
- Serology – immunodiffusion test is useful for diagnosis but remains positive after successful treatment. Complement fixation tests allow an assessment of response to treatment but cross-react with *H. capsulatum*. More-specific tests have been developed.
- Skin testing – not useful for diagnosis.

Treatment

- It is important that specific therapy is combined with measures to improve general health.
- Imidazoles – oral ketoconazole results in major improvement in around 90% of cases, with 10% relapse after 5 years. It should be given for at least 6–18 months as guided by response. Patients should be monitored for liver impairment. Itraconazole appears to be even more effective, requiring shorter treatment courses with fewer side-effects and lower relapse (3–5%).
- Sulfadiazine and amphotericin B – paracoccidiomycosis is the only mycosis sensitive to sulfa drugs. Treatment takes weeks or months and can be reduced only when improvement is apparent. Courses often last as long as 5 years to avoid relapse (up to 25% of cases). Mortality can be as high as 25%. In severe cases it is used in combination with amphotericin B, which may also be added to imidazole-based regimes in patients unresponsive to a single agent.

Penicillium marneffei

A thermally dimorphic fungus (exists in a yeast or mould form depending on the temperature) that causes life-threatening disseminated infection.

Epidemiology

Its distribution is limited to South-East Asia and southern China. Humans and bamboo rats are the only known hosts. The exact route of transmission is unknown but is thought to be inhalation or, rarely, inoculation. Infection is commonly seen in young adults with HIV, but cases are seen in immunocompetent children and adults. Occupational exposure to soil is a risk factor.

Clinical features

Patients present with around a one-month history of illness with low-grade fever, weight loss, and skin lesions (pustules, papules, ulcers, or abscesses of the face, upper trunk, or extremities). Pharyngeal and palatal lesions are common in those with HIV. Most have anaemia and weight loss, with around half presenting with fungaemia or lymphadenopathy. Hepatomegaly, splenomegaly, haemoptysis (secondary to cavitating lung lesions), joint infections, and pericarditis may occur. The diagnosis should be considered in those with immunocompromise and history of travel to an affected area. Infection may present many years after travel.

Diagnosis

Diagnosis may be made on smear (skin lesion, sputum), biopsy (lymph node, bone marrow), or culture. Microscopic examination may reveal yeast forms both extracellularly and within phagocytes. Histology may demonstrate either a granulomatous (commonly seen in the immunocompetent), suppurative, or necrotizing (commonly seen in the immunocompromised) response. Culture at 30°C produces a mould with sporulating structures that may be converted to yeast form by culture at 37°C. This dimorphism is not seen in other members of the genus *Penicillium*. In vitro *P. marneffei* is highly sensitive to itraconazole, voriconazole, terbinafire and flucytosine. It is intermediately sensitive to amphitericin. There are no randomised controlled trials on the acute treatment of penicilliosis.

Treatment[1]

Disseminated infection has been most successfully treated with 2 weeks IV amphotericin B followed by 10 weeks PO itraconazole. Patients with mild disease can be treated with oral itraconazole 400mg/day for 8 weeks. All HIV infected patients who complete treatment for penicilliosis should be given secondary prophylaxis with oral itraconazole 200mg/day. Secondary prophylaxis may be found in patients on antiretroviral therapy who have a CD4 count of >100 cells/mm^3 for at least 6 months.

Reference

1. Guidelines for the prevention and treatment of opportunistic infection in HIV infected adults and adolescents (2008). Available from http://AIDSinfo.nih.gov

Plasmodium species (malaria)

Malaria, an infection caused by *Plasmodium* species, has affected mankind for millennia. The word malaria means 'bad air', and refers to the association between the illness and the marshes where *Anopheles* mosquitoes breed. Although malaria has virtually disappeared from Europe and the USA (apart from imported cases), it remains a major problem in tropical countries where it causes 3–500 million cases and 2–3 million deaths per year.

Plasmodium species

Five *Plasmodium* species cause human infection:

- *P. falciparum* can invade red blood cells of all ages, may be drug-resistant and is responsible for most severe, life-threatening infections. It does not produce dormant liver stages (hypnozoites) or cause relapse
- *P. vivax* and *P. ovale* cause clinically similar, milder infections. They produce hypnozoites and may cause relapse months after the initial infection
- *P. malariae* rarely causes acute illness in normal hosts, does not produce hypnozoites, but may persist in the bloodstream for years.
- *Plasmodium knowlesi* causes malaria in macaques and has recently been recognised as a cause of human malaria in SE Asia. Microscopically it resembles *P. malariae* but can cause fatal disease (like *P. falciparum*).

Mixed infections may occur in 5–7% of patients.

Life cycle

Humans acquire malaria from sporozoites transmitted by the bite of the female *Anopheles* mosquito. Sporozoites travel through the bloodstream and enter hepatocytes. Here they mature into tissue schizonts which rupture and release merozoites into the bloodstream. These invade red blood cells and mature into ring forms, then trophozoites, and finally schizonts before rupturing to release merozoites. Alternatively, some erythrocytic parasites develop into gametocytes (sexual forms), which are ingested by the mosquito and complete the sexual life cycle. In *P. vivax* and *P. ovale* infections, some parasites remain dormant in the liver as hypnozoites for months before they mature into tissue schizonts.

Epidemiology

The epidemiology of malaria varies and depends on a number of factors: climate, plasmodium species and life cycle, efficiency of transmission by vectors, and drug resistance. Thus, in sub-Saharan Africa, *P. falciparum* can survive as a result of the year-round presence and efficient transmission by its mosquito vectors (*A. gambiae* and *A. funestus*). In contrast, *P. vivax,* which is found in more-temperate zones, requires hypnozoites to sustain its transmission.

Pathogenesis

The following mechanisms contribute to the pathogenesis of severe falciparum malaria:

- **cytoadherence** – adherence of parasitized red blood cells to the vascular endothelium is mediated by *P. falciparum*-infected erythrocyte membrane protein 1 (PfEMP1), which binds to specific endothelial receptors, e.g. thrombospondin, CD36, ICAM-1, VCAM1, and ELAM1.

This results in peripheral sequestration of parasites, which protects them from removal from the circulation as they pass through the spleen, and oxidant damage as they pass through the lungs

- **rosetting** – *PfEMP1* also binds to complement receptor-1, resulting in clustering of unparasitized red cells around parasitized red cells
- **hyperparasitaemia** (>5%) is associated with a greater risk of death, particularly in non-immune patients. Reasons for this include more-severe metabolic effects, e.g. hypoglycaemia and lactic acidosis.

Clinical features

- Fevers (cyclical or continuous with intermittent spikes)
- Malarial paroxysm – chills, high fever, sweats
- Complications – cerebral malaria, pulmonary oedema, severe anaemia, hyperparasitaemia, hypoglycaemia, uraemia, lactic acidosis

Laboratory diagnosis

- Thick and thin blood smears, stained with Field's stain or Giemsa stain, examined under light microscopy. Giemsa is better for species identification.
- Malaria dipstick tests, e.g. OptiMAL-IT and Paracheck-Pf®. These detect plasmodial lactate dehydrogenase and offer a rapid diagnostic near-patient test. Some tests enable distinction of species.
- Laboratory findings – haemolytic anaemia, thrombocytopaenia (common), uraemia, hyperbilirubinaemia, abnormal LFTs, coagulopathy.

Treatment

- **Antimalarials** (📖 p.148) – remain the mainstay of therapy, but successful treatment is threatened by increasing drug resistance. The main classes of drugs are:
  - quinoline derivatives (chloroquine, quinine, mefloquine, halofantrine)
  - antifolates (pyrimethamine, sulfonamides)
  - ribosomal inhibitors (tetracycline, doxycycline, clindamycin)
  - artemisinin derivates (artemisinin, artemether, arteether, artesunate).
  - Combination therapy with artemisinin derivatives shows rapid parasite clearance, low toxicity, and no clinical reports of resistance. Despite superior efficacy to quinine, most artemisinin derivatives are not available in the UK.
- **Supportive therapy** – good supportive therapy with careful management of seizures, pulmonary oedema, acute renal failure, and lactic acidosis is essential in severe malaria. Exchange transfusion may be helpful in hyperparasitaemia.
- **Adjunctive therapies** – adjunctive therapies for severe malaria have proved disappointing. Monoclonal antibodies directed against TNF-α reduced fever but showed no effect on mortality and may have increased morbidity. Dexamethasone has been shown to increase the duration of coma and was associated with poorer outcome in cerebral malaria.

Reference

Warell DA, Hooareeoyran S, Warrell MJ et al. Dexamethasone proves deleterious in cerebral malaria. A double-blind trial in 100 comatose patients. *New Engl J Med* 1981;306:313–319.

Prevention

- **Insecticide-treated bed nets** have been shown to reduce intradomilcilary vector populations and protect against infection
- **Insect repellents** such as diethyltoluamide (DEET) reduce the risk of transmission in areas where mosquitoes are active before bedtime.
- **Chemoprophylaxis**, taken rigorously, is efficacious in reducing the incidence of malaria in travellers.
- **Vaccines –** a number of candidate vaccines using various antigens have been developed. Intense efforts to produce a vaccine have so far failed to yield a good candidate, but studies are ongoing.

Box 4.21 Malaria treatment and prevention guidelines

Treatment

- UK guidelines:
  - Lalloo D, Shingadia D, Pasvol G et al. UK malaria treatment guidelines. J Infect 2007;54(2):111–121.
  - See also section on treatment of malaria in British National Formulary www.bnf.org
- US guidelines: CDC Treatment of malaria (Guidelines for Clinicians) 2009. http://www.cdc.gov/malaria/diagnosis_treatment/treatment.html
- WHO guidelines *for the Treatment of Malaria* (2006). http://www.who.int/malaria/publications/atoz/9241546948/en/index.html

Prophylaxis

- UK guidelines
  - Chiodini P, Hill D, Lalloo D et al. Guidelines for malaria prevention in travellers from the United Kingdom. London, Health Protection Agency, January 2007. http://www.hpa.org.uk/webw/HPAweb&HPAwebStandard/HPAweb_C/1203496943315?p=1249920576136.
  - HPA. Malaria. www.hpa.org.uk/infections/topics_az/malaria/menu.htm.
  - See also section on prophylaxis against malaria in British National Formulary www.bnf.org
- US guidelines: National Center for Infectious Diseases Travellers' Health; The Yellow Book – Health Information for International travel 2008. Atlanta: Centers for Disease Control, 2008. http://wwwn.cdc.gov/travel/ybToc.aspx
- WHO guidelines: World Health Organization. *International travel and health: vaccination requirements and health advice*. Geneva: WHO, 2004. www.who.int/ith

Babesia

Babesiosis is a zoonotic infection caused by *Babesia* spp., a malaria-like parasite that parasitizes erythrocytes of animals and causes fever, haemolysis, and haemoglobinuria. It typically causes mild illness in humans, but fulminant disease may occur in asplenic or immunosuppressed patients.

The parasite

There are more than 70 *Babesia* species worldwide that infect a wide range of mammals and birds. The rodent strain *B. microti* (USA) and the cattle strains *B. divergens* and *B. bovis* (Europe) are the main causes of human disease. *Babesia* spp. vary in length from 1–5 micrometre, and are pear-shaped, oval, or round; their ring conformation and peripheral location in erythrocytes may lead to their misidentification as *P. falciparum*. *Babesia* spp. are transmitted from their animal reservoir to humans via a tick vector *Ixodes scapularis* (USA) or *Ixodes ricinus* (Europe). The tick has three developmental stages (larva, nymph, and adult), and requires a blood meal, often from different mammalian species (e.g. deer, rodent) to mature to the next stage.

Epidemiology

The first fatal human case of babesiosis was reported in 1996. Since then more than 100 cases have been reported worldwide, most from the north-eastern coastal regions of the USA. Based on seroprevalence data, most infections appear to be subclinical. Transfusion-associated, transplacental and perinatal transmission may occur. The clinical features of babesiosis vary markedly between regions. Virtually all of the European cases have been caused by *B. bovis* or *B. divergens*, have occurred in splenectomized patients, and have had a fulminant and usually fatal course. In contrast, epidemiological data from the USA suggest that most infections are caused by *B. microti* and are mild or subclinical; clinical infections are more likely in the asplenic, immunosuppressed, elderly, or patients with concomitant Lyme disease.

Clinical features

- Clinical features include fever, chills, malaise, fatigue, anorexia, headache, myalgia, arthralgia, nausea, vomiting, abdominal pain, dark urine, depression, and emotional lability. Photophobia, conjunctival injection, sore throat, cough, and adult respiratory distress syndrome have also been described.
- Laboratory abnormalities include haemolytic anaemia, reticulocytosis, normal or low WCC, thrombocytopaenia, raised eryhtrocyte sedimentation rate (ESR), positive direct Coombs' test, mild elevations in LFTs, renal impairment, reduced serum haptoglobin levels.
- Urinalysis reveals haemoglobinuria and proteinuria.

Laboratory diagnosis

- Microscopy – examination of thick and thin blood smears stained with Giemsa or Wright stains show parasitized erythrocytes, sometimes with diagnostic tetrads of merozoites. *Babesia* spp. can be distinguished

from *P. falciparum* by its lack of haemozoin, and absence of schizonts and gametocytes.
- Serology – indirect immunofluorescent antibody titre for *B. microti* is available from CDC, Atlanta. A titre of ≥1:256 is considered diagnostic for acute *B. microti* infection.
- Molecular methods – a PCR-based assay may be used for the detection of low levels of parasitaemia.

Treatment
- Most patients with *B. microti* infection have mild, self-limited illness that does not require specific treatment.
- In patients with severe infections, a combination of quinine and clindamycin for 7–10 days appears to be effective.
- Atovaquone and azithromycin is an alternative regimen – less rapidly effective but fewer adverse effects.
- Exchange transfusion is used in critically ill patients to reduce parasitaemia.

Prevention
- Avoid exposure to ticks in endemic areas between May and September.
- Wear light-coloured, long-sleeved clothing, and tuck trousers into socks or boots.
- Use insect repellent (e.g. diethyltoluanide) on skin and clothes.
- Carefully remove any ticks.
- Discourage blood donations from donors in endemic areas between May and September, from donors with fevers 2 months prior to donation, and from donors with a history of tick bite

Toxoplasma gondii

Toxoplasmosis is a zoonotic infection caused by *Toxoplasma gondii*, a coccidian parasite of cats that affects humans and other mammals as intermediate hosts. Although infection with *T. gondii* is common, it rarely causes disease apart from in congenitally acquired infection and in patients with cell-mediated immunodeficiency, especially AIDS.

Classification

T. gondii belongs to subphylum *Apicomplexa*, class *Sporozoa* and exists in three forms: the occyst (which releases sporozoites), the tissue cyst (which contains bradyzoites), and the tachyzoite.

Life cycle

Oocysts are produced in the cat's intestine and shed in its faeces. Once outside the cat, the oocysts sporulate and develop sporozoites. Oocysts are ingested by other animals and release sporozoites, which develop into tachyzoites. These infect a wide variety of cells, multiply rapidly to form rosettes, lyse the cells, and spread to other cells or parts of the body. In the tissues, formation of tissue cysts may occur, with slowly replicating bradyzoites inside them.

Epidemiology

Toxoplasmosis is a worldwide zoonosis infecting a wide variety of mammals. Human infection occurs through the ingestion of:

- tissue cysts in raw or undercooked meat
- food or water contaminated with oocysts
- transplacental transmission from mother to fetus
- rarely, through organ transplantation from a seropositive donor, contaminated blood transfusion, needlestick injury.

In HIV-infected patients, cerebral toxoplasmosis is usually due to reactivation of latent infection due to advanced immunosuppression (CD4 count <100 cells/mm^3.) The incidence of toxoplasmosis among HIV-infected individuals is directly related to the seroprevalence of *T. gondii* antibodies in the general HIV population. Thus rates are higher in Western Europe and Africa than in the USA. However the introduction of HAART and use of co-trimoxazole prophylaxis for PCP have resulted in a dramatic fall in the incidence of toxoplasmosis in the developed world.

Pathogenesis

T. gondii penetrates intestinal epithelial cells and multiplies intracellularly. Organisms spread to the regional lymph nodes before being carried to distant organs in the lymphatics and blood. Infection with *T. gondii* induces both humoral and cellular immune responses, which are important for the early clearance of organisms from the blood and limit the parasite burden in other organs. Cyst formation is responsible for persistent or latent infection; the main sites are the brain, skeletal and cardiac muscle, and the eye. In immunocompetent individuals, initial infection is often asymptomatic, chronic/latent infection is not clinically significant, and immunity is lifelong. In immunosuppressed patients, toxoplasmosis may be caused by primary infection but is usually due to reactivation of latent infection and uncontrolled proliferation of organisms. The histological features of cerebral toxoplasmosis include: focal (or diffuse) necrotizing encephalitis, microglial nodules, multiple brain abscesses, hydrocephalus.

Clinical features

- **Immunocompetent patients** – 10–20% of infections are symptomatic. Clinical features include lymphadenopathy and/or an infectious mononucleosis-like syndrome. Rarely, severe, disseminated disease (myocarditis, pneumonitis, hepatitis, encephalitis) may occur.
- **Immunodeficient patients** – toxoplasmosis in HIV-negative patients is associated with organ transplants and lymphoma, and presents with CNS, myocardial, or pulmonary involvement. In AIDS patients, cerebral toxoplasmosis is the most common diagnosis, and presents subacutely with focal neurological symptoms. Other manifestations include spinal cord involvement, pneumonitis, chorioretinitis, pituitary abnormalities, oorchitis, and GI involvement.
- **Ocular toxoplamosis** –*T. gondii* is an important cause of chorioretinitis. Congenitally acquired infection usually presents in the 2nd or 3rd decade of life, with bilateral disease, macular involvement, and old retinal scars. Postnatal infection usually presents in the 4th to 6th decade of life with unilateral involvement and macular sparing. Ocular toxoplasmosis has also been reported in HIV-infected patients, especially from Brazil.
- **Congenital toxoplasmosis** – the incidence of fetal infection varies with trimester: 10–25% in 1st trimester, 30–54% in 2nd trimester, 60–65% in 3rd trimester. The risk of severe congenital infection is highest in the 1st and 2nd trimesters (weeks 10–24). Clinical features include: chorioretinitis, strabismus, blindness, seizures, microcephaly, intracranial calcification, hydrocephalus, anaemia, jaundice, rash, encephalitis, pneumonitis, diarrhoea, hypothermia. In contrast, infants who acquire infection in the 3rd trimester may be born with subclinical infection but, if untreated, go on to develop disease, e.g. chorioretinitis, developmental delay.

Laboratory diagnosis

- Serology – this remains the mainstay of diagnosis. The main problem with serological tests is that antibodies are present in many healthy individuals and persist at high levels for years. Different tests measure different antibodies and there is no single test that can be used to differentiate acute from chronic infection. A combination of tests is often used:
  - IgG antibodies usually appear within 1–2 weeks, peak at 1–2 months and persist for life. The most widely used tests are: ELISA, indirect fluorescent antibody (IFA) test, modified direct agglutination test, IgG avidity test, and Sabin–Feldman dye test (gold standard)
  - IgM antibodies may appear and decline more rapidly than IgG antibodies. However, high IgM levels may persist for years, limiting its use as the sole marker of acute infection. Various tests exist: ELISA (false positives with antinuclear antibody (ANA) and rheumatoid factor); IFA; and IgM immunosorbent agglutination assay (ISAGA)
  - IgA antibodies have higher sensitivity the IgM assays for the diagnosis of congenital toxoplasmosis
  - IgE antibodies are present for a shorter duration than IgM or IgA and may be useful for diagnosing recently acquired infection.

- PCR – the detection of *T. gondii* DNA in body fluids and tissues has been used to diagnose all forms of toxoplasma infection. It has been used for the diagnosis of intrauterine infection and disseminated disease. Sensitivity is 15–85% in blood/buffy coat, and 11–77% in CSF.
- Isolation – isolation of *T. gondii* from blood, body fluids, placenta, or fetal tissues is diagnostic of acute infection. Isolation may be performed by tissue culture (3–6 days) or mouse inoculation.
- Histology – demonstration of tachyzoites in tissues or body fluids is also diagnostic of acute infection. Various methods may be used to demonstrate organisms: fluorescent antibody staining, immunoperoxidase, ELISA, fluorescein-labelled monoclonal antibodies, electron microscopy, Wright–Giemsa staining of centrifuged deposit/smear.
- Antigen-specific lymphocyte transformation and typing – lymphocyte proliferation in response to *T. gondii* antigens is a sensitive and specific indicator of previous infection in adults and has been used to diagnose congenital infection. An increase in CD8+ T cells may occur with acute infection in immunocompetent adults.

Radiological features

Radiological imaging is helpful in patients with CNS disease. In neonates with congenital toxoplasmosis, ultrasound or CT may demonstrate intracranial calcification and ventricular dilatation. In immunodeficient adults with cerebral toxoplasmosis, CT typically shows multiple ring-enhancing lesions. However, scans may be normal, or show solitary lesions or cortical atrophy. MRI appears to be more sensitive than CT and is the imaging modality of choice.

Treatment[1]

Currently recommended drugs act primarily against the tachyzoite form and do not eradicate the encysted form. Pyrimethamine is the most effective agent and should be given with folinic acid to prevent bone marrow suppression. A second drug, sulfadiazine or clindamycin is also given. Alternative agents include co-trimoxazole or pyrimethamine plus one of azithromycin, clarithromycin, atovaquone, or dapsone.

- **Immunocompetent adults** do not usually require treatment unless symptoms are severe and persistent, visceral disease is overt, or infection is parenterally acquired.
- **Immunodeficient patients** – acute/primary therapy is recommended for 3–6 weeks followed by lifelong maintenance therapy/secondary prophylaxis. AIDS patients with cerebral toxoxplasmosis usually respond clinically within 2 weeks; those who do not should be investigated for other alternative diagnoses, e.g. CNS lymphoma.
- **Ocular toxoplamosis** – treatment may not be required for small peripheral retinal lesions in immunocompetent adults, but is generally indicated for lesions that threaten or cause visual loss.
- **Toxoplasmosis in pregnancy** – patients with suspected acute toxoplasmosis in pregnancy should be referred to a specialist unit for further investigation and management. Treatment with spiramycin reduces the risk of transmission to the fetus. As spiramycin does not cross the placenta, if fetal infection occurs treatment should be changed to pyrimethamine (not in 1st trimester) and sulfadiazine.

- **Congenital toxoplasmosis** – infants with congenital toxoplasmosis should be referred to a specialist unit. Treatment is with pyrimethamine and sulfadiazine for up to 12 months.

Prevention[1]

- Prevention of primary infection in susceptible individuals e.g. pregnant women and immunosuppressed patients is by education:
  - avoid contact with cat faeces in gardens and cat litters
  - avoid ingestion of undercooked meat.
- For HIV-infected patients with CD4 count <200 cells/mm^3, primary prophylaxis with co-trimoxazole has been shown to reduce the incidence of cerebral toxoplasmosis. Prophylaxis may be discontinued when CD4 count remains persistently >200 cells/mm^3.
- Some countries, e.g. France and Austria, advocate monthly screening of seronegative pregnant women during pregnancy.

Reference

1 Guidelines for prevention and treatment of opportunistic infections in HIV-infected adults and adolescents (2008). Available from http://AIDSinfo.nih.gov.

Cryptosporidium

Cryptosporidium is an intracellular protozoan, first described in 1907 in mice, and thought to be rare and clinically insignificant for 50 years. It has since been recognized as a common enteric pathogen and is associated with waterborne outbreaks and diarrhoea in AIDS patients. It infects and replicates in epithelial cells of the digestive and respiratory tracts of most vertebrates. Twenty species are recognized; *Cryptosporidium parvum* is the most important species.

Life cycle

Ingestion of oocysts is followed by encystation, usually following exposure to digestive enzymes or bile acids, then release of four sporozoites which attach to the epithelial cell wall. Sporozoites mature asexually into meronts and release merozoites intraluminally. Some of these re-invade the host cells (autoinfection), while others mature sexually into oocysts which are excreted in the faeces

Epidemiology

Cryptosporidium is a ubiquitous enteric pathogen of all age groups.
Transmission occurs by person-to-person, animal-to-person, waterborne or, less commonly, foodborne spread. The prevalence of faecal oocyst excretion varies from 1–3% in industrialized countries to 5–10% in Asia and Africa. Seroprevalence data indicate that cryptosporiodiosis is more common than surveys of faecal oocyst excretion demonstrate, e.g. 25–35% seroprevalence in Europe and N America. Crytosporidiosis has been recognized as an important cause of diarrhoea in HIV/AIDS patients. Other risk factors: malnutrition, immunoglobulin deficiencies, intercurrent viral infections, diabetes mellitus, organ transplantation, haematological malignancies.

Clinical features

- Symptoms usually develop 7–10 days after ingestion of oocysts.
  - GI symptoms: – diarrhoea (may be copious), cramping abdominal pains, anorexia, nausea, vomiting, toxic megacolon (rare)
  - Other symptoms – low-grade fever, weakness, malaise, fatigue, cholecystitis (especially HIV patients), hepatitis, pancreatitis, reactive arthritis, respiratory symptoms, disseminated disease. Recovery depends on the immune status of the patient: immunocompetent have self-limited disease

Diagnosis

- Stool microscopy – examination of several specimens may be required (intermittent shedding). Most laboratories use a faecal concentration method followed by microscopy using a modified acid-fast stain: oocysts stain red/pink (carbol fuschin) against a blue (methylene blue) or green (malachite green) counterstain. Other stains include safranin-methylene blue, methenamine silver-nigrosin acridine orange, auramine-rhodamine, and auramine-carbolfuschin.

- Antigen detection – immunofluorescent antibody or ELISA tests may be used to detect cryptosporidial antigens in clinical or water specimens.
- Serology is primarily used as an epidemiological tool rather than acute diagnosis.
- PCR-based assays have been developed but are not in routine use.
- Histology has low sensitivity and is now rarely used.

Treatment

- Disease is usually self-limited in immunocompetent patients, and supportive therapy (hydration, parenteral nutrition) is key. Nitazoxanide has been used in immunocompetent patients. Antiretroviral therapy has been associated with improvement in HIV/AIDS patients.

Prevention

- Prevention of exposure in 'at-risk' individuals, e.g. water filters, avoid exposure to human and animal faeces, boiling water during outbreaks.

Isospora

Isospora belli, a coccidian parasite first described in 1915, is the only species that infects humans. It is found predominantly in tropical and subtropical climates. It causes a self-limited diarrhoeal illness in immunocompetent patients. It has been recognized as an important cause of chronic or severe diarrhoea in immunocompromised individuals especially HIV/AIDS patients. Rare presentations include disseminated disease, cholecystitis, and reactive arthritis. Diagnosis is by identification of oocysts in stool in wet mounts or acid-fast smears of faecal concentrates. Ultraviolet autofluorescence microscopy is a rapid, simple, and sensitive diagnostic test that is based on detection of oocyst autofluorescence when a 330–380nm UV filter is used. Treatment is with co-trimoxazole (960 mg bd for 10 days for immunocompetent patients, 960 mg qds for 10 days for immunocompromised patients, followed by 960 mg 3 times a week (long-term suppressive therapy).

Cyclospora

Cyclospora cayetanensis was first described humans in Papua New Guinea in 1977. Since then it has emerged as a worldwide cause of diarrhoea in travellers, children, and HIV-infected patients. Transmission is by contaminated food and water. Most of the early cases were described in Nepal, Peru, and Haiti. The epidemiology of the disease varies according to the type of patient. In endemic areas, the duration of illness is short and there are many asymptomatic carriers. In travellers, diarrhoea may last for over a month. In immuncompromised patients symptoms may be severe and last longer. Diagnosis is by microscopic detection of oocysts in stool using modified acid-fast or safranin stains. Ultraviolet autofluorescence of occysts is both rapid and sensitive but not specific. Treatment is with co-trimoxazole (960 mg bd for 7 days for immunocompetent patients, 960 mg qds for 10 days followed by lifelong suppressive therapy in HIV patients, until CD4 count consistently >200 cells/mm^3.

Trypanosoma

The genus *Trypanosoma* consists of approximately 20 species of protozoa. They are common animal pathogens causing severe disease in domestic animals. Three species infect humans:

- *Trypanosoma cruzi* which causes Chagas' disease
- *Trypanosoma brucei gambiense* causes W. African sleeping sickness
- *Trypanosoma brucei rhodesiense*, causes E African sleeping sickness.

Trypanosoma cruzi

Chagas' disease is a zoonosis caused by *T. cruzi*. The disease is endemic in wild and domestic animals in Central and South America. It is transmitted by blood-sucking insects called triatome or 'kissing' bugs which transmit the parasite between and among many mammalian species. Humans are considered accidental hosts. Transmission of *T. cruzi* may also occur through blood transfusions, organ transplantation, or vertically from mother to child.

Life cycle

The parasites multiply in the mid-gut of the insects as promastigotes. In the hindgut they transform into trypomastigotes which pass out in the faeces during blood meals. Transmission to a second mammalian host occurs when breaks in the skin, mucous membranes, or conjunctivae are contaminated with bug faeces. The parasites enter host cells, transform into amastigotes, and multiply and differentiate into trypomastogotes. The cell ruptures releasing the parasites which invade local tissue and spread haematogenously.

Pathogenesis

In acute Chagas' disease, the inflammatory lesion that develops at the site of entry is called the chagoma. Trypomastigotes released by cell rupture may be detected my microscopic examination of the blood. Muscles are the most heavily parasitized tissues. The heart is the most commonly affected organ in chronic Chagas' disease – clinical features include myocarditis and conduction defects. The gastrointestinal tract is also affected with dilatation and muscular hypertrophy of the oesophagus and the colon.

Clinical features

There are four main clinical syndromes:

- **acute Chagas' disease** is usually an illness of children. An inflammatory lesion called a chagoma develops at the site of entry. Romaña's sign (painless periorbital oedema) may be seen if the site of entry is the conjunctivae. Localized signs may be followed by fever, malaise, anorexia, oedema, lymphadeonpathy, and hepatosplenomegaly. CNS involvement is rare but carries a poor prognosis. Severe myocarditis with congestive cardiac failure may also occur. The acute illness is usually resolved and the patient enters the asymptomatic phase

- **cardiac disease** – chronic disease becomes symptomatic many years after the primary infection. The heart is the most common site and clinical features are dizziness, syncope, arrhythmias, seizures, congestive cardiac failure, and thromboembolism. Death occurs within months of developing cardiac failure
- **gastrointestinal disease** – patients with megaoesophagus present with symptoms of achalasia, e.g. dysphagia, odynophagia, cough, chest pain, and regurgitation. Aspiration pneumonitis is common and may be fatal. An increased incidence of oesophageal cancer has been reported. Patients with megacolon present with constipation, abdominal pain, intestinal obstruction, or bowel perforation
- **disease in immunosuppressed patients** – reactivation of *T. cruzi* may occur in immunosuppressed patients, e.g. solid organ transplantation, HIV. The clinical presentation is similar to acute Chagas' disease but may be more severe, with CNS involvement.

Diagnosis

Diagnosis of Chagas' disease is based on:

- history of exposure to *T. cruzi*, e.g. residence or travel to an endemic area, blood transfusion in an endemic area
- acute Chagas' disease – wet prep or Giemsa smear for detection of circulating parasites. In immunocompromised patients other specimens may need to be examined, e.g. lymph node, bone marrow aspirates, pericardial fluid, CSF. If the smear is negative, culture of blood or specimens in liquid media or xenodiagnosis may be attempted
- chronic Chagas' disease is diagnosed by detection of IgG antibodies to the parasite. Many serological assays are available but their performance is variable
- PCR assays have been developed over the past 20 years. Sensitivity is usually >90% (range 47–100%).

Treatment

The current treatment of Chagas' disease is far from ideal.

- Nifurtimox, a nitrofuran derivative, has been used to treat acute and congenital Chagas' disease. Although it reduces the duration and severity of symptoms, parasitological cure rates are only ~70%. Side-effects include nausea, vomiting, abdominal pain, weight loss, and neurological symptoms.
- Benznidazole, a nitroimidazole derivative, is the alternative agent. Its efficacy is similar to nitrofurimox. Side-effects include rash, peripheral neuropathy, and neutropenia.
- Chronic infection – treatment is supportive. Patients have an ECG every 6 months. Pacemakers are helpful in patients with bradyarrhythmias. Megaoesophagus may be treated with balloon dilatation/myomectomy of the lower oesophageal sphincter. Megacolon is managed with high-fibre diet and laxatives/enemas. Surgery may be required for complications.

Trypanosoma brucei complex

The organisms that cause African sleeping sickness are morphologically indistinguishable and belong to the *T. brucei* complex:

- *T. brucei brucei* (animal pathogen)
- *T. brucei rhodesiense* (East African trypanosomiasis)
- *T. brucei gambiense* (West African trypanosomiasis).

They are transmitted by the blood-sucking tsetse flies in Africa.

Life cycle

Tsetse flies ingest trypomastigotes during a blood meal from an infected mammalian host. Once in the mid-gut, the short, stumpy trypomastigotes transform into long, slender, procyclic trypomastigotes. After several cycles of replication they migrate to the salivary glands where they differentiate into epimastigotes and continue to multiply. The epimastigotes transform into infective trypomastigotes which are inoculated into a second mammalian host at the next blood meal. African trypanosomes differ from *T. cruzi* in that they exhibit antigenic variation and are thus able to evade the host immune response.

Pathogenesis

The pathogenesis of African sleeping sickness is complex and incompletely understood. An acute inflammatory lesion (trypanosomal chancre) develops at the site of the tsetse fly bite. Multipliaction of the parasite occurs in this lesion resulting in inflammation, oedema, and local tissue destruction. The parasites spread to the local lymph nodes and then disseminate in the bloodstream. In stage 1 disease (haemolymphatic), there is widespread lymphadenopathy and histiocytic infiltration followed by fibrosis. The heart may be involved. Stage 2 disease (meningoencephalitic) is characterized by CNS invasion.

Clinical features

- West African trypanosomiasis is caused by *T. brucei gambiense*. Infected humans are the main reservoir of infection. A trypanosomal chancre develops 1–2 weeks after the tsetse fly bite and resolves within several weeks. Stage 1 disease is marked by the onset of intermittent high fevers, posterior cervical lymphadenopathy (Winterbottom's sign), hepatosplenomegaly, transient oedema, pruritis, and rash. Stage 2 disease is characterized by the insidious onset of neurological symptoms (headache, somnolence, listless gaze, extrapyramidal signs), accompanied by CSF abnormalities.
- East African trypanosomiasis is caused by *T. brucei rhodesiense*. Infected wild animals are the main reservoir of infection. The illness is more acute than the West African disease, with onset of symptoms a few days after the insect bite. Intermittent fever and rash are common features – lymphadenopathy is less prominent than in West African disease. Cardiac manifestations such as arrhythmias and congestive cardiac failure may result in death prior to the onset of CNS disease. Untreated, this condition is fatal in weeks to months.

Diagnosis

The diagnosis of African trypanosomiasis is based on:

- history of exposure e.g. residence in or travel to an endemic area
- compatible clinical features
- examination of chancre fluid or lymph nodes for trypanosmomes (wet prep and Giemsa stain)
- examination of blood for trypanosomes (wet prep and Giemsa stain). This is more likely to be positive in haemolymphatic stage and in East African trypanosomiasis (higher parasitaemia). Serial specimens should be examined
- examination of Buffy coat for trypansomes (if blood film is non-diagnostic).
- CSF examination – this shows increased cell count, increased CSF pressure, elevated IgM and total protein concentrations. A patient with any CSF abormalities should be regarded as having CNS disease
- bone marrow aspiration may be helpful in patients whose other tests are negative
- serology – assays are available but limited by variable performance.

Treatment

A number of drugs are available to treat African trypanosmiasis (Table 4.26):

- melarsoprol (📖 p.158)
- pentamidine (📖 p.159)
- suramin (📖 p.160)
- eflornithine (📖 p.157).

The treatment of African trypanosomiasis depends on the infecting species, drug resistance patterns, and stage of disease.

Table 4.26 Treatment of African trypanosomiasis

	Stage 1 disease	**Stage 2 disease**
T. brucei gambiense	Pentamidine or eflornithine	Eflornithine or melarsoprol
T. brucei rhodesiense	Suramin or pentamidine	Melarsoprol

Prevention

Control programmes involving vector eradication and drug treatment of animals and humans have had limited success. Chemoprophylaxis is not recommended because of drug toxicity.

Leishmania

Leishmaniasis is caused by various *Leishmania* spp. that vary in their geographical distribution and clinical features. There are three clinical syndromes, each of which may be caused by several species:
- visceral leishmaniasis (kala-azar)
- cutaneous leishmaniasis
- mucosal leishmaniasis (espundia)
- post kala-azar dermal leishmaniasis.

The parasite

Leishmania spp. have a dimorphic life cycle and live in macrophages as intracellular amastigotes in mammalian hosts and extracellular promastigotes in the gut of their sandfly vectors. *Leishmania* spp. cannot be differentiated on the basis of morphology. Speciation was initially based on epidemiological and clinical features; several molecular assays are now used.

Pathogenesis

Cell-mediated immune responses are responsible for controlling leishmanial infections. The clinical manifestations depend on complex interactions between the parasite's invasiveness, tropism, and pathogenicity and the host's genetically determined immune response e.g. NRAMP-1 polymorphisms. The resolution of leishmanial infections is associated with expansion of leishmania-specific $CD4^+$ T cells of the TH1 type, which secrete IFN-γ and IL-2 in response to leishmanial antigens. Macrophages and their products are also important in controlling infection. *Leishmania* spp. may also elicit disease-enhancing immune responses, e.g. expansion of $CD4^+$ T cells of the TH2 type which produce IL-4, IL-5 and IL-10.

Epidemiology

- Visceral leishmaniasis has a wide geographical distribution. It is caused by *L. donovani* spp. (Indian, Pakistan, Nepal, E Africa, E China), *L. infantum* (Middle East, Mediterranean, Balkans, central and SW Asia, N and W China, N and sub-saharan Africa) or *L. chagasi* (Latin America). Rarely, *L. amazonensis* or *L. tropica* may cause visceral leishmaniasis.
- Cutaneous leishmaniasis is also widely distributed. The classic form of Old World cutaneous leshmaniasis, the oriental sore, is found in the Middle East, Mediterranean, Africa, India, and Asia. It is usually caused by *L. major*, *L. tropica*, *L. aethiopica*, and occasionally by *L. donovani* and *L. infantum*. New World cutaneous leishmaniasis is endemic in Latin America. It is caused by *L. brazilensis*, *L. mexicana*, *L. panamensis*, and occasionally by *L. chagasi*.
- Mucosal leishmaniasis (espundia) mainly occurs in Latin America and is usually caused by *L. braziliensis*

Clinical features

- **Visceral leishmaniasis –** incubation period 3–8 months. Onset may be acute or gradual. Symptoms include abdominal enlargement, fever, weakness, anorexia, weight loss. Examination shows pallor, hepatosplenomegaly ± lymphadenopathy (Sudan). The skin becomes dry, thin, scaly and discoloured (kala-azar = black fever). Haemorrhage may occur at various sites. Secondary infections are common in advanced disease and may lead to death. Laboratory findings include anaemia, leucopaenia, and hypergammaglobulinaemia. Visceral leishmaniasis may the presenting feature of HIV infection.

- **Cutaneous leishmaniasis** – incubation period 2 weeks to several months. A wide variety of skin lesions may occur from small dry crusted lesions (usually *L. tropica*) to large, deep ulcers with a granulating base and overlying exudate (usually *L. braziliensis*). Lesions may be single or multiple and tend to occur on exposed areas. Secondary bacterial infections and lymphadenopathy may occur.
- **Mucosal leishmaniasis** – a small proportion of patients with cutaneous leishmaniasis develop mucous membrane involvement of the nose, oral cavity, pharynx, and larynx months to years after their skin lesions have healed. Symptoms include nasal stuffiness, discharge, or epistaxis. The nasal septum may be destroyed resulting in nasal collapse. Perforation may occur through the nose or soft palate. Occassionally, patients may be unable to eat, or develop aspiration penumonia.

Diagnosis

- **Visceral leishmaniasis** – splenic (most sensitive), bone marrow, lymph node, or liver biopsy may confirm the diagnosis. Amastigotes may be seen in Wright- or Giemsa-stained smears. Specimens should be inoculated into special media (e.g. Novy, McNeal and Nicoll medium, Schneider insect medium) and cultured at 22–26°C. Motile promastigotes develop after days to weeks. Anti-leishmanial antibodies may be present at high titre in immunocompetent patients but may be absent or low titre in HIV-infected patients. False-positive reactions occur with leprosy, Chagas' disease, malaria, schistosomiasis, toxoplasmosis, or cutaneous leishmaniasis. An antigen test has been developed.
- **Cutaneous leishmaniasis** – skin biopsies taken from the edge of a lesion may show on amastigotes on Wright or Giemsa staining. Lesions may also be injected and aspirated with saline and examined for amastigotes. Samples may be cultured using special media (see above). Anti-leishmanial antibodies may be present in some patients. The leishmanin (Montenegro) skin test becomes positive during the course of the disease, but is no longer used. A *Leishmania* PCR test has recently been developed and evaluated.
- **Mucosal leishmaniasis** – a definitive diagnosis is made by the identification of amastigotes in tissue biopsies, or isolation of promastigotes in culture. However, the diagnosis is often presumptive and based on the presence of a characteristic scar and positive leishmanin skin test or antileishmanial antibodies.

Treatment

- Pentavalent antimony compounds (sodium stiboglutamate or meglumine antimoniate) have been used for decades but drug resistance, treatment failures, and relapses are becoming more common. The recommended dose is 20 mg/kg/day pentavalent antimony for 20–28 days, depending on clinical syndrome. Side-effects include abdominal pain, anorexia, nausea, vomiting, myalgia, arthralgia, headache, malaise, raised amylase and lipase, renal failure, ECG abnormalities, cardiac arrhythmias, and sudden death (high doses).

- Liposomal amphotericin B is licensed for the treatment of visceral leishmaniasis. It is as effective as, but less toxic than, pentavalent antimony. Conventional amphotericin B deoxycholate and pentamidine isetionate are effective but more-toxic alternatives.
- Fluconazole and itraconzole have been shown to be effective in cutaneous leishmaniasis.
- Miltefosine has been used to treat both cutaneous and visceral leishmaniasis in the developing world.
- Topical therapy with paromycin and methylbenzethonium has been used in cutaneous leishmaniasis.
- Local heat therapy or cryotherapy have been used in cutaneous leishmaniasis.

Prevention

Prevention strategies include:

- controlling sandfly vectors (insecticides, bed nets)
- controlling animal reservoirs (difficult)
- treating infected humans.

Although there is no effective form of immunoprophylaxis, there are ongoing efforts to produce a vaccine.

Giardia lamblia

Giardia lamblia, a flagellated intestinal protozoan, is a common cause of diarrhoea throughout the world.

The pathogen

The differentiation of *Giardia* species has traditionally relied on the morphological features and the identity of the host. However, they may now be classified on the basis of antigen, isoezyme, and genetic analysis. *G. lamblia* is the only species that infects humans. The life cycle consists of two stages: trophozoite and cyst.

Epidemiology

G. lamblia has a worldwide distribution and is the most commonly identified intestinal parasite. It is usually acquired by ingestion of contaminated water but may also be spread by person-to person (children in day care centres, institutionalized people, sexual) or by foodborne transmission. Natural or experimental infections with *Giardia* spp. have been documented for many mammalian species; whether these act as reservoirs for transmission to humans is less clear.

Pathogenesis

Infection occurs after ingestion of as few as 10–25 cysts. After encystation, trophozoites colonize and multiply in the small bowel. Several pathogenic mechanisms have been postulated: disruption of the brush border, mucosal invasion (rare), stimulation of inflammatory infiltration leading to fluid and electrolyte secretion, and villous changes. The production of gastrointestinal secretory IgA antibodies appears to be key in preventing and clearing infection. The cellular immune response is also important in clearing infection, by coordinating IgA secretion and cellular cytotoxicity. Susceptibility to giardiasis has been seen in patients with common variable immune deficiency, X-linked agammagobulinaemia, previous gastric surgery, or reduced gastric acidity.

Clinical features

- Incubation period – symptoms develop 1–2 week after ingestion of cysts; detection of cysts in the stool may take longer.
- Clinical features include asymptomatic cyst passers (5–15%), diarrhoeal syndrome (25–50%), subclinical infection (35–70%).
- Symptomatic giardiasis is characterized by diarrhoea, abdominal cramps, bloating, flatulence, malaise, nausea, anorexia, weight loss. Initially stools may be profuse and watery but later may become greasy, foul smelling and may float. Vomiting, fever, and tenesmus are less common.
- Unusual features include urticaria, reactive arthritis, biliary disease, and gastric infection (if achlorhydria).
- Severe volume depletion may occur in young children and pregnant women, necessitating hospital admission.

Diagnosis

The diagnosis should be considered in all patients with chronic diarrhoea, particularly if associated with malabsorption or weight loss.

- Stool examination – a wet mount of fresh liquid stool may show motile trophozoites; iodine staining may reveal cysts. Formol-ether concentration techniques may increase the yield.
- Antigen detection assays detect *G. lamblia* by immunofluorescence or ELISA. They are most useful in investigating outbreaks and testing patients after treatment.
- Duodenal string test (Entero-Test) may be helpful in difficult cases.
- Duodenal aspirate and biopsy is more invasive but may help to exclude other diagnoses.
- Antibody tests are not widely available but are useful in distinguishing acute from past infection, and in epidemiological surveys.
- *In vitro* culture and molecular assays are available in research settings.

Treatment

- Metronidazole is the drug of choice and has an efficacy of 80–95%. Drug resistance can be induced *in vitro* and may occur *in vivo*. Side-effects: metallic taste, nausea, dizziness, headache disulfiram reaction (with alcohol), neutropenia (rare). Concerns about teratogenicity mean that it is contraindicated in pregnancy (1st trimester) and not recommended in children.
- Tinidazole (📖 p.161), another nitroimidazole, is given in a single dose, and has an efficacy of approx 90%.
- Quinacrine is an alternative and has an efficacy of 90%. Side-effects: nausea, vomiting, abdominal cramps, yellow discolouration of skin, urine and sclera (rare), exfoliative dermatitis (rare).
- Furazolidine (see 📖 Nitrofurans, p.102), a nitrofuran, has a lower efficacy rate (80%) but is available as a liquid suspension. Side effects: GI symptoms, brown discolouration of urine, mild haemolysis (in G6PD deficiency).
- Benzimidazoles (see 📖 Antihelminthic drugs 1, p.162). Mebendazole has proved disappointing, but albendazole looks promising in several studies.
- Paromomycin (see 📖 Aminoglycosides, p.78), an oral aminoglycoside, has been used in pregnancy but has a lower efficacy rate (60–70%) than other drugs.

Prevention

- Good sanitation with proper treatment of public water supplies
- Boiling or purificaton of water with chlorine- or iodine-based preparations in endemic areas
- Prevention of person-to-person spread by good personal hygiene/handwashing/avoidance of orogenital or oro-anal sex.

Trichomonas vaginalis

T. vaginalis, a flagellated protozoan, was initially thought to be a harmless commensal, but has since become recognized as an important cause of genital infection.

The pathogen

On microscopic examination of genital specimens, *T. vaginalis* is a pear-shaped organism (10 x 7 micrometre) with twitching motility. There are four anterior flagellae that arise from single stalk and a fifth flagellum which is embedded in the undulating membrane. The organism only exists as a vegetative cell, reproduces by binary fission, and generates energy with unique organelles called hydrogenosomes.

Epidemiology

The incidence appears to be declining in W Europe and the USA. Trichomoniasis is usually sexually transmitted and its incidence is highest in women with multiple partners, in patients with other STIs, and in HIV-infected patients. Trichomoniasis is occasionally acquired non-venereally (e.g. in institutionalized patients), or by vertical transmission during delivery.

Pathogenesis

All areas of the cell surface are capable of phagocytosis and can ingest bacteria, leucocytes, erythrocytes, and epithelial cells. Trichomonads appear to damage genital epithelium by direct contact, which is mediated by surface proteins, and cause micro-ulceration. Specific virulence factors have not been defined and the immune response is incompletely understood. *T. vaginalis* activates the alternative complement pathway and attracts neutrophils which may kill the protozoan.

Clinical features

- The incubation period is 5–28 days.
- Symptoms often begin or worsen during periods and include vaginal discharge (may be smelly or itchy), dyspareunia, dysuria, and lower abdominal discomfort.
- Signs include vulvar erythema, yellow/green or frothy vaginal discharge, vaginal inflammation, punctuate haemorrhages on the cervix ('strawberry cervix').
- Most infected men are asymptomatic but those that are symptomatic may have urethritis that is clinically indistinguishable from other causes of non-gonococcal urethritis.
- Complications of vaginal trichomoniasis include vaginitis emphysematosa (gas-filled blebs in the vaginal wall), vaginal cuff cellulitis after hysterectomy, premature labour and low birth weight infants.

Diagnosis

- Diagnosis relies on identification of the organism in genital specimens.
- The wet mount will identify organisms in 48–80% of infected women and 50–90% of infected men.

- Various staining methods (e.g. Gram, Giemsa, Pappenheim and acridine orange) are less sensitive than the wet mount.
- Other methods (e.g. direct fluorescent antibody staining, latex agglutination, ELISA, DNA probe, and PCR-based assays) are more sensitive than wet prep but less sensitive than culture.
- Culture remains the most sensitive technique, and trichomonads can be cultured on a variety of media: modified Diamond's media is the best.
- Serological diagnosis is hampered by low sensitivity and poor specificity, particularly in high-risk populations.

Treatment

- Metronidazole (📖 p.100) is the treatment of choice and can be given as a single 2 g dose or in divided doses for 7 days. The main disadvantage of a single dose is the risk of re-infection if the partner is not treated simultaneously.
- Alternative drugs, for metronidazole intolerance or resistance (which appears to be increasing), include nimorazole, tinidazole or ornidazole.

Prevention

- General advice about prevention of STIs
- Use of barrier contraceptive methods if sexually active

Entamoeba histolytica

E. histolytica is a common cause of diarrhoea worldwide, particularly in the tropics. It can also cause extraintestinal disease sites, e.g. abscessses in the liver, lung, brain, or genitourinary tract.

The parasite

E. histolytica is one of several *Entamoeba* species that infect man. Other species are non-pathogenic and include *E. dispar* (morphologically identical), *E. hartmanii* and *E. coli*. The organism exists in two forms: the trophozoite (10–60 micrometre with single nucleus ± ingested erythrocytes) and the cyst (5–20 micrometre with four nuclei). Ingestion of the cyst results in excystation in the small bowel and trophozoite infection of the colon resulting in symptoms. When conditions are no longer favourable, the trophozoite encysts and is passed out in the faeces. Cysts remain viable for weeks or months in moist environments.

Epidemiology

Ten per cent of the world's population is estimated to be infected with *E. histolytica*. There is a wide geographical variation in prevalence ranging from ≤5% in developed countries, to 20–30% in the tropics. Risk factors for amoebiasis in endemic areas include low socioeconomic status, poor sanitation, and overcrowding. In low-prevalence countries, certain groups are at higher risk: immigrants or travellers from endemic regions, institutionalized individuals, and promiscuous male homosexuals. Factors associated with severe disease: neonates, pregnancy, corticosteroid therapy, and malnutrition.

Pathogenesis

The pathogenesis of invasive amoebiasis requires adherence of trophozoites by galactose-inhibitable lectin, direct cytolytic and proteolytic effects that damage tissue, and resistance of the parasite to the host immune response.

Clinical features

The clinical features of amoebiasis can be divided into intestinal and extraintestinal syndromes:

- intestinal manifestations include asymptomatic infection, symptomatic non-invasive infection, amoebic dysentery (gradual onset, abdominal pain/tenderness, bloody diarrhoea), fulminant colitis (rare but carries a high mortality), toxic megacolon, chronic colitis, amoeboma (annular lesion of the colon), perianal ulceration
- extra-intestinal manifestations include amoebic liver abscess (complications: empyema, pericarditis, peritonitis), lung abscess, brain abscess, genitourinary disease; 50% of liver abscess patients have no history of dysentery.

Diagnosis

- Stool microscopy remains the cornerstone of diagnosis but sensitivity is poor and multiple specimens may need to be examined. A fresh liquid stool should be examined by wet mount for motile trophozoites. A formol-ether concentrate with examination of iodine-stained deposit increases the likelihood of seeing cysts.
- Stool culture is more sensitive than microscopy but not routinely available.
- Colonoscopy and biopsy may be helpful in confirming the diagnosis in patient with colitis. Endoscopic features include punctuate haemorrhages and ulcers, but may appear normal in early disease.
- Serological tests, e.g. indirect haemagglutination assays are helpful in the diagnosis of invasive intestinal amoebiasis, but titres may be negative in early disease and remain high for years. Other less-sensitive assays (e.g. counterimmunoelectrophoresis or gel diffusion precipitation) wane more rapidly and may be helpful in diagnosing acute disease.
- Molecular methods include serum and faecal antigen detection tests and PCR-based assays
- Imaging studies, e.g. ultrasound, CT or MRI scans are useful in assessing patients with suspected amoebic liver abscess. Aspiration of the abscess yields a brown, odourless, sterile liquid, which may show trophozoites. The main risk of aspiration is peritoneal spillage/peritonitis. Many cases may be diagnosed and treated with aspiration.

Treatment

The treatment of amoebiasis is complicated by a number of factors including a variety of clinical syndromes, varying sites of action of different drugs, and the availability of different drugs in different countries.

- Intraluminal carriage (e.g. asymptomatic cyst passers) should be treated because of the risk of invasive disease. Possible regimens include: diloxanide furoate, paromomycin, tetracycline + iodoquinol.

- Invasive disease (e.g. dysentery, colitis) should be treated with metronidazole or tinidazole followed by an intraluminal agent (see above).
- Extraintestinal amoebiasis (e.g. liver abscess) should be treated by metronidazole followed by an intraluminal agent (see above). In severely ill patients, emetine or dehydroemetine (less toxic) may be added for the first few days. Aspiration or percutaneous drainage is usually only required for large cysts or to confirm the diagnosis. Surgical attempts to correct amoebic bowel perforation or peritonitis should be avoided.

Prevention

- Avoid ingestion of contaminated water and food
- In endemic areas vegetables should be treated with a detergent and soaked in acetic acid or vinegar
- In endemic areas water should be boiled, as purification with chlorine or iodine may not be sufficient to kill cysts.
- Avoid sexual practices that involve faeco-oral contact

Free-living amoebae

Human infection with free-living amoebae is infrequent, but may be severe and life-threatening. Three clinical syndromes occur:

- primary amoebic meningoencephalitis, caused by *Naegleria fowleri*
- granulomatous amoebic encephalitis, caused by *Acanthamoeba* spp. and *Ballamuthia mandrillaris*
- amoebic keratitis, caused by *Acanthamoeba* spp.

Naegleria fowleri

- **Epidemiology** – found throughout the world in soil, rivers, lakes, and thermally polluted water; grows well in temperatures up to 45°C. Causes primary amoebic encephalitis in children and young adults who have recently swum in warm freshwater lakes or ponds.
- **Pathogenesis** – amoebae penetrate the olfactory mucosa and enter the CNS through the cribriform plate resulting in a diffuse meningoencephalitis, purulent leptomeningitis, and cortical haemorrhages.
- **Clinical features** – symptoms occur 2–5 days after exposure. Patients may initially report changes in smell or taste followed by an abrupt onset of fever, anorexia, nausea, vomiting, headache, meningism, and altered mental status. Patients rapidly progress to coma and death within a week. Myocarditis is found in 7–16% of patients at autopsy but patients do not appear to develop arrhythmias or heart failure.
- **Diagnosis** is based on clinical suspicion and confirmed by the demonstration of trophozoites in the CSF. A variety of molecular assays have been developed. Serological tests are not helpful as the majority of adults tested in endemic areas, e.g. Florida, have antibody.
- **Treatment** – six patients are known to have survived primary amoebic encephalitis. Most were treated with systemic and intrathecal amphotericin B. One patient was treated with miconazole, riafmpicin,

and sulfisoxazole. Various other agents have been tried in animal models, including passive immunotherapy.

Acanthamoeba spp.

- **Epidemiology** – *Acanthamoeba* spp. have been isolated from soil, water, and air. Serological surveys indicate that exposure is common and the organism may be isolated in pharyngeal swabs from healthy people. Encephalitis tends to occur in debilitated or immunosuppressed patients, e.g. HIV, liver disease, diabetes mellitus, organ transplantation, corticosteroid therapy, chemotherapy. In contrast, keratitis occurs in healthy patients.
- **Pathogenesis** – granulomatous amoebic encephalitis is characterized by necrotizing granulomatous lesions containing perivascular trophozoites and cysts located in the cerebellum, midbrain, and brainstem. Keratitis is characterized by cysts and trophozoites in the cornea, with an acute or mixed inflammatory infiltrate.
- **Granulomatous amoebic encephalitis** has an insidious onset and presents with focal neurological deficits. Clinical features include altered mental status, seizures, fever, headache, hemiparesis, meningism, visual disturbance, and ataxia. The duration of CNS illness until death is 7 – 120 days. Other clinical syndromes include skin lesions, pneumonitis, adrenalitis, leucocytoclastic vasculitis, osteomyelitis.
- **Amoebic keratitis** is associated with minor corneal trauma or the use of soft contact lenses. Clinical features include a foreign body sensation followed by severe pain, photophobia, tearing, blepharospasm, conjunctivitis, and blurred vision. The diagnosis is often delayed because of initial misdiagnosis or periods of temporary remission.
- **Diagnosis** – granulomatous amoebic encephalitis is usually diagnosed post mortem, but may be diagnosed ante mortem by brain biopsy. CT brain scans have shown multiple lucent non-enhancing lesions. Lumbar puncture is contraindicated because of the risk of herniation. When performed, CSF examination has been non-diagnostic with elevated white cell counts and protein and decreased glucose levels. The diagnosis of amoebic keratitis depends on the demonstration of *Acanthamoeba* in corneal scrapings, contact lenses, or contact lens fluid by histology or culture. Corneal scrapings may be examined by wet mount for motile trophozoites, or fixed and stained using a variety of stains. A non-nutrient agar overlaid with *E. coli* is most commonly used for culture. Molecular techniques include PCR and DNA probes.
- **Treatment** – little is known about the treatment of granulomatous amoebic encephalitis. Drugs that are active *in vitro* include propamidine, pentamidine, ketoconazole, miconazole, paromomycin, neomycin, 5-flucytosine and, to a lesser extent, amphotericin B. One case has been successfully treated with co-trimoxazole. The treatment of amoebic keratitis involves aggressive surgical debridement and topical therapy with miconazole, propamidine isetionate, and neosporin for 4–6 weeks.

Ballamuthia mandrillaris

- **Epidemiology** – *B. mandrillaris* is a soil inhabitant which contaminates fresh water. It causes granulomatous amoebic encephalitis in both immunocompetent and immunocompromised hosts.
- **Pathogenesis** – CNS lesions are characterized by a chronic inflammatory infiltrate with or without granulomas. Cysts and trophozoites occur in perivascular pattern and associated with angiitis and haemorrhagic necrosis of the meninges and underlying brain tissue.
- **Clinical features** are of a subacute or chronic meningoencephalitis with fever, headache, nausea, vomiting, seizures, and focal neurological signs. Death occurs one week to several months after onset of symptoms.
- **Diagnosis** – CT brain scan shows multiple hypodense lesions with mass effect. CSF abnormalities include a mononuclear pleocystosis (10–500 cells), raised protein, and low glucose. Brain biopsy specimens may demonstrate the cyst and trophozoite. Previously, *B. mandrillaris* was difficult to distinguish from *Acanthamoeba* spp., but an immunofluorescence assay and a cell-free growth medium have been developed.
- **Treatment** – there is no known effective treatment for *B. mandrillaris* encephalitis. Studies have found some efficacy with pentamidine isetionate.

Microsporidia

First identified in 1857, the microsporidia are a diverse group of obligate intracellular, spore-forming protozoa that belong to phylum *Microspora*, order *Microsporida*. Although eukaryotic, they are unusual in having 70S ribosomes, no mitochondria, and simple vesicular Golgi membranes. The microsporidial spore is a highly specialized structure that varies in size and shape according to species. Over 1000 species exist and they infect a wide range of vertebrate and invertebrate hosts. At least 11 species have been implicated in human disease, e.g. *Enterocytozoon bieneusi* and *Encephalitozoon* spp., being the most common.

Epidemiology

Human infection has been identified worldwide (except in Antarctica). Most severe infections are associated with immunocompromise, e.g HIV infection, organ transplantation corticosteroid therapy. However, infections are becoming increasingly recognized in immunocompetent patients, e.g. residents of or travellers from tropical countries. Routes of transmission include waterborne, person-to person spread, inhalation/aerosol or by zoonotic spread.

Pathology

Microsporidia can infect many different organs:

- eye – punctuate epithelial keratopathy
- respiratory tract – rhinitis, sinusitis, nasal polyposis, tracheitis, bronchitis, bronchiolitis ± pneumonia

- genitourinary tract – chronic and granulomatous interstitial nephritis, acute tubular necrosis, microabscesses, granulomas, necrotizing ureteritis and cystitis, prostatic abscess
- gastrointestinal and hepatobiliary tract – enteritis, ulceration, mucosal invasion, granulomatous hepatitis, and cholecystitis
- central nervous system – ring-enhancing lesion with central areas of necrosis filled with spores/macrophages, surrounded by microsporidia-filled astrocytes
- musculoskeletal system – myositis, muscle fibrosis.

Clinical features

The clinical manifestations of microsporidiosis can be divided into two groups, according to the host's immune status.

Immunocompetent patients

- Intestinal infections. Most common manifestation. Caused by *Ent. bieneusi* or *Enc. intestinalis*. Presents with watery diarrhoea, nausea, abdominal pain, fever, and is usually self-limited.
- Ocular infections are rare and may present with corneal stromal infection or keratoconjunctivitis.
- Cerebral or disseminated infections are extremely rare.

HIV-infected patients

- *Ent. bieneusi* typically causes intestinal infections with chronic diarrhoea, anorexia, weight loss, and malabsorption. CD4 counts are typically <100 cells/mm^3. Patients may also develop cholecystitis or cholangitis with fever, nausea, vomiting, and abdominal pain.
- *Enc. intestinalis* causes intestinal and systemic infections that appear similar to those caused by *Ent. bieneusi*. Disseminated disease, particularly to the kidneys, may occur.
- *Enc. hellem* and *Enc. cuniculi* can both cause keratoconjunctivitis sicca. Patients often have laboratory evidence of disseminated infection, and may present with bronchiolitis, sinusitis, nephritis, cystitis, urethritis, prostatitis, hepatitis, peritonitis, cerebral infection, or nodular skin infections.
- Myostis may be caused by various microsporidial species, e.g. *Pleistophora* spp., *Trachipleistophora hominis, Brachiola vesicularum*.
- Systemic infections due to other microsporidial species have been described in case reports.

Diagnosis

- Stool examination is the easiest and most practical method for diagnosing intestinal infections. Several chromotrope-based stains, e.g. Weber stain may be used which stain the microsporidial wall bright pink. Chemifluoresecnt stains may also be used.
- Cytology is used for diagnosis of microsporidiosis in other organs. Various stains may be used, e.g. Weber, Gram, Giemsa, Steiner silver, trichrome blue, chemifluorescent stains.

- Histology remains important in the diagnosis of microsporidiosis. Various stains may be used, e.g. modified Gram, Giemsa, periodic-acid–Schiff and Steiner silver stains.
- Electron microscopy may be useful to identify microsporidia to genus or species level.
- Nucleic acid amplification assays – several PCR-based assay have been developed for species-specific diagnosis. They are usually restricted to a few research laboratories.
- Immunofluorescent detection methods using polyclonal antisera can detect microsporidia (except *Ent. bieneusi*) in most clinical specimens. Sensitivity is poor in stool specimens.
- Serology is unhelpful in the diagnosis of microsporidiosis.
- Tissue culture is only available in a few specialist laboratories.

Treatment

There are limited data on the therapy of human microsporidiosis:

- albendazole appears to improve symptoms in HIV-associated *Enc. intestinalis* infections.
- Fumagillin[1]
- HAART appears to improve symptoms, normalize intestinal architecture, and clear parasites from the stool. However, recurrent diarrhoea and parasitological relapse can occur.

Prevention

Prevention strategies for environmental or zoonotic exposure have not been established, but meticulous handwashing and adherence to existing guidelines for the general prevention of opportunistic infections in HIV-infected patients may be pertinent. As yet there are no clinical trial data to support antimicrobial prophylaxis. HAART may be important in preventing microsporidiosis.

Reference

1 Molina JM, Toumeur M, Sarfati C, Cherret S et al. Fumagillin treatment of intestinal microspondiosis. *N Engl J Med* 2002;346:1963–9.

Nematodes

There are >60 species of nematodes or roundworms that infect humans, some of which are shown in Table 4.27. They are the most common human parasites and are estimated to infect 3–4 billion people worldwide. Helminth infections are a major public health burden in the developing world. All nematodes are elongated, cylindrical, non-segmented organisms with a smooth cuticle and body cavity containing a digestive tract and reproductive organs.

Intestinal nematodes

Intestinal nematodes are the largest group of human helminths. The most-common intestinal nematodes (*Ascaris lumbricoides*, *Ancylostoma duodenale*, *Necator americanus*, and *Trichuris trichuria*) cannot reproduce in humans and are referred to as geohelminths, as their eggs have to develop in the soil. The exceptions are *Stronglyoides stercoralis* and *Enterobius vermicularis*, which can be transmitted from person to person.

Tissue nematodes

The tissue-dwelling roundworms are also a major public health problem, particularly in the tropics. Some affect humans only, while others have an animal reservoir. All of the parasites have a complex life cycle involving intermediate hosts, except *Trichinella* spp. Adult worms do not multiply in humans, so the worm load and severity of disease depend on intensity of exposure.

Table 4.27 Medically important nematodes

Type	Disease	Species
Intestinal	Ascariasis	*Ascaris lumbricoides*
	Trichuriasis	*Trichuris trichiura*
	Hookworm	*Ancylostoma duodemale*, *Necator americanus*
	Strongyloidiasis	*Strongyloides stercoralis*
	Pin worm	*Enterobius vermicularis*
Tissue	Trichinosis	*Trichinella spiralis*
	Dracunculiasis	*Dracunculus medinensis*
	Filiariasis	*Wuchereria bancrofti*, *Brugia malaya*, *Brugia timori*
	Onchocerciasis	*Onchocerca volvulus*

Ascaris lumbricoides

Ascariasis is the most common helminthic infection of humans, with an estimated prevalence of >1 billion. It is caused by *Ascaris lumbricoides* (roundworm) and is found worldwide, most commonly in the tropics.

The parasite

The adult worms (white or reddish yellow, 15–35 cm in length) live in the small intestine and have a lifespan of 10–24 months. Each female produces up to 200,000 ova/day, which pass out in the faeces. When ingested, the eggs hatch in the small intestine, penetrate the intestinal wall, migrate through the venous system to the lungs where they break into the alveoli, migrate up the bronchial tree before they are swallowed, and develop into mature worms in the intestine.

Epidemiology

Ascaris infection is most common in young children, but can occur at any age. Transmission is by the faeco-oral route and is enhanced by the high output of ova and their ability to survive unfavourable environmental conditions. In endemic areas, most people have light-to-moderate worm burdens.

Clinical features

Most infected patients are asymptomatic. Clinical features depend on the site and intensity of infection:

- pulmonary manifestations occur during larval migration through the lungs. Patients may present with Löffler's syndrome (respiratory symptoms, pulmonary infiltration, and peripheral eosinophilia)
- gastointestinal manifestations include malnutrition, malabsorption, steatorrhoea, and intestinal obstruction
- biliary obstruction may cause abdominal pain, cholangitis, pancreatitis, and obstructive jaundice
- ectopic infections occur rarely, e.g. umbilical or hernial fistulae, fallopian tubes, bladder, lungs, and heart.

Diagnosis

Stool examination by wet prep usually confirms the diagnosis. The eggs are oval shaped with a thick mamillated shell and measure 45–70 micrometre (length) by 35–50 micrometre (breadth) (Fig 4.3).

Treatment

- Mebendazole (see 🕮 Antihelminthic drugs 1, p.162) is the treatment of choice.
- Pyrantel pamoate can also be used.
- Piperazine citrate syrup via nasogastric tube is recommended for intestinal or biliary obstruction; piperazine narcotizes the worms and helps to relieve symptoms.

Trichuris trichiuria

Trichuriasis is one of the most prevalent helminthic infections, with an estimated 800 million cases worldwide. Infection is mainly asymptomatic, but heavy infection may cause anaemia, bloody diarrhoea, growth retardation, or rectal prolapse.

The parasite

T. trichiura principally infects humans, residing in the caecum and ascending colon. The mean life span of adult worms is 1 year, and each female worm produces 5–20,000 eggs per day. After excretion, embryonic development occurs over 2–4 weeks. The embryonated egg is ingested and the larva escapes its shell, penetrating the small intestinal mucosa, before migrating down into the caecum or colon. The anterior whip-like portion remains embedded in the mucosa, while the shorter posterior end is free in the lumen.

Epidemiology

T. trichiura has a worldwide prevalence but is more common in most tropical environments, particularly in rural communities with poor sanitation. Infection results from ingestion of embryonated eggs by direct contamination of hands, food, or drink, or indirectly through flies or other insects. The intensity of infections is usually light; heavy infection is more common in children.

Clinical features

Infection is mainly asymptomatic, but heavy infection may present with a variety of symptoms:

- iron deficiency anaemia
- abdominal symptoms and signs
- acute bloody diarrhoea
- chronic colitis with growth retardation
- rectal prolapse.

Diagnosis

Stool examination. The diagnosis is confirmed by detection of characteristic lemon-shaped ova (52 x 22 micrometre) in the stool (Fig 4.3).

Treatment

Albendazole is the treatment of choice. Alternatives: mebendazole or ivermectin.

Prevention

Improved sanitation and meticulous handwashing may help to prevent infection.

Ancylostoma duodenale and *Necator americanus*

Human hookworm infection is estimated to affect 25% of the world's population and is caused by two species, *Ancylostoma duodenale* and *Necator americanus*.

The parasite

Adult hookworms are small, cylindrical (1 cm long), and greyish-white in colour. They live in the upper small intestine, attached to the mucosa. Adult worms produce about 7000 eggs per day. They pass out in the stool and, under suitable conditions, hatch into larvae that moult once to become infective to humans. Skin penetration requires contact-contaminated soil for 5–10 minutes. The larvae are carried by the venous circulation to the lungs where they migrate up the respiratory tree to be swallowed and carried to the small intestine.

Epidemiology

The distribution and prevalence of hookworm infections are limited by environmental conditions – ova fail to develop at temperatures below 13°C larvae are destroyed by drying or direct sunlight. Transmision requires walking barefoot through faecally contaminated topsoil. Transmission is also though to occur from mother to child, either transplacentally or during breastfeeding.

Clinical features

- **Skin rash** – patients may present early in the disease with 'ground itch', intense pruritis, erythema, and a papular/vesicular rash at the site of larval penetration.
- **Pulmonary manifestations** – patients may present with Löffler's syndrome (respiratory symptoms, pulmonary infiltration and eosinophilia) caused by migration of larvae through the lungs.
- **Iron deficiency anaemia** is the most common manifestation. The average daily blood loss is 0.2 mL for *A. duodenale*, and 0.03 mL for *N. americanus*.
- **Protein energy malnutrition** is a common complication. Patients may also have abdominal pain, diarrhoea, weight loss, and malabsorption.

Diagnosis

Stool examination by wet mount readily identifies the ova in clinically significant hookworm infections. The ova are ovoid, thin-shelled and measure 58 x 36 micrometre.

Treatment

- Mebendazole is the treatment of choice.
- Albendazole is an alternative.
- Iron deficiency anaemia should be treated.

Strongyloides stercoralis

Strongyloidiasis, although uncommon compared with other intestinal nematode infection, is important because of its potential to cause overwhelming infection in immunocompromised hosts.

The parasite

S. stercoralis worms can survive as parasitic forms in humans or free-living forms in the soil. The female worm is 2.2 mm, whereas the male is 0.7 mm. Adult worms inhabit the small intestine where the females deposit ova. Eggs hatch in the mucosa, releasing larvae and enter intestinal lumen where they pass out in the faeces. The usual route of infection is through skin contact with contaminated soil. Humans can also be infected via the faeco-oral route. The larvae migrate through the bloodstream to the lungs where they migrate up the respiratory tree to be swallowed to the small intestine.

Epidemiology

S. stercoralis infection is most common in the tropics where transmission depends on climatic conditions, soil conditions, and sanitation. In temperate zones it is a disease of institutions. The patient's worm burden depends on the size of the inoculum and the degree of autoinfection.

Clinical features

- Asymptomatic infection occurs in about one-third of infected people.
- Skin rash – patients may present with a pruritic papulo-vesicular rash at the site of larval penetration; 5–22% of patients develop an urticarial rash which starts perianally and extends to the buttocks, thighs, and abdomen.
- Pulmonary manifestations – patients may present with Löffler's syndrome (respiratory symptoms, pulmonary infiltration, and eosinophilia) caused by migration of larvae through the lungs.
- Abdominal symptoms are common and include colicky abdominal pain, diarrhoea, passage of mucus, nausea, vomiting, weight loss, malabsorption, protein-losing entropathy. Eosinophilia is common.
- Hyperinfection syndrome – massive larval invasion may occur with autoinfection, particularly in immunocompromised hosts, e.g. patients with leukaemia, lymphoma, lepromatous leprosy, receiving corticosteroids, or HIV infection. Clinical features include shock, severe abdominal pain, ileus, pulmonary infiltrates, and Gram-negative bacillary meningitis or septicaemia. Mortality is high.

Diagnosis

Diagnosis depends on demonstration of the larvae in stool or duodenal aspirates. Serological or molecular assays may be used to support the diagnosis.

Treatment

- Tiabendazole (see 📖 Antihelminthic drugs 2, p.165) is the treatment of choice.
- Albendazole (see 📖 Antihelminthic drugs 1, p.162) is an alternative.

Enterobius vermicularis

Infection with *Enterobius vermicularis*, or pinworm, is highly prevalent, particularly in temperate climates. Pinworm infection is most common in young children and institutionalized populations.

The parasite

E. vermicularis is a small, white thread-like worm which inhabits the caecum and ascending colon. Female worms contain about 11,000 ova and live for 11–35 days. The gravid females migrate at night to the perianal/perineal region where they deposit their eggs. The eggs embryonate within hours and are transferred from the perianal region to night clothes, bedding, dust, and air. The most common route of transmission, however, is via the patient's hands.

Epidemiology

The prevalence of pinworm is highest in children aged 5 to 14 years. Pinworm is primarily a family or institutional infection, with no particular socioeconomic associations. As the lifespan of the worm is relatively brief, and eggs can only survive out of the body for 20 days, longstanding infections must be due to continuous re-infection.

Clinical features

- Most infected patients are asymptomatic.
- Perianal/perineal pruritis and disturbed sleep are the most common symptoms.
- Occasionally, migration of the worms may cause ectopic disease, e.g. appendicitis, salpingitis, ulcerative bowel lesions.
- Serum eosinophilia or raised IgE are uncommon.

Diagnosis

- A sellotape slide is used to collect worms from the perianal region.
- The ova are oval shaped but flattened on one side and measure 56 x 27 micrometre.
- The number of examinations is correlated with rate of detection, e.g. 50% for a single examination; 90% for three examinations.
- All family members of an affected individual should be screened for infection.

Treatment

- Mebendazole (see 🕮 Antihelminthic drugs 1, p.163) should be given to all family members.
- Albendazole is an alternative drug.

Prevention

Although good personal hygiene is a useful general principle, its role in the management of enterobiasis is debatable.

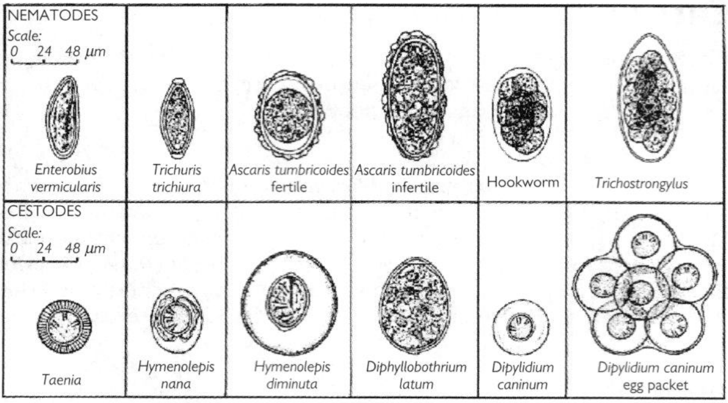

Fig. 4.3 Identification of helminth ova (nematodes and cestodes).

Cutaneous larva migrans

Cutaneous larva migrans is characterized by an erythematous, pruritic, serpinginous skin lesion. It is usually caused by *Ancylostoma braziliense*, the dog and cat hookworm. Other animal and human hookworms may cause similar findings e.g. *Ancylostoma caninum*, *Unicinaria stenocephala, Bunostomum phlebotomum, Strongyloides stercoralis* and *Gnathostoma spinigerum*.

The parasite

The larvae infect dogs or cats by burrowing through the skin. The adults live in the host's intestine and shed eggs in the faeces, which develop into larvae in the sandy soil.

Epidemiology

Infections are most common in warmer climates, especially in holidaymakers who visit tropical sandy beaches. Infection is more common in children than adults.

Clinical features

- The larvae penetrate the skin of humans (an accidental host) causing tingling, itching, and vesicle formation. They then migrate through the skin, causing a characteristic raised, erythematous, pruritic serpinginous track.
- In severe infections many tracks may be seen.
- Systemic symptoms are rare, although pulmonary symptoms and lung infiltrates have been reported.

Diagnosis

- The diagnosis is usually made clinically.
- Skin biopsy may show an eosinophilic inflammatory infiltrate but the migrating parasite is rarely found.

Treatment

- Without treatment, the skin lesions will gradually disappear.
- Both topical and oral tiabendazole are effective.
- Oral albendazole (for 3 to 7 days) or ivermectin (one dose) are also highly effective.

Visceral larva migrans

Visceral larva migrans is characterized by fever, hepatomegaly, and eosinophilia. It is usually caused by *Toxocara canis* but may be caused by *Toxocari cati* or other helminths.

The parasite

T. canis infects dogs and other animals by a variety of mechanisms. Usually, ingested eggs hatch in the small intestine and migrate to the liver, lung, and trachea. They are then swallowed and mature in the lumen of the small intestine where eggs are shed. Other larvae migrate to the muscles and remain dormant but remain capable of development months or years after infection, particularly in pregnancy. During or post-pregnancy, pups may become infected transplacentally or by breast milk and shed larvae in their faeces.

Epidemiology

Toxocariasis has a worldwide distribution. The prevalence of infection is unknown but seroepidemiological surveys show prevalence rates ranging from 3% to 54%, depending on the community selected. Most seropositive people are asymptomatic. Visceral larva migrans occurs most commonly in children <6 years.

Clinical features

- Most infections are asymptomatic. Disease manifestations vary from mild symptoms to fulminant disease and death.
- Clinical features include fever, cough, wheeze, hepatomegaly, splenomegaly, lymphadenopathy, urticaria, skin nodules, seizures, and eye involvement (ocular larva migrans).
- Laboratory abnormalities include eosinophilia, leucocytosis, hypergammaglobulinaemia, and elevated isohaemagglutinin titres to blood group A and B antigens (due host immune response to cross-reacting antigens on *T. canis*).

Diagnosis

- The diagnosis is usually made clinically in a young child with typical clinical features and a history of exposure to puppies.
- A definitive diagnosis is made by finding larvae in the tissues by histological examination.
- Serological tests, e.g. ELISA, may help to confirm the diagnosis (but may also be positive in asymptomatic patients).

Treatment

- There is no proven effective therapy and most patients recover without specific treatment.
- Anti-helminthic drugs (e.g. albendazole, mebendazole, tiabendazole, diethylcarbamazine) may be helpful in patients with severe disease affecting the lungs, heart, or brain. However, treatment may provoke a severe inflammatory response and worsen the clinical condition.
- Corticosteroids have been used, with or without anti-helminthics, with some reports of success.

Prevention

Prevention measures include preventing dogs from contaminating the environment, and preventing children from ingesting eggs.

Trichinella species

Trichinosis occurs after ingestion of *Trichinella* larvae. Most infections are asymptomatic, but heavy exposure may cause fever, diarrhoea, periorbital oedema, and myositis.

The parasite

The genus *Trichinella* comprises five species: *T. spiralis* (the most common cause of human infection), *T. nativa*, *T. brotovi*, *T. pseudospiralis*, and *T. nelsoni*. The cysts are ingested in undercooked meat and the larvae liberated by acid-pepsin digestion of the cysts in the stomach. The larvae invade enterocytes where they develop into adult worms. The adult worms may disseminate in the bloodstream and seed the skeletal muscles where they encyst.

Epidemiology

Trichinella spp. have a worldwide distribution, infecting a wide range of animals, e.g. pigs, rats, horses, bears, foxes, wild boar, and big cats. Humans are incidental hosts.

Clinical features

- Most infections are subclinical.
- Clinical features include fever, myalgia, malaise, periorbital oedema, headache, rash, oedoema, diarrhoea, nausea, subconjunctival haemorrhages, splinter haemorrhages, cough, and vomiting.

Diagnosis

- The diagnosis should be suspected in any patient who presents with fever, periorbital oedema, and myositis, particularly if there is a history of ingestion of undercooked meat.
- Routine laboratory tests show eosinophilia, raised ESR, elevated CPK and lactate dehydrogenase (LDH) levels.
- Antibodies are detectable 3 weeks after infection. Various assays may be used, e.g. ELISA, immunofluorescence, indirect haemagglutination, precipitin, and bentonite flocculation assays.
- Molecular DNA detection assays have been developed.

Treatment

- There is no satisfactory treatment; patients are treated symptomatically with bed rest and salicylates.
- Corticosteroids have been given for severe disease but the evidence to support this is equivocal.
- When a patient is known to have recently ingested trichinous meat, oral tiabendazole may be given. This is active against intestinal worms but not tissue larvae, and does not alter the course of established disease.
- Albendazole has similar efficacy to tiabendazole and may be better tolerated.

Prevention

The most effective way to prevent trichinosis is to cook meat properly.

Dracunculus medinensis

Dracuculiasis (guinea worm infection) occurs after drinking water containing crustaceans infected with *Dracunculus medinensis*. It is characterized by a chronic ulcer from which the worm protrudes.

The parasite

After ingestion of crustaceans containing *D. medinensis*, the larvae are released in the stomach, pass into the small intestine, penetrate the mucosa, and reach the retroperitoneum where they mature and mate. About a year later, the female worm migrates to the subcutaneous tissues of the legs. The overlying skin ulcerates, and a portion of the worm protrudes. On contact with water, large numbers of larvae are released. These are ingested by crustaceans where they undergo further development before the cycle is repeated.

Epidemiology

D. medinensis is found predominantly in Africa where water supplies are used both for drinking water and for bathing.

Clinical features

- There are often no clinical signs until the worm reaches the skin surface.
- Initially a stinging papule develops on the lower leg.
- Some patients may develop generalized symptoms such as urticaria, nausea, vomiting, diarrhoea, and dyspnoea.
- Over the next few days, the lesion vesiculates and ruptures and forms a painful ulcer. If the area is rinsed with water a milky fluid containing larvae wells up.
- Discharge continues intermittently over weeks and the worm is slowly absorbed or extruded, after which the ulcer heals.
- Multiple ulcers may occur and secondary infection is common.

Diagnosis

The diagnosis is mainly clinical, but larvae may be seen on examination of the discharge fluid.

Treatment

- Tiabendazole (see 🕮 Antihelminthic drugs 2, p.165) plus metronidazole (see 🕮 Nitroimidazoles, p.100) have no effect on the worms but produce resolution of the inflammation within days. This permits gradual removal of the worm by rolling it around a small stick.
- Corticosteroid ointments shorten the time to healing, and topical antibiotics may reduce the risk of secondary infection.
- Unerupted worms may be removed by minor surgery under local anaesthesia.
- Secondary bacterial infections should be treated with antibiotics.

Prevention

Boiling, chlorinating, or sieving drinking water prevents guinea worm infection. In W Africa, major advances in prevention have dramatically reduced infection rates.

Filiariasis

Filiariasis is caused by three species: *Wuchereria bancrofti, Brugia malay* and *Brugia timori*.

The parasites

After the bite of an infected mosquito, larvae enter the lymphatics and lymph nodes where they mature into white, thread-like adult worms. The adults live for 5 years, and females discharge microfilariae into the bloodstream, usually around midnight. In the south Pacific, the peak is less pronounced and occurs during the day.

Epidemiology

It is estimated that 120 million people are infected with these parasites.

- *W. bancrofti* occurs throughout the tropics and subtropics.
- *B. malaya* occurs in south and south-east Asia.
- *B. timori* is restricted to eastern Indonesia.

Humans are the only host for *W. bancrofti* but *B. malaya* has been found in felines and primates. Only a small proportion of people who are bitten by infected mosquitoes develop clinical disease.

Clinical features

- Most patients are asymptomatic despite microfiliaraemia.
- Acute infection may present with lymphangitis, lymphadenitis, fever, headache, backache, nausea, epididymitis, or orchitis.
- Chronic hydrocoele is the most common manifestation.
- Chronic lymphadenopathy is common and may progress to lymphoedema or elephantiasis of the lower limb. Ulceration and secondray infection may occur.
- Chyluria may occur if lymphatics burst into the urinary tract.

Diagnosis

- Definitive diagnosis depends on demonstrating microfilariae in the peripheral blood. A sample should be taken at midnight (apart from if the patient is from the south Pacific), and a smear made, stained and examined.
- Micofilariae are occasionally seen in hydrocoele fluid, chylous urine, or lymph node aspirates.
- Serological tests may be positive but do not distinguish the different species, or current from past infection. Immunoassays and PCR-based assays have been developed.
- Ultrasonography of the lymphatic vessels in the spermatic cord may show motile adult worms.

Treatment

- Diethylcarbamazine (DEC) or DEC–albendazole are used. Ivermectin is an alternative. Doxycycline has been shown to be effective in reducing microfilaraemia and adverse effects.

Prevention

Avoid mosquito bites (protective clothing, insect repellent).

Loa loa

Loiasis is caused by *Loa loa* and characterized by transient subcutaneous swellings. Occasionally worms can migrate through the subconjunctiva, causing conjunctivitis.

The parasite

The white, thread-like worms measure 30 to 70 x 0.3 mm and migrate through the connective tissues. The microfilariae measure 300 x 8 micrometre, and appear in the blood during the day.

Epidemiology

Loa loa is endemic in west and central Africa. It is transmitted to humans by tabanid flies (*Chrysops* spp).

Clinical features

- Many patients are asymptomatic but may have eosinophilia.
- The characteristic feature is transient oedematous swellings (Calabar swellings), which are caused by worms migrating through the subcutaneous tissues. They are usually preceded by localized pain and itching, solitary, commonly found around joints, and may last for days to weeks.
- Occasionally a worm may migrate across the subconjunctiva causing conjunctivitis.
- Complications include worms in the penis or breast tissue, endomyocardial fibrosis, retinopathy, encephalopathy, peripheral neuropathy, arthritis, pleural effusion.

Diagnosis

- The diagnosis is often clinical, based on finding typical clinical features in a patient from west or central Africa.
- The diagnosis is confirmed by demostrating microfilariae in a peripheral blood smear taken during the daytime.
- A PCR-based assay has been developed but is not widely available.
- Occasionally an adult worm may be extracted from the eye.

Treatment

- Diethylcarbamazine (see 📖 Antihelminthic drugs 1, p.164) eliminates microfilariae from the blood but does not kill adult worms. Treatment may precipitate encephalopathy in patients with high microfilarial loads.
- Ivermectin decreases microfilarial loads in the peripheral blood. Side-effects include fever, pruritis, headache, arthralgia.
- Albendazole slowly reduces microfilarial loads.

Prevention

- Avoid insect bites (protective clothing, insect repellent) in endemic areas.
- Mass treatment with diethylcarbamazine or ivermectin interrupts transmission in endemic areas.
- Temporary visitors to endemic areas may take prophylactic diethylcarbamazine.

Onchocerca volvulus

Onchocerciasis (river blindness) is caused by *Onchocerca volvulus* and characterized by dermatitis, subcutaneous nodules, keratitis, and choroidoretinitis.

The parasite

O. volvulus is transmited to humans by the *Simulium* blackfly. After a bite, the larvae pentrate the skin and migrate into the connective tissues where they develop into filiform adults. The worms are often found tangled in nodules of subcutaneous tissue. Each female produces large numbers of microfilariae that migrate through the skin and connective tissues.

Epidemiology

O. volvulus occurs in west, central, and east Africa with scattered foci in central and south America. There is no known animal reservoir.

Clinical features

- Skin lesions – initially there is an itchy, erythematous, papular rash. In severe infections, thickening of the skin, depigmentation, lymphoedema, lymphadenopathy. Fibrous nodules containing adult worms may develop over bony prominences. Systemic features include fever, weight loss, and musculoskeletal pains.
- Eye manifestations include punctuate keratitis, pannus formation, and corneal fibrosis. Iridocyclitis, glaucoma, choroiditis, and optic atrophy may occur.

Diagnosis

- The diagnosis is confirmed by detecting microfilariae in skin snips or in the cornea or anterior chamber of the eye on slit lamp examination. Microfilariae are sometimes found in the urine.
- Reliable immunodiagnostic tests are not generally available, but molecular tests are being developed.
- If the diagnosis is strongly suspected, but the parasite cannot be found, a test dose of diethylcarbamazine may exacerbate the rash (see Treatment).

Treatment

- Diethylcarbamazine (see 📖 Antihelminthic drugs 1, p.164) kills microfilariae but not the adult worms. Severe Mazotti reactions may occur, e.g. fever, rash, generalized body pains, keratitis, and iritis.
- Ivermetin has been shown to be safer and more effective. It also kills microfilariae but not adult worms. Side-effects include fever, pruritis, headache, arthralgia.
- Other active drugs include suramin (not recommended), albendazole, and amocarzine.

Prevention

- Avoid insect bites (protective clothing, insect repellents).
- Vector control with larvicides is being used in west Africa.

Cestodes

Human cestode infections occur in one of two forms: mature tapeworms residing in the gut, or larval cysts in the tissues. The form that the infection takes depends on the species. The medically important cestodes are summarised in Table 4.28.

Parasite structure

The parasitic cestodes are flatworms (platyhelminths). The worms consist of several parts: a head (scolex, which has suckers and sometimes hooks), a short neck, and a strobila (a segmented tail made of proglottids). The proglottid has both male and female sexual organs and is responsible for egg production. They become gravid and eventually break free of the tapeworm, releasing eggs in the stool or outside the body.

Parasite life cycle

Cestodes divide their life cycle between two animal hosts: the definitive carnivorous host and the intermediate herbivorous/omnivorous host. Mature tapeworms reside in the intestinal tract of the definitive host and shed eggs into the stool. The eggs may be embryonated (can immediately infect the intermediate host) or non-embryonated (require development outside the body). The intermediate host is infected by the ingestion of eggs that hatch in the intestine releasing an oncosphere. This penetrates the gut mucosa and spreads through the circulation to the tissues where it forms a larval cyst. The life cycle is completed when the carnivorous host ingests the cyst-infected tissues of the intermediate host.

Table 4.28 Medically important cestodes

Type of cestode	Disease	Species
Intestinal	Tapeworm	*Taenia saginata*
		Taenia solium
		Diphyllobothrium latum
		Hymenolepis nana
Invasive	Cysticercosis, echinococcosis	*Taenia solium*
		Echinococcus granulosus
		Echinococcus vogeli
		Echinococcus multilocularis

Taenia saginata (beef tapeworm)

- The parasite – adult worms are long (up to 10 m in length) and contain more than 1000 proglottids, each capable of producing thousands of eggs.
- Epidemiology – transmitted to humans by ingestion of larval cysts in rare or undercooked meat from infected cattle. Common in cattle breeding areas of the world such as central Asia, the near East, and central and eastern Africa where prevalence of infection may be >10%. Areas of lower prevalence (<1%) include Europe, southeast Asia, central America,

- Clinical features – patients are usually asymptomatic but a minority complain of abdominal cramps, malaise. The proglotids are motile and may be seen on the perineum or clothing.
- Diagnosis is confirmed by examination of proglottids (with 15–20 lateral uterine branches). The eggs are morphologically indistinguishable from those of *T. solium*.
- Treatment is with praziquantel (5–10 mg/kg) or niclosamide (2 g) PO stat.

Taenia solium (pork tapeworm)

- The parasite – *T. solium* tapeworms may live for 10–20 years and grow up to 8 m in length.
- Epidemiology – humans can be definitive or intermediate hosts for *T. solium*. Individuals who ingest larval cysts in raw or undercooked pork acquire pork tapeworm; those who ingest eggs develop tissue infection with cysts (cysticercosis, p.605). *T. solium* infection is endemic in Mexico, central and south America, southeast Asia, India, the Philippines, and southern Europe.
- Clinical features – most patients are asymptomatic unless autoinfection with parasite eggs occurs.
- Diagnosis – infection is readily diagnosed by detection of eggs in the stool (morphologically indistinguishable from those of *T. saginata*). Definitive diagnosis is by examination of the proglottids (with 7–13 lateral uterine branches).
- Treatment is with praziquantel (5–10 mg/kg) or niclosamide (2 g) PO stat.

Diphyllobothrium latum (fish tapeworm)

- The parasite – *D. latum* tapeworms may grow up to 25 m in length. The tapeworm takes 3–6 weeks to mature and may survive for more than 30 years.
- Epidemiology – human infection is acquired by uncooked freshwater fish containing cysts. Areas of endemic infection (>2% prevalence) include Siberia, Scandanavia, and other Baltic countries, north America, Japan, and Chile where there is stable zoonotic transmission through other animal hosts, e.g. seals, cats, bears, foxes, and wolves.
- Clinical features – infection is usually asymptomatic but patients may report weakness, dizziness, salt cravings, diarrhoea, or intermittent abdominal discomfort. Prolonged/heavy infection may lead to megaloblastic anaemia caused by vitamin B_{12} deficiency ± folate deficiency.
- Diagnosis – stool examination shows operculated eggs (45–65 micrometre). Recovery of proglottids (with a characteristic central uterus) also confirms the diagnosis.
- Treatment is with praziquantel (5–10 mg/kg) or niclosamide (2 g) PO stat. Mild vitamin B_{12} deficiency resolves with eradication of the tapeworm; severe deficiency requires parenteral treatment.

Hymenolepis nana (dwarf tapeworm)

- The parasite – *H. nana* is the only tapeworm that can be transmitted directly from human to human. Adult tapeworms measure 15–50 mm.
- Epidemiology – areas of endemic infection (up to 26%) include Asia, southern and eastern Europe, central and south America, and Africa. Infection is more common in children and institutionalized patients.
- Clinical features – heavy infection may be associated with abdominal cramps, nausea, diarrhoea, and dizziness.
- Diagnosis is made by identification of eggs (30–47 micrometre, with a characteristic double membrane) in the stool.
- Treatment is with praziquantel (25 mg/kg PO stat, repeated after 1 week) or niclosamide (2 g PO daily for 1 week).

Cysticercosis

- The parasite – cysticercosis is an infection with larval cysts of the cestode *Taenia solium* (see above).
- Epidemiology – infection is acquired by consumption of *T. solium* eggs and occurs wherever *T. solium* infection is prevalent (see above). The cumulative risk of infection increases with age, frequency of pork consumption, and poor household hygiene.
- Clinical features – infected individuals may harbour multiple cysts throughout the body but are often asymptomatic. Symptoms may develop because of local inflammation at the site of infection. Serious disease is rare but occurs with cardiac or CNS involvement (neurocysticercosis). The latter may present with focal symptoms, seizures, chronic meningitis, or spinal cord compression. Neurocysticercosis is the most common cause of seizures in central America. Racemose cysticerocosis is an aggressive form of basilar neurocysticercosis, resulting in coma and death.
- Diagnosis – asymptomatic patients may be diagnosed incidentally by detection of calcified cysts on plain radiographs. Neurocysticercosis may be diagnosed by CT or MRI scan, which show multiple enhancing and non-enhancing unilocular cysts. The diagnosis may be supported by a positive ELISA indicating prior exposure to *T. solium* antigens. However, patients infected with other helminths may have cross-reactive antibodies. Immunoblotting techniques using purified glycoprotein from cyst fluid may be more-sensitive/specific.
- Treatment – surgical resection is the treatment of choice for symptomatic cysts outside the CNS. The drug therapy of neurocysticercosis remains controversial. For complicated neuocysticercosis, treatment may be given with high-dose praziquantel (50 mg/kg/day for 1–21 days) or albendazole (10–15 mg/kg/day for 7–30 days). Evidence favours albendazole over praziquantel and suggests that longer courses may be needed in patients with multiple lesions. CNS inflammation may be reduced by concurrent administration of dexamethasone, but this may reduce the efficacy of praziquantel. Seizures should be treated with anti-epileptic medication, and hydrocephalus with shunting.

Echinococcosis

- The parasite – the canine tapeworms *Echinococcus* spp. inadvertently infect humans causing visceral cysts. Hydatid cyst disease is caused by *Echinococcus granulosus* or *Echinococcus vogeli*, whereas alveolar cyst disease is caused by *Echinococcus granulosus*.
- Epidemiology – infection is acquired by ingestion of parasite eggs excreted by tapeworm-infected animals. *E. granulosus* is prevalent worldwide and transmitted by sheep, goats, horses, camels, and domestic dogs (livestock rearing areas). *E. multilocularis* is transmitted by wild canines and found in northern forest areas of Europe, Asia, north America, and the Arctic. *E. vogeli* is found in south America.
- Clinical features – the hydatic cysts of *E. granulosus* usually affect the liver (50–70%) or lungs (20–30%) but may affect any organ of the body. They are often asymptomatic and found incidentally on radiological imaging. Symptoms may occur as a result of expansion or rupture into adjacent organs. Cyst rupture may cause a severe allergic reaction or seeding to distant organs. Alveolar cyst disease caused by *E. multilocularis* is more aggressive, with invasion of adjacent tissue and metastasis to distant organs. Complications include biliary disease, portal hypertension, and Budd–Chiari syndrome.
- Diagnosis – infection detected by radiological imaging (ultrasound, CT or MRI scan) may be confirmed serologically by ELISA or Western blot assay. This confirms exposure to the parasite and is most sensitive for liver cysts.
- Treatment – asymptomatic cysts may be monitored whereas symptomatic cysts should be treated. The optimal treatment is by surgical resection of the whole cyst 30 min after instillation of a cysticidal agent, e.g. 30% saline, iodophor, or 95% ethanol. Peri-operative antihelminthic agents (e.g. albendazole 400mg bd or mebendazole 40–50mg/kg/day in 3 divided doses) may be given, and care must be taken to prevent cyst rupture or spillage during surgery. For inoperable cysts, medical therapy improves symptoms (55–79%), although cure rates are low (29%). In patients treated surgically, an antiparasitic agent, preferably albendazole, is continued for at least 2 years. In inoperable patients, treatment is considered for years, and possibly for life. The PAIR procedure (puncture, aspiration, injection and re-aspiration) is becoming more popular, as it is less invasive than surgery.

Trematodes (flukes)

Flukes are parasitic worms of the class *Trematoda*. They are usually oval shaped and vary in length (1 mm to several cm). Structurally, they have an oral sucker, a ventral sucker (usually), a blind bifurcate intestinal tract, and prominent reproductive organs. The human flukes belong to the digenetic group in which sexual reproduction is followed by asexual multiplication. Most human parasites are hermaphrodites, except *Schistosoma* spp.

Table 4.29 Medically important trematodes

Type of fluke	Disease	Species
Blood	Schistosomiasis	*Schistosoma haematobium*
		Schistosoma japonicum
		Schistosoma mansoni
		Schistosoma mekongi
		Schistosoma intercalatum
Liver	Clonorchiasis	*Clonorchis sinensis*
	Opisthorchiasis	*Opisthorcis felineus*
		Opisthorcis viverrini
	Fascioliasis	*Fasciola hepatica*
Intestinal	Fasciolopsiasis	*Fasciolopsis buski*
	Heterophyiasis	*Heterophyes heterphyes*
Lung	Paragonimiasis	*Paragonimus westermani*

Schistosoma spp.

- Humans are the principal host of the five *Schistosoma* species (see Table 4.29). Adult worms live in the venous plexus of the urinary bladder (*S. haematobium*) or the portal venous system (*S. mansoni*, *S. japonicum*) where they mate and shed their eggs. Eggs are passed out in the urine or faeces and hatch in fresh water, releasing miracidia that enter the snail (intermediate host). The miracidia multiply asexually in the snail and eventually release cercariae. These infective forms penetrate human skin, and migrate through the lungs and the liver, before passing to their final habitat.
- **Epidemiology** – 200 million people worldwide are estimated to be infected with *Schistosoma* spp. Each species has a specific geographical location: *S. haematobium* (Africa, Middle East), *S. mansoni* (Arabia, Africa, South America, Caribbean), *S. japonicum* (Far East), *S. mekongi* (southeast Asia), *S. intercalatum* (West and Central Africa). Two factors are responsible for endemicity – the presence of the snail vector and contamination of fresh water by human waste.

- **Pathogenesis** – the disease syndromes that characterize schistosomiasis coincide with the three stages of parasite development:
  - cercariae penetrate the skin to cause a rash
  - some weeks after infection the mature worms deposit their eggs; this may be accompanied by acute schistosomiasis (Katayama fever)
  - production of large numbers of eggs results in chronic granulomatous inflammation and fibrosis of the urinary tract or portal venous system.
- **Clinical features** of schistosomiasis include:
  - swimmer's itch – a papular, pruritic dermatitis that occasionally occurs 24 h after penetration of the skin by cercariae. It is appears to be a sensitization phenomenon as it rarely occurs on primary exposure
  - acute schistosomiasis or Katayama fever – occurs 4–8 weeks after infection and is characterized by fever, chills, sweating, headache, cough, hepatosplenomegaly, and lymphadenopathy. Peripheral eosinophilia is common. Symptoms usually resolve within a few weeks but, rarely, death may occur
  - chronic schistosomiasis – occurs in patients with heavy infestation
  - intestinal shistosomiasis, caused by *S. mansoni*, *S. japonicum*, or *S. mekongi*, may present with fatigue, colicky abdominal pain, diarrhoea, dysentery, chronic granulomatous bowel lesions, mucosal ulceration or anaemia. *S. intercalatum* infection may also present with symptoms
  - hepatic schistosomiasis – also caused by *S. mansoni*, *S. japonicum*, or *S. mekongi*, may present with hepatomegaly, portal hypertension, splenomegaly, oesophageal varices, or decompensated liver disease. *S. mekongi* infection may also present with hepatomegaly
  - urinary schistosomiasis – caused by *S. haematobium*, causes granulomatous inflammation in the bladder and ureters. Patients may complain of dysuria and terminal haematuria. Haematospermia is common. Progresion of disease may cause urinary obstruction with hydronephrosis and hydroureter
  - central nervous system schistosomiasis – rare, but complicates 3% of S. japonicum infections. It may present with a space-occupying lesion, encephalopathy, or seizures. *S. haematobium* and *S. mansoni* may cause spinal cord lesions and present with a transverse myelitis.
- **Diagnosis** – the diagnosis should be suspected in any patient with compatible symptoms and an appropriate travel history. The diagnosis is confirmed by detection of eggs in a terminal urine specimen collected between 12 and 2 pm (*S. haematobium*), or in faeces examined by the Kato thick smear procedure (other species). Serodiagnostic tests may be useful in returning travellers (p.628).

- **Management** – praziquantel is the treatment of choice for all species. The dose is 20 mg/kg bd for 1 day (*S. haematobium* and *S. mansoni*), or 20 mg/kg tds for 1 day (*S. japonicum* and *S. mekongi*). Side-effects are mild: abdominal discomfort, fever, and headache. Drug resistance may become a problem in endemic areas with mass treatment. Alternative drugs are metrifonate 7.5 mg/kg 2 doses, 2 weeks apart (*S. haematobium*), and oxamniquine 15–20 mg/kg stat (*S. mansoni*).

Clonorchiasis

- *Clonorchis sinensis* (Chinese or oriental liver fluke) is a parasite of fish-eating mammals in the Far East. The adult flukes are flat elongated worms (15 x 3 mm) that inhabit the distal biliary capillaries where they deposit small yellow operculated eggs (30 x 14 micrometre). The eggs pass out in the stool and are ingested by snails, inside which they hatch into miracidia. Miracidia multiply into cercariae that pass into the water and penetrate freshwater where they encyst as metacercariae. Humans are infected by ingestion of raw or undercooked fish. Once ingested, the metacercariae excyst in the duodenum and migrate to the bile ducts.
- **Epidemiology** – millions of humans are estimated to be incidentally infected, mainly in China, Hong Kong, Korea, and Vietnam.
- **Clinical features** – most infected people are asymptomatic. Heavy infection may result in cholangitis and cholangiohepatitis. Infection has been associated with an increased risk of cholangiocarcinoma.
- **Diagnosis** – infection is confirmed by demonstration of characteristic oprculated, embryonated eggs in the stool (Fig 4.4).
- **Management** – praziquantel (see 🕮 Antihelminthic drugs 2, p.164) 25 mg/kg tds for 2 days, or albendazole (see 🕮 Antihelminthic drugs 1, p.162) 10 mg/kg/day for 7 days. Surgery is needed, rarely, to relieve biliary obstruction.

Opisthorciasis

- *Opisthorcis felineus* and *Opisthorcis viverrini* are common liver flukes of cats and dogs that are occasionally transmitted to humans. The life cycle is similar to *Clonorchis sinensis* (see above).
- **Epidemiology** – *O. felineus* is endemic in south-east Asia and eastern Europe, whereas *O. viverrini* is found in Thailand.
- **Clinical features** – mild or moderate infection is usually asymptomatic. Biliary tract symptoms and ultrasonographic signs are more common in patients aged 20–40 years with heavy infection. An association between *O. viverrini* infection and cholangiocarcinoma has been reported in Thailand.
- **Diagnosis** – this is confirmed by the detection of eggs in the stool (fig 4.4).
- **Management** – praziquantel is the drug of choice.

Fascioliasis

- *Fasciola hepatica* is a liver fluke of sheep and cattle that can infect humans. The adult worms are large, flat, brown, and leaf-shaped, (2.5 x 1cm) and live in the biliary tract of their mammalian host. The large, oval, yellow-brown, operculated eggs (140 x 75 micrometre) pass out in the faeces and complete their development in water.

The miracidia hatch and enter the snail intermediate host where they multiply into unforked-tail cercariae. These emerge and undergo encystment into metacercariae on aquatic plants, grasses, and sometimes soil. After ingestion, the metacercariae excyst releasing larvae that penetrate the intestinal wall, peritoneum, and liver capsule to migrate to the biliary tract.

- **Epidemiology** – infection is more common in sheep- and cattle-rearing areas, e.g. south America, Europe, Africa, China, and Australia.
- **Clinical features** – *F. hepatica* infection has two distinct clinical phases:
  - acute hepatic migratory phase, characterized by fever, right upper quadrant pain, hepatomegaly, and peripheral eosinophilia. Nodules or linear tracks may be seen on ultrasound, CT, or MRI scan
  - chronic biliary phase which may be asymptomatic or present with biliary obstruction or cirrhosis.
- **Diagnosis** – this is confirmed by detection of characteristic ova in the stool or bile (Fig 4.4). Serological tests may also be helpful.
- **Management** – triclabendazole 10mg/kg PO 1 dose is the treatment of choice. Alternatives: nitazoxanide (500 mg bd for 7 days) or bithionol.

Fasciolopsiasis

- *Fasciolopsis buski* is a large intestinal fluke (2–7.5 cm in length) which is endemic in SE Asia and the Far East. It inhabits the duodenum and jejunum producing large operculated eggs (135 x 80 micrometre), which are excreted and hatch into miracidia in fresh water. These enter the snail intermediate host where they multiply and develop into cercariae which encyst into metacercariae on aquatic plants. Humans are infected by ingestion of contaminated plants. The metacercariae excyst in the intestine and develop into adult worms.
- **Clinical features** – fasciolopsiasis is usually asymptomatic, but heavy infection may present with diarrhoea, abdominal pain, or malabsorption.
- **Diagnosis** – this is confirmed by detection of eggs in the stool (Fig 4.4).
- **Management** – praziquantel 25 mg/kg tds for 1 day.

Heterophyiasis

- *Heterophyes heterophyes* is a tiny intestinal fluke (<2 mm in length) which is endemic in the Nile Delta, southeast Asia and the Far East. The lifecycle is similar to *Fasciolopsis buski* (see above), except that the metacerceriae encyst in fish. Humans are infected by consumption of undercooked fish.
- **Clinical features** – infection may present with abdominal pain and diarrhoea.
- **Diagnosis** – adult worms produce small operculated eggs (30 x 15 micrometre), which may be detected in the stool (Fig 4.4).
- **Management** – praziquantel 25 mg/kg tds for 2 days.

Paragonimiasis

- *Paragonimus westermani* is a lung fluke that is widely distributed (West Africa, Indian subcontinent, Far East, Central, and South America).
- **Life cycle** – adult worms inhabit the lungs and produce golden brown operculated eggs that pass into the bronchioles and are coughed up or are swallowed and pass out into the faeces. In fresh water the

eggs mature and release miracidia that infect the snail intermediate host. After 3–5 months, cerceriae are released and infect freshwater crustaceans (crayfish and crabs) where they encyst in the muscles. Humans are infected by ingestion of raw or pickled crustaceans. The metacercariae excyst in the intestine, penetrate the intestinal wall, enter the peritoneal cavity, and migrate through the diaphragm and pleural cavities to the lungs. Worms may also lodge in the peritoneal cavity or the brain.

- **Clinical features** – infection may be asymptomatic or present with cough, brown sputum, intermittent haemoptysis, pleuritic chest pain, and peripheral eosinophilia. Complications include lung abscess, empleural effusion. Ectopic infection is rare and may present with abdominal masses, epilepsy, or focal neurological signs.
- **Diagnosis** – this is confirmed by detection of characteristic eggs in the sputum or faeces (fig 4.4). Serology may be helpful in ectopic infections.
- **Management** – praziquantel 25 mg/kg tds for 2 days, or biothionol 30-50mg/kg PO every other day for 10 days.

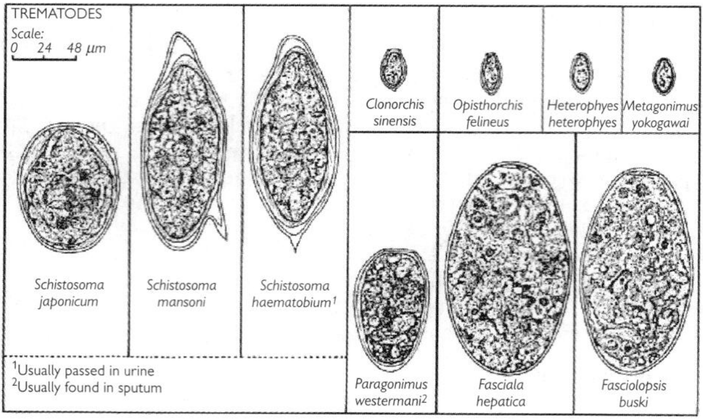

Fig. 4.4 Identification of trematode eggs.

Abdominal angiostrongyliasis

The parasite *Angiostrongylus cantonensis*, a rodent parasite, may cause abdominal symptoms or eosinophilic meningitis (see Chronic meningitis, p.762). The definitive host is the rat, where the parasite lives in the arteries and arterioles of the ileocaecum. Eggs hatch in the intestinal tissue and are excreted in the faeces before being ingested by a slug intermediate where a similar cycle occurs. Humans may be infected by accidental ingestion of foods contaminated by larvae or slugs.

Epidemiology

Infection usually affects children in Central and South America and, rarely, Africa.

Clinical features

Patients usually complain of abdominal pain, fever, and vomiting. Physical findings include fever, abdominal tenderness, and a right lower quadrant mass (50%).

Diagnosis

The syndrome resembles appendicitis apart from the presence of eosinophilia. Radiological features are non-specific. Serology and PCR detection of DNA exist but are not routinely available.

Treatment

Most patients undergo laparatomy and removal of infected tissue. Some patients have been successfully treated with diethycarbamazine and tiabendazole. Alternative: mebendazole.

Anisakiasis

Anisakis and *Phocanema* are parasites of marine mammals, e.g. dolphins, seals and whales. The eggs are excreted in the faeces and hatch as free-swimming larvae which are ingested by crustanceans and then by fish and squid. Humans are accidentally infected following ingestion of raw or poorly cooked seafood.

Epidemiology

Infection occurs most frequently in countries where raw fish is consumed, e.g. Japan and the Netherlands.

Clinical features

Symptoms usually occur 48 h after ingestion, and are caused by penetration of the worms into the gastrointestinal tract. Gastric anisakiasis is usually caused by *Phocanema* and characterized by abdominal pain, nausea, and vomiting. Small intestinal involvement is usually caused by *Anisakis* and characterized by lower abdominal pain and signs of obstruction. Symptoms may become chronic with development of abdominal masses. Occasionally, acute allergic symptoms, e.g. urticaria and anaphylaxis, may also occur with *Anisakis*.

Diagnosis

The diagnosis should be suspected in any patient with a history of ingestion of raw fish and suggestive abdominal symptoms. Leucocytosis commonly occurs with intestinal involvement but eosinophilia is rare. Gastric aniskiasis may be confirmed by radiological features, endoscopy, and histological examination of biopsy specimens. Intestinal anisakiasis is confirmed by radiological features and detection of eosinophils in aspirated ascites. Serological tests are not routinely available.

Treatment

Symptoms usually improve without specific therapy but may resolve more quickly if gastric worms are removed by endoscopy. Occasionally removal of an intestinal mass may be required.

Prevention

Infection may be prevented by cooking or freezing fish for 24 h prior to ingestion.

Capilliariasis

Although the life cycle of *Capillaria philippinensis* is incompletely understood, its larvae have been found in freshwater fish and are known to be infectious for human and birds. After ingestion of raw fish the larvae invade the jejunum and ileum and produce both eggs and larvae. The parasite multiplies in the gut and may cause autoinfection and overwhelming infection (similar to *Strongyloides stercoralis*, 📖 p.594).

Epidemiology

Infections usually occur in the Philippines and Thailand, but two cases have been reported in the Middle East.

Clinical features

These are consistent with malabsorption and protein-losing enteropathy, and include abdominal pain, vomiting, diarrhoea, abdominal distension, borborygmi, malaise, weight loss, peripheral oedema.

Diagnosis

This is confirmed by detecting ova or larvae in the stool. There are no serological tests.

Treatment

Treatment with mebendazole or albendazole is therapeutic and lifesaving and has superseded tiabendazole (25 mg/kg/day for 30 days). Mortality rates of up to 30% have been reported in untreated patients.

Ectoparasites

An ectoparasite is an organism that survives through interaction with the cutaneous surface of the host (e.g. obtaining a blood meal or living in the skin). Most ectoparasites belong to the phylum *Arthropoda*. Two classes are important in human disease: *Hexapoda* (six-legged insects, e.g. lice, bugs, flies, mosquitoes) and *Arachnida* (eight-legged mites, spiders, and ticks). Ectoparasitic diseases are a common health problem in non-industrialized tropical countries.

Lice (pediculosis)

Aetiology

Three species of sucking lice affect humans: *Pediculus humanus* var. *capitis* (head louse), *Pediculus humanus* var. *corporis* (body louse), and *Phthirus pubis* (pubic or crab louse). The first two species are morphologically similar with small, flat, elongated bodies, and pointed heads. The pubic louse is shorter, wider, and resembles a crab. Small ovoid eggs (nits) are laid by the adult female and adhere to hair and clothing; 7–10 days later the nymphs emerge and, after three successive moults, the adult lice develop and mate. The females produce up to 300 eggs per day for 3–4 weeks until they die. Lice pierce the skin, inject saliva, and defaecate while feeding.

Epidemiology

Lice infestations occur worldwide, are transmitted by direct contact, and are associated with poor hygiene and overcrowding. Lice cause skin disease and can also act as vectors for other infectious diseases, e.g. epidemic typhus (*Rickettsia prowazekii*), trench fever (*Bartonella quintana*), and relapsing fever (*Borrelia recurrentis*).

Clinical features

- *Pediculus capitis* affects the scalp and causes pruritis. Complications include secondary bacterial infection and regional adenopathy
- *Pediculus corporis* is usually found in the seams of clothing. Symptoms include pruritis, erythematous macules, papules, and excoriations, usually on the trunk. Complications include impetigo, hyperpigmentation, and hyperkeratosis
- *Phthirus pubis* resides in the pubic hair but may also be found in eyebrows, eyelashes, axillary and chest hair. Symptoms include pruritis, erythematous macules, papules, and excoriations but are usually less-severe than with other species. Small greyish-blue macules (maculae cerulae), caused by injection of an anticoagulant, may be seen. Eyelash infestation may be associated with nits at the base of the eyelashes and crusting of the eyelids.

Diagnosis

Diagnosis is usually clinical but may be confirmed by microscopic examination of the organism.

Management

- *Pediculus capitis* – 0.5% malathion lotion has been shown to be the most effective treatment but requires prolonged application (8–10 h) and has an unpleasant odour. Other pediculicides are comparable in efficacy and require only a 10-min application. These include 1% lindane, gamma

benzene hexachloride shampoo, pyrethrins, 1% permethrin cream rinse. Nits can be removed by combing the hair with a fine-toothed comb. Permethrin-resistant organisms have been reported.
- *Pediculus corporis* – body lice can be eradicated by discarding the clothing, washing clothes in a hot cycle, and ironing the seams or dusting clothing with 1% malathion powder or 10% DDT powder.
- *Phthirus pubis* may be treated with the same agents as head lice (see above). Eyelid infestation may be treated by applying a thick layer of petrolatum bd for 8 days, or 1% yellow oxide of mercury qds for 2 weeks.
- Pruritis is treated symptomatically with antihistamines and topical corticosteroids. Secondary bacterial infection should be treated with an oral anti-staphylococcal agent, e.g. flucloxacillin.

Scabies

Aetiology

Sarcoptes scabiei var. *hominis* is an eight-legged mite that resides in human skin. The adult female lays 2–3 eggs per day which burrow into the skin. After 72–84 h the larvae emerge and, after several moults, develop into adults and mate. The males die shortly afterwards but the gravid female lives for 4–6 weeks.

Epidemiology

Scabies occurs worldwide and epidemics are associated with poverty, malnutrition, overcrowding, and poor hygiene. Scabies is transmitted by direct contact (often sexual) or by fomites.

Clinical features

- Human scabies – symptoms include intense pruritis (more severe at night). Signs include linear burrows, erythematous papules, excoriations and, occasionally, vesicles. Complications include secondary bacterial infection and hypersensitivity reactions, e.g. eczematous eruption, nodular scabies.
- Norwegian scabies is a severe variant which may occur in institutionalized, debilitated, or immunosuppressed patients. Cutaneous lesions are hyperkeratotic, crusted nodules, or plaques, and nail involvement may occur. Complications include secondary bacterial infection, septicaemia, and even death.
- Animal scabies – humans may occasionally be infected by *Sarcoptes scabiei* var. *canis* from their pet dog. Skin lesions are pruritic, papular, or urticarial.

Diagnosis

Diagnosis is usually clinical but may be confirmed by microscopic examination of skin scrapings for the organism, egg, or faeces.

Management

- Permethrin 5% cream applied for 8–10 h is the most effective treatment.
- Lindane 1% lotion is also effective but is absorbed through the skin and may cause toxicity, e.g. irritability and seizures in infants.

- Pregnant women and children may be treated with 6–10% precipitated sulfur in petrolatum daily for 3 days.
- Ivermectin (see 📖 Antihelminthic drugs 1, p.162) 200–250 microgram/kg single dose has been used experimentally in HIV-infected patients and institutional outbreaks.
- Secondary bacterial infection should be treated with an anti-staphylococcal agent, e.g. oral flucloxacillin.
- Household members and close contacts should be treated simultaneously. Infected clothing and bed linen should be washed and dried on a hot cycle.

Myiasis

Epidemiology

Myiasis is an infestation caused by the larvae (maggots) of dipterous (two-winged) flies. It occurs more commonly in tropical climates and is an important veterinary problem. Human disease occurs as a result of travel to an endemic area or exposure to infected animals.

Aetiology

A number of species may cause myiasis:

- *Dermatobia hominis* (human or tropical botfly)
- *Cordylobia anthrophaga* (tumbu fly)
- *Cordylobia rodhaini*
- *Oestrus ovis* (sheep botfly)
- *Gasterophilus* spp. (horse botfly)
- *Hypoderma bovis* (cattle botfly)
- *Cuterebra* spp. (N American botfly)
- *Cochliomyia hominivorax* (New World screwworm)
- *Chrysomyia bezziania* (Old World screwworm).

Clinical features

- Furunculoid myiasis is usually caused by *D. hominis*, *C. anthrophaga*, or *C. rhodhaini*, and occurs in travellers returning from Latin America or Africa. It is characterized by single or multiple cutaneous nodules, each containing a larva. A central punctum develops and may exude serosanguineous or purulent fluid. *D. hominis* infestations occur in the scalp, face, and extremities, and may be associated with local pain. *C. anthrophaga* usually affects the trunk, buttocks, and thighs. *C. rodhaini* is similar but lesions are larger and more painful.
- Subcutaneous infestation is caused by *Gastrophilus* spp. and characterized by migratory integumomiasis (creeping eruption), which is due to migration of larvae through the skin. *H. bovis* may cause similar lesions, which become furunculoid as the larvae mature.
- Wound myiasis may be caused by *C. hominovorax* or *C. bezziana*, and is characterized by local tissue destruction and secondary bacterial infection.
- Ophthalmomyiasis is caused by *O. ovis* and may be superficial (ophthalmomyiasis externa), with conjunctivitis, lid oedema, and punctuate keratopathy, or deep (opthalmomyiasis interna) with invasion of the globe.

Management

- Cutaneous myiasis – removal of the larvae from the affected tissue may be achieved by occlusion of the punctum, e.g. with Vaseline or clingfilm which encourages the larva to partially emerge from the punctum to avoid asphyxiation. It can then be removed with forceps. Surgical excision may be required.
- Wound myiasis – treatment requires removal of larvae and wound debridement.
- Opthalmomyiasis – external infection is managed by removal of larvae under local anaesthetic using forceps and slit lamp examination. With internal infection, dead larvae (without associated inflammation) may be left *in situ*. Inflammation requires topical corticosteroids and mydriatics. Surgical intervention is indicated for live larvae or involvement of critical structures.

Mites

Aetiology

Mites belong to the class *Arachnida*. They occur worldwide and may be free living or parasitize plants, insects, animals, and humans. The mites that infect humans include:

- chiggers (harvest mite or red bug)
- animal mites, e.g. *Sarcoptes scabiei* var. *canis, Cheyletiella* spp., *Liponyssoides sanguineus*, *Ornithonysus bacoti*
- bird mites, e.g. *Dermanyssus gallinae*
- food, grain, and straw mites
- follicle mites, e.g. *Demodex folliculorum*, *Demodex brevis*
- house dust mites, e.g. *Dermatophagoides pteronyssinus*, *Dermatophagoides farinae*
- scabies (see above).

Clinical features

Mites may cause cutaneous disease in humans or act as vectors for infectious diseases, e.g. rickettsial diseases (📖 p.397) Q fever (📖 p.398), tularaemia (📖 p.366), and plague (📖 p.361).

Management

Treatment of mite bites is symptomatic, with oral antihistamines or topical corticosteroids. Most lesions resolve within a week. Secondary bacterial infection should be treated with an antistaphylococcal antibiotic.

Ticks

Ticks are bloodsucking arthropods of the class *Arachnida*. There are three classes: *Ixodidiae* (hard ticks), *Argasidae* (soft ticks), and *Nuttalliellidae* (with characteristics of both).

Epidemiology

Ticks occur worldwide and are important vectors of infectious diseases, e.g. Lyme disease (📖 p.390), babesiosis (📖 p.565), erlichiosis, rickettsial diseases (📖 p.397), Q fever (📖 p.398), tularaemia (📖 p.366), and relapsing fever (📖 p.389).

Clinical features and management

- Tick bites – most bites are asymptomatic. Attached ticks should be removed with forceps to prevent disease transmission. After removal, a pruritic, erythematous, papule or plaque may persist for 1–2 weeks. Sometimes a tick bite granuloma may develop.
- Tick paralysis is rare complication of prolonged attachment of certain tick species. Clinical features are of an ascending paralysis caused by a neurotoxin in the tick salivary gland. Symptoms usually resolve with removal of the tick. A hyperimmune globulin against Ixodes holocyclus is effective against tick paralysis caused by this species.

Chapter 5

Clinical syndromes

Fever: introduction

Fever has been recognized as a clinical syndrome since the 6th century BC. Several centuries later, Hippocratic physicians proposed that body temperature was a balance between the four corporal humors – blood, phlegm, black bile, and yellow bile. Devices to measure body temperature have been around since the 1st century BC. Thermometry became a part of clinical practice in 1868 when Wunderlich declared 37.4°C (98.6°F) to be normal body temperature and described the diurnal variation of body temperature.

Definitions

Different terminologies exist for separate but overlapping conditions.

- **Fever** is defined as 'a state of elevated core temperature which is often, but not necessarily, part of the defensive responses of a multicellular organism (the host) to the invasion of live (microorganisms) or inanimate matter recognized as pathogenic or alien to the host'.
- **Infection** – the presence of organisms in a normally sterile site, usually accompanied by a host inflammatory response.
- **Bacteraemia** – the presence of bacteria in the blood; may be transient.
- **Septicaemia** – similar to bacteraemia but more severe.
- **Systemic inflammatory response syndrome (SIRS)** – response to a wide variety of clinical insults which include infectious and non-infectious causes.
- **Sepsis** – clinical evidence of infection plus evidence of systemic response to infection, e.g. hypoxia, lactic acidosis, oliguria, altered mentation.
- **Sepsis syndrome** – sepsis plus evidence of altered organ perfusion.
- **Severe sepsis** – sepsis associated with organ dysfunction, hypoperfusion, or hypotension
- **Septic shock** – sepsis with hypotension despite adequate fluid resuscitation along with the presence of perfusion abnormalities.
- **Refractory septic shock** – septic shock that lasts for >1 h and does not respond to interventions.

Sepsis syndrome

Sepsis and sepsis syndrome are clinical definitions and are not directly related to microbiological data. However, information on bloodstream infections is available from passive surveillance systems, e.g. UK national bacteraemia surveys. These show that the rate of bloodstream infections has increased over the past two decades, with a dramatic increase in staphylococcal bacteraemias, notably meticillin-resistant *Staphylococcus aureus* (MRSA).

Pathophysiology

The usual setting for the sepsis syndrome is bacterial invasion of the host or toxin production. The best studied example is systemic disease caused by Gram-negative bacteria which have bacterial endotoxin or lipopolysaccharide (LPS). LPS triggers humoral enzymatic mechanisms including complement, clotting fibrinolytic, and kinin pathways. Fever and inflammation are mediated by cytokines e.g. tumour necrosis factor (TNF)-α, interleukin (IL)-1, IL-6, IL-8 and interferon (IFN)-γ. Disseminated intravascular coagulation (DIC) may occur, also mediated by cytokines. Systemic activation of coagulation results in the deposition of fibrin and microvascular thrombosis in critical target organs. Conversely, consumption of clotting factors may lead to bleeding. Infection or shock may trigger injury to the pulmonary vascular endothelium, leading to acute respiratory distress syndrome (ARDS). In models of acute lung injury, various stimuli, e.g. LPS, thrombin, complement, platelet-activating factor, or arachidonate metabolites, may trigger this.

Clinical features

The symptoms and signs suggestive of severe bacterial infection as a cause of sepsis syndrome include:

- fever, chills, hypothermia
- hypotension, cardiac failure
- hyperventilation, respiratory failure, cyanosis
- jaundice, liver failure
- oliguria, anuria, renal failure
- skin lesions
- bleeding
- altered mental status.

Laboratory diagnosis

- Routine investigations may show leucopaenia, thrombocytopaenia, and lactic acidosis.
- At least two sets of blood cultures should be taken from different sites.
- Cultures of potential sources of infection should also be taken.
- Lumbar puncture and cerebrospinal fluid (CSF) examination should be performed for all patients with altered mental status.

Management

- **Antimicrobials** – combination antimicrobial therapy is usually given empirically until the results of cultures are available. The choice of agents depends on the clinical presentation, i.e. community or hospital acquired, neutropenic sepsis, colonization with MRSA or other drug-resistant pathogens
- **Volume replacement** – careful management of fluid and electrolyte balance is crucial. Haemodynamic monitoring may help to guide therapy. Depending on the clinical situation, crystalloids, colloids, or blood products may be used.
- **Inotropes** – sympathomimetic agents, e.g. epinephrine (adrenaline) and norepinephrine (noradrenaline) have been widely used to treat shock but there are no controlled trials of comparative efficacy. Alternative agents, e.g. dopamine, dobutamine, and isoprenaline (isoproterenol), are being increasingly used of instead noradrenaline.
- **Corticosteroids** – the use of corticosteroids for the treatment of septic shock, in the absence of adrenal sufficiency, remains controversial. Early small studies suggested benefit but this was not confirmed in larger controlled trials.
- **Anticoagulation** – the routine use of anticoagulation to treat septic shock associated with DIC has not been shown to be beneficial. If there is bleeding secondary to a coagulopathy this should be corrected with appropriate replacement therapy, e.g. platelets, cryoprecipitate, or fresh-frozen plasma. For patients with refractory shock and coagulopathy despite therapeutic measures, heparin may be beneficial in terminating DIC. Activated protein C is sometimes given in severe sepsis.
- **Diuretics** – despite lack of evidence, diuretics are commonly used in the early oliguric or anuric phases of septic shock with a view to preventing acute renal failure.
- **Cytokine inhibition** – the use of antibodies directed against LPS have failed to show benefit in clinical trials of Gram-negative sepsis syndrome. Similarly, results from trials using recombinant IL-receptor (IL-1Ra) antagonists, monoclonal antibodies to TNF-α and TNF-α receptors have proved disappointing.

Prevention

A number of measures may be implemented to try to prevent sepsis:

- infection control measures in hospitals
- prophylactic antimicrobials, e.g. polymixin sprays, nebulized colistin, selective digestive decontamination
- management of high-risk patients, e.g. bone marrow transplant patients in a protective environment
- active or passive immunoprophylaxis with type-specific or cross-reactive antibodies
- the use of granulocyte transfusions or colony-stimulating factors.

However, these approaches are controversial as no consistent benefit has been demonstrated and concern remains about selection of drug-resistant organisms.

Pyrexia of unknown origin

Definition

The first definition of pyrexia of unknown origin (PUO) was proposed by Petersdorf and Beeson in 1961: 'fever of >38.3°C (101°F) on several occasions persisting without diagnosis for at least 3 weeks despite at least 1 week's investigation in hospital'.

Since then the definition has been modified to reflect changes in medical practice and there are now four different subtypes:

- classic PUO (>38°C for > 3 weeks, >2 visits or 3 days in hospital)
- nosocomial PUO (>38°C for 3 days, not present or incubating on admission)
- immune-deficient PUO (>38°C for >3 days, negative cultures after 48 h)
- HIV-related PUO (>38°C for >3 weeks for outpatients or >3 days for inpatients).

Causes of PUO

- **Classic PUO** – although a wide variety of conditions can cause classic PUO, most fall into five categories:
  - infections (27–50%), e.g. abscesses, endocarditis, tuberculosis, complicated urinary tract infections (UTIs). Some causes show distinct geographical variation, e.g. visceral leishmaniasis (Spain), melioidosis (southeast Asia)
  - neoplasms (13–25%), e.g. lymphoma
  - connective tissue disorders (9–17%), e.g. Still's disease, systemic lupus erythematosus (SLE), rheumatoid arthritis, temporal arteritis, polymyalgia rheumatica
  - miscellaneous disorders (15–21%)
  - undiagnosed conditions (5–23%).

The relative frequency of disorders within these five categories varies according to the era in which the study was conducted, geographical region, age of patient, type of hospital.

- **Nosocomial PUO** presents as fever after hospitalization for at least 24 h. Risk factors include intravascular devices, urinary or respiratory tract instrumentation, surgical procedures, immobility, and drug therapy. However, knowledge is limited due to lack of published data.
- **Immune-deficient PUO** occurs in patients receiving cytotoxic therapy or with haematological malignancies. Because of impaired immune function, signs of inflammation may be modest, leading to atypical presentations. During episodes of neutropenia, infections caused by pyogenic bacteria are most common. In patients with impaired cell-mediated immunity, viral infections are more common.
- **HIV-related PUO** may occur with primary infection or in advanced disease, where it is due to opportunistic infections (e.g. mycobacteria, visceral leishmaniasis, *Pneumocystis jiroveci*, bacterial infections, cytomegalovirus (CMV), toxoplasmosis, cryptococcosis) or malignancies.

Evaluation of PUO

- **History and examination** – a comprehensive history should include details of recent travel, contact with persons who have a similar illness, exposure to animals, work environment, past medical history, drug history, family history for hereditary causes of fever. A careful physical examination may reveal clues as to the aetiology, e.g. stigmata of endocarditis. The presence of fever should be verified although fever patterns are neither sensitive nor specific enough to be diagnostically reliable.

- **Laboratory investigations** include simple blood tests (e.g. full blood count, erythrocyte sedimentation rate (ESR), biochemistry) and urinalysis. Blood cultures (at least three separate specimens) should be taken prior to initiation of antimicrobial therapy). Serology may be helpful for viral infections (e.g. Monospot, CMV IgM, HIV test) and autoimmune disorders (e.g. antinuclear antibody (ANA), antineutrophil cytoplasmic antibody (ANCA), rheumatoid factor).
- **Imaging studies** – all patients should have a chest radiograph. Further radiological imaging should be guided by the clinical presentation, e.g. abdominal computerized tomography (CT) or radiolabelled leucocyte scans for suspected abscesses. Venous duplex scans of the lower extremities may reveal deep vein thromboses.
- **Invasive procedures** – diagnostic biopsies (e.g. needle biopsy, excision biopsy, open biopsy) may be helpful. With the exception of temporal artery biopsies, the diagnostic yield of bedside biopsies is low compared with CT-guided or operative biopsies.
- **Therapeutic trials** – there are a number of limitations and risks assocaited with empirical therapy. The underlying condition may remit spontaneously. Empirical treatment may not be specific, e.g. rifampicin is active against a range of pyogenic bacteria as well as *M. tuberculosis*. Therapeutic trials may confound or delay the diagnosis of PUO. Thus therapeutic trials should be reserved for those few patients in whom all other approaches have failed, or the occasional patient who is too ill for therapy to be withheld.

Management

A fundamental principle of management of classic PUO is that therapy should be delayed until the cause has been identified so that it can be targeted appropriately. However this ideal is frequently ignored in clinical practice and may confound or delay the diagnosis of PUO. In contrast, for neutropenic sepsis, which carries a high risk of serious bacterial infections, empiric broad-spectrum antibiotic therapy should be started immediately after appropriate cultures are taken.

Prognosis

The prognosis of PUO depends on the cause of the fever and the underlying disease. Elderly patients and those with malignant disease have the poorest prognosis. Diagnostic delay adversely affects outcome in intra-abdominal infections, disseminated tuberculosis and fungal infections, and recurrent pulmonary emboli. Patients who have undiagnosed PUO after extensive evaluation generally have a favourable outcome (3.2% 5-year mortality rate).

Reference

1 Durack DT. Fever of unknown origin. In: Mackowiak PA (ed). *Fever. Basic mechanisms and management*, 2nd edn. Philadelphia: Lippincott-Raven, 1997: 237–49.

Imported fever

Fever in returning travellers is estimated to affect 20–70% of the 50 million people who travel from industrialized countries to the developing world each year. Although most illnesses are mild, 1–5% of people are ill enough to seek medical attention, 0.1% require medical evacuation, and 1 in 100,000 dies. An increased risk of travel-associated infections is seen in people who visit family and friends abroad and adventure travellers, often because they underestimate the risks of illness. Potential problems include exposure to 'new' pathogens (via contaminated water and food, poor hygiene, or sanitation), inexperience of clinician in diagnosis, and possible transmission of 'common' resistant organisms between countries. For an excellent review see Ryan et al.[1]

Clinical features

- Patients may present with a wide spectrum of disease, ranging from asymptomatic carriage to fulminant disease (see table 5.1).
- Three main syndromes are seen: fever, diarrhoea, and skin conditions.
- Fever can be differentiated into diagnostic groups according to
  - duration of symptoms (<14 days, 14 days to 6 weeks, and >6 weeks)
  - symptoms, e.g. undifferentiated fever, respiratory symptoms, central nervous system (CNS) symptoms, or haemorrhage.
- Diarrhoea may be acute (<2 weeks) or chronic (2 to 4 weeks).
- Skin lesions may be divided into four categories, e.g. papules, subcutaneous swellings/nodules, ulcers, or linear/migratory lesions.
- Don't forget to consider non-travel-related infections or conditions.

History and examination

The following are essential in the assessment of a returning traveller:

- travel history – where, how long, urban or rural
- exposure history – animals and insects
- incubation period
- duration of illness
- symptoms and physical findings
- immunization status
- antimalarial chemoprophylaxis: regimen, compliance.

Laboratory investigations

- Full blood count, white cell count (WCC) and differential, thick and thin films
- Urea, creatinine, electrolytes, liver function tests, C-reactive protein (CRP)
- Urinalysis, urine microscopy and culture
- Stool microscopy for microscopy, ova, cysts, parasites, and culture
- Skin scrapings or biopsy of lesions
- Sputum microscopy and culture
- Blood cultures
- Serology – arboviruses, rickettsia, schistosomiasis, leptospirosis, viral hepatitis, HIV
- Imaging, e.g. chest x-ray (CXR), abdominal ultrasound
- Bone marrow examination may be helpful in certain conditions, e.g. typhoid, leishmaniasis

Treatment

- Supportive treatment (mild infections or those with no treatment)
- Specific treatment according to the causative organism

Prevention

- Pre-travel vaccination, e.g. hepatitis A (HAV), tetanus, typhoid, diphtheria, meningitis, rabies, Japanese B encephalitis, cholera, yellow fever
- Chemoprophylaxis, e.g. antimalarials
- Advice regarding risk avoidance, e.g. water purification, avoid uncooked food, barrier contraception

Table 5.1 Causes of imported fever

Incubation	Syndrome	Causes
<14 days	Undifferentiated fever	Malaria, dengue, rickettsial spotted fevers, scrub typhus, leptospirosis, bacterial gastroenteritis, typhoid, acute HIV
	Fever with respiratory symptoms	Influenza, legionellosis, Q fever, acute histoplasmosis, acute coccidioidomycosis
	Fever with CNS symptoms	Bacterial meningitis, viral meningitis, encephalitis, cerebral malaria, typhoid, typhus, rabies, arboviral encephalitis, Angiostrongylus cantonensis eosinophilic meningitis, polio, E African trypansomiasis
	Fever with haemorrhage	Meningococcaemia, leptopirosis, Streptococcus suis, malaria, viral haemorrhagic fevers
14 days to 6 weeks		Malaria, typhoid, hepatitis A, hepatitis E, acute schistosomiasis, amoebic liver abscess, leptospirosis, acute HIV infection, E African trypansomiasis, viral haemorrhagic fevers, Q fever
>6 weeks		Malaria, tuberculosis, hepatitis B, hepatitis E, visceral leishmaniasis, lymphatic filiariasis, schistosomiasis, amoebic liver abscess, chronic mycosis, rabies, African trypanosomiasis, HIV

Adapted from Ryan et al.[1]

Reference

1 Ryan ET, Wilson ME, Kain KC. Illness after international travel. *N Engl J Med* 2002;347(7):505–16.

Common cold

The term applied to the acute minor coryzal illness caused by viruses belonging to a number of different families.

Aetiology

- Common causes – members of the myxovirus, paramyxovirus, adenovirus, picornavirus, and coronavirus families – mostly the rhinoviruses (40%) and the coronaviruses (10%). Some of these have several different antigenic types (e.g. over 100 in the case of rhinovirus).
- Other causes – parainfluenza, respiratory syncytial virus (RSV), adenovirus, certain enteroviruses, streptococcal pharyngitis (cannot be differentiated on clinical grounds alone and is considered a cause of colds).
- Re-infections may occur with the same virus; ~25% of cases are attributed to agents as yet unidentified. Influenza can produce a cold-like syndrome but generally produces more-severe lower respiratory tract infection (LRTI).

Epidemiology

- Epidemics occur in the winter months in temperate areas and during the rainy season in tropical regions, reflecting the seasonal nature of viral circulation (perhaps due to the effects of humidity on viral survival).
- In developed countries adults experience on average 2–4, and children 6–8 colds a year. Smokers tend to experience more-significant symptoms.
- Prime reservoir for cold viruses is the upper airway of young children and spread occurs in schools and the home. Mothers have higher secondary attack rates than fathers. Spread is probably by a mix of aerosol and contact with contaminated skin and surfaces.

Pathogenesis

- The pathological mechanisms of different viruses differ. All invade the mucosa of the upper respiratory tract, and sloughed nasal columnar epithelial cells may be identified in the nasal secretions.
- Chemical mediators (cytokines, histamine, prostaglandin) and activation of parasympathetic nervous pathways cause inflammation and engorging of blood vessels of the nasal turbinates with congestion and discharge.
- Peak of rhinovirus cold symptoms coincides with maximal viral shedding.
- Cold viruses may affect the bacterial flora of the respiratory tract and facilitate secondary bacterial infections.

Clinical features

- Incubation 12 h to 3 days. Symptoms: nasal discharge, congestion, sneezing, cough, sore throat. Fever is mild and commoner in children.

- Symptoms reach peak severity by day 3. Most resolve by 1 week, some last up to 2 weeks. Conjunctivitis may be seen in adenovirus and enterovirus infection.

Diagnosis

- Identification of the causative agent is unhelpful in uncomplicated cold.
- Important to recognize secondary bacterial sinusitis (2% of cases) and otitis media (2% of cases). Marked pharyngeal inflammation or exudates raises the possibility of streptococcal, adenovirus, herpes simplex virus (HSV) or Epstein–Barr virus (EBV) infection.
- Rapid antigen detection for group A streptococci may be useful in those with prominent pharyngeal symptoms.

Treatment

Patients should be encouraged to wash their hands and take measures to avoid contamination of others at the peak of symptoms.

- Sedating antihistamines provide relief from sneezing, discharge, cough.
- Non-steroidal anti-inflammatory drugs (NSAIDs) such as naproxen reduce cough.
- Decongestants may be given topically or orally. Rebound nasal congestion may follow withdrawal of topical agents after prolonged use.
- Topical anaesthetic-containing lozenges may relieve sore throat.
- Large doses of vitamin C may have a modest therapeutic effect but have no demonstrable preventative effect in controlled trials.
- There is no role for antibiotics in uncomplicated cold.

Pharyngitis

Pharyngitis is an infection/irritation of the pharynx and/or tonsils.

Aetiology

- 40–60% viral (mostly rhinovirus and adenovirus)
- 5–40% bacterial (up to ≤15% caused by group A streptococci, group A *Streptococcus* (GAS))
- Other causes – allergy, trauma, toxins, malignancy
- It is important to distinguish pharyngitis from more-serious conditions such as epiglottitis and para/retropharyngeal abscess

Pathogenesis

- Bacteria or viruses invade the pharyngeal mucosa, causing a local inflammatory response. Certain viruses (rhinovirus) promote nasal secretions which cause secondary pharyngeal irritation.
- Streptococcal protein/toxins facilitate local invasion and may lead to complications such as rheumatic fever and post-streptococcal glomerulonephritis.

Epidemiology

- Commonest in children. Cases of both viral and bacterial aetiology peak among school-aged children (4–7 years).
- Group A streptococci account for 30% of childhood pharyngitis and 15% of adult cases. *Mycoplasma pneumoniae* and *Chlamydiophila pneumoniae* are common causes among teenagers and young adults.
- Consider gonococcal pharyngitis if there is a history of orogenital contact.

Clinical features

- Viral and bacterial causes are not easily distinguished clinically.
- General features – fever, malaise, sore throat, myalgia. On examination, erythema and oedema of the tonsils and pharyngeal mucosa. Purulent tonsillar exudate suggests streptococcal infection or EBV. Conjunctivitis suggests adenovirus, vesicles suggest HSV stomatitis or Coxsackie A infection. Document murmurs to monitor for potential rheumatic fever. Chest signs may indicate LRTIs by *M. pneumoniae* or *C. pneumoniae*. Hepatosplenomegaly may be seen in EBV infection.
- Features of specific bacterial infections:
  - group A streptococcal infection – 15% of cases. Occur winter to early spring. Consider if: abrupt onset, recent contact with others diagnosed with GAS, headache, vomiting, no cough, swollen tender cervical lymphadenopathy, high WCC. Scarlet fever is also caused by GAS infection (see 🕮 Group A *Streptococcus*, p.290). Complications: rheumatic fever (1 in 400 untreated GAS infections), glomerulonephritis, abscess, toxic shock syndrome, airway obstruction
  - group C, G, and F streptococci – 10% of cases. Resemble GAS but without the immunological sequelae

    - other – *Arcanobacterium haemolyticus* (5% – common in young adults, causing outbreaks of a scarlet fever-like rash), *M. pneumoniae* (young adults with headache, pharyngitis, chest symptoms, cough), *C. pneumoniae* (similar to *M. pneumoniae* – pharyngitis may precede pulmonary infection), *N. gonorrhoeae* (rare – usually follows orogenital contact), *Corynebacterium diphtheriae* (rare in the developed world – risk of airway obstruction), *Borrelia* species.
- Viral – pharyngitis following several days of coughing or rhinorrhoea:
    - adenovirus – 5% of cases and the commonest cause in children under 3 years. Causes an associated conjunctivitis
    - HSV – vesicular lesions are common in children but may be absent in older patients (may be indistinguishable from GAS infection)
    - other – Coxsackie viruses A and B (similar presentation to HSV), rhinovirus, coronavirus, EBV (difficult to distinguish from GAS – p.446), CMV, HIV (pharyngeal oedema and erythema, mouth ulcers).
- Other causes – candidal infection (particularly in immunocompromised), irritation (dry air, postnasal drip, allergy, oesophageal reflux, smoking). Differential – diphtheria, mycoplasmal pneumonia, parapharyngeal abscess, malignancy.

Investigations

- Serology – a rise in ASO titre confirms GAS infection retrospectively.
- Throat swab – antigen tests and culture are sensitive for the presence of GAS and are widely used in the USA. High rates of asymptomatic carriage and positivity do not indicate causality. Antigen tests do not detect group C or G streptococci or other bacterial pathogens. Culture for *N. gonorrhoea* if indicated.
- Consider specific tests for other causes: EBV, *C. pneumoniae, M. pneumoniae* etc, viral culture. Viral serology may be useful in retrospect.

Management

- General – consider airway, sepsis, exclude abscess, assess hydration.
- Most cases are viral, do not need antibiotic treatment, and resolve spontaneously within 10 days. Early treatment of GAS infection reduces the incidence of immunological sequelae but these are in any case very rare in the UK. Antibiotic therapy may prevent suppurative complications.
- Although their value might be questioned, most people would treat suspected GAS infection and all patients who appear unwell, e.g. penicillin V for 10 days. Avoid ampicillin or amoxicillin in those in whom EBV infection has not been excluded. Rates of recurrence are higher in those who do not complete the 10-day course.
- Recurrent tonsillitis – treat with β-lactamase stable agents such as clindamycin. Consider ear, nose and throat (ENT) referral – tonsillectomy may be appropriate.

Complications

LRTI, suppurative complications (abscess, sinusitis, otitis media, epiglottitis, mastoiditis), non-suppurative sequelae of GAS infection. Scarlet fever (see 📖 Bacterial causes of childhood illness, p.820) may be seen in association with a streptococcal infection at any site.

Retropharyngeal abscess

Infection causing abscess in one of the deep spaces of the neck. Potentially life-threatening due to risk of airway compromise but rare in developed countries where antibiotic usage in the treatment of upper respiratory tract infection (URTI) is widespread. Generally a paediatric diagnosis. Other pharyngeal abscesses (lateral pharyngeal, peritonsillar) are commoner in older children/adults.

Pathogenesis

- The retropharyngeal space is located posterior to the pharynx, anterior to the cervical vertebrae, and extends from the base of the skull to the level of the tracheal bifurcation.
- It may become infected by contiguous spread (e.g. an URTI that has spread to the retropharyngeal lymph nodes), or through direct inoculation by penetrating trauma or instrumentation (feeding tube insertion, head and neck surgery, fish bone).

Clinical features

- Patients may have experienced previous URTI (perhaps weeks previously) and may not recall specific trauma (e.g. running and falling with a lollipop in the mouth).
- Symptoms: fever, chills, malaise, voice change (classically a duck quack: 'cri du canard'), sore throat, dysphagia, neck stiffness, sensation of a lump in the throat, pain in back and shoulders on swallowing. Difficulty breathing is a serious sign that may herald airway obstruction.
- On examination – fever, septic, unilateral lymphadenopathy, neck mass, mass in posterior pharyngeal wall on oral examination (30% cases), signs of vascular complications (jugular vein thrombophlebitis, carotid rupture – bleeding from the ear, nose, or mouth and ecchymosis in the neck).
- Consider the diagnosis in those with fever, neck stiffness, and normal lumbar puncture.
- Complications – mass effect (e.g. expansion against the trachea causing airway compression), abscess rupture (pus aspiration may lead to pneumonia), spread of infection (inflammation and destruction of adjacent tissues – mediastinitis, pericarditis, empyema, jugular vein thrombosis, carotid artery rupture, osteomyelitis of the cervical vertebrae leading to subluxation and spinal cord damage). The infection itself may rarely develop into necrotizing fasciitis and sepsis.

Diagnosis

- Imaging – lateral soft tissue x-ray of the neck (inspiration with the neck in normal extension) may show widening of the prevertebral tissue and gas. Contrast CT is more useful. Magnetic resonance imaging (MRI) produces more-detailed images but these are usually unnecessary and children often require sedation.

- Microbiology – usually polymicrobial members of the oropharyngeal flora: Gram-positives (group A streptococci, *S. aureus*) and anaerobes (*Bacteroides* species) predominate. Gram-negatives may also be found.
- Consider tuberculosis (TB) and coccidiosis in patients with risk factors who fail to respond to standard antibiotic therapy.
- Differential – foreign body, epiglottitis, cystic hygroma, other forms of pharyngeal collection.

Management

- Urgent ENT referral for incision and drainage.
- Intravenous (IV) antibiotic therapy with benzylpenicillin and clindamycin, or cephalosporin and metronidazole.
- 30% may require a surgical airway.

Quinsy (peritonsillar abscess)

- Relatively common infection of the peritonsillar space (between the capsule of the palatine tonsil and the pharyngeal muscles), which usually follows bacterial pharyngitis but may arise *de novo*.
- One-third of cases occur in children. Peak incidence age 20–30 years.
- Usually polymicrobial in origin (GAS, aerobes such as *S. milleri* and anaerobes such as *Peptostreptococcus*, *Streptococcus sanguis*, and *Fusobacterium* species).
- Symptoms – drooling, fever, an abrupt increase in pain and dysphagia.
- Examination – asymmetrical tonsillar enlargement and neck swelling, may be a palpable fluctuant mass and bad breath. Trismus may render examination impossible. Ultrasound scan (USS) or CT may be useful in such patients.
- Management – ENT referral for consideration of surgical drainage and IV benzyl-penicillin or co-amoxiclav. Perimucosal needle aspiration may be appropriate in patients who can tolerate it with no evidence of deep neck tissue extension, septicaemia, or toxicity. Material can also be obtained for culture. Complications: airway obstruction.
- Treatment success is 90–95% for appropriately treated uncomplicated abscesses, with a 10–15% recurrence rate (the majority very shortly after the initial episode, suggesting ongoing infection). Such patients may benefit from interval tonsillectomy.

Lemierre's disease

- Oropharyngeal infection ('anaerobic tonsillitis') spreading to cause internal jugular thrombosis. Usually seen in young, healthy adults.
- Caused by *Fusobacterium necrophorum*. Severe sore throat progresses to sepsis with secondary abscesses (liver, lungs, pleura, bones, joints, brain) and local spread (Quinsy and vascular thrombosis).
- US, CT and MRI confirm the diagnosis. The organism may be grown from blood or pus – some cases are mixed infections (oropharyngeal flora, e.g. *S. milleri*). Prolonged antibiotic therapy is required (e.g. 6 weeks' co-amoxiclav).

Croup

An acute laryngotracheitis or laryngotracheobronchitis of viral aetiology occurring in young children and resulting in breathlessness with stridor-like inspiration and a barking, seal-like cough.

Epidemiology

- A common illness in young children (around 10% of LRTI), peaking during the second year of life although it can occur in children as old as 15 years.
- Autumn cases tend to be due to parainfluenza types 1 and 2, winter cases to influenza and RSV, sporadic cases to parainfluenza type 3, and several less-frequently identified agents e.g. adenovirus, rhinovirus.
- Measles is a cause of severe croup and remains so in areas of poor vaccination coverage. *M. pneumoniae* can cause a croup-like syndrome.

Pathophysiology

- Viral infection of the upper airway spreads down and may involve the entire respiratory tract. Inflammation of the larynx and trachea causes stridor, hoarseness and cough, and is greatest at the subglottic level near the cricoid. The compliant airways of young children further restrict airflow on inspiration at this narrow point.
- Decreased tidal volume leads to an increase in respiratory rate, and if the obstruction is severe, exhausation, hypercapnia, and hypoxia may follow. Hypoxia is exacerbated by parenchymal inflammation.
- The predilection of croup for young children is probably a combination of their anatomy (small compliant airways) and the fact that many children are experiencing primary infections. Immune mechanisms may play a part – not every child experiencing a parainfluenza infection gets croup, and immune defects similar to those found in atopic patients may contribute to the mechanics of disease in certain cases.

Clinical features

- It is vital to distinguish croup from epiglottitis – see 🕮 p.638.
- Children usually have a short history of URTI with sore throat, mild cough, and fever. The onset of croup often occurs at night with barking, 'seal-like' cough, and hoarseness. Stridor and dyspnoea may develop. The child may sit forward to aid breathing. Ausculation can reveal rhonchi and wheeze in severely affected children. Some develop pneumonia. Respiratory rates of 40/min are not unusual. Symptoms fluctuate, improving or worsening within an hour, tending to be more severe in the evening.
- In severe cases, exhaustion or severe obstruction may necessitate respiratory support or intubation.
- Indicators of impending respiratory failure – stridor at rest (may be quiet), sternal wall retractions, lethargy, or decreased level of consciousness, paradoxical breathing, quiet breath sounds.
- Some children experience repeated episodes of croup (spasmodic croup) perhaps related to airway hyper-reactivity and allergic disease.

Diagnosis

Diagnosis is clinical and croup must be differentiated from other causes of stridor (e.g. foreign body) and bacterial epiglottitis. Epiglottitis has a more-rapid course and children tend to be more toxic with dysphagia and drooling. Croup's distinctive cough is also absent.

- General laboratory features – WCC may be normal or raised, hypoxia is seen in most hospitalized children and hypercapnia in half.
- Viral identification – immunofluorescence and reverse transcription polymerase chain reaction (RT-PCR) may allow rapid identification of a causative agent.
- Serology is not useful in the diagnosis of croup.

Management

- Steam inhalation providing humidification of the upper airways is a mainstay of home therapy. There is no evidence to demonstrate its efficacy.
- Patients unwell enough to require hospitalization require close observation to identify signs of airway obstruction, exhaustion, and respiratory failure requiring intubation and ventilation. Interventions such as blood sampling should be kept to the minimum necessary to avoid exacerbating anxiety and breathlessness.
  - Respiratory rate is the best indicator of hypoxia. Cyanosis may not be present and the severity of stridor indicates the degree of subglottic obstruction and does not necessarily relate to oxygenation.
  - Pulse oximetry is sufficient a measure of oxygenation in most, but the more severely ill patient should have arterial CO_2 measured. Children who are hypoxic with normal CO_2 should respond to low concentrations of supplemental oxygen. Humidification has not been shown to have any benefit – in fact dry air may be preferable.
- Pharmacological interventions:
  - nebulized budesonide is associated with faster clinical improvement compared to placebo in moderate to severe cases
  - corticosteroid therapy – a single dose of dexamethasone (oral or intramuscular (IM) 0.15–0.6 mg/kg) reduces the proportion of those with mild disease requiring further medical attention. It is beneficial in cases of moderate and severe croup, equal to or more efficacious than nebulized steroids, easier to give, and less distressing to the anxious child when given orally
  - nebulized racemic adrenaline (epinephrine) 2.25% – diminishes subglottic swelling and produces clinical improvement in those children with severe stridor. Fast acting but improvement is transitory – around 2 h. It does not improve oxygenation but results in less-frequent need for intubation due to exhaustion
  - there is no role for antibiotics in the management of croup in the absence of concomitant bacterial infection.

Epiglottitis

A potentially life-threatening condition of inflammation, oedema, and obstruction of the epiglottis and surrounding structures. Classically a disease of children (aged 3–7 years) due to infection by *Haemophilus influenzae* type B. Since the introduction of *H. influenzae* type B (Hib) vaccination, recent epidemiology suggests that it is now more common in adults and no one organism is predominant. Other causes: group A, B, and C *Streptococcus*, *S. pneumoniae*, *K. pneumoniae*, *Candida albicans*, *S. aureus*, *Haemophilus parainfluenzae*, *N. meningitides*, and certain viruses. Hib epiglottitis has occurred in vaccinated children.

Clinical features

- Abrupt onset of severe sore throat and fever with stridor, drooling, anxiety, and refusal to eat. Children may adopt the typical posture: sitting up, leaning forward, and generally looking seriously unwell.
- Attempting to examine the throat (e.g. use of a tongue depressor) may result in total airway obstruction, as may IV cannulation. These should be deferred until anaesthetic support is present.
- Complications – bacteraemia, pneumonia, meningitis, arthritis, cellulitus.

Management

- Securing the airway is the absolute priority. Once the airway is secured, mortality is less than 1%. Outcome is significantly better with elective intubation than emergency intubation.
- Antibiotics – IV cephalosporin usually for 7–10 days. However, consider *C. albicans* if there are white patches on examination.
- Prophylaxis – rifampicin should be given to the patient and all household/day care contacts including adults if there are other susceptible (unvaccinated or immunocompromised) children in the family.
- Mortality – around 0% with a quick diagnosis in specialist centres, 9–18% where diagnosis is delayed, 6% where patients are managed without intubation.

Investigations

All investigations should follow securing the airway.

- WCC and inflammatory markers are raised
- Cultures – blood and epiglottic swabs
- Radiology – lateral neck XR may show an enlarged epiglottitis protruding from the anterior wall of the hypopharynx ('thumb sign')
- Differential – croup, diphtheria, inhaled foreign body

Bacterial tracheitis

An atypical form of croup, its clinical picture having more in common with epiglottitis. It is uncommon, tending to affect older children. Those with a history of either recent intubation or viral illness are at greater risk. Presentation is with an abrupt-onset of fever, stridor, and breathlessness with large amounts of purulent sputum. Progression can be rapid, necessitating intubation. Obstruction is subglottic, the epiglottis itself being only minimally inflamed. Organisms that may be recovered include *S. aureus*, group A, β-haemolytic streptococci, and (prior to vaccination) *H. influenzae* type B. Antibiotics should be given promptly and direct laryngscopy can confirm the diagnosis and provide local secretions for culture.

Laryngitis

Aetiology

An inflammation of the larynx that may be caused by any of the major respiratory viruses including rhinovirus, influenza, parainfluenza, adenovirus, and coronavirus. It can be a feature of streptococcal sore throat. Other common bacterial agents associated with laryngitis include *M. catarrhalis*, *H. influenzae*, *Mycoplasma pneumoniae*, and *Chlamydia pneumoniae*. Worldwide diphtheria continues to be an important cause. Uncommon causes include *Candida*, *Coccidioides immitis*, *Cryptococcus neoformans*, TB, and blastomycosis.

Clinical features

Most cases occur in adults between 18 and 40 years of age. It is a feature in 38% of cases of pneumonia, 24% of cases of children with sore throat, and 75% of toddlers with croup. It presents as a hoarse or harsh voice with lowered pitch, episodes of aphonia, and sore throat. Duration: 3–8 days.

Treatment

Management is symptomatic (analgesia and voice rest). Routine antibiotics are not recommended. Treatment should be directed at the underlying cause.

Sinusitis

Inflammation of the paranasal sinuses usually due to viral, bacterial, or fungal infection, or non-infectious causes. Acute sinusitis lasts <4 weeks and chronic sinusitis lasts >4 weeks.

Aetiology

- Bacterial sinusitis:
  - community acquired – *Streptococcus pneumoniae*, *Haemophilus influenzae*, α-haemolytic streptococci, *Moraxella catarrhalis*
  - nosocomial – anaerobes, *S. aureus*, *P. aeruginosa*, other enterobacteria
- Viral sinusitis – rhinovirus, influenza, parainfluenza, adenovirus
- Fungal sinusitis:
  - community acquired – *Aspergillus* spp., *Rhizopus* spp. *Mucor* spp., *Sporothrix schenkii*, *Scedosporium apiospermum*
  - immunocompromised patients – *Candida* spp., *Cryptococcus neoformans*
- Non-infectious causes – chemical irritants, tumours, foreign bodies, Wegener's granulomatosis

Epidemiology

- 1–5% of European adults are diagnosed with acute sinusitis each year.
- Viral sinusitis is seen as part of the common cold from autumn to spring. Cases of acute community-acquired bacterial sinusitis peak in association with these (occur in around 0.5% of colds).
- Non-viral cases occur throughout the year in association with allergy, swimming, nasal polyps, foreign bodies, tumour, immunodeficiency, cystic fibrosis.
- Nosocomial cases (often polymicrobial) may occur secondary to head trauma, prolonged nasotracheal/nasogastric intubation, neutropenia, diabetic ketoacidosis, corticosteroids, or broad-spectrum antibiotic use.

Pathogenesis

- Viral – 90% of patients with a cold develop viscous discharge and reduced clearance of secretions from sinuses, which may be due to local viral infection or the effect of inflammatory mediators.
- Bacterial – the majority of cases probably occur by spread of nasopharyngeal organisms to the usually sterile sinuses. Sinus obstruction prevents effective clearance, and bacterial growth leads to destruction of epithelial cells, inflammatory infiltration, and the formation of mucus/pus.
- Fungal sinusitis may be non-invasive (two forms: allergic fungal sinusitis or sinus mycetoma) or invasive (occurs in hospitalized or immunocompromised patients). There are three forms of invasive sinusitis – acute fulminant (high mortality rate), chronic, and granulomatous.

Clinical features

- Viral infections – cough, sneeze, nasal discharge (clear or purulent) and obstruction, headache, and facial pressure may occur in viral infections.

- Bacterial infections – high fever and facial pain are characteristic. Sphenoid sinus infection may cause severe headache and sensory changes. Advanced frontal sinusitis may cause swelling and oedema of the forehead as pus collects under the periosteum.
- Nosocomial cases – infection may not be apparent if patient is unwell or unconscious (on intensive care unit (ICU)).
- Allergic fungal sinusitis – consider in patients with intractable sinusitis, and a history of allergic rhinitis and nasal polyposis.
- Sinus mycetoma – symptoms of sinusitis plus gravel-like material from the nose. Often found incidentally on CT scan.
- Acute invasive fungal sinusitis – fever, cough, nasal discharge, headache, confusion, dark ulcers on the septum/turbinates/palate.
- Chronic invasive fungal sinusitis – long history; may have reduced vision or ocular mobility (mass in the superior orbit).
- Granulomatous invasive fungal sinusitis – chronic sinusitis and proptosis, bone erosion. Often rapidly progressive.

Complications

Meningitis, brain abscess, subdural empyema, cavernous sinus and cortical vein thrombosis, orbital cellulitus and abscess, Pott's puffy tumour (osteomyelitis, usually staphylococcal, of the frontal bone).

Diagnosis

- Most cases are diagnosed clinically. Consider a bacterial aetiology in those whose symptoms fail to improve or worsen after 1 week.
- Nosocomial cases present during the second week of hospitalization.
- Radiology – an air-fluid level on skull x-ray correlates well with a positive bacterial aspirate culture, but sensitivity is low.
- Microbiology – sinus cavity specimens must be collected by sinus puncture and aspiration via the antrum below the inferior turbinate; 60% of aspirates are positive in suspected bacterial sinusitis.

Treatment

- Two-thirds of cases resolve spontaneously, reflecting the fact that many of these have a viral aetiology.
- Decongestants may relieve symptoms of obstruction but have no effect on sinus drainage.
- Patients who do not respond to antimicrobial therapy should be considered for sinus puncture and lavage to avoid progression.
- Bacterial disease – treatment is usually given empirically. Co-amoxiclav, cefuroxime, and quinolones such as levofloxacin have been shown to be effective, and treatment should be continued for 10 days. Those with severe infection, or complications such as intracranial extension should receive intravenous therapy with broad-spectrum agents until culture results are available. These groups require CT or MRI imaging and may need diagnostic lumbar puncture or surgical interventions. Nosocomial cases may require broader therapy than community-acquired disease.
- Fungal infection (community acquired) is effectively treated by surgical debridement. Complicated cases/immunocompromised patients are likely to need a combination of surgical and antifungal therapy.

Mastoiditis

Pathogenesis

The mastoid bone is penetrated by airspaces which are lined with modified respiratory mucosa and are connected via the antrum to the middle ear. It is clinically important as it is adjacent to many important structures: the posterior and middle cranial fossae, the sigmoid and lateral sinuses, the facial nerve canal, and the semicircular canals. The advent of antibiotics has brought the incidence of clinically significant mastoiditis to very low levels. Infection of the mastoid leads to the collection of purulent exudates within the air cells. The thin bony septa necrose, and pus coalesces into cavities.

Clinical features

Acute mastoiditis is accompanied by acute middle ear infection. Shortly after the onset of the signs of acute otitis media, pain, swelling and erythema develop over the mastoid. The pinna may be displaced downwards and tympanic perforation is followed by a purulent discharge. Chronic disease may lead to erosion through the roof of the antrum (causing a temporal lobe abscess) or extend posteriorly (where it may cause thrombosis of the lateral sinus).

Diagnosis

Plain x-ray may show something of the destruction of bony septa and inflammation of the air cells. CT demonstrates the extent of the disease. It is useful to obtain fresh pus as it exudes from the tympanic membrane to enable culture of material from the middle ear.

Treatment

- Treatment is with systemic antibiotics providing cover for *S. pneumoniae* and *H. influenzae*, as well as *S. aureus* and Gram-negative organisms in cases that have had a prolonged course. Therapy can be modified once culture results are available.
- Mastoidectomy is indicated in those cases where a mastoid abscess has formed, and should be performed once sepsis has been adequately controlled

Otitis externa

Bacterial and, less commonly, fungal infection of the external auditory canal.

Pathogenesis

The external auditory canal is around 2.5 cm long. The lateral half is cartilaginous and the medial half runs through the temporal bone with a narrowing at the junction between the two. The bacterial flora is that of skin elsewhere: *S. epidermidis*, *S. aureus*, and some anaerobes. In situations in which the skin becomes damaged these organisms may proliferate resulting in infection and inflammation of the skin itself. In addition to native organisms, invasive disease may be caused by Gram-negative species such as *Pseudomonas aeruginosa*. Fungal species such as *Aspergillus* and *Candida albicans* may also cause otitis externa.

Clinical features and therapy

- Acute localized otitis externa – may be due to a pustule (*S. aureus*) or erysipelas (group A streptococci) involving the canal. There may be regional lymphadenopathy. Systemic antibiotics are usually curative – surgical drainage is rarely necessary.
- Acute diffuse otitis externa – a large portion of the canal skin becomes oedematous, red, itchy, and painful. Gram-negative bacteria (e.g. *P. aeruginosa*) are important. Cases occur in humid weather and may be associated with swimming. Irrigation with saline and alcohol/acetic acid mixes may help. Neomycin/hydrocortisone ear drops or oral antibiotics with topical hydrocortisone reduce inflammation and speed resolution.
- Chronic otitis externa – due to discharge through a perforated tympanic membrane secondary to chronic suppurative otitis media which results in irritation of the external canal. Rare causes: TB, leprosy, sarcoid.
- Invasive otitis externa – a severe necrotizing infection usually caused by *P. aeruginosa*. It is seen in the elderly, diabetic, and immunocompromised, and characterized by infection spreading from the skin of the external auditory canal to the tissue, vessels, and bone beneath. Symptoms: pain and tenderness of the tissue around the ear, pus discharging from the canal. Disease may be life-threatening in those cases where the sigmoid sinus, skull base, and meninges become involved. Cranial nerves 7, 9, 10, and 12 may be affected (sometimes permanently). Diagnosis can be confirmed and the extent of tissue damage ascertained on CT or MRI. Necrotic tissue may need removing, and topical steroids and antipseudomonal antibiotics combined with systemic therapy for 4–6 weeks. Any underlying disease should be identified and treated

Otitis media

An acute inflammatory condition characterized by fluid in the middle ear. It is a common cause of fever, pain, and hearing impairment in children and may cause sequelae in adults.

Epidemiology

- The most frequent diagnosis made by general practitioners in those <15 years old. By the age of 3 years, two-thirds of children have had at least one episode. Complicates one-third of respiratory infections and occurs at higher rates in children attending day care centres.
- Most children with recurrent or severe disease have no obvious predisposition. Increased risk in those with anatomical abnormalities (e.g. cleft uvula), and immune deficiency (e.g. HIV infection).
- Commoner in males and tends to be more severe in Native Americans and Australian aborigines.

Pathogenesis

- Dysfunction of the Eustachian tube, e.g. due to mucosal congestion, leads to accumulation of middle ear secretions which provide a hospitable environment for bacteria.
- *S. pneumoniae* is the most frequent bacterial cause of otitis media. *H. influenzae* is second commonest cause, of which only 10% are caused by type B. Other bacterial causes include group A streptococci and *Moraxella catarrhalis*. Causes of chronic suppurative otitis media: *P. aeruginosa*, *S. aureus*, *Corynebacterium* spp., and *Klebsiella pneumoniae*. Uncommon causes: *Chlamydia trachomatis* (in infants <6 months of age), diphtheria, TB.
- Respiratory viruses (e.g. rhinovirus, RSV, influenza) cause up to 25% of clinical cases and may co-infect with bacteria.

Clinical features

- Acute otitis media – fluid in the middle ear with features of acute illness such as ear pain, discharge, hearing loss, fever, and lethargy. Vertigo, nystagmus, and tinnitus may occur. Tympanic erythema is an early feature but is not specific to middle ear infection. Otitis media with effusion ('glue ear') is seen in 10% of children (90% of those with cleft palates). The presence of fluid is associated with hearing impairment, and children with a history of recurrent episodes of acute otitis media show delay in speech and language abilities.
- Chronic suppurative otitis media – inflammation of the middle ear lasting ≥6 weeks and associated with discharge. Some authorities consider a single infection that resolves leaving a persistent effusion as 'chronic otitis media'.

Diagnosis

- Detecting fluid – use techniques that assess the mobility of the tympanic membrane (e.g. pneumatic otoscopy, tympanometry) or the difference in acoustic reflectivity between a fluid-filled or air filled middle ear.

- Culture – not usually attempted. Causative organisms are well defined and routinely covered in antibiotic therapy. Take blood cultures and local samples if patients are severely unwell or have infection elsewhere.
- Even after therapy and the resolution of acute features, fluid persists for some weeks (10% of children still have fluid at 3 months after infection). Tympanocentesis may be warranted in those with immune impairment, those unresponsive to antibiotic therapy, or the critically ill.

Management

- Acute otitis media – uncomplicated cases with no systemic features usually resolve without antibiotic therapy (~25% are viral). Consider antibiotics after 72 h if no improvement or if symptoms worsen. Agents must achieve good middle ear penetration, e.g. oral amoxycillin (erythromycin if allergic). Co-amoxiclav or ceftriaxone should be considered if there is no improvement after 24–48 h. Ensure Gram-negative coverage in newborn infants, the immunosuppressed, or suppurative chronic otitis media. If the tympanic membrane has perforated, culture and sensitivity testing of discharge may guide antibiotic choice.
- Recurrent episodes of acute infection:
  - antibiotic prophylaxis (e.g. oral amoxicillin) may benefit those experiencing well-documented recurrent episodes. Consider it in those experiencing two episodes in the first 6 months of life or three episodes in 6 months for older children. It reduces the number of febrile episodes attributable to otitis media. Risk of side-effects and selection of resistant organisms. Patients should be reviewed monthly and assessed for the presence of an asymptomatic effusion
  - xylitol is a polyol sugar alcohol which inhibits bacterial colonization and, used by children in chewing gum form (5 times a day!), has been shown to reduce the number of episodes of acute otitis media
  - pneumococcal vaccination prevents otitis media caused by *S. pneumoniae* and is not very effective in children under 2 years of age due to poor immunological response. Hib vaccination only prevents 10% of cases caused by *H. influenzae* (90% are non-type B).
- Persistent effusion – systemic antibiotics are not usually indicated. Refer to ENT if 'glue ear' persists for >1–2 months. Interventions include:
  - myringotomy – once commonplace, now used only in cases of intractable pain, for drainage of a persistent effusion unresponsive to medical therapy or to speed the resolution of mastoid infection
  - adenoidectomy – may improve Eustachian tube function in selected children and reduce the time spent with effusion
  - tympanostomy tubes – placed in the tympanic membrane to allow ventilation and drainage in cases of persistent effusions unresponsive to medical therapy over 3 months.
- Chronic suppurative otitis media – thorough cleaning with microsuction may resolve long-standing infection. Acute exacerbations of chronic infection will require systemic therapy with broad cover (e.g. amoxicillin and metronidazole) and may need to be given parenterally.

Dental infections

Pathogenesis

- Fermentation of dietary carbohydrates by bacteria (e.g. *S. mutans* and lactobacilli) leads to acid production and dental caries.
- Dental caries erodes protective enamel allowing bacteria access to the dentin and pulp. In the pulp, infection may track through the root, reaching the medullar cavity of the maxilla or mandible.
- Advanced infection perforates the bony cortex, draining into tissues of the oral cavity or deep fascial planes. Neutropenic patients are at particular risk of sepsis and airway compromise.

Clinical features

- Local infections – local pain, oedema and sensitivity to percussion and temperature. Severe local infections may be associated with abscess formation.
- Mandibular infection – pain and local swelling (e.g. infection of the mandibular incisors causing submental space infection may produce midline swelling beneath the chin, sublingual space infection may cause swelling of the floor of the mouth and tongue elevation). Retropharyngeal infection may follow molar infection (📖 p.634). Horner's syndrome or cranial nerve palsies may follow involvement of deep areas of the neck.
- Buccal space infection due to infection of posterior teeth may cause facial oedema. Masticator space infection causes trismus, is typically indicated by cheek oedema, and is due to infection of posterior teeth, usually premolar or molar.
- Ludwig's angina – a severe cellulitus of the floor of the mouth; 75% of odontogenic cases follow infection of the 2nd or 3rd mandibular molars. Features: elevation/swelling of the tongue, drooling, airway obstruction, sensation of choking/suffocating (from which it gets its name). It is polymicrobial in nature and can cause widespread infection.
- Vincent's angina – an acute necrotizing ulcerative gingivitis which presents with oral pain, bleeding gums, fetid breath, fever, anorexia, and lymphadenopathy. It is caused by invasive fusiform bacteria and spirochetes. On examination patients have ulcerated interdental papillae, with necrosis and pseudomembrane formation over the tonsils and gums. Differential includes candidiasis, HSV stomatitis, diphtheria. Material swabbed from the affected area should be cultured. Treatment: penicillin V and metronidazole or co-amoxiclav. Risk factors: poor dental hygiene, smoking, severe intercurrent illness. A rare complication in the immuncompromised is progression to noma, a gangrenous stomatitis

Management

- Airway protection is paramount.
- Dental x-ray, facial series or soft tissue x-ray of the neck may help localize the region involved. A CT scan is necessary in more-advanced infections.
- Localized infections may respond to oral antibiotics. Patients who appear unwell should be given IV antibiotics initially. More-extensive infection and abscesses may require surgical intervention.

Lateral pharyngeal abscess

Infection of lateral pharyngeal space may complicate pharyngitis, tonsillitis, mastoiditis, and dental abscesses.

- Caused by mixed organisms
- Anterior infection may present with fever, pain, trimus, swelling below the mandible, dysphagia, and displacement of the tonsil towards the midline. Posterior infection may present with sepsis and little pain
- Complications – systemic sepsis, respiratory obstruction (laryngeal oedema), thrombosis of internal jugular, erosion of the internal carotid
- Management – airway protection, parenteral antibiotic, and surgical drainage

Acute bronchitis

A syndrome of tracheal and bronchial inflammation associated with respiratory infection. Diagnoses peak in the winter months and are made most frequently in children less than 5 year of age.

Aetiology

- Viral causes commonest – rhinoviruses, coronaviruses, adenovirus, and influenza (acute bronchitis cases are common during flu outbreaks). Other viruses (e.g. measles) are uncommon but produce severe disease.
- Less-common bacterial causes include *Mycoplasma pneumoniae*, *Chlamydophila pneumoniae*, and *Bordetella pertussis*.

Pathogenesis

- The exact nature of pathogenesis varies between viruses. Some (e.g. influenza) invade the LRT. Others (e.g. rhinovirus) do not, and symptoms may be secondary to inflammatory mediators. Either way the outcome is an inflamed oedematous tracheobronchial tree with increased secretions. The extent of epithelial damage varies with the aetiological agent.
- Attack severity may be increased by exposure to irritants such as cigarette smoke, and may lead to long-term airway damage. Patients with acute bronchitis are more likely to have a history of atopic disease which may be associated with airway hyper-reactivity. Some progress to develop adult-onset asthma.

Clinical features

- Severe, prolonged cough distinguishes acute bronchitis from other respiratory syndromes such as the common cold. It lasts over 4 weeks in over 40% of patients. May be productive early on but typically dry.
- Cough and respiration may be associated with retrosternal pain in those cases with severe tracheal inflammation. Dyspnoea and more-severe respiratory symptoms are seen only in those with underlying chest disease. Fever is seen in some cases – most frequently with agents such as influenza or *M. pneumoniae*.

Diagnosis

- Bronchitis is a diagnosis of exclusion, and a complete history and examination should be performed seeking any more serious cause of cough.
- Vaccination and exposure histories may point to a specific aetiological agent, e.g. influenza, pertussis infection (a common cause of bronchitis in adults whose childhood vaccine immunity is waning).
- Cultures of respiratory secretions may be useful in looking for specific viruses or agents such as *M. pneumonia* or *B. pertussis*, but routine bacteriological culture is unhelpful in defining a cause of bronchitis.
- Those in whom cough persists beyond a reasonable duration of illness should be investigated for other causes (e.g. foreign body, tuberculosis, malignancy).

Treatment

- Healthy patients do not usually require any therapy other than management of cough.
- Those with severe bronchitis (particularly those with underlying heart or lung disease) may develop respiratory impairment and require oxygen or ventilatory assistance.
- Cough suppression has traditionally been attempted with codeine-based agents such as dextromethorphan and pholcodine (which have fewer side-effects than codeine itself). However randomized controlled trials (RCTs) (albeit small) have found no significant difference in cough severity between such agents and placebo in children or adults with acute bronchitis. They should not be given to those under 1 year of age.
- NSAIDs, sedating antihistamines such as diphenhydramine (the cough suppressant component of many commercial 'cough mixtures'), and short courses of inhaled steroids are widely used but their efficacy remains unproven.[1]
- Antibiotics should be used only in those cases in which a bacterial cause is suspected. There is no evidence that extended-spectrum agents are more effective than amoxicillin or doxycycline. Specific therapies for viral and bacterial causes are covered in the appropriate sections.
- Smoking should be actively discouraged.

Reference

1 Schroeder K, Fahey T. Over the counter medications for acute cough in children and adults in ambulatory settings (Cochrane review). *The Cochrane Library*, Issue 1. Oxford: Update Software, 2003.

Chronic bronchitis

Defined as a cough productive of sputum on most days during at least 3 months of two successive years which cannot be attributed to other specific diseases (e.g. TB, bronchiectasis). Where airflow obstruction exists the patient is considered to have chronic obstructive pulmonary disease (COPD), which may coexist with emphysema.

Epidemiology

- Common, affecting 10–25% of the adult population. Men are affected more than women and it is more common in those >40 years of age.
- Associations – cigarette smoking (although only 15% of smokers develop chronic bronchitis), pollution, and exposure to allergens.

Pathogenesis

- Even modest smoking is associated with increased alveolar macrophages, inflammation of the respiratory bronchioles, epithelial hyperplasia, and fibrosis of the bronchiolar and alveolar walls.
- The inflammation and oedema seen in patients with chronic bronchitis results from the interaction between exogenous irritants and the pathological response, including: an increase in bronchial mucus-secreting cells, granulocytic infiltration in response to chemokines produced by epithelial cells, increased airway secretions, and the production of neuropeptides promoting bronchospasm.
- Acute exacerbations of disease may be related to bacterial infections. Pathogenic bacteria can be cultured from the bronchi of most chronic bronchitics (as well as those with other lung pathologies such as TB), and the development of purulent sputum is not associated with the appearance of specific organisms but does correlate with an increase in number. *H. influenza* and *S. pneumoniae* are found in half of chronic bronchitics. Organisms such as *M. pneumoniae*, *Staphylococcus aureus* and Gram-negative organisms are identified infrequently. One-third of acute exacerbations are though to be due to viral infections.

Clinical features

- Frequent productive cough, most severe in the morning when patients may produce large amounts of sputum which may be mucoid and white or obviously purulent in appearance.
- Patients may be incapacitated only when they develop acute infections; however, most patients have some degree of airflow limitation. COPD exists as a spectrum of clinical disease with emphysema predominant at one end (breathless, less sputum, fewer infections, barrel-chest with hyper-expanded clear lungs) and bronchitis predominant at the other (productive coughing, frequent infections, wheeze, widespread crepitations, and right heart failure in severe cases).

Diagnosis

- Acute exacerbations of chronic bronchitis can be difficult to identify. Most patients do not develop fever or leucocytosis. Diagnosis is made on symptoms such as an increase in sputum production, or a change in colour with increasing cough and breathlessness.
- CXR is useful only in the exclusion of other illnesses.

Treatment

General measures

- Exclude other causes of recurrent chest infections and have a high index of suspicion for lung malignancy.
- Smoking cessation, weight control, avoidance of environmental irritants, and assessment for allergic disease.
- Record of baseline spirometry, arterial blood gas (ABG), and oxygen saturations.
- Pneumococcal and influenza vaccinations.
- Pulmonary rehabilitation programmes and postural drainage where appropriate.

Maintenance therapy to improve airflow obstruction symptoms

- For example, regular inhaled steroids, β_2-agonists, anticholinergic agents. Oral steroids as required.
- Prophylactic antibiotics may be useful in selected patients with very frequent exacerbations (four or more per year) – most specialists believe they do not have a role in routine treatment due to concerns over the development of resistance. RCTs prior to 1970 demonstrated that chronic bronchitis patients using prophylactic antibiotics had a small but significant reduction in exacerbations compared to placebo, and antibiotics significantly reduced the number of days of disability per person per month treated. There are no contemporary studies.

Intensive therapy for acute exacerbations

- Intensification of normal therapy (e.g. course of oral steroids).
- Antibiotic therapy – infection can cause respiratory decompensation (a common cause of death these patients) and despite difficulties in assessing their efficacy, antibiotic therapy does improve clinical outcome. They are usually given orally for 7–10 days. Specific agents are best determined locally and with reference to previous microbial sensitivities.

Bronchiolitis

An acute infection of the lower respiratory tract characterized by the acute onset of wheeze, and associated with cough, nasal discharge, breathlessness, and respiratory distress.

Aetiology

- RSV (see 📖 Respiratory synctial virus, p.426) accounts for up to 75% of cases.
- Other viruses, e.g. parainfluenza types 1–3, adenoviruses, rhinoviruses, influenza, and *Mycoplasma pneumoniae* have been implicated.

Epidemiology

- In temperate regions cases peak in winter and early spring (RSV) with small peaks in autumn (parainfluenza). Most cases occur in children aged between 2 and 10 months.
- Estimated incidence varies widely; 80% of children have evidence of previous RSV infection by their second birthday. A UK study estimated the total mean annual incidence of hospital admissions of children aged < 1 year attributable to RSV at 28.3 per 1000.
- Factors associated with high rates of hospitalization with bronchiolitis – young age, young maternal age, living in crowded/polluted areas, large number of siblings, airway hyper-reactivity, RSV identified as cause.

Pathogenesis

- Virus infects the upper respiratory mucosa and spreads to the lower airways. Bronchial and bronchiolar inflammation and necrosis follow with oedema and peribronchiolar mononuclear cell infiltration. In severe cases, interstitial pneumonitis may develop.
- Inflammation and oedema reduce airway caliber. Necrotic material may block small airways. Distally trapped air is later absorbed, resulting in multiple areas of atelectasis and a low ventilation/perfusion ratio. Children become breathless and tachypnoeic, developing respiratory distress and respiratory failure in severe cases.
- Infants who develop wheeze with respiratory virus infection in early life are more likely to have some form of recurrent lower respiratory tract disease. Whether this is related to a propensity to atopy, or pre-exisiting lung dysfunction, or has a viral aetiology is not clear.

Clinical features

- Mild fever and signs of URTI. Progresses after 2–3 days to lower respiratory tract features, with cough, a raised respiratory rate, anorexia, lethargy, and wheeze. Fever may resolve.
- Severe cases develop tachypnoea, tachycardia, and signs of increased breathing work (nasal flaring, chest wall retraction, grunting). Cyanosis is rare even in the presence of hypoxia. Apnoea is relatively common in young infants hospitalized with RSV infection.
- Auscultatory findings are variable – wheeze, crepitations, decreased breath sounds in severe cases. Other findings: dehydration, otitis media, diarrhoea. Symptoms begin to settle after 2–3 days with recovery taking 2 weeks or more.

Diagnosis

- General features – the white cell count may be elevated in severe cases. CXR findings include hyperinflation, hyperlucent parenchyma, multiple areas of atelectasis, and pneumonia may be present. Findings do not correlate with clinical severity.
- Diagnosis is usually clinical in the setting of a seasonal outbreak. Other causes of wheeze and dyspnoea must be considered (e.g. a first episode of asthma, gastric reflux and aspiration, congestive cardiac failure, or airway obstruction by a foreign body).
- The specific agent may be identified from respiratory secretions (preferably a nasopharyngeal aspirate) by tissue culture or PCR. Positive predictive value diminishes outside the setting of an epidemic. Serology is rarely helpful.

Treatment

- General measure – oxygen to maintain the saturation above 92% and ventilation if indicated. Bronchodilators are often given but are of limited benefit. Corticosteroids are often used and have been reported to bring about a modest improvement in symptoms but with no effect on the duration of hospitalization.
- Ribavirin may be given to hospitalized infants at risk of severe disease by aerosol for 8–12 h a day for 2–5 days. Such children include those with underlying cardiac or respiratory disease, or the premature. Although it has been shown to speed the improvement in oxygenation, its use does not result in shorter hospital stays.

Prevention

- Palivizumab is a humanized monoclonal antibody directed against the F glycoprotein of RSV. It is indicated for the prevention of respiratory syncytial virus infection in infants at high risk of severe disease. It may also be given to 'at-risk' exposed paediatric inpatients.
- In the UK, children at risk are considered to be those less than 6 months of age born at less than 35 weeks' gestation, those under 2 years treated within the previous 6 months for bronchopulmonary dysplasia, or those with with haemodynamically significant heart disease.
- It is given IM monthly during the RSV season, the first dose being given prior to the start of the season.

Complications

- Infants with underlying cardiac or pulmonary disease or immunodeficiency are at greatest risk of severe disease with respiratory failure and prolonged hypoxia.
- Although up to 75% of infants requiring hospital admission have recurrent episodes of bronchospasm within the first 2 years after recovery, the number continuing to experience such episodes drops year by year. The majority with long-term problems tend to either have a predisposition to atopy, or have reduced lung function at birth.

Community-acquired pneumonia

Epidemiology

- Incidence ~1 per 100 people per year; 20–40% of cases require hospital admission. Mortality varies with the patient group (overall 5–10%, 50% in those requiring ICU admission).
- Peak age 50–70 years and onset midwinter and early spring; 58–89% have underlying disease (e.g. COPD, diabetes, cardiovascular disease, immunosuppression etc).
- Some organisms are acquired by person-to-person spread or are existing commensals (*S. pneumoniae, H. influenzae*). Others are acquired from the environment (*L. pneumophilia*) or animals (*C. psittaci*).
- ~80% of community-acquired pneumonia (CAP) is managed in primary care – viruses, *S. pneumoniae*, and *M. pneumoniae* are the commonest causes.
- Risk factors – extremes of age, smoking, COPD, diabetes, cardiovascular disease, severe intercurrent illness, recent anaesthetic/ intubation, immunosuppression

Aetiology

- Organisms vary with country, study, age, and patient group (e.g. *Moraxella* and *H. influenzae* are commoner in COPD).
- Pneumonia in childhood is usually viral.
- Common bacterial isolates vary with age:
  - 0–1 month – *E. coli*, group B *Streptococcus, Listeria monocytogenes*
  - 1–6 months – *Chlamydia trachomatis, S. aureus*, RSV
  - 6 months to 5 years – RSV, parainfluenza viruses
  - 5–15 years – *Mycoplasma pneumoniae*, influenza
  - 16–30 years – *M. pneumoniae, S. pneumoniae*
  - older adults – *Streptococcus pneumoniae, Haemophilus influenzae*.
- Some infections, e.g. *M. pneumoniae*, are associated with epidemics.
- *S. pneumoniae* infection is associated with viral illness, e.g. influenza.
- Mixed infections are commoner in the elderly – *S. aureus* and Gram-negatives are seen more frequently among those in residential care.
- Severe disease occurs with *S. pneumoniae*, MRSA, *L. pneumophilia*, and Gram-negative organisms. Mortality is 20–53%.
- Rare causes of pneumonia include anthrax, plague, and melioidosis.
- *Pneumocystis jiroveci* (📖 p.530) is an important cause in HIV-infected patients, who are also at increased risk of infection with mycobacteria, *C. neoformans*, and viruses (e.g. CMV).

Clinical features

- History – most patients present with sudden-onset chills, fever, cough, mucopurulent sputum, pleuritic chest pain, fatigue, anorexia, sweats, and nausea; 20% do not have cough. Ask about predisposing conditions, travel, and exposure to animals.
- Symptoms and signs – fever, tachypnoea, tachycardia, postural blood pressure drop (may indicate dehydration), and consolidation (absent in ~70%). Signs of respiratory distress in severe cases. All findings are less pronounced in the elderly in whom presentation may be insidious (confusion, abdominal pain). Other findings: herpes labialis (40% of

pneumococcal pneumonia patients), bullous myringitis (mycoplasma pneumonia).

Investigations

- Blood tests – full blood count (FBC) shows neutrophilia, or neutropenia if very unwell. Biochemical abnormalities include raised urea, hyponatraemia (especially elderly due to syndrome of inappropriate antidiuretic hormone secretion (SIADH)), abnormal liver function tests, (especailly legionella), raised CRP.
- CXR changes are non-specific:
  - lobar consolidation, cavitation, effusions suggest a bacterial cause
  - CXR worse than examination findings suggest mycoplasmal or viral pneumonia
  - diffuse bilateral involvement may suggest pneumocystis pneumonia (PCP), *Legionella* infection, or primary viral pneumonia.
  - Pneumatocoeles may be seen in pneumonia caused by *S. aureus*, *K. pneumoniae*, *H. influenzae*, and *S. pneumoniae*.
- CT is helpful in recurrent pneumonias or those unresponsive to therapy (e.g. to identify a tumour). In immunocompromsed patients, some pathogens (e.g. *Aspergillus*) have typical CT appearances which may aid diagnosis.

Microbiological investigations

Microbiological tests are not routinely recommended for patients managed in the community – consider in those who do not respond to empirical antibiotic therapy or in whom unusual pathogens are suspected.

- Blood culture – positive in only 1–16% of hospitalized patients with CAP (≤25% of cases of pneumococcal pneumonia).
- Sputum examination has low sensitivity; ~50% of sputum cultures are negative even in proven bacterial cases. Sputum examination is however useful in the identification of organisms such as *Legionella* spp., *Pseudomonas* spp., *Burkholderia* spp., *M. tuberculosis*, and *P. carinii*.
- Serology – severely ill patients who have had symptoms for 5–7 days should have serum tested for antibodies to atypical pathogens (*Legionella*, *Mycoplasma*, and *Chlamydia* spp.) A second specimen is taken 7–10 days later.
- Antigen testing – urinary legionella antigen testing should be performed on all patients with severe disease and at least 5 days of symptoms. Some authorities also recommend pneumococcal antigen testing.
- Bronchoalveolar lavage (BAL) – a bronchoscope is used to instill sterile fluid into a segment of lung and the fluid examined microscopically and cultured. A threshold of 10^4 colony-forming units (CFU)/mL is used to define significant isolates. It is particularly useful in the diagnosis of *M. tuberculosis*, *Pneumocystis jiroveci*, CMV, and ventilator-associated pneumonia (VAP).
- Pleural fluid sampling – where positive, pleural fluid cultures are specific for the organism causing the underlying pneumonia. Fluid analysis helps differentiate other causes of lung disease (e.g. TB, tumour).
- Other tests – immunofluorescence or PCR for respiratory viruses; immunofluorescnce for *Chlamydia* spp.; cold agglutinins for *M. pneumoniae* (<25% positive), lung biopsy (e.g. immunosuppressed patients with no diagnosis).

Severity assessment

- The CURB-65 score (**C**onfusion, **U**rea, **R**espiratory rate, **B**lood pressure, age **65** or over) enables rapid assessment of severity and guides initial management (Fig. 5.1).
- Additional adverse features – hypoxia regardless of oxygen therapy (arterial oxygen saturation (SaO_2) <92% or partial pressure of arterial oxygen (PaO_2) <8 kPa), bilateral or multilobe involvement on CXR, positive blood cultures, WCC <4 × 10^9/L or >20 × 10^9/L.
- Severity should be reassessed regularly during the course of the illness.

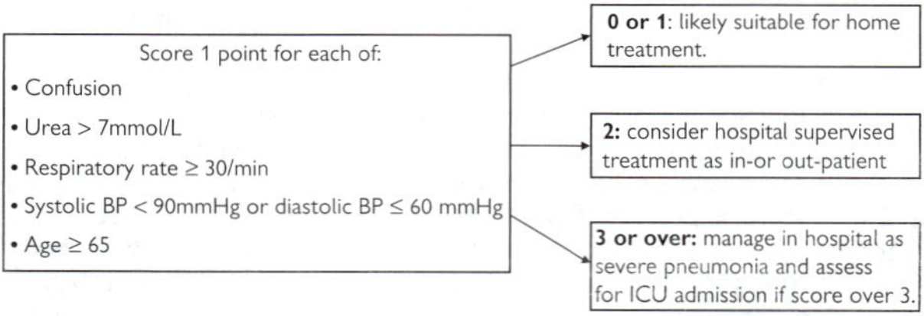

Fig. 5.1 The CURB-65 severity-assessment tool. SBP: systolic blod pressure; DBP: diastolic blood pressure.

Management

See British Thoracic Society (BTS) guidelines for community acquired pneumonia 2010 update.[1]

- General – IV fluids, appropriate oxygen therapy (with repeated ABG in those with COPD), fluids, frequent reassessment of progress and severity particularly aimed at the early identification of those who may require ICU support.
- Antibiotics – empirical therapy should be started as soon as possible:
  - CURB-65 score 0 or 1 – amoxicillin 500 mg to 1 g tds PO. Alternatives: erythromycin 500 mg qds (four times a day) PO, or clarithromycin 500 mg bd (twice a day) PO.
  - CURB-65 score 2 – amoxicillin 500 mg tds (three times a day) PO and erythromycin 500 mg qds PO (or clarithromycin 500 mg bd PO)
  - CURB-65 score ≥3 – co-amoxiclav 1.2 g tds IV plus erythromycin 500 mg qds PO.
- Older fluoroquinolones (e.g. ciprofloxacin) are not recommended for empirical treatment due to poor activity against *S. pneumoniae*. Newer agents (e.g. levofloxacin and moxifloxacin) may be used.
- Antibiotic therapy should be tailored to the causative organism in the light of microbiological data:
  - *S. pneumoniae* – amoxicillin or benzylpenicillin
  - *M. pneumoniae* – erythromycin or clarithromycin
  - *C. pneumoniae* – erythromycin or clarithromycin
  - *C. psittaci* – tetracycline
  - *C. burnetii* – tetracycline
  - *Legionella* spp. – clarithromycin ± rifampicin

1 Available from http://www.brit-thoracic.org.uk/clinicalinformation/pneumonia/pneumonia-guidelines.aspx.

- *H. influenzae* – amoxicillin (non-β-lactamase producer) or co-amoxiclav (β-lactamase producer)
- Gram-negative enteric bacilli – cefuroxime, cefotaxime, or ceftriaxone
- *P. aeruginosa* – ceftazidime and gentamicin or tobramycin
- *S. aureus* (methicillin sensitive) – flucloxacillin ± rifampicin
- MRSA – vancomycin.

- IV antibiotics should be switched to oral therapy as soon as there is evidence of clinical improvement (preferably within 48 h). However, IV treatment may be continued in patients with severe infections caused by *Legionella*, *S. aureus*, or aerobic Gram-negatives.
- 3 to 5 days' therapy is usually sufficient in uncomplicated cases. Treatment duration may be prolonged to 7–21 days in severe disease.
- CXR should be repeated at 6 weeks if symptoms persist or in those at increased risk of lung cancer (e.g. smokers).
- Failure to respond – adequately treated pneumonia will resolve clinically over 7–10 days. Older patients and those with underlying disease may take longer. CXR findings should normalize within 4 weeks but may take longer in the elderly, those with multilobe involvement, and those with pre-existing pulmonary disease. Consider infection with resistant organisms, underlying malignancy, empyema, or lung abscess. Additionally consider IV catheter infection or antibiotic-associated diarrhoea in those with prolonged fever.

Complications

- Parapneumonic effusion
- Empyema (📖 p.664)
- Adult respiratory distress syndrome (ARDS)
- Sepsis syndrome
- Metastatic infection (meningitis, arthritis, endocarditis)
- Rare neurological sequelae may follow *M. pneumoniae* infection, e.g. meningoencephalitis, cranial nerve palsies, and Guillain–Barré syndrome (GBS)

Prevention

There is a pneumococcal polysaccharide vaccine available, made from the 23 capsular serotypes that cause over 90% of invasive infections.

Reference

British Thoracic Society (BTS) guidelines for community-acquired pneumonia 2010 update. Available from www.brit-thoracic.org.uk

Atypical pneumonias

These account for 7–28% of CAP. The term was originally used to describe cases that failed to respond to penicillin or sulpha drugs, and in which no organism was identified. More recently it has meant those cases that start with an apparently mild respiratory tract illness (which may last up to 10 days) followed by pneumonia with dyspnoea and cough with or without sputum. Clinical signs tend to be milder than the CXR would suggest. It usually involves the lower lobes and may be unilateral or bilateral. The clinical course is usually benign although certain organisms may cause extrapulmonary symptoms (e.g. mycoplasmal infection and neurological sequeale) or severe disease (e.g. *Legionella*). CXR tends to improve faster than typical pneumonia. It is good practice to avoid the term 'atypical pneumonia' although it is still useful to refer to 'atypical organisms' as they have certain features in common: they tend not to be respiratory tract colonizers, they affect healthy individuals of all age groups, they occur in epidemics, and they do not respond to penicillin.

Mycoplasma pneumoniae (📖 p.402)

- Causes autumn epidemics every 4–8 years (more frequently within closed populations such as prisons). Most common in children >4 years and young adults.
- Certain respiratory viruses cause a similar clinical picture (e.g. influenza, parainfluenza, adenovirus, RSV).
- Resolves without complications in the majority of cases although the illness may last a few weeks, with a protracted cough.
- Difficult to culture – diagnosis is usually retrospective by serology.
- Serum cold agglutination is a non-specific test (positive in 50–70% of patients after 7–10 days of infection).
- CXR appearance is of basal atelectasis or involvement of a single lower lobe, sometimes a nodular infiltration resembling that associated with other diseases with granulomatous pathology, such as tuberculosis, mycoses, and sarcoidosis.
- Treatment is with erythromycin or clarithromycin.
- Rare complications – pericarditis, arthritis, Stevens–Johnson syndrome, haemolytic anemia, thrombocytopenia, CNS infections, GBS, peripheral neuropathy, other neurological manifestations, and ocular complications.

Legionella pneumophila (📖 p.368)

- Accounts for 1–20% of CAP cases. More common in summer.
- The organism colonizes water piping systems, and outbreaks are associated with acquisition from contaminated water sources including cooling systems, showers, decorative fountains, humidifiers, respiratory therapy equipment, whirlpool spas.
- Risk factors – smoking, diabetes, malignancy, AIDS, end-stage renal disease, alcohol abuse.
- Clinical features – 1–2-day prodrome (mild headache and myalgias) is followed by high fever, chills, and rigors, cough (non-productive becoming productive as the disease progresses), dyspnoea, pleuritic chest pain, haemoptysis, nausea, vomiting, diarrhoea, abdominal pain, altered mental status, arthralgias, myalgias. Consider it in those with a

high fever, multilobar involvement, a need for ICU and rapidly evolving gastrointestinal (GI), neurological, and radiographic abnormalities. Laboratory abnormalities include DIC, SIADH, abnormal liver function tests (LFTs), renal impairment.

- Treatment – clarithromycin ± rifampicin.
- 25% mortality (may be related to comorbidities).

Chlamydophila pneumoniae (📖 p.404)

- Accounts for 3–10% of CAP cases in adults.
- Causes mild pneumonia or bronchitis in adolescents and young adults.
- Incidence is highest in the elderly who may experience more-severe disease and repeated infections.
- Mortality rate ~9%.

Chlamydophila psittaci (📖 p.404)

- Usually associated with exposure to birds – pet shop employees and poultry industry workers are at risk.
- Clinical spectrum ranges from an asymptomatic infection to fulminant toxicity.
- Consider in those patients with pneumonia, splenomegaly, and a history of bird exposure (especially sick birds).
- May develop rash, hepatitis, haemolytic anemia, DIC, meningoencephalitis, or reactive arthritis.
- Treatment – tetracycline.
- Mortality rate <1%.

Coxiella burnetii (📖 p.398)

- An intracellular pathogen found worldwide, with the exception of New Zealand. Highly prevalent in parts of Spain and France (2nd-commonest cause of CAP in some regions).
- Reservoir primarily farm animals (e.g. cattle, goats, sheep). Excreted in urine, milk, and faeces, and attains high concentration in birth products.
- Rare human-to-human transmission from exposure to the placenta of an infected woman and from blood transfusions.
- Acute Q fever may cause a febrile illness, a pneumonia, or hepatitis.
- Most cases of acute Q fever resolve spontaneously with 2 weeks, but 14–21 days of treatment with doxycycline reduces symptom duration.

Aspiration pneumonia

Elderly patients, those with neurological impairment (e.g. acute phase of stroke), and others with altered consciousness (e.g. alcoholics) and abnormal swallow/gag reflexes are at risk of aspiration. Acid aspiration results in the release of proinflammatory cytokines which recruit neutrophils into the lung. These are thought to be the key mediators of acute lung injury, with bacterial pneumonia developing several days later, perhaps facilitated by bronchial obstruction by inhaled debris (e.g. peanut). Abscess and empyema are not uncommon. Anaerobes on their own or mixed with aerobes are the commonest bacteriological findings (e.g. *Bacteroides, Fusobacterium*). Bronchoscopy may be indicated to provide material for culture and exclude foreign bodies.

Hospital-acquired pneumonia

- Hospital-acquired pneumonia (HAP) is the leading cause of infection-related deaths in hospital. Defined as pneumonia developing more than 48 h after admission.
- 60% of cases are caused by aerobic Gram-negatives, the majority being enterobacteria and *Pseudomonas* spp. Other causes include *S. aureus* and *S. penumoniae* and anaerobes.
- Nosocomial outbreaks of viral pneumonia are not uncommon.
- Risk factors:
  - patient related – age >70 years, severe underlying disease, malnutrition, coma, metabolic acidosis, and possibly sinusitis
  - Infection control related – poor healthcare worker hand hygiene, contaminated respiratory equipment
  - Intervention related – sedatives, corticosteroids and cytotoxic drugs, prolonged antibiotic use, ventilation (risk of acquiring pneumonia 20 times that of unventilated patients).

Ventilator-associated pneumonia

- Ventilator-associated pneumonia (VAP) complicates the course of 8–28% of patients receiving mechanical ventilation.[1]
- **Causes** – the predominant organisms responsible for infection are *P. aeruginosa*, *S. aureus*, enterobacteria, *Haemophilus* spp., and *Acinetobacter* spp., but aetiological agents vary according to ICU, duration of inpatient stay, and prior antibiotic use. Polymicrobial infections are common.
- **Pathogenesis** – intubation compromises the natural barrier between the orphraynx and the trachea and also facilitates entry of bacteria into the lung by polling and leakage of containminated secretions around the endotracheal tube cuff.

- **Risk factors** – hypoalbuminaemia, age ≥60 years, ARDS, COPD, coma, burns, trauma, organ failure, gastric aspiration, gastric colonization and pH, upper respiratory tract colonization, sinusitis, H_2 receptor antagonists, paralytic agents, prior antibiotics, continuous sedation, mechanical ventilation >2 days, re-intubation, tracheostomy, nasogastric tube, supine position.
- **Clinical features** – the diagnosis of VAP is based on three features: systemic signs of infection, new or worsening pulmonary infiltrates, bacteriological evidence of parencyhmal infection. Present with purulent secretions, CXR changes, neutrophilia, fever and increased ventilatory requirements. Differential includes: pulmonary embolism (PE), ARDS, aspiration (chemical pneumonitis).
- **Diagnosis** – can be difficult. Quantitative cultures of endotracheal aspirates may be misleading as they may only reflect upper respiratory tract colonizers. Bronchoscopic sampling provides lower respiratory tract samples for microbiological analysis. This may be directed (i.e. in area of maximal pulmonary infiltrate) or non-directed. Quantitative BAL cultures indicate the likelihood of infection with $>10^4$ organisms/mL being diagnostic of infection. Some authorities recommend surveillance BALs in patients on ICU.
- **Treatment**[2] – this should be targeted against the likely organism and antimicrobial susceptibility pattern. In early-onset VAP (≤4 days after hospital admission), the organisms are community acquired and unlikely to be multi-drug resistant. Treatment should be with ceftriaxone or fluoroquninolone or ertapenem. In late-onset VAP (>4 days after hospital admission) organisms are likely to be hospital flora and multi-drug resistant. Treatment is with an antipseudomonal cephalosporin or Tazocin® or carbapenem plus a fluoroquinolone or an aminoglycoside. Vancomycin should be added if there is a possibility of MRSA.
- **Prognosis** – mortality ranges from 24% to 76%.

References

1 Chastre J, Fagon JY. Ventilator associated pneumonia. *Am J Respir Crit Care Med* 2002;165:867–903.

2 American Thoracic Society guidelines for hospital acquired, ventilator associated and health care associated pneumonia. *Am J Respir Crit Care Med* 2005;171:388–416.

Pulmonary infiltrates with eosinophilia

The differential diagnosis is broad and includes:

- tropical eosinophilia
- parasitic infections e.g. *Ascaris* or *Strongyloides*
- tuberculosis
- brucellosis
- psittacosis
- coccidioidomycosis
- histoplasmosis
- bronchopulmonary *Aspergillus*
- drug allergy
- sarcoidosis,
- Churg–Strauss syndrome
- eosinophilic leukaemia
- hypersensitivity pneumonitis.

Empyema

Microbial infection of the pleural space – prognosis can be poor in those cases where diagnosis is missed or therapy inadequate.

Aetiology

- Usually secondary to pneumonia. Cases may occur with no evidence of pneumonia (primary empyema). Other precipitants: surgery, trauma, oseophageal perforation, chest drains.
- Common organisms in cases secondary to pneumonia: *S. aureus*, *S. pneumoniae*, and *S. pyogenes*. *H. influenzae* has declined since the introduction of the Hib vaccine.
- Cases associated with aspiration, or arising from gastrointestinal sites are more likely to be due to anaerobic organisms. Those associated with subdiaphragmatic disease are often polymicrobial. Aerobic Gram-negative organisms are common in cases complicating trauma or surgery, and those associated with serous effusions.
- The immunocompromised have higher rates of Gram-negative and fungal empyema generally in the context of disseminated disease.

Clinical features

- Chest pain, breathlessness, weight loss, night sweats, fever. Examination reveals only the signs of the effusion in most cases.
- Consider in patients with persistent fever already receiving antibiotics for pneumonia.
- Consider oesophageal rupture in those who develop a pleural effusion soon after significant retching or vomiting.

Diagnosis

See BTS guidelines for the management of pleural infection.[1]

- Radiology – CXR shows pleural effusion. Ultrasound permits diagnostic aspiration and the identification of loculated effusions. Contrast CT distinguishes empyema from most lung abscesses and is used to monitor treatment.
- Diagnostic sampling – should be performed in all patients with a pleural effusion in the context of pneumonia or sepsis. Samples should be kept tightly sealed on ice to prevent changes in pH and glucose. The presence of pus cells or high numbers of microorganisms on Gram stain confirms diagnosis. Cultures may be negative in patients receiving antibiotics.
- Biochemical tests – empyema is confirmed by measurement of pH (<7.2), glucose (<2.2 mmol/L) and lactate dehydrogenase (LDH) (>1000 IU/L). Low pH values (pH 6–6.7) should raise the suspicion of oesophageal rupture or chronic empyema. See Table 5.2.
- Culture-negative cases – consider blood culture, urine antigen testing for legionella or histoplasmosis, pleural biopsy in suspected TB (95% positive on histology compared to 23% by pleural fluid culture/microscopy), serology for *E. histolytica* (positive in 98% of patients with pleural amoebiasis), microscopy/culture of empyema pus for acid-fast bacilli (AFB) in those at risk of nocardiosis, examination of stool/sputum for eggs in cases with pleural or blood eosinophilia suggestive of paragonimiasis

Management

Empyema requires prompt treatment to prevent complications and the need for surgical drainage. Involve a respiratory specialist.

- Drainage – indications for chest tube insertion are (1) purulent pleural fluid; (2) pleural fluid pH <7.2; (3) positive pleural fluid Gram stain or culture; (4) loculated pleural collections; (5) large non-purulent pleural effusions; and (6) poor response to antimicrobial therapy alone.
- Antibiotics – the regimen should be guided by culture results. Culture-negative cases should receive antibiotics covering community-acquired and anerobic organisms (e.g. ceftriaxone and metronidazole). Broader-spectrum cover should be initiated in cases of hospital-acquired empyema. Consider adding a macrolide in cases of suspected *Legionella* infections. Once the fever has settled, convert to oral antibiotics and continue treatment for at least 3 weeks.
- Intrapleural thrombolytic therapy (e.g. urokinase 100,000 IU od for 3 days) has been recommended although it was not associated with improved radiological appearance demonstrated no improvement in mortality rate, rate of surgical intervention, or length of hospital stay[2].
- Consider bronchoscopy if there is a high index of suspicion of bronchial obstruction.
- In patients with persistent sepsis and/or residual pleural effusion, review the diagnosis and perform a CT chest to confirm chest tube position, effusion anatomy, and look for obstructing lesions etc.
- Surgery – patients should be considered for surgery if they have ongoing sepsis with a persistent pleural collection by day 7 despite chest tube drainage and antibiotics. Modalities include video-assisted thoracoscopic surgery (VATS), open thoracic drainage, or thoracotamy and decortication.

Table 5.2 Pleural fluid charactersics in empyema

Stage	Pleural fluid findings	Comments
Simple parapneumonic effusion	Clear fluid; pH > 7.2; LDH < 1000 IU/L, glucose > 2.2 mmol/L; no organisms on culture or Gram stain	Will usually resolve with antibiotics alone. Perform chest tube drainage for symptom relief if required
Complicated parapneumonic effusion	Clear or cloudy fluid; pH < 7.2, LDH > 1000 IU/L, glucose > 2.2 mmol/L; may be Gram stain- or culture-positive	Requires chest tube drainage
Empyema	Purulent fluid; may be Gram stain- or culture-positive	Requires chest tube drainage

References

1 British Thoracic Society. Guidelines for the Management of Pleural Infection. *Thorax* 2003;58(Suppl II):ii18–ii28.

2 Maskell NA, Danei CWH, Numa AJ et al. UK controlled trial of intrapleural strep for pleural infection. New Engl J Med 2005; 1352:865–74.

Lung abscess

A suppurative lung infection that destroys lung parenchyma producing a pus-filled cavity with an air-fluid level.

Aetiology

- The risk of pneumonia progressing to lung abscess in the absence of effective therapy relates to the size of the inoculum and state of host defence mechanisms. Anaerobic organisms are the commonest causes.
- Some cases follow primary pneumonia (e.g. *S. aureus*, *Klebsiella pneumoniae*, *Pseudomonas aeruginosa*).
- Most follow aspiration in the context of states of altered consciousness (e.g. alcoholism, stroke) or dysphagia (e.g. neurological disease).
- Other associations – intestinal obstruction, periodontal disease or gingivitis, septic embolization, bronchiectasis, immunosuppression, and pharyngeal instrumentation (e.g. endotracheal intubation). Up to 5% of pulmonary emboli may become secondarily infected.

Pathogenesis

- Bacteria are usually endogenously acquired from the flora of the upper respiratory tract. Some may be acquired nosocomially.
- Lung regions that are dependent when lying flat are commonly affected – the posterior segment of the right upper lobe and apical segments of the lower lobes. Bases may be affected in cases of subdiaphragmatic extension (e.g. amoebic liver abscess).
- Multiple abscesses follow septic embolization (e.g. *S. aureus* right heart endocarditis) or bacteraemia (enteric Gram-negatives and anaerobes).

Microbiology

- Community-acquired aspiration – mixed anaerobic infections: Gram-negatives (*Prevotella oralis*, *Bacteroides fragilis*, *Fusobacterium nucleatum*) and Gram-positives (*Peptostreptococcus* and *Clostridium perfringens*). α-haemolytic streptococci are the commonest aerobes.
- Hospital-acquired infection: *S. aureus*, *Klebsiella*, *Pseudomonas*, and *Proteus* in combination with anaerobes.
- Necrotizing infection – *S.aureus*, *S. pyogenes*, *K. pneumonia*, *P. aeruginosa*, *E. coli*, *Legionella pneumonphilia*, and certain mycobacteria and fungal infections in combination with anaerobes. Uncommon causes of cavitating pneumonia include: *Nocardia*, *Burkholderia*, and melioidosis.

Clinical features

- Cases diagnosed late may present with several weeks or months of cough, low-grade fever, weight loss, anaemia, and clubbing. Sputum is copious and may be foul smelling. Findings are those of a severe pneumonia with or without effusion.
- In cases of secondary lung abscess, the primary lesion may also be apparent (endocarditis, subphrenic infection etc) and lung lesions multiple (e.g. *S. aureus* in intravenous drug users (IVDUs)) with pain and haemoptysis.

- Necrotizing pneumonia – seen in severe cases of anaerobic infection and may affect a single segment or extend to involve one or both lungs with associated empyema. Disease rapidly spreads destroying large volumes of parenchyma. Patients appear ill with a pronounced leucocytosis. Pulmonary actinomycosis may present similarly.
- Amoebic lung abscesses – features of the co-existent liver abscess and presents with cough productive of brown-red (anchovy sauce) sputum.
- Complications – empyema (one-third of cases), brain abscess, localized bronchiectasis. Tuberculosis should be considered in the differential diagnosis.

Diagnosis

- Radiology – CXR may reveal a cavity with an air-fluid level. CT scanning facilitates the detection of smaller lesions.
- Microbiology – sputum culture is helpful only in the diagnosis of amoebic infection. Culture of empyema fluid or percutaneous transtracheal aspiration (CT guided) is useful where possible. Quantitative culture of bronchoscopic sampling can provide good results. It is essential samples are placed in anaerobic conditions for transport to the laboratory. Blood cultures may be positive but may not reveal the entire infecting flora.

Treatment

- Empirical therapy whilst awaiting culture – community-acquired cases, e.g. ceftriaxone and metronidazole; nosocomially acquired infection (cover for *S. aureus*), e.g. piperacillin/tazobactam. Vancomycin should be considered if local rates of MRSA warrant it.
- Antibiotics should be given for 2–4 months. Patients should be monitored carefully for relapse.
- Bronchoscopy and postural physiotherapy may facilitate drainage.
- Surgical resection is rarely required outside the context of malignancy.

Prognosis

- Overall mortality is <15% for anaerobic lung abscesses and ~25% for anaerobic necrotizing pneumonia.
- Mortality is higher in acute pneumonias caused by organisms such as *S. aureus*.

Cystic fibrosis

A recessive genetic disorder with complex pathogenesis and a variable clinical syndrome. One-third of identified patients are adults.

Pathogenesis

The *CF* (cystic fibrosis) gene (chromosome 7) codes for the cystic fibrosis transmembrane regulator (CFTR), a plasma membrane cAMP-regulated chloride channel. Defective CFTR chloride channel function has been identified in epithelial cells of the airways, sweat ducts, and small intestine in CF patients, resulting in viscous secretions.

Microbiology

The colonizing organisms and antimicrobial resistance patterns change over time:

- *S. aureus* – colonizes in childhood
- *P. aeruginosa* – colonizes in childhood or early adolescence, >80% are infected by adulthood. Early non-mucoid isolates can be eradicated. Later isolates produce large amounts of mucoid polysaccharide (alginate), are difficult to eradicate, and are associated with greater mortality than non-mucoid strains. Chronic infection is associated with rapid decline in lung function, and increased mortality
- *Burkholderia cepacia* complex – an important and highly transmissible group of pathogens, intrinsically resistant to aminoglycosides and polymixins. May be difficult to identify, requiring specific isolation media ± referral to reference laboratory. Colonized patients should be separated from the non-colonized. Infection with *B. cenocepacia* can lead to a rapid deterioration in pulmonary function, bacteraemia, and even death among adolescents and young adults (cepacia syndrome)
- other – *H. influenzae*, non-tuberculous mycobacteria (clinical significance unknown).

Clinical features

Reflect obstruction of organs by viscous secretions, and the presence of chronic bacterial lung infection.

- General features – chronic cough, wheeze, recurrent pneumonia, sinusitis, clubbing, haemoptysis, pneumothorax, signs of respiratory impairment. Hypoxia and CO_2 retention are uncommon. The CXR may show airway thickening, retained secretions, and bronchiectasis
- Acute respiratory infections – patients often produce a large amount of purulent sputum even when well. Episodes of deterioration are associated with increased volume and purulence of sputum, dyspnoea, wheeze, chest ache, anorexia, and malaise. High fever or sepsis is unusual despite the large number of organisms in the secretions (10^8 organisms/mL of sputum). CXR may be unchanged from the patient's normal film. Forced expiratory volume in 1 s (FEV_1) falls, returning to pre-infection levels with successful antibiotic therapy.
- Other – diabetes mellitus, pancreatic insufficiency, urogential, and gastrointestinal features.

Management

The aims are to slow lung damage by removing viscous airway secretions, control bacterial infection, and monitor the appearance of highly transmissible or antibiotic-resistant organisms. Antibiotics need to be given to CF patients for longer and at frequent, higher doses than non-CF patients. Outpatient IV antibiotic therapy may be given via a long line. Those requiring very frequent antibiotics may require insertion of a Porto-cath.

- **General measures** – postural drainage, deep breathing, coughing, exercise, aerosolized DNase I (reduces mucus viscosity, clears airway secretions), inhaled steroids, bronchodilators (helpful in some patients). Pneumococcal and annual influenza vaccinations are recommended. Lung transplantation should be considered if life expectancy is less than 2 years and quality of life is severely impaired despite medical therapy.
- **Antimicrobial prophylaxis** – controversial. A Cochrane review[1] found that antibiotic prophylaxis against *S. aureus* infection reduced isolation of *S. aureus* from sputum but had no impact on lung function. Whereas the US guidelines do not recommend it, the UK Cystic Fibrosis Trust recommends oral flucloxacillin from diagnosis until age 2 years and some UK clinics promote lifelong prophylaxis[2].
- **Eradication of *Pseudomonas* colonization** – the first isolation of a non-mucoid *Pseudomonas* strain should be treated with the aim of eradication (e.g. 6 weeks of PO ciprofloxacin 750 mg bd with nebulized colistin 1 mega unit bd).
- **Long-term management of *Pseudomonas* colonization** – patients with established colonization may benefit from long-term nebulized antibiotics (e.g. colistin or tobramycin) to reduce the bacterial burden – studies have shown improved lung function and reduced episodes of respiratory illness in such patients. Some centres advocate elective courses of IV anti-pseudomonal therapy every 3 months to reduce the frequency of exacerbations and consequent lung damage. There is no evidence to support this.
- **Treatment of acute exacerbations** – the patient's most recent sputum culture result should be used to guide therapy. Broad-spectrum oral agents are beneficial despite the presence of resistant *P. aeruginosa*. High doses and prolonged therapy (3–4 weeks) are recommended. Aggressive IV therapy is indicated in those patients that do not respond to oral treatment. Such therapy is usually directed at *P. aeruginosa*, e.g. ceftazidime and an aminoglycoside. Once-daily aminoglycoside dosing is as effective as three times a day, and associated with reduced toxicity in children. *Haemophilus influenzae* and *S. aureus* should be treated if isolated even if the patient is asymptomatic. *Burkholderia cepacia* is very resistant and should be treated with a combination of two or three agents such as ceftazidime and an aminoglycoside. Nebulized vancomycin can be used to treat MRSA colonization of sputum, but IV therapy is required for exacerbations. Parenteral therapy may be given on an outpatient basis and should continue for 10–14 days or longer. *Aspergillus* is frequently cultured from sputum. Treatment (steroids ± antifungals) is indicated only for allergic bronchopulmonary aspergillosis, if present.

Reference

1. Smyth A, Walteo S. Phophylatic anti-stapphylococcal antibities for cystic fibrosis. Cochrane Database of Systematic Reviews 2003, Issue 3. Art. No: CDøø1912. DOI: 10.1002/14651858.CDDøø1912.

2. Antibitc treatment for cystic fibrasis. Second Edition. September 2002. http://www.cytrust.org.uk/aboutof/publication/consensusdor

Infective endocarditis

Infective endocarditis (IE) is characterized by infection of the endocardial surface of the heart. It may be classified as acute, subacute, or chronic, depending on the time course of the infection. It is now more-commonly classified according to the type of valve (native or prosthetic) and the aetiological agent (e.g. staphylococcal, streptococcal, enterococcal, fungal, culture negative, etc).

Epidemiology

The incidence of IE is estimated to be 0.16–5.4 cases per 1000 hospital admissions. Most patients are aged 30–60 years, M>F. The disease is uncommon in children in the absence of a predisposing condition. Risk factors include congenital heart disease, rheumatic heart disease, degenerative heart disease, prosthetic valves, intravenous catheters, intravenous drug use, mitral valve prolapse.

Pathogenesis

The development of IE requires the simultaneous occurrence of a number of events: alteration of the valvular surface, deposition of platelets and fibrin, colonization by bacteria, bacterial multiplication, and development of a vegetation.

Aetiology

- 80% of cases of native valve endocarditis are due to streptococci (viridans group, *S. bovis*) or staphylococci.
- *S. aureus* is the most common isolate in IVDUs and tricuspid valve IE.
- *S. epidermidis* is the most common isolate in early (<2 months) prosthetic valve endocarditis (PVE).
- Enterococcal endocarditis is usually associated with malignancy or manipulation of the genitourinary (GU) or GI tracts.
- Other organisms, e.g. *Corynebacteria*, *Listeria*, *Bacillus*, *Salmonella*, *E. coli*, *Enterobacter*, *Citrobacter*, and *Pseudomonas* spp. are uncommon.
- HACEK organisms or fungi are associated with large vegetations.
- Culture-negative endocarditis (~5%) may be caused by *Coxiella burnetii*, *Chlamydia* spp., *Legionella* spp., *Mycoplasma pneumoniae*, *Bartonella* spp., *Brucella* spp.
- Polymicrobial infections occur in 1–2%.

Clinical features

- The incubation period may vary from days to weeks.
- Symptoms are protean and include fever, chills, weakness, dyspnoea, sweats, anorexia, weight loss, malaise, cough, skin lesions, stroke, nausea, vomiting, headache, myalgia, arthralgia, oedema, chest pain, abdominal pain, delirium, coma, haemopytsis, and back pain
- Physical findings include fever, cardiac murmur, Roth spots, clubbing, splinter haemorrhages, Osler's nodes, Janeway lesions, petechiae, peripheral emboli, splenomegaly, septic complications (pneumonia, meningitis, mycotic aneurysms, see 📖 Endovascular infections, p.676).

Laboratory diagnosis

- Blood cultures are positive in ~2/3 of cases. Three blood culture sets should be obtained in the 1st 24 h and incubated for 3 weeks.
- Blood tests may show elevated ESR (90–100%), anaemia (70–90%), leucocytosis (20–30%), leucopaenia (5–15%), and thrombocytopaenia (5–15%). Hypergammaglobulinaemia (20–30%) may result in false-positive results for rheumatoid factor and Venereal Disease Research laboratory (VDRL). Renal impairment and hypocomplementaemia occur in 5–15%.
- Urinalysis is frequently abnormal with proteinuria (50–60%), microscopic haematuria (30–60%), gross haematuria, pyuria, bacteriuria, red cells casts, and white cell casts.
- Serology is useful for diagnosis of culture-negative endocarditis.
- Echocardiography – transthoracic echocardiography (TTE) allows visualization of vegetations in 60–75% of cases compared with >95% with transoesophageal echocardiography (TOE).
- Electrocardiography – lengthening of the PR interval in aortic valve endocarditis indicates aortic root involvement.

Duke criteria

This schema stratifies patients with suspected IE into three categories:

- definite – identified by histopathologically or clinical criteria. Clinical diagnosis requires the presence of two major criteria, one major and two minor criteria, or five minor criteria:
  - major criteria – ≥2 positive blood cultures (or single positive culture for *C. burnetii*), echocardiographic evidence for endocardial involvement
  - minor criteria – predisposing condition (heart condition, IVDU), temp > 38°C, vascular phenomena, immunological phenomena, microbiological evidence (not satisfying major criteria)
- possible – one major and one minor criterion, or three minor criteria
- rejected – firm alternative diagnosis, rapid resolution with no or short course antibiotics, no pathological evidence of IE.

Management

- Antimicrobial therapy is targeted at the causative organism
- UK endocarditis guidelines[1]
- US endocarditis guidelines[2]
- Surgery is indicated in patients with life-threatening congestive cardiac failure or cardiogenic shock due to surgically treatable valvular disease, if the patient has a reasonable prospect of recovery. Surgery is recommended for annular or aortic abscesses, heart block, recurrent emboli on therapy, antibiotic-resistant infections, and fungal endocarditis.

Prevention

There are few data to support antimicrobial prophylaxis for the prevention of IE. For detailed recommendations see National Institute for Health and Clinical Excellence (NICE) guidelines for prevention of endocarditis,[3] and American Heart Association guideline for prevention of endocarditis.[4]

References

1 Elliott TS, Foweraker J, Gould FK et al: Working Party of the British Society for Antimicrobial Chemotherapy. Guidelines for the antibiotic treatment of endocarditis in adults: report of the Working Party of the British Society for Antimicrobial Chemotherapy. *J Antimicrob Chemother* 2004;54:971–81.

2 Bonow RO, Carabello BA, Chatterjee K et al. ACC/AHA 2006 guidelines for the management of patients with valvular heart disease: a report of the American College of Cardiology/American Heart Association Task Force on Practice Guidelines (writing Committee to Revise the 1998 guidelines for the management of patients with valvular heart disease) developed in collaboration with the Society of Cardiovascular Anesthesiologists endorsed by the Society for Cardiovascular Angiography and Interventions and the Society of Thoracic Surgeons. *Circulation* 2006; 114: e84

3 NICE. Antimicrobial prophylaxis against infective endocarditis. London: NICE, 2008. http://www.nice.org.uk/guidance/index.jsp?action=byID&o=11938 (accessed 11 August 2008).

4 Wilson W, Taubert KA, Gemitz M et al. Prevention of Infective Endocantitis: Guidelines from the American Heart Association. *Circulation* 2007; 116:1736–1794.

Intravascular catheter-related infections

Definitions

- **Catheter colonization** – significant growth of organism in quantitative or semi-quantitative culture from catheter tip, subcutaneous segment, or catheter hub
- **Phlebitis** – induration, erythema, pain, or tenderness around exit site
- **Exit site infection** – exudate at exit site yielding microorganism or phlebitis <2 cm from exit site + signs of infection (fever, pus) ± bloodstream infection
- **Tunnel infection** – phlebitis ≥2 cm from exit site, along subcutaneous tract of catheter ± bloodstream infection
- **Pocket infection** – infected fluid in subcutaneous pocket of implanted intravascular device, often associated with local erythema, induration, tenderness, rupture and drainage, necrosis of skin, ± bloodstream infection
- **Bloodstream infection** – bacteraemia or fungaemia in a patient who has an intravascular device and ≥1 positive blood culture obtained from a peripheral vein and no obvious source (apart from the device)

Epidemiology

In the USA, >200,000 nosocomial bloodstream infections occur per year; most of these are related to intravascular devices. Risk factors for intravenous catheter (IVC)-related infections include type of catheter, site of catheter, duration of placement, and hospital demographics.

Aetiology

- Staphylococci, e.g. coagulase-negative staphylococci, *S. aureus*
- Aerobic Gram-negative bacilli, e.g. *E. coli, Klebsiella* spp., *Pseudomonas* spp., *Enterobacter* spp., *Serratia* spp., *Acinetobacter* spp.
- Fungi, e.g. *Candida albicans, Candida* spp., *Malassezia furfur*

Clinical features

The clinical features are unreliable. The most-sensitive clinical features, e.g. fever and chills, lack specificity whereas inflammation and purulence at the catheter site are specific but not sensitive.

Diagnosis

- Rapid diagnostic techniques – e.g. Gram stain or acridine orange stain may be used for diagnosis of exit site infections but have poor sensitivity.
- Cultures of IVC samples – semi-quantitative (roll plate method) or quantitative (flush, vortex or sonication methods) have greater specificity than qualitative methods.
- Paired blood cultures drawn through IVC and percutaneously – all patients with suspected IVC-related infections should have two sets of blood cultures drawn, at least one peripherally. A positive culture from a line requires clinical interpretation, whereas a negative culture virtually excludes catheter-related bloodstream infection.

- Quantitative cultures of CVC and peripheral blood samples – a 5–10-fold difference in colony count between the central and peripheral culture, or an absolute colony count of 100 CFU/mL from a central culture supports the diagnosis of catheter-related bloodstream infection.
- Differential time to positivity for CVC and peripheral cultures – this method takes advantage of continuous blood culture monitoring. e.g. by radiometric methods to compare differential times to positivity between central and peripheral cultures. It correlates well with quantitative methods and is suitable for use in routine labs.

Management

See Infectious Diseases Society of America (IDSA) guidelines for management of intravascular catheter-related infections.[1]

- Peripheral venous catheters – remove IVC, swab exit site if pus present, and take two sets of blood cultures before starting antimicrobial therapy.
- Non-tunnelled CVCs – if there are local or systemic signs or positive blood cultures the CVC should be removed, antimicrobial therapy started and the CVC replaced at a new site:
  - complicated infections (septic thrombosis, endocarditis, osteomyelitis): remove CVC and treat with systemic antimicrobials for 4–6 weeks
  - uncomplicated coagulase-negative staphylococcal infection: remove CVC and treat with 5–7 days of systemic antibiotics
  - uncomplicated *S. aureus* bacteraemia: remove the CVC and treat with 14 days systemic antibiotics (if negative TOE) or 4–6 weeks antibiotics (if positive TOE)
  - uncomplicated Gram-negative bacteraemia: remove CVC and treat with 10–14 days of systemic antibiotics
  - uncomplicated candidaemia: remove CVC and treat with antifungals for 14 days after last positive blood culture.
- Tunnelled CVCs and implanted devices (IDs) – investigations should be performed to establish the CVC or ID as the source of infection:
  - tunnel infection or port abscess: remove the CVC/ID and treat with systemic antibiotics for 10–14 days
  - complicated infections (septic thrombosis, endocarditis, osteomyelitis): remove CVC/ID and treat with systemic antibiotics for 4–6 weeks
  - uncomplicated coagulase-negative staphylococcal infection: retain CVC/ID and treat with 7 days of systemic antibiotics plus antibiotic lock therapy for 10–14 days. Remove CVC/ID if persistent bacteraemia or clinical deterioration
  - uncomplicated *S. aureus* bacteraemia: remove the CVC/ID and treat with 14 days of systemic antibiotics (if negative TOE) or 4–6 weeks of antibiotics (if positive TOE). For salvage therapy see IDSA guidelines[1]
  - uncomplicated Gram-negative bacteraemia: remove CVC/ID and treat with 10–14 days of systemic antibiotics.

Prevention

See IDSA guidelines for the prevention of intravascular catheter-related infections.[2]

References

1 Mermel LA, Farr BM, Sherertz RJ et al. Guidelines for management of intravascular catheter-related infections. *Clin Infect Dis* 2001;32:1249–72.

2 O'Grady NP, Alexander M, Dellinger EP et al. Guidelines for the prevention of intravascular catheter-related infections. *Clin Infect Dis* 2002;35:1281–307.

Endovascular infections

Persistent bacteraemia (i.e. multiple blood cultures taken on different occasions which are positive for the same isolate) suggest endovascular infection. These include endocarditis, IVC-related infections, mycotic aneurysms, pacemaker infections, and vascular graft infections.

Mycotic aneurysm

- **Definition** – a mycotic aneurysm may be any extra- or intra-cardiac aneurysm of infectious origin, except syphilitic aortitis.
- **Aetiology** – in the pre-antibiotic era, mycotic aneurysms were usually associated with infective endocarditis and caused by streptococci and staphylococci. Today mycotic aneurysms are usually due to haematogenous seeding of atherosclerotic vessels or trauma. Pathogens include *S. aureus*, *Salmonella* spp., aerobic Gram-negative bacilli, *L. monocytogenes*, *B. fragilis*, group A and C streptococci, *C. septicum*, enterococci, and pneumococci.
- **Clinical features** – symptoms and signs of infective endocarditis (p.668) may be present. Intracranial mycotic aneurysms are usually silent but may present with headache, homonymous hemianopia, or focal neurological symptoms and signs. Symptomatic intracranial haemorrhages carry a high mortality. Visceral mycotic aneurysms are uncommon. The most common location is the superior mesenteric artery but other sites include the hepatic artery, coeliac artery, external iliac artery, femoral, peripheral, and carotid arteries. Aortic aneurysms are usually associated with infected atherosclerotic lesions and present with fever, back or abdominal pain ± draining cutaneous sinus.
- **Diagnosis** – blood cultures may identify the causative organism. Echocardiography is useful to visualize aortic valve mycotic aneurysms. CT brain scan may show intracerebral haemorrhages although cerebral angiography is the investigation of choice for intracranial mycotic aneurysms. MR angiography is less invasive but less sensitive. Abdominal x-ray may show calcified abdominal aorta. CT or MRI abdomen are less sensitive than angiography for intra-abdominal aneurysms.
- **Management** – intracranial mycotic aneurysms should be treated with antimicrobial therapy, monitored by angiography, and excised if they enlarge or bleed. Infected atherosclerotic aneurysms should be operated on before rupture occurs, and systemic antibiotics continued for 6–8 weeks postoperatively. Peripheral vessel mycotic aneurysms are managed by surgical resection/reconstruction and antibiotic therapy.

Pacemaker infections

- Pacemaker infections affect 1–7% of procedures. Superficial infections involve the generator pocket and/or subcutaneous electrode. Deep infections involve the transvenous intravascular electrode ± generator.
- **Aetiology** – coagulase-negative staphylococci and S. aureus are the most common isolates. Other organisms include *Enterobacteriaceae* spp., *P. aeruginosa*, *C. albicans*, streptococci, enterococci, *Corynebacteria* spp., *Listeria* spp., *Aspergillus* spp.
- **Clinical features** – infections confined to the generator pocket usually present with local swelling, erythema, tenderness ± discharge through incision site or fistula ± systemic symptoms. Infection of the epicardial electrodes may be associated with pericarditis (p.678), mediastinitis (p.680), bacteraemia, and systemic symptoms. Infection of the intravascular portion presents with clinical features of endocarditis (p.668).
- **Management** – superficial infection confined to the subcutaneous elements may be treated by a one-stage exchange under antimicrobial cover. For pacemaker endocarditis, the generator and electrodes should be removed transcutaneously, if possible, but surgical extraction may be required. A temporary pacing system is inserted and at least 2 weeks' systemic antimicrobial therapy is given before insertion of a permanent system. A full course of endocarditis therapy should be given; this may need to be extended if there is evidence of metastatic infection.

Vascular graft infections

- **Epidemiology and pathogenesis** – the incidence of vascular graft infection is 1–5%. Three mechanisms are thought to be responsible for infection: intra-operative contamination (most common), extension from adjacent infected tissue, haematogenous seeding.
- *Aetiology* – *S. aureus* is the most common cause. However, a wide range of organisms may cause infection, e.g. *Enterobacteriaceae* spp., coagulase-negative staphylococci, enterococci, streptococci, *P. aeruginosa*, *Bacteroides* spp., corynebacteria.
- Clinical features – these depend on the site of the graft infection:
  - inguinal graft infections present with an inguinal mass ± pain, erythema, fever, sinus formation
  - abdominal graft infections present with fever, abdominal pain or mass, retroperitoneal bleeding, lower extremity emboli, GI bleeding due to erosion into the GI tract.
- **Diagnosis** – superficial graft infections may be readily diagnosed clinically. Deep grafts may require radiological imaging (e.g. CT or MRI abdomen) to confirm the infection. Blood cultures are often negative, unless infection involves the graft lumen.
- **Management** – surgical resection of the infected graft and revascularization (preferably through an extra-anatomic, uninfected route) is the treatment of choice. Systemic antimicrobial therapy is given for 4–6 weeks postoperatively. If the arterial stump is found to be infected at the time of surgery, culture-specific antimicrobial therapy is given for 6 months postoperatively. Salvage therapy with long-term suppressive antibiotic therapy is sometimes given for vascular graft infections that are not surgically resectable.

Myocarditis

An inflammation of the myocardium. Most cases are viral. Up to 4% of unselected post-mortems reveal unsuspected myocarditis, especially young people dying abruptly. There is histological evidence of myocarditis in up to 20% of cases of 'idiopathic' dilated cardiomyopathy.

Aetiology

Myocardial injury may be a consequence of direct cell damage by an infectious agent, by a circulating toxin, or by immune reactions following infection. The cause is not identified in most cases.

- Viruses are the commonest agents in the developed world, e.g. measles, influenza, polio, mumps, adenovirus, and the group B Coxsackie viruses.
- Bacterial infection may cause myocarditis through immunoloigcal mechanisms (e.g. Lyme disease, acute rheumatic fever), or through direct myocardial infection with associated inflammation (e.g. brucellosis, meningococcal, streptococcal and staphylococcal sepsis, *Legionella* spp., *M. pneumoniae*, and *C. psittici* infection). Toxin-mediated damage is seen in infection by *C. diptheriae* (blocks cellular protein synthesis) and *C. perfringens*.
- Parasitic causes include trypanosomal disease e.g. *T. cruzi* is a common cause in South America.
- Disseminated infection in the immunocompromised may lead to myocarditis (e.g. *Toxoplasma*, *Aspergillus*, and *Cryptococcus* spp.). Cardiac dysfunction may be clinically apparent in up to 20% of AIDS patients, with echocardiographic abnormalities in up to 65% of AIDS patients.

Pathogenesis

The pathological process varies according to the mechanism of the injury and whether it is acute or chronic. All lead to an inflammatory infiltrate and damage to adjacent myocardial cells. In addition some agents damage vascular endothelial cells. Routine histology rarely allows a definitive aetiological diagnosis. Where normal cardiac function is regained, histological abnormalities may lag behind clinical improvement. Cases that leave permanent damage are marked by interstitial fibrosis and a loss of muscle fibres.

Clinical presentation

- Myocarditis may be asymptomatic or result in severe heart failure or sudden death.
- Myocarditis should be considered in a young person developing cardiac abnormalities in the context of a recognized systemic illness or in an otherwise well individual developing unexpected heart failure or arrhythmias (e.g. supraventricular tachycardia (SVT) or extrasystoles).
- Fever, malaise, upper respiratory tract symptoms, tachycardia, dyspnoea, and chest pain may precede Coxsackie virus myocarditis.
- On examination there may be cardiomegaly, murmurs, and signs of cardiac failure.
- Pericarditis may co-exist.

Diagnosis

- Diagnosis can be difficult in those cases occurring in the context of fulminant systemic infection.
- Electrocardiogram (ECG) changes are non-specific (e.g. sequential ST elevation and T wave inversion).
- Around one-third of patients with biopsy-proven myocarditis have raised troponin levels.
- Echocardiography may demonstrate systolic dysfunction and can be used to track the progression of disease.
- MRI scanning is a sensitive means of detecting myocardial inflammation.
- Endomyocardial biopsy is considered the gold standard for diagnosis but can miss cases in which disease is focally distributed, and timing is critical – most experts believe it is not warranted in most cases.
- Demonstrating a viral cause requires isolation of the virus or viral material from myocardium. Generally diagnosis is inferred by serology (e.g. Lyme disease) or detection of the organism in other specimens (e.g. stool or blood).

Treatment

- Therapy should be directed at the causative agent where possible.
- General measures include bed rest (exercise is associated with increased death in mouse models), management of heart failure and arrhythmias. Most patients recover completely.
- Severe cases may require cardiac assist devices.
- Steroids are of no benefit and are probably deleterious overall. Immunoglobulin administration may help certain subgroups (e.g. CMV myocarditis). Antivirals may be of benefit in the future.

Differential diagnosis

Pericarditis, idiopathic congestive cardiomyopathy, acute rheumatic fever, non-infectious myocarditis (collagen vascular disease, thyrotoxicosis, drug or radiation induced).

Pericarditis

An inflammation of the pericardium which may be acute or chronic

Aetiology

- 'Idiopathic' cases account for up to 86% of cases and are probably viral.
- Viruses – enteroviruses (particularly Coxsackie viruses) are the commonest cause. Others include echoviruses, adenovirus, mumps, influenza, EBV, varicella, CMV, HSV and hepatitis B virus.
- Bacteria – before antibiotics became widely available, bacterial pericarditis (e.g. *S. aureus* and *S. pneumoniae*) was a recognized complication of pneumonia. Bacterial pericarditis is now uncommon and tends to occur following Gram-negative infection in older people with predisposing conditions, e.g. oesophageal perforation, head/neck infections (usually anaerobes), and in cases of meningococcal septicaemia. Other bacterial causes include: *M. pneumoniae*, *Legionella pneumophilia*, *Haemophilus influenzae*. Tuberculous pericarditis is a major cause of heart failure in sub-Saharan Africa – chronic disease is associated with a constrictive pericarditis.
- Fungi – these include *H. capsulatum*, *C. immits*, *Aspergillus* spp., *C. neoformans*, and *Candida* spp.
- Parasites – these include *T. gondii*, *E. histolytica*, and *T. canis*.

Pathogenesis

- Viruses reach the pericardium haematogenously and infection results in inflammation of both the visceral and parietal pericardium, with or without a pericardial effusion. Most patients recover – some may experience episodes of relapse, a phenomenon that is probably related to immune mechanisms rather than persistent viral infection. It is rare that viral pericarditis leads to constriction.
- Bacterial infection may occur as a result of direct inoculation (trauma or surgery), contiguous spread (e.g. endocarditis or untreated pneumonia), or bacteraemia. Fluid is usually grossly purulent and subsequent organization with adhesions may lead to constriction.
- TB pericarditis may arise from haematogenous spread (during primary infection), lymphatic spread (from the regional lymph nodes), or contiguous spread (from infected lung or pleura). Initial fibrin deposition, granuloma formation, and polymorphonuclear cell infiltration is followed by the development of a serous/serosanguinous effusion in with lymphocytes and plasma cells. Later, the pericardium is thickened by fibrin deposition and granulomas. In late disease the pericardial space is taken up with adhesions and fibrous tissue, leading to constriction.

Clinical presentation

- Idiopathic or viral pericarditis – retrosternal chest pain, radiating to the shoulder/neck and aggravated by breathing or lying flat. Fever may be present along with flu-like features.
- Bacterial pericarditis is usually seen in the context of severe systemic infection with an acutely ill patient. Chest pain and pericardial rubs may

be reported in less than one-third of patients. Bacterial pericarditis may be recognized late, after the onset of haemodynamic complications.

- TB pericarditis has an insidious onset with chest pain, weight loss, night sweats, cough, and breathlessness. The classic clinical finding is a pericardial rub. Where the effusion is significant, there may be jugular venous distension, pulsus paradoxus.

Diagnosis

- Diagnosis is often made clinically and depends on the history.
- The ECG is abnormal in 90% of cases (due to diffuse subepicardial inflammation), with 50% showing the classic findings of early ST-elevation in multiple leads, resolving over a few days to be replaced with T wave flattening/inversion.
- Echocardiography is useful is diagnosing and assessing effusions and the extent of any compromise.
- Virus isolation from throat swabs or stool, or acute and convalescent viral serology, may lead to a diagnosis but rarely affect management.
- Diagnostic sampling may be indicated if the effusion persists for >3 weeks, when TB, fungal infection, malignancy, or connective tissue disorders should be considered.
- Pericardiotomy with biopsy is preferable to pericardiocentesis as it has a higher diagnostic yield and fewer complications.

Treatment

- Viral/idiopathic – bed rest, analgesia, and monitoring for haemodynamic complications. Steroids should be avoided in the acute phase due to the frequent concomitant myocarditis in which steroids are contraindicated. They may, however, have a role in preventing recurrent pericarditis once a patient has recovered. Colchicine may also be useful.
- Purulent bacterial – surgical drainage and appropriate antibiotic therapy are essential. Early pericardiocentesis may be life saving, but fluid often reaccumulates. Overall mortality, however, remains at around 30% – particularly in those cases associated with endocarditis or following surgery.
- TB – antituberculous therapy should be intiated. Constrictive pericarditis may develop in up to half of patients despite this. Steroids (prednisolone 60 mg for 4 weeks, then reducing over 7 weeks) in addition to antituberculous therapy reduce the need for repeat drainage, as well as resulting in a modest reduction in those developing constriction. Those developing haemodynamic compromise secondary to effusion reaccumulation or progressive pericardial thickening, benefit from early surgery (pericardectomy).

Mediastinitis

Acute mediastinitis is an uncommon but potentially devastating infection involving the mediastinal structures.

Epidemiology and pathogenesis

Primary infection is rare. Almost all cases are secondary to:

- oesophageal perforation, e.g. iatrogenic, trauma, spontaneous
- head and neck infections, e.g. odontogenic, Ludwig's angina, phayngitis, tonsillitis, epiglottitis, parotitis
- spread from other infections, e.g. pneumonia, empyema, subphrenic abscess, pancreatitis, skin or soft tissue infections of chest wall, osteomyelitis of sternum, clavicle, ribs or vertebrae, haematogenous seeding
- cardiothoracic surgery, now the most-common cause.

Aetiology

The spectrum of organisms causing infection varies strikingly according to the underlying cause.

- Oesophageal perforation or head and neck infections are usually polymicrobial and caused by oral streptococci (e.g. viridans streptococci, peptococci, peptostreptococci) and anaerobic Gram-negative bacilli (e.g. *Bacteroides* spp., *Fusobacterium* spp., *Prevotella* spp., *Porphyromonas* spp.).
- Cardiothoracic-surgery related infections are primarily due to Gram-positive cocci (e.g. *S. aureus, S. epidermidis*, enterococci, streptococci), with lesser contributions from Gram-negative bacilli (e.g. *E. coli, Enterobacter* spp., *Klebsiella* spp., *Proteus* spp. *Pseudomonas* spp.).

Clinical features

The clinical manifestations also depend on the underlying cause:

- head and neck infections usually present with fever, pain, and swelling of the affected site
- oesophageal perforation may be obvious or clinically inapparent
- symptoms include chest pain (site depends on location of infection), respiratory distress and dysphagia
- physical signs include fever, tachcycardia, crepitus, and oedema of the head and neck. Hamman's sign (a crunching sound heard over the praecordium synchronous with the cardiac rhythm) is due to emphysema of the mediastinum
- post-cardiothoracic mediastinitis usually presents within 2 weeks of surgery, with fever, wound erythema/discharge, and chest pain (often pleuritic). Sternal instability, wound dehiscence and chest wall emphysema may occur.

Diagnosis

- Blood tests show leucocytosis and raised inflammatory markers. Blood cultures may yield the causative organism(s).
- Chest x-ray may show mediastinal widening, air-fluid levels, and subcutaneous or mediastinal emphysema.

- CT thorax is particularly useful in postoperative mediastinitis to distinguish superficial wound infections from deep retrosternal infections.

Management

- Prompt surgical intervention is required with drainage, debridement, and repair (in cases due to oesophageal perforation). Postoperative mediastinitis may be managed by the open technique (wound debrided and left open to heal by secondary intention) or the closed technique (debridment, primary closure and irrigation through drains).
- Appropriate parenteral antibiotic therapy should be initiated as soon as the diagnosis is made. Empiric therapy should cover the most likely organisms (e.g. penicillin and metronidazole or clindamycin for head and neck/oesophageal infections, and vancomycin and meropenem for postoperative mediastinitis).

Complications

- Pericardial effusion/cardiac tamponade
- Pleural effusions/empyema
- Peritonitis
- Sternal osteomyelitis (postoperative mediastinitis)

Oesophagitis

Inflammation of the oesophagus, generally non-infectious (e.g. gastrooesophageal reflux) but may also be caused by a variety of infectious agents, usually in the context of impaired immunity (HIV, transplant recipients, or those receiving cancer chemotherapy).

Aetiology

- **Candida** – *C. albicans* is the commonest cause of oesophagitis. Other *Candida* species are less commonly isolated. Colonization is seen in 20% of the population (particularly those receiving antacid therapy). Infection follows when breakdown of local and systemic defences permit invasion to the deeper epithelial layers. Endoscopy reveals yellow-white plaques adhering to a hyperaemic oeosphagus (usually the distal third). Removing these reveals an inflamed friable surface. Perforation occurs rarely. Predisposing factors: acute or advanced HIV infection, diabetes mellitus, haematological malignancy, broad-spectrum antibiotic therapy, corticosteroid therapy, conditions that impair oesophageal motility (systemic sclerosis, achalasia), reflux oesophagitis.
- **Cytomegalovirus** – seen usually in AIDS patients (the cause in around 30% of such patients reporting oesophageal symptoms) or the severely immunosuppressed. Endoscopy may demonstrate large (10 cm^2) shallow 'punched-out' ulcers. Diagnosis is best made by histopathological examination of biopsies obtained from the ulcer edge and base, which show enlarged endothelial cells with large intranuclear inclusions. Isolation of CMV in culture is not reliable due to contamination from blood or saliva. Co-infection with HSV or candida is common.
- **Herpes simplex virus** – usually seen in those with significant immunosuppression – rare in healthy adults. HSV-1 is commoner than HSV-2 which is rarely implicated. Accounts for up to 16% of HIV patients with oesophageal symptoms. Presentation may be with odynophagia, chest pain, fever, nausea, and vomiting; <25% may develop clinically significant GI bleeding. Oral/labial or cutaneous HSV infection may be apparent (<38% of cases). Endoscopy reveals multiple small superficial ulcers in the distal third of the oesophagus. Large confluent ulcers and denuded epithelium may be seen as infection progresses. Viral culture of brushing or biopsies is the most-sensitive means of diagnosis.
- **Idiopathic ulceration** – extensive ulceration may occur in those with acute or advanced HIV, or in mild form in the otherwise healthy. These may be attributed to unrecognised infectious agents. A course of prednisolone (40 mg a day for 14 days then tapered to stop) improves symptoms in the majority of HIV-related aphthous ulceration.

Clinical features

Patients present with difficulty in, or pain on, swallowing. Liquids may be better tolerated than solids. Pain may be worse with acidic substances. Severe ulcerative oesophagitis can cause such severe pain that oral intake is limited to the point of weight loss and dehydration. Gastrointestinal bleeding can occur. Oesophagitis can exist in the absence of symptoms (<41%).

Fever may be seen in those with CMV or mycobacterial infection. Vomiting is more common with CMV than other causes. The presence of oral lesions may be indicative of the cause of oesophagitis (e.g. oral thrush, or herpetic ulceration).

Diagnosis and treatment

Accurate diagnosis of oesophageal candidiasis requires endoscopic brushing (with a sheathed cytology brush) and biopsy. The gross appearance can mislead: white lesions may be seen with HSV, CMV and candidal infection. Histopathological examination and viral culture may identify viral causes. Fungal culture is useful only in the management of refractory cases – e.g. to identify the species and sensitivities.

Management

- Diagnostic endoscopy may not always be feasible (bleeding, severe pain, critical illness), particularly in those patients developing oesophagitis secondary to cancer chemotherapy. Empirical treatment for *Candida* and HSV infection may be appropriate (e.g. intravenous amphotericin B and aciclovir), particularly if symptoms are very severe, or oral thrush/ HSV stomatitis are apparent.
- Patients receiving immunosuppressant therapy may need drug level monitoring if treated with antifungals such as fluconazole.
- Candida – fluconazole PO or IV for 14–21 days.
- HSV – aciclovir IV for 7–14 days or PO for 14–21 days.
- CMV – ganciclovir IV for 14–21 days.

Oesophagitis in AIDS patients

Oesophageal symptoms are seen in 40–50% of patients with AIDS at some point in their illness, and affect nutritional status and morbidity.

- *Candida* oesophagitis is commonest and can be treated empirically with fluconazole in mild cases if oral thrush is observed in a symptomatic patient (70% of such cases will have oesophageal involvement); 5% of patients with proven *Candida* oesophagitis do not respond to fluconazole therapy due to resistance. This is commoner in those with previous exposures to fluconazole. Some may response to itraconazole (see 🕮 Triazoles, p.124) or amphotericin B (see 🕮 Polyenes, p.120).
- Viruses cause one-third of cases (often in association with candidiasis). Three-quarters will have a partial or complete response to induction therapy with antiviral drugs but relapses are common without maintenance treatment; 70% of HSV oesophagitis responds to aciclovir but relapse is seen in 15% within 4 months.

Other rare cause of oesophagitis in HIV: EBV, *Mycobacterium avium* complex, *Cryptococcus neoformans, Cryptosporidium, Actinomyces.*

Peptic ulcer disease

Heliobacter pylori is a motile, curved, Gram-negative rod that lives within the mucus layer overlying the gastric (and occasionally duodenal or oesophageal) mucosa. It is present in most people with peptic ulcer disease, increasing the risk of several inflammatory and neoplastic processes. All clinical isolates of *H. pylori* produce urease. It has been isolated from people in all parts of the world – humans appear to be the major reservoir with transmission likely to be via faeco-oral and possibly oral-oral routes. Rates of colonization are equal between men and women. Carriage is near universal by age 20 years in developing countries. Prevalence is over 50% by age 50 years in the UK – 1–3% of those who remain free of the organism by adulthood acquire the bacteria each year.

Clinical features

- Acute acquisition – may cause an acute upper GI illness with nausea and abdominal discomfort with vomiting, burping, and fever lasting 3–14 days. However, infection is clinically silent in most individuals. There have been some documented cases of acute self-limited infection.
- Persistent colonization – *H. pylori* persists for decades in the majority of people. Acute symptoms do not recur in most although the incidence of non-ulcer dyspepsia is slightly higher in colonized individuals.
- Duodenal ulceration – 90% of those with such ulceration carry *H. pylori* and it is usually associated with cases occurring in the absence of aspirin or NSAID use. The organism is found only in areas of metaplastic gastric-type epithelium, and its presence is associated with an over 50 times greater risk of duodenal ulceration.
- Gastric ulceration – 50–80% of patients with benign gastric ulceration are colonized. This is less than duodenal ulceration as a greater proportion of gastric ulcers are due to NSAID or aspirin use.
- Gastric carcinoma – the presence of *H. pylori* has been identified as a risk factor for gastric carcinoma. It induces a chronic gastritis which is thought to lead to atrophic and metaplastic changes over decades. However, *H. pylori* is neither necessary nor sufficient for oncogenesis.
- Gastric lymphoma – colonization is strongly associated with mucosa-associated lymphoid tumours (MALT – lymphomas arising from B lymphocytes). There is evidence to suggest that eradication may lead to improvement in tumour histology.
- Oesophageal disease – as the incidence of *H. pylori* colonization falls it appears the incidence of gastro-oesophageal reflux disease (GORD), Barrett's oesophagus, and oesophageal adenocarcinoma are on the rise. Certain *H. pylori* strains may have an inverse association with Barrett's oesophagus. It has been shown that eradication of *H. pylori* in those with duodenal ulceration doubles the rate of GORD development, and patients with GORD are less likely to be colonized with *H. pylori* than controls.

Pathogenesis

Several features help the organism survive in the hostile gastric environment: its microaerophilic characteristics (enabling its survival within the mucus layer), its motility, and its urease activity (allowing it to generate ammonium ions to buffer the acidity). It colonizes gastric mucosa (stomach or ectopic cells in the duodenum/oesophagus) and induces a cellular infiltrate (lymphocytes, monocytes, plasma cells, neutrophils). It does not appear to invade tissue, and damage is due to either contact with the organism or its extracellular products (ammonia and proteins). The presence of *H. pylori* induces the production of proinflammatory cytokines, and colonized persons have higher gastrin levels than those who are not colonized.

Diagnosis

- Endoscopy with biopsy – expensive and invasive but allows the examination of pathology (e.g. if a neoplasm is considered in the differential). Specimens may be cultured on selective media (e.g. Skirrow's media) in microaerobic conditions for up to 5 days at 37°C. This allows a determination of antibiotic sensitivities. Histology is generally more sensitive than culture – the organism may be visualized in biopsy specimens with Gram or silver stains, or by immunofluorescence.
- Antibody tests – IgG is positive in nearly all colonized patients and may be more sensitive than biopsy – the organism may be present in focal regions of the stomach. Antibody levels decline 3–6 months after successful eradication.
- Urease breath test – patients fast and are then given a meal containing ^{13}C- or ^{14}C-urea. Over the next hour their breath is examined for $^{13}CO_2/^{14}CO_2$. Results may be falsely negative after therapy that suppresses but fails to eradicate the organism, and it should not be performed within 4 weeks of treatment with an antibacterial agent, or within 2 weeks of treatment with anti-secretory drug. Commercial test kits are available – collected breath samples are sent to an appropriate laboratory for analysis.

Treatment

- Antibiotic therapy aimed at eradicating *H. pylori* is associated with significantly lower recurrence rates than acid suppression alone in the treatment of duodenal and gastric ulcers associated with the organism. Eradication is associated with tumour regression in patients with gastric MALTomas.
- Treatment is with a combination of antimicrobials – single agents are rarely effective. They are given in combination with an acid suppressant – proton pump inhibitors are directly inhibitory to *H. pylori* and appear to be potent urease inhibitors. Example regimes: 7 days of either omeprazole 20 mg bd, amoxicillin 1g bd, clarithromycin 500 mg bd, or ranitidine bismuth citrate 400 mg bd, clarithromycin 500 mg bd and metronidazole 400 mg bd. (see *BNF* section 1.3 for detailed regimes and cost).
- Acquired resistance may develop after therapy with certain agents (e.g. quinolone, imidazoles, macrolides).
- True eradication is considered to have been achieved if the biopsy or breath test is negative at 1 month, or serology at 6 months.

Infectious diarrhoea

Definitions

- Gastroenteritis is inflammation of the stomach and intestinal epithelium.
- Diarrhoea is the passage of ≥3 loose/liquid stools within 24 h.
- Food poisoning is vomiting and/or diarrhoea caused by eating food contaminated with microorganisms or toxins (bacterial or otherwise, e.g. poisonous mushrooms).
- Dysentery is bloody diarrhoea with mucus, tenesmus, pain, and fever usually caused by bacterial, parasitic, or protozoan infection.

Aetiology

Some individuals are at higher risk (e.g. previous gastric surgery, immunodeficiency) and recent antibiotic use predisposes to antibiotic–associated diarrhoea. Bacterial infections are more common in the tropics. In the UK causes of gastroenteritis include:

- general patients – viruses (50–70% cases e.g. rotavirus, norovirus), *Campylobacter*, *Shigella*, *Salmonella*, *C. perfringens*, *S. aureus*, *B. cereus*, *E. coli*, *C. difficile*, parasites (10–15% e.g.*Giardia*, *Cryptosporidium*)
- immunosuppressed patients – general causes plus increased *E. coli*, *Cryptosporidium*, mycobacteria, *Microsporidia*, CMV and HSV (especially HIV patients with CD4 count <200/mm^3)
- returning travellers – enterotoxigenic *E. coli* (30–70%), *Shigella* spp. (5–20%), *Salmonella* spp. (5%), *Campylobacter* (5–20%), *V. parahaemolyticus* (shellfish), viral (10–20%), protozoal (5–10%).

Clinical features

- History:
  - nature of the diarrhoea – blood, mucus or pus? Is it painful? Watery diarrhoea with blood or mucus implies large bowel pathology; fatty or smelly diarrhoea suggest small bowel involvement
  - timing (acute or chronic) – an abrupt onset suggest bacterial and viral infections. Chronic diarrhoea is more characteristic of parasites and may also occur with non-infectious causes, e.g. inflammatory bowel disease or malignancy
  - food history – specific restaurant, reheated food, unusual diets, fish
  - are other people affected – is it an outbreak and if so what was the source?
  - recent antibiotic use (in community or hospital)
  - foreign travel – country, city or rural, with reference to timing of possible exposures, e.g. food from a street vendor
  - risk factors for immunosuppression.
- Examination:
  - is the patient, febrile, systemically unwell, and shocked? Consider infection as well as surgical causes, e.g. diverticulitis
  - wasting implies a longer-standing problem, e.g. small bowel malabsorption, immunosuppression, malignancy
  - rectal examination – blood, mucus, faecal occult blood, impacted faeces causing overflow diarrhoea, rectal carcinoma.
- When gastrointestinal symptoms are followed by neurological signs think of *C. botulinum* (nausea, dry mouth, cranial nerve palsies, and descending

weakness with respiratory and autonomic dysfunction, or *C. jejuni* infection associated GBS (occurs 1–3 weeks after GI symptoms).
- Differential diagnosis – non-infectious causes of food poisoning include mushrooms and metal poisoning. Non-infectious causes of diarrhoea include perforation, appendicitis, diverticulosis, inflammatory bowel disease (IBD), colonic malignancy, ischaemic colitis, malabsorption, irritable bowel syndrome, constipation with overflow, thyrotoxicosis, drugs, autonomic neuropathy.

Investigations
- Blood tests – anaemia or macrocytosis may be due to malabsorption. Renal failure may occur with dehydration or haemolytic uraemic syndrome (HUS). Blood film shows red cell fragmentation in HUS.
- Sigmoidoscopy may show inflamed colonic mucosa ± pseudomembranes (*C. difficile* colitis). Biopsies may be taken to exclude IBD.
- Abdominal x-ray or CT abdomen may be necessary to exclude surgical causes.
- Stool samples should be sent to the laboratory for:
  - microscopy – blood and pus cells indicate infectious diarrhoea (e.g. *Salmonella*, *Shigella*, or *Campylobacter* spp.) or IBD. Ova, cysts, and parasites may be diagnostic in patients with a history of foreign travel. Modified Ziehl–Neelsen (ZN) stain for *Cryptosporidium* (preschool children and the immunocompromised)
  - culture – detects specific pathogens such as *Salmonella* spp., *Shigella* spp., *Campylobacter jejuni, E. coli* O157, *Yersinia* spp., *Vibrio* spp. Special media are required
  - toxin detection – either the toxin itself within stool (e.g. *C. difficile*), or the toxin gene in isolated organisms (e.g. *E. coli* O157).

Management
- Oral rehydration is sufficient in mild cases. Oral rehydration salts are commercially available. Patients with moderate or severe dehydration require IV replacement or fluid and electrolytes.
- Antibiotics are not generally indicated in immunocompetent patients. The exceptions are: early *C. jejuni* enteritis, *Y. enterocolitica* infections (children and immunocompromised), *S. dysenteriae*, severe *S. enteritidis*, and *S. typhimurium* infections (e.g. bacteraemia), *Giardia lamblia* and *Entamoeba histolytica*.
- Antispasmodic agents – useful in mild diarrhoea without blood. Do not use if there is a suggestion of dysentery.

Public health aspects
All cases of suspected food poisoning or dysentery should be notified to Public Health (consultant in communicable disease control (CCDC)). The UK HPA has issued guidelines for the prevention of person-to-person spread following gastrointestinal infections.[1]

Reference
1 Working group of the former PHLS Advisory Committee on Gastrointestinal Infections. Preventing person-to-person spread following gastrointestinal infections: guidelines for public health physicians and environmental health officers. *Commun Dis Public Health* 2004;7:362–84. http://www.hpa.org.uk/cdph/issues/CDPHvol7/No4/guidelines2_4_04.pdf (accessed 11 August 2008).

Enteric fever

The clinical and pathological features of typhoid fever were first described in the 19th century by Louis (1829) and then by Jenner (1850). The term enteric fever was proposed by Wilson in 1869. The species that cause enteric fever are *Salmonella enterica* serovar *typhi* (typhoid fever)[1] and *Salmonella enterica* serovar *paratyphi* A, B, or C (paratyphoid fever)[2].

Epidemiology

Enteric fever is a global health problem, affecting an estimated 12 to 33 million people per year. The disease is endemic in many developing countries, e.g. Indian subcontinent, Asia, Africa, and Central and South America. The organisms are usually spread by ingestion of faecally contaminated food or water. Direct person-to-person spread is rare, and laboratory transmission has been reported. Outbreaks in developing countries may result in high morbidity and mortality, especially when caused by multi-drug-resistant strains. In developed countries infection is usually associated with international travel, although food-borne outbreaks do occur.

Pathogenesis

Data from volunteer studies and outbreak investigations suggest that inoculum size and decreased gastric acidity are important determinants of disease severity. The ability of the organisms to survive within macrophages is essential to disease pathogenesis and spread. Organisms multiply in Peyer's patches then enter the bloodstream and re-invade the small bowel causing bleeding and peritonitis. The Vi antigen of *S. typhi* prevents antibody-mediated opsonization, increases resitance to peroxide, and confers resistance to complement-mediated lysis.

Clinical features

- The incubation period ranges from 5 to 21 days, depending on the inoculum size and host immune status.
- Abdominal symptoms (e.g. diarrhoea, constipation, or abdominal pain) may initially develop then resolve. This is followed by non-specific symptoms (e.g. chills, diaphoresis, headache, anorexia, cough, weakness, sore throat, muscle pains, dizziness, delirium, or psychosis) prior to the onset of fever.
- On examination, patients are acutely ill with fever, relative bradycardia (<50%), rose spots, abdominal tenderness, hepatosplenomegaly. Cervical lymphadenopathy, respiratory crepitations, cholecystitis, pancreatitis, seizures, or coma may also occur.
- Complications occur in the 3rd or 4th week of illness, and include intestinal perforation or haemorrhage, endocarditis, pericarditis, hepatic or splenic abscesses, oorchitis.
- Mortality rates are <1% in developed countries but may be as high as 10–30% in developing countries, as a result of delayed treatment or multi-drug-resistant strains.
- Long-term carriage (presence of salmonellae in stool or urine for >1 year) occurs in 1–4% of patients with *S. typhi*. It is associated with biliary abnormalities, concurrent infection with *Schistosoma haematobium*, and an increased risk of developing cholangiocarcinoma.

Diagnosis

- Definitive diagnosis requires isolation of the organism from cultures of blood (50–70% sensitivity) or bone marrow (>90% sensitivity). Other diagnostic samples include stool, urine, rose spots, peripheral blood mononuclear cells (PBMCs), gastric, or intestinal secretions. A combination of specimens increases the likelihood of diagnosis.
- A number of serological tests exist (e.g. Widal test) but none are sensitive, specific, or rapid enough for clinical use.
- DNA probes for *S. typhi* have been developed but these are not as sensitive as culture or commercially available.

Management

The management of enteric fever depends on the clinical severity and drug susceptibility of the isolate:

- uncomplicated typhoid (community to outpatient)
  - fully susceptible – fluoroquinolone 15 mg/kg/day for 5–7 days
  - multi-drug resistant – fluoroquinolone (ciprofloxacin, ofloxacin or plefloxacin) 15 mg/kg/day for 5–7 days
  - nalidixic acid resistant – azithromycin 8–10 mg/kg/day for 10 days
- severe typhoid (hospitalized)
  - fully susceptible – fluoroquinolone 15 mg/kg/day for 10–14 days
  - multi-drug resistant – fluoroquinolone (ciprofloxacin, ofloracin or plefloxacin) 15 mg/kg/day for 10–14 days
  - nalidixic acid resistant – ceftriaxone 60 mg/kg/day or cefotaxime 80 mg/kg/day for 10–14 days.

Prevention

- Immunization may be used to prevent enteric fever in travellers to endemic areas, in laboratory personnel who work with *S. typhi*, and in household contacts. The available vaccines include: heat-killed or inactivated whole-cell vaccines the Vi capsular polysaccharide vaccine (Typhim Vi) and the live attenuated oral vaccine Ty21a.
- Food handlers – may return to work after symptoms have resolved. Two negative stool samples are recommended for food handlers who prepare foods that are consumed raw.

While Typhoid Mary Mallon worked as a cook in New York in the early 1900s, she infected almost 50 people with typhoid. Three of them died from the disease. In the end, she was forcibly quarantined, and ultimately died. Some commentators say the problem was exacerbated by the prejudice against working-class Irish immigrants at that time. Typhoid Mary is probably the most famous disease-carrier of all time. Her name is often used as a generic term for a disease-carrier who is a danger to the public because they refuse to take appropriate precautions.

Reference

1 Parry CM, Hien TT, Dongan G et al. Typhoid Fever. *New Eng J Med* 2002; 347(322):1770–82 et al. *N Engl J Med* 2002;347(22):1770–82.

2 Bhan MK, Bahl R, Bhatnagar S. Typhoid and paratyphoid fever. *Lancet* 2005; 366:749–62

Cholera

Cholera is an epidemic diarrhoeal disease caused by *Vibrio cholerae*. *V. cholerae* is a curved Gram-negative rod that belongs to the family *Vibrionaeceae*. The bacterium has a single polar flagellum that causes erratic movement, visible on microscopy. It has a flagellar H antigen and a somatic O antigen; the latter enables differentiation of the two types that cause cholera (O1 and O139). *V. cholerae* O1 can be classified into three serotypes (Inaba, Ogawa and Hikojima) based on the presence of somatic antigens, and two biotypes (classic and El Tor) based on phenotypic characteristics.

Epidemiology

- Cholera has the ability to cause both epidemics with pandemic potential and to remain endemic in affected areas.
- Epidemic cholera affects non-immune individuals of all ages, occurs after a single introduction, spreads by the faeco-oral route, and has high secondary spread.
- Endemic cholera affects children aged 2–15 years, has an aquatic or asymptomatic human reservoir, and is spread by water, food or faeco-orally. Immunity increases with age, and secondary spread is variable.
- Seven pandemics have occurred between 1817 and 1923. The first six were caused by *V. cholerae* O1 classic biotype, originated from the Indian subcontinent, and spread to Europe and the Americas.
- The seventh pandemic originated in Indonesia in 1961 and was caused by *V. cholerae* O1 El Tor biotype. This pandemic was the most extensive and is still ongoing in some countries.
- Finally in 1992 a new epidemic caused by *V. cholerae* O139 occurred in India and Bangladesh.

Pathogenesis

V. cholerae causes disease by secreting an enterotoxin that causes secretion of fluid and electrolytes by the small intestine. The toxin consists of five B subunits (which bind to a ganglioside receptor on the mucosal surface) and two A subunits. Activation of the A1 subunit by adenylate cyclase results in an increase in cyclic AMP which blocks absorption of sodium and chloride by the microvilli and promotes secretion of chloride and water by crypt cells. The infectious dose varies with route of transmission: 10^2–10^6 organisms. Reduced gastric acidity is associated with an increased severity of disease.

Clinical features

- Cholera is characterized by the sudden onset of profuse watery diarrhoea ('rice water stool') and vomiting, accompanied by varying degrees of dehydration.
- Fever is uncommon (<5% of patients) but the pulse may be rapid and weak and the blood pressure unrecordable.
- Patients may be anxious, restless, or obtunded, with sunken eyes, dry mucous membranes and loss of skin elasticity.

- Hypoglycaemia, seizures, coma, and fever are more common in children. Hypoglycaemia in children carries a higher risk of death.
- Severe disease is more common in pregnancy and associated with fetal loss in up to 50%

Laboratory diagnosis

- Microscopy and culture – dark-field microscopy shows large numbers of vibrios moving chaotically; their movement may be blocked by specific antisera. The organism is cultured on selective media (thiosulphate citrate bile salts sucrose agar, or tellurite taurocholate gelatin agar) and its identity confirmed using specific antisera.
- Laboratory abnormalities – these include raised urea and creatinine, normal or low serum sodium and potassium levels, a metabolic acidosis with high anion gap, raised white cell count, packed cell volume, serum specific gravity total protein. Hyperglycaemia is more common than hypoglycaemia.

Management[1,2,3]

- The goal of therapy is to restore fluid losses rapidly and safely.
- Evaluate the patient for degree of dehydration: mild (<5% fluid loss), moderate (5–10% fluid loss), or severe (>10% fluid loss).
- Rehydrate the patient in two phases: intensive phase (2–4 h); maintenance phase (until diarrhoea resolves).
- Use intravenous fluids only for: severely dehydrated patients in rehydration phase (50–100 mL/kg/h); moderately dehydrated patients who cannot tolerate oral fluids; high stool volumes (>10 mL/kg/h) in maintenance phase.
- Use oral hydration salts (OHS) for patients in maintenance phase (800–1000 mL/h) matching input with output.
- Discharge patients when all the following criteria are fulfilled: oral intake ≥1000 mL/h; urine volume ≥400 mL/h; stool volume ≤400 mL/h.
- Antimicrobial agents play a secondary role in treatment and have been shown to reduce duration and volume of diarrhoea. Oral tetracycline (500 mg qds (four times a day) PO for 3 days) or doxycycline (300 mg stat po) are the agents of choice in adults with sensitive strains. Alternative agents include: furazolidone (100 mg qds PO for 3 days), co-trimoxazole (960 mg bd PO for 3 days), ciprofloxacin (20 mg/kg stat), or azithromycin (20 mg/kg stat).

Prevention

- Provision of uncontaminated water and good sanitation prevents cholera transmission.
- Active surveillance for *Vibrios* organisms in the environment and at-risk populations has been shown to predict the onset of an epidemic.
- Vaccines – parenteral cholera vaccines have proved disappointing. Oral cholera vaccines (e.g. oral inactivated whole cell plus B subunit (WCBS) or live attenuated oral vaccines) look more promising.

References

1 Seas C, Du Pont HL, Valdez LM et al. Practical guidelines for the treatment of cholera. *Drugs* 1996;51:966–73.

2 Morris JG. Cholera and other types of vibriosis: a story of human pandemics and oysters on the half shell. *Clin Infect Dis* 2003; 37: 272–80.

3 Sack DA, Sack RB, Nair GB, Siddique AK. Cholera. *Lancet* 2004; 363: 223–33.

Antibiotic-associated colitis

Diarrhoea is the most common complication of antibiotic therapy, occuring in up to 15% of those receiving β-lactams, and 25% of those receiving clindamycin. Predisposing factors include increased age, underlying illness, recent surgery, the recent use of bowel-motility-altering medications (e.g. GI stimulants, enemas).

Aetiology

- *Clostridium difficile* is a frequent cause (20–30% of antibiotic-associated diarrhoea (*C. difficile*-associated diarrhoea (CDAD)), 50–75% of antibiotic-associated colitis), and is associated with higher death rates and an increase in hospital cost and length of stay.
- Other causes: *Candida*, enterotoxigenic *Clostridium perfringens*, and salmonellosis. The aetiology of diarrhoea/colitis not caused by *C. difficile* is not well understood.

Epidemiology

- Incidence varies with antibiotic and epidemiological setting. Toxigenic *C. difficile* is the commonest cause of hospital-acquired diarrhoea. Rates vary with prescribing patterns, endemic strains, and the clinical setting; 3% of healthy adults carry *C. difficile*. Asymptomatic carriage rates rise to 20% among hospitalized adults receiving antibiotic therapy. Such patients rarely harbour significant quantities of toxin in their stool.
- Most frequently implicated inciting agents – clindamycin, cephalosporins (particularly 3rd generation), penicillins, fluoroqunolones, and proton pump inhbitors. Certain anti-neoplastic agents have also been associated with *C. difficile* diarrhoea, including doxorubicin, cyclophosphamide, and methotrexate.
- Most disease-causing organisms are thought to be acquired from the environment. *C. difficile* may be cultured from surfaces, the hands of hospital personnel, soil, swimming pools, beaches, and tap water. Many outbreaks feature a predominant strain, or related group of strains.

Clinical features

- Infection with toxigenic *C. difficile* may be asymptomatic (particularly neonates), mildly symptomatic, or cause fulminant and occasionally fatal colitis. Symptoms commonly start 5–10 days after antibiotic therapy (<10 weeks after therapy has finished). Features include fever, abdominal pain, nausea, dehydration, and leucocytosis.
- Severe cases – perforation, toxic megacolon (acute dilatation of the colon to a diameter >6 cm with systemic toxicity with no mechanical obstruction – mortality >64%) in the absence of diarrhoea.
- Extraintestinal complications (e.g. splenic abscess, osteomyelitis, Reiter's syndrome) are rare.
- Differential diagnosis – other infectious causes of diarrhoea, adverse drug reaction, ischaemic colitis, inflammatory bowel disease.

Pathogenesis

The complex ecosystem of the normal gut flora confers resistance to *C. difficile*. Antibiotics or antineoplastic agents disrupt this and render the colon more susceptible to *C. difficile* colonization. The organism is acquired faeco-orally, spores surviving the acidic environment of the stomach. They convert to the vegetative form in the small intestine, and begin to produce two toxins, A and B. These cause intestinal fluid secretion, mucosal injury, and inflammation. The resulting neutrophilic infiltrate with mucin, fibrin, and nuclear debris produces raised white/yellow plaques, giving rise to the name 'pseudomembranous colitis'. The severity of disease seen in the UK has increased dramatically with the introduction of a strain originally identified in the US (Ribotype 027).

Diagnosis

- Toxin detection from stool samples is the most widely used means of diagnosis. Enzyme-linked immunosorbent assay (ELISA)-based tests for toxins A or B are rapid and specific but not as sensitive as the slower, more cumbersome cytotoxicity tests (in which filtered stool is incubated with a mammalian tissue culture cell line and assessed for its cytotoxic effect). Rare isolates produce toxin B alone, therefore testing for A only will miss some. Combined with strict clinical criteria and positive culture, ELISA sensitivity ranges from 63% to 94%, with specificity from 75% to 100%.
- Culture is sensitive and essential in the assessment of an outbreak, but most hospital microbiology labs cannot distinguish between toxigenic and non-toxigenic strains.
- PCR-based assays for toxins A and B are available but not yet in widespread use.
- Endoscopy is usually reserved for specific situations – when a rapid diagnosis is required, if a patient has ileus and cannot produce stool, or when the differential diagnosis includes disease that could be confirmed endoscopically.

Management

- General measures – isolate the patient, implement infection control measures, discontinue the precipitating drug, replace fluid/electrolyte losses, and avoid antimotility agents.
- Specific treatment – oral metronidazole (500 mg PO qds for 10 days), or oral vancomycin (125 mg qds for 10 days). Patients unable to take oral antibiotics should receive antibiotics via a nasogastric (NG) tube or intravenously. Other antibiotics with activity against *C. difficile* infection include teicoplanin, fusidic acid, and nitazoxanide. Note that ribotype 027 may be less sensitive to metronidazole.
- Recurrent disease – CDAD recurs in 8–50% of cases. Re-treat with the original drug, e.g. metronidazole – if this fails try the alternative, e.g. vancomcyin. In cases of multiple recurrence or refractory disease, consider the use of probiotics, immunoglobulins, or corticosteroids.
- Surgery is rarely necessary but is life-saving in cases of toxic megacolon or perforation. Mortality in such cases is around 35%.
- Asymptomatic carriers should not be treated – vancomycin suppresses the organism but may cause prolonged carriage; metronidazole is ineffective.

Prevention

Limit the use of inciting agents, e.g. antibiotics and proton pump inhibitors. Infection control measures such as handwashing, universal precautions, and phenolic disinfectants for environmental cleaning are of proven benefit in reducing *C. difficile* transmission in healthcare settings.

Acute cholecystitis

Inflammation of the gall bladder, usually secondary to cystic duct obstruction. Acute cholecystitis usually occurs on the background of chronic cholecystitis – most gall bladders removed at cholecystectomy exhibit fibrosis and other histological changes indicative of chronic inflammation.

Pathogenesis

Ninety per cent of patients have gallstones impacted in the cystic duct. The consequent increase in intraductal pressure impairs blood supply and lymphatic drainage. Tissue necrosis and bacterial proliferation follow within the gall bladder. Complications occur in 10–15% of cases: gall bladder empyema, emphysematous cholecystitis (elderly diabetic men), gall bladder perforation and peritonitis, pericholecystic abscess, intraperitoneal abscess, cholangitis, liver abscess, pancreatitis, bacteraemia. The differential diagnosis includes: myocardial infarction, ulcer perforation, intestinal obstruction, right lower lobe pneumonia.

Clinical features

Early obstruction may cause only mild epigastric pain and nausea. Transient cases may settle in 1–2 h. Persistent obstruction sees the symptoms localize to the right upper quadrant and increase in severity with signs of peritoneal irritation (shoulder tip pain). The gall bladder may be palpable in 30–40% of cases. Fever may occur. Most patients settle within 4 days, with 25% requiring surgery or developing complications. Severe fever with jaundice or hypotension should raise the suspicion of suppurative cholangitis (and obstruction of the CBD).

Diagnosis

- Blood tests – WCC is usually raised. 50% of patients have mild elevations of the bilirubin, 40% have raised aspartate transaminase (AST), 25% raised alkaline phosphatase (ALP).
- Microbiology – bacteria may be isolated from bile even in asymptomatic cases of cholecystitis. Rates of bile infection rise with the duration of symptoms, the age of the patient, and in jaundiced patients (particularly with CBD obstruction). The organisms isolated are those of the intestinal flora: enteric Gram-negative bacilli (*E. coli*, *Klebsiella*, *Enterobacter*, *Proteus* spp.), enterococci, and anaerobes (*Bacteroides*, *Clostridia*, *Fusobacterium* spp.). Anaerobic organisms may be found in polymicrobial infection and are often isolated following biliary tract procedures.
- Radiology – CXR is of limited use. Gas in the gall bladder wall or lumen is diagnostic of emphysematous cholecystitis. USS is the diagnostic study of choice, showing a sensitivity of around 90% (presence of stones, thickened gall bladder wall, dilated gall bladder lumen, pericholecystic collection). Other: nuclear medicine hepatobiliary scanning, CT, MRI.

Treatment

- Antibiotics – elderly, or severely ill patients with signs of infection or complications should be treated empirically, e.g. an aminoglycoside (gentamicin) with ampicillin and metronidazole. Therapy should be altered when culture results are available. Antibiotics do not affect the outcome or incidence of infectious complications (e.g. empyema) in uncomplicated cases – perhaps due to poor penetration into gall bladder bile due to obstruction. Perioperative antibiotics do reduce postoperative infectious complications such as wound infection, and are recommended in the elderly, those with jaundice, CBD obstruction, or fever (cases associated with significant bactobilia).

- Surgery:
  - acute uncomplicated cholecystitis – some surgeons advocate delayed surgery after acute infection has settled. However, there is no difference in morbidity between early and delayed surgery and presentations can be deceptively benign, leading to urgent surgery under less-favourable conditions
  - gangrenous/emphysematous cholecystitis, perforation, percholecytstic abscess: all require urgent surgery.

Acute cholangitis

Obstruction of the common bile duct (by gallstones, tumour, parasite etc) leads to inflammation and infection of the biliary tree. Patients usually have a history of gall bladder disease.

- **Aetiology** – similar to that seen in acute cholecystitis
- **Clinical features** – onset is acute with high fever, sweats, jaundice (Charcot's triad – seen in 85% of cases) and diffuse right upper quadrant pain. Gram-negative sepsis with shock is common
- **Diagnosis** – similar to cholecystitis. WCC, bilirubin and ALP are high. However patients with cholangitis have a high rate (50%) of bacteraemia – *E. coli*, then anaerobes are the most-frequently isolated organisms. Those with biliary stents are more likely to grow *Pseudomonas* spp.
- **Treatment** – antibiotic therapy, e.g. co-amoxiclav and gentamicin, should be commenced immediately as there is a high incidence of septic shock. Therapy is aimed at managing the bacteraemia. Prompt decompression of the common bile duct by endoscopic retrograde cholangiopancreatography (ERCP) or surgery is essential. Intraoperative cholangiography should be performed

Primary peritonitis

Inflammation of the peritoneum due to infection not directly related to other intra-abdominal abnormalities. It occurs in all ages groups.

Aetiology

- Children – associated with post-necrotic cirrhosis, nephrotic syndrome, and urinary tract infection but can occur in those with no predisposing condition. Its incidence has fallen in children with the use of antibiotics.
- Adults – at-risk groups: those with ascites associated with alcoholic cirrhosis, chronic active hepatitis, congestive cardiac failure (CCF), metastatic malignancy, SLE, HIV. A cause of decompensation in those with previously stable chronic liver disease.

Pathogenesis

- Infection is acquired via the lymph, the blood (particularly in those with portosystemic shunting in association with cirrhosis which may increase the rates and duration of bacteraemia), by bacterial transmural migration from the gut lumen or via the Fallopian tubes in women (e.g. gonococcal or chlamydial perihepatitis).
- Enteric organisms account for nearly 70% of infections in cirrhotic patients (*E. coli*, *Klebsiella pneumoniae*, enterococci, and other streptococci). *Staph aureus* and anaerobes are less commonly isolated. Bacteraemia may occur in up to 75% of those with aerobic organisms. Unusual causes of peritonitis include *M. tuberculosis* and *Coccidioides immitis* – such organisms are usually found in disseminated infection. *S. pneumoniae* is the most common cause in HIV patients.

Clinical features

- An acute febrile illness with fever, diffuse abdominal pain, nausea/vomiting, diarrhoea, and rebound tenderness on examination. Resembles acute appendicitis. The onset may be insidious and patients can present with signs of infection/sepsis and no localizing features.
- Cirrhotic patients may have other features of chronic liver disease.
- Tuberculous peritonitis is gradual in onset with fever, weight loss, night sweats, and abdominal distension.
- Gonococcal or chlamydial perihepatitis is usually seen in women. Presents with pain, guarding, and tenderness in the right upper quadrant.

Diagnosis

- Abdominal paracentesis is indicated in all cirrhotic patients with ascites. Peritoneal fluid should be sent for cell count, Gram stain, culture, and protein concentration. Culture yield is improved by the direct inoculation of 10 mL fluid into blood culture bottles at the bedside. A positive Gram stain is diagnostic but is negative in 60% of cirrhotics with infection. Ascitic fluid neutrophil count of over 500/microlitre is the best single predictor of peritonitis (86% sensitivie, 98% specific) – generally a threshold of 250/microlitre is used (93% sensitive, 94% specific). Improved diagnostic accuracy is achieved by combining cell

counts and the ascitic fluid pH (neutrophil count > 500/microlitre and an ascitic fluid pH < 7.35 gives 100% sensitivity and 96% specificity).
- Blood cultures may be positive in ~one-third of patients. The diagnosis of primary peritonitis can be made only after other potential primary sources of infection have been excluded.
- Contrast-enhanced CT can help identify intra-abdominal sources of infection.
- Some surgeons will exclude appendicitis in children only at operation. Tuberculous peritonitis may be confirmed at operation or histology/culture of peritoneal biopsies.

Treatment

- Treat those patients with positive cultures or Gram stain regardless of the cell count (nearly 40% of those with positive cultures and normal cell counts go on to develop peritonitis) and all culture-negative patients with raised cell counts.
- Initial treatment is empirical while culture results are awaited – ampicillin in combination with an aminoglycoside, or a third-generation cephalosporin (avoids the risks of nephrotoxicity). Patients with primary peritonitis respond within 48 h to appropriate antibiotic therapy. Antibiotics are usually given for 10–14 days.
- Follow-up peritoneal fluid cell counts are useful but not essential.
- In those who do not respond, another primary source of infection should be considered (e.g. perforation, intra-abdominal abscess).

Prevention

Antimicrobial prophylaxis (e.g. ciprofloxacin) decreases the frequency of primary peritonitis in certain high-risk groups (patients with ascites admitted with GI bleeds, those awaiting liver transplantation, those with ascitic protein levels <1 g/dL), but does not confer a survival advantage.

Secondary peritonitis

Secondary peritonitis occurs as a result of a breach in the mucosal barrier resulting in spillage of organisms from the GI or GU tracts into the peritoneal cavity. This normally occurs in the context of intra-abdominal infections (e.g. appendicitis, diverticulitis) or surgery (abdominal, gynaecological, or obstetric).

Aetiology

- Most cases are due to infection by the commensal flora of the mucous membranes within the abdominal cavity. Peritonitis also complicates an exogenously acquired visceral infection (e.g. *S. aureus*, *M. tuberculosis*).
- Infection is usually polymicrobial. The commonest isolates are *E. coli*, *B. fragilis*, enterococci, other *Bacteroides* spp., *Fusobacterium*, *C. perfringens*, *Peptococcus*, and *Peptostreptococcus*. Antibiotic-resistant organisms are more likely to be found among those patients who acquire peritonitis while receiving antibiotics in hospital (e.g. *Candida*, enterococci, *Enterobacter*, *Serratia*, *Acinetobacter* spp.). Vaginal flora, e.g. group B streptococci, may be present after vaginal surgery or labour.

Pathogenesis

- Many anaerobic infections are synergisitic – e.g. facultative anaerobes providing a sufficiently reduced environment for the establishment of obligate anaerobic organisms.
- Leaking bile or acid may cause a chemical peritonitis that leads to inflammation, necrosis, and further intra-abdominal damage, facilitating the establishment of bacterial infection.
- Local response – local inflammatory response of peritoneum leads to fluid production, and granulocyte entry to the peritoneal cavity. The exudate contains fibrinogen which forms plaques around inflamed surfaces aimed at localizing infection and may later lead to adhesions. Some instances of infection may be contained and resolve. Others may lead to local abscess formation. If localization fails completely, diffuse peritonitis may result.

Clinical features

- Symptoms – initial features are those of the primary disease process (e.g. appendicitis). Moderate abdominal pain, aggravated by movement, becomes more severe and diffuse as infection spreads throughout the abdomen. Pain may reduce in intensity and become more focal if localization strategies are effective. Other: vomiting, fever, distension, anorexia, unable to pass flatus, thirst.
- Signs – patient lying still, alert, and restless at first, later becoming listless. Fever is usually present. Hypothermia may be noted in early chemical peritonitis and is a severe sign late in the course of patients presenting with sepsis. Tachycardia, hypotension, tachyopnoea, abdominal tenderness (maximal over primarily affected organ) with rebound and guarding, bowel sounds present initially later disappear. Some of these features may be masked in patients receiving glucocorticoids, or whose abscess has been localized away from the anterior abdominal wall (e.g. subphrenic).

Diagnosis

- Lab tests – peripheral WCC 17,000–25,000 cells/mL with a left shift (in some situations massive peritoneal inflammation may lead to low peripheral WCC with an extreme shift to immature forms), haemoconcentration, elevated amylase, acidosis in late disease, features of underlying condition (diabetic ketoacidosis (DKA), haematuria, pyuria, pancreatitis).
- Radiology – contrast CT is the preferred initial study; plain XR: signs of inflammation, free air, distended loops of adynamic bowel, signs of the underlying condition (obstruction, volvulus, instussusception, gall bladder calcification); USS may be limited by the presence of air-filled loops of bowel.
- Blood cultures.
- Peritoneal lavage or aspiration may be appropriate in some situations.
- Differential diagnosis – pneumonia, sickle cell anaemia, herpes zoster, DKA, porphyria, familial Mediterranean fever, SLE, uraemia.

Treatment

- General measures – fluid resuscitation, circulatory and respiratory support, appropriate surgical interventions.
- Antimicrobial therapy – broad spectrum antimicrobial therapy should be started immediately after taking blood cultures. Combinations of two or three antibiotics are used to provide good activity against aerobes and anaerobes, e.g. IV cephalosporin, metronidazole ± gentamicin, or IV co-amoxiclav ± gentamicin. Detailed culture and sensitivity results may take several days as cultures are often mixed and some organisms are slow growing. Antibiotics may not need to be active against every organism isolated – the elimination of the majority may allow host defences to eliminate the remainder. Antifungal therapy (e.g. amphotericin B) should be used if *Candida* spp. are isolated; 5 to 7 days of treatment should be sufficient after adequate surgical intervention depending on the severity of infection and clinical response. Conversion to oral therapy may be indicated in those patients with a good response.

Prognosis

- Survival depends on age, comorbid conditions, duration of peritoneal contamination, the primary process, and microorganisms involved.
- Mortality ranges from 3.5% in those with early infection caused by penetrating trauma to 60% in those with established infection and secondary organ failure. Death is thought to follow uncontrolled cytokine release.

Prevention

Pre/perioperative antibiotics reduce infections in clean, contaminated surgery (e.g. appendectomy for appendicitis without rupture, penetrating wounds of the abdomen, vaginal hysterectomy in premenopausal women). Postoperative infection rates fall from 20–30% to 4–8% with prophylactic antibiotic use in such infections.

Continuous ambulatory peritoneal dialysis peritonitis

Peritonitis was a common complication of peritoneal dialysis until Tenckhoff introduced his improved catheter in 1968. Rates fell further as techniques and bag adapters etc improved. However it still occurs at around one episode per patient year, with up to 70% of patients experiencing an episode of infection in their first year of dialysis. Recurrent infection is one of the commonest reasons for discontinuing continuous ambulatory peritoneal dialysis (CAPD) (20–30% of patients). Prognosis is good – mortality is less than 1%.

Pathogenesis

Infection is commonly acquired by contamination of the catheter by skin organisms. Enteric organisms may be cultured from the skin of some CAPD patients. Organisms can also enter via the catheter exit site and through contamination of the dialysate delivery system, as well as trans-murally in the manner seen in some cases of primary peritonitis. Host defences are impaired in CAPD patients by the low pH, high osmolality, and low IgG and C3 levels of dialysate (the addition of IgG to the fluid has been shown to have a prophylactic effect). Gram-positive organisms account for over 60% of isolates (CoNS, *S. aureus*, streptococcal species and diphtheroids). Other organisms include Gram negatives (*E. coli*, *Klebsiella*, *Enterobacter*, and *Pseudomonas*). Anaerobic organisms and yeasts are uncommon isolates. *Pseudomonas aeruginosa* peritonitis is associated with high treatment failure rates, and organisms such as *S. aureus*, *Streptococcus*, and fungi are associated with more-severe disease.

Clinical features

Abdominal pain, tenderness (up to 80% of patients), nausea and vomiting, fever (10%), and diarrhoea.

Diagnosis

Fluid should be taken from the catheter under sterile technique. It is usually cloudy in appearance. A white cell count above 100/microlitre is indicative of infection (neutrophils dominate). Peritoneal eosinophilia may be seen after tube placement (represents an allergic reaction to the tubing) and in some cases of fungal infection. Gram stain is positive in less than half the cases. Blood cultures are rarely positive. Fluid should be inoculated into blood culture bottles. Yield can be improved by culturing the sediment of 50 mL of centrifuged fluid.

Treatment

- Bacterial – intraperitoneal antibiotics are preferred; there is no advantage in IV therapy. This allows for ambulatory treatment. Therapy should be started empirically, then guided by the Gram stain and culture results. Treatment should continue for between 10 and 21 days, or for one week after catheter removal. Most patients improve within 2–4 days. Those who do not should be re-evaluated and unusual (e.g. fungi) or resistant organisms considered, as well as alternative diagnoses.

- Fungal – most cases can be treated with amphotericin B which can be given intraperitoneally but may cause abdominal pain. Most patients with fungal infections will require catheter removal and IV therapy. Flucytosine may be used but levels must be carefully monitored. Some *Fusarium* species are resistant to amphotericin.
- Catheter removal – up to 20% of patients require catheter removal due to persistent skin or tunnel infection, intractable infections (fungal, mycobacterial, or *P. aeruginosa* infection), recurrent infections with the same organism, or catheter failure.

Prevention

Good technique helps reduce infection rates, e.g. exit site care, connection methods, patient training. Antibiotic prophylaxis may be of some benefit in those patients undergoing extensive dental procedures or lower GI endoscopy. Further details are available in the guidelines produced by the International Society for Peritoneal Dialysis.[1]

Reference

1 ISPD Guidelines. www.ispd.org/lang-en/treatmentguidelines/guidelines (accessed 12 August 2008).

Diverticulitis

Pathogenesis

Diverticulae are small mucosal herniations that protrude through the intestinal layers and smooth muscle along openings created by the nutrient vessels in the wall of the colon. They are associated with a low-fibre diet, constipation, and obesity. Diverticulosis affects >10% of those over 45 years of age, and 80% of those over 85 years. Inflammation occurs in 20% of patients with diverticulae. It is commoner in the elderly and those with extensive disease. Inflammation may remain confined to the bowel wall or be complicated by the formation of fistulae or perforation.

Presentation

Uncomplicated sigmoid diverticulitis can resemble appendicitis, but with left-side findings: fever, tenderness and guarding in the lower left quadrant of the abdomen, a mass, raised WCC. Urinary symptoms may reflect inflammation close to the bladder, and fistulae may form (pneumaturia, faecaluria). Severe disease may present with peritonitis.

Diagnosis

Often made clinically acutely – patients should be managed medically with investigations following symptom resolution. CT has replaced barium enema for the diagnosis of diverticulosis and is also useful in guiding the drainage of any abscesses. Sigmoidoscopy may help exclude other pathology. Differential diagnosis: malignancy, IBD, pelvic inflammatory disease, mesenteric ischaemia.

Management

- First attacks of acute uncomplicated diverticulitis – manage medically if tolerated: oral antibiotics (e.g. co-amoxiclav) and liquid diet. Otherwise patients require admission and IV antibiotics.
- Complicated disease – intravenous antibiotics (as for secondary peritonitis), nasogastric tube and surgery may be indicated. Small abscesses may resolve with IV antibiotics alone, or be amenable to guided percutaneous drainage. Abscesses containing gross faecal material require surgical intervention. Fistuale may need repairing.
- Perforated diverticulitis – a two-stage procedure is commonly performed with resection of the diseased bowel and proximal colostomy to decompress the bowel and divert the faecal stream, followed by reanastomosis a few months later. A single-stage procedure may be appropriate in certain patients. Intravenous antibiotics should be started preoperatively.
- Consider surgery for those patients failing to respond to conservative therapy within 72 h, with persistent obstruction, or if malignancy is suspected.

Intra-abdominal abscess

Intra-abdominal abscesses may complicate peritonitis of any cause. Primary abscesses develop following primary peritonitis. Secondary abscesses may follow appendicitis, diverticulitis, biliary tract lesions, pancreatitis, IBD, perforated peptic ulcers, trauma, and surgery.

Pathogenesis

Infections are usually polymicrobial, with anaerobes isolated in up to 70% of cases. Other organisms include enterococci, enterobacteriaceae, *P. aeruginosa*, *S. aureus*. Abscess location is related to that of the primary disease and the direction of peritoneal drainage (e.g. most appendicitis-related abscesess occur in the right lower quadrant or pelvis).

Clinical features

These include high/fluctuating fever, rigors, abdominal pain, and tenderness over the affected area. Specific features will vary with the location, e.g. subphrenic abscesses may cause costal tenderness and chest signs on examination. Presentation can be acute or chronic (particularly subphrenic abscesses where the patient has been receiving antibiotics), and may follow primary abdominal disease (e.g. pancreatitis) or abdominal surgery with a prolonged recuperation.

Diagnosis

Ultrasound, CT, or MRI scans are the most effective means of identifying abscesses. A pleural effusion on chest x-ray may indicate a subphrenic abscess. The diagnosis is confirmed by radiologically guided diagnostic aspiration. Samples should be sent to the lab for microscopy, culture, and sensitivity (MC&S) testing.

Treatment

- Drainage is key – percutaneous drainage (radiologically guided) is suitable for unilocular collections that are readily accessible, are not vascular, and are likely to drain easily by simple dependent drainage. Repeat scanning should be used to confirm resolution following adequate drainage. Surgical drainage is usually required for multiple or loculated abscesses, or those with very viscous pus.
- Antibiotics – agents should be directed against the most-likely organisms, e.g. *Enterobacteriaceae* and anerobes. Therapy should be started immediately after blood cultures have been taken. The antimicrobial regimen should be tailored to culture results. Repeat samples may be required in patients with prolonged antimicrobial therapy.

Retroperitoneal abscess

- Abscesses may form in the retroperitoneal space following direct extension of infection from a retroperitoneal structure (e.g. pyelonephritis, spinal osteomyelitis), intra-abdominal sepsis, traumatic haemorrhage, or bacteraemia.
- Common organisms include *S. aureus* and coliforms. Anaerobes and polymicrobial infections are less common. *M. tuberculosis* may be seen in endemic areas

- Clinical features – patients present with fever, abdominal/flank/lumbar pain, and a palpable mass. If the psoas sheath is involved there may be pain on hip flexion.
- Diagnosis is made by CT or MRI scan. Blood cultures and aspirated pus should be sent to the lab for culture.
- Management is by surgical drainage and empirical broad-spectrum IV antibiotic therapy while awaiting results of cultures. Treatment should be tailored to culture results.

Pancreatic abscess

- Up to 9% of patients with acute pancreatitis develop a pancreatic abscess. Abscesses may also develop following penetration by a peptic ulcer or secondary infection of a pancreatic pseudocyst. Up to 50% of abscesses are polymicrobial. Haematogenous seeding of bacteria may explain those abscesses caused by single organisms.
- Clinical features – presents with failure to improve, or abrupt deterioration following initial recovery from acute pancreatitis. Most patients experience abdominal pain radiating to the back, with fever and vomiting. Rarer manifestations include jaundice, distension, peritonitis, abdominal mass. Serum amylase may be elevated.
- Diagnosis – USS and CT demonstrate the abscess in the majority of cases but distinguishing abscess from pseudocyst may require guided diagnostic needle aspiration.
- Treatment – surgical drainage/debidridement is essential (53–86% survival). Percutaneous drainage may be helpful in some patients requiring stabilization prior to surgery but is rarely sufficient. Initial antibiotic therapy needs to be broad and can later be adjusted according to sensitivity testing.
- Complications – retroperitoneal extension of infection; fistula formation between the abscess and stomach, duodenum or colon; erosion of major blood vessels causing intra-abdominal haemorrhage.

Splenic abscess

- Uncommon and may be due to bacteraemic seeding of infection (e.g. bacterial endocarditis, IVDU), splenic infarction (e.g. blunt trauma, sickle cell disease), or direct extension of intrabdominal infection. They are usually multiple.
- Aetiology – causes include *S. aureus*, streptococci, enterobactericeae, anaerobes, and *Candida* spp. Around a quarter are polymicrobial.
- Clinical features – left upper quadrant pain with shoulder tip discomfort and fever. Multiple small abscesses may not cause spleen enlargement.
- Diagnosis – chest x-ray may demonstrate an elevated hemidiaphragm, basal pulmonary infiltrates, or pleural effusion. Ultrasound, CT, or MRI scanning confirms the diagnosis.
- Treatment – initial antibiotic therapy must be broad spectrum (e.g. cephalosporin and metronidazole) and modified following culture results. Multiple or large single abscesses may necessitate splenectomy. Incision and drainage may be preferred in cases where the spleen is held by extensive adhesions.

Hepatic abscess

Aetiology

- Bacterial liver abscesses – relatively uncommon. Associated with liver transplantation, chronic granulomatous disease, and sickle cell anaemia. Infection may be acquired by the biliary tract (may cause multiple abscesses), the portal vein (e.g. in appendicitis or inflammatory bowel disease), from an adjacent structure (e.g. gall bladder), haematogenous seeding of infection from elsewhere in the body, and trauma to the liver. Bacterial abscesses are usually polymicrobial e.g. enterobacteriaceae and anaerobes. *S. aureus* is found in <20% of cases, many of these young children. Other organisms identified in the context of liver abscess include *Y. enterocolitica* and *Candida* spp. (patients are often immunosuppressed).
- Amoebic liver abscess – caused by *Entamoeba histolytica*. Complicates <10% of cases of amoebic colitis. Occurs in males more than females. Abscesses are usually solitary, and more common in the right lobe.

Presentation

- Bacterial abscesses – present with fever over days or weeks. Ascending cholangitis causing multiple abscesses presents with characteristic spiking temperatures. Abscesses in the upper right lobe may cause cough, pleuritic and shoulder pain; <70% of patients have hepatomegaly. Jaundice is unusual out of the context of ascending cholangitis or in cases of extensive hepatic involvement (e.g. multiple abscesses).
- Amoebic – particularly suggested by a history of diarrhoea with point tenderness over the right chest wall.

Diagnosis

- Radiology – plain CXR may reveal elevation of the right diaphragm, with a right pleural effusion or gas in the abscess cavity. USS or CT are most useful and may be used to guide diagnostic aspiration.
- Blood tests – WCC, CRP and ALP may be raised.
- Blood cultures – positive in 50% of patients with bacterial abscesses.
- Amoebic serology is useful outside endemic areas and is positive in 90% of those with amoebic abscess Aspirated material should be cultured aerobically and anaerobically and examined for *E. histolytica* on direct microscopy. Brown sterile fluid without a foul odour suggests an amoebic abscess, although secondary infection with bacteria is common.

Treatment

- Bacterial abscesses **–** broad-spectrum IV antibiotic therapy should be started as soon as diagnosis is suspected. Although there are reports of successful resolution in the absence of drainage, most experts recommend percutaneous drainage. Material should be cultured to guide specific therapy. Antibiotics should be continued for at least 1 month, and often as long as 4 months in the case of multiple abscesses. Most fevers resolve within 2 weeks – some may take up to 4 weeks. The rate of resolution may be assessed with serial scans. Surgical drainage should be considered in those cases that fail to respond, loculated or viscous abscesses, and those resulting from biliary obstruction. Cure rates may be >90% in cases where the diagnosis is made early and the abscess is not associated with serious underlying disease.

- Amoebic abscesses – metronidazole is active against both the intestinal and hepatic stages of the organism. Aspiration is probably not necessary unless the lesion is very large, threatens to rupture, or fails to respond to medical therapy. Large lesions may require repeated aspiration. The mortality rate of uncomplicated ameobic abscesses is under 1%. Higher mortalities are associated with those abscesses that rupture into the peritoneum (18%), pericardium (30%), or pleura/bronchi (6%).

Acute hepatitis

A self-limiting inflammation of the liver, which may be caused by a broad range of infectious or non-infectious agents.

Aetiology

- Hepatitis viruses – e.g. hepatitis A (📖 p.457), hepatitis B (📖 p.459), hepatitis C (📖 p.462), hepatitis D (📖 p.461), hepatitis E (📖 p.467). Novel candidate viruses include hepatitis GB virus C (HGBV-C), hepatitis G (HGV), transfusion transmissible virus (TTV).
- Other viruses – Epstein–Barr virus (📖 p.447), cytomegalovirus (📖 p.451), herpes simplex virus (📖 p.441), varicella zoster virus (📖 p.456), measles (📖 p.444), rubella (📖 p.430), adenovirus (📖 p.434), Coxsackie B, Entesvirus virus, and yellow fever virus (📖 p.501)
- Non-viral infectious diseases – spiochaetes (📖 p.386), coxiella (📖 p.398), legierella, M. enterineosis, ponicella, fancicella, yersinea, syphilis, (📖 p.752), leptospirosis (📖 p.393), Q fever (📖 p.398), sepsis (📖 p.624), legionellosis (📖 p.368), tuberculosis (📖 p.407), brucellosis (📖 p.359), tularaemia (📖 p.366), plague (📖 p.361)
- Drug-induced hepatitis may be caused by a variety of drugs. Common culprits include aspirin, paracetamol, isoniazid, rifampicin, pyrazinamide, phenytoin, and halothane
- Alcoholic hepatitis may mimic acute viral hepatitis
- Anoxic liver injury may be caused by hypotension, cardiac failure, or cardiopulmonary arrest
- Other liver diseases – Wilson's disease, Budd–Chiari syndrome

Clinical features

- There are no clinical features that distinguish the various causes.
- Acute viral hepatitis can be divided into four clinical stages: incubation period, pre-icteric phase, icteric phase, and convalescence.
- Clinical features may range from asymptomatic disease to anorexia, malaise, abdominal pain, and jaundice to fulminant hepatic failure.
- Hepatitis B and C may cause immune complex-mediated diseases, e.g. serum sickness, polyarteritis nodosa (HBV), glomerulonephritis, mixed cryoglobulinaemia.
- Fulminant viral hepatitis, characterized by liver failure and hepatic encephalopathy, occurs within 8 weeks after onset of symptoms.

Diagnosis

- Routine blood tests – AST and alanine aminotransferase (ALT) are usually dramatically elevated and bilirubin may be variably elevated. A prolonged prothrombin time (PT) is rare and suggests severe hepatic necrosis.
- Serology – anti-HAV IgM, HBsAg and anti-HBc IgM, anti-HCV should be performed initially. If these are negative, other diagnoses should be considered (see above).
- Liver ultrasound is usually normal in acute viral hepatitis. Abnormalities, e.g. hepatic lesions, cirrhosis, portal hypertension, or ascites suggest alternative diagnoses (see above).
- Liver biopsy may be performed to establish the diagnosis in acute hepatitis with negative serology.

Management

- Supportive care – most patients with acute viral hepatitis do not require hospitalization unless they are at risk of dehydration, have clinical evidence of liver failure, or a rising bilirubin or PT. Bed rest and alcohol avoidance are recommended while patients are symptomatic. Most medications should be avoided but symptomatic therapy for nausea or pain may be required. Vitamin K may be given if the prothrombin time is prolonged.
- Treatment – there is no specific treatment for acute viral hepatitis. Corticosteroids have been recommended for cholestatic hepatitis and fulminant hepatic failure, although clinical trials have failed to show benefit. Interferon-α has also been used in fulminant HBV but the evidence is poor. There is some evidence that treatment of acute HCV infection with interferon-β may prevent chronic infection.
- Monitoring – inpatients should be monitored regularly for signs of liver failure and with blood tests (bilirubin, AST, ALT, and PT). Hepatitis serology should be rechecked after 6 months to determine chronicity.
- Liver biopsy may be performed for various reasons, e.g. diagnostic uncertainity, if more than one cause is a possibility, or if specific treatment is being considered
- Liver transplantation is the only available treatment for fulminant hepatic failure, and patients should be promptly referred for consideration of transplantation.

Chronic hepatitis

This is a descriptive term used to denote ongoing inflammation (>6 months) of the liver.

Causes

- Chronic viral hepatitis: hepatitis B virus (HBV, 📖 p.459), hepatitis C virus (HCV, 📖 p.462), hepatitis D virus (HDV) hepatitis E issue (HEV, 📖 p.467).
- Autoimmune hepatitis
- Hereditary haemochromatosis
- Wilson's disease
- α_1-antitrypsin deficiency
- Fatty liver and non-alcoholic steatohepatitis (NASH)
- Alcoholic liver disease
- Drug-induced liver disease
- Hepatic granulomas – infectious, drug induced, neoplastic, idiopathic

Clinical features

There are no specific clinical features and most patients remain asymptomatic until they develop end-stage liver disease. Non-specific features (e.g. fatigue and right upper quadrant discomfort) are common. Symptoms such as jaundice, weight loss, abdominal distension, or confusion suggest decompensation. Examination may show signs of chronic liver disease (e.g. palmar erythema, Dupytren's contractures, jaundice, spider naevi, hepatosplenomegaly, caput medusa, ascites). Clinical features of hepatic encephalopathy include confusion, drowsiness, asterixis, ophthalmoplegia, and ataxia.

Diagnosis

- Routine blood tests – AST and ALT are usually dramatically elevated and bilirubin may be variably elevated. A prolonged PT suggests hepatic failure. A low albumin occurs in cirrhosis.
- Serology – HBsAg and anti-HCV should be performed for patients with suspected chronic viral hepatitis. If these are negative other diagnoses should be considered (see above).
- Liver ultrasound may show hepatomegaly or cirrhosis, portal hypertension, or ascites. Hepatic lesions may be due to hepatocellular carcinoma.
- Liver biopsy may be performed to establish the diagnosis in chronic hepatitis with negative serology, or to determine the degree of fibrosis in patients with suspected cirrhosis. In patients with deranged clotting this may be performed by the transjugular route.

Management

- Chronic HBV infection is usually treated with antiviral agents (see 📖 Antivirals for chronic viral hepatitis, p.136), e.g. interferon-α lamivudine, adefovir dipivoxil, and entecavir. Tenofovir (TDF, also has anti-HBV activity and should be used to treat HIV/HBV co-infected patients. Liver transplantation may be performed for patients with end-stage liver disease. However, the risk of re-infection is 20% even with prophylaxis (lamivudine and polyclonal anti-hepatitis B immunoglobulin, HBIG).

- Chronic HCV infection is also be treated with antiviral therapy, e.g. pegylated interferon-α and ribavirin (📖 Antivirals for chronic viral hepatitis, p.136). End-stage liver disease secondary to chronic HCV is the leading indication for hepatic transplantation. HCV recurrence occurs in >90% by 1 year after transplantation. The best strategy to prevent or treat re-infection has not yet been established. At present combination therapy with interferon-α and ribavirin may be given in the following situations: before transplantation to prevent complications, e.g. hepatoma; prophylactically just before transplantation to prevent recurrence; pre-emptively just after transplantation; when HCV recurs after transplantation.

Other gastrointestinal infections

Mesenteric adenitis

- Inflammation of the mesenteric lymph nodes – may be acute or chronic depending on the infecting agent. Organisms are thought to pass through intestinal lymphatics to the lymph nodes where they produce inflammation and sometimes suppuration. Most common in children <15 years of age.
- Causes – These include *Yersinia* species, *Staphylococcus* spp., *E. coli, Streptococcus* spp., *M. tuberculosis, Giardia lamblia*, non-typhoidal salmonellae and viruses (e.g. Coxsackie and adenovirus).
- Clinical features – fever, abdominal pain, and tenderness. It may be difficult to distinguish clinically from appendicitis.
- Diagnosis – CT scan may demonstrate enlarged mesenteric lymph nodes. The key feature in diagnosis is to recognize appendicitis and other problems requiring surgical intervention; <20% of appendectomies may reveal evidence of non-specific mesenteric adenitis. Blood cultures may be positive in those bacterial cases that progress to sepsis. Serological tests may demonstrate evidence of *Y. enterocolitica* infection.
- Treatment – patients with mild symptoms need only supportive care. Ill patients with more-obvious evidence of infection require antibiotic treatment.
- Complications – abscess formation, sepsis, peritonitis.

Typhlitis

- Inflammation of the caecum. May occur in patients with HIV or severe neutropenia. It is thought that bacteria from the lumen invade ulcerations in the bowel wall during periods of neutropenia, proliferate, and produce exotoxins causing damage to the gut wall.
- Clinical features – resembles acute appendicitis with fever, pain, and rebound tenderness in the right iliac fossa. Rapid progression to an acute abdomen may occur.
- Treatment – broad-spectrum IV antibiotic therapy (aerobic and anaerobic cover) with surgical resection of necrotic bowel is recommended as the mortality rate of severe cases of neutropenic enterocolitis is >50%.

Tropical sprue/enteropathy

- A syndrome of acute or chronic diarrhoea, weight loss, and malabsorption of at least two nutrients, which is believed to follow an intestinal microbial infection that causes enterocyte injury and bacterial overgrowth. Villous destruction and demonstrable nutrient malabsorption occur in varying degrees. It has been described in tropical climates throughout the world but primarily southeast Asia and the Caribbean.
- Clinical features – symptoms develop over months after several years resident in an affected area. It presents with weight loss, fatigue, and the features of the loss of specific nutrients: most commonly folate, vitamin B_{12}, and iron.

- Diagnosis – there are no specific tests. Laboratory studies reveal the features of the specific nutritional deficiencies (e.g. megaloblastic anaemia). Stool studies may demonstrate fat malabsorption, and small bowel biopsy may show villous atrophy. The diagnosis is one of exclusion.
- Treatment – management is by nutritional support and antibiotic therapy (e.g. tetracycline or metronidazole for 6–12 months).

Whipple's disease

- A rare chronic and systemic infectious disorder caused by a Gram-positive bacterium, *Tropheryma whippelii*. Clinical manifestations probably follow a disordered host response to the organism's infiltration of various body tissues. The organism is taken up into tissue macrophages which may be seen with periodic acid–Schiff (PAS) staining.
- Predominantly affects middle-aged Caucasian men. Patients with HIV infection do not develop the disease, and one small study found *T. whippelii* DNA in 35% of healthy volunteers.
- Clinical features – presents with arthritis, fever, diarrhoea, abdominal pain, and weight loss (90%). Malabsorption follows disruption of the villous architecture. Cardiovascular (endocarditis), respiratory, and central nervous system (supranuclear ophthalmoplegia, cerebellar ataxia, disinhibition, meningoencephalitis) involvement may occur.
- Diagnosis – biopsy of affected organs (small bowel, synovium, brain, endocardium) to demonstrate the typical histopathology (not pathognomic and may be seen with mycobacterium avium intracellulare (MAI), cryptococcosis, or other parasitic infections usually observed in patients who are immunosuppressed with HIV disease) and DNA testing for *T whippelii*. Almost universally fatal within 12 months if untreated.
- Treatment – prolonged course of antibiotics, e.g. 2 weeks of IV ceftriaxone followed by 1 year of co-trimoxazole. PCR may be the best way to demonstrate remission – therapy should be continued if patients remain positive, perhaps with an alternative regime. Malnourished patients will need nutritional support and vitamin supplementation.

Urinary tract infections: introduction

Definitions

- The term 'urinary tract infection' (UTI) covers the whole spectrum of infection from asymptomatic bacteriuria to severe pyelonephritis.
- Uncomplicated UTI is considered to be infection of a structurally and functionally normal urinary tract, e.g. acute cystitis in women.
- Complicated UTI is infection of a urinary tract that is abnormal, either structurally (obstruction, stents, medullary scars), functionally (vesicoureteral reflux, incomplete bladder emptying), or neurologically.
- All UTIs in men, pregnant women, and children are considered complicated. Any patient with a complicated UTI should be referred to a specialist for assessment and follow-up.

Epidemiology

- Asymptomatic bacteriuria occurs in all age groups and does not necessarily result in clinical infection.
- Infants – incidence of UTI 1–2%, more common in males.
- Children – asymptomatic bacteriuria and UTI more common in girls. UTI is rare in boys and suggests a structural abnormality.
- Women – asymptomatic bacteriuria occurs in 1–3% of non-pregnant women and 2–9.5% of pregnant women. 10–20% of women experience symptomatic UTI during their life. Risk factors: frequent sexual intercourse, diaphragm use, spermicide use, lack of urination after intercourse. and history of recurrent infections; 20–30% of pregnant women with untreated asymptomatic bacteriuria go onto develop acute pyelonephritis.
- Men – asymptomatic bacteriuria <0.1%. Circumcision is associated with a decreased risk of UTI.
- Elderly – 10% of men and 20% of women aged >65 years have bacteriuria. Risk factors: prostatic disease, poor bladder emptying, perineal soiling, and urinary tract instrumentation.
- Hospitalized patients – high rates of bacteriuria; 10% of catheterized patients develop UTI. Other risk factors: female diabetics, pregnant black women with sickle cell trait, those with interstitial renal disease, renal transplant recipients.
- Renal transplant patients – asymptomatic bacteriuria is associated with a high incidence of pyelonephritis and the risk of graft loss.

Aetiology

- 70–95% of acute community infections are due to *E. coli*.
- Recurrent UTIs or nosocomial infections are associated with *Proteus*, *Pseudomonas*, *Klebsiella*, *Enterobacter*, enterococci and staphylococci.
- Polymicrobial infections are common in those with structural abnormalities, as are antibiotic-resistant organisms (secondary to antibiotic exposure and instrumentation).
- Fungi (particularly *Candida*) occur in patients with indwelling catheters receiving antimicrobial therapy.
- Coagulase-negative staphylococci (CoNS) are a common cause among young sexually active women.

- *S. aureus* infection is usually haematogenous; associated with renal/perinephric abscesses.
- Adenoviruses (especially type 11) are implicated in acute haemorrhagic cystitis in children and allogenic bone marrow transplant recipients.

Pathogenesis

- Host factors – the urinary tract is normally sterile and fairly resistant to bacterial colonization (except the urethra). Host antibacterial mechanisms include urinary flow and pH, bactericidal cytokines, inhibitors of bacterial adherence, local immune responses, and the inhibitory effect of prostatic fluid secretions. Individuals who are particularly susceptible to UTIs may have defects in these mechanisms. Abnormalities of the urinary tract may also undermine the effectiveness of the host response.
- Organism factors – infection may be caused by many bacterial species, each of which has its own virulence factors, e.g. *E. coli* has P fimbriae which facilitate adherence to the uroepithelium. Bacteria enter the urethra tract by (a) ascending the urethra, or (b) haematogenously. The urethra is normally colonized with bacteria which may be forced into the bladder by sexual intercourse or catheterization. Women have shorter urethras and higher rates of UTI. Haematogenous spread is less common but occurs in the context of staphylococcal bacteraemia/endocarditis or candidaemia.

Diagnosis

- Specimen collection – urine may be collected by midstream clean catch (to reduce the number of urethral organisms collected), by catherization, or by suprapubic aspiration of the bladder.
- Urine dipstick analysis – pyuria (10–50 white cells/mm^3 urine) is nonspecific and does not necessarily indicate infection, but most patients with UTI have pyuria. The leucocyte esterase test is sensitive (75–96%) and specific (94–98%) for detecting >10 white cells/mm^3 urine. Haematuria may be seen in certain infections, but calculi, tumour, vasculitis, glomerulonephritis, and renal TB should be considered. Proteinuria is common in UTI and should be <2 g/24 h. The dipstick nitrite test detects the products of bacterial nitrate reduction. It may be falsely negative in the presence of diuretic use, low dietary nitrate, or organisms that do not produce nitrate reductase (eg, *Enterococcus*, *Pseudomonas*, and *Staphylococcus*). The combined sensitivity/specificity of dipstick leucocyte/nitrite testing is 79.2% and 81% respectively.
- Urine culture – patients with UTI usually have $\geq 10^5$ organisms/mL urine in properly collected specimens. Patients without infection will have counts of $<10^4$ organisms/mL urine. However symptomatic infections can result with counts of 10^4–10^5 organisms/mL urine. These criteria apply to Gram-negative organisms. Infection caused by Gram-positive organisms, fastidious bacteria, and fungi rarely reach over 10^4 bacteria/mL.
- Blood cultures – should be taken if systemic infection is a possibility.
- Urological assessment – exclude anatomical abnormlaities, stones, tumours in patients with recurrent UTIs, complicated UTIs, or UTIs in men, pregnant women, infants and children.

Cystitis

A superficial mucosal infection confined to the lower urinary tract and characterized by dysuria, frequency, and urgency. These symptoms may also be related to urethritis or inflammation without infection. Non-bacterial causes of cystits include infectious agents (viral, mycobacterial, chlamydial, and fungal species), and non-infectious precipitants (radiation, chemical, autoimmune, hypersensitivity, and interstitial cystitis). Consider these non-bacterial causes in cases of cystitis that are culture negative and fail to respond to antibiotic therapy.

Clinical features

- These include dysuria, urgency, hesitancy, polyuria, incomplete voiding, urinary incontinence, haematuria, and suprapubic or low back pain.
- Elderly patients may present with confusion and no localizing features.
- Constitutional symptoms such as fever are mild or absent.

Management[1]

- General measures – these include hydration, management of diabetes, investigation, and management of obstruction or structural abnormalities.
- Antibiotic therapy – all symptomatic infections should be treated, and empirical regimes based on local resistance data. The resolution of bacteriuria is related to the concentration of the antimicrobial agent achieved in the urine – dosage modifications are necessary in patients with renal insufficiency for agents excreted primarily by the kidney:
  - uncomplicated lower urinary tract infections in women: 3-day treatment courses are as effective as 7-day schedules and have as few side-effects as the less-effective 1-day regimens. e.g. nitrofurantoin 60 mg qds
  - lower urinary tract infection in other groups – short courses have not been evaluated in men, and are not suitable for children or women with a history of previous urinary infection caused by antibiotic-resistant organisms. These groups should receive 7–10 days of treatment.
  - complicated infection – 7–14 days treatment.

Recurrent infection

May follow relapse (bacteriuria with the same organism that was present when treatment was started) or re-infection (bacteriuria with a different organism from that before treatment). Re-infection with the same organism may occur if it has persisted in nearby areas, e.g. vagina.

- Relapse – consider renal involvement (necessitating a longer course of therapy 2–6 weeks), a structural abnormality (e.g. calculi, obstruction – consider urological investigation) or chronic prostatitis (📖 p.728). Certain patients experiencing repeated relapses in whom surgical correction is not indicated or is not feasible may be appropriate for long-term antibiotic therapy. Such patients should have regular urine cultures (looking for antibiotic resistance), assessment of renal function, and renal imaging. There is no consensus on how long prophylaxis should last. Rates of infection return to pre-treatment levels once therapy is stopped. Cranberry juice does reduce the frequency of episodes compared to placebo.

- Re-infection – certain patients experience repeated re-infections (with successful clearance following appropriate therapy between each episode). Those cases related to sexual intercourse may benefit from post-coital voiding. One RCT demonstrated reduced recurrence rates on taking a single dose of antibiotic (co-trimoxazole) up to 2 h after intercourse.[2] Where no associated precipitating event can be identified, long-term chemoprophylaxis may be appropriate, particularly in children who may be at risk of renal damage.

Prognosis

- Children – those without obstruction (e.g. urethral valves) or vesico-ureteric reflux (VUR) have a good prognosis. Obstruction can lead to severe destruction of the renal parenchyma. VUR is seen in 30–50% of children with bacteriuria and can lead to renal scarring. Infants and preschool children are at the greatest risk. Severe reflux may lead to repeated infection and renal impairment. Reflux alone, particularly intrarenal reflux, may be capable of causing renal scarring even in the absence of infection. Infection exacerbates reflux which reduces with the elimination of bacteriuria.
- Adults – once a woman has had a UTI she is more likely to go on have further episodes. There may be a role for prophylactic antibiotic therapy in women with recurrent uncomplicated UTIs.

Urethral syndrome

Seen in women with acute onset of urinary symptoms (dysuria etc) but with $<10^5$ bacteria/mL. Studies have shown that the majority of these patients have genuine infection with a low number of organisms confined to the lower urinary tract. Others may represent patients with sterile pyuria and urethritis secondary to infection with *Chlamydia trachomatis* or *N. gonorrhoeae*. However, some patients with the syndrome have no pyuria and persistently sterile cultures – the cause of their symptoms is not clear but vaginitis and genital herpes should be excluded.

References

1 Guidelines for Antimicrobial Treatment of Uncomplicated Acute Bacterial Cystitis and Acute Pyelonephritis in Women. *Clin Inf Dis* 1999 29:745–58.
2 www.clinicalevidence.com (accessed 12 August 2008).

Acute pyelonephritis

Infection and inflammation of the renal parenchyma characterized by flank pain and fever with or without dysuria, frequency and urgency. Potentially organ-threatening, and each episode of infection may scar the kidney, impairing renal function. Complications: renal failure, abscess formation (nephric, perinephric), sepsis, or shock and multi-organ failure.

Clinical features

Symptoms develop over hours to (rarely) days and vary greatly in severity. Symptoms of lower UTI may be present: dysuria, frequency, hesitancy, lower abdominal pain, urgency, haematuria (haemorrhagic cystitis), suprapubic pain. Symptoms of pyelonephritis: flank pain, back pain, fever, rigors, chills, weakness, anorexia. Signs: fever, tachycardia, hypotensive if shock, suprapubic tenderness, flank tenderness.

Diagnosis

Usually easily diagnosed in women, but may be less obvious in men, the elderly, and the hospitalized in whom infection may develop insidiously. See general points on diagnosis of UTI (📖 p.716).

- Urinalysis – macroscopic haematuria is unusual in pyelonephritis, and is seen more commonly in lower UTI (haemorrhagic cystitis). Consider calculi, cancer, glomerulonephritis, tuberculosis, trauma, vasculitis.
- Urine culture – all patients with presumed pyelonephritis should be tested because of the possibility of antibiotic resistance.
- Blood cultures – indicated in the hospitalized. Up to 20% are positive.
- Imaging – rarely indicated in the diagnosis of those presenting with typical signs. It is useful in those with atypical features, and those who deteriorate or do not respond to therapy (e.g. fever and positive blood cultures more than 48 h after therapy initiated, complicated UTIs) for the purposes of identifying organ- or life-threatening complications. Contrast-enhanced spiral CT is the study of choice (more sensitive than USS). Dimercaptosuccinic acid (DMSA) scanning is used in children to lessen radiation exposure. Other patients in whom early imaging may be useful are those at increased risk of complications: AIDS patients and the immunosuppressed, those with poorly controlled diabetes, sepsis, shock.

Management[1]

- General – rehydration, antipyretics, analgesics.
- Antibiotic therapy – start empirically as guided by local resistance patterns while awaiting culture and sensitivity testing. Most community-acquired cases are due to *E. coli* or other enterobacteriaceae. Appropriate therapy would include IV co-amoxiclav or ceftriaxone ± a single dose of gentamicin. Treatment duration is usually 14 days; 7 days' ciprofloxacin is sufficient in healthy young women with uncomplicated disease. Consider the addition of vancomycin in those at risk of MRSA infection (e.g. recent instrumentation, previous MRSA isolation in urine), and the hospitalized or those from residential institutions may require enterococcal coverage.
- Surgery may be required in some patients with predisposing conditions who fail to respond to therapy, and in those developing certain

complications, e.g. renal cortical abscess, corticomedullary abscess, emphysematous pyelonephritis.
- Patients with complicated infection: obtain follow-up urine cultures, consider for follow-up urinary tract imaging, and refer for specialist management. Such cases would include the first episode in a child, those presenting with renal impairment, the pregnant, patients with unusual organisms, and the immunocompromised.

Prevention

- Cases with an obvious precipitant – practice changes (e.g. different means of contraception, administration of prophylactic antibiotics, early identification and treatment of lower UTIs) may prevent them. If they do not, consider an underlying structural abnormality.
- Long-term catheter-related infections – ensure a closed system, consider intermittent or suprapubic catheterization.
- Renal transplant recipients – antibiotic prophylaxis may be given to some groups of patients in the first 6–12 months after transplantation.

Prognosis

- Acute renal failure (ARF) is rare in children, healthy adults, and pregnant women outside the context of hypovolaemia, obstruction, or sepsis. ARF may follow papillary necrosis (sloughing necrotic renal papillae cause ureteric obstruction), which may be seen in those with diabetes mellitus, sickle cell disease, or urinary tract obstruction.
- Renal scarring:
  - children – seen in 6–15% of children after a febrile UTI. This is often associated with a degree of VUR (thought to be congenital), and patients with scarring are at risk of hypertension and renal impairment in later life. This risk increases with delayed treatment in those experiencing recurrent infection
  - adults – a single episode of acute pyelonephritis in an adult woman leads to renal scarring in 46%, as demonstrated by Tc99m-labelled DMSA scanning 10 years later. Acute pyelonephritis in pregnancy may lead to acute renal impairment, ARF, ARDS, low birth weight children, preterm delivery, and sepsis. Renal scarring is four times more likely after pyelonephritis in pregnant women than in non-pregnant women. Renal impairment is seen particularly in infections causing severe papillary necrosis.
- Pyelonephritis becomes potentially fatal when secondary conditions develop, such as emphysematous pyelonephritis (20–80% mortality rate), perinephric abscess (20–50% mortality rate), or sepsis. Severe sepsis mortality is higher in those with chronic renal disease, acute renal impairment, and in the elderly.
- Acute renal transplant pyelonephritis occurring in the first 3 months after transplant has a significant association with graft loss (>40%) by 96 months as compared to all renal transplant cases with or without the occurrence of pyelonephritis at any time after the transplant up to 96 months (25–30%).

Reference

1 Warren JW, Abrutyn E, Hebel JR, et al. Guidelines for Antimicrobial Treatment of Uncomplicated Acute Bacterial Cystitis and Acute Pyelonephritis in Women. *Clin Inf Dis* 1999; 29:745–58.

Chronic pyelonephritis

Not so well defined as acute pyelonephritis but generally considered to refer to the pathological changes of diffuse interstitial inflammation which can be caused by several conditions including obstruction, calculi, analgesic nephropathy, hypokalaemic nephropathy, renal TB and following acute pyelonephritis in childhood in the context of VUR.

Chronic pyelonephritis secondary to VUR

VUR is congenital incompetence of the ureterovesical valve due to an abnormally short intramural segment of the ureter. The condition is present in 30–40% of young children with symptomatic UTIs, and in almost all children with renal scars. VUR may also be acquired by patients with a flaccid bladder due to spinal cord injury. This may lead to impaired renal function (reflux nephropathy). Sometimes this diagnosis is established based on radiological evidence obtained during an evaluation for recurrent urinary tract infection (UTI) in young children. Infection without reflux is less likely to produce injury.

- **Symptoms** – patients with chronic pyelonephritis present with fever, lethargy, nausea and vomiting, flank pain, dysuria, and children may fail to thrive. Hypertension may be noted.
- **Investigations** – proteinuria (negative prognostic feature), urine cultures (negative cultures do not exclude the diagnosis), demonstration of renal stones/dilatation (intravenous urogram, renal USS), reflux (voiding cystourethrogram, cystoscopy), and renal scarring (radioisotopic scanning with technetium DMSA).
- **Management** – infection should be treated and underlying structural abnormalities corrected (e.g. ureteric re-implantation).
- **Complications** – proteinuria, focal glomerulosclerosis, renal impairment secondary to scarring (rate of progress of scars can be slowed by speedy institution of appropriate antibiotic therapy), pyonephrosis (if obstructed), nephrosis (may occur in cases of obstruction), hypertension (increases rate of decline in renal function), xanthogranulomatous pyelonephritis.

Emphysematous pyelonephritis

- A severe, necrotizing acute multifocal bacterial nephritis with extension of the infection through the renal capsule. Gas is found in the renal substance and perinephric space.
- 85–100% cases are seen in patients with diabetes.
- Most cases due to enterobacteriaceae.
- Patients present with fever, chills, pain, flank mass (50%), crepitation (over thigh or flank), and urinary symptoms.
- Diagnosis is best confirmed by CT.
- Treatment – antibiotics, drainage, nephrectomy.
- Mortality – 60% in those with gas within the kidney alone and managed with antibiotics and drainage, 80% in those with gas extending to the perinephric space and managed by antibiotic therapy alone, 20% in those managed by nephrectomy.

Xanthogranulomatous pyelonephritis

- A rare serious, debilitating illness characterized by a chronic inflammatory mass originating in the renal parenchyma. Gross appearance: mass of yellow tissue composed of lipid-laden macrophages and inflammatory cells (perhaps with an abscess cavity), regional necrosis, and haemorrhage.
- Causes – often associated with infection by *Proteus*, *E. coli*, or *Pseudomonas* spp. in the context of chronic obstruction (stones are seen in 75% of patients, e.g. staghorn calculus).
- Patients are often immunocompromised or diabetic and it is four times commoner in women than men.
- Clinical features – patients appear chronically ill, with dull persistent flank pain, fever, weight loss, fistulae (pyelocutaneous and ureterocutaneous fistulae have been described). Can present acutely with fever and flank pain. Renal function is reduced in almost all cases.
- Diagnosis – CT scan helps diagnosis. Resembles a neoplastic lesion in its radiographic appearance and tendency to involve adjacent structures including the psoas muscle and peri-renal space. Renal pelvis is dilated. Many are confirmed only at operation. Bacteria are not typically cultured from urine, but if culture is positive the commonest organisms are *Proteus mirabilis*, *E. coli*, and *Pseudomonas* spp.
- Treatment – appropriate antibiotic therapy may be important in initial stabilization but definitive therapy is always surgical, usually nephrectomy. Other factors complicating response to therapy: obstructing calculus, renal papillary necrosis.

Renal abscess

Perinephric abscess

- Follows chronic or recurrent pyelonephritis, rupture, or extension of suppuration within the kidney, or dissemination/direct extension from another site. Located between the renal capsule and surrounding fascia and may extend to involve the GI tract, groin, lung (pleuritic pain, raised hemidiaphragm, pleural effusion), and psoas (may be signs of psoas irritation, e.g. scoliosis, pain on hip flexion).
- Clinical features – presentation is insidious with fever, chills, unilateral flank pain (70%), dysuria (40%), nausea, vomiting, weight loss (25%), flank tenderness, abdominal tenderness (60%), referred pain (i.e. hip, thigh, or knee), flank or abdominal mass (<50%), pyuria (70%), sterile urine (40%), and bacteraemia (40%).
- Diagnosis is often not apparent – one-third of patients are diagnosed at autopsy. CT helps confirm the diagnosis. USS may be falsely negative. MRI defines extension.
- Treatment – drainage is always required. Specific agents providing pseudomonal or enterococcal coverage may be indicated. Other organisms that have been reported include tuberculosis and fungi. Nephrectomy may be necessary.
- Mortality <50%, less if recognized early and managed appropriately (e.g. surgery and aminoglycoside with anti-staphylococcal agent).

Renal corticomedullary abscess

- Usually associated with urinary tract abnormalities and commonly caused by the *Enterobacteriaceae*. Disease is part of a spectrum that ranges from acute focal bacterial nephritis affecting a single lobe, to severe emphysematous pyelonephritis. Males and females are equally affected in most cases.
- Clinical features – fever, chills, flank pain, nausea, vomiting (usually absent in cortical abscesses), flank mass, hepatomegaly. Urinary symptoms may be absent (but seen more frequently than with cortical abscesses), and urinalysis is normal in 30%.
- Diagnosis is best confirmed by CT. Microbiology: midstream urine (MSU), blood cultures, culture of pus obtained by CT/USS-guided aspiration or drainage.
- Treatment is by antibiotic therapy (e.g. ciprofloxacin and gentamicin) and drainage or surgical intervention. Structural abnormalities should be corrected, e.g. obstruction relieved.

Renal cortical abscess

- Uncommon and usually due to the haematogenous spread of *S. aureus*, most commonly from a skin infection. Risk factors include IVDU, diabetes mellitus, and haemodialysis. Microabscesses forming in the cortex coalesce to form a circumscribed abscess over days to months. Commoner in men than women.
- Clinical features – the onset is often insidious. Symptoms include fever, chills, back pain, abdominal pain, flank mass, and rarely urinary symptoms (if the abscess communicates with and involves the collecting system).

- Diagnosis best confirmed by CT and this or USS may be used to guide aspiration or drainage.
- Microbiology – MSU (culture is usually normal), blood cultures (often negative), culture of aspirated pus.
- Treatment – IV antibiotics (e.g. high-dose flucloxacillin for 4 weeks) and drainage (for all but the smallest abscesses) which may be successfully achieved percutaeously. Nephrectomy rarely required.

Catheter-associated UTI

Urinary tract infection is the most common nosocomial infection, accounting for 40% of hospital-acquired infections, 80% of which occur in association with the use of catheters and related devices.

Pathogenesis of UTI

- Catheterization thwarts a number of the defence mechanisms that reduce the incidence of UTI in healthy individuals.
- Organisms may be introduced from the perineum or urethra at the time of insertion, contaminate the collecting device (greatly reduced following the introduction of closed drainage systems in the 1950s), or enter via the space between the catheter and the urethral mucosa.
- Once in the urinary tract, organisms are not eliminated as efficiently as is usual and can reach large numbers within a couple of days. Some are capable of producing biofilms which facilitate their growth. An inflammatory response may result in cystitis and pyuria. Organisms may ascend and cause upper urinary tract infection.
- Risk factors associated with catheter-associated bacteriuria include the duration of catheterization, the absence of a drip chamber, microbial colonization of the collection system, diabetes mellitus, quality of catheter care, and being female.

Short-term catheterization

Up to 25% of patients have a catheter in at some point during a hospital stay, most for less than 4 days; 10–30% of catheterized patients develop bacteriuria. Common organisms: *E. coli* (24%), *Candida* spp. (26%), *Pseudomonas aeruginosa*, *Klebsiella pneumoniae*, *Proteus mirabilis*, enterococci, and CoNS. Most bacteriuric episodes in this group are caused by a single organism. Organisms isolated from the catheter itself may not be found in the urine (due to sequestration of the bacteria within catheter-related biofilm). Most episodes of bacteriuria are asymptomatic but 5% of such patients develop fever. Bacteraemia is uncommon but occurs at higher rates in bacteriuric patients undergoing instrumentation (e.g. prostatectomy).

Long-term catheterization

The two most-frequent indications are urinary incontinence (women) and outflow obstruction (men). Such patients may be catheterized for months to years. All develop bacteriuria at some point, and certain species possess adhesins that enable them to persist in the catheterized urinary tract. Polymicrobial bacteriuria is seen in 95% of long-term catheterized patients. Mildly symptomatic UTIs occur fairly regularly, most lasting only a day and resolving without treatment. Bacteraemia occurs in 4–10% of institutionalized patients undergoing catheter removals or replacements, often following the development of acute pyelonephritis.

Other complications of long-term catheterization: symptomatic UTI, catheter obstruction (by bacteria, crystals – particularly in association with *Proteus* infection, protein, glycocalyx), urinary stones, chronic renal inflammation, peri-urinary infection, bladder metaplasia, malignancy (in very long-term patients). Some of the complications of long-term catheterization once seen in spinal-injury patients are now seen much less frequently, as such individuals manage themselves with intermittent catheterization.

Prevention

- Patients should be catheterized for clear indications only. Incontinence in particular may be more appropriately managed by other means.
- When urethral catheterization cannot be avoided, carers should be meticulous in maintaining a closed collection system and the catheter should be used for as short a period as possible.
- Alternatives to indwelling urethral catheterization: conveens (lower incidence of bacteriuria but have infection risks and other complications of their own), intermittent catheterization, suprapubic catheterization (cleaner skin region is associated with lower rate of infection).
- A single dose of gentamicin at insertion may reduce infection.

Treatment

- Asymptomatic bacteriuria – no evidence that treating catheterized patients with bacteriuria in the absence of symptoms significantly reduces the number of people who go on to develop symptoms. Long-term catheterized patients treated with antibiotics for bacteriuria regardless of symptoms showed no difference in the number of febrile episodes. Certain situations do warrant treatment: identification of organisms with a high incidence of bacteraemia (e.g. *Serratia marcescens*), control of organisms associated with an outbreak of UTI in an institution, and bacteriuria in those patients at high risk of serious complications (e.g. pregnant women, immunosuppresed patients), or patients undergoing urological surgery.
- Symptomatic catheter-associated UTI – cultures of blood and urine should be taken and most patients treated with empirical parenteral antibiotics based on the locally known organisms and previous infections the patient might have experienced. These can be modified once culture results are available, and given orally once afebrile. Bacteria may persist in catheter biofilm and it is sensible to replace or remove the catheter prior to treatment. Antiseptic/antibiotic irrigation is of no benefit. Treatment duration is usually 7–10 days. Rule out obstruction and renal stones.
- Candiduria – seen in many catheterized patients, and particularly related to hospitalization and previous antibiotic exposure. It is usually asymptomatic. Catheter removal resolves it in 40%, changing the catheter resolves it in 20%. Patients who must remain catheterized and continue to have candiduria may benefit from a course of fluconazole if they have a non-krusei candidal cystitis. Systemic therapy with IV amphotericin B or fluconazole and possibly surgery may be indicated. Complications: fever, renal/perirenal abscess, fungus balls in the bladder and renal pelvis, dissemination.

Prostatitis

Up to 50% of men will experience symptoms of prostatitis at sometime in their lives. Key clinical issue: to distinguish patients with bacteriuria and bacterial prostatitis from the larger number of patients without bacteriuria.

Acute bacterial prostatitis

- Clinical features – symptoms are those of a lower UTI (dysuria, frequency) and possibly obstruction (due to prostatic oedema), and fever. On examination: lower abdominal/suprapubic discomfort, extremely tender, firm prostate on PR (per rectum).
- Investigations – urinalysis shows pyuria and cultures are positive; blood cultures may be positive either spontaneously or following vigorous PR.
- Management – response to antimicrobial therapy is usually rapid; agents should provide good coverage of pseudomonads, enterococci and the enterobacteriaceae. Urinary retention is best managed by suprapubic catheterization to avoid obstructing the drainage of prostatic secretions.
- Complications – prostatic abscess, prostatic infarction, chronic prostatitis.

Chronic bacterial prostatitis

- An important cause of bacterial persistence in the lower urinary tract. Gram-negative organisms are the most important causes.
- Patients often experience repeated infections with the same organism, and are asymptomatic between episodes with a normal prostate on examination.
- Long treatment courses fail in one-third, cure one-third, and bring about resolution while on treatment with subsequent relapse in one-third. These poor results may be a consequence of poor drug penetration into the prostatic parenchyma, or perhaps infected calculi serving as persistent foci for infection. Those not cured may remain asymptomatic on long-term low-dose suppressive antibiotic therapy despite the persistence of prostatic bacteria.

Chronic prostatitis/chronic pelvic pain syndrome

- The largest subset of patients with symptoms of prostatitis. There is no history of bacteriuria or evidence of infection.
- Symptoms – difficulty voiding, erectile dysfunction, and a dull aching pain which may be pelvic, perineal, suprapubic, scrotal, or inguinal and is exacerbated by ejaculation. Examination is unremarkable.
- Some patients may have leucocytes in semen or prostatic secretions (expressed by digital massage), whereas others have no evidence of inflammation. Although the reason for this is not known, patients with leucocytes are more likely to have bacteria in their prostatic parenchyma and it is suggested that those without are experiencing a non-infectious disease.

Asymptomatic inflammatory prostatitis

Prostate inflammation with no symptoms. Such patients may be identified in working-up the cause of a raised prostate-specific antigen (PSA) with prostate biopsy showing a simple inflammatory process.

Granulomatous prostatitis

A histological reaction that may follow acute bacterial prostatitis, tuberculous prostatitis (and that following BCG (bacillus Calmette–Guérin) therapy for transitional cell carcinoma of the bladder), systemic mycoses. It may cause an indurated, firm, or nodular prostate, clinically indistinguishable from that caused by malignancy.

Prostatic abscess

- A rare complication of acute bacterial prostatitis. Patients most commonly affected: those with urinary tract obstruction or foreign bodies, those with diabes, the immunocompromised, and those not adequately treated for their acute episode.
- Most cases are caused by the common uropathogens acquired by the ascending route and, rarely, organisms such as *S. aureus*.
- Symptoms resemble those of acute bacterial prostatitis: fever, dysuria and signs of urinary sepsis. A fluctuant area of the prostate may be apparent on PR.
- Definitive diagnosis can be made by US, CT, or MR of the pelvis.
- Treatment – drainage and appropriate antibiotics.

Epididymitis

An inflammatory reaction of the epididymis to infection or trauma. There are two distinct patterns of infective epididymitis: sexually transmitted and non-specific (non-sexually transmitted) bacterial epididymitis. Underlying genitourinary tract abnormalities are common only in the latter group.

General features

- Symptoms – painful swelling of the scrotum which may be acute (over 1–2 days) or more gradual in onset, dysuria with or without urethral discharge. Fever may be present, particularly in hospitalized patients who develop the condition following urinary tract manipulation.
- Examination – tender swelling and erythema of the scrotum, usually unilateral. Early in disease, swelling may be localized to one portion of the epididymis. Consequent involvement of the associated testis is common, producing epididymo-orchitis. Secretion of inflammatory fluid can lead to the development of a hydrocoele.

Non-specific bacterial epididymitis

- The most common pathogens in men >35 years are coliforms and *Pseudomonas* spp. Other infectious agents: *M. tuberculosis* (tuberculous epididymitis is the commonest form of male genital TB), systemic mycoses (e.g. blastomycetes).
- Patients often have underlying urinary tract pathology or a history of recent genitourinary tract manipulation (cases may occur weeks or months after the intervention), particularly if bacteriuric at the time. Bacterial prostatitis or long-term urethral catheters are other important predisposing factors.
- Complications – testicular infarction, scrotal abscess, pyocoele, scrotal sinus, infertility, chronic epididymitis.
- Management – empirical antibiotics aimed at covering Gram-negative rods and Gram-positive cocci while awaiting urinary cultures. Bed rest, scrotal elevation, analgesics. Some complications may require surgical intervention.

Sexually transmitted epididymitis

- The commonest form in young men
- Major pathogens – *C. trachomatis*, *N. gonorrheae*
- Many patients do not complain of discharge. *Chlamydia* spp. may be carried for prolonged periods (≥1 month) before developing symptoms
- Diagnosis requires high index of suspicion and appropriate cultures. Patient should be evaluated for the presence of other sexually transmitted infections (STIs), and sexual partners followed up. Underlying genitourinary abnormalities are uncommon in this group
- Treatment – specific therapy covering both chlamydial and gonococcal infections. If symptoms do not subside within 3 days of therapy, or tenderness/swelling persists on completion, the diagnosis should be reviewed. Consider abscess, infarction, malignancy, tuberculous/fungal epididymitis
- Complications – abscess, testicular infarction, infertility, chronic epididymitis

Orchitis

Less common than epididymitis or prostatitis. Blood-borne dissemination is the major route of infection.

Viral orchitis

Viruses are by far the commonest cause: e.g. mumps, Coxsackie B virus. Mumps rarely causes orchitis in pre-pubescent males but is seen in 20% of post-pubertal patients. Testicular pain and swelling follows 4–6 days after parotitis, and may be seen even the absence of parotitis; 70% of cases are unilateral. Contralateral involvement may occur a few days after the first testicle. Symptoms range from mild discomfort to severe pain with nausea, vomiting, prostration, fever, and constitutional symptoms. Mild cases resolve within 4–5 days, severe ones may take 3–4 weeks; 50% of patients experience some degree of testicular atrophy. Contrary to previous belief, mumps orchitis rarely results in infertility.

Bacterial orchitis

Isolated bacterial orchitis is extremely rare. It usually follows from contiguous spread from an infected epididymis. Most cases of pyogenic orchitis are caused by: *E. coli*, *K. pneumoniae*, *P. aeruginosa*, staphylococci, and streptococci. Patients are acutely ill with a high fever, marked discomfort, testicular swelling, and nausea and vomiting. Pain radiates to the inguinal canal. There is usually an acute hydrocoele, and the testis is swollen and tender. Overlying skin may be erythematous and oedematous. Treatment is as for bacterial epididymitis. Complications (e.g. infarction, abscess) may require surgery.

Sexually transmitted infections: introduction

Sexually transmitted infections (STIs) have been on the rise in the UK and many other western countries in recent years, fuelled by a decline in the practice of 'safer sex'. The most severely affected groups are teenage females and homosexual men. The number of new STI diagnoses made in UK genito-urinary medicine clinics continues to rise with an increase of 2–3% each year since 2003.

Risk factors

The risk factors that influence an individual's chance of acquiring a particular STI are broadly the same for all STIs. This means that patients with one STI should be assessed for the presence of others including syphilis and HIV. Risk factors include: the number of sexual partners an individual has, failure to use barrier contraception, frequency of partner change, lower socioeconomic status, age <25 years, residence in an inner city, symptomatic partner, sexual orientation (syphilis, gonorrhoea, HIV, and hepatitis B are more prevalent among men who have sex with men in the UK), sexual practices (orogenital and anogenital contact).

Contact tracing

During the Second World War, fears of a UK STI epidemic led to laws enabling the compulsory treatment of a sexual contact named by more than one person with a diagnosed STI. These laws were repealed after the war and led to the concept of partner notification. Partner notification aims to prevent re-infection of treated persons and break any chain of onward STI transmission. Patients are encouraged to notify their sexual partners of any infection risk with the help and advice of trained health advisers. This process should be carried out by a genitourinary medicine (GUM) clinic and it is essential that individuals experiencing such infections are referred. It may be appropriate to treat asymptomatic contacts presumptively. Partner notification is voluntary in the UK, but a legal requirement in some states of the USA and Sweden.

Patient assessment

- History – last intercourse, contraceptive method, nature of sexual contacts and number, frequency of partner change, sexual orientation, sexual practices, previous history of STI, previous treatments received, menstrual history, drug use, foreign travel.
- Examination – skin (rashes, lesions), lymphadenopathy, hair loss, jaundice, mucosal lesions, conjunctivits, urethritis, arthritis, detailed examination of the genitalia including a speculum examination of women and the subpreputal space and male urethra in men. A rectal examination and proctoscopy may be indicated.
- Tests – samples of genital secretions should be taken – in practice this will usually done by an experienced individual in the GUM clinic. In men: urethral swab (*Chlamydia*) and smear for Gram stain and culture (*N. gonorrhoea*) in women: high vaginal swab in Stuart's media for microscopy and culture (*Candida*, *G. vaginalis*, anaerobes, *Trichomonas*), endocervical swab (*C. trachomatis*). Other tests: swab ulcers for HSV

culture and dark microscopy for syphilis, blood serology for syphilis, hepatitis and HIV.

Differential diagnoses

The following list is reproduced from Pattman, R, Snow, M, Hardy, P, Sarkar, KN, Elawad, B. et al. *Oxford Handbook of Genitourinary Medicine, HIV and AIDS*. Oxford: Oxford University Press, 2005, with permission from Oxford University Press.

- Men with urethritis – *N. gonorrhoeae,* non-gonococcal urethritis (chlamydia, trichomoniasis, UTI)
- Balanitis – if associated with ulcers or blisters consider causes of genital ulceration. If associated with erythema or excoriation consider: chlamydia, causes of urethritis, trichomoniasis, candida, bacterial infection. Non-STI causes: consider dermatological causes such as dermatitis, lichen simplex, lichen planus etc
- Vulval irritation/pain – if associated with ulcers/blisters consider causes of genital ulceration. Otherwise consider candidiasis (especially if pregnant, diabetic, discharge or recent antibiotics), trichomoniasis, bacterial vaginosis. Non-STI causes: dermatological conditions – especially atopic vulvitis and consider vulval intraepithelial neoplasia
- Abnormal vaginal discharge – watery white/grey with fishy smell consider bacterial vaginosis, white curdy discharge with vulval rash consider candidiasis, malodourous green/yellow discharge consider trichomoniasis. Other: gonorrhoea, chlamydia, cervical herpes simplex. Non-STI causes: retained foreign body (e.g. tampon)
- Anogenital ulceration – herpes (preceded by vesicles), syphilis, tropical. Non-STI causes: neoplasia, drug reactions, Behcet's disease, trauma
- Genital lumps – genital warts, molluscum contagiosum, condylomata lata. Non-STI causes: folliculitis, lichen planus, keratoacanthoma, carcinoma

Infestations that may be transmitted sexually include pubic lice and scabies.

Bacterial vaginosis

Bacterial vaginosis (BV) causes vaginal discharge in ≤50% of symptomatic women (the other causes are vulvovaginal candidiasis and trichomoniasis) and is termed 'vaginosis' rather than 'vaginitis' due to the absence of inflammation. Rather than being due to a single organism, BV is caused by complex changes in the balance of the microbiological flora.

Epidemiology

- Worldwide prevalence ranges from 11% to 48% in women of child-bearing age.
- Risk factors for acquisition – new or multiple sexual partners, vaginal douching, smoking. It can occur in women who have never had vaginal intercourse.

Pathology

- A reduction in the normally dominant lactobacilli and increase in other organisms, especially anaerobes such as *G. vaginalis* and *Bacteroides* spp. The mechanism by which this change occurs in not certain.
- Lactobacilli produce hydrogen peroxide which lowers the pH – the loss of these organisms permits an increase in pH and overgrowth of vaginal anaerobes. These produce proteolytic enzymes which degrade vaginal peptides into offensive-smelling products and promote discharge and exfoliation of the epithelial layers.

Clinical features

- <75% of cases are asymptomatic
- Thin, white, fishy smelling discharge, most noticeable after intercourse
- May be associated with cervicitis which may or may not occur in the presence of simultaneous chlamydial or gonococcal infection
- Vaginal pain or vulval irritation is uncommon
- Complications – pregnant women with BV have a higher rate of pre-term delivery; BV is associated with endometritis, post-partum fever, and infections following gynaecological surgery; it is a risk factor for HIV acquisition and transmission, and acquisition of HSV-2, chlamydia and gonorrhoea

Diagnosis

- The Amsel criteria – sensitivity is 90%, specificity 77% if three of the four criteria are present. Remember that trichomonal infection may cause the first three findings:
  - homogeneous, watery white-grey discharge coating the vaginal walls
  - vaginal pH > 4.5
  - positive amine test – add 10% KOH to sample of discharge: positive if produces a fishy odour
  - the presence of 'clue cells' (epithelial cells studded with adherent coccobacilli) on a saline wet mount – the single best predictor of BV. At least 20% of epithelial cells should be clue cells in those women with BV. Gram staining is most sensitive but impractical in standard clinical practice.

- No bacteria are specific for BV and bacterial culture is not useful.
- Diagnostic cards tests for pH indicate the presence of amines.

Differential diagnosis

Trichomoniasis, atrophic vaginitis (dyspareunia and inflammation are present in these cases).

Management

- Infection resolves spontaneously in one-third of cases.
- Treatment may reduce the risk of acquiring other STDs.
- Who to treat:
    - all women presenting with symptoms – oral treatment is safe in pregnancy and not associated with adverse fetal effects
    - asymptomatic women proceeding to abortion or hysterectomy – reduces the risk of postoperative infection
    - asymptomatic pregnant women with previous pre-term delivery may also benefit from treatment. BV is associated with a higher rate of pre-term birth (perhaps due to chorioamnionitis) but studies have not demonstrated that treating it brings about a significant reduction. However, treating BV in those women with a history of pre-term delivery is associated with reduced rates of pre-term pre-labour rupture of membranes and low-birth-weight babies. Consider screening those women with a history of pre-term labour for BV.
- Regimes:
    - metronidazole – 500 mg bd PO for 7 days (single 2 g dose has lower efficacy and is no longer recommended), or 5 g od PV (per vagina) 0.75% metronidazole gel for 5 days. Early cure rates >90%, 80% at 4 weeks
    - clindamycin – 300 mg bd PO for 7 days, or 100 mg ovules od PV for 3 days. The use of clindamycin may be associated with the acquisition of clindamycin-resistant anaerobes. No resistance to metronidazole has been demonstrated.
- 30% of patients experience recurrence within 3 months. A prolonged (e.g. 14 days) or alternative treatment course should be used in such patients. Those who experience multiple relapses may benefit from a long term maintenance regime of twice weekly PV metronidazole gel. Clindamycin should not be used for this purpose.
- Treating partners does not appear to reduce recurrence.

Vulvovaginal candidiasis

Vulvovaginal candidiasis accounts for one-third of cases of vaginitis.

Epidemiology

- *Candida* species may be found in the lower genital tract of up to 50% of asymptomatic women.
- It is common with <75% of pre-menopausal women reporting at least one episode. It is less common in post-menopausal women.
- Candidal infection is uncommon in pre-pubertal women but does occur in children who have had recent antibiotic therapy, wear nappies, or are immunosuppressed.
- There is an increase in incidence at the time at which most women begin regular sexual activity.

Pathology

- *Candida albicans* is the cause of around 90% of cases, but the incidence of other candida species such as *C. glabrata* may be increasing as a result of increasing use of over-the-counter drugs.
- Sporadic episodes usually occur with no identifiable predisposing factor. Risk factors include: diabetes mellitus, immunosuppression, recent antibiotic use, oral contraceptive use, or oestrogen therapy,

Clinical features

- Pruritus, dysuria, dyspareunia, soreness
- There may be discharge which might be white and clumpy, or thin and watery but it is often absent with only vulvar and vaginal erythema on examination
- Recurrent infection – defined ≥4episodes a year and seen in 5–8% of women. Predisposing factors such as diabetes are seen in a minority and susceptibility seems to be largely determined genetically. Behavioural factors seem to play a part – a two-fold increase in risk has been associated with the consumption of cranberry juice, the use of sanitary towels, and sexual lubricants

Diagnosis

- Self-diagnosis unreliable – one study demonstrated that only 34% of those women self-diagnosing candidal infection actually had it.
- A wet mount of the discharge with 10% KOH may allow recognition of yeast and hyphae but microscopy is negative in around 50%.
- Vaginal pH is around 4–4.5 (unlike trichomonal infection or bacterial vaginosis).
- Perform culture in patients with persistent discharge or recurrent symptoms unresponsive to azole treatment – they may have non-albicans *Candida* infection. Routine culture is unhelpful.
- As well as other infective causes consider allergic reactions, and contact dermatitis in the differential.

Management

- Treatment is indicated for symptoms. Asymptomatic carriage does not require therapy.
- 90% of cases represent uncomplicated infection (healthy non-pregnant women with mild/moderate symptoms, infrequent episodes and infection with *C. albicans*). Oral and topical treatments are similarly effective with topical therapy relieving symptoms more rapidly but oral being preferred by women, e.g. PO fluconazole.
- 10% of cases are complicated (infection with non-albicans species, severe symptoms, four or more episodes per year, and those cases occurring in the pregnant, uncontrolled diabetics or the immunosuppressed).
    - The immunosuppressed and those with severe symptoms are unlikely to respond to short treatment courses – 7–14 days of topical therapy, or two doses of oral fluconazole (150 mg) 72 h apart is indicated.
    - *C. glabrata* infection – 50% of women infected with this species fail treatment with azoles. Moderate success may be seen with intravaginal boric acid, and greater than 90% cure may be seen with topical flucytosine cream.
    - Pregnancy – treat only for symptoms using a topical imidazole for 7–14 days (e.g. clotrimazole). Oral azoles are contraindicated in pregnancy. Vaginal candidiasis is not associated with adverse outcomes in pregnancy.
    - Recurrent infection (four or more episodes per year) – aim to eliminate risk factors (e.g. better glucose control, lower-oestrogen-containing contraceptives, behavioural changes where appropriate). After the initial treatment course, long-term suppressive therapy (e.g. fluconazole 150 mg PO weekly for 6 months) is effective at preventing relapses. However more than half are likely to experience further infections in the months following cessation of suppressive therapy. These patients should be treated acutely once again, and then given a year of suppressive treatment after culture confirmation of relapse. Development of azole resistance has not yet been associated with long-term therapy in this setting.
- Most experts do not recommend treatment of asymptomatic sexual partners.

Genital warts

Anogenital warts are one of the most common sexually transmitted viral infections. They are caused by human papilloma virus, a highly infectious double-stranded DNA virus of which there are over 100 serotypes (see 📖 p.485); 90% of cases are related to serotypes 6 and 11 (the least likely to exhibit malignant potential). Serotypes 16 and 18 have a strong association with malignancy.

Epidemiology

- Exposure is usually sexual, and incubation is from a few weeks to several months. The risk of disease increases with the number of sexual partners. Women tend to be affected more than men in most settings.
- Anal disease can occur in women as a result of extension of perineal infection, or receptive anal intercourse. Men usually experience lesions on the shaft of the penis or the preputial cavity. Anal lesions are commoner among men who have sex with men, but also occur among heterosexual men.
- The prevalence of anogenital warts is higher among those who are HIV-positive or have other sexually transmitted disease. The risk increases with lower CD4 counts and decreases with antiretroviral treatment.
- Most infections are cleared within 2 years, but persistent infections can occur and are associated with the development of squamous cell carcinoma.

Clinical features

- Those with a small number of lesions may experience no symptoms.
- A larger number of lesions may be associated with pruritus, bleeding, dysuria, PV discharge, pain, and tenderness.
- Rarely, warts may form larger exophytic masses that can interfere mechanically with intercourse, defaecation, and even child birth.
- Anal disease may cause strictures.

Diagnosis

- Usually made visually. Lesions are pink, and can take the form of flattened papules or more-classic verrucous papilliform warts. Application of 5% acetic acid causes lesions to turn white. This is not specific for anogenital warts however.
- Anoscopy, colposcopy, etc allows the extent of disease to be assessed.
- Biopsy should be performed where the diagnosis is in doubt, in immunocompromised patients (higher risk of malignancy), and in cases that do not respond to therapy.
- The differential includes: condyloma lata (flat, velvety lesions of secondary syphilis), anogenital squamous cell carcinoma (may co-exist with genital warts), vulvar intraepithelial neoplasia, skin tags, molluscum contagiosum.

Management

Spontaneous regression is seen in up to 30% of immunocompetent cases by 3 months. The choice of therapy where indicated is governed by the number and extent of the lesions. All modalities have high rates of recurrence. Women should have a pap smear. Small external lesions can be managed by the application of a topical treatment either in clinic or by the patient where appropriate. Large, multiple, or internal lesions should be referred to a surgeon or gynaecologist, and pathological studies undertaken where indicated.

- Chemical agents:
  - podophyllin contains an anti-mitotic agent which stops the cell cycle in metaphase causing cell death; 25% solution is administered 1–2 times per week – usually at a clinic – and achieves clearance rates of 20–50% at 3 months. It should be applied to small areas of skin, allowed to dry, and washed off 6 h later. It should not be used on the cervix or vaginal epithelium (burns). It is contraindicated in pregnancy (teratogenic)
  - podophyllotoxin is a related agent that can be self-administered to external warts with similar rates of success
  - trichloroacetic acid acts by protein coagulation and has similar rates of success and similar side-effects to podophyllin. It can be used on internal lesions and during pregnancy. Neighbouring skin can be protected from its caustic effects by the application of petroleum jelly prior to use
  - 5-fluorouracil/adrenaline gel injected intralesionally can achieve an initial cure rate of up to 60%, but half relapse by 3 months.
- Immuno-modulation:
  - imiquimod is applied topically as a cream to external lesions only and acts by cytokine induction. It achieves high rates of clearance (over 80%) and low rates of recurrence (under 20%)
  - interferon-α given systemically achieves similar rates of clearance as thermocoagulation but has higher rates of recurrence.
- Surgery – ablation or excision should be performed where medical therapies have failed or are not indicated (e.g. due to size). Cryotherapy with liquid nitrogen or a cryoprobe is safe in pregnancy and achieves clearance rates of over 90% at 3 months – repeated applications are required. Laser therapy is expensive, requires anaesthesia, and places the operator at risk of developing warts. It achieves cure rates of almost 100% at 1 year. Surgical excision has clearance rates of 36% at 3 months. As well as the standard risks associated with anaesthesia and surgery, patients may develop strictures. Excised lesions should be examined pathologically for signs of malignancy.
- Newer therapies include topical cidofovir, topical BCG, infra-red coagulation.

Tropical genital ulceration

Genital ulceration is much commoner in patients presenting with STI in the developing world and an important factor in the spread of HIV. The common causes of genital ulcers in the developed world (HSV and syphilis) remain common in developing regions (e.g. HSV remains the top cause in Jamaica and South Africa) but may be pushed out of top place by certain other infections (e.g. chancroid in Rwanda). Diagnosing lesions clinically can be difficult – syphilis classically causes a single painless ulcer but so may HSV and lymphogranuloma venereum (LGV). Where facilities allow, investigations should include: serologic testing for syphilis, a diagnostic evaluation for herpes, and where appropriate Gram-stain and culture on selective media (for *H. ducreyi*).

Chancroid

- Caused by *Haemophilus ducreyi*, a fastidious Gram-negative rod.
- It produces a potent 'cytolethal distending toxin' which is likely to contribute to both the formation of ulcers and their slow healing.
- Incubation after infection is around 1 week, following which painful erythematous papules develop on the external genitalia (prepuce, corona, or glans in men; the labia, vagina, and perianal areas in women), develop into pustules and then erode into sloughy, non-indurated haemorrhagic ulcers. Lesions are usually multiple, often developing on adjacent skin surfaces (thigh, scrotum), and suppurative inguinal lymphadenopathy is common (sometimes forming fluctuant buboes). Co-infection with HIV may result in atypical presentations with multiple lesions, extra-genital involvement, and delayed response to treatment.
- Diagnosis is by clinical appearance and culture and Gram stain (organisms clump in parallel strands – 'school of fish' appearance) of material from the ulcer or aspirated lymph nodes. Enriched culture media are required. PCR-based tests are in development.
- Treatment is with erythromycin 500 mg qds for 1 week or azithromycin 1 g PO stat or ceftriaxone 250 mg IM stat. Fluctuant buboes should be aspirated (risk of fistulas).

Lymphogranuloma venereum (LGV)

- Genital ulcer disease caused by the L1, L2 and L3 serovars of *C. trachomatis*. Endemic in areas of East and West Africa, India, SE Asia and the Caribbean. Since 2003 a series of outbreaks of LGV have been reported in men who have sex with men (MSM). In 2004 the Health Protection Agency launched an enhanced surveillance programme.
- Asymptomatic infection in women is common and may serve as a reservoir. Incubation is 3–12 days. Primary infection is characterized by a transient painless genital ulcer. Direct local extension leads to a secondary lesion 2–6 weeks later – an inflammatory reaction in the inguinal lymph nodes with fever, headache, weight loss, ± pneumonia, meningoencephalitis, and arthritis. Lymphadenopathy may be so severe as to bulge each side of the inguinal ligament ('groove sign'). An inflammatory mass may form in the rectum leading to pain, constipation, tenesmus, and rectal discharge. LGV proctitis may be confused with inflammatory bowel disease. Late disease may lead to

fibrosis and strictures in the anogenital tract, genital elephantiasis, anal fistulae, frozen pelvis, and infertility.
- Diagnosis is based on clinical features. Culture of pus from lesions and nodes is possible but rarely practical in regions where the disease occurs. Chlamydial serology is useful but not specific for serovars. PCR-based tests are available.
- Treatment is with doxycycline or erythromycin for 3 weeks.

Granuloma inguinale *(Klebsiella granulomatis)*

- Endemic in Western New Guinea, the Caribbean, Southern India, South Africa, SE Asia, Australia, and Brazil, granuloma inguinale (or donovaniasis) is a primarily sexually transmitted infection causing indolent, painless non-purulent ulceration. Infection may also be acquired faecally and by passage through an infected birth canal.
- After an incubation of 1–3 months the 'beefy red' ulcers appear on the prepuce or labia and enlarge over months to 5 cm diameter or more. Auto-inoculation may see ulcers forming on adjacent skin. Local extension and fibrosis occur and late lesions may cause elephantiasis-like swelling of the external genitalia. Regional lymphadenopathy is rare but metastatic spread to the bones, joints, and liver has been reported.
- Diagnosis is clinical and by microscopy of Giemsa-stained material from ulcers which may demonstrate bipolar intracellular bacteria ('Donovan bodies' – a characteristic safety-pin appearance). Culture is extremely difficult and rarely performed.
- Treatment is with co-trimoxazole or doxycycline for 3 weeks. Alternatives include azithromycin 1 g weekly or erythromycin (again for 3 weeks). An aminoglycoside may be added if there is no initial response (which may be seen in HIV-positive patients).

Genital herpes

Genital herpes simplex infections are a major public health problem across the world. Like all herpes viruses, herpes simplex establishes a latent state following primary infection and may reactivate causing episodic local disease.

Epidemiology

- HSV-2 is the commonest cause of genital herpes, but an increasing number of cases are due to HSV-1 infection.
- Asymptomatic HSV-2 infection is more likely in those previously infected with HSV-1 and vice versa.
- The incidence of genital herpes has been increasing in the UK and the presence of HSV-related ulcers is associated with an increased risk of HIV transmmission.

Clinical features

- Incubation is usually 3–7 days. Primary infection is characterized by local burning followed by a painful genital vesicular eruption. These vesicles then rupture forming ulcers. Other symptoms: fever, dysuria, tender inguinal lymphadenopathy, headache, herpetic proctitis. New lesions appear for around a week. Resolution over 1–3 weeks. Up to 60% of primary cases are asymptomatic, thus the first clinical attack may actually represent the first reactivation.
- Recurrent attacks tend to be less severe with a shorter duration of symptoms and infrequent systemic features. Up to half of patients with reactivation experience prodromal symptoms (local tingling, shooting pains). The majority of patients developing primary infection will experience a recurrence within the first year. Prolonged first episodes are associated with earlier and more-frequent relapses. Recurrence rates are much higher with HSV-2 infection and in immunosuppressed patients.
- Rare extragenital features of primary infection include: meningitis, urinary retention due to autonomic dysfunction, distant skin lesions.
- Subclinical viral shedding can occur in the absence of lesions. This is of importance as it leads to unrecognized transmission to neonates and sexual partners. It is commoner with HSV-2.

Diagnosis

- Type-specific antibodies to HSV develop in the first few weeks of infection and are maintained indefinitely. Commercial tests are available but a positive test does not allow one to distinguish present from previous infection.
- Viral culture from lesions allows definitive diagnosis – it is more likely to be positive if the fluid is taken from vesicles that have not yet ruptured.
- Viral-antigen-detection tests are available and allow rapid, type-specific diagnosis.
- PCR-based viral detection is rapid, specific and allows recognition of asymptomatic viral shedding.

Management

- Antiviral treatment – aciclovir, valaciclovir, and famciclovir reduce the severity of episodes but do not alter the course of the disease. Treatment should be commenced as soon as possible – ideally with 24 h – for maximum benefit. Systemic antivirals are less useful in recurrent attacks which are in any case shorter and milder. Topical aciclovir is of little benefit; IV aciclovir is indicated in neonatal infection or severe disease. Drug treatment reduces but does not eliminate viral shedding.
- Recurrent disease – patients with frequent recurrences (more than four per year) may benefit from prophylactic aciclovir. HIV-positive patients developing severe genital herpes should be given prophylactic therapy if their CD4 count is under 100 cells/mm^3.
- If catheterization is required (e.g. autonomic disturbance or pain), the suprapubic route may be better to reduce pain, aid recognition of return of normal micturition, and reduce the risk of ascending infection.

Pelvic inflammatory disease

An acute infection of the female upper genital tract which may involve the uterus, Fallopian tubes, ovaries, and even adjacent pelvic structures. It occurs when organisms, either transmitted sexually or constituting part of the normal vaginal flora, breach the barrier of the endocervical canal and gain access to the upper genital organs. Such a breach may be precipitated by infection with organisms such as *N. gonorrheae* or *C. trachomatis*.

Epidemiology

- The majority of cases present within 1 week of menses, which is thought to enhance the ascension of vaginal organisms.
- Those at greatest risk of pelvic inflammatory disease (PID) are those with multiple sexual partners. It is rarely seen in celibate women and those in longstanding monogamous relationships.
- Other risk factors include: age (highest incidence in those aged 15–25 years), the presence of symptomatic STI in the partner, previous PID, and possibly vaginal douching.

Microbiology

PID is a polymicrobial infection. Species identified include: streptococci, *E. coli*, *Klebsiella* spp, *Proteus mirabilis*, *Haemophilus*, *Bacteroides*, *Peptococcus*, and *Peptostreptococcus* spp. Around 15% of those acquiring endocervical gonorrhoea go on to develop PID, and it accounts for one-third of PID presentations worldwide (less in Europe). *C. trachomatis* serovars D–K account for another one-third of cases overall (much more in Western Europe), and as with gonococcus around 15% of endocervical chlamydia infections produce PID. Studies have demonstrating that screening for chlamydial infection reduces the rate of PID (see 📖 p.750).

Clinical features

- Infection of the upper genital structures may precipitate any or all of endometritis, salpingitis, oophoritis, peritonitis, and perihepatitis.
- Symptoms – lower abdominal pain which may start during or shortly after menses, vaginal discharge, abnormal uterine bleeding. Rebound tenderness is common in the lower quadrants, and fever is seen in half of patients. Those with perihepatitis may also develop upper-abdominal pain. The uterus and andnexae will be tender on pelvic examination.
- PID can be a subclinical disease and cause of infertility – one-third of women with no history of PID were found to have *C. trachomatis* in the upper genital tract with no clinical findings except infertility.

Diagnosis

- No test for PID achieves high sensitivity or specificity. It is appropriate to initiate empirical antibiotic therapy when clinical suspicion in strong.
- Consider the diagnosis in patients with abdominal pain *and* one of: cervical or uterine/adnexal tenderness, fever >38°C, raised WCC, abnormal cervical or vaginal discharge, or raised inflammatory markers. Have a low threshold for treating such individuals empirically.

- Investigations include:
  - microscopy of vaginal discharge – if this demonstrates Gram-negative intracellular diplococci the probability of PID is high
  - laparoscopy – although specificity approaches 100%, laparoscopy has been found to be only around 50% sensitive in the diagnosis of PID. It should be considered in patients who do not respond to empirical therapy within 72 h (less if acutely ill) and those in whom there is a high suspicion of an alternative diagnosis (e.g. appendicitis)
  - endometrial biopsy – the demonstration of plasma cell endometritis is a common finding in cases of clinical PID, but it is also found in asymptomatic women with no other evidence of PID
  - other tests – transvaginal ultrasound has a low specificity and sensitivity for PID but is useful in the identification of pelvic abscesses; positive DNA testing for gonococcus and chlamydia increases clinical probability.
- Confirmed cases are considered to be those with pelvic pain/tenderness *and* one of: endometritis/salpingitis on biopsy, *N. gonorrhoeae or C. trachomatis* in the genital tract, salpingitis seen on laparoscopy or laparotomy, isolation of pathogenic bacteria from the upper genital tract, or inflammatory pelvic peritoneal fluid with no other cause.
- All patients should have a pregnancy test and urinalysis.
- Differential diagnosis – appendicitis, cholecystitis, IBD, UTI, dysmenorrhoea, ectopic pregnancy, ovarian cyst/torsion/tumour.

Treatment

- Most people can be treated as outpatients. Consider admitting pregnant women, those failing to respond to oral medications, those with severe clinical features (high fever, vomiting, severe pain), those with abscesses or likely to require surgical intervention.
- Selected antibiotics should cover *N. gonorrhoea*, *C. trachomatis*, group A and B streptococci, anaerobes and the common Gram-negative enterics. Avoid fluoroquinolones (increasing *N. gonorrhoea* resistance). Suitable regimes include:
  - inpatient therapy – cefoxitin 2 g IV qds *and* doxycycline 100 mg PO or IV bd; or clindamycin 900 mg IV tds *and* gentamicin 5–7 mg/kg IV od. Patients should be moved to an oral regime within 24–48 h and complete 14 days of therapy (as below but omitting the IM stat dose)
  - outpatient therapy – single dose ceftriaxone 250 mg IM stat then doxycycline 100 mg PO bd 14 days *and* metronidazole 500 mg tds PO 14 days
  - penicillin-allergic patients in whom it is considered too risky to use ceftriaxone should be admitted and given the clindamycin regime described above, for 48 h. If improving they can then complete treatment on metronidazole 500 mg tds PO *and* clindamycin 450 mg PO qds *or* doxycycline 100 mg PO bd for 14 days.
- As with all STIs, contacts should be traced and patients should be counselled and screened for HIV, hepatitis, etc where indicated.

Toxic shock syndrome

A syndrome of fever, skin rash, and shock due to toxins produced by certain strains of infecting *S. aureus*. Classically associated with menstruation and the use of highly-absorbent tampons, around half of cases are now non-menstrual (e.g. surgical and wound infections, osteomyelitis, septic arthritis, burns, etc).

Pathogenesis

S. aureus establishes infection and produces toxins. The first of these to be identified, toxic shock syndrome toxin-1 (TSST-1) was first isolated in the 1970s when a series of cases of toxic shock syndrome (TSS) occurred associated with the use of highly absorbent tampons. It remains commonly associated with menstrual-related cases. Non-menstrual cases are more likely to be associated with other toxins, e.g. staphylococcal enterotoxin B. These exotoxins are superantigens, capable of activating a large number of T cells simultaneously, resulting in massive cytokine production leading to fever, muscle proteolysis, and shock.

Epidemiology

Menstrual cases of TSS fell rapidly following the withdrawal of certain types of highly absorbent tampons in the 1970s, but still occur. Women who develop the syndrome are more likely to have used tampons of high absorbancy, and kept individual tampons in place for longer periods. Non-menstrual TSS accounts for half of reported cases. Patients tend to be slightly older, and studies imply an equal incidence in men and women if vaginal and post-partum cases are excluded.

Clinical features

- Symptoms (develop rapidly, 1–2 days post-menstruation or post-surgery) – fever, hypotension (may be unresponsive to fluid therapy), skin lesions (erythroderma resembling sunburn, conjunctival bleeds, petechiae, mucosal ulceration and vesicles), myalgia (due to cytokine-mediated muscle damage), headache, sore throat, vomiting, diarrhoea, and abdominal pain.
- Certain skin manifestations may occur late in disease with itchy maculopapular rash appearing 1–2 weeks later, desquamation up to 3 weeks later, and even nail loss 1–2 months later.
- Multiple organ systems may be involved – renal failure (of both pre-renal and renal aetiology), gastrointestinal symptoms (diarrhoea), encephalopathy (due to cerebral oedema), cardiac depression, liver function impairment, and anaemia/thrombocytopenia.
- Recurrent disease can occur if an episode is not adequately treated with anti-staphylococcal antibiotics. A subacute form of TSS may occur in patients with HIV, with symptoms occurring over weeks.

Diagnosis

Isolating *S. aureus* is not a prerequisite for the diagnosis of TSS. It is rarely isolated in blood cultures (<5%) but frequently grown from swabs of wounds or mucosal sites – the toxin produced may be identified by specialist laboratories. The Centers for Disease Control (CDC) have produced clinical criteria for the diagnosis of TSS. These were designed for epidemiological surveillance and should not be used to exclude the diagnosis in a patient in whom TSS is considered likely.

- Confirmed case – a patient with fever of 39°C or more, hypotension, diffuse erythroderma, involvement of three organ systems, and desquamation (if the patient survivies).
- Probable case – lacks one of the above.

Differential diagnosis includes: streptococcal TSS (important to recognize as urgent surgical debridement may be required), meningococcal disease, leptospirosis, dengue, typhoid fever.

Management

- Supportive – hypotension can be difficult to manage and may not respond to fluid therapy alone, necessitating the use of vasopressors such as dopamine.
- Surgical – foreign bodies should be removed in the case of menstrual TSS, and wound debridement may be required in post-surgical cases.
- Antibiotics – while it is not clear whether antibiotics are required in the acute management of TSS, they are important in preventing recurrent disease. Clindamycin *may* be more effective than antibiotics acting solely on the bacterial cell wall, as it suppresses protein synthesis (and therefore potentially toxin production). MRSA should be covered in any regime. Recommended empirical treatment is clindamycin 600 mg IV tds and vancomycin 30 mg/kg/day in two divided doses IV. Treatment should be for 10–14 days.
- Intravenous immunoglobulin – there are no trials to demonstrate the effectiveness of immunoglobulin in the management of staphylococcal TSS. Case reports suggest it may be of benefit in severe cases that fail to respond to fluids and vasopressors.
- Corticosteroids – there is no evidence to suggest the use of steroids affects outcome.

Prognosis

Death usually occurs within a few of days of presentation but can occur up to 2 weeks later. Causes of death include cardiac arrhythmias, respiratory failure, and bleeding. Mortality in menstrual cases is around 1.8%, non-menstrual cases 6%.

Gonorrhoea

A purulent infection of mucous membranes (e.g. urethra, rectum, cervix, conjunctiva, pharynx) caused by sexually transmitted *Neisseria gonorrhoeae.*

Epidemiology

- Infection is common across the world. In the developing world, perinatal transmission and neonatal eye infections remains a significant problem.
- It is the 2nd-commonest STI in the UK affecting predominantly young people (peaking in males aged 20–24 years and females aged 16–19 years), with the highest rates in urban areas. Infection is concentrated among homosexual/bisexual men and black ethnic minority populations.
- The recent increase in incidence (probably related to an increase in unsafe sexual practices) and growing prevalence of antimicrobial resistance have made it a major public health concern.
- Resistance to first-line antibiotic treatment is related to an increased risk of treatment failure (with consequent disease complications) and onward transmission within a community. In 2005, 21% of isolates were resistant to ciprofloxacin overall (42.4% among homosexual men and 11.3% among heterosexual men), and 17.9% of isolates were resistant to penicillin. Therefore first-line therapy should be ceftriaxone or cefixime.

Clinical features

- Incubation is 2–5 days. Lower genital tract infection may be asymptomatic or cause urethritis with purulent discharge and dysuria in men and endocervicitis with PV discharge, itch, and dysuria in women. Although infection of the female urethra, pharynx, and rectum (common in homosexual men, uncommon otherwise and causing discharge and tenesmus) are probably common, they are usually asymptomatic.
- Retrograde spread may occur, causing salpingitis/endometritis, pelvic inflammatory disease, and tubo-ovarian abscesses in up to 20% of women with cervicitis. In rare cases frank peritonitis or peri-hepatitis (Fitz-Hugh–Curtis syndrome) are seen. Men with gonococcal urethritis can develop epididymitis or epididymo-orchitis.
- Disseminated gonococcal infection may follow around 1% of genital infection; 75% of such cases occur in women who are at increased risk if mucosal infection occurs during menstruation or pregnancy. Features include: rash, fever, arthralgias, migratory polyarthritis, septic arthritis, endocarditis, meningitis.
- Neonates acquiring infection intrapartum present with ophthalmia neonatorum and disseminated infection. Conjunctivitis can also occur in adults following direct inoculation of organisms, and may lead to blindness.

Diagnosis

- Gram stain – highly sensitive from swabs of urethral discharge (the presence of Gram-negative diplococci within leucocytes is 95% sensitive for gonococcal infection). Sensitivity and specificity are less good for endocervical or pharyngeal specimens.
- Culture – the gold standard. All infected areas should be swabbed and cultured both to confirm diagnosis and to provide sensitivity data.
- Disseminated infection – joint effusions, blood and CSF should be sent for culture and Gram stain where appropriate. Negative cultures do not rule out disseminated infection.

Management

- Antibiotics – first-line therapy is with a cephalosporin: ceftriaxone 250 mg IM single dose, or cefixime 400 mg PO single dose for uncomplicated infection.
- Co-infection with other STIs is common and both patients and their partners should be screened. Some physicians treat patients empirically for associated chlamydial infection – chlamydial urethritis can present as apparent treatment failure after confirmed gonorrhoea.
- Pregnancy test should be performed on all women with suspected gonococcal infection.

Chlamydia

Epidemiology

The commonest sexually transmitted infection in the UK with rates highest in the under 25s. A significant number of cases are asymptomatic, and 10–40% of untreated infected women develop pelvic inflammatory disease making it an important reproductive health issue. The number of cases in the UK has been rising steadily since the mid-1990s, prompting the initiation of a national screening programme. The responsible organisms are *Chlamydia trachomatis* serovars D–K (📖 p.404).

Clinical features

- Incubation period is 1–3 weeks.
- Around 50% of infected males and 80% of infected females are asymptomatic – such infection may persist for many years if untreated.
- Symptoms – mucopurulent cervicitis in females and urethritis with dysuria and discharge in males. Ascending genital tract infection may lead to pelvic inflammatory disease in women and is the commonest cause of epididymitis in men under 35 years. Proctitis and pharyngitis occur in men and women.
- Other presentations – LGV (the cause of 10% of genital ulcers in tropical countries – see 📖 p.740), neonatal conjunctivitis (📖 p.780), and neonatal pneumonia may occur in children born to infected mothers.
- Complications: PID (📖 p.744), Reiter's disease (urethritis, conjunctivitis, reactive arthritis), perihepatitis, conjunctivitis.
- Co-infection of chlamydia and gonorrhoea is common (40% of women and 20% of men with chlamydia also have gonorrhoea).

Diagnosis

- Culture is not performed routinely.
- Enzyme immunoassay is available and cheap but is only 40–60% sensitive.
- DNA testing – widely available and highly sensitive and specific. May be performed on cervical swabs, urine, vulvovaginal swabs (best on urine in men and endocervical swab in women), and is the first-line investigation for all specimens.

Management

- Antibiotics – azithromycin 1 g single dose (the drug of choice for reasons of compliance) or doxycycline 100 mg bd for 7 days. Alternative regimens are erythromycin 500 mg bd for 10-14 days or ofloxacin 200 mg bd or 400 mg od for 7 days. First-time cure rates of over 95%.
- Patients should be thoroughly worked up for other STIs and contacts traced.
- Recurrent infection is usually due to re-infection by untreated partners. Female partners of men with urethritis should be treated whether or not there is evidence of infection, given the high risk of asymptomatic disease.

Trichmoniasis

An infection caused by the flagellated protozoan, *Trichomomas vaginalis* (p.581), which may be asymptomatic (particularly in men) or lead to vaginal discharge, dysuria, and lower abdominal pain in women, or urethritis in men.

Epidemiology

- Transmission is by sexual contact and its incidence in highest in women with multiple sexual partners, and those with other STIs including HIV. Vertical transmission may take place during delivery.
- Non-sexual transmission (e.g. by contact with contaminated linen in institutions) occurs but is very rare.

Clinical features

- Incubation is 5–28 days.
- Infection is asymptomatic in most men and 50% of women.
- Symptoms tend to develop during menstruation or pregnancy (higher vaginal pH provides a favourable environment for parasite replication) and include: yellow vaginal discharge (may be itchy and smelly), dyspareunia, dysuria, lower abdominal pain.
- On examination, the vulva may be erythematous with obvious discharge, vaginal inflammation, and punctuate haemorrhages on the cervix ('strawberry cervix'). Symptomatic men experience a urethritis indistinguishable from other causes of non-gonococcal urethritis.
- Complications – vaginitis emphysematosa (gas-filled blebs in the vaginal wall), vaginal cuff cellulitis after hysterectomy, premature labour, low-birth-weight infants.

Diagnosis

- Phase-contrast or dark-ground microscopy of wet preparation of genital specimens will demonstrate the motile flagellated protozoans in 48–80% of infected women and 50–90% of infected men.
- Culture is the most-sensitive technique – modified Diamond's media produces the best results.
- Other methods such as ELISA, DNA probe, PCR, and fluorescent antibody staining are more sensitive than wet prep microscopy but less sensitive than culture.

Management

Metronidazole 2 g stat dose. The main disadvantage of single-dose treatment is the risk of re-infection should the partner not be treated simultaneously. Metronidazole is relatively contraindicated in the first trimester of pregnancy – treatment should be deferred until the start of the second trimester if symptoms are severe. Partners and asymptomatic individuals should be treated.

Syphilis

Caused by *Treponema pallidum* subspecies *pallidum* (see 📖 *Treponema* species, p.387).

Epidemiology

- Generally transmitted by sexual contact; can also be transmitted vertically and via blood transfusions. Highest rates are seen in adults.
- Cases of syphilis have risen dramatically in the UK since the late 1990s, driven by outbreaks in London and Manchester.
- HIV infection is associated with treatment failures and more-frequent, earlier neurological disease.

Clinical features

- **Primary syphilis** – after an incubation period of 14 days to 3 months, a painless, erythematous papule develops. This ulcerates forming a painless 'punched out' chancre on the genitalia (rarely on the mouth, hands, anus). Associated with regional lymphadenopathy. Multiple chancres can occur, particularly in HIV-infected patients. They are highly infectious and heal spontaneously after 1–2 months.
- **Secondary syphilis** – organisms disseminate from the chancre causing symptoms 1–6 months later:
  - rash – localized or diffuse mucocutaneous rash may be macular, papular, pustular, or mixed. Involves the trunk, limbs, palms, and soles. Mucosal ulcers may occur. Condylomata lata occur in warm moist areas (e.g. skin folds) and are highly infectious
  - early neurosyphilis (more common in HIV) – may be asymptomatic (CSF findings: pleocytosis, raised protein, decreased glucose, reactive CSF VDRL test), present with syphilitic meningitis (chronic basal meningitis with headache and cranial nerve palsies – fever is usually absent), or cause meningovascular syphilis (headache, fits, limb paralysis). May also present with stroke, cervical myelopathy, and hemiplegia
  - other features – fever, sore throat, 'snail-track ulcers' in mouth, lymphadenopathy, malaise, hepatitis, periostitis, iritis, arthritis, glomerulonephritis.
- **Latent infection** – spontaneous resolution of secondary syphilis occurs at 3–12 weeks. During the latent period infectivity is low but up to one-quarter of patients experience recrudescence of disease.
- **Tertiary syphilis** – rare, follows a latent period of up to 20 years. Characterized by chronic inflammation:
  - gummatous syphilis – granulomatous lesions usually affecting skin, mucous membranes, and bone or organs causing local destruction (e.g. saddle nose). Gummata may be indurated, nodular, or ulcerated and can be painful
  - cardiovascular syphilis – endarteritis of the aorta leads aortic regurgitation (may present with angina and left ventricular failure (LVF)) and aneurysm formation (ascending aorta). Other large arteries may be affected. VDRL can be negative
  - late neurosyphilis – two forms: (1) general paresis of the insane (presents with gradual confusion, hallucinations, delusions, fits, cognitive impairment, tremor of lips and tongue, brisk reflexes,

extensor plantars, Argyll–Robertson pupils), and (2) tabes dorsalis (atrophy of the dorsal columns of the spinal cord with autonomic neuropathy and cranial nerve lesions). Presents with ataxia, sensory loss, sphincter disturbance, shooting pains, sensory loss, arreflexia.

Diagnosis

- Microscopy – detection of organisms on dark-field microscopy, or immunofluorescence of samples taken from chancre exudates.
- PCR-based tests can be used to confirm the diagnosis or to test samples taken from oral lesions, which may be contaminated by commensal spirochaetes, e.g. *T. macrodentium* and *T. microdentium*.
- Serology[1] – tests fall into two groups, non-treponemal (e.g. VDRL) and treponemal (e.g. treponema pallidum haemagglutination assay (TPHA)). May be negative in HIV (see 📖 p.387).
- CSF findings – in asymptomatic neurosyphilis include pleocytosis, low glucose, raised protein, and a positive VDRL test (may be negative in HIV). Symptomatic patients have more severe CSF changes and the CSF VDRL is almost always positive. CSF changes occur in general paresis (elevated lymphocytes, raised protein and positive CSF VDRL), but are variable in tabes (may be normal. 25% of CSF VDRL tests non-reactive).
- Primary syphilis – dark-field or immunofluorescent microscopy of samples taken from chancre exudates. PCR-based tests for oral lesions or to confirm microscopy findings. VDRL may be positive in around 75%, TPHA in around 90%.
- Secondary syphilis – VDRL present at high titre in almost 100%, TPHA positive in 100%. CSF-VDRL is usually positive in early neurosyphilis.
- Latent infection – VDRL falls with time, and following treatment so a negative test does not rule out infection. TPHA remains positive.
- Tertiary syphilis – in gummatous syphilis both VDRL and TPHA are positive. In contrast in syphilitic aortitis and late neurosyphilis, VDRL may be only weakly positive or even negative. TPHA is positive.

Management

- All patients should be tested for HIV infection.
- Early syphilis – benzathine penicillin G 2.4 million IU IM stat as two injections into separate sites, or doxycycline 100 mg bd for 14 days or erthromycin 500 mg qds for 14 days. Penicillin treatment may be complicated by the Jarisch–Herxheimer reaction (📖 p.390).
- Late syphilis – benzathine penicillin G 2.4 million IU IM as two injections into separate sites weekly for 3 weeks, or doxycycline 100 mg bd PO for 28 days.
- Neurosyphilis – benzylpenicillin G 3–4 million IU IV 4-hourly for 14 days, or procaine penicillin G 2.4 million IU IM daily with probenecid 500 mg PO qds for 14 days, or ceftriaxone 2 g IV od for 14 days, or doxycycline 200 mg PO bd for 28 days.
- Treatment success is assessed by symptoms and repeat VDRL. Repeat lumbar puncture in neurosyphilis at 3–6 months and 3-monthly thereafter until CSF normal and CSF VDRL non-reactive. Failure to achieve resolution by 2 years should prompt retreatment.

Reference

1 Egglestone SI, Turner AJ. Serological diagnosis of syphilis. PHLS Syphilis Serology Working Group *Commun Dis Public Health* 2003; 3:158–62.

Acute meningitis

Definition

Acute meningitis is defined as a syndrome characterized by the onset of meningeal symptoms (headache, neck stiffness, vomiting, photophobia) and cerebral dysfunction (confusion, coma) over hours to days. It is identified by an abnormal number of white blood cells in the CSF. Table 5.3 summarizes the causes.

Table 5.3 Causes of acute meningitis

Category	Causes
Bacteria	Group B streptococcus, *E. coli*, *L. monocytogenes*, *K. pneumoniae*, *H. influenzae*, *S. pneumoniae*, *N. meningitidis*, *E. coli*, *Klebsiella spp.*, *Salmonella spp.*, *Serratia marcescens*, *Pseudomonas aeruginosa*, *Enterobacter spp.*, *S. aureus*, *S. epidermidis*, *P. acnes*
Viruses	Enteroviruses, mumps virus, measles virus, herpes viruses, influenza and parainfluenza viruses, HIV, arboviruses, lymphocytic choriomeningitis virus
Rickettsia	*R. rickettsii*, *R. conorii*, *R. prowazekii*, *R. typhi*, *R. tsutsugamushi*, *Erlichia* spp.
Protozoa	*Naegleria fowleri*, *Acanthamoeba* spp., *Angiostrongylus cantonensis*
Helminths	*Strongyloides stercoralis*
Other infectious diseases	Infective endocarditis, para-meninigeal foci of infection, viral post-infectious syndromes, post-vaccination
Medications	Antimicrobials, non-steroidals, azathioprine, OKT-3, cytosine arabinoside, carbamazepine, immune globulin, ranitidine
Systemic diseases	Systemic lupus erythematosus
Procedure related	Post-neurosurgery, spinal anaesthesia, intrathecal injections
Miscellaneous	Seizures, migraine, Mollaret's meningitis

Bacterial meningitis

The cause of acute bacterial meningitis depends on the age, immune status, and whether there has been recent head trauma or neurosurgery (see Table 5.4).The initiation of infection usually begins with nasopharyngeal colonization by a new organism, followed by systemic invasion. Important bacterial virulence factors include fimbriae, bacterial capsule, and production of IgA proteases. Host factors that predispose to meningitis include splenectomy and complement deficiencies.

Table 5.4 Causes of bacterial meningitis

Age/condition	Common organisms
0 to 4 weeks	Group B streptococcus, *E. coli*, *L. monocytogenes*, *K. pneumoniae*, *Enterococcus spp.*, *Salmonella* spp.
4 to 12 weeks	Group B streptococcus, *E. coli*, *L. monocytogenes*, *K. pneumoniae*, *H. influenzae*, *S. pneumoniae*, *N. meningitidis*
3 months to 18 years	*H. influenzae*, *N. meningitidis*, *S. pneumoniae*
18 to 50 years	*N. meningitidis*, *S, pneumoniae*, *S. suis*
>50 years	*S. pneumoniae*, *N. meningitidis*, *L. monocytogenes*, *aerobic Gram-negative bacilli*, *S. suis*
Immunocompromised	*S. pneumoniae*, *N. meningitis*, *L. monocytogenes*, aerobic Gram-negative bacilli *(e.g. E.coli*, *Klebsiella* spp., *Salmonella* spp., *Serratia marcescens*, *Pseudomonas aeruginosa)*
Basal skull fracture	*S. pneumoniae*, *H. influenzae*, group A streptococci
Head trauma, post-neurosurgery	*S. aureus*, *S. epidermidis*, aerobic Gram-negative bacilli
CSF shunt	*S. aureus*, *S. epidermidis*, *P. acnes*, aerobic Gram-negative bacilli

Clinical features

- Classical features include fever, headache, meningism (neck stiffness, photophobia, positive Kernig's and Brudzinski's signs), and cerebral dysfunction (confusion and/or reduced conscious level).
- Seizures occur in 30% of patients. Cranial nerve palsies (especially III, IV, VI, and VI) and focal signs are seen in 10–20% of cases. Hemiparesis may be due to a subdural effusion.
- Papilloedema is rare (<1%).
- Skin rash (initially macular then petechial) occurs in patients with meningococcal septicaemia but can occur in pneumococcal, *H. influenzae* or *S. suis* septicaemia.
- Rhinorrhoea or otorrhoea suggests basal skull fracture.
- Patients with *L. monocytogenes* have an increased risk of seizures and focal signs; some patients present with ataxia, cranial nerve palsies, nystagmus caused by rhomboencephalitis.

- Neonates may present with non-specific symptoms e.g. temperature instability, listlessness, poor feeding, irritability, vomiting, diarrhoea, jaundice, respiratory distress. Seizures occur in 40%, and a bulging fontanelle is a late sign.
- Elderly patients may present insidiously with confusion, lethargy, obtundation, no fever, and variable signs of meningeal inflammation.

Diagnosis

The diagnosis is confirmed by examination and culture of the CSF. In bacterial meningitis the following are typically seen:

- opening pressure >18 mm CSF
- CSF white cell count 1000–5000 cells/mm^3 (range 100–10,000)
- CSF neutrophils ≥80%
- CSF protein 0.1–0.5 g/dL
- CSF glucose ≤40 mg/dL or ≤2.2mmol/L
- CSF lactate ≥35 mg/dL or ≥1.9mmol/L
- Gram stain positive in 60–90%
- culture positive in 70–85%
- bacterial antigen detection positive in 50–100%
- bacterial PCR positive in 90%.

Management

- For acute management see flowchart in back cover
- Empirical antimicrobial therapy should be commenced immediately, pending investigations (Table 5.5).

Table 5.5 Empirical antibiotic therapy

Age/condition	Empiric therapy
Age 0–4 weeks	Ampicillin + cefotaxime or aminoglycoside
Age 4–12 weeks	Ampicillin + cefotaxone or ceftriaxone
Age 3 months – 18 years	Cefotaxime or ceftriaxone
Age 18–50 years	Cefotaxime or ceftriaxone
Age >50 years	Ampicillin + cefotaxime or ceftriaxone
Immunocompromised	Ampicillin + ceftazidime ± vancomycin
Basal skull fracture	Cefotaxime or ceftriaxone
Head trauma/neurosurgery	Vancomycin + ceftazidime
CSF shunt	Vancomycin + ceftazidime

- If the CSF Gram stain or culture is positive treatment should be tailored to the infecting organism (Table 5.6).
- Adjunctive corticosteroids have been recommended for the treatment of acute bacterial meningitis. The recommended regimen was dexamethasone 10 mg qds for 4 days, administered before or with the first dose of antibiotic. However, more-recent data from the developing world do not support this recommendation.[2,3]

Table 5.6 Specific antibiotic therapy

Organism	Antimicrobial therapy
H. influenzae type b	Ceftriaxone or cefotaxime for 7 days
N. meningitidis	Penicillin G or ampicillin or ceftriaxone for 7 days
S. pneumoniae	Ceftriaxone ± vancomycin for 10–14 days
L. monocytogenes	Ampicillin or penicillin G for 21 days
Group B streptococcus	Ampicillin or penicillin G for 14–21 days
E. coli	Ceftriaxone or cefotaxime for 21 days

- Reduction of raised intracranial pressure (ICP) may be achieved by various methods – elevating the head of the bed to 30 degrees to maximase venous drainage, hyperventilation to cause cerebral vasoconstriction, and use of hyperosmolar agents, e.g. mannitol.
- Neurosurgery may be required in certain circumstances – persistent CSF leak after basal skull fracture, congenital defects leading to recurrent meningitis, subdural empyema.

Prevention

- Vaccination – *H. influenzae* type B and meningitis C conjugate vaccine are part of the routine childhood immunization schedule in the UK. The quadrivalent meningitis vaccine (ACYW135) is recommended for patients with complement or properdin deficiency, asplenic patients, travellers to endemic areas, medical or laboratory personnel routinely exposed to *N. meningitidis*. *S. pneumoniae* vaccination is recommended in certain high-risk groups, e.g. age >65 years, chronic cardiovascular, pulmonary, renal, or liver disease, diabetes mellitus, alcoholism, CSF leak, asplenia, HIV, haematological and other malignancies, bone marrow transplant patients, immunosuppressive therapy.
- Chemoprophylaxis should be given within 24 h to household contacts, kissing contacts, and medical personnel involved in resuscitation of the index case. Rifampicin is the agent of choice for *H. influenzae type B* meningitis. For *N. meningitidis* the agents used are rifampicin (600 mg bd for 2 days), ciprofloxacin (500 mg stat) or ceftriaxone (250 mg IM). NB Rifampicin interacts with the oral contraceptive pill and may reduce its efficacy. Penicillin is not recommended to prevent secondary cases of *S. pneumoniae* meningitis but is recommended for children with sickle cell disease, although the optimum duration is unknown. Intravenous ampicillin, penicillin, clindamycin, or erythromycin is recommended for pregnant women colonized with group B streptococci or with obstetric risk factors for invasive disease. Meningitis C vaccination should be given for unvaccinated close contacts of meningitis C cases.

References

1. de Gans J, van de Beck D. Dexamethadone in adults with bacterial meningitis. *New Engl J Med* 2002; 347:1549–56.
2. Mai NTH, Chau TTH, Thuastas G et al. Dexamethadone in Vietnamase Adolescents and Adults with Bacterial Meningits. *New Engl J Med* 2007;357:2431–40.
3. Scarborough M, Gardan SB, whitty CJM et al. Corticosterords for bacterial meningitis in adults in sub-saharan Africa. *New Engl J Med* 2007; 357: 2441–50.757

Viral meningitis

Viruses are the major cause of the aseptic meningitis syndrome. This is usually characterized by lymphocytic pleocytosis in the CSF and sterile bacterial cultures.

Causes

- Enteroviruses are the leading cause of viral meningitis, e.g. echoviruses, Coxsackie viruses, enteroviruses 70 and 71.
- Arboviruses that cause meningitis include St Louis encephalitis virus, California, Eastern equine, Western equine, Venezuelan equine, and Colorado tick fever.
- Mumps virus is a common cause in unimmunized populations. CNS disease may occur in the absence of parotitis and is usually a benign self-limited disease.
- Herpes viruses include herpes simplex viruses (HSV-1 and HSV-2), varicella zoster virus (VZV), cytomegalovirus (CMV), Epstein–Barr virus (EBV), human herpes viruses. Although all of these can cause meningitis, herpes simplex viruses are the most-common cause and are often associated with primary genital HSV-2 infection.
- Human immunodeficiency virus (HIV) may cause meningitis as part of primary infection.
- Lymphocytic choriomeningitis virus (LCMV) is a rare cause of aseptic meningitis. It usually occurs in laboratory personnel, pet owners, or persons living in unsanitary conditions.

Pathogenesis

After colonization of mucosal surfaces, the virus invades and replicates prior to haematogenous dissemination. CNS invasion may occur by several mechanisms: via the cerebral microvascular endothelial cells, via the choroid plexus epithelium, or by spread along the olfactory nerve. Once CNS invasion occurs, inflammatory cells accumulate leading to the release of inflammatory cytokines, e.g. IL-6, IFN-γ, IL-1β, and synthesis of immunoglobulins, e.g. oligoclonal IgG.

Clinical features

- Enterovirus – in neonates, fever is accompanied by vomiting, anorexia, rash, upper respiratory tract symptoms. Meningeal signs (nuchal rigidity, bulging anterior fontanelle) may be present or absent, and focal signs are uncommon. A severe form may occur in the early neonatal period with hepatic necrosis, myocarditis, necrotizing enterocolitis, and encephalitis. In older children and adults, symptoms are milder with fever, headache, neck stiffness, and photophobia. There may be non-specific symptoms, e.g. anorexia, vomiting, rash, diarrhoea, cough, phayngitis, and myalgia. Other clues include community enteroviral epidemics, maculopapular or pustular rashes, conjunctivitis, pleurodynia, pericarditis, herpangina.
- Mumps virus – CNS symptoms usually occur 5 days after the onset of parotitis. Other findings include salivary gland enlargement (50%), neck stiffness, lethargy, abdominal pain.

- Herpesviruses – HSV-2 meningitis presents with classical symptoms. Complications include urinary retention, dysaesthesia, paraesthesia, neuralgia, motor weakness, paraparesis, difficulties in concentration, impaired hearing; these usually resolve within 3–6 months. EBV meningitis is associated with pharyngitis, lymphadenopathy, and splenomegaly. VZV meningitis is associated with a characteristic, diffuse vesicular rash.
- HIV – HIV-infected patients may present with a typical aseptic meningitis syndrome associated with acute primary HIV infection.
- LCMV – this is usually a biphasic illness that starts with non-specific viral symptoms, followed by improvement; 15% of patients develop severe headache, photophobia, lightheadedness, myalgia, and pharyngitis. Occasionally, arthritis, oorchitis, myopericarditis, and alopecia may occur.

Diagnosis

- CSF examination – CSF pleocytosis (100–1000 cells/mm^3) usually occurs. This may show a neutrophil predominance initially but becomes lymphocytic over 6–48 h. CSF protein level may be normal or mildly elevated. CSF glucose level is normal or mildly reduced.
- Viral culture – enteroviral meningitis may be identified by tissue culture although sensitivity is only 65–75%. Prolonged or asymptomatic viral shedding may occur. LCV is diagnosed by viral culture of blood or CSF (early infection) or urine (later infection). HSV-2 has been cultured from CSF and buffy coat of some patients. HIV has been isolated from the CSF of some patients with neurological disease. Arboviruses may be cultured from blood and CSF.
- Serology – rapid diagnosis of enteroviral infections is possible by detection of enteroviral IgM antibodies; the specificity of some tests is unsatisfactory. A four-fold rise in mumps antibody titres confirms the diagnosis of mumps meningitis. A salivary antibody test has been developed that looks promising. LCV and arboviral infections are usually diagnosed serologically.
- Molecular methods – cDNA nucleic acid probes for enteroviruses have been developed but have poor specificity (≤33%). PCR-based assays for enteroviruses are more promising with higher sensitivity and specificity than tissue culture. PCR-based assays are the diagnostic test of choice for herpes virus infections, e.g. HSV-2, CMV, VZV. HIV-RNA has been isolated from the CSF of some patients with meningitis.

Management

- Treatment of viral meningitis is mainly supportive, e.g. analgesics, antipyretics.
- Pleconaril has been used for enteroviral meningitis.
- Intravenous aciclovir is used for meningitis associated with HSV infection.
- No specific antiviral therapy exists for arboviruses, mumps virus, or LCV meningitis
- Antiretroviral therapy may be indicated for HIV infection, but is not given in primary infection (unless as part of a clinical trial).

Chronic meningitis

Chronic meningitis is a syndrome characterized by the subacute onset of meningoencephalitic symptoms (fever, headache, nausea, vomiting, neck stiffness, lethargy, and confusion) and CSF abnormalities which persist for at least 4 weeks There are a large number of infectious and non-infectious causes (Table 5.7).

Table 5.7 Causes of chronic meningitis

	Syndrome	Causes
Infectious	Meningitis	*Acanthamoeba* spp., *Angiostrongylus cantonensis*, brucellosis, candidiasis, coccidioidomycosis, cryptococcosis, histoplasmosis, Lyme disease, sporotrichosis, syphilis, tuberculosis
	Focal lesions	Actinomycosis, blastomycosis, cysticercosis, aspergillosis, nocardiosis, schistosomiasis, toxoplasmosis
	Encephalitis	African trypanosomiasis, cytomegalovirus, enterovirus (hypogammaglobulinaemia), measles (subacute sclerosing panencephalitis (SSPE)), rabies
Non-infectious	Meningitis	Behçet's disease, benign lymphocytic meningitis, granulomatous angiitis, malignancy, sarcoidosis

Clinical features

- History – an exposure history may suggest certain infections, e.g. tuberculosis, brucellosis, cysticerocosis, coccidiodomycosis, histoplasmosis, Lyme disease, syphilis, or HIV infection. In non-infectious cases, there may be a history of pre-existing systemic disease.
- Examination – diagnostic physical findings are rare. Skin lesions may be found in cryptococcosis, sarcoidosis, *Acanthamoeba* infection, coccidioidomycosis, blastomycosis, and secondary syphilis. Subcutaneous nodules may be found in cysticercosis and metastatic carcinoma. Lymphadenopathy and hepatomegaly suggest systemic disease. Eye examination may show choroidal tubercles, sarcoid granulomas, papilloedema, iritis, or uveitis. Neurological examination is non-discriminatory: focal signs indicate a cerebral mass lesion; hydrocephalus and cranial nerve palsies indicate basal meningitis; peripheral neuropathy suggests sarcoidosis or Lyme disease.

Laboratory diagnosis

- Blood tests – in addition to routine blood tests (FBC, ESR, CRP, creatinine, liver function tests) the following may be indicated: Mantoux test, blood culture for fungi and mycobacteria, serology for HIV and syphilis, serum cryptococcal antigen, ANA, and ANCA. Depending on the patient's exposure history the following tests may be indicated: serology for *Brucella*, *Borrelia burgdorferi*, *Histoplasma*, *Coccidiodes*.

- Radiology – a chest x-ray and CT or MRI brain scan should be performed in all cases. Meningeal enhancement and hydrocephalus are common findings.
- CSF examination should be performed in all cases (unless contraindicated by scan findings). The CSF should be analysed for cell count and differential, protein, glucose, lactate (Table 5.8). Diagnostic tests include Gram stain and culture, Ziehl–Neelsen stain for mycobacteria, India ink and cryptococcal antigen, syphilis serology. Depending on the patient's exposure history the following tests may be indicated: CSF antibodies to *Histoplasma*, *Coccidiodes*, *Blastomycosis*, *Taenia solium*, *Brucella*, and measles virus.

Table 5.8 CSF findings in chronic meningitis

CSF characteristic	Causes
Lymphocytic pleocytosis	Viral causes, TB meningitis
Neutrophilic pleocytosis	Actinomycosis, nocardiosis, HIV-associated CMV, early *M. tuberculosis* infection, aspergillosis, candidiasis
Eosinophilic pleocytosis	*Angiostrongylus*, *coccidioides*, cysticercosis, schistosomiasis, lymphoma, chemical
Pleocytosis <50 cells/microlitre	Behçet's disease, benign lymphocytic meningitis, carcinoma, HIV-associated cryptococcosis, sarcoidosis, vasculitis
Low CSF glucose	Actinomycosis, nocardiosis, carcinoma, cysticercosis, fungi, tuberculosis, syphilis, toxoplasmosis, chronic enterovirus, HIV-associated CMV, sarcoidosis, subarchnoid haemorrhage

In cases where the diagnosis remains obscure, the following tests may be helpful: repeat Mantoux test or TB interferon-gamma release assay; immunoglobulins, serum ACE; CSF antibody for *Sporothrix schenkii*, enteroviral culture and PCR. Biopsy of the brain or other tissues may also be indicated.

Management

- Specific therapy is tailored according to the cause of chronic meningitis (for further details see section on causative organism).
- Therapeutic trials may be indicated when a specific cause is not found despite comprehensive evaluation. Response to treatment may be slow, making interpretation difficult. Attempts to establish a diagnosis should be continued during therapeutic trial. In areas where tuberculosis is endemic, tuberculous meningitis (see next section) is the most-common cause of chronic meningitis, and empirical therapy is often initiated if the clinical presentation and CSF indices are compatible. Positive cultures or a clinical response to treatment are indications for continuing therapy. In areas where TB is not endemic, chronic meningitis is usually not infectious.

Tuberculous meningitis

- Caused by *M. tuberculosis* (see 📖 Mycobacterium tuberculosis, p.407).
- Clinical features – non-specific with gradual onset of meningeal symptoms, cranial nerve palsies (III, IV and VI), hemiplegia, or paraplegia, urinary retention. CXR is abnormal in 50% of cases and may show pulmonary or miliary tuberculosis. CT or MRI brain scan may show hydocephalus, basal meningeal enhancement, infarcts or tuberculomas.
- Laboratory diagnosis – CSF findings include a lymphocytic pleocytosis (100–500 cells/mm^3), increased CSF protein and decreased CSF glucose levels. Neutrophils may predominate in early disease and in HIV-infected patients. Diagnosis confirmed by detection of *M. tuberculosis* by CSF Ziehl–Neelsen smear or culture. Smear positivity rates are generally low (10–22%) but may be increased to >50% if the spun deposit of a large volume of CSF (5–10 mL) is examined meticulously. PCR detection of mycobacterial DNA shows a sensitivity of 27–85% and a specificity of 95–100%. CSF cultures are positive in 38–88% of cases.
- Management – the optimum drug choice and duration of treatment has not been established in clinical trials. The UK and US TB treatment guidelines both recommend a four-drug initiation phase (rifampicin, isonazid, pyrazinamide, and ethambutol) for 2 months, followed by a two-drug continuation phase (rifampicin and isoniazid) for 10 months. As isoniazid and pyrazinamide are the only two drugs that have good CSF penetration, some experts recommend continuing pyrazinamide during the continuation phase. Studies from India and South Africa suggest that 6 months of therapy may be adequate. Adjunctive dexamethasone has been shown to reduce mortality by 30%.

Cryptococcal meningitis

- Caused by *Cryptococcus neoformans* (see 📖 Cryptococcus neoformans, p.527). *C. neoformans* var. *neoformans* has a worldwide distribution and tends to cause disease in immunocompromised patients. *C. neoformans* var. *gatii* occurs in tropical and subtropical climates and tends to affect non-immunocompromised patients.
- Clinical features – subacute presentation with fever, meningoencephalitis, visual loss, and focal signs (<30%).
- Laboratory diagnosis – CSF findings include raised CSF pressure, lymphocytic pleocytosis (40–400 cells/mm^3), low CSF glucose (55%), and positive India ink stain (≤50%). CSF findings may be normal in HIV patients. Serum and CSF crypococcal antigen tests can increase the diagnostic rate to ≥90%. CSF cultures are positive in 75% of patients. Cultures of blood, urine, and sputum may increase the diagnostic rate.
- Management – immunocompetent patients are treated with amphotericin (0.4 mg/kg/day) and flucytosine (150 mg/kg/day) for 6 weeks. Alternatives include amphotericin alone (0.5–0.7 mg/kg/day) or fluconazole (200–800 mg/day). Treatment of HIV-associated cryptococcal meningitis is with amphotericin ± flucytosine for 2 weeks followed by fluconazole or itraconazole 400 mg/day for 8 weeks. Higher doses of fluconazole (800–2000 mg/day) have been tried, with or without flucytosine. Lifelong

maintenance therapy with fluconazole 200 mg is indicated in all patients. Raised intracranial pressure may benefit from serial lumbar punctures, diuretics (e.g. acetazolamide), or venticuloperitoneal shunts, although comparative evidence is lacking.

Coccidioidal meningitis

- *Coccidiodes immitis* (see 📖 Coccidioides immitis, p.556) is endemic in the arid and semi-arid regions of the western hemisphere, e.g. southwest USA, Mexico, central and south America.
- Clinical features – CNS involvement may be part of generalized coccidiodidomycosis or may be the only site of extra-pulmonary disease. Meningitis usually occurs within 6 months of pulmonary infection, but can occur up to 12 years after primary infection. Headache is the most-prominent symptom but the clinical syndrome is indistinguishable from other causes of chronic meningitis.
- Laboratory diagnosis – CSF findings resemble those of cryptococcal meningitis but eosinophilia may be seen. CSF cultures are positive in 30% and spherules are sometimes seen on CSF smear. CSF antibodies are positive in 55–95% of patients; ELISA to spherule is more sensitive than complement fixation tests (CFTs). A serum CFT titre of 1:16 is supportive of the diagnosis. Skin tests with spherulin are positive in 33–55% of patients.
- Management – fluconazole (400 mg/day) is associated with a 70% response rate. Higher fluconazole doses may be used in patients who do not respond initially. Ketoconazole is an alternative. Intrathecal amphotericin B is used in patients who do not respond to oral azole therapy. Hydrocephalus may require a ventriculoperitoneal shunt.

Histoplasma meningitis

- Caused by *Histoplasma capsulatum* (see 📖 Histoplasma capsulatum, p.548).
- Clinical features are non-specific with fever and gradual onset of meningitic symptoms over weeks or months. Oral mucosal lesions occur in 16% of patients and are more common than skin lesions.
- Laboratory diagnosis – CSF cultures are positive in 27–65% of cases. Blood cultures should also be done. Detection of serum and CSF antibodies is the most-sensitive test, but problems occur with cross-reactivity to other fungi. Histoplasma polysaccharide antigen may be detected in blood, urine, or CSF in 61% of patients.
- Management – treatment is with amphotericin (0.7–1.0 mg/kg/day). Lumbar punctures should be performed weekly for 6 weeks then every 2 weeks to assess response. Initial response rates are good but relapse rates are high resulting in an overall cure rate of 50%.

Neuroborreliosis

- Caused by *Borrelia burgdorferi*, transmitted to humans by the asymptomatic bite of the deer tick *Ixodes scapularis*.
- Clinical features – early infection is characterized by flu-like symptoms and a characteristic rash (erythema chronicum migrans). Neuroborreliosis is characterized by fever, headache, cranial nerve palsies, peripheral neuropathy ~4 weeks after primary infection.
- Laboratory diagnosis – CSF examination shows a lymphocytic pleocytosis. Diagnosis is confirmed by positive serology in the context of an appropriate exposure history. Serology is insensitive in early disease and false-positive and -negative results are a considerable problem. The most specific test is detection of *B. burgdorferi* antibodies in the CSF and comparison of CSF and serum antiboidy levels by an immunocapture assay. PCR detection of *B. burgdorferi* has poor sensitivity (25–38%) but high specificity.
- Management – IV ceftriaxone 2 g/day for 2–4 weeks. Alternatives: cefotaxime 2 g tds or penicillin G 3 g IV qds.

Neurocysticercosis

- Caused by *Taenia solium* (p.604), the most-common parasitic disease of the CNS. Infection is endemic in Mexico, central and south America, the Carribean, sub-Saharan Africa, India, and China.
- Pathogenesis – infection is acquired by the ingestion of *T. solium* eggs. Larvae hatch in the small intestine, invade the bloodstream, and disseminate to the muscle, eye, and brain where they encyst. Symptoms occur when the larvae die causing an inflammatory response.
- Clinical features – seizures, focal neurological deficits, chronic basilar meningitis, and hydrocephalus. CT or MRI scans show multiple cystic and calcified lesions in the brain paranechyma. Skeletal muscle calcification may also be seen (<10%).
- Laboratory diagnosis – CSF examination shows a lymphocytic or eosinophilic pleocytosis, low CSF glucose (25%), and elevated CSF protein. Positive serology is supportive of the diagnosis, but there is considerable cross-reactivity with other helminth infections. Immunoblotting techniques, using the purified glycoprotein fraction of cyst fluid, appear sensitive and specific. Sensitivity of antibody testing appears higher in patients with multiple cysts.
- Management – remains controversial as the benefit of antihelminthic therapy has not been firmly established. If the decision is made to treat albendazole (15 mg/kg/day for 7–30 days) or praziquantel (50 mg/kg/day for 1–21 days) may be given. More recent studies have recommended a single day regimen of praziquantel. Seizures should be controlled with anticonvulsants and symptomatic hydrocephalus relived by shunting. Corticosteroids may be given to reduce CNS inflammation, but can reduce praziquantel levels.

Encephalitis

Encephalitis is an inflammation of the brain which is usually caused by viral infections or non-infectious agents (Table 5.9). It is often accompanied by meningeal inflammation, and is therefore referred to as encephalomyelitis.

Table 5.9 Viral cause of encephalitis

Family	Viruses
Togaviridae	Eastern equine, Westerrn equine Venezuelan equine viruses
Flaviviridae	St Louis, Murray Valley, West Nile, Japanese B encephalitis, dengue
Bunyaviridae	La Crosse, Rift Valley, Toscana
Paramyxoviridae	Mumps, measles, Hendra, Nipah
Arenaviridae	Lymphocytic choriomeningitis, Machupo, Lassa, Benin
Picornaviridae	Polio, Coxsackie, echovirus, hepatits A
Reoviridae	Colorado tick fever
Rhabdoviridae	Lyssavirus, rabies
Filoviridae	Ebola, Marburg
Retroviridae	HIV
Herpesviridae	HSV-1, HSV-2, VZV, HBV, CMV, HHV-6, HHV-7, EBV
Adenoviridae	Adenovirus

Pathogenesis

Infectious agents cause clinical symptoms and signs in the CNS, either by direct invasion or indirectly, without invading the parenchyma. Viruses enter the CNS in two of ways: via the bloodstream (most viruses) or via the peripheral nerves (e.g. HSV, VZV, polio, rabies). Once the infectious agent enters the brain, only certain cells will become infected – this results in variable clinical manifestations, e.g. seizures, demyelination, impaired consciousness, coma, respiratory failure. In fatal viral encephalitis an inflmammatory reaction is usually prominent in the meninges and in a perivascular distribution in the brain. Neural cells may show degenerative changes and apparent phagocytosis of neural cells by macrophages. Intranuclear inclusion bodies are seen in herpesvirus and adenovirus infections. When acute demyelinating disease complicates viral infections outside the brain, damage is though to be related to induction of an immune response against CNS myelin rather than invasion on the brain.

Clinical features

- Patients usually present with signs and symptoms of meningeal irritation – headache, neck stiffness, and a CSF pleocytosis.
- Viral encephalitis is characterized by alterations in consciousness, progressing from mild lethargy to confusion, to stupor and coma.
- Focal neurological signs frequently develop and seizures are common.

- Motor weakness, attenuation of reflexes, and extensor plantar responses may be seen.
- Some viruses may cause CNS symptoms as part of a postinfectious encephalomyelitis, e.g. mumps, measles, rubella, influenza.
- Certain diseases are associated with characteristic symptoms or signs:
  - HSV encephalitis – personality change, hallucinations, aphasia
  - VZV encephalitis – acute contralateral hemiparesis after herpes zoster ophlamicus (due to frontal infarct)
  - Japanese B encephalitis – Parkinsonian syndrome
  - rabies – local paraesthesia at the site of the bite
  - polio – flaccid paralysis
  - Rocky Mountain spotted fever – rash on palms and soles
  - Enteroviral infections – viral exanthems
  - tick-borne encephalitis – history of tick bite
  - rabies – history of bat bite.

Diagnosis

- CSF examination is essential.
- CSF pleocytosis is variable (10–2000 cells/mm^3) and lymphocytes usually predominate. However, in early disease there may be no cells in the CSF or neutophils.
- Red cells may be found in HSV encephalitis.
- CSF protein is usually increased.
- CSF glucose is usually normal or slightly low.
- Detection of viral nucleic acids in the CSF by PCR is useful for the diagnosis of number of infections, e.g. herpesviruses and enteroviruses. PCR may be negative in early disease so should be repeated after 48–72 h.
- Acute and convalescent serology may help to confirm the diagnosis.
- Electroencephalogram (EEG) is often abnormal.

Treatment

Treatment is mainly supportive. Specific treatment is only available for some of the causes of viral encephalitis:

- aciclovir for HSV encephalitis
- ganciclovir and foscarnet for CMV disease
- antiretroviral therapy for HIV disease.

Prevention

Some diseases may be prevented by vaccination, e.g. mumps, measles, polio, Japanese B encephalitis.

Cerebral abscess

A focal intracerebral infection that begins as a local area of cerebritis and develops into a collection of pus surrounded by a well-vascularized capsule. The introduction of surgical drainage and antimicrobial theapy has transformed this condition from an almost uniformly fatal disease to a treatable condition.

Epidemiology

Brain abscesses are an uncommon but severe disease. They tend to occur more frequently in males with a median age of presentation of 30–40 years. If the original site of infection is the paranasal sinuses, most of the patients are 10–30 years old. If the initial site of infection is the ear then patients are generally <20 or >40 years of age; 75% of cerebral abscesses occur in adults. Case fatality rates range from 0% to 24%.

Aetiology

Cerebral abscesses may be caused by a broad spectrum of pathogens. Some pathogens are associated with certain predisposing conditions (Table 5.10).

Table 5.10 Factors Predisposing to cerebral abscess

Predisposing condition	Microorganisms
Otitis media/mastoiditis	Streptococci, *Bacteroides* spp., *Prevotella* spp., *Enterobacteriaceae*
Sinusitis	Streptococci, *Bacteroides* spp., *Enterobacteriaceae*, *S. aureus*, *Haemophilus* spp.
Dental sepsis	*Fusobacterium*, *Prevotella*, *Bacteroides* spp., streptococci
Trauma or neurosurgery	*S. aureus*, streptococci, *Enterobacteriaceae*, *Clostridium* spp.
Pulmonary/pleural sepsis	*Fusobacterium*, *Actinomyces*, *Bacteroides*, *Prevotella* spp., streptococci, *Nocardia* spp.
Endocarditis	*S. aureus*, streptococci
Congenital heart disease	Streptococci, *Haemophilus* spp.
Neutropenia	Aerobic Gram-negative bacilli, *Aspergillus* pp., *Mucorales*, *Candida* spp.
Transplantation	*Candida* spp., *Aspergillus* pp., Mucorales, *Enterobacteriaceae*
HIV	*T. gondii*, *Nocardia* spp., *Mycobacterium* spp., *L.monocytogenes*, *C. neoformans*

Pathogenesis

Bacteria reach the brain by several different mechanisms:

- contiguous spread, e.g. from the ear or sinuses
- haematogenous spread, e.g. bloodstream infections, pulmonary sepsis
- direct inoculation, e.g. trauma or surgery.

The brain may be more susceptible to infection than other tissues such as the skin. It may also have different susceptibilities to infection by different organisms. The role of bacterial virulence factors in brain abscess formation has not been adequately studied.

Clinical features

Cerebral abscesses may present with a gradual decline or a rapid deterioration. The presenting symptoms and signs include:

- headache
- altered mental status
- focal neurological deficits
- fever
- seizures
- nausea and vomiting
- neck stiffness
- papilloedema.

The location of the abscess defines the clinical symptoms:

- frontal – headache, drowsiness, inattention, altered mental state, hemiparesis, speech disorder
- temporal – ipsilateral headache, aphasia, visual field defect
- cerebellar – ataxia, nausea, vomiting dysmetria.

Certain pathogens may have specific charactersistics:

- *Nocardia* spp. – chronic pulmonary condition and skin nodules
- rhinocerebral mucormycosis – eye and sinus pain, nasal stuffiness.

Diagnosis

- An urgent CT scan of the brain enables rapid diagnosis – cerebral abscesses appear as ring-enhancing lesions surrounded by oedema. MRI is more sensitive than CT, enabling the early detection of cerebritis.
- If single or multiple ring-enhancing lesions are seen then the patient should be referred to neurosurgery for a diagnostic aspirate.

Treatment[1]

- A diagnostic aspirate should be obtained and the patient commenced on broad-spectrum empiric therapy (e.g. ceftriaxone and metronidazole).
- Once culture results are available, treatment can be rationalized according to antimicrobial sensitivities. Antimicrobial therapy is given for 2–4 weeks intravenously followed by 2–6 months orally.
- Adjunctive corticosteroids should be given to patients with significant oedema and mass effect.

Reference

1 de Louvois J, Brown EM, Bayston Rn et al. The rational use of antibiotics in the treatment of brain abscess. *British Journal of Neurosurgery* 2000; 14(6):525–30.

Subdural empyema

A collection of pus in the space between the dura and the arachnoid.

Epidemiology

Accounts for 15–20% of localized intracranial infections. Risk factors: sinusitis, otitis media, mastoiditis, skull trauma, neurosurgery, infection of pre-existing subdural haematoma, cranial traction devices, nasal surgery, ethmoidectomy, or polypectomy. Metastatic infection accounts for ~5%. A complication of meningitis in infants. Spinal subdural empyema is rare and occurs secondary to metastatic infection.

Aetiology

Causative organisms include streptococci, staphylococci, aerobic Gram-negative bacilli, anaerobes. Polymicrobial infections are common. Post-operative/traumatic empyemas are usually caused by staphylococci or aerobic Gram-negative bacilli. Unusual causes include *Salmonella* spp., *Propionobacterium acnes*, *M. tuberculosis*, and *Candida* spp.

Clinical features

Acute onset of fever, headache (may be localized to infected sinus or ear), vomiting, altered mental state (disorientation, drowsiness, coma), and focal neurological signs (hemiparesis, cranial nerve palsies, dysphasia, homonymous hemianopia, cerebellar signs); ~80% of patients have meningeal symptoms/signs. Seizures occur in ≤50% of cases. There may be rapid neurological deterioration with signs of raised intracranial pressure and cerebral herniation. Complications: septic venous thrombosis, cerebritis, cerebral abscess. In infants with subdural empyema, persistent fever, decline in neurological status, and seizures are seen. Spinal epidural abscess presents with radicular pain and signs of spinal cord compression.

Diagnosis

Consider the diagnosis in any patient with meningism and focal neurological signs. Lumbar puncture is contraindicated because of the risk of cerebral herniation. CT or MRI brain shows a crescentic or elliptical area of hypodensity below the cranial vault adjacent to the falx cerebri. After administration of contrast, a fine line of enhancement is seen between the subdural collection and the cerebral cortex. MRI is more sensitive than CT and is considered the investigation of choice.

Management

Subdural empyema is an emergency and requires immediate surgical and medical management. Samples should be sent for urgent microscopy and culture. Commence IV antibiotics immediately after aspiration, based on the likely infecting organisisms, e.g. ceftriaxone and metronidazole. Vancomycin should be added for suspected staphylococcal infection. Tailor treatment to culture results once available. Outcome is related to conscious level at presentation (>90% for patients who are awake/alert and <50% in patients who are unresponsive to pain); 10–44% of survivors experience permanent neurological sequelae.

Epidural abscess

A localized collection of pus between the dura mater and the overlying skull or vertebral column. May be complicated by subdural empyema.

Epidemiology

The epidemiology of cranial epidural abscess is similar to subdural empyema. Spinal epidural abscess usually occurs following haematogenous spread from another site of infection (e.g. bacteraemia) or by extension of vertebral osteomyelitis. Risk factors: bacteraemia, diabetes mellitus, skin infections, spinal trauma/surgery, decubitus ulcers, lumbar puncture, epidural anaesthesia/analgesia.

Aetiology

The causes of cranial epidural abscess are similar to subdural empyema. *S. aureus* is the most common cause of spinal epidural abscess. Others include aerobic and anaerobic streptococci, aerobic Gram-negative bacilli (especially *E. coli* and *P. aeruginosa*); 5–10% are polymicrobial. Unusual causes include *Nocardia* spp., *M. tuberculosis*, and fungi.

Clinical features

- The presentation of cranial epidural abscess may be insidious, masked by the primary focus of infection, e.g. sinusitis, otitis media. Headache is common and focal neurological signs and seizures eventually develop followed by signs of raised intracranial pressure.
- Gradenigo's syndrome, characterized by unilateral facial pain and V and VI cranial nerve palsies, may occur if the abscess is close to the petrous bone.
- Spinal epidural abscess may present acutely (hours to days with haematogenous seeding) or chronically (weeks to months with vertebral osteomyelitis). Pain is the most common symptom (70–90%) followed by fever (60–70%). There are four clinical stages: (1) back pain and tenderness; (2) nerve root pain; (3) spinal cord symptoms, e.g. motor or sensory deficits, sphincter dysfunction; and (4) paralysis.

Diagnosis

Gadolinium-enhanced MRI is the diagnostic investigation of choice; abscesses appear as low-density lesions with linear enhancement.

Management

Epidural absces is an emergency and requires immediate surgical and medical management. The management of cranial epidural abscess is similar to subdural empyema – surgical drainage and antibiotics (3–6 weeks after drainage). The management of spinal epidural abscess is urgent surgical decompression (laminectomy) and antibiotics. Empirical therapy should cover staphylococci (e.g. vancomcyin) and aerobic Gram-negative bacilli (e.g. ceftazidime or meropenem). The outcome of spinal epidural abscess depends on the level of neurological deficit before decompression. Complete recovery is possible if neurological signs have been present for <24 h.

CSF shunt infections

Infection is a frequent complication of neurosurgical procedures used to treat hydrocephalus, occurring in approximately 10% of cases. The types of device that may become infected are:

- ventriculo-atrial (VA), ventriculoperitoneal (VP) or ventriculopleural shunt
- Ommaya drains
- external ventricular drains (EVDs)
- lumbar-peritoneal or lumbar-pleural shunt.
- Shunt infections may be classified as:
- internal (associated with CSF abnormalities), or
- external (associated with soft tissue abnormalities)

Aetiology

- *S. epidermidis* is the most common isolate
- *S. aureus* including MRSA
- Streptococci, enterococci
- Corynebacteria e.g. *Propionobacterium acnes*
- Gram-negative organisms
- Mycobacteria
- Fungi

Pathogenesis

The mechanisms of infection are:

- contamination (at implantation of device)
- externalization (erosion of shunt through skin)
- retrograde (perforation of VPS through bowel)
- haematogenous (rare).

Clinical features

- These depend on age of the patient, site of infection, and whether there is raised intracranial pressure.
- Symptoms include fever, headache, vomiting, neck stiffness, impaired conscious level, endocarditis (VA shunts), abdominal pain (VP or lumbar-peritoneal shunts).
- External shunt infections are associated with erythema, abscess, or shunt erosion.

Laboratory diagnosis

- Blood tests – full blood count, differential white cell count, urea and electrolytes (U&Es), glucose, liver function tests, ESR, CRP. NB a normal white cell count, ESR, and C-reactive protein do not exclude shunt infection
- Blood cultures – 90% positive with VA shunt infections
- CT/MRI brain to look for raised intracranial pressure
- Urine dipstick for haematuria and proteinuria – VA shunts may be associated with shunt nephritis

- CSF examination should be performed by neurosurgical team. CSF samples should be taken for urgent MC&S, protein, and glucose. All abnormal results should, ideally be confirmed by second sample within 24 h, unless clinical condition mandates immediate treatment
- Chest x-ray, if VA or ventriculo-pleural shunt
- Consider abdominal ultrasound or CT if abdominal symptoms or signs and VP or lumbar peritoneal shunt
- Consider echocardiogram if VA shunt

Management[1,2]

CSF shunt infections should be managed by neurosurgeons with infectious disease/microbiology input.

- External shunt infections should be managed by drainage of pus, removal of infected device and bone flap if present, soft tissue closure if possible, insertion of temporary device at a new site, interval antibiotics for 7–14 days, followed by replacement with new permanent device.
- Internal shunt infections should ideally be managed by shunt removal, external ventricular drainage placement or ventricular taps, interval antibiotics for 7–14 days, followed by insertion of a new device when CSF sterility is achieved.
- Empiric antibiotic therapy should be with vancomycin intravenously (1 g bd IV, monitor levels) and intrathecally (10 mg od intrathecal). Intravenous meropenem (2 g tds) should be added if there are abdominal symptoms or Gram-negative organisms seen in the CSF. Consider gentamicin intravenously (5 mg/kg od IV) and intrathecally (1–5 mg od, monitor levels) if there is evidence of endocarditis.
- Specific antibiotic therapy should be tailored in the light of culture results and clinical response.
- Salvage therapy – in some cases e.g. coagulase-negative staphylcoccal infections where shunt removal is not possible, salvage therapy may be attempted with vancomycin and rifampicin (but there are insufficient data to support this strategy).

Prognosis

The prognosis of internal shunt infections varies with management strategy:

- intravenous + intrathecal antibiotic therapy with two-stage exchange – 90% cure
- intravenous antibiotic therapy with one-stage exchange – 70% cure
- Intravenous + intrathecal antibiotic therapy – 40% cure
- intravenous antibiotic therapy alone – 20% cure
- salvage therapy (for coagulase-negative staphylococcal infection only) without shunt exchange (but with an Ommaya reservoir) using intrathecal vancomycin and oral rifampicin – 40–99% cure.

References

1 Treatment of infections associated with shunting for hydrocephalus. Working party on the use of antibiotics in Neurosurgery of the British Society of Antimicrobial Chemotherapy. *Br J Hosp Med* 1995;53:368–73.

2 Management of neurosurgical patients with post operative bacterial or aseptic meningitis or external ventricular drain-associated ventriculitis. Infection in Neurosurgery Working Party of the British Society of Antimicrobial Chemotherapy. *Br J Neurosurg* 2000;14(1):7–12.

Peri-orbital infections

Blepharitis

Inflammation of the lid margins. Bacterial infection is usually secondary to minor trauma, and often occurs in association with seborrhoeic dermatitis, acne rosacea, and pubic lice infestations.

Anterior blepharitis

Affects the lid where eyelashes attach and usually caused by bacteria colonizing the base of the eyelashes (e.g. *S. aureus*). Infection of the pilosebaceous glands of Zeiss and Moll may result in an abscess (a 'stye'). Anaerobic infection may follow certain injuries, e.g. bites. Symptoms: erythema, pruritus, and crusting of lid margins. Chronic infections are caused by infection with *S. aureus*, CoNS, and more rarely *Pseudomonas* spp., *Proteus mirabilis*, or *Capnocytophaga ochracea*. Clinical features: hyperaemia, crusted exudates around the base of the lashes, lash loss. Exclude the presence of lice and their eggs. Cell-mediated immunologic mechanisms have been implicated in the pathogenesis of chronic blepharitis.

Posterior blepharitis

Affects the inner portion of the eyelid where it contacts the eye and due to Meibomian gland dysfunction and infection. May present acutely as an 'internal stye' or hordeolum (pain and swelling are usually apparent on the conjunctival surface of the lid) or chronically as a painless cysts (chalazion). Symptoms: eye watering, foreign body or burning sensation. An internal stye may rarely progress to cause preseptal cellulitus.

Treatment

Eyelid hygiene may be sufficient in most cases. Blepharitis thought to be infectious in nature should be treated with a topical antibiotic, frequency and duration of treatment determined by severity (chloramphenicol, bd for up to 2 weeks). Chalazion may require incision and drainage. Cases following trauma may require oral therapy with anaerobic cover, e.g. animal bites. Predisposing conditions should be treated, e.g. lice (malathion), acne rosacea (oral tetracycline), seborrhoeic dermatitis (topical antifungal/ steroid combinations).

Other causes of lid inflammation: cosmetic contact allergy, molluscum contagiosum, louse infestation (e.g. *Phthirus pubis*), dermatoblepharitis secondary to HSV infection or spread of adjacent impetigo.

Infections of the lacrimal apparatus

The lacrimal gland is found at the lateral upper lid margin. It produces around 10 mL of tears a day, the act of blinking serving to smear the tear film from the lateral to the medial edge of the eye surface. Drainage is via the puncta at the inner canthus into the canaliculi, and from here to the lacrimal sac, the nasolacrimal duct, and out into the nose.

Canaliculitis

Low-grade inflammation of the canaliculi usually chronic and due to infection by *Propionibacterium* spp. or *Actinomyces*. Form gritty casts that obstruct the lacrimal duct leading to eye-watering, chronic conjunctivitis and nasal lid swelling. Treatment: antibiotic irrigation with canaliculotomy and curettage where necessary.

Dacryocystitis

Inflammation of the lacrimal sac usually in the setting of obstruction of the sac or duct (congenital, secondary to infection, tumor, or trauma). Common in infants, resolving spontaneously by 12 months of age. Organisms include *S. penumoniae*, *S. aureus*, *P. aeruginosa*. Recurrent cases may be seen with sarcoidosis or *Chlamydia trachomatis*. The only symptom may be eye-watering. Acute cases follow obstruction of both the proximal and distal ends of the drainage system, e.g. sarcoidosis, trauma. The main symptom is pain in the region of the tear sac. It may be possible to express purulent material through the lacrimal puncta. Cases may require dacryocystorhinostomy. Orbital cellulitus is a serious complication of acute dacrocystitis. Treatment: in newborn lacrimal sac massage may be sufficient to resolve the blockage and most cases will resolve with time – if not, probing may resolve the problem. Adults: systemic antibiotics.

Dacryoadenitis

Inflammation of the lacrimal gland. Symptoms: localized tenderness/swelling of the outer upper eyelid, with conjunctivitis and peri-orbital oedema. Pyogenic bacteria are the usual causes. Viral infections may be seen in children, e.g. mumps. Ocular motility defects may occur. Chronic infections may be caused by TB, syphilis, leprosy, and fungi. Treatment: systemic antibiotics. Drainage if a collection develops.

Mikulicz syndrome: dacryoadenitis associated with inflammation and swelling of the salivary glands (of any aetiology).

Orbital infections

The orbital septum is a fibrous sheet lying beneath orbicularis oculi. It extends from the periosteum of the orbit and fuses to the levator aponeurosis in the upper lids and orbital retractor in the lower lids. It acts as a physical barrier to infection. Orbital cellulitus (infection within the septum) is an ophthalmic emergency and must be differentiated from the less-devastating preseptal cellulitus. Early involvement of an ophthalmologist is essential. Children with preseptal infection are at high risk of progressing to orbital cellulitus due to the undeveloped nature of the orbital septum, and should be managed as orbital cellulitus.

Preseptal cellulitus

An infection of the superficial skin around the eyes anterior to the orbital septum. It may follow infection of adjacent structures (e.g. dacryocystitis) or trauma.

- **Aetiology** – *S. aureus*, group A streptococci, *H. influenzae* (if unvaccinated).
- **Clinical features** – hyperaemia of eyelid skin, soft tissue distention, low-grade fever. Proptosis and impairment of ocular motility are not seen and their presence suggests orbital cellulitus. Optic nerve function is normal. Complications include CNS infections and progression to orbital cellulitus.
- **Investigations** – Gram stain and culture of any discharge, CT/MRI if any question of orbital involvement.
- **Management** – oral antibiotics are sufficient in simple cases (e.g. flucloxacillin 500 mg qds with metronidazole 400 mg tds, both for 7 days). Intravenous antibiotics may be required, particularly if *H. influenzae* is suspected, e.g. a cephalosporin. Ensure anaerobic cover if infection developed following trauma (e.g. bite).

Orbital (postseptal) cellulitus

Acute infection of orbital contents. An ophthalmic emergency with the risk of visual loss and posterior extension to the cavernous sinus (leading to possible thrombosis and death). Most cases result from contiguous spread from infected sinuses, e.g. paranasal, ethmoid, frontal. Other causes: trauma, otitis media, dental infection.

- **Aetiology** – *S. aureus, group A streptococci, S. pneumoniae*, anaerobes and Gram-negatives (particularly in those cases following chronic sinus infection). Rare causes: TB, fungi (aspergillosis, mucormycosis – causes a rare but very aggressive form of orbital cellulitus. See p.536).
- **Clinical features** – fever, lid oedema, rhinorrhoea, pain and tenderness over the eye, headache, limited and painful eye movements. Proptosis (risk of exposure keratopathy) and ophthalmoplegia develop as infection progresses. Late signs: increased orbital pressure, reduced corneal sensation, and congestion of retinal veins. Cases of very posterior infection (orbital apex syndrome) manifest as visual loss and ophthalmoplegia with very limited external features. The most serious complication is cavernous sinus thrombosis manifest by headache, eye pain, neck stiffness, swelling over forehead and eyelids, with papilloedema, ophthalmoplegia, and altered level of consciousness. Other complications: meningitis, cerebral abscess.

- **Investigations** – Gram stain and culture of material if possible (external drainage is rare and aspiration of fluid from the orbit is usually contraindicated except as part of a surgical procedure), blood culture, lumbar puncture if meningism, sinus x-rays, CT/MRI (MRI preferred to diagnose cavernous sinus thrombosis). If surgical intervention (e.g. ethmoidectomy) is performed, the organism may be recovered from material despite empirical antibiotic therapy.
- **Treatment** – urgent antibiotics (e.g. IV ceftriaxone and PO metronidazole), ophthalmology opinion, ENT review (sinus surgery may be required). Continuing deterioration on therapy suggests the development of an abscess, and a repeat CT should be performed with a view to surgical drainage if necessary. The management of fungal orbital cellulitus is a complex mix of surgical debridement and antifungal therapy.

Table 5.11 Orbital versus preseptal cellulitis

	Preseptal	Orbital
Proptosis	Absent	Present
Ocular motility	Normal	Painful and restricted
Visual acuity	Normal	Reduced in severe cases
Colour vision	Normal	Reduced in severe cases
Relative afferent pupillary defect	Normal	Present in severe cases

Reproduced from Denniston A and Murray P *Oxford Handbook of Ophthalmology* . Oxford: Oxford University Press, 2005, with permission from Oxford University Press.

Conjunctivitis

Conjunctivitis is the most common ocular inflammation and may be a primary/ local infection, or part of a systemic infection (e.g. leptospirosis, measles). Some organisms (e.g. *Chlamydia trachomatis*) cause very specific syndromes, but most cannot be distinguished clinically. Viruses are the most common cause. Acute conjunctivitis resolves within 4 weeks, chronic conjunctivitis persists for ≥4 weeks. Conjunctivitis is typically self-limiting but can progress to potentially sight-threatening infections.

Aetiology

- Viruses – the most common cause, e.g. adenovirus, Coxsackie A24, enterovirus 70, herpes simplex virus, varicella zoster virus, smallpox, vaccinia, rubella, rubeola, mumps, influenza, EBV
- Chlamydia – *C. trachomatis, C. pneumoniae*
- Bacterial – *S. aureus, S. pneumoniae, H. influenzae, Moraxella* spp., C, *diphtheriae, Neisseria* spp., and enteric Gram-negative rods
- Parasitic – *Leishmania* spp., *Trypanosoma* spp., microsporidia, cryptosporidia, fly larvae, loa loa, *Phthirus pubis* (pubic lice), *Demodex* (mites)
- Fungal – *Candida* spp., *Blastomyces* spp., *Sporothrix schenkii*
- Allergic or toxic – cosmetics, soaps, detergents, medications

Clinical features

- Irritation and itching are the most common symptoms. Ocular pain is unusual unless there is ulceration, e.g. HSV or corneal involvement.
- Visual acuity is normal or slightly reduced (unless the cornea is involved).
- Skin lesions are seen with HSV, VZV, pox viruses, immune-mediated diseases, e.g. Stevens–Johnson syndrome.
- Conjunctival hyperaemia is worse in the periphery than in the limbal region. Saccular aneurysms, petechiae, subconjunctival or intraconjunctival haemorrhages may be present.
- Ocular secretion may be due to increased lacrimal flow or impaired drainage.
- Conjunctival oedema (chemosis) may be marked, resulting in an inability to close the eyelids. Chronic chemosis may lead to conjunctivochalasis (laxity of the conjunctiva).
- Conjunctival papillae – conjunctival inflammation may result in dilated subepithelial blood vessels which become surrounded by an inflammatory infiltrate to form mounds, called papillae. More common in bacterial and allergic conjunctivitis.
- Conjunctival follicles – small elevated clusters of lymphocytes, similar to papillae but with no central vascular core. Most commonly associated with viral, chlamydial, or toxic conjunctivitis.
- Membrane and pseudomembranes – inflammatory exudate may coalesce forming a yellow-white membrane overlying the palpebral conjunctiva. Membranes are adherent and cause bleeding when removed; pseudomembranes are not. More common in viral and bacterial conjunctivitis.

- Conjunctival phlyctenules – a phlyctenule is a whitish, nodular collection located at or near the limbus, often in the centre of a hyeraemic area. It is a delayed-type hypersensitivity reaction and is associated with *S. aureus* and *M. tuberculosis*. Occassionally seen in fungal or parasitic conjunctivitis.
- Conjunctival granuloma – a granulomatous nodule of inflammatory cells. Seen in Parinaud's oculoglandular conjunctivitis, foreign body, tuberculosis, and sarcoidosis, and sometimes in chlamydial or fungal conjunctivitis.
- Corneal involvement – may be mild (superficial epithelial erosions) or severe (ulceration or perforation). Corneal dendritic ulceration is a feature of HSV conjunctivitis. Symptoms include foreign body sensation, pain, decreased visual acuity, and photophobia.
- Lymphadenopathy – preauricular adenopathy is a non-specific finding associated with viral, chlamydial, and gonococcal causes of conjunctivitis. Submandibular and submental adenopathy are uncommon but may be present in Parinaud's oculoglandular conjunctivitis.

Diagnosis

- Laboratory investigations are not usually performed for most cases of conjunctivitis, especially if a viral aetiology is suspected.
- All cases of ophthalmia neonatorum (conjunctivitis occurring within the first month of life) should be investigated with smears and cultures for bacteria and viruses. The most common causes are *C. trachomatis* and *N. gonorrheae*. Others causes include *S. aureus*, *S. pneumoniae*, *Pseudomonas* spp., *Shigella flexneri*, *M. catarrhalis*. and HSV.
- Swabs and conjunctival scrapings should be taken for Gram stain, culture (on blood, chocolate and Sabouraud agar), and chlamydia and viral diagnostics (e.g. immunofluorescent staining, PCR etc).

Management

- Treatment should be directed at the cause.
- Viral conjunctivitis usually resolves spontaneously and is usually treated supportively, e.g. artificial tears and cold compresses.
- HSV and VZV conjunctivitis – no role for antivirals or corticosteroids but a prophylactic antibacterial ointment is sometimes given to prevent secondary bacterial infection.
- Chlamydial conjunctivitis – trachoma is treated with azithromycin 20 mg/kg stat or doxycycline for 21 days, or erythromycin for 14 days. Adult inclusion conjunctivitis is treated with doxycycline or erythromycin for 3 weeks; sexual partners should be treated simultaneously.
- Acute bacterial conjunctivitis – topical antibiotic eyedrops, e.g. chloramphenicol for 7–10 days.
- *N. gonorrheae* or *N. meningitis* – IV ceftriaxone for 1 day (3 days if corneal involvement). Patients with gonococcal conjunctivitis should be screened and treated for other STIs.
- Chronic bacterial conjunctivitis –treat with appropriate antibiotic therapy (e.g. against *S. aureus*) and aggressive lid hygiene.
- Microsporidia – oral albendazole or topical fumagillin.

Keratitis

Keratitis is an nflammation of the cornea that may be caused by infectious or non-infectious agents. Any corneal inflammation should be considered potentially sight-threatening and requires prompt investigation and management. Corneal perforaton can occur within 24 h with certain organisms. Subsequent endophthalmitis may lead to loss of vision or even loss of the eye.

Aetiology

- Microbial agents do not usually cause keratitis in immunocompetent patients with an intact corneal epithelium. Exceptions include *N. gonorrheae*, *L. monocytogenes*, *Shigella* spp. and *Corynebacterium* spp.
- Risk factors – trauma, contact lens use, contaminated cleaning fluids, immunological impairment secondary to malnutrition, alcoholism, or diabetes, recent or pre-existing eye disease (e.g. sicca syndrome, recent topical steroid use)
- Bacteria – the most common cause of keratitis. Causes include *Staphylococcus* spp., *Streptococcus* spp., *Corynebacterium* spp., *Bacillus* spp., *Propionobacterium* spp., *Pseudomonas* spp., *Haemophilus* spp., *Moraxella* spp., *N. gonorrheae*, and *Enterobacteriaceae*
- Myobacteria – *M. tuberculosis*, *M. chelonae*, *M. gordonae*, *M. avium intracellulare*
- *Chlamydia trachomatis*
- Spirochaetes – *Treponema pallidum, Borrelia burgdorferi*
- Viruses – HSV, VZV, vaccinia, adenovirus, enterovirus, molluscum contagiosum, EBV, Coxsackie virus, measles
- Fungi – *Fusarium* (most common), *Aspergillus*, *Curvularia*, *Paecilomyces*, *Phialophora*, *Blastomyces*, *Sporothrix*, *Exophiala*, *Pseudallescheria*, *Scedosporium*, and *Alternaria* spp.
- Parasites – *Acanthamoeba*, *Onchocerca volvulus*, *Leishmania*, *Microsporidia*, and*Trypanosoma* spp.

Clinical features

- Rapid onset of eye pain is characteristic and may hinder physical examination. Topical anaesthesia may facilitate eye examination but can result in further epithelial damage.
- Eye pain is accompanied by conjunctival injection, tearing, photophobia, blepharospasm, and decreased visual acuity.
- Other features include corneal infiltrate, epithelial defects (visualized by fluorescein stain under cobalt blue light), stromal suppuration, corneal oedema, corneal neovascularization, intraocular inflammation (white cells or protein flare in the anterior chamber, hypopyon, synechiae, glaucoma), and loss of corneal tissue (keratolysis).

Diagnosis

- Because of the limited amount of tissue available, extreme care must be taken in the collection, transport and processing of specimens – discuss with the microbiology laboratory.
- Corneal scrapings (or biopsies) should be taken using sterile technique and transferred to glass slides and appropriate culture media. It may

also be helpful to culture material from the conjuntiva, eyelids, contact lenses/solutions/storage cases.

- For viruses, samples should be collected into viral transport media and inoculated into cell culture the same day. PCR assays enable rapid diagnosis of certain viruses, e.g. HSV and VZV.

Management

Patients may need to be admitted to hospital for management in order to give immediate therapy and to monitor them, particularly if there is evidence of corneal thinning.

- Bacterial keratitis – broad-spectrum topical therapy, e.g. cephalosporin (or vancomycin) and an aminoglycoside (gentamicin or tobramycin) are given for severe keratitis. Topical fluoroquinolones are increasingly being used. The use of topical corticoteroids is controversial. Supportive measures: topical cycloplegics, temporary soft contact lens for corneal ulceration.
- Chlamydial keratitis – systemic antimicrobials, e.g. oral tetracycline, erythromycin or azithromycin. Sexual partners should be treated simultaneously.
- Interstitial keratitis – an immune phenomenon associated with syphilis and Lyme disease. Specific therapy may be indicated for the primary disease but has liitle impact on the cornea. Topical corticocosteroids may be helpful.
- HSV keratitis – acute infection requires topical (trifluridine 1% eye drops for ≥7 days), or systemic therapy (aciclovir, famciclovir, or valaciclovir for 14–21 days).
- VZV keratitis – acute herpes zoster ophthalmicus requires antiviral therapy (aciclovir, famciclovir or valaciclovir), pain management (e.g. amitriptyline). Exposure keratopathy – topical antibiotic ointment (to prevent secondary bacterial infection). Dendritiform keratopathy – 3% vidarabine ointment or 1% trifluridine drops or oral antivirals. Immune keratopathy – topical corticosteroids. General measures – non-steroidal analgesia and cycloplegics.
- Ocular vaccinia – topical 1% trifluridine drops or 3% vidarabine ointment. Intravenous immunoglobulin (IVIG) not indicated unless required for other reasons, e.g. eczema vaccinatum, progressive vaccinia.
- Viral keratoconjunctivitis – no specific treatment required; supportive treatment only, e.g. artificial tears ± cycloplegics. If severe symptoms, topical steroids and cycloplegics may be helpful.
- Fungal keratitis – topical agents (e.g. amphotericin, flucytosine, fluconazole, and itraconazole) have poor corneal penetration. May require combined topical and systemic therapy for months.
- Parasitic keratitis – the optimal treatment for *Acanthamoeba* keratitis is unknown and various agents have been used e.g. diamidines, biguanides, aminoglycosides, and azoles. Onchocerciasis is treated with ivermectin. Microsporidia may be treated with albendazole.

Uveitis

Uveitis is an inflammation of the uveal tract (iris, ciliary body, choroid) or adjacent ocular structures such as the retina. Inflammation may occur in different anatomical regions of the eye e.g. anterior (most common), intermediate, posterior, or panuveitis. Uveitis may be caused by infections, autoimmune conditions, or, rarely, trauma; 50% are idiopathic. Some infectious causes may affect particular locations. Aspiration of aqueous or vitreous material may allow identification of the causative organism. Involve an ophthalmologist early in management.

Classification

- **Anterior uveitis** – inflammation affects the iris (iritis), anterior ciliary body (cyclitis) or both (iridocyclitis). It presents with a unilateral red eye, deep ocular pain, a tender eyeball, irregular/constricted pupil, photophobia and eye-watering. Ocular findings include white blood cells in the aqueous humor, keratitic precipitates, and iris nodules. Most cases are associated with autoimmune conditions (45%) or are idiopathic (40%). Infectious causes include HSV, syphilis, TB, Lyme disease and leprosy.
- **Intermediate uveitis** – inflammation involving the anterior vitreous, and adjacent portion of the retina. Ocular findings include white cells which may be clumped as 'snow balls' in the vitreous, and pars plana white exudate or 'snow bank'. Most cases have an unknown aetiology (69%) or are due to sarcoidosis or multiple sclerosis. Lyme disease causes <1%.
- **Posterior uveitis** – inflammation involving the choroid (choroiditis), retina (retinitis), or both (choroidoretinitis). Ocular findings include lesions in the choroid, retina, or both. Vitritis sometimes occurs; >40% of cases are due to infection e.g. *Toxoplasma*, CMV, acute retinal necrosis (HSV), *Toxocara*, syphilis and *Candida*.
- **Panuveitis** – inflammation involving all parts of the uvea. Ocular findings include white cells in the aqueous and vitreous. Causes: mostly autoimmune, idiopathic (25%), and infections (10%) e.g. syphilis, TB and *Candida*.

Aetiology

- **Tuberculosis** – ocular involvement occurs in ~1% of TB cases and may present without evidence of systemic disease; 50% have a normal CXR. TB can involve any part of the eye but the most common finding is choroiditis. Other ocular findings include chronic anterior uveitis (usually granulomatous), retinal vasculitis/periphlebitis, scleritis, interstitial keratitis, and optic neuritis.
- **Syphilis** – ocular findings in congenital disease include interstitial keratitis and 'salt and pepper' fundi. In secondary syphilis the most common ocular finding is iritis (>70%). In tertiary syphilis the main symptom is progressive visual loss. All patients with ocular syphilis should have a lumbar puncture to exclude concomitant neurosyphilis. HIV-infected patients have higher rates of ocular syphilis and neurosyphilis.

- **Acute retinal necrosis** – a rapidly progressive necrotizing retinitis due to human herpesviruses, e.g. HSV, VZV, and, rarely, CMV. It presents with eye pain, photophobia and reduced vision. Fundoscopy shows one or more foci of retinal necrosis in the peripheral retina. These extend circumferentially and posteriorly over 3–21 days. Although initially unilateral, bilateral involvement occurs in 70% of untreated patients.
- **CMV retinitis** – prior to highly active antiretroviral therapy (HAART), CMV disease occurred in 25–40% of all AIDS patients and its introduction has decreased the incidence of CMV retinitis by 55–83%. Blindness follows the development of retinitis in areas affecting central vision, or retinal detachment, as a result of breaks in the peripheral, necrotic retina. Antiviral therapy (for CMV and HIV) has improved outcome over the last decade see 📖 Antivirals for CMV, p.130, and 📖 HIV treatment, p.138–147
- ***Toxoplasma*** – may be asymptomatic. Bilateral in 40%. Ocular findings include vitritis, retinitis, retinal vasculitis, anterior uveitis (often with raised intra-ocular pressure), scleritis, punctate outer retinitis, neuroretinitis, retinal detachment, pigmented retinopathy, endophthalmitis, and systemic features of infection. Complications: glaucoma, cataract.
- ***Toxocara*** – disease is usually unilateral and presentation varies with age. Ocular findings include diffuse chronic endophthalmitis, chronic anterior uveitis, posterior synechiae, granulomas, retinal detachment, cataract, isolated anterior uveitis, intermediate uveitis, optic papillitis, systemic features of toxocariasis.
- ***Microsporidia*** – affects the immunocompromised and presents with irritation, photophobia, punctate keratopathy, and a follicular conjunctivitis, anterior uveitis.
- **Onchocerchiasis** – causes sclerosing keratitis, chorioretinitis, sclerosis of the retinal vessels, optic neuritis, and optic atrophy.
- ***Candida*** – findings include decreased acuity, floaters, pain, multifocal retinitis (yellow-white fluffy lesions), vitritis (cotton ball colonies which may be joined forming a string of pearls). Complications include retinal necrosis, tractional retinal detachment.
- ***Aspergillus*** – usually occurs in immunosuppressed patients with chronic pulmonary disease. Clinical findings: subretinal hypopyon, intraretinal haemorrhages, dense vitritis, anterior chamber hypopyon.
- **Rare causes** – leptospirosis, Lyme disease, brucellosis, Whipple's disease, leprosy, *Bartonella henselae*.

Diagnosis

- The diagnosis of uveitis is almost always presumptive as the uvea cannot be biopsied without risking sight. The aqueous and vitreous humors may be sampled but these samples rarely yield a diagnosis.
- Molecular diagnostic techniques may be helpful, e.g. PCR for HSV, VZV and CMV. Serology is unhelpful apart from in the diagnosis of syphilis.

Treatment

The treatment of the infectious causes of uveitis is the same as the treatment for CNS infection caused by the same pathogen. Systemic corticosteroids may be given in some conditions, e.g. ocular syphilis.

Endophthalmitis

Inflammation of the ocular cavity, e.g. aqueous and vitreous humours. It is usually caused by bacteria or fungi, which may be introduced into the eye from an external (exogenous) source (e.g. trauma, surgery, keratitis, bleb-related), or enter the eye haematogenously from a distant (endogenous) site of infection, e.g. endocarditis. Panophthalmitis refers to inflammation of all ocular tissue.

General features

- Symptoms – eye pain, redness, lid swelling, decreased visual acuity, headache, photophobia, discharge. Fungal endophthalmitis may have a more indolent course with symptoms developing over days to weeks. Consider in anyone with a history of penetrating injury with a plant substance or soil-contaminated foreign body. Symptoms of primary source of infection may be seen in endogenous cases (e.g. fever, meningism).
- Signs – lid swelling/erythema, inflamed conjunctiva, hypopyon, chemosis, corneal oedema, discharge, reduced/absent red reflex, papillitis, cotton-wool spots, vitritis, the fluffy yellow-white retinal or vitreo-retinal lesions of growing fungi, fever and, late in panophthalmitis, proptosis.

Aetiology

- Bacterial – the most common infectious cause. Onset is abrupt and progression rapid. Most cases are seen after intraocular surgery. Slit-lamp examination is necessary to confirm the diagnosis and detect early signs of infection. Two types:
  - exogenous – symptoms develop 24–48 h after eye trauma, later in patients undergoing extracapsular cataract extraction (<5 days postop). Ocular surface flora are responsible for the majority of infections, and preoperative conjunctival sterilization may reduce the incidence. Common postoperative causes: CoNS, coryneforms, *S. aureus*. Common post-traumatic causes: *S. aureus*, *Bacillus cereus*, group A streptococci, coliforms, anaerobes, *Pseudomonas* spp. Delayed endophthalmitis after cataract extraction in patients with an intraocular lens may run a chronic course – associated organisms include *P. acnes*, *Corynebacterium* spp., CoNS, non-tuberculous mycobacteria. Visual outcome after recovery is poor
  - endogenous – tends to affect the posterior segment of the eye. Patients are usually very unwell and often immunocompromised. Foci of primary infection include meningitis, endocarditis, pneumonia, abdominal infection, and dental procedures. Causes: *S. aureus*, *S. pneumoniae*, and group A streptococci.
- Fungal – increasing in incidence with the use of immunosuppressive agents and antibiotics. Haematogenous cases are most commonly caused by *Candida* species. Treatment is difficult and the risk of permanent damage high. Cases have occurred in healthy patients following injections of contaminated anaesthetic. Ocular features may arise many days after *Candida* being recovered from the blood. Patients are often sick (e.g. ventilated on ICU) and may not report early visual symptoms. Have a low threshold for ocular examination

of such patients. *Aspergillus* endophthalmitis may occur in seriously immunosuppressed patients and IVDUs. Other causes: *Cryptococcus neoformans*, *Blastomyces dermatitidis*, *Histoplasma capsulatum*, *Nocardia* spp.

- Viral – HSV, VZV, and CMV cause a spectrum of eye disease from acute retinal necrosis (usually healthy patients) to progressive outer retinal necrosis (severely immunocompromised patients). PCR from vitreal samples may allow identification of the causative virus. CMV retinitis is seen in AIDS patients and those receiving chemotherapy for acute leukaemia and malignant lymphomas Measles retinopathy may be seen 6–12 days after skin rash, and chorioretinitis can be a complication of SSPE.
- Parasitic – causes include *Toxoplasma gondii* and *Toxocara canis*.

Diagnosis

Early recognition and prompt microbiological investigations are essential if functional vision is to be salvaged in bacterial endophthalmitis. Samples from the vitreous humour have the greatest yield. It may also be appropriate to obtain material from the anterior chamber, and any wound. Surgical specimens should be examined. Gram, Giemsa and PAS stains should be performed, and samples cultured for aerobic and anaerobic bacteria, mycobacteria, and fungi. ELISA or PCR testing of samples may also be appropriate. Blood cultures should be taken if the patient is systemically unwell. Viral retinities may have a characteristic appearance on fundoscopy and urgent specialist examination is indicated.

Treatment

Successful outcome is dependent upon a low threshold of clinical suspicion for diagnosing infectious endophthalmitis. Urgent specialist referral is indicated.

- Bacterial – broad-spectrum intravitreal antibiotics (vancomycin, amikacin, ceftazidime) should be started immediately after urgent diagnostic aspirates. Modify as guided by cultures. Those with visual acuity of light perception or worse benefit from immediate vitrectomy; those with better vision than this do no better with vitrectomy and intravitreal antibiotics than with biopsy and intravitreal antibiotics (unless perhaps they are diabetic). Outcome is influenced by the time to diagnosis and appropriate treatment, and the virulence of the organism: *P. aeruginosa* and *S. aureus* can destroy the eye within 24 h of presentation. There may be a role for early corticosteroids – much of the damage to the eye is done by the immune response.
- Fungal – *Candida*: systemic and intravitreal amphotericin B with vitrectomy in those cases with intravitreal abscess. *Aspergillus*: vitrectomy and high-dose intravenous amphotericin B. 5-FC (5-flucytosine) may be used in combination. Prognosis is poor. Steroids are contraindicated.
- Viral – a combination of systemic and in some cases, intraocular antiviral agents is indicated. See 📖 Cytomegalovirus, p.451 for details regarding the treatment of CMV retinitis.

Skin and soft tissue infections[1]

Impetigo

- Caused by *S. aureus* and/or group A streptococci. Commonly affects children in tropical/subtropical regions; also prevalent in temperate regions in the summer months.
- Clinical features – occurs on the face and extremities. Lesions start as small vesicles that develop into flaccid bullae which rupture releasing a yellow discharge that forms thick crusts.
- Treatment – mupirocin is the best topical agent (although resistance has been described). Patients who have numerous lesions or who do not respond to topical treatment should receive oral antibiotics, e.g. penicillinase-resistant penicillin or a first-generation cephalosporin.

Folliculitis

- A superficial infection of the hair follicles and apocrine structures.
- Aetiology – *S. aureus* (most common), *P. aeruginosa* ('hot tub' folliculitis), *Enterobacteriaceae* (complication of acne), *Candida* spp. and *Malassezia furfur* (in patients taking corticosteroids). Eosinophilic pustular folliculitis occurs in AIDS patients.
- Clinical features – lesions consist of small, erythematous, pruritic papules, often with a central pustule.
- Treatment – empiric treatment is with oral flucloxacillin. If the clinical response is slow consider other pathogens.

Cutaneous abscesses

- Collections of pus within the dermis and deeper skin structures.
- Usually polymicrobial containing skin/mucous membrane flora; *S. aureus* is the sole pathogen in ~25% of cases.
- Clinical features – painful, tender, fluctuant nodules, usually with an overlying pustule and surrounded by a rim of erythematous swelling.
- Treatment is with incision and drainage. Antibiotics are rarely necessary unless there is extensive infection, systemic toxicity, or the patient is immunocompromised.

Furuncles and carbuncles

- A furuncle (boil) is a deep inflammatory nodule that usually develops from preceding folliculitis. Furuncles usually occur in areas of the hairy skin, e.g. face, neck, axillae, and buttocks.
- A carbuncle is a larger, deeper lesion made of multiple abscesses extending into the subcutaneous fat. Usually occur at the nape of the neck, on the back or on the thighs. Patients may be systemically unwell.
- Outbreaks of furunculosis caused by meticillin-sensitive *S. aureus* (MSSA) and MRSA have been described in groups of individuals with close contact, e.g. families, prisons, and sports teams.
- Most furuncles may be treated with application of moist heat which promotes localization and spontaneous drainage. Large lesions require surgical drainage. Systemic antibiotics are indicated if fever, cellulites, or lesions are located near the nose or lip. Control of outbreaks may require washing with chlorhexidine soaps, no sharing of cloths or towels, laundering of clothing, towels and bedclothes, and eradication of staphylococcal carriage in colonized persons.

Ecthyma

- Punched-out ulcers surrounded by raised violaceous margins.
- Caused by group A streptococci. Similar lesions (ecthyma gangrenosum) may occur with *P. aeruginosa* in neutropaenic patients.
- Empiric treatment is with oral penicillin or clindamycin. Antipseudomonal agents, e.g. Tazocin®, should be given for *P. aeruginosa* infections.

Erysipelas

- An acute spreading skin infection with prominent lymphatic involvement. Usually affects children, infants, and the elderly. Predisposing factors include skins lesions, venous stasis, paraparesis, diabetes mellitus, and alcohol abuse.
- Causes – group A streptococci (most common), group C and G streptocococci, *S. aureus*, or group B streptococci.
- Clinical features – painful, erythematous, oedematous lesion with an elevated, sharply demarcated border. Usually occurs on the face or the legs. Systemic symptoms are common; 5% are bacteraemic.
- Treatment is with penicillin. If *S. aureus* is suspected treatment should be with a flucloxacillin, clindamycin, or erythromycin.

Cellulitis

- An acute spreading infection of the skin that extends into the subcutaneous tissues. *S. aureus* is the main cause if cellulitis is associated with skin lesions. Other causes group A, C, G, or B streptococci. Clinical clues to other causes include physical activities, trauma, water contact, and animal, insect, or human bites and immunosuppression. Examples include *Enterobacteriaceae, Legionella pneumophila, Aeromonas hydrophila, Vibrio vulnificus, Erysipelothrix rhusiopathiae*, and *Cryptococcus neoformans.*
- Clinical features – spreading erythematous, hot, tender lesion, usually accompanied by systemic symptoms.
- The diagnosis is usually clinical; cultures should only be obtained in patients who do not respond to first-line treatment. If unusual organisms are suspected or in immunocompromised hosts, appropriate culture material should also be obtained. Unfortunately, aspiration of skin is not helpful in 75–80% of cases of cellulitis, and results of blood cultures are rarely positive (<5% of cases).
- Treatment – empiric treatment of cellulitis is with flucloxacillin, clindamycin, or erythromycin (unless infections resistant to these agents are common in the community). Vancomycin, teicoplanin, linezolid, or daptomycin are indicated for MRSA cellulitis. If a Gram-negative organism is suspected, an intravenous cephalosporin, e.g. cefuroxime, may be used. Oral metronidazole should be added for patients with diabetic or vascular ulcers. The affected limb should also be immobilized and elevated and cool sterile saline dressings used to remove any exudate.

Reference

1 Stevens DL, Pislo AL, Chambers HI et al. Practice Guidelines for the management of skin and soft tissue infections. *Clin Infec Dis* 2005;41:1373–406.

2 Smartz MN. Cellulitis. *New Engl J Med* 2004;350:904–12

Bite infections

Animal bites

- **Aetiology** – animals bites are usually caused by domestic pets (e.g. dogs or cats) but may be caused by exotic pets or wild animals. Most infections are polymicrobial. The predominant pathogens are the oral flora of the biting animal, e.g. *Pasteurella multocida*, *Capnocytophaga canimorsus*, *Bacteroides* spp., *Fusobacterium* spp., *Prevotella* spp., *Porphyromonas* spp., *Propionobacterium* spp., and peptostreptococci. Secondary bacterial infection with *S. aureus* or group A streptococci may occur.
- **Clinical features** – patients who present >8 h after injury usually have established infection which may be non-purulent, purulent, or abscesses.
- **Diagnosis** – this is clinical but samples may be taken to identify the causative organisms. Complications include septic arthritis, osteomyelitis, subcutaneous abscesses, tendonitis, and bacteraemia.
- **Management** – wounds should be irrigated copiously with sterile saline and any debris removed; debridement is rarely necessary. Wounds should steri-stripped but not sutured (except facial wounds by a plastic surgeon). Empiric antibiotic therapy is with PO amoxicillin/clavulanic acid. Alternatives include PO doxycline or IV piperacillin/tazobactam or carbapenems. A tetanus booster should be given to those whose vaccination status in unknown. Rabies vaccination should be considered for animal bites in endemic regions. Prophylactic valaciclovir should be considered for monkey bites (simian herpes virus).

Human bites

- **Aetiology** – human bites result from aggressive behaviour and are often more serious than animal bites. The causative organisms are usually the oral flora of the biter, e.g. oral streptococci, staphylococci, *Haemophilus* spp., *Eikenella corrodens*, *Fusobacterium* spp., *Prevotella* spp., *Porphyromonas* spp., and, rarely, *Bacteroides* spp. Human bites may also potentially transmit viral infections e.g. HBV, HCV, and HIV.
- **Clinical features** – bite wounds may be may be occlusive injuries (where teeth bite the body part) or clenched fist injuries (where one person's fist hits the other person's teeth). Complications of closed fist injuries include tendon or nerve damage, fractures, septic arthritis, or osteomyelitis.
- **Management** – the principles are the same as for animal bites, e.g. wound irrigation and prophylactic antimicrobial therapy (see above). Hand injuries should be evaluated for complications by a hand surgeon. Post-exposure prophylaxis (📖 p.194–5) of hepatitis B and HIV should be considered if the source is potentially infected.

Surgical site infections

Infections of surgical wounds are common adverse events following surgery. The frequency of surgical site infections (SSIs) is related to the category of operation and is highest with contaminated or high-risk surgical procedures. There are three categories of SSI:

- superficial incisional SSI – involves subcutaneous tissue, occurs within 30 days of operation
- deep incisional SSI – involves muscle and fascia, occurs with 30 days of operation (or 1 year if prosthesis inserted)
- organ/space SSI – involves any part of the anatomy (organs or spaces) other than the incisional site

Aetiology and pathogenesis

The most common organisms are *S. aureus* and MRSA. Others include coagulase-negative staphylococci, aerobic Gram-negative bacilli, *Bacillus* spp., and corynebacteria. SSIs that occur after an operation on the GI tract or female genitalia have a high probability of having mixed flora. The presence of prosthetic material greatly reduces the number of organisms that are required to initiate infection.

Clinical features

- Most SSIs have no clinical manifestations for at least 5 days after the operation, and many may not become apparent for up to 2 weeks.
- Local signs of pain, swelling, erythema, and purulent drainage are usually present. Fever may not be present until a few days later.
- In morbidly obese patients or in patients with deep, multilayer wounds, e.g. thoracotomy, the external signs of SSIs may appear late.

Diagnosis

The diagnosis is usually clinical but samples of fluid or tissue should be sent to the laboratory for Gram stain and culture.

Management

- The primary therapy for SSI is to open the incision, debride the infected material, and continue dressing changes until the wound heals by secondary intention.
- Although patients commonly receive antibiotics for SSIs, there is little or no evidence supporting this practice.
- A common practice, endorsed by expert opinion, is to open all infected wounds. If there is minimal evidence of invasive infection (<5 cm of erythema), and if the patient has minimal systemic signs of infection (temperature <38.5°C, pulse rate of <100/min), antibiotics are unnecessary. For patients with a temperature of >38.5°C or a pulse rate of >100 beats/min, a short course of antibiotics (24–48 h) may be indicated.
- In the UK, the Health Protection Agency performs surveillance of SSIs.[2]

References

1 Stevens DL, Bisno AL, Chambers HF, et al. Practice guidelines for the diagnosis and management of skin and soft tissue infections. *Clin Infect Dis* 2005;41:1373–406.

2 Health Protection Agency. Surgical Site Infection Surveillance Service (SSISS). http://www.hpa.org.uk/infections/topics_az/surgical_site_infection/default.htm (accessed 15 Aug 2008).

Gas gangrene

This is a rapidly progressive, life-threatening, skeletal muscle infection caused by *Clostridia* spp. (clostridial myonecrosis).

Aetiology and pathogenesis

- Gas gangrene usually occurs following muscle injury and contamination of the wound by soil or foreign material containing clostridial spores. *Clostridium perfringens* (see 📖 Other clostridia, p.312) is the predominant cause (80–95%). The pathological effects are mediated by production of α and θ toxins.
- Spontaneous or non-traumatic gas gangrene may occur in the absence of an obvious wound. This form is usually caused by *C. septicum* and associated with intestinal abnormalities, e.g. colonic cancer, diverticulitis, bowel infarction, necrotizing enterocolitis.
- Other organisms include *C. novyi*, *C. bifermentans*, *C. histolyticum*, and *C. fallux*. Organisms such as *E.coli*, *Enterobacter* spp., or enterococci may be isolated, reflecting contamination of the wound.

Clinical features

- The incubation period is usually 2–3 days, but may be shorter.
- Patients present with acute onset of excruciating pain and signs of shock (fever, tachycardia, hypotension, jaundice, renal failure).
- Local oedema and tenderness may be the only early signs or there may be an open wound, herniation of muscle, a serosanguinous, foul-smelling discharge, crepitus, skin discoloration, and necrosis.
- Progression is rapid and death may occur within hours.

Diagnosis

- The diagnosis is usually clinical but may be confirmed by Gram stain of the wound or aspirate.
- Liquid anaerobic cultures may be positive within 6 h.
- Plain radiographs may show gas in the affected tissues.

Management

- Emergency surgical exploration and debridement of the affected area may be life saving.
- Antimicrobial therapy with intravenous benzylpenicillin and clindamycin is widely recommended (based on animal experimental data).
- Additional antibiotics (e.g. ciprofloxacin, ceftriaxone, or chloramphenicol) may be given if Gram-negative bacilli are seen in the initial Gram smear.
- The role of adjunctive hyperbaric oxygen therapy remains controversial – it should never delay surgery.

Necrotizing fasciitis

A severe acute infection involving the superficial and deep fascia.

Aetiology and pathogenesis

- Type I necrotizing fasciitis involves at least one anaerobic species (e.g. *Bacteroides* or *Peptostreptococcus* spp.) as well as one or more facultative anaerobic species (e.g. non-group A streptococci, *E. coli, Enterobacter, Klebsiella, Proteus* spp.).
- Type II necrotizing fasciitis is caused by group A streptococci alone or in combination with other species (e.g. *S. aureus*). This form usually occurs after trauma, wounds, or surgery in patients with predisposing conditions (e.g. diabetes mellitus, peripheral vascular disease, hepatic cirrhosis, corticosteroid therapy, IVDU). Recently there has been an increase in cases associated with streptococcal toxic shock syndrome.

Clinical features

- Necrotizing fasciitis most commonly affects the lower limbs but may affect any part of the body, e.g. wound sites, abdominal wall, groin, or perianal area. The affected area is usually red, hot, swollen, and exquisitely tender/painful. There is rapid progression with skin discolouration, bulla formation, and cutaneous gangrene. The affected area becomes anaesthetic as result of small vessel thrombosis and destruction of superficial nerves. Systemic toxicity is common.
- In the newborn, necrotizing fasciitis may complicate omphalitis and spread to involve the abdominal wall, flanks, and chest wall.
- Fournier's gangrene is a form of necrotizing fasciitis that affects the male genitals and is usually polymicrobial.
- Craniofacial necrotizing fasciitis is usually associated with trauma and caused by A streptococci.
- Cervical necrotizing fasciitis is usually associated with dental or pharyngeal infections and is polymicrobial.

Diagnosis

Diagnosis is usually clinical but may be confirmed by Gram stain of the exudate. Blood cultures are frequently positive. CT or MRI scans may demonstrate subcutaneous and fascial oedema but should never delay surgical exploration.

Management

Emergency surgical exploration and debridement confirms the diagnosis and is the mainstay of treatment.
Empiric antimicrobial therapy should cover the most likely organisms, e.g. intravenous cefuroxime or co-amoxiclav, clindamycin, and gentamicin.
Adjunctive therapies (e.g. intravenous immunoglobulin for streptococcal toxic shock syndrome or hyperbaric oxygen) remain unproven.

References

1 Bisno AL, Stevens DL. Streptococcal infections of skin and soft tissues. *N Engl J Med* 1996; 334: 240–5.

2 Stevens DL, Bisno AL, Chambers HF, et al. Practice guidelines for the diagnosis and management of skin and soft tissue infections. *Clin Infect Dis* 2005;41:1373–406.

Pyomyositis

Pyomyositis (primary muscle abscess) is an acute bacterial infection of skeletal muscle.

Epidemiology

Most cases occur in the tropics, where it may affect any age group, and accounts for 1–4% of hospital admissions. In temperate climates pyomyositis usually affects adults or the elderly, 40% of whom have a predisposing condition (e.g. diabetes mellitus, alcoholic liver disease, corticosteroid therapy, haematological malignancies, and HIV infection).

Aetiology

S. aureus accounts for 95% of cases in the tropics and 66% of cases in N America. Group A streptococci account for 1–5% of cases. Uncommon causes include groups B, C, and G streptococci, *S. pneumoniae*, and *S. anginosus*. Rare causes include *Enterobacteriaceae*, *Y. enterocolitica*, *N. gonorrhoeae*, *H. influenzae*, *Aeromonas hydrophila*, anaerobes, *Burkholderia mallei*, *Burkholderia pseudomallei*, *Aspergillus fumigatus*, *Candida* spp., *M. tuberculosis*, *M. avium* complex.

Clinical features

Between 20% and 50% of cases have had recent blunt trauma or vigorous exercise of the affected area. There are three clinical stages:

- invasive stage – subacute onset of fever, local swelling ± erythema, mild pain and minimal tenderness
- suppurative stage (2–3 weeks later) – fever, distinct muscle swelling, and tenderness; pus may be aspirated from muscle
- systemic stage – sepsis syndrome, erythema, exquisite tenderness, fluctuance. May progress to metastatic abscesses, shock, and renal failure. Complications include compartment syndrome.

Diagnosis

- Early pyomyositis may be difficult to differentiate from a number of other conditions, e.g. thrombophlebitis, muscle haematoma, muscle rupture, PUO, osteomyelitis. Iliacus pyomyositis may mimic septic arthritis of the hip, and iliopsoas pyomyositis may mimic appendicitis.
- CT or MRI scans are useful both to delineate intramuscular abscesses and to aid diagnostic aspiration.

Management

- Percutaneous or open drainage of all abscesses is the essential. Fasciotomies and debridement may be required for compartment syndrome.
- Empiric antimicrobial therapy should cover *S. aureus* and group A streptococci, e.g. intravenous flucloxacillin.
- Antibiotic therapy should be modified in the light of Gram stain or culture results. Ongoing fever despite appropriate therapy should prompt a search for further foci of infection.

Fungal skin infections

Fungi may cause primary infection of the skin or present with cutaneous manifestations of systemic disease.

Candida (📖 p.518)

- Localized skin infections – 'erosio interdigitalis blastomycetica' (between fingers and toes), folliculitis, intertrigo, nappy rash, paronychia, onychomycosis, balanitis.
- Generalized cutaneous candidiasis – in which lesions spread and become confluent, affecting widespread areas of the trunk, thorax, and extremities (uncommon).
- Chronic mucocutaneous candidiasis – a group of candidal infections that fail to respond to normally adequate therapy, resulting in complications such as oesophageal stenosis, alopecia, and disfigurement of the face, scalp and hands. These failures seem to be associated with immunological abnormalities. Most cases present in infancy or by the age of 20 years. There is a wide spectrum of severity. Up to half of patients subsequently develop certain endocrinopathies (e.g. hypoparathyroidism). Most patients have good life expectancies. The commonest cause of death is bacterial sepsis rather than disseminated candidiasis. Chronic mucocutaneous disease is very difficult to treat. Intravenous amphotericin is initially effective but most patients relapse on its cessation. Months or years of treatment with azoles may be necessary.

Malessezia furfur (📖 p.526)

Causes pityriasis versicolor (a superficial skin infection characterized by hypopigmented lesions usually confined to trunk and proximal limbs), folliculitis, as well as IV catheter infections. *Malessezia* spp. have been implicated in the pathogenesis of seborrhoeic dermatitis.

Aspergillus (📖 p.532)

Cutaneous infection is rare, usually occuring in burn wounds or neutropenic patients at the site of IV catheter insertion. More common is otomycosis, caused by *A. niger* in those with chronic otitis externa. Cleaning and topical therapy with an agent such as amphotericin B 3% or clotrimazole is curative.

Mucormycosis (📖 p.537)

Infection by fungi belonging to the order Mucorales. Risk factors: immunsuppression, transplantation, diabetes mellitus, trauma. Presents with chronic ulcer or cellulitus – if unrecognized the organism penetrates deeper into the skin, with vascular invasion, necrosis, and possible dissemination.

Eumycetoma (📖 p.539)

Chronic, slow-growing destructive fungal infection of the hands or feet. Found worldwide in tropical regions but rare in temperate areas.

Pseudallescheria boydii (📖 p.544)

May cause eumycetoma, skin, and soft tissue infection, abscesses.

Scedosporium prolificans (📖 p.544)

Extremely rare. Focal (e.g. osteoarticular) disease in the immunocompetent, disseminated infection (including skin) in the immunocompromised (e.g. those undergoing bone marrow transplant (BMT)).

Fusarium species (📖 p.544)

Rare in immunocompetent people. Skin lesions start as macules and progress to necrotic papules. Systemic infection is seen in patients with acute leukaemia with prolonged neutropenia, and those undergoing BMT.

Sporothrix schenckii (📖 p.545)

Inoculated into skin at sites of minor trauma. May cause either a fixed plaque, or painless smooth or verrucous erythematous nodular papules with secondary lesions that follow the routes of lymphatic vessels.

Chromomycosis (📖 p.547)

An itchy small pink papule is followed by crops of either warty, violaceous nodules, or firm tumours, which may enlarge forming groups with ulceration and dark haemopurulent material on the surface. Satellite lesions may occur. Some people develop annular, papular lesions with active edges and healing in the centre which can become scarred, or form keloid. Fibrosis and oedema of the affected limb may occur in severe cases.

Dermatophytes (📖 p.541)

A group of fungi capable of invading the dead keratin of skin, hair, and nails. Clinical classification is by the body area involved: tinea capitis (scalp hair and the commonest in children), tinea corporis (trunk and limbs), tinea manuum and pedis (palms, soles and the commonest overall worldwide), tinea cruris (groin), tinea barbae (beard area and neck), tinea faciale (face), tinea unguium (nail – also known as onychomycosis).

Cutaneous manifestations of systemic fungal infection

Systemic fungal infections that present with cutaneous disease include:

- disseminated candidiasis
- cryptococcosis (📖 p.527)
- *Penicillium marneffei* (📖 p.561)
- *Blastomyces dermatitidis* (📖 p.554)
- *Coccidioides immitis* (📖 p.556)
- *Paracoccidioides brasiliensis* (📖 p.559)
- *Fusarium* spp. (📖 p.544)
- *Scedosporium prolificans* (📖 p.544)
- mucormycosis (📖 p.537).

Viral skin infections

Herpes simplex virus (📖 p.441)

Cutaneous manifestations of HSV infection include:

- pharyngitis/gingivostomatitis – the commonest presentation of primary HSV-1, generally seen in children and young adults. General features: fever, malaise, difficulty chewing, cervical lymphadenopathy. Ulcers and exudative lesions are found on the posterior pharynx and sometimes the tongue, buccal mucosa, and gums. Patients with eczema may develop severe orofacial disease (eczema herpeticum) which may disseminate requiring systemic therapy. HSV has been associated with up to 75% of cases of erythema multiforme
- recurrent herpes labialis – the most frequent manifestation of HSV-1 reactivation. May be asymptomatic (viral secretion). Reactivation is usually localized with symptoms that are milder and of shorter duration than primary infection. Mild prodromal tingling is followed by the development of lesions within 48 h – resolution is quick, usually within 5 days. Immunosuppressed patients may experience severe mucositis with spread to skin surrounding the mouth. AIDS patients can develop persistent HSV ulceration
- herpetic whitlow – HSV infection of the finger which may result from auto-inoculation (existing oral or genital infection), or by direct inoculation from some other environmental exposure – e.g. healthcare workers. Oedema, tenderness, and erythema are followed by the development of vesicular lesions. Regional lymphadenopathy may occur. Diagnosis is important if only to prevent unnecessary surgical intervention and onward transmission
- *Herpes gladiatorum* – mucocutaneous infection of surfaces such as chest, ears, face, and hands seen in rugby players and wrestlers.

Varicella zoster virus (📖 p.444)

- Chickenpox – the illness associated with primary varicella virus infection; 90% of cases occur in children under 13 years of age. Incubation is 10–14 days and may be followed by a 1–2-day febrile prodrome before the onset of constitutional symptoms (malaise, itch, anorexia) and rash. Lesions start as maculopapules (<5 mm across), progressing to vesicles which quickly pustulate and form scabs which fall off 1–2 weeks after infection. They appear in successive crops over 2–4 days starting on the trunk and face and spreading centripetally. May rarely involve the mucosa of the oropharynx and vagina. Complications include secondary bacterial infection, pneumonitis, and encephalitis. Disease may be severe in pregnancy and the immunocompromised.
- Shingles (herpes zoster, – the localized recurrence of varicella virus. Causes a unilateral vesicular eruption in a dermatomal distribution (most commonly thoracic and lumbar), often preceded by 2–3 days of pain in the affected area. Maculopapular lesions evolve into vesicles, with new crops forming over 3–5 days. Resolution may take 2–4 weeks. Complications include keratitis (herpes zoster opthalmicus), Ramsay–Hunt syndrome (VIII cranial nerve palsy), encephalitis, and paralysis (anterior horn cell involvement).

Smallpox (📖 p.490)

- Smallpox is caused by variola virus, an orthopoxvirus. There are two strains: variola major (mortality 20–50%) and variola minor (mortality <1%).
- The last reported case was in Somalia in 1977 and the virus was declared eradicated by the World Health Organization (WHO) in 1980. Virus stocks exist in two laboratories and there are concerns about its potential use as a bioterrorism agent (p.876).
- The incubation period is 10–12 days and is followed by a prodromal period of 1–2 days. The centrifugal rash is initially maculopapular and progresses to vesicles, pustules, and scabs over 1–2 weeks. Death may occur with fulminant disease.
- Diagnosis may be confirmed by electron microscopy or PCR (to differentiate it from other pox viruses).
- There is no specific treatment. Management is by isolation of cases to prevent transmission.

Monkeypox (📖 p.490)

- Monkeypox caused by an orthopox virus.
- It causes a vesicular illness in monkeys and rodents in west and central Africa. Infection may sporadically be transmitted to humans. A large outbreak in humans in the USA was traced to importation and sale of exotic pets. The disease is similar but less severe than smallpox.
- Diagnosis is by electron microscopy or PCR (to differentiate it from other pox viruses).
- Management is symptomatic as there is no specific treatment.

Orf (📖 p.490)

- Orf is caused by a parapox virus and primarily affects sheep and cattle. Humans are infected following direct exposure to infected animals.
- It presents with vesicular rash on sites of contact, e.g. hands and arms. This progresses to pustules which coalesce, scab, and gradually resolve over weeks.
- Diagnosis is by virus culture and identification of the virus by electron microscopy.
- Management is symptomatic as there is no specific treatment.

Molluscum contagiosum (📖 p.491)

- Molluscum contagiosum is caused by a pox virus.
- It is spread by close human contact and may cause severe, generalized disease in HIV-infected patients.
- The lesions are small, firm, umbilicated papules which occur on exposed epithelial surfaces or the genitalia. Lesions may resolve spontaneously or persist for months or years.
- Diagnosis is confirmed by histology or electron microscopy.
- Management is with local therapy, e.g. laser, cryotherapy, or incision and curettage. One small uncontrolled study of three patients has shown benefit with cidofovir.

Reference

1 Toutous-Trellu L, Hirschel B, Piguet V, et al. Treatment of cutaneous human papilloma virus, poxvirus and herpes simplex virus infections with topical cidofovir in HIV positive patients. *Ann Dermatol Venereol.* 2004;131(5):445–9.

Miscellaneous skin infections

Cutaneous anthrax

- Cutaneous anthrax is caused by *Bacillus anthracis* (📖 p.298). It usually affects humans who are in direct contact with infected animals (e.g. cattle and sheep) or animal products.
- The lesion begins with a pruritic papule that enlarges to form an ulcer surrounded by vesicles, and then develops into an eschar surrounded by oedema. There may be regional lymphangitis, lymphadenopathy, and systemic symptoms.
- Diagnosis is confirmed by microscopy and culture of vesicle fluid. If the patient has received antibiotics, or cultures are negative, a punch biopsy may be taken for immunohistochemistry or PCR.
- Antimicrobial therapy does not accelerate healing of the skin lesion but may reduce oedema and systemic symptoms. Empiric therapy is with oral ciprofloxacin for 5–9 days; the duration of treatment should be increased to 60 days if inhalational anthrax is a possibility.

Erysipeloid

- Erysipeloid is caused by *Erysipelothrix rhusiopathiae* (📖 p.306). It usually affects people who handle fish, marine mammals, poultry, or swine.
- After exposure, a red maculopapular lesion develops, usually on the fingers or hands. Erythema spreads centrifugally with central clearing. A blue ring with a peripheral red halo may appear. Regional lymphangitis/lymphadenopathy occurs in ~1/3 cases. A severe, generalized cutaneous infection may also occur.
- Diagnosis is confirmed by culture of a lesion aspirate and/or biopsy specimen; blood cultures are rarely positive.
- Untreated erysipeloid resolves during a period of 3–4 weeks, but treatment probably hastens healing and perhaps reduces systemic complications. Treatment is with oral pencillin or amoxicillin for 7–10 days.

Cat scratch disease

- Cat scratch disease is mainly caused by *Bartonella henselae* (📖 p.400).
- A papule or pustule develops 3–30 days after a scratch or a bite. Regional adenopathy occurs ~3 weeks after inoculation and ~10% of nodes suppurate. Extranodal disease (e.g. CNS, liver, spleen, bone, and lung) occurs in 2% of cases.
- Diagnosis is by serology (poor specificity), PCR, or histology (Warthin–Starry silver stain).
- Treatment is with oral azithromycin for 5 days. Clinical response is rarely dramatic but lymphadenopathy usually resolves by 6 months.

Bacillary angiomatosis

- Bacillary angiomatosis may be caused by *Bartonella henselae* or *Bartonella quintana*. It usually occurs in immunosuppressed patients, especially AIDS.

Reference

1 Stevens DL, Bisno AL, Chambers HF, et al. Practice guidelines for the diagnosis and management of skin and soft tissue infections. *Clin Infect Dis* 2005;41:1373–406.

Septic arthritis

An inflammatory reaction of the joint space caused by an infectious agent.

Aetiology

- *Staphylococcus aureus*
- Streptococci, e.g. groups A, B, C, and G streptococci, *S. pneumoniae*,
- Coagulase-negative staphylococci
- *E. coli*
- *H. influenzae*
- *N. gonorrhoeae*
- *N. menigitidis*
- *P. aeruginosa*
- *Salmonella* spp.
- Others, e.g. *Pasteurella multocida*, *Capnocytophaga canimorsis*, *Eikenella corrodens*, *Streptobacillus moniliformis*, *Brucella* spp., *Burkholderia pseudomallei*, *Clostridium* spp.
- Polymicrobial infections

Epidemiology

The reported incidence of septic arthritis varies from 2–5/100,000 per year in the general population. Risk factors for septic arthritis include rheumatoid arthritis, intra-articular injections, trauma, diabetes mellitus, immunosuppression, intravenous drug use, human and animal bites.

Pathogenesis

Septic arthritis usually occurs after haematogenous seeding of pathogenic microorganisms, but may occur following trauma.

Clinical features

- Children and adults with acute septic arthritis usually present with fever (60–80%) and monoarticular involvement (90%).
- The knee is the most commonly affected joint, followed by the hip. Clinical features include pain, swelling, and reduced mobility in the joint.
- Polyarticular infections occur in 10–20% of patients, especially those with rheumatoid arthritis and viral causes.
- Infections with mycobacteria or fungi usually have an insidious onset.

Differential diagnosis

- Inflammatory arthritides
- Post-infectious arthritis
- Chronic bacterial arthritis, e.g. *Borrelia burgdorferi*, *Brucella* spp., *Tropheryma whippelii*, *Nocardia asteroides*
- Viruses, e.g. parvovirus B19, hepatitis B, mumps, rubella, HTLV-1, HIV-1, lymphocytic choriomengitis virus, Chikungunya virus, and Ross River virus
- Mycobacteria, e.g *M. tuberculosis* (most common), *M. kansasii*, *M. marinum*, *M. avium intracellulare*, *M. fortuitum*, *M. haemophilum*, *M. leprae*
- Fungi, e.g. *Sporothrix schenkii*, *Blastomyces dermatitidis*, *Coccidioides immitis*, *Paracoccidiodes braziliensis*, *Candida albicans*, *Pseudallescheria boydii*
- Parasites, e.g. filarial infections, schistosomiasis

Diagnosis

- Laboratory investigations frequently show a raised white cell count and inflammatory markers. Aspiration of the joint reveals a purulent synovial fluid with high neutrophil count. Gram stain is positive in 50%, and culture is positive in 80–90% of cases. False-positive Gram stains may occur with artefacts from stain, mucin, and cellular debris. Direct inoculation of the synovial fluid into blood culture bottles may improve recovery of pathogens. Samples should also be sent for microscopy for crystals.
- Radiological investigations – plain radiographs may show periarticular soft tissue swelling, fat pad oedema, periarticular osteoporosis, loss of joint space, periosteal reactions, erosions, and loss of subchondral bone. Ultrasound can be used to confirm an effusion and guide aspiration. CT and MRI are highly sensitive for imaging early septic arthritis. CT is better for imaging bone lesions. MRI may not distinguish septic arthritis from inflammatory arthropathies.

Management

- Drainage of the joint, either by closed aspiration or arthroscopic washout, should be performed urgently. Open drainage may be required either when repeated drainage has failed to control the infection, or for drainage of hip joints. Prosthetic joint infections usually require removal of the prosthesis.
- Antimicrobial therapy should be guided by the initial Gram stain findings. If the Gram stain is negative then intravenous ceftriaxone ± vancomycin is a reasonable choice. Definitive therapy should be tailored to the organism isolated and its antimicrobial susceptibility pattern. Treatment is usually for 3 weeks.
- Adjunctive therapy – short-course systemic corticosteroid treatment has been shown to be of benefit in children with haematogenous bacterial arthritis.

Septic bursitis

- This common condition usually affects the olecranon bursa or the prepatellar bursa. Bacteria are usually introduced following minor trauma, but may rarely be inoculated during intrabursal injection of corticosteroids. Infection of the deep bursa is rare but may occur in association with septic arthritis.
- *S. aureus* is the cause in >80% of cases; the remainder are caused by streptococci, various Gram-negative bacteria and fungi.
- Diagnosis is made by aspiration of the affected bursa and examination of the synovial fluid for cells, organisms and crystals.
- Treatment is with antibiotics (guided by Gram stain results) and aspiration. Treatment should be tailored to the organism isolated and its antimicrobial susceptibility pattern. The duration of treatment is 14 days.

Reactive arthritis

Also known as Reiter's syndrome, this condition is characterized by arthritis, urethritis, and uveitis and, often, lesions of the skin and mucous mebranes. It complicates 1–2% of cases of non-gonococcal urethritis (NGU), and is the most common inflammatory arthritis in young men.

Aetiology

Reiter's syndrome is associated with sexually transmitted infections and gastrointestinal infections:

- *C. trachomatis*
- *N. gonorrhoeae*
- *Salmonella* spp.
- *Shigella* spp.
- *Campylobacter* spp.
- *Yersinia* spp.

It is also reported after antibiotic-associated colitis, cryptosporidiosis, after bladder instillation of BCG, and after respiratory infections with *C. psittaci* and *C. pneumoniae*.

Pathogenesis

- The pathogenesis of the condition is not fully understood; it probably represents an abnormal host response to infectious agents.
- Associated with the presence of HLA-B27 which is found in >90% of patients with Reiter's syndrome.

Clinical features

- Urethritis is the first symptom and usually occurs 14 days after sexual intercourse. Cystitis has been reported in women.
- The other features occur after 1–5 weeks.
- Arthritis most frequently affects the knees, but other joints may be involved, e.g. ankles, small joints of feet, sacroiliac joints. Complications: ankylosing spondylitis, calcaneal spurring, dactylitis.
- Ocular findings include conjunctivitis, iritis, keratitis, or uveitis.
- Skin manifestations include waxy papules, keratoderma blenorrhagica, circinate balanitis.
- Initial episode lasts 2–6 months but the disease may recur.
- Rare complications – pericarditis, myocarditis, first-degree heart block, aortic regurgitation.

Diagnosis

- There is no diagnostic test.
- Anaemia and raised ESR are common.
- Synovial fluid shows a raised WCC with a neutrophil predominance and low glucose level. Histology shows non-specific changes.

Management

- Treatment is controversial.
- Because of the association with STIs, some authorities recommend treatment with antibiotics, but there are limited data to support this.
- NSAIDS are given for symptomatic relief.
- Sulfsalazine, methotrexate, or ciclosporin may be used in refractory cases.

Osteomyelitis

Osteomyelitis is an infection of the bone characterized by progressive bone destruction and formation of sequestra. It may be due to haematogenous seeding, contiguous spread from adjacent infected tissues, or traumatic or surgical inoculation of microorganisms.

Classification

Two classification systems exist:

- The Cierny–Mader system is functional classification, based on the affected portion of bone and physiological status of the host, and is useful in guiding therapy:
  - anatomic type – stage 1 = medullary osteomyelitis, stage 2 = superficial osteomyelitis, stage 3 = localized osteomyelitis, stage 4 = diffuse osteomyelitis
  - physiologic class – A = normal host, B = host with local (BL) or systemic (Bs) compromise, C = treatment worse than disease
- the Lee and Waldvogel system is essentially an aetiological classification based on duration of illness (acute or chronic), mechanism of illness (contiguous or haematogenous), presence or absence of vascular insufficiency. It is less helpful in terms of treatment.

Pathogenesis

In experimental models, normal bone is highly resistant to infection; osteomyleitis only occurs after inoculation of large inocula, as a result of bone trauma, or in the presence of foreign material. When digested by osteoclasts, *S. aureus* can survive in a dormant state for a long time, making it difficult to treat with antimicrobials, and resulting in high relapse rates if short courses of antibiotics are given. Biofilm formation associated with prosthetic material can also make these infections difficult to treat.

Aetiology

- *S. aureus* and coagulase-negative staphylococci are the most common cause of osteomyelitis, accounting for >50% of cases.
- Less-common causes (>25%) include streptococci, enterococci, *Pseudomonas* spp., *Enterobacter* spp., *Proteus* spp., *E. coli*, *Serratia* spp., anaerobes.
- Rare causes (<5%) include *M. tuberculosis*, *M. avium complex*, rapidly growing mycobacteria, dimorphic fungi, *Candida* spp., *Aspergillus* spp., *Mycoplasma* spp., *Tropheryma whipplei*, *Brucella* spp., *Salmonella* spp., and *Actinomyces* spp.
- In haematogenous long bone osteomyelitis, the infection is usually monomicrobial, whereas contiguous infection is often polymicrobial.

Clinical features

- Osteomyelitis usually presents with subacute to chronic onset of pain around the affected site.
- Systemic symptoms and signs are frequently absent and local erythema and swelling is unusual.
- In chronic osteomyelitis a draining sinus may be present.

Diagnosis

Diagnosis is often suspected clinically and may be confirmed by a combination of radiological, microbiological, and histopathological investigations.

- Blood tests – the peripheral WCC may be normal or raised; inflammatory markers (ESR and CRP) are often elevated.
- Radiology – although insensitive, a plain radiograph is readily available, inexpensive, and may show changes after 10–14 days. Bone scans are sensitive but non-specific. CT or MRI scans are expensive but highly sensitive and specific, and have become the investigations of choice. MRI is contraindicated in patients with metalware; these may also cause artefacts on CT.
- Biopsy – radiologically guided or surgical biopsy, preferably taken prior to antibiotic therapy, is essential to identify the causative organism(s). Samples should be sent for both microbiology and histology. Sinus tract swabs are of dubious value as they may represent colonizing flora.

Management

- **General principles** – owing to the lack of good clinical trial data, most of the recommendations for the management of osteomyelitis come from animal models, retrospective cohort studies, and expert opinion. The goal of therapy is to eradicate infection and restore/preserve function. Osteomyelitis in adults is usually treated with a combination of surgical debridement and antibiotic therapy, preferably started after surgical debridement.
- **Surgery** – the principles of surgical therapy are debridement of infected tissue, removal of metalware, management of dead space (using a flap), wound closure and stabilization of infected fractures.
- **Antimicrobial therapy** – choice of antimicrobial therapy depends on the organism isolated and its drug-susceptibility results. The optimal duration of treatment is not known but most experts advocate 4–6 weeks of intravenous therapy. The addition of rifampicin to β-lactams has been shown to be effective in animal models of staphylococcal osteomyelitis, and is often used in infections, particularly those involving prosthetic material. Once patients are clinically stable they may be discharged from hospital and treated with outpatient antimicrobial therapy via a long-term intravenous catheter, if appropriate.
- **Adjunctive therapy** – hyperbaric oxygen has been shown to be effective in animal studies but there are inadequate data to support this approach in humans.

Special situations

- Osteomyelitis after open fracture
- Vertebral osteomyelitis, discitis, epidural abscess
- Osteomyelitis in patients with diabetes or vascular insufficiency
- Acute haematogenous osteomyelitis
- SAPHO (synovitis, acne, plantar pustolsis, hyperosteosis, osteitis) syndrome

Reference

1 Mader JT, Shirtleff M, Cellous JH. Staging and Staging Application in Oxyesmyelitis. *Clin Inf Dis* 1997;25:1303–9.

Prosthetic joint infections

Prosthetic joint infections complicate 1–5% of implants, resulting in considerable morbidity. The treatment of prosthetic joint infections is difficult and expensive (estimated cost £25,000–30,000).

Epidemiology

Risk factors for prosthetic joint infection include: previous surgery at same site, rheumatoid arthritis, diabetes mellitus, immunosuppression, poor nutrition or obesity, psoriasis, elderly patients.

Pathogenesis

Infection occurs by one of two mechanisms:
- local infection (60–80%), e.g. peri-operative contamination or local wound infection
- haematogenous seeding (20–40%), e.g. *S. aureus* bacteraemia, dental, genitourinary, or gastrointestinal tract procedures.

As foreign bodies, the metallic implant and the cement contribute by reducing the number of organisms required to establish infection, and permitting pathogens to establish/survive within biofilms.

Aetiology

- Prosthetic joint infections are usually monomicrobial, but may be mixed.
- Common causes include coagulase-negative staphylococci (22%), *S. aureus* (22%), α-haemolytic streptococci (9%), groups A, B, G streptococci (5%), enterococci (7%), Gram-negative bacilli (25%), and anaerobes (10%).
- Rarer causes include corynebacteria, propionobacteria, *Bacillus* spp., mycobacteria, and fungi.

Clinical features

- Cardinal symptoms include joint pain (95%), fever (43%), periarticular swelling (38%), and wound or cutaneous sinus drainage (32%).
- Constant joint pain is suggestive of infection, whereas pain that only occurs on motion or weight bearing is more suggestive of prosthetic loosening.
- The majority of patients present with gradual onset of pain ± discharging sinus.
- Some patients present acutely with high fever, severe joint pain, swelling, erythema, and systemic toxicity.
- The pattern of clinical presentation is determined by the virulence of the infecting pathogen, route of infection, and nature of host tissue.

Diagnosis

- Blood tests are non-specific, and usually show a raised blood white cell count, ESR, and CRP.

- Plain x-rays may show abnormal lucency (>2 mm) at the bone cement surface, periosteal reaction, cement fractures, changes in position of the prosthetic components, and movement of components on stress views. However, these changes are evident in 50% of cases and may also occur with prosthetic loosening.
- Radioisotope scans are generally unhelpful. Technetium diphosphonate scans may show increased uptake in infected joints, but this may also occur in normal joints for up to 6 months after arthroplasty. Sequential technetium gallium bone scans also have poor sensitivity (66%) and specificity (81%).
- Arthrocentesis (aspiration of joint fluid) may confirm the diagnosis. Synovial fluid should be aspirated using aseptic technique, and sent for cell count, microscopy, and culture. The diagnostic sensitivity of arthrocentesis is 86–92%, and the specificity is 82–97% in larger studies. The Gram stain is only positive in 32%.
- Operative samples (arthroscopic or open surgical) are used to make a definitive diagnosis. Several (3 to 6) operative samples of tissue and fluid should be taken, ideally prior to antibiotic therapy, and sent for microbiology and histology. Three positive cultures for the same organism give a 94.8% probability of prosthetic joint infection.

Management

- The most successful approach is a two-stage surgical procedure – removal of the prosthesis and cement, then 6 weeks of intravenous antimicrobial therapy, followed by insertion of a new prosthesis. Success rates of 90–97% have been reported. The use of antibiotic-impregnated spacers (prior to re-implantation) and antibiotic-impregnated cement (at re-implantation) is common, although clinical trial data are lacking.
- An alternative approach is a one-stage surgical procedure – removal of the infected prosthesis and cement, immediate implantation of a new prosthesis, and antimicrobial therapy. This method is often used in elderly or infirm patients. Success rates of 80–83% have been reported.
- Simple surgical drainage with retention of the prosthesis and antibiotic therapy is only successful in 20–36% of cases. Treatment without prosthesis removal is more likely to be successful in early postoperative (<1 month), early haematogenous seeding (<1 month symptoms), and with certain pathogens.
- Long-term suppressive antibiotic therapy is given in special circumstances, e.g. prosthesis removal impossible, prosthesis not loose, pathogen relatively avirulent, pathogen highly sensitive to oral antibiotic, patient able to tolerate long-term oral antibiotics. If all five criteria are met, a 63% success rate is seen with retained hip arthroplasty and variable success rates with total knee replacements.

Prevention

Prosthetic joint infections may be prevented by perioperative antimicrobial prophylaxis, meticulous surgical technique, filtered laminar airflow systems in operating theatres, early recognition, and prompt treatment of wound infections.

References

1 Lentino JR. Prosthetic joint infections: bane of orthopedists, challenge for infectious diseases specialists. *Clin Infect Dis* 2003;36:1157–61.

2 Atkins BL et al. Prospective evaluation of criteria for microbiologic diagnosis of prosthetic joint infections. *J Clin Microbiol* 1998;36:2932–9.

Diabetic foot infections

A diabetic foot infection may be defined as any inframalleolar infection in a patient with diabetes mellitus. The most common lesion is an infected diabetic ulcer, but the spectrum of infections is broad and may include paronychia, cellulitis, myositis, abscesses, necrotizing fasciitis, septic arthritis, tendonitis, and osteomyelitis.

Epidemiology

Foot infections in diabetic patients are common, debilitating, and difficult to manage. Risk factors include: peripheral sensory, motor and/or autom-nomic neuropathy, neuro-osteopathic deformity (e.g. Charcot joint), vascular insufficiency, hyperglycaemia leading to poor immune function and wound healing, patient disabilities (e.g. poor vision, limited mobility, previous amputations), maladaptive patient behaviours (inadequate foot care or footwear), health system failure (inadequate education and management of diabetes and footcare).

Aetiology

A number of organisms may be associated with various syndromes (Table 5.12).

Table 5.12 Aetiology of diabetic foot infections

Foot infection syndrome	Pathogens
Cellulitis	β-haemolytic streptococci (groups A, B, C, and G), *S. aureus*,
Infected ulcer, antibiotic naive	Often monomicrobial: *S. aureus* or β-haemolytic streptococci (groups A, B, C, and G)
Infected ulcer, chronic, previous antibiotic therapy	Usually polymicrobial: *S. aureus*, β-haemolytic streptococci (groups A, B, C, and G), *Enterobacteriaceae*
Macerated ulcer	*Pseudomonas aeruginosa* ± other organisms as above
Longstanding, non-healing wound, prolonged antibiotic therapy	Usually polymicrobial with antibiotic-resistant organisms: aerobic Gram-positive cocci (*S. aureus* , coagulase-negative staphylococci, enterococci), diphtheroids, *Enterobacteriaceae*, *Pseudomonas* spp., non-fermentative Gram-negative rods, fungi
'Fetid foot': extensive necrosis or gangrene	Mixed aerobic Gram-positive cocci (*S. aureus* , coagulase-negative staphylococci, enterococci), *Enterobacteriaceae*, non-fermentative Gram-negative rods, obligate anaerobes

Pathogenesis

Neuropathy plays the central role, leading to ulceration due to trauma or excessive pressure on a deformed foot that lacks protective sensation. Once the protective layer of skin is breached, underlying tissues are exposed to bacterial colonization. This wound may progress to become actively infected, and, by contiguous extension, the infection can involve deeper tissues.

Clinical features

These can range from mild to severe and life threatening:

- foot ulcer with no signs of infection
- foot ulcer with surrounding inflammation or cellulitis <2 cm from edge of wound
- local complications – cellulitis >2 cm from edge of wound, lymphangitis, spread beneath superficial fascia, deep tissue abscess, gas gangrene, involvement of muscle, tendon, or bone
- systemic toxicity or metabolic instability – fever, chills, tachycardia, hypotension, confusion, vomiting, leucocytosis, acidosis, hyperglycaemia, uraemia.

Diagnosis

- Diagnosis is based on clinical features (above). It is important to assess perfusion (peripheral pulses) as well as sensation (using a monofilament). If the peripheral pulses are not palpable, use a Doppler ultrasound to determine ankle brachial pressure indices (ABPIs).
- Imaging, e.g. MRI scan, may be helpful to determine the extent of infection, e.g. deep collections osteomyelitis.
- Deep tissue specimens (not superficial swabs) should be taken and sent to the laboratory for microscopy and culture prior to commencing antimicrobial therapy.

Management

See IDSA guidelines for management of diabetic foot infections.[1]

- Determine the need for hospitalization.
- Stablize the patient if systemically unwell and correct any metabolic abnormalities.
- Antibiotics – do not give antibiotics for uninfected ulcers. Initial empiric therapy should be based on severity of infection and available microbiological data, e.g. Gram stain. If there are no data, initial therapy should be broad spectrum, e.g. IV co-amoxiclav.
- Assess the need for surgery – patients with severe infections, e.g. necrotizing fasciitis, gas gangrene, extensive tissue loss, critical limb ischaemia, should be referred for urgent surgical review.
- Wound care plan – the wound should be dressed in a way that permits daily inspection. Special aids may be available to offload pressure on the wound.
- Adjunctive therapies – the use of granulocyte colony-stimulating factors has not been shown to be beneficial. Hyperbaric oxygen may have a role.

Reference

1 Lipsky BA, Berendt AR, Deery HG et al. Diagnosis and management of diabetic foot infections. *Clin Infect Dis* 2004;39:885–910.

Neonatal sepsis

This is defined as sepsis occurring within 4 weeks of birth:

- early onset – within 7 days, associated with microbes acquired from the mother either transplancentally or intrapartum; 85% of early-onset cases present within 24 h of delivery
- late onset – after 7 days, associated with organisms acquired from the environment (e.g. caregivers or urinary or vascular devices)
- premature infants experience the most rapid onset
- certain viral infections may cause an indistinguishable clinical picture.

Epidemiology

- Incidence of culture-proven sepsis is 2/1000 live births in the US, but up to 7–13% of neonates may be evaluated for sepsis due to the non-specific nature of the early signs.
- Neonatal sepsis contributes up to 15% of all neonatal deaths from meningitis and 4% of all neonatal deaths.
- Risk factors:
  - early-onset sepsis – maternal colonization with group B streptococci, premature rupture of membranes, prolonged rupture of membranes, prematurity, maternal UTI, chorioamnionitis, maternal fever >38°C at delivery, poor maternal nutrition, recurrent abortion, meconium staining, congenital abnormalities
  - late-onset sepsis – prematurity, central venous catheterization (duration >10 days), continuous positive pressure nasal cannula, H_2 antagonist/proton pump inhibitor use, GI tract pathology.

Aetiology

- Early onset – group B streptococci, *E. coli*, *H. influenzae*, *L. monocytogenes*
- Late onset – coagulase-negative staphylococci, *S. aureus*, *Klebsiella* spp., *E. coli*, *Pseudomonas* spp., *Candida* spp., *Enterobacter* spp., *Serratia* spp., *Acinetobacter* spp., group B streptococci, anaerobes.

Clinical features

- Pneumonia – neonates may aspirate organisms during delivery or have developed intrauterine pneumonia following aspiration of amniotic fluid. Signs: tachypnoea, cyanosis, grunting, apnoea, costal/sternal retractions, nasal flaring. CXR may show bilateral consolidation and pleural effusions. *Klebsiella* spp. and *S. aureus* may generate severe lung damage with abscesses and empyema. Early-onset group B streptococcal pneumonia may be fulminant with significant mortality.
- Cardiac features – overwhelming sepsis may be associated with pulmonary hypertension, decreased cardiac output, and hypoxia. Late features: overt shock, pallor, poor capillary perfusion, oedema.
- Metabolic features of sepsis – hypo/hyperglycaemia, acidosis, jaundice.
- Neurological signs – ventriculitis, meningitis (36% group B streptococci, 31% *E. coli*, 5–10% *Listeria*), cerebral vasculitis, cerebral oedema, cerebral infarction. Meningitis in early-onset sepsis occurs within 24–48 h –signs of meningitis are present in only 30% and CSF WCC may be normal. In late-onset disease, 80–90% have neurological features and

CSF changes may be markedly abnormal, especially with Gram-negative organisms. Neonates with meningitis are likely to be hypothermic.

- Haematological abnormalities – thrombocytopenia, DIC, high or low WCC (50% normal). The immature-to-total neutrophil ratio is a more useful marker of infection.
- Gastrointestinal – necrotizing enterocolitis has been associated with the presence of a number of species in immature gut.

Investigations

- Blood tests – FBC, U&E, LFTs, CRP
- Microbiological – blood, CSF, and urine cultures. Gram stain may provide early identification. Cultures may be negative if the mother received intrapartum antibiotics. If CSF is culture-positive a follow-up lumbar puncture is often performed at 24–36 h after initiation of antibiotic therapy to document CSF sterility
- Other tests – infection markers such as IL-6, IL-8 and CD64 have been used in the evaluation of sepsis in neonates
- Radiology – CXR may show lobar changes but more usually resembles respiratory distress syndrome, with a diffuse reticulogranular pattern. Cranial ultrasound may show evidence of ventriculitis and chronic changes. CT may be required in complex meningitis with obstruction and abscesses

Management

- Medical emergency – IV antimicrobials should be commenced as soon as cultures are taken. Antibiotic choice should be guided by maternal history, and local drug resistance patterns. A 2004 Cochrane review found no significant difference in outcome between various antibiotic regimes.[1] Generally, a glycopeptide is combined with an aminoglycoside. Treatment for 7–10 days may be appropriate even in the absence of positive cultures.
- Cardiovascular, respiratory, and nutritional support may be required.
- Infants with bacterial meningitis require antibiotics capable of penetrating the blood–brain barrier to achieve therapeutic concentrations in the CSF, and longer courses of treatment (up to 3 weeks). If CSF is not sterile on a follow-up lumbar puncture 24–36 h after the initiation of therapy, consider modification of therapy.
- Surgical interventions – the development of hydrocephalus may require the placement of a ventriculoperitoneal shunt. Abscesses may require surgical drainage.
- With early diagnosis and treatment prognosis is good, although residual neurological damage is seen in 15–30% of neonates with septic meningitis.
- Follow-up – hearing assessments before discharge and at 3 months if aminoglycosides have been given, follow-up for those at risk of developing neurological sequelae (with a paediatric neurologist).

Prevention

Some authorities recommend antibiotic prophylaxis should be given to certain groups of women at risk of carriage of group B streptococci.

[1] Mtitimila CL, Cooke RWI. Antibiotic regimes for suspected early neonatal sepsis. Cochane Database of Systamatic Reviews 2004, Issue 4, Article No:CD ØØ4495

Viral causes of childhood illness

Table 5.13 Viral causes of childhood illness

Virus	Typical age	Features
Herpes simplex virus	Neonate: 90% of infections acquired perinatally, 5% congenitally and the remainder postnatally (e.g. from an adult with herpes labialis).	Nearly all infected infants manifest disease. Incubation: 3–14 days. May be localized to the eye or CNS. 70% of untreated cases disseminate (hepatomegaly, jaundice, pneumonitis, encephalitis, vesicular rash). Neonates have the highest rates of visceral and/or CNS infection of any patient group.
		70% of infected infants are born to mothers with no apparent disease. 70% of cases are due to HSV-2. 50% of babies delivered via an infected birth canal become infected. Most cases of HSV-1 follow maternal acquisition of genital HSV-1 late in pregnancy, with consequent neonatal contact with infectious secretions during delivery. Untreated the death rate is 65%. Less than 20% of those with CNS infection develop normally. CNS morbidity is less severe with HSV-1 than HSV-2. See p.824 for prevention. Systemic aciclovir is essential and has reduced the death rate of neonatal herpes to < 25%.
	Childhood (highest incidence HSV-1 infection is seen in children aged 6 months to 3 years).	>80% of primary HSV infections are asymptomatic. Symptoms: fever, anorexia, sore mouth (an ulcerative gingivostomatitis), local lymphadenopathy. Contamination of skin by infectious saliva may lead to secondary lesions on the peri-oral skin, eye, fingers and vulva. Those with disseminated infection should be isolated. Topical aciclovir is of no benefit in acute primary infection of children. Systemic treatment can decrease healing time and is important in the immunocompromised.
Varicella zoster virus	90% of cases occur in those under 13 years of age	Primary infection causes chickenpox, a maculopapular rash that forms pustulating vesicles. Complications: bacterial superinfection, cerebellar ataxia, and encephalitis. See p.823 and 827 for the prevention and management of neonatal disease.
Measles ('first disease')	Uncommon in those populations with vaccination	Acute highly infectious disease characterized by cough, coryza, fever, and rash. Severe manifestations and complications include pneumonia, encephalitis, bacterial superinfection and subacute sclerosing panencephalitis.

Table 5.13 Viral causes of childhood illness (*Continued*)

Virus	Typical age	Features
Rubella ('third disease')	Prior to vaccination incidence was highest in the spring among children aged 5–9 years	Acute mild exanthematous viral infection of children and adults resembling mild measles but with the potential to cause fetal infection and birth defects.
Parvovirus B19 (slapped cheek disease, erythema infectiosum, 'fifth disease')	Infection common in childhood – 50% are IgG-positive by 15 years	20% asymptomatic. Prodrome (5–7 days) of myalgia, arthralgia, malaise, rhinorrhoea, and fever, then a bright red rash on the cheeks followed 1–2 days later by a maculopapular rash on the trunk, legs, arms, and buttocks. This clears after a few days leaving a characterisitic lacey pattern which fades/reappears over the following 3 weeks.
Human herpesvirus 6 (roseola, exanthem subitum, 'sixth disease')	Most children acquire infection between 4 months and 3 years of age	Abrupt onset of fever (±peri-orbital oedema) is followed 3–5 days later by a rash (rose-pink papules which are mildly elevated, non-pruritic, and blanche on pressure) on the back and neck and spread to the chest and limbs sparing the feet and face. Lasts ~2 days. Other features: malaise, vomiting, diarrhoea, cough, pharyngitis and lymphadenopathy, febrile convulsions (10% of primary infections). Meningitis and encephalitis are less-commonly seen.
Mumps	90% of cases occured in those under 15 years prior to the introduction of vaccination. Many cases now occur in older children and those at university	Acute generalized viral infection of children and adolescents causing swelling and tenderness of the salivary glands and, rarely, epididymo-orchitis.
Enteroviral infections	All age groups but commoner in younger children	Accounts for the majority of childhood fever-rash syndromes as well as meningitis, myocarditis, sepsis, hand-foot-and-mouth disease, and herpangina.
Epstein–Barr virus (the cause of 90% of infectious mononucleosis)	Approximately 50% of UK children are infected by age 5 years, around 90% by age 25 years	Primary infection in childhood is asymptomatic. Infection in adolescence may present with an acute infectious mononucleosis syndrome.
Other common viral infections of childhood	Adenovirus, molluscum, RSV, metapneumovirus, rhinovirus, rotavirus	

Enteroviral infections

The non-polio enteroviruses (including Coxsackie virus, enterovirus, echoviruses) cause a large number of different clinical syndromes accounting for the majority of childhood fever-rash syndromes as well as being important causes of meningitis, myocarditis, and even neonatal sepsis. Of the many clinical syndromes they cause, only hand-foot-and-mouth disease and herpangina have a clinical presentation distinct enough to allow identification.

Epidemiology

- Transmission is faeco-oral or via contact with discharging skin lesions. Respiratory and oral-to-oral routes occur in crowded conditions. Viruses can survive at room temperature for several days and tolerate the acidic pH of the gastrointestinal tract.
- Found worldwide, affects all age groups but highest infection rates are seen in children (secondary to exposure, hygiene, and immune status). Infection course is benign in older children, more serious in neonates.
- Neonatal infections are probably acquired after birth.

Clinical features

Incubation is 3–10 days, and virus may be excreted in stool for weeks. Infection may be asymptomatic (90% of infections), or cause an undifferentiated flu-like illness, or a more-characteristic syndrome.

- Undifferentiated illness – low-grade fever of sudden onset with or without upper respiratory and gastrointestinal symptoms, e.g. flu-type syndrome of malaise, myalgias, sore throat, headache, conjunctivitis, nausea, vomiting, and diarrhoea. Orchitis and epididymitis can occur. Symptoms last 3–7 days.
- Herpangina (enteroviral vesicular pharyngitis) – typically seen during the summer in children aged 3–10 years (may occur in young adults). Painful vesicles (usually 3–6) and ulcers of the posterior pharynx and tonsils with fever and a sore throat. There are no exudates present. Pain may make the child reluctant to eat. Symptoms last 3–7 days. Organism: Coxsackie virus group A and sometimes group B.
- Hand, foot, and mouth disease (HFM, enteroviral vesicular stomatitis with exanthema) – fever and vesicular eruption in the anterior pharynx, palms, and soles of toddlers/school-aged children. Oral vesicles not initially painful but later burst leaving painful ulcers. Cutaneous vesicles heal by resorption of fluid and do not crust. May develop characteristic rash. Organism: Coxsackie virus group A, serotype 16 (among others). Complications: HFM caused by enterovirus-71 has been associated with a higher incidence of neurological complications (polio-like syndrome, aseptic meningitis, encephalitis, encephalomyelitis, acute cerebellar ataxia, acute transverse myelitis, Guillain–Barré syndrome, opso-myoclonus syndrome, and benign intracranial hypertension), rare cases of myocarditis, interstitial pneumonitis, and pulmonary oedema.
- Other – viral exanthems (pink, maculopapular, blanching rash, rarely urticarial, vesicular or petechial – can mimic rubella and roseola but with no significant adenopathy), aseptic meningitis, myocarditis/

pericarditis, pleurodynia (lancinating chest pain attacks with fever and headache, also seen with Coxsackie B infection), neonatal sepsis.

Diagnosis and management

- Diagnosis of herpangina and HFM is clinical (both are mild self-limiting illnesses that do not warrant laboratory diagnosis) and management supportive (e.g. soft food for those with painful mouth ulcers, antipyretics, topical analgesics). See individual sections for the diagnosis and management of the more severe clinical presentations.
- The virus can be identified from respiratory secretions, cutaneous lesions, and stool. Paired serology will confirm recent infection. PCR tests are available.
- Hygiene – to prevent continued faeco-oral spread (handwashing etc).

Bacterial causes of childhood illness

Scarlet fever

A syndrome of exudative pharyngitis, fever, and scarlatiniform rash caused by infection with an erythrogenic exotoxin-producing group A beta-haemolytic streptococcus. Once known as 'second disease' – the second exanthematous disease of childhood.

Clinical features

- Normal inhabitants of the nasopharynx, group A streptococci may cause pharyngitis, skin infections, pneumonia, and bacteraemia. Scarlet fever usually occurs in association with pharyngeal infection.
- Usually seen in children aged 5–15 years.
- Rash appears 1–2 days after onset of sore throat with a diffuse red blush with scattered points of deeper red. First noticed on the chest, it spreads to involve the trunk, neck, and extremities, sparing the palms, soles, and face. The face may, however, be flushed with circumoral pallor. Skin folds in the neck, axillae, groin, elbows, and knees may appear as lines of deeper red. There maybe petechiae and a sandpaper texture to the skin due to sweat gland occlusion. Examination of the oropharynx may reveal exudative pharyngitis, tonsillitis, and small red haemorrhagic spots on the palate. The tongue may be coated in early disease but then becomes beefy red ('strawberry tongue'). The skin rash fades over a week and is followed by desquamation which may last several weeks.
- Although common and fatal in the 19th and early 20th century, it is rarely considered serious today. Severe scarlet fever may be due to haematogenous spread or systemic toxicity with high fever which may be complicated by arthritis and jaundice.
- Complications – suppurative complications (e.g. abscess), rheumatic fever, post-streptococcal glomerulonephritis, erythema nodosum.

Diagnosis and management

- Throat swab for culture and anti-streptolysin O titre (ASOT)
- Treatment – penicillin V PO or IV benzylpenicillin.
- Differential – consider measles, infectious mononucleosis, other viral infections with rash, Kawasaki's disease, staphylococcal infection.

Staphylococcal epidermal necrolysis

Also known as (staphylococcal) scalded skin syndrome (SSSS), this condition is caused by an exotoxin produced by *S. aureus* which leads to exfoliation of the upper layers of the epidermis; 98% of cases occur in children under 6 years of age due to lack of immunity and immature renal clearance capability. Mortality is low in children (1–5%) but can be higher in adults who are usually immunocompromised or have renal failure.

Clinical features

- An infection commonly occurs at a site such as the oral or nasal cavities, throat, or umbilicus. Epidermolytic toxins are produced locally and act at a remote site leading to the abrupt onset of generalized skin erythema.
- The epidermis beneath the granular cell layer separates due to binding of the toxins to desmoglein 1 in desmosomes. Bullae form, and diffuse sheet-like desquamation may occur 1–2 days later (Nikolsky sign positive). This leaves a raw and tender exposed surface.
- There may be associated conjunctivits, stomatitis, and urethritis.
- Most patients do not appear very ill but significant dehydration can develop. Healing occurs over 1–2 weeks.

Diagnosis and management

- *S. aureus* can usually be cultured from the site of remote infection, WCC is usually normal but inflammatory markers may be elevated, a PCR test for the toxin is available. Blood cultures are usually negative in children (but may be positive in adults).
- Differential diagnosis – toxic epidermal necrolysis (part of the disease spectrum that contains bullous erythema multiforme and Stevens–Johnson syndrome and associated with a deeper epidermal detachment than that of SSSS), erythema multiforme, burns.

Management

- Antibiotics – flucloxacillin, or erythromycin if penicillin allergic.
- Fluids – patients can leak a lot of proteinaceous fluid through the skin and may require IV supplementation. Wound care is similar to that given for burns, and very severe cases may require specialist burns unit input. The skin damage can make patients vulnerable to secondary infection.

Pertussis

A highly contagious bacterial infection of the respiratory tract, spread by droplets and characterized by paroxysmal cough ('whooping cough'). Caused by *Bordetella pertussis* (📖 p.357), a Gram-negative pleomorphic bacillus of which humans are the sole reservoir, and less-commonly *B. parapertussis*.

Epidemiology

- Infection is worldwide but unusual in the UK and other countries with widespread vaccination. Neither infection nor vaccination provides complete or lifelong immunity. Protection against typical disease lasts 3–5 years and immunity is not detectable after 12 years. The UK introduced a pre-school pertussis booster in 2001 which has seen morbidity at the lowest levels yet in both vaccinated groups and infants too young to receive the vaccine.
- Most cases occur in infants/children (the majority infected by coughing adults and older children). Adults (10% of cases) experience milder disease. Children <1 year of age are most likely to require hospitalization.
- Worldwide it remains a major cause of death. Around 50 million cases and 600,000 estimated deaths each year.

- Those at risk of severe disease (pneumonia, encephalopathy, death) include premature infants, and those patients with underlying cardiac, pulmonary, neuromuscular/neurological disease.

Clinical features

- Incubation – 3–12 days. Patients are infectious from the onset of illness until towards the end of the paroxysmal phase.
- Pertussis is a 6-week illness of three stages, each lasting around 2–4 weeks. Older children and adults may not exhibit these distinct stages.
  - Stage 1 (catarrhal phase) – indistinguishable from the common cold: congestion, sneezing, mild fever, and rhinorrhoea. Patients are at their most infectious during this phase.
  - Stage 2 (paroxysmal phase) – paroxysms of intense coughing which can last several minutes may be followed by a loud whoop in older infants and toddlers. Infants <6 months may have apnoeic episodes but do not whoop. Vomiting is common after coughing. Subconjunctival haemorrhages and facial petechiae may occur. Most deaths occur in infants (coughing leading to choking and apnoea).
  - Stage 3 (convalescent phase) – chronic cough, which may last for weeks, triggered by intercurrent viral infections.
- Differential diagnosis – bronchiolitis, mycoplasmal pneumonia, chlamydial pneumonia, inhaled foreign body.
- Complications – pneumonia, secondary bacterial infection, pneumothorax, diaphragmatic rupture, surgical emphysema, neurological deficits secondary to hypoxia.

Diagnosis

- Lab confirmation is usually delayed. Diagnosis should be made clinically.
- General – leucocytosis (WCC > 100,000 is associated with an increased risk of death), CXR may be normal or show peribronchial thickening, consolidation (secondary bacterial infection, rarely pertussis pneumonia), pneumothorax, pneumomediastinum, or air in the soft tissues.
- Microbiological culture – requires special media (e.g. Regan–Lowe or Bordet–Gengou agar). Culture specimens are best obtained by flexible swab or deep nasopharyngeal aspiration during the catarrhal or early paroxysmal phase. Culture for 7 days. Usually negative in those previously immunized or given antibiotics.
- Serology is useful to confirm the diagnosis retrospectively. PCR-based tests are available.

Management

- General – supportive care is the mainstay. Consider admitting patients at risk of severe disease and complications (see above, plus those younger than 3 months, or 3–6 months with severe paroxysms); 50% of infants require hospitalization. Infection control measures should be taken for those patients in the contagious phase of the disease.
- Antimicrobial therapy – erythromycin given early in the catarrhal phase shortens the duration of the paroxysmal stage. Once cough is established, antimicrobial agents do not alter the course of the illness but serve to limit the spread of disease. Treatment duration: 14 days. Patients

should be isolated. Consider treating close contacts of pertussis cases (including children and staff at day centres) who are particularly vulnerable, unvaccinated, partially vaccinated, or under 5 years of age.
- Other agents – there is no evidence for any benefit from corticosteroids or β_2-adrenergic agents. Pertussis-specific immunoglobulin is an experimental therapy that may be effective in decreasing paroxysms of cough.

Prevention
- Vaccination is recommended for all babies at 2, 3, and 4 months as part of the DTP (diphtheria, tetanus, polio) vaccine. It may not prevent the illness entirely, but lessens disease severity and duration.
- There is no transfer of protective maternal antibody even from mothers with a documented history of infection or vaccination. Nearly all cases of fatal pertussis in developed countries occur in infants too young to be immunized. In the UK and US, children are given boosters at 3–4 or 11–12 years of age respectively, with the aim of reducing transmission to pre-vaccination infants.

Other common causes of bacterial infection in childhood
- Bacterial meningitis, including *N. meningitidis* – 📖 p.756
- Bacterial causes of pneumonia – 📖 p.654
- Infectious diarrhoea – 📖 p.688
- Urinary tract infections – 📖 p.716
- Upper respiratory tract infections – 📖 p.630–669
- Superficial bacterial infections of the skin (e.g. erysipelas) – 📖 p.788

Congenital infections

Congenital infections may be acquired *in utero* or perinatally. The acronym TORCH has been used to describe a group of infections which generally cause a mild or inapparent infections in the mother yet may cause severe disease in the infant, e.g.

- **T**oxoplamosis
- **O**thers (e.g. syphilis, parvovirus and VZV)
- **R**ubella
- **C**ytomegalovirus
- **H**SV (mother may be asymptomatic).

Long-term consequences include growth retardation, microcephaly, congenital defects with long-term sequelae, and progressive disease in childhood. They may also present with unusual exanthemata, organomegaly, or thrombocytopenia in the neonatal period. Any infant with features of TORCH illness should undergo thorough serological evaluation for these agents as well as others, e.g. HIV.

Toxoplasmosis

- Maternal – may be subclinical or present with an infectious mononucleosis-like illness with lymphadenopathy.
- Fetal – risk of transmission is lowest in 1st trimester (but associated with more-severe abnormalities) and highest in last trimester. Clinical features include chorioretinitis, hydrocephaly, microcephaly, aqueductal stenosis, agenesis of corpus callosum, cerebral calcifications, nonimmune hydrops. Three-quarters of infants are asymptomatic at birth
- Diagnosis – IgM typically turns positive after 1 week and may remain positive for years. IgG follows the same course, but remains positive for life.
- Treatment – maternal infection is treated with spiramycin which reduces fetal infection rate. Refer to a specialist centre for treatment and ultrasound monitoring of fetus. Congenital infection – treat with pyrimethamine, sulfadiazine, and folinic acid. Seek specialist advice.

Syphilis

- Maternal – risk of congenital infection is 50% in primary and secondary syphilis, 40% in latent infection, and 10% in tertiary syphilis. Maternal disease is detected by routine antenatal serological screening.
- Fetal – clinical features include stillbirth, intrauterine growth restriction (IUGR), non-immune hydrops, rhinitis, skin rash, hepatosplenomegaly, 'mulberry molars', 'saber shins', saddle nose deformity, interstitial keratitis, eigth nerve deafness, peg-shaped incisors.
- Diagnosis – based on serology or positive dark-field microscopy or staining for treponemes in samples from the placenta or umbilical cord.
- Treatment – maternal infection treated with penicillin. Neonates should be treated for neurosyphilis (cannot be excluded).

Parvovirus B19

- Maternal – presents with fever, malaise, polarthralgia, coryza, and rash. May be mistaken for rubella. See 📖 Management of rash contact in pregnancy, p.828.
- Fetal – anaemia leading to non-immune hydrops. Fetal loss 9%.
- Diagnosis – serology in the mother. Fetal infection can be diagnosed by amniotic fluid sampling, foetal blood sampling or post mortem.
- Treatment – intrauterine blood transfusion. One-third resolve without transfuson.

Varicella zoster

- Maternal – presents with a vesicular rash. Increased risk of complications, e.g. pneumonitis. See 📖 Management of rash illness in pregnancy, p.830.
- Fetal – Primary VZV infection during the first half of pregnancy has been associated with limb hypoplasia, cicatricial lesions, psychomotor retardation, cutaneous scars, chorioretinitis, cataracts, cortical atrophy, microcephaly, microphthalmus and IUGR. The risk of developing the syndrome is 1% if less than 20 weeks and 2% at 13–20 weeks.
- Diagnosis – usually clinical. May be confirmed by VZV PCR of skin lesions in mother.
- Treatment – aciclovir PO or IV for the mother. Varicella zoster immune globulin (VZIG) for the neonate if mother develops chickenpox within 5 days of delivery.

Rubella

Congenital rubella syndrome is rare since the introduction of vaccination.

- Maternal – presents with rash illness. See 📖 Management of rash illness in pregnancy, p.830.
- Fetal – infection is more severe in early pregnancy with >85% chance of being affected by either multiple defects or spontaneous abortion in first 2 months. Clinical features include 'blueberry muffin skin',cataracts, glaucoma, microphthalmia, sensorineural deafness, patent ductus arteriosus (PDA), atrioventricular (AV) septal defects, pulmonary artery stenosis, microcephaly, meningoencephalitis, IUGR, hepatosplenomegaly, interstitial pneumonitis, thrombocytopenia.
- Diagnosis – based on maternal serology.
- Treatment – none.

Cytomegalovirus

- Maternal – infection is usually asymptomatic but may present with an infectious mononucleosis-like syndrome. Occurs in 1–2% of seronegative women during pregnancy; a small number of cases are due to viral reactivation.
- Fetal – infection of the neonate may occur *in utero* or, more commonly, perinatally. Clinical features include UGR, microcephaly, periventricular calcifications, sensorineural deafness, blindness with chorioretinitis, mental retardation, hepatosplenomegaly, thrombocytopenic purpura, haemolytic anemia. Incidence of sequelae is 25% for primary infection, and 8% with recurrent infection; ~1% die at or soon after birth.

15% appear normal but hearing defects or mental retardation become apparent in later life.

- Diagnosis – detection of CMV in the urine within the first week of life confirms the diagnosis.
- Treatment – there is is little information regarding the use of ganciclovir in the setting of congenital CMV infection.

Herpes simplex

- Maternal – primary infection presents with a vesicular genital rash. Approximately 15% of all pregnant women with a history of genital HSV infection experience recurrent lesions at delivery. Around 2% of pregnant women with a history of recurrent HSV infection are asymptomatically shedding at the time of delivery.
- Fetal – usually acquired intrapartum but may be acquired *in utero*. 75–90% of infants with neonatal HSV are born to infected asymptomatic mothers who have no known history of genital HSV. Oral-labial herpes presents a greater risk of postnatal HSV acquisition than genital HSV. Clinical features include skin lesions, chorioretinitis, microcephaly, hydranencephaly, and microphthalmia. While primary HSV infections in the first trimester are associated with higher rates of spontaneous abortion and stillbirth, infection later in pregnancy appears more likely to be associated with preterm labour or growth restriction. Of greatest concern is the risk of primary infection acquired at birth which could lead to herpetic meningitis.
- Diagnosis – usually clinical in the mother. May be confirmed by serology or PCR.
- Treatment – 1st and 2nd trimesters: treat mother with PO or IV aciclovir. 3rd trimester or genital lesions at time of delivery – consider Caesarean section. Infants born to women with active lesions should be isolated and the infant closely observed during the first month of life for features of neonatal HSV infection.

Maternal infections associated with neonatal morbidity

Pelvic inflammatory disease

Pelvic inflammatory diesease (PID) is associated with chlamydial infection in over 50% of cases, and gonorrhoea in around 14%. Many cases are asymptomatic. Pregnant women with PID should be treated with IV antibiotics as it is associated with an increase in pre-term delivery, maternal and fetal morbidity (e.g. ophthalmia neonatorum).

Listeria monocytogenes

Pregnant women are 20 times more likely to contract listeriosis than other adults; ~33% of all cases of listeriosis occur during pregnancy. Acquisition is mainly by the ingestion of contaminated food. Pregnant patients are often asymptomatic or present with a flu-like febrile illness. These mild symptoms notwithstanding, listeriosis can still lead to premature delivery, neonatal sepsis, and stillbirth. Placental transfer of the organism can cause amnionitis with spontaneous septic abortion or premature labour with delivery of an infected baby. Fetal infection may cause sepsis, meningo-encephalitis, or disseminated infection with microabscesses. Neonatal infection has a mortality of around 50%, particularly in early-onset sepsis (🕮 p.814). Late-onset infection typically presents as meningitis at 3–4 weeks of age.

Management

Prevention is by avoidance of potentially contaminated foods. Treatment is with ampicillin (or co-trimoxazole in those with serious penicillin allergies. It has not been approved for use in pregnancy).

Other

- Malaria (🕮 p.562)
- Varicella zoster and herpes simplex (🕮 p.441 and 444)
- Hepatitis B (🕮 p.459).
- Group B streptococcus (🕮 p.292)
- Chorioamnionitis (🕮 p.834)
- Leptospirosis (🕮 p.393)
- *Ureaplasma urealyticum* (🕮 p.402)
- *Mycoplasma hominis* (🕮 p.402)

Management of rash contact in pregnancy

See also the Health Protection Agency (HPA) guidance on rashes in pregnancy and the Department of Health (DH) *Green Book*.[1,2]

Contact with a non-vesicular rash

The important infective causes are measles, rubella, and parvovirus B19. Intervention is indicated for other possible causes in the absence of the development of symptoms in the pregnant woman. *All* pregnant women with contact with a non-vesicular rash illness should be investigated for asymptomatic parvovirus B19 infection, and for asymptomatic rubella infection unless there is satisfactory evidence of past rubella infection (vaccine or natural infection). A significant contact is defined as being in the same room for a significant period of time (>15 min) or face-to-face contact.

- **Rubella** – a mother is extremely unlikely to be susceptible to rubella if she has had at least two previous positive rubella antibody tests or at least two doses of rubella vaccine documented, or one vaccine dose followed by one positive rubella antibody test. She should be reassured, but told to reattend if a rash develops. If rubella susceptibility is possible a serum sample should be taken and tested for rubella-specific IgG and IgM.
  - If IgG-positive and IgM-negative, the mother can be considered immune with no evidence of recent primary infection. If IgG levels are low it is worth repeating the test – there are rare occasions when IgG may precede IgM positivity in primary infection.
  - If IgM and IgG are negative the patient is susceptible.
  - If IgM is detected further advice should be sought – the control of rubella in the UK means that *most* rubella-specific IgM-positive results don't reflect recent rubella. Unless seroconversion has been demonstrated further specialist testing is required.
- **Parvovirus B19** – all women should be investigated for asymptomatic infection. This should not be delayed pending the development of symptoms, as asymptomatic infection is just as likely as symptomatic to infect and damage the fetus, and active management of the infected fetus reduces the risk of poor outcome. Maternal serum should be taken as soon after rash contact as possible and tested for parvovirus B19-specific IgG and IgM.
  - If IgM but not IgG is detected, tests should be repeated on another fresh serum sample.
  - If IgG is detected and IgM is not present the mother can be reassured.
  - If both IgG and IgM are negative, another sample should be taken one month after the last contact. If these remain negative the mother can be reassured she has not been infected but informed that she is susceptible. If the mother is found to have developed asymptomatic infection she should be managed as detailed in 📖 Management of rash illness in pregnancy, p.828.

- **Measles** – if epidemiological and clinical features suggest the source patient has measles, passive prophylaxis with IM human normal immunoglobulin should be considered. This should be given as soon as possible after exposure and certainly within 6 days – it attenuates maternal illness, but does not confer any benefit on the fetus. If the mother has received two doses of measles vaccine the probability of becoming infected is very low. If vaccination history is negative or uncertain, serum should be taken for an urgent measles IgG and results awaited before giving human normal immunoglobulin (HNIG). If IgG-positive within 10 days of contact no further action is required. If measles IgG is not present HNIG should be given and serological tests repeated at 3 weeks. There is no point in giving HNIG if the contact was more than 10 days prior to presentation.

Contact with a vesicular rash

Pregnant women who are exposed to varicella or herpes zoster in pregnancy should seek medical attention as soon as possible. Contact is considered significant if the mother was in the same room for 15 min or more, or had face-to-face contact of any duration.

- They can be reassured that they are protected if they themselves have a history of varicella or herpes zoster.
- If there is an uncertain or absent history of infection, the mother's susceptibility should be determined urgently by VZV IgG testing. VZIG should be offered to VZV IgG-negative women within 10 days of exposure (or within 10 days of rash onset in cases of continuous household exposure – e.g. an infected child in the house). The 10-day administration window allows ample time for antibody testing before proceeding with administration of VZIG. Around 50% of women who receive VZIG following household exposure will develop chickenpox; another 25% are infected subclinically. However, disease is attenuated (the risk of fatal varicella is estimated to be about five times higher in pregnant than non-pregnant adults, with fatal cases concentrated late in the second or early third trimester) and the risk of fetal infection reduced. NB: if contact was before the infectious period (i.e. more than 48 h before chickenpox rash onset or before the appearance of shingles vesicles), VZIG is not indicated.

References

1 Health Protection Agency. *Rashes in pregnancy – HPA Guidelines – information and advice*. http://www.hpa.org.uk/infections/topics_az/pregnancy/rashes/default.htm (accessed 19 August 2008).

2 Department of Health. *Green Book*. http://www.dh.gov.uk/PolicyAndGuidance/HealthAndSocialCareTopics/GreenBook/fs/en (accessed 19 August 2008).

Management of rash illness in pregnancy[1,2]

Rubella, parvovirus B19, and varicella-zoster virus are the infections of most relevance because of their potential impact on the fetus and neonate. With the exception of varicella, these infections do not have a specific impact on the fetus beyond 20 weeks' gestation. Investigation is recommended at any gestation: age calculation may not be accurate, and achieving an accurate diagnosis is helpful in guiding the advice given to the mother regarding contact with other pregnant women and neonates (e.g. antenatal clinics). Other infections (enterovirus, infectious mononucleosis, syphilis, streptococcus, meningococcus) are managed as normal in the mother and the neonate should be followed up. *NB: if investigation is commenced some weeks after rash or contact, it may not be possible to confirm or refute a possible diagnosis.*

Patients presenting with a non-vesicular rash

All pregnant women with a non-vesicular rash illness compatible with rubella or parvovirus B19 should be investigated simultaneously for both infections *regardless* of previous history, immunization, or prior testing.

- Rubella infection – around 1–2% of young pregnant women in the UK are susceptible and the rates of infection following contact are high. Maternal infection is extremely rare in the UK today.
  - Risk of adverse fetal outcome – 90% at a gestational age <11 weeks, 20% at 11–16 weeks, falling to a minimal risk of deafness at 16–20 weeks and no increased risk after this time.
  - Management of pregnant women with proven primary or symptomatic re-infection with rubella varies with the gestation at which infection took place and the individual circumstances of the woman. The only available intervention is termination of pregnancy.
  - There is a low but significant risk to the fetus in maternal asymptomatic re-infection within the first 16 weeks of gestation. It may be that further fetal investigation by virus detection (to ascertain whether infection has occurred) is warranted. Such investigations are, however, invasive and risk adverse outcome.
- Parvovirus B19 – 40–50% of young pregnant mothers are susceptible and the UK rates of infection are around 1 in 400 infections at present.
  - Fetal infection at <20 weeks' gestation is associated with a 9% excess fetal loss and 3% rate of hydrops fetalis (of which ~50% die).
  - Maternal parvovirus B19 infection diagnosed during pregnancy – the fetus should be scanned by USS 4 weeks after the onset of illness onset or date of seroconversion, and then at 1–2-weekly intervals until 30 weeks' gestation. If findings suggest the development of hydrops fetalis, the patient should be referred to a fetal medicine unit for consideration of fetal blood sampling and intrauterine transfusion (improves outcome). Termination is not recommended.

Patients presenting with a vesicular rash

~10% of young pregnant women are susceptible to varicella zoster virus infection and are at risk of severe disease, particularly in the late second and early third trimester. The case fatality rate for women developing varicella in pregnancy is 1 in 1000. They must be advised to consult their general practitioner (GP) at the first sign of chickenpox, and those with suspected chickenpox should avoid contact with others who might be at risk (other pregnant women and neonates).

- **Diagnosis** – if clinical diagnosis cannot be made with some certainty confirm varicella infection by virus, antigen, or virus detection in vesicle fluid and urgent serological testing for VZV IgM.
- **Management** – women presenting within 24 h of onset of the first observable lesion should be offered 7 days' aciclovir or valaciclovir. Antivirals are not recommenced in those presenting over 24 h after rash onset (no evidence that they alter the clinical course in uncomplicated cases).
  - Uncomplicated cases can be managed at home with daily review, but those in whom fever persists and fresh vesicles are appearing 6 or more days after initial presentation should be referred to hospital. Also consider hospitalization in those patients who are approaching term, have a bad obstetric history, are smokers, have chronic lung disease, have poor social circumstances, or if the GP is unable to monitor the patient closely.
  - Severe disease – pneumonitis, neurological symptoms other than headache, haemorrhagic rash/bleeding, severe extensive rash or numerous mucosal lesions, significant immunosuppression. Urgent hospital review is indicated. Those with severe disease should be referred to specialist isolation facilities under the joint care of an obstetrician, infectious disease specialist, and paediatrician. Treatment is with IV aciclovir for at least 5 days.
- The fetus – consequences of primary maternal varicella in the first 20 weeks include spontaneous miscarriage in the first trimester and the risk of congenital varicella syndrome (~1% in the first 12 weeks, and 2% between weeks 13 and 20 of pregnancy). Features: dermatomal skin scarring, eye defects (chorioretinitis, cataract, microophthalmia), limb hypoplasia and neurological abnormalities (microcephaly, cortical atrophy, mental retardation, bladder and bowel dysfunction).
  - Infection before 20 weeks' gestation – perform a specialist ultrasound at 5 weeks post-infection (or 16–20 weeks' gestation) looking for polyhydramnios, microcephaly, hyperechogenic liver foci, hydrops fetalis). A neonatal eye examination should be performed at birth.
  - Infection after 20 weeks' gestation – congenital varicella syndrome does not occur but maternal infection up to one week from delivery may lead to herpes zoster in an otherwise healthy infant. Occasional reports of mild fetal damage up to 28 weeks' gestation.
  - Infection from one week before to one week after delivery – may lead to severe neonatal varicella. Such infants should be given prophylactic varicella-zoster immunoglobin (VZIG), with aciclovir if maternal disease onset was 4 days before to 2 days after delivery. See p.830 for management of disease exposure in neonates.

References

1 Health Protection Agency. *Rashes in pregnancy – HPA Guidelines – information and advice*. http://www.hpa.org.uk/infections/topics_az/pregnancy/rashes/default.htm (accessed 19 August 2008).

2 Department of Health. *Green Book*. http://www.dh.gov.uk/PolicyAndGuidance/HealthAndSocialCareTopics/GreenBook/fs/en (accessed 19 August 2008).

Prevention of congenital/perinatal infection

General points

- Any strategy aimed at preventing congenital/perinatal infection should start well before conception with education and general public health measures, e.g. vaccination against diseases such as rubella which are associated with congenital/perinatal disease, and in some countries screening for group B *Streptococcus* (GBS) carriage.
- Infection history and rash contact advice – midwives should ask all pregnant women at booking whether they have previously had chickenpox or shingles, and if they have not advise that they make urgent contact if they develop a rash in pregnancy, or have contact in pregnancy with someone with a rash.
- In the UK all maternal booking bloods are screened as a matter of course for evidence of immunity to:
  - rubella – <2% pregnant women are non-immune and should be vaccinated post partum. Vaccination in pregnancy is not recommended
  - syphilis – to prevent congenital syphilis in the neonate
  - hepatitis B – identification of maternal infection allows a course of active and passive immunization to be undertaken in at-risk neonates after birth, in an attempt to prevent infection.
- Routine maternal HIV screening is recommended in the UK so measures can be taken to prevent vertical transmission.
- Pregnancies complicated by congenital infection should be referred to regional fetomaternal medicine centres. Amniocentesis is the method of choice for fetal sampling in those cases of possible congenital infection in which such invasive investigations may be warranted. Maternal investigations for possible infective causes in cases of fetal hydrops, fetal brain lesions, unexplained severe growth restriction, or *in utero* demise are recommended.
- Certain interventions may preven or reduce neonatal acquisition, or treat infections in neonates exposed *in utero*. These are detailed below.
- Infants in whom congenital infection is suspected and those born preterm, where infection may have played a role, need follow-up with a paediatric neurologist.

Preventing neonatal varicella

- Give VZIG to the following infants:
  - those whose mothers develop varicella from 7 days before to 7 days after delivery as they will not be protected by maternal antibody
  - those who are VZV antibody-negative (i.e. born to susceptible uninfected mothers) and are exposed to varicella or herpes zoster in the first 7 days of life
  - those born before 28 weeks' gestation or weighing less than 1 kg exposed to varicella or herpes zoster, as transfer of maternal IgG antibodies may be inadequate. Some infants beyond 28 weeks' gestation at birth may become VZV antibody-negative if they are more than 60 days old or have had repeated blood samples, despite

a maternal history of varicella or zoster – serological testing is recommended.
- Intravenous aciclovir should be:
  - given urgently to those infants developing varicella despite VZIG
  - considered as prophylactic treatment in those infants whose mothers develop varicella from 4 days before to 2 days after delivery (high risk of fatal outcome despite VZIG prophylaxis).
- Mothers with varicella can breast feed, but if they have lesions close to the nipple they should express milk from the affected breast until the lesions have crusted. This milk can be fed to the baby if they are covered by VZIG and/or aciclovir.
- If other children in the family have varicella, and the mother has had varicella (or is VZV antibody-positive) there is no reason to prevent a new baby going home. If the mother is susceptible, contact with siblings with varicella should be delayed until the new baby has reached seven days of age.

Preventing neonatal HIV

Neonatal HIV transmission is a significant problem, particularly in resource-poor countries with a high prevalence of maternal HIV. Maternal transmission can occur *in utero* by passage of virus across the placenta, during delivery from blood and placental fluids, and through breast milk. Transmission can be reduced by delivering through Caesarean section, avoidance of breast feeding (advice that may be impractical in certain developing countries), and maternal treatment with antiretrovirals. Untreated, most maternally infected children die by 10 years of age. See 📖 HIV prevention, p.870 for more details.

Preventing neonatal hepatitis B

In contrast to adult infection, most (90%) neonates infected with hepatitis B perinatally go on to become chronic carriers. All infants born to HBsAg-positive mothers should receive IM hepatitis B immune globulin within 12 h of birth, along with their first dose of hepatitis B vaccine (into the *other* thigh). This strategy is effective in 90% of cases; a Cochrane review reported the relative risk of developing hepatitis B infection at 0.08 compared to no intervention.[1] The second and third vaccine doses are given at 1 and 6 months with testing to confirm immunity at 1 year.

Prevention and treatment of other neonatal infections

Preventing other infections requires a combination of good maternal health (aiming to reduce the chance of uterine or intrapartum transmission) and a low threshold for investigating and treating at-risk neonates when infection develops if severe disease is to be avoided. Specific interventions are indicated in certain cases of group B streptococcal infection (📖 p.292), neonatal HSV infection (📖 p.441), toxoplasmosis (📖 p.567), and neonatal CMV infection (📖 p.451).

Reference

1 Lee C, Gong Y, Brok J, Boxall EH, Gluud C, Hepatitis B immunisation for newborn infants of hepatitis B surface antigen-positive mothers. *The Cochrane Database of Systematic Reviews* 2006; Issue 4. Oxford: Update Software.

Chorioamnionitis

Inflammation of the chorion and amnion, the membranes surrounding the fetus. The early recognition of maternal chorioamnionitis is important as it is associated with early-onset bacterial infections of the neonate with consequent neonatal morbidity and mortality.

Aetiology

Usually chorioamnionitis is associated with a bacterial infection. Organisms residing in the vagina and cervix ascend into the uterus initiating infection of the fetal membranes and amniotic fluid. Ascending infection may be facilitated by poor urogenital hygiene and certain sexual practices. Villitis is seen in around 6% of placentas after delivery (not necessarily due to infection). Conservative estimates place the incidence of infective chorioamnionitis at around 1% of all deliveries. Chorioamnionitis (which may be clinically silent) greatly increases the risk of preterm labour. The risk of developing chorioamnionitis is highest following premature rupture of the membranes (PROM) or prolonged labour. Maternal mortality is rare. More significant is the impact on the neonate. The risk of neonatal infection increases with the time from membrane rupture.

Although bacterial vaginosis is an important cause of premature labour, overt neonatal infection is uncommon.

Clinical features

Fever (>37.8°C), maternal tachycardia, less commonly fetal tachycardia (>160 beats/min), purulent amniotic fluid or vaginal discharge, uterine tenderness, raised maternal WCC. The presence of at least two of these features is associated with an increased risk of neonatal sepsis. Remember that mothers with genuine chorioamnionitis may be asymptomatic. Epidural anaesthesia during labour may be associated with a low-grade fever that can prompt suspicion of maternal chorioamnionitis. The reasons for this are not clear.

Diagnosis

Usually made clinically by the above criteria in the intrapartum period. Antenatal screening examinations may have detected GBS carriage, itself associated with an increased risk of chorioamnionitis. Asymptomatic mothers presenting with premature labor or PROM should have silent chorioamnionitis excluded (e.g. amniotic fluid examination).

- Amniotic fluid examination – amniotic fluid may be obtained by amniocentesis if appropriate (risk of rupturing fetal membranes if intact). It can be examined for WCC, pH, glucose, cytokine levels, and microscopy and microbiological culture performed. Certain centres perform PCR to detect common causes of infection.
- Blood tests – WCC (may be raised in mothers who have been given steroids), CRP, blood cultures if febrile.
- Histology – diagnosis may be confirmed or refuted only on histological examination of the placenta, fetal membranes, and umbilical cord for evidence of inflammation and infection.
- Neonates born to mothers with suspected chorioamnionitis should be assessed for evidence of sepsis.

Management

- Mothers presenting with PROM and no obvious infection – a balance needs to be struck between avoiding the complications of prematurity and those of chorioamnionitis. Subclinical infection may be a precipitant of PROM. Mothers are usually given steroids to promote fetal maturation prior to delivery. In the absence of clinical infection there is no evidence that prophylactic antibiotics improves outcome but they are normally given along with steroids. Mother and fetus must be assessed regularly for signs of distress or the onset of chorioamnionitis.
- Mother presenting with acute chorioamnionitis – delivery must be expedited. If signs of fetal distress develop, emergency delivery may be necessary. Antibiotics should *not* be withheld with the intention of obtaining neonatal cultures.
- Mothers in preterm labour or with PROM at less than 36 weeks' gestation should receive prophylactic antibiotics as should mothers in labour at term with risk factors for fetal GBS infection (📖 p.292).
- The standard drug treatment of the mother with chorioamnionitis includes clindamycin and an aminoglycoside. Ampicillin or penicillin may have already been given to some mothers as prophylaxis against GBS infection of the neonate. Ampicillin covers GBS, *Haemophilus* species, most enterococci strains, and *Listeria* species.
- The infant should be assessed and treated for any evidence of infection.

Puerperal sepsis

Any infection following delivery is classified as postpartum or puerperal infection. Puerperal pyrexia is defined as a maternal temperature of >38°C on more than one occasion on the first 14 days after delivery; 90% of infections are genital or urinary in origin.

Aetiology

- Sources of infection include endometritis (commonest), wound infections, perineal cellulitis (usually seen around day 2 after delivery), mastitis, pneumonia (a complication of anaesthesia), retained products of conception, UTIs, and septic pelvic phlebitis (pregnant women are at increased risk of thrombosis).
- Risk factors include: Caesarean delivery, PROM, frequent cervical examination, internal fetal monitoring, pre-existing pelvic infection, diabetes, obesity.
- The uterine cavity is normally sterile until rupture of the amniotic sac, and the organisms isolated in endometritis are those normally present in the bowel, vagina, perineum, and cervix. Uterine infections are most likely following prolonged rupture of the membranes and after instrumental delivery. Genital tract infections may be polymicrobial and include *E. coli*, GAS, GBS, *Bacteroides,* and *Clostridium* species.

Clinical features

- The source of infection may be indicated by the history. Was the delivery vaginal (with or without instruments?) or Caesarean? Did premature rupture of the membranes occur? Was there intrapartum fever?
- Patients may be febrile and shocked and may have symptoms and signs indicative of the causative infection. Look for signs of UTI, DVT, wound infection, respiratory symptoms (pneumonia or septic pulmonary embolus), abdominal pain and tenderness on bimanual examination with foul-smelling vaginal discharge (suggestive of endometritis – although GAS infections are associated with odourless lochia), evidence of mastitis.

Diagnosis and management

- **Investigations** – FBC, U&E, blood cultures, urine cultures, swab and culture any wounds or discharges, swab for *Chlamydia* from the cervix and lochia. Pelvic USS may help detect pelvic abscesses or infected haematomas. Contrast abdominal CT may be required if non-pregnancy-related abdominal sources of infection are suspected.
- **Management** – fluid resuscitation and respiratory support if required. Antibiotic therapy should be guided by the likely source of infection. Avoid tetracyclines if breast feeding. Mild cases of endometritis may be managed by broad-spectrum PO antibiotics (e.g. co-amoxiclav); moderate-to-severe cases require intravenous therapy. Mastitis should respond to PO flucloxacillin – mothers should continue to express milk to prevent blockage and breast engorgement. Check for abscess development. Treat UTI/pyelonephritis as indicated. Septic pelvic thrombosis requires anticoagulation and broad-spectrum antibiotics.

Infected wounds may need surgical debridement or drainage in combination with antibiotic therapy.

- If the patient fails to respond, check the culture results and appropriateness of antibiotic therapy; exclude pelvic/abdominal collections, abscesses (wound, breast). If sensitivities are not available consider adding gentamicin and changing to a third-generation cephalosporin. Early surgical referral is essential if there is evidence of spreading skin infection despite antibiotic therapy – consider synergistic gangrene. Urgent surgical debridement may be required.
- If GBS, *Chlamydia*, or *N. gonorrhoea* are cultured, inform the paediatrician or family GP so infection can be excluded in the child.

Primary immunodeficiency

Most of these rare conditions are inherited as single gene disorders and present in early infancy or childhood. They can be divided into three groups:

- antibody-deficiency syndromes
- selective T-cell deficiencies
- mixed T- and B-cell defects.

Antibody-deficiency syndromes

The lifetime prevalence of severe antibody deficiency syndromes is ~16/million of the population the West. Partial antibody deficiency occurs in about 1/700 Caucasians, most of whom are healthy.

X-linked agammaglobulinaemia

Presents with recurrent infections during the first 2 years of life. Affected children have very few B cells and low levels of circulating IgG. Prone to infections with the following pathogens: *H. influenzae*, *S. pneumoniae*, *Mycoplasma* spp., *Ureaplasma* spp., *Campylobacter jejuni*, *Giardia lamblia*, enteroviruses. Treatment is with intravenous immunoglobulin (IVIG). Prognosis is relatively good with >90% survival at 30 years.

Common variable immunodeficiency (CVID)

Peak incidence in early childhood and late adolescence. Serum Ig levels are variable but IgA is virtually absent, IgG is <2 g/L and IgM is <0.2 g/L (but may be normal or raised). There is often a family history of selective IgA deficiency and/or autoimmune disease. Associated with the major histocompatibility complex (MHC) haplotype HLA A1, B8, C4A*QO,DR3 in 50% of patients. One-third of patients are severely lymphopaenic with CD4+ T-cell counts of $<0.4 \times 10^9$/L, low numbers of B cells, and a relative increase in CD8+ T cells; 70% of patients have features such as inflammatory bowel disease, splenomegaly, lymphadenopathy, autoimmune diseases, and malignancies. Treatment is with IVIG. Prognosis is relatively good with ~70% survival at 30 years.

Thymoma with hypogammaglobulinaemia

Thymoma occurs in patients >40 years and is associated with or followed by hypogammaglobulinaemia. Clinical features are similar to CVID but prognosis is poorer; patients usually die within 15 years of symptom onset.

Selective IgA deficiency

Complete absence of IgA occurs in ~1/700 Caucasians. Most people are healthy and only ~5% suffer from recurrent respiratory tract infections. May be a mild variant of CVID and is also associated with the MHC haplotype HLA A1, B8, C4A*QO, DR3. There is a small increase of IgA deficiency in coeliac disease, Still's disease, rheumatoid arthritis, and epilepsy patients but this may be drug related, e.g. sulphasalazine, gold, penicillamine, or anti-epileptics.

IgG subclass deficiencies

The clinical significance of IgG subclass deficiencies is controversial. IgG2 deficiency is the most common and may be associated with recurrent respiratory tract infections or be asymptomatic.

Selective IgM deficiency

Complete deficiency is very rare. Most cases are associated with rare conditions e.g. Wiskott–Aldrich or Bloom's syndromes, or are secondary to lymphoma.

Functional immunoglobulin deficiency

This is defined as the complete or partial failure to produce antibodies to specific proteins, peptides or polysaccharides in the presence of normal total Ig levels. Mechanism is unknown. Only functional IgG deficiency is clinically important.

Selective T-cell deficiency

These conditions are very rare. Infections associated with T-cell deficiency include herpes simplex virus (HSV), varicella zoster virus (VZV), cytomegalovirus (CMV), adenovirus, papilloma virus, rotavirus, *Mycobacterium* spp, *Cryptococcus neoformans*, *Toxoplasma gondii*, *Candida* spp., *Aspergillus* spp., *Pneumocytis jiroveci*, *Cryptosporidium* spp., and *Strongyloides* spp. Children remain relatively healthy unless the condition is also associated with macrophage dysfunction or antibody deficiency.

Thymic aplasia (Di George syndrome)

Caused by fetal malformation of the 3rd and 4th branchial arch during gestation. Severely affected infants usually die from associated cardiac abnormalities. Those that survive have a few circulating T cells (<10%) and remain healthy. Severely affected infants are prone to recurrent infections especially life-threatening CMV and VZV disease. Treatment is with a thymus graft.

Purine nucleoside phosphorylase (PNP) deficiency

Autosomal recessive condition characterized by T-cell deficiency, susceptibility to severe CMV and VZV infections, autoimmune blood dyscrasias, lymphoma, and neurological disease. Most patients present in infancy. Bone marrow transplantation is the best treatment.

Severe combined immunodeficiency (SCID)

This is due to rare mutations in genes that influence the maturation of lymphocytes, e.g. adenosine deaminase deficiency, X-linked SCID, lymphocyte MHC class II deficiency, reticular dysgenesis; and multiple interleukin deficiency. Infections are usually much more severe than those seen in primary hypogammaglobulinaemia and selective T-cell deficiency, probably because macrophage function is affected. Infants present with failure to thrive, diarrhoea, *Pneumocystis jiroveci* pneumonia, mucocutanoeus or systemic candidiasis. Immunization with live vaccines, e.g. polio or BCG may cause fatal infection. Most patients die within 2 years unless they undergo bone marrow transplantation.

Secondary immunodeficiency

Secondary immunodeficiency is defined as a defect in the components or function of the immune system, occurring as a result of another disease or condition. It may affect humoral immunity, cell-mediated immunity or both. HIV infection, drugs and lymphoreticular malignancies are the most important causes (Table 5.14).

Table 5.14 Causes of secondary immunodeficiency

Cause	Examples	Humoral immunity	Cell-mediated immunity
Viruses	HIV	√	√
	Rubella	√	
Drugs	Corticosteroids		√
	Cyclophosphamide		√
	Azathioprine		√
	Ciclosporin		√
	Mycophenylate		√
	Anti-T-cell antibodies		√
	Gold	√	
	Penicillamine	√	
	Sulphasalazine	√	
	Phenytoin	√	
	Methotrexate		√
	Bleomycin		√
	Vincristine	√	√
	Cis-platinum		√
Malignancy	Chronic lymphocytic leukaemia	√	
	Myeloma	√	
Metabolic	Renal failure	√	√
	Liver failure	√	√
	Trauma	√	√
	Vitamin A deficiency	√	
	Vitamin B_{12} deficiency	√	
	Zinc deficiency		√
Ig loss	Nephrotic syndrome	√	
	Protein-losing enteropathy	√	
	Dystophia myotonica	√	

Infections in asplenic patients

The spleen is the largest lymphoid organ in the body and performs a wide range of immunological functions that protect it from severe infections. The relationship between an absent or hypofunctioning spleen and severe infection has long been recognized and is termed post-splenectomy sepsis (PSS) or overwhelming post-splenectomy infection (OPSI). Asplenia is usually acquired (due to surgical removal of the spleen for traumatic or therapeutic reasons) but may rarely be congenital. A number of conditions may also lead to functional hyposplenism.

Aetiology

The following organisms are associated with PSS:

- *S pneumoniae*
- *H. influenzae*
- *N. meningitidis*
- *Capnocytophaga canimorsis*
- *Salmonella* spp.
- *Plasmodium falciparum*
- Babesiosis
- Erlichiois
- *Bartonella bacilliformis*

Clinical features

- The risk of PSS is highest in the first few years after splenectomy.
- PSS has a short prodrome with fever, chills, pharyngitis, muscle aches, vomiting, and diarrhoea. In adults there is usually no obvious site of infection, whereas in children meningitis is common.
- Deterioration is usually rapid and occurs over hours with septic shock, DIC, seizures, and coma.

Diagnosis[1]

- The diagnosis is clinical and all asplenic patients with fever should be considered to have and managed as PSS.
- The presence Howell–Jolly bodies (nuclear remnants) on the blood film confirms hyposplenism, but they may not always be present.

Management[1]

- Asplenic patients should be given a supply of prophylactic antibiotics for self-administration at the first sign of serious illness.
- If the patient presents acutely with PSS they should receive immediate treatment with IV antibiotics, e.g. ceftriaxone + vancomycin.

Prevention[1]

- Prophylactic antibiotics – lifelong oral penicillin V should be given to patients with an absent or dysfunctional spleen.
- Immunization against *S. pneumoniae*, *N. meningitis*, *H. influenzae*, and influenza virus is recommended for all apslenic patients.

Reference

1 Davies JM. Updated guideline. The prevention and treatment of infection in patients with an absent or dysfunctional spleen. *Clin Med* 2002;2:440–3.

Neutropaenic sepsis

Neutropenia is associated with an increased risk of bacteraemia and severe infection. Although neutropenia is defined as an absolute neutrophil count <0.5 × 10^9 cells/L, many experts believe that the risk of infection increases when the neutrophil count falls below 1 x 10^9 cells/L. In very immunocompromised patients, signs of infection may be absent and the first and only sign of infection may be fever (>38°C). Febrile neutropenia is considered a medical emergency, and appropriate antimicrobial therapy should be started immediately as the mortality of neutroapenic patients with untreated Gram-negative sepsis approaches 40%.

Classification

Febrile neutropenia can be divided into four categories:
- microbiological documented infections (MDI) with bacteraemia
- MDI with isolation of a significant pathogen from a well-defined site of infection, e.g. urine, respiratory, abscess
- clinically documented infection without microbiological proof
- unexplained fever without clinical or micrbiological proof.

Aetiology

- In the 1970s Gram-negative rods were the predominant pathogen.
- In the 1980s Gram-negative bacteria became the predominant pathogens, mainly as a result of fluoroquinolone prophylaxis.
- In the late 1990s an increase in Gram-negative pathogens was again seen; the reasons for this are not clear.

Evaluation of the febrile neutropaenic patient

- The following factors should be considered:
- underlying disease (e.g. solid tumour or haematological malignancy) and stage
- previous history of infection
- clinical features – fever, symptoms/signs of infection, systemic upset, metabolic instability
- neutrophil count – at time of onset and previous 30 days, expected neutrophil count kinetics and duration of neutropenia
- lymphocyte count – at time of onset and previous 30 days
- platelet and fibrinogen levels.

A number of studies have been performed to try to stratify patients into into high- and low-risk patients using clinical and laboratory parameters.

Management

- Initial studies used a combination β-lactam/aminoglycoside regimen.
- More-recent studies have shown that monotherapy with β-lactam active against *Pseudomomas* spp., (e.g. ceftazidime, Tazocin®, or meropenem) was associated with a lower rate of clinical failures and adverse events, and a trend towards better survival.
- The choice of antimicrobial regimen should follow local antimicrobial policy and be based on local microbiological data.

- If clinical findings suggest a possible Gram-positive infection, e.g. line infection, then a glycopeptide should also be started.
- Micobiologically documented infections should be treated with appropriately tailored antimicrobial therapy.
- Neutropaenic patients who remain febrile after 24–48 h of therapy should be evaluated for treatment failure:
  - persistence of fever >39°C and chills 24–48 h after starting therapy
  - relapse of fever >38°C after at least 24 h of defervescence
  - progression of sepsis syndrome
  - development of DIC, ARDS, multiorgan failure
  - persistence of positive cultures despite 24 h of therapy
  - relapse of primary infection
  - appearance of a new infection.
- Consider alternative causes of fever:
  - drug-resistant organisms (bacteria, viruses, fungi, protozoa)
  - drug reactions
  - infusion of blood products
  - graft versus host disease.
- Empirical anti-Gram-positive therapy – despite a lack of evidence of benefit, many centres add empirical Gram-positive cover after 48 h of antibiotic therapy. The IATG-EORTC (International Antimicrobial Therapy Group of the European Organization for Research and Treatment of Cancer) trial XIV showed that there was no difference in time to defervescence in patients receiving Tazocin® who were randomized to vancomycin or placebo[1].
- Empirical antifungal therapy – this has been shown to be of benefit in patients with persistent febrile neutropenia. The IATG-EORTC showed that although there were no differences in defervescence and survival between patients who received and did not receive antifungal therapy, there were no deaths from fungal infection in patients receiving amphotericin, and the number of documented fungal infections was higher in the placebo group[2].
- Alternative antifungal regimens:
  - liposomal amphotericin preparations (p.120) are as effective and less toxic but considerably more expensive than conventional amphotericin B
  - voriconazole has recently been evaluated and did not meet criteria for non-inferiority to liposomal amphotericin B
  - caspofungin has recently been evaluated and found to be non-inferior to liposomal amphotericin B.
- Adjunctive therapies include:
  - granulocyte-colony-stimulating factor (G-CSF)
  - granulocyte-macrophage-colony-stimulating factor (GM-CSF)
  - granulocyte infusions
  - infusion of immunoglobulins.

Reference

1 Cometta A, Ken WV, deBock R et al. International Antimicrobial Therapy Group of the European Organisation for Research Treatment of Cancer. *Clin Infect Dis.* 2003;37(3):382–9.

2 Hughes WT, Armstrong D, Baley GP et al. 2002 guidelines for the use of antimicrobial agents in neutropaenic patients with cancer. *Clin Infect Dis.* 2002;34(6):730–51

Infections in transplant recipients

The advent of immunosuppression has resulted in a marked growth in organ transplantation over the last 30 years. Apart from medical and surgical issues related to function and rejection of transplanted organs, infections are the most important problem. Most infections occur during the first 4 to 6 months after transplantation. The following risk factors have been identified:

- pre-transplant factors (underlying medical conditions, lack of immunity, prior latent infections, colonization with nosocomial organisms, prior medications)
- transplantation factors (type of organ, trauma of surgery)
- immunosuppression (medication, chemotherapy, irradiation)
- allograft reactions (graft versus host disease).

Sources of infection

The sources of infection can be divided into three categories:

- host factors (endogenous)
  - reactivation of viruses (HSV, VZV, CMV)
  - barrier disruption causing invasive disease (mucositis, IV catheters)
  - colonization with resistant flora (Gram-negatives, vancomycin-resistant enterococci (VRE), yeasts)
  - reactivation of fungi (*Aspergillus*)
  - reactivation of parasites (*Toxoplasma*, *Strongyloides*)
- environmental factors (exogenous)
  - importance of positive pressure ventilation (*Aspergillus* spores)
  - opportunistic pathogens (*Legionella*, *P. jirovecii*, *Listeria*)
- organ transplant/blood products
  - viral (CMV, HBV, HCV, HIV, HHV-6, HHV-7, TTV, parvo, HTLV-1)
  - bacterial (organ contamination, TB)
  - Unknown (variant Creutzfeldt–Jakob disease (vCJD), pig retroviruses).

Clinical features

- The clinical manifestations are variable and depend on a number of factors including prior immune status of the host, reason for transplant (e.g. viral hepatitis, diabetic nephropathy), type of transplant (solid organ or bone marrow), preconditioning treatment, time after transplantation, degree of immunosuppression, likelihood of exposure, infecting pathogen.
- **Bone marrow transplant-associated infections**:
  - pre-engraftment, profound neutropenia is associated with ablation of humoral and cell-mediated immunity. Infections during this period may include bacteraemia (Gram-positive and Gram-negative) HSV, candidaemia, invasive fungal disease
  - during the post-engraftment period up until 100 days post-BMT, B- and T-cell immunity starts to recover. The commonest infections are caused by Gram-positive and Gram-negative bacteria, fungi, CMV, PCP, other viruses (RSV, parainfluenza, adenovirus, JC, BK)
  - From 100 days to 1 year post-BMT, B- and T-cell function continue to recover but may take 18–36 months to fully recover. The commonest infections are caused by VZV and encapsulated bacteria.

- **Solid organ transplant-associated infections**:
  - in the early post-operative period, standard post-operative infections (VAP, line infection, wound infection, UTI) with nosocomial pathogens predominate
  - between the 2nd and 6th post-operative months, immunosuppression is established and a wide range of bacterial, viral, fungal, and protozoal infections may occur
  - After 6 months, patients with good allograft function on a stable immunosuppressive regime have the same risk of bacterial infections as a minimally immunosuppressed patient in the community. In contrast, patients with rejection are at increased risk of opportunistic infections such as those seen prior to 6 months.

Pre-transplantation screening

Pre-transplantation screening of the donor, recipient and/or blood products is performed in order to try and prevent or predict transplant-related infections:

- recipient screening:
  - ongoing or active infection
  - serological testing for HBV, HCV, HIV, HSV, VZV, EBV, CMV, *T pallidum*, *T gondii*
  - in endemic areas consider *T cruzi*, *Histoplasma*, *Strongyloides*
- donor screening:
  - serological testing for HBV, HCV, HIV, *T. pallidum*, *T. gondii* (cardiac)
  - culture of cadaveric organs perfusates, transport medium
  - if living donor, take clinical and epidemiological history and consider tuberculin testing and fungal serology
  - in endemic areas, consider screening for malaria, *T. cruzi*
- blood products
  - screening for HBV, HCV, HIV, *T. pallidum*
  - leucodepleted blood reduces the risk of CMV.

Post-transplantation surveillance

Post-transplantation surveillance is performed in order to guide pre-emptive therapy and monitor response to treatment:

- CMV disease by PCR
- *Candida* or *Aspergillus* infection by antigen tests
- surveillance cultures for multi-drug-resistant pathogens.

Prevention of infection

- Routine immunizations for people with chronic diseases, e.g. pneumococcal, influenza immunization, should be given prior to transplantation
- Prophylatic antimicrobials are commonly given in the first few months following transplantation, e.g. co-trimoxazole, antivirals (aciclovir, valaciclovir, ganciclovir or valganciclovir), and antifungals (nystatin, fluconazole, or itraconazole). Protocols differ between different transplant centres.

HIV epidemiology

There are an estimated 33 million people living with HIV/AIDS, with 2.7 million new infections and 2.1 million deaths in 2007. The majority of new infections occur in young adults (aged 15–24 years old) in developing countries. Sub-Saharan Africa bears the brunt of the epidemic but there are growing epidemics in Asia, notably in India and China, which may eclipse the African epidemic in the next decade. Although the introduction of highly active antiretroviral therapy (HAART) has dramatically reduced morbidity and mortality in north America and western Europe, the majority of patients who require HAART in the developing world do not have access to it.

HIV transmission

HIV may be transmitted via a number of routes:
- sexual transmission
- perinatal transmission – intrapartum, peripartum, breast feeding
- blood transfusion
- intravenous drug use/sharing needles
- occupational transmission – needlestick injury or mucocutaneous exposure.

The risk of transmission differs with the route of infection (Table 5.15).

Table 5.15 HIV transmission rates

Exposure	Risk/10,000 exposures
Blood transfusion	9,000
Intravenous drug use	67
Receptive anal intercourse	50
Needlestick injury	30
Receptive vaginal intercourse	10
Insertive anal intercourse	6–7
Insertive vaginal intercourse	5

HIV natural history

The natural history of HIV infection is divided into the following stages (Fig. 5.2):

- **primary infection** – diagnosis is based on a plasma HIV RNA level >10,000 copies/mL + indeterminate or negative serology or recent seroconversion
- **acute retroviral syndrome** (2–3 weeks) – clinical features include fever (96%), adenopathy (74%), pharyngitis (70%), rash (70%), myalgia (54%), diarrhoea (32%), headache (32%), nausea and vomiting (27%), hepatosplenomegaly (14%), weight loss (13%), thrush (12%), and neurological symptoms (12%). This is accompanied by rapid decline in CD4 T-lymphocyte count and high concentrations of HIV RNA in the plasma
- **recovery and seroconversion** (2–4 weeks) – characterized by recovery of CD4 cell count and reduction in plasma HIV viral load to a set point
- **asymptomatic chronic HIV infection** (average 8 years) – associated with gradual decline in CD4 cell count
- **symptomatic HIV infection/AIDS** (average 1–3 years) – occurs when the CD4 count declines to <200/mm^3 and the viral load begins to rise
- death usually occurs 10–11 years after infection in untreated individuals.

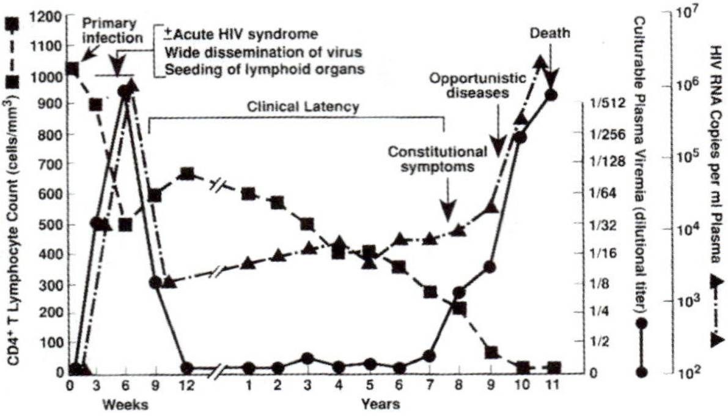

Fig. 5.2 Typical course of HIV infection. Reproduced from Fauci AS Patales G, Stanley S, Weuoman D. Immunopathogenic Mechanisms of HIV infection. *Ann Intern Med* 1996;124(7):654–63.

Complications of HIV infection

As the CD4 count declines, the complications shown in Table 5.16 may occur.

Table 5.16 CD4 count and HIV complicaitons

CD4 count	Complications
>500/mm^3	Acute retroviral syndrome, candida vaginitis, persistent generalized lymphadenopathy (PGL), Guillain–Barré syndrome, myopathy, aseptic meningitis
200–500/mm^3	Pneumococcal and other bacterial pneumonia, pulmonary tuberculosis, herpes zoster, oropharyngeal candidiasis, cryptosporidiosis, Kaposi's sarcoma, oral hairy leucoplakia, cervical intraepithelial neoplasia, cervical cancer, B-cell lymphoma, anaemia, mononeuritis multiplex, idiopathic thrombocytopaenic purpura (ITP), Hodgkin's lymphoma, lymphocytic interstitial pneumonitis (LIP)
<200/mm^3	*Pneumocystis jiroveci* pneumonia (PCP), disseminated histoplasmosis, disseminated cocciodiodomycosis, miliary/extrapulmonary TB, progressive multifocal leucoencephalopathy (PML), wasting syndrome, peripheral neuropathy, HIV-associated dementia, cardiomyopathy, vacuolar myopathy, progressive radiculopathy, non-Hodgkin's lymphoma (NHL)
<100/mm^3	Disseminated herpes simplex, toxoplasmosis, cryptococcosis, chronic cryptosporidiosis, microsporidiosis, oesophageal candidiasis
<50/mm^3	Disseminated cytomegalovirus (CMV), disseminated *Mycobacterium avium* complex (MAC), primary central nervous system lymphoma (PCNSL)

HIV classification

There are two commonly used classification systems

CDC 1993 revised classification

This categorizes patients according to clinical categories (A, B, C) and CD4 count categories (1,2,3). Patients in categories A3, B3, and C1 to C3 are defined as having AIDS (Table 5.17).

Table 5.17 CDC 1993 revised classification of HIV

	A (asymptomatic, acute HIV, PGL)	B (symptomatic)	C (AIDS indicator conditions)
CD4 ≥500/mm^3	A1	B1	C1
CD4 200–499/mm^3	A2	B2	C2
CD4 <200/mm^3	A3	B3	C3

World Health Organization classification

This categorizes patients into four clinical stages, regardless of CD4 cell count. Patients with clinical stage 4 are defined as having AIDS (Table 5.18).

Table 5.18 WHO classification of HIV

Stage	Symptoms
1	Asymptomatic
	Persistent generalized lymphadenopathy (PGL)
	Performance scale 1: asymptomatic, normal activity
2	Weight loss, <10% of body weight
	Minor mucocutaneous manifestations (seborrhoeic dermatitis, prurigo
	Fungal nail infections, recurrent oral ulcerations, angular cheilitis)
	Herpes zoster, within the last 5 years
	Recurrent upper respiratory tract infections (e.g. bacterial sinusitis)
	And/or performance scale 2: symptomatic, normal activity
3	Weight loss, >10% of body weight
	Unexplained chronic diarrhoea, >1 month
	Unexplained prolonged fever (intermittent or constant), >1 month
	Oral candidiasis (thrush)
	Oral hairy leucoplakia
	Pulmonary tuberculosis, within the past year
	Severe bacterial infections (e.g. pneumonia, pyomyositis)
	And/or performance scale 3: bedridden, <50% of the day during the last month

Table 5.18 WHO classification of HIV (*Continued*)

Stage	Symptoms
4	HIV wasting syndrome, as defined by CDC[a]
	Pneumocystis jiroveci pneumonia
	Toxoplasmosis of the brain
	Cryptosporidiosis with diarrhoea, >1 month
	Cryptococcosis, extra-pulmonary
	Cytomegalovirus (CMV) disease of an organ other than liver, spleen, or lymph nodes
	Herpes simplex virus (HSV) infection, mucocutaneous >1 month, or visceral any duration
	Progressive multifocal leucoencephalopathy (PML)
	Any disseminated endemic mycosis (e.g. histoplasmosis, coccidioidomycosis)
	Candidiasis of the oesophagus, trachea, bronchi, or lungs
	Atypical mycobacteriosis, disseminated
	Non-typhoid *Salmonella* septicaemia
	Extra-pulmonary tuberculosis
	Lymphoma
	Kaposi's sarcoma (KS)
	HIV encephalopathy, as defined by CDC[b]
	And/or performance scale 4: bedridden, >50% of the day during the last month

[a] HIV wasting syndrome: weight loss of >10% of body weight, plus either unexplained chronic diarrhoea (>1 month), or chronic weakness and unexplained prolonged fever (>1 month).

[b] HIV encephalopathy: clinical finding of disabling cognitive and/or motor dysfunction interfering with activities of daily living, progressing over weeks to months, in the absence of a concurrent illness or condition other than HIV infection that could explain the findings.

Initial evaluation of HIV patient

All newly diagnosed HIV patients should be carefully evaluated to determine the clinical stage of disease, co-existent infections, and laboratory abnormalities:

- full blood count
- biochemistry profile – renal and liver function tests
- fasting blood glucose and serum lipids
- serology for HIV (if laboratory confirmation not available), hepatitis A, B, and C, CMV, syphilis, and *T. gondii*
- urinalysis
- cervical smear in women
- testing for *C. trachomatis* and *N. gonorrhoeae* is optional but should be considered for those at high risk
- tuberculin skin test (unless previous history of TB or positive test)
- chest x-ray
- **CD4 T-cell count** – this is an indicator of immunocompetence and is a string predictor of progression and survival. It is checked at baseline to assess the need for antiretroviral therapy. Once treatment starts, the CD4 count usually rises by 100–150 cells/mm^3 per year, with an accelerated response in the first 3 months. During treatment it is monitored every 3–6 months
- **plasma HIV RNA** – numerous studies have shown an association between decrease in plasma viraemia and improved survival. Baseline viral load may be a consideration in when to start treatment. It should be measured immediately before and 2–8 weeks after initiation of treatment. Its main role is in monitoring the response to therapy. HIV viral load should be checked every 3–4 months in patients on a stable antiretroviral regimen, or earlier if clinically indicated
- **HIV drug resistance testing** – for patients with HIV-RNA >100,000 copies/mL, genotypic drug resisatnce testing is recommended, regardless of whether the patient starts antiretroviral therapy. If treatment is deferred this should be repeated prior to commencing therapy. In antiretroviral naive patients a genotypic assay is generally preferred. Drug resistance testing should also be performed in the setting of virological failure to assist with choosing new drugs
- **HLA-B*5701 screening** – the abacavir hypersensitivity reaction (ABC HSR) is a multi-organ clinical syndrome which occurs in 5–8% of patients within 6 weeks of starting abacavir therapy. Several studies have shown an association between the HLA allele HLA-B*5701 and ABC HSR. For this reason all patients should be screened for this allele prior to commencing abacavir therapy and those who are HLA-B*5701-positive should not be given the drug
- **co-receptor tropism assays** – during acute/recent infection most patients harbour a CCR5 tropic (R5) virus. In untreated patients the virus shifts to using CXCR4 (R4 tropic) or both (dual/mixed tropic). The CCR5 inhibitors (maraviroc and vicriviroc) are a new class of drugs that prevent HIV entry into target cells by binding to CCR5. Co-receptor tropism assays should be performed prior to considering therapy with CCR5 receptor antagonists.

HIV skin complications (1)

Skin manifestations are common in HIV-infected patients and may be caused by bacteria, fungi, viruses, parasites, or drugs.

Bacillary angiomatosis

- Caused by *Bartonella henselae* and *Bartonella quintana* (see 🕮 *Bartonella* species, p.400).
- Lesions start as red or purple papules that expand into nodules or pedunculated masses. They appear vascular and may bleed with trauma.
- Skin biopsy shows vascular proliferation, inflammation and typical organisms on Warthin–Starry silver stain. Serology (immunofluorescence assay (IFA) or enzyme immunosorbent assay (EIA)) may be used to support the diagnosis.
- Treatment is with oral erythromycin or doxycycline for >3 months.

Cutaneous candidiasis

- Usually caused by *Candida albicans*.
- Lesions are moist, red, scaly, and may have satellite lesions. May also cause intertrigo, balanitis, glossitis, angular cheilitis, paronyhchia, and nail dystrophies.
- Diagnosis is usually clinical but may be confirmed by KOH preparation or wet mount.
- Treatment may be topical (e.g. ketoconazole, miconazole, clotrimazole, or nystatin) or systemic (e.g. ketoconazole or fluconazole).

Cryptococcosis

- Caused by *Cryptococcus neoformans* (see 🕮 *Cryptococcus neoformans*, p.527) as part of disseminated disease.
- Lesions are nodular, popular, follicular, or ulcerated, and may resemble molluscum. Usually occurs on face, neck, and scalp.
- Skin biopsy with Gomori methenamine silver stain shows typical budding yeasts and positive culture. Serum cryptococcal antigen is usually positive. Lumbar puncture should be performed in patients with positive cultures or cryptococcal antigen, to exclude cryptococcal meningitis (see 🕮 *Cryptococcus neoformans*, p.527).
- Treatment is with oral fluconazole 400 mg/day for 8 weeks followed by 200 mg/day.

Dermatophyte infections

- Caused by a variety of fungi, e.g. *T. rubrum*, *T. mentagrophytes*, *T. tonsurans*, *T. soudanense*, *M. canis*, *E. floccosum*.
- May present with fungal nail infections (onychomychosis), athlete's foot (tinea pedis), or ringworm (tinea corporis, tinea cruris, tinea capitis).
- Diagnosis is confirmed by skin scrapings for KOH preparation and culture.
- Treatment is with oral terbinafine or itraconazole for nail infections, or topical agents (e.g. clotrimazole, ketoconazole, miconazole, terbinafine) for skin infections.

Drug eruptions

- Common causes include antibiotics (e.g. co-trimoxazole and beta-lactams), anticonvulsants and non-nucleoside reverse transcriptase inhibitors (NNRTIs) (see 📖 Non-nucleoside reverse transcriptase inhibitors, p.142).
- Lesions occur within 2 weeks of a new drug and present as an itchy, morbilliform, exanthematous eruption ± fever. More-severe forms include urticaria, toxic epidermal necrolysis, Stevens–Johnson syndrome, hypersensitivity reactions (especially abacavir and nevirapine), and anaphylaxis.
- Treatment of uncomplicated cases is with antihistamines and topical antipruritics and topical corticosteroids. For severe reactions, discontinue the drug and administer supportive care.

Folliculitis

- Usually caused by *S. aureus*. Other causes include *Pityrosporum ovale* (intrafollicular yeast), *Demodex folliculorum* (intrafollicular mite), and eosinophilic folliculitis (unknown cause).
- Lesions are itchy, follicular papules and pustules on the face, trunk, and extremities. Occurs at CD4 count 50–250 cells/mm^3. Often relapses and remits. May recur with immune reconstitution.
- Diagnosis is clinical and confirmed by skin biopsy.
- Treatment is of the underlying cause.

Herpes simplex

- Caused by herpes simplex virus (see 📖 Herpes simplex virus, p.439).
- Lesions begin as papules which develop into vesicles and ulcerate and crust. They are usually found on the lip, in the mouth, or in the genital region. Recurrences are common.
- Diagnosis is usually clinical but may be confirmed by PCR, HSV antigen detection, viral culture or Tzanck prep.
- Treatment is with oral valaciclovir for 7–10 days. Severe or disseminated disease may require intravenous aciclovir (see 📖 Antivirals for HSV and VZV, p.129).

Herpes zoster (shingles)

- Caused by reactivation of varicella zoster virus (see 📖 Varicella zoster virus, p.444). This affects 5% of healthy adults but is 15–25 times more common in HIV-infected adults.
- The rash is usually preceded by painful prodrome in the region of a dermatome. This is followed by the development of a dermatomal vesicular rash.
- Diagnosis is usually clinical but may be confirmed by PCR, HSV antigen detection, viral culture or Tzanck prep.
- Treatment is with oral valaciclovir for 7–10 days. Severe or disseminated disease may require intravenous aciclovir (see 📖 Antivirals for HSV and VZV, p.129).

HIV skin complications (2)

Kaposi's sarcoma

- Caused by human herpes virus 8 (HHV-8, see Human herpes virus type 8, p.483). Can occur in the general population but is 20,000 times more common in HIV-infected individuals.
- Lesions are purple to brown-black macules, papules, nodules, and patches which occur on the legs, face, mouth, and genitalia. Complications include lymphoedema and visceral involvement.
- Diagnosis is clinical but may be confirmed by biopsy.
- Treatment – HAART is associated with lesion regression, decreased incidence and prolonged survival. Local injection of vinblastine reduces lesion size. Extensive disease is treated with chemotherapy, e.g. anthracyclines or paclitaxel (more toxic).

Prurigo nodularis

- Cause unknown.
- Lesions are hyperpigmented, hyperkeratotic papules and nodules which usually occur on the chest. They are usually intensely itchy, resulting in excoriation, ulceration, and scars.
- Diagnosis is usually clinical but may be confirmed by biopsy.
- Treatment is with topical steroids, occlusive dressings, antihistamines, or phototherapy. Refractory cases may benefit from thalidomide.

Scabies

- Caused by *Sarcoptes scabiei*.
- Lesions are small red papules that are intensely itchy; sometimes there are 'burrows'. Occur in the finger webs, wrist, periumbilical area, axilla, thighs, buttocks, genitalia, legs, and feet. A severe, crusted form (Norwegian scabies) may occur in immunocompromised patients.
- Diagnosis is clinical and confirmed by microscopic examination of the mite.
- Treatment – the patient and all close contacts should be treated simultaneously with permethrin cream 5%, or lindane 1%, or ivermectin (severe or refractory cases). Pruritis may respond to antihistamines. Bedding and clothing must be washed in hot water or dry cleaned.

Seborrhoeic dermatitis

- Cause unknown, but *Pityrosporum* is often recovered from lesions.
- Lesions are erythematous plaques with greasy scales and indistinct margins. Usually occur on scalp, face, behind ears, sternum, axillae, and sometimes pubic area.
- Diagnosis is clinical.
- Treatment is with a weak topical steroid, e.g. hydrocortisone 2.5% ± ketoconazole 2% cream. Scalp lesions may be treated with various shampoos, e.g. tar-based, selenium sulfide or zinc pyrithione containing, or ketoconazole shampoo.

HIV oral complications

A number of oral complications may occur in HIV-infected patients.

Apthous ulcers

- Cause unknown. Differential diagnosis includes HSV, CMV (📖 Cytomegalovrius, p.451), drug-induced ulcers.
- Diagnosis is clinical. Biopsy recommended for non-healing ulcers.
- Treatment is with topical lidocaine solution, Orabase®, Aphthasol® (amlexanox oral paste 5%). Refractory cases may require oral prednisolone, colchicine, dapsone, pentoxifylline, or thalidomide.

Oral candidiasis (thrush)

- Caused by *Candida* species, usually *C. albicans* (see 📖 *Candida* species, p.513).
- Lesions are white, painless plaques on the buccal and pharyngeal mucosa or the tongue. Risk factors include CD4 <250 cells/mm^3, antibiotics, corticoteroids.
- Diagnosis is clinical but may be confirmed by KOH preparation. Culture is only indicated for speciation and drug suscepitibility testing, e.g. for refractory cases.
- Treatment is with oral clotrimazole, nystatin, or fluconazole. Most cases respond in 7–14 days, unless prior azole exposure and CD4 count <50 cells/mm^3. Refractory cases may require itraconazole suspension, amphotericin suspension, or intravenous amphotericin.

Oral hairy leucoplakia

- Caused by Epstein–Barr virus (see 📖 Epstein–Barr virus, p.447) and found almost exclusively in HIV-infected patients with low CD4 count.
- Presents as unilateral or bilateral adherent white frond-like patches on lateral margins of the tongue.
- Diagnosis is usually clinical; biopsy is rarely required.
- Treatment is not usually required as is usually asymptomatic. Responds to HAART. Occasionally treated for pain or cosmetic reasons, with topical podophyllin, surgical excision, cryotherapy, or antivirals.

Salivary gland enlargement

- May be due to HIV-related lymphoid proliferation.
- Presents with unilateral or bilateral parotid enlargement. Usually asymptomatic but may present with pain or xerostomia.
- Diagnosis is by fine needle aspiration (FNA) for microbiology, cytology, and decompression. Occasionally biopsy is required to exclude a tumour.
- Treatment is by FNA for decompression of fluid-filled cysts.

Other conditions

- The oral cavity may be involved in a number of conditions, e.g.:
- herpes simplex virus infections
- Kaposi's sarcoma
- leishmaniasis (📖 p.577)
- syphilis (📖 p.387).

HIV cardiovascular complications

The main concern is the increased risk of cardiovascular disease due to metabolic complications of HAART, e.g. hyperlipidaemia.

Dilated cardiomyopathy

- Cause unknown, but hypotheses include mitochondrial toxicity from zidovudine, HIV infection of myocardial cells, l-carnitine deficiency, selenium deficiency
- Incidence declining – previously 30–40% AIDS patients (pre-HAART), now 3–15%
- Clinical features include congestive cardiac failure, arrhythmias, cyanosis, syncope, sudden death
- Diagnosis is by echocardiogram which shows ejection fraction ≤50% ± arrhythmias on ECG
- Treatment is with HAART and an ACE inhibitor. Diuretics, e.g. furosemide or spironolactone are given for persistent symptoms. Digoxin may be given for refractory cases. Other options: treat hypertension and hyperlipidaemia, discontinue alcohol and cocaine, discontinue azathioprine, some recommend supplements of l-carnitine or selenium (if deficient)

Endocarditis

HIV has no effect on the incidence of endocarditis, apart from in intravenous drug users in whom it is more common. Associated with increased mortality in AIDS patients (30%).

Myocarditis

The cause is unclear and most cases. HIV may have a direct effect. 20% associated with infections e.g. cryptococcis, CMV, EBV, HSV, TB and *T. gondii*.

Pericardial effusion

Cause unknown. Occurs in ~10% of untreated HIV patients. Incidence declining in the HAART era. Most are small. High mortality in AIDS patients.

Pericarditis

Causes include mycobacteria, pyogenic bacteria, lymphoma, Kaposi's sarcoma, viruses and fungi. Aspiration and pericardial biopsy may yield a diagnosis. Effect of HAART unknown.

Pulmonary hypertension

- Cause unknown but human herpes virus 8 (HHV-8) has been implicated.
- Clinical features are similar to primary pulmonary hypertension with extertional dyspnoea, fatigue, cough, haemoptysis, chest pain, and syncope.

- Diagnosis – CXR shows enlarged pulmonary vessels, right ventricular and right atrial hypertrophy. Echocardiography shows a dilated right atrium and ventricle ± tricuspid regurgitation. Cardiac catheterization shows increased pulmonary artery pressure, increased right atrial pressure, and normal pulmonary capillary pressure.
- Treatment is difficult as the condition is usually progressive. Some studies report improvement with HAART; others show no benefit. Other options include epoprostenol, diuretics, oral anticoagulant, sildenafil, antiviral therapy (controversial).

HIV pulmonary complications

HIV infection may be associated with a number of pulmonary complications, some of which are considered AIDS-defining illnesses.

Pneumocystis jiroveci pneumonia (PCP)

- This fungus, formerly called *Pneumocystis carinii*, is the most common cause of infection in HIV-infected patients. It is a ubiquitous environmental organism, transmitted by inhalation. PCP may be due to acquisition of a new infection or reactivation of previous infection.
- Associated with CD4 count <200 cells/mm^3. Incidence in the West is low because of HAART and PCP prophylaxis. Remains a common AIDS presentation in developing countries.
- Clinical features – progressive dynspoea, fever, chills, malaise, chest pain, weight loss. Examination reveals crepitations or wheeze but may be normal in 50%. Hypoxia is common. May desaturate on exercise.
- Chest x-ray shows diffuse interstitial/alveolar infiltrates; 25% are normal in early disease. CXR signs may be highly variable.
- Induced sputum is diagnostic in ~60%. Bronchoalveolar lavage is diagnostic in >95%. Molecular diagnostic tests look promising.
- Treatment – co-trimoxazole is the treatment of choice. Alternatives include: clindamycin and primaquine; dapsone, atovaquone, intravenous pentamidine. If partial pressure of arterial oxygen (PaO_2) <7 kPa or alveolar–arterial (A–a) gradient >3.5 KPa, give prednisolone.
- Secondary prophylaxis should be given to all patients who have had PCP and continued until CD4 count is >200 cells/mm^3 for ≥3 months.

Pneumonia

- An acute infection of the lung which presents with fever, dyspnoea, cough ± sputum production.
- May be caused by a wide variety of pathogens: pyogenic bacteria (*S. pneumoniae*, *H. influenzae*, *P. aeruginosa*, *S. aureus*), PCP, TB, cryptococcosis, cytomegalovirus, or *Aspergillus* spp.
- Onset and duration of symptoms – short in influenza and bacterial pneumonia; longer in PCP and TB.
- CD4 count – a high CD4 count is usually correlated with 'normal' pathogens, e.g. *S. pneumoniae*, TB, *S. aureus*, and influenza. A CD4 count <200 cells/mm^3 is associated with opportunistic pathogens, e.g. *Pneumocystis jiroveci*, *C. neoformans*, histoplasmosis, coccidiodomycosis, *Nocardia* spp., *Rhodococcus equi*.
- CXR may appear show typical or atypical appearances or may even be normal in patients, e.g. PCP and TB. Intrathoracic lymphadenopathy suggests TB, atypical mycobacteria, lymphoma, Kaposi's sarcoma.
- Injection drug use is associated with *S. aureus* pneumonia.
- Prophylaxis – co-trimoxazole reduces the incidence of PCP and bacterial pneumonia. Influenza vaccination appears to reduce the risk of influenza. The effects of pneumococcal vaccination are variable.

- Laboratory diagnosis – expectorated sputum is used for TB diagnosis. Induced sputum is better for PCP. Bronchoscopy has a yield of ~95% for PCP. Tests to consider in patients who are not responding to treatment are *Legionella* urinary antigen, serum cryptococcal antigen, *H. capsulatum* serum or urinary antigen. Also consider bronchoscopy and CT thorax.
- Treatment is of the underlying cause.
- Secondary prophylaxis – patients who present with AIDS-defining pulmonary infections, e.g. PCP, should be treated with co-trimoxazole until their CD4 count is consistently >200 cells/mm^3.

Lymphoid interstitial pneumonitis (LIP)

- This is a diffuse interstitial lung disease, characterized by a polyclonal lymphoid cell infiltrate in the alveolar septae.
- It occurs in <1% of HIV-infected adults but up to 40% of HIV-infected children.
- Clinical features are fever, dyspnoea, weight loss, pleuritic pain, arthralgia. Adults may be asymptomatic. Chest examination shows bibasal creptitations. Children may have clubbing and adenopathy.
- CXR shows bilateral reticulonodular shadowing. High-resolution CT scan shows ground-glass shadowing and pulmonary nodules.
- Diagnosis is established by lung biopsy in adults. In children, persistence of abnormalities for >2 months and exclusion of other infectious causes is considered diagnostic.
- Treatment is with oral prednisolone. Optimal duration of treatment is unknown – 6–12 months is usually given. The condition may improve with institution of HAART.
- Resolves after 6–12 months in some patients. Others may require lifelong low-dose steroids.

Pneumothorax

- In HIV-infected patients the most common cause is *Pneumocystis jiroveci* (PCP), and a spontaneous pneumothorax should prompt investigation for PCP. May also occur in TB, pulmonary cryptococcosis, and lymphoid interstitial pneumonitis. Iatrogenic causes include central line insertion, thoracocentesis, and bronchoscopy. Risk factors include male sex, smoking, patients on inhaled pentamidine, patients with bullae on CXR, ventilated patients, and injection drug users.
- Presents with pleuritic chest pain, dyspnoea, and cough. On examination there is hyperresonance to percussion and reduced breath sounds.
- The diagnosis is confirmed on CXR which shows a rim of air around the lung.
- Treatment is with aspiration or chest drain insertion.
- PCP-associated pneumothorax is associated with higher mortality.

References

1 Guidelines for the prevention and treatment of opportunistic infections in HIV-infected adults and adolescents (2008) available from http://aidsinfo.nih.gov/Guidelines/Default.aspx?MenuItem=Guidelines.

HIV gastrointestinal disease

Gingivitis/periodontitis

- Gum inflammation and bleeding, caused by oral anaerobic bacteria.
- Four phases – linear gingival erythema, necrotizing gingivitis, necrotizing periodontis, necrotizing stomatitis.
- Diagnosis is clinical. Orthopantogram may show bony loss.
- Treatment – routine dental care plus antiseptic mouth wash. A dental opinion should be sought re curettage and debridement. Antibiotics, e.g. metronidazole are given for necrotizing stomatitis.

Oropharyngeal candidiasis

Usually caused by *Candida albicans*. Diagnosis is clinical. Treatment is with oral fluconazole.

Oesophageal candidiasis

- Caused by a number of *Candida* spp. Usually occurs in patients with a CD4 count <100 cells/mm^3.
- Clinical features – dysphagia, odynophagia, and retrosternal pain.
- Diagnosis is clinical but upper GI endoscopy may performed if atypical presentation or poor response to treatment.
- Treatment – fluconazole or itraconazole for 14–21 days.
- Secondary prophylaxis may be considered in patients with >3 episodes in one year.

Nausea and vomiting

- Often due to medications (especially antiretrovirals, antibiotics, opiates). May also be due to abacavir hypersensitivity, nevirapine hepatotoxocity, lactic acidosis. Other causes: adrenal insufficiency, ureamia, hyoercalcaemia, CNS lesions, GI disease.
- Investigations – lactate level, ultrasound/CT abdomen, CT brain.
- Treat the underlying cause. If due to drugs, consider changing them. Otherwise symptomatic treatment.

Diarrhoea

- Usually caused by drugs (especially protease inhibitors. But may be due a number of pathogens e.g. *Salmonella* spp., *Shigella* spp, *Campylobacter* spp., *E. coli*, *S. aureus*, viruses.
- Chronic diarrhoea may be due to *Cryptosporidium*, *Microsporidium*, *M. avium* complex, *Cyclospora*, *Isospora*, *G. lamblia*, *E. histolytica*, or HIV enteropathy.
- Clinical features – fever, abdominal cramps, diarrhoea, tenesmus, PR bleeding.
- Diagnosis – send blood cultures and stool to the lab for microscopy, ova, cysts, and parasites. *C. difficile* toxin. Take blood cultures for MAC and *Salmonella* spp. CT abdomen may be helpful to determine the cause of an acute abdomen. Colonscopy is helpful for the diagnosis of CMV colitis – also helps to rule out Kaposi's sarcoma and lymphoma.
- Treatment – treat the underlying cause. If due to drugs, consider chaging them. Otherwise symptomatic treatment.

HIV liver disease

Deranged liver function tests

These are common in HIV-infected patients and may be due to a variety of causes:

- drug toxicity e.g. NNRTIs (especially nevirapine), protease inhibitors, co-trimoxazole, TB drugs, statins, paracetamol
- alcohol toxicity or substance abuse
- viral hepatitis (acute infection, flare of chronic disease, drug resistance)
- opportunistic infections, e.g. MAC, CMV, TB.

Lactic acidosis/hepatic steatosis

- Hyperlactataemia is defined as a venous lactate level >2 mmol/L.
- It can occur with any nucleoside reverse transcriptase inhibitor (NRTI) but is most common caused by d4T, ddI or both.
- The mechanism is NRTI-mediated inhibition of DNA polymerase gamma, leading to depletion of mitochondrial DNA.
- It can be asymptomatic or associated with a symptomatic and sometimes fatal acidosis.
- Asymptomatic elevations in lactate occur in 8–20% of patients on prolonged HAART.
- Symptomatic hyperlactataemia occurs in 0.5–1/100 patient-years of NRTI exposure.
- Diagnosis – patients who have symptoms compatible with lactic acidosis (nausea, vomiting, abdominal pain) or abnormal liver function tests/amylase should have their lactate levels measured. A lactate level >5 mmol/L is considered diagnostic.
- Treatment – stop the NRTIs and switch to another drug class. Substitution of abacavir, 3TC (lamivudine), FTC (emtricitabine) or TDF (tenofozir) may be possible in patients who are not seriously ill.
- Prognosis is related to lactate level: 7% if lactate 5–10 mmol/L; >30% if lactate 10–15 mmol/L, and >60% if lactate >15 mmol/L.

Viral hepatitis

- All patients who develop deranged liver function tests should be screened for the hepatitis viruses.
- Patients may acquire an acute hepatitis (e.g. HAV or HEV) transmitted by the faeco-oral route.
- HIV patients may be chronically infected with other blood-borne viruses such as HBV or HCV.
- All patients should be advised to avoid or limit their intake of alcohol. They should also be vaccinated against HAV and HBV, if non-immune when CD4 count >200 cells/mm^3.
- Treatment of HBV – In patients who do not require HAART, treatment is with pegylated inferon-α; alternatives entecavir or adefovir. In patients who require HAART, treatment with two agents active against HBV is recommended, e.g. tenofovir and emtricitabine
- Treatment of HCV – Patients who require treatment should be given pegylated interferon and ribavirin. Sustained virological response rates are lower for genotype 1 (14–29%) compared with genotypes 2 and 3 (44–73%)

Cholangiopathy

- Syndrome of biliary obstruction caused by infection-associated strictures. Four types: papillary stricture/stenosis, sclerosing cholangitis-like, papillary stenosis + sclerosing cholangitis, extrahepatic strictures.
- Causes – *Cryptosporidium parvum* (most common), *Microsporidia, Cyclospora cayetanensis,* CMV; 20 – 40% idiopathic.
- CD4 count usually <100 cells/mm^3.
- Clinical features – right upper quadrant pain, fever, diarrhoea (if intestinal involvement).
- Liver function tests show cholestatic derangement. Ultrasound shows dilated bile ducts. Diagnosis confirmed by ERCP.
- Treatment – sphincterotomy for papillary stenosis. Endoscopic stenting for isolated bile duct stricture. Ursodeoxycholic acid for has ben used experimentally in cholangiopathy without stenosis. Pathogen-specific therapy if possible.
- Outcome – average survival in HAART era is 9 months – worse if ALP >1000 IU/L.

Pancreatitis

- Pancreatitis is a well recognized complication in HIV-infected patients
- There are a number of causes: drugs (especailly ddl or ddl plus d4T). May also occur as a complication of protease inhibitor (PI)-associated hypertriglyceridaemia or NRTI-induced lactic acidosis. Opportunistic infections such as CMV, MAC, TB, crypotosporidiosis, toxoplasmosis, and cryptococcosis have also been implicated.
- Diagnosis is based on an amylase >3 times the upper limit of normal. A CT scan should be performed to stage the disease, detect complications, and exclude other diagnosis.
- Treatment is supportive – fluids, antibiotics, analgesia.
- Prognosis is related to the APACHE II score.

HIV-associated nephropathy

- This is the leading cause of end-stage renal disease (ESRD) in HIV-infected patients.
- Risk factors – black race, male, family history of renal disease.
- Usually a late manifestation of HIV disease (CD4 count <200 cells/mm^3).
- Proteinuria may be massive and predates renal insufficiency.
- Rapid progression to ESRD if not treated.
- Diagnosis – protenuria (nephrotic range) is common. Renal ultrasound shows normal to large echogenic kidneys. Renal biopsy is diagnostic
- Treatment – HAART improves renal survival. Blood pressure should be maintained <125 mmHg systolic using an ACE inhibitor. Corticosteroids may be given as rescue therapy (or a bridge to HAART).

HIV neurological complications

Peripheral neuropathy

Peripheral neuropathy is another relatively common complication of HIV therapy. There are a number of possible causes:

- distal sensory neuropathy (DSN) – pain and numbness in glove and stocking distribution. CD4 count usually <200 cells/mm^3
- antiretroviral toxic neuropathy (ATN) – same as DSN but associated with ddl, ddC and d4T. More common in older patients with diabetes. Can occur at any CD4 count
- tarsal tunnel syndrome – pain and numbness in anterior part of soles of feet
- HIV-associated neuromuscular weakness syndrome – ascending paralysis with arreflexia ± cranial nerve or sensory involvement. Usually associated with d4T. Poor survival
- HIV-associated myopathy (AZT myopathy) – pain, aching, and weakness of proximal muscles. Associated with AZT and raised creatinine kinase (CK). Can occur at any CD4 count
- polyradiculitis – rapidly evolving weakness and numbness in legs with bladder and bowel incontinence. May be caused by CMV or HSV. CD4 count <50 or >500 cells/mm^3
- vacuolar myopathy – stiffness, weakness and numbness in legs followed by bowel and bladder incontinence. Need to exclude vitamin B_{12} deficiency and HTLV-1 infection. CD4 count <200 cells/mm^3. Physiotherapy, methionine, or HAART may be helpful
- inflammatory demyelinating polyneuropathies – weakness in arms and legs with minor sensory component. Can occur at any CD4 count. Treatment: plasmapheresis, IVIG and/or HAART
- mononeuritis/mononeuritis multiplex – asymmetrical mix of motor and sensory defects occurring over weeks. CD4 count variable. Treat with steroids if CD4 count >200 cells/mm^3. Treat for CMV if CD4 count <50 cells/mm^3.

Central nervous system manifestations

Central nervous system involvement may be due to HIV itself, opportunistic infections, or malignancies:

- **Cryptococcal meningitis** (8–10% of all AIDS patients). Presents with fever, headache, visual changes, stiff neck, cranial nerve deficits, seizures. Progresses over 2 weeks. CD4 count <100 cells/mm^3. CSF India ink-positive in 60–80%; CSF culture positive in 95–100%. Cryptococcal antigen >95% sensitive and specific
- **HIV dementia** (HAD, 7%) – presents with a triad of cognitive, motor, and behavioural dysfunction over weeks to months. Afebrile. CD4 count <200 cells/mm^3. Elevated β_2 microglobulin. Neuropsychological tests show subcortical dementia
- **Toxoplasmosis** (2–4%) – presents with fever, reduced conscious level, focal neurological deficits, and seizures. CD4 count <200 cells/mm^3. MRI scan shows ring-enhancing lesions. Toxoplasma IgG falsely negative in 5%; 85% respond to empiric therapy in 7 days. Brain biopsy makes the definitive diagnosis
- **Primary CNS lymphoma** (2%) – clinical presentation afebrile with altered mental status, focal neurological deficits, seizures with progression over 2–8 weeks. CD4 count <200 cells/mm^3. Suspect if patient has no response to anti-toxoplasma therapy. Thallium 201 SPECT scan 90% sensitive and specific
- **Progressive multifocal leucoencephalopathy** (PML, 1–2%) – presents with impaired speech, vision, and motor function. No fever or headache. CD4 count usually <100 cells/mm^3. MRI shows multifocal lesions in the subcortical white matter. CSF or brain biopsy positive for JC virus (p.488)
- **CNS tuberculosis** (0.5–1%) – fever, headache, meningism, impaired conscious level, focal neurological deficits. CD4 count <100 cells/mm^3. CT/MRI scan shows meningeal enhancement, hydrocephalus, tuberculomas. ZN smear positive in 20%. CXR shows active TB in 50%. Gold standard for diagnosis is positive CSF culture for *M. tuberculosis*
- **CMV encephalitis** (>0.5%) – presents with fever, lethargy, delirium, disorientation, headache, neck stiffness, photophobia, cranial nerve deficits. CD4 count <100 cells/mm^3. May have CMV retinitis. CSF CMV PCR positive. Definitive diagnosis is made by brain biopsy
- **Neurosyphilis** (0.5%) – various clinical presentations: meningitis-like, tabes dorsalis, general paresis of the insane, meningovascular strokes/myelitis; ocular manifestations (iritis, uveitis, optic neuritis). Occurs at any CD4 count. Diagnosis positive CSF VDRL (60–70%).

HIV-associated malignancies

Malignancies are generally more common in HIV-infected compared with HIV-uninfected patients. Certain malignacies are particularly assocated with HIV infection:

- **Kaposi's sarcoma** – caused by human herpes virus 8 (HHV-8). Most common HIV-associated malignancy. Rate is 20,000 fold higher in HIV-positive than HIV-negative individuals, and 300-fold higher than in other immunosuppressed patients. It is more common in men who have sex with men (MSM) and more common in women. Presents with purple/brown/black macules, nodules and papules on the face, mouth, legs, and genitals. Visceral involvement affects the lungs and GI tract. HAART associated with decreased incidence and regression of lesions. Treatment may be local, e.g. vinblastine injections, or systemic, e.g. liposomal anthracycline
- **non-Hodgkin's lymphoma** – 50–80% are EBV-positive. CD4 count <100cells/mm^3; 200–600 times more common in HIV than in the general population. The majority are high-grade diffuse large-cell or Burkitt-like lymphomas. Usually present with advanced disease, sparse lymph nodes, and constitutional 'B' symptoms. Diagnosis is made by biopsy of the brain, lymph nodes, or bone marrow. CT better than endoscopy for assessing GI involvement. Treatment: HAART plus chemotherapy. Initial response rates 60–80% but median survival <1 year
- **primary CNS lymphoma** (see 📖 HIV neurological complications, p.866)
- **primary effusion lymphoma** – caused by HHV-8 and EBV. Very rare, accounting for <0.14% of non-Hodgkin's lymphomas in AIDS. Presents with serous effusions (pleural, pericardial, peritoneal, joint spaces). Diagnosis is made by examining the effusions. Treatment: HAART plus chemotherapy
- **cervical cancer** is associated with human papilloma virus (HPV) types 16, 18, 31 33 and 35). Cervical intraepithelial neoplasis (CIN) and invasive cervical cancer are both more common in HIV-positive than HIV-negative women. The frequency and severity of cervical dysplasia increases with progressive immune compromise. The US CDC recommends a gynaecological examination and Pap smear at baseline at 6 months and then yearly in HIV-infected women.

HIV prevention

Almost three decades after the start of the HIV epidemic, it has reached every country of the world. Although HIV was initially described in MSM, 80% of infections are now transmitted heterosexually and >50% of of HIV-infected people are now women. Mother-to-child transmission accounts for >90% of HIV-infected children worldwide. Intravenous drug use is also fuelling transmission in central and eastern Europe and certain countries in Asia. The epidemic continues unabated, especially in sub-Saharan Africa and southeast Asia. In developed countries, advances in antiretroviral therapy have resulted in significant reductions in morbidity and mortality. However, complacency and recurrence of high-risk behaviour among some populations has resulted in a resurgence of sexually transmitted diseases and a recent increase in HIV incidence in these populations.

Prevention of perinatal transmission

The three major factors associated with perinatal HIV transmission are: high maternal plasma HIV viral load, prolonged rupture of membranes, and breast feeding. A number of studies have been performed in order to try to prevent perinatal transmission, and have shown:

- antiretroviral therapy of the mother during pregnancy and labour reduces maternal viraemia and HIV transmission
- antiretroviral therapy of the child (*in utero* and after birth) reduces HIV transmission
- delivery by Caesarean section has led to a decline in HIV transmission
- continuation of antiretroviral therapy in mothers who choose to exclusively breast feed is beneficial
- there are concerns about the use of single-dose nevirapine in the developing world because of the potential for development of NNRTI resistance.

Prevention of sexual transmission

A number of factors have been identified to be important in sexual transmission: plasma HIV RNA level, genital HIV RNA level, acute infection versus advanced disease, degree of immunosuppression, genital ulcers, inflammatory STIs, cervical ectopy, uncircumcised status, host genetics, levels of cytokines and chemokines. Interventions to reduce sexual transmission include:

- reduction of HIV RNA level – studies in Africa have shown that antiretroviral therapy reduces the risk of HIV transmission between sero-discordant couples by >80%
- HSV-2 suppression – treatment with valaciclovir was associated with reduction in plasma and genital HIV-1 levels and reduced incidence of HIV transmission. However, use of aciclovir was not associated with reduced HIV acquisition in two other studies[1]
- male circumcision – several ecological studies and three randomized controlled trials have shown that male circumcision is associated a 60–70% reduction in rates of HIV acquisition[1]. Circumcision also reduced the frequency of genital ulcer disease and HIV acquisition by female sexual partners. Modelling studies now suggest that male circumcision in sub-Saharan Africa could potentially prevent

5.7 million infections and 2 million deaths over the next 20 years, if the intervention could be delivered safely and cost-effectively

- microbicides – whereas circumcision is a method that can protect men from HIV, there is an urgent need to develop female-controlled methods of protection. Numerous studies have investigated the effectiveness of female condoms, diaphragms, and microbicides but have failed to show benefit. In some studies, the use of microbicides has been associated with an increased risk of HIV acquisition.
- post-exposure prophylaxis following sexual exposure (PEPSE)[2]. This is recommended where the individual presents within 72 hours after anal and or vaginal intercourse with a known HIV positive source, or a source from a group or area of high HIV prevalence.

HIV vaccines

Despite billions of pounds and over 20 years of research, an effective vaccine for the prevention of HIV remains elusive. Specific characteristics of the virus that hinder vaccine cebelopment include the extreme genetic variability in circulating viral isolates worldwide, biological properties of HIV that impede immune attack and a high mutation rate that allows for rapid escape from adaptive immune responses.

Reference

1. Padian NS, Buvé A, Balkus J et al. Biomedical interventions to prevent HIV infection: evidence, challenges and the way forward. *Lancet* 2008;372:585–99.

2. Fisher M, Benn P, Evans B et al. UK guideline for the use of post-exposure prophylaxis for HIV following sexual exposure. *Int Jour STD and AIDS* 2006;17:81–92.

Immunisations

Routine childhood immunisations

All children in the UK are entitled to free immunisations to protect them from childhood illnesses. The introduction of immunisation has resulted in dramatic declines in certain diseases e.g. meningitis caused by *H. influenzae* type B and *N. meningitidis* serogroup C. Routine childhood immunisation schedules vary from country to country. The UK routine childhood immunisation schedule is summarized in Table 5.19

In September 2008 the human papillomavirus vaccine (HPV) was introduced into the routine immunization schedule for girls. Unlike the other childhood immunisations, the HPV vaccine is used to prevent the development of cervical cancer, rather than a childhood infectious disease.

Table 5.19 Routine childhood immunization schedule.

Child's age	Vaccine(s) given	Diseases protected against
2 months	DTaP IPV Hib PCV	Diphtheria, tetanus, pertussis Polio *H. influenzae* type b (Hib) Pneumococcal infection
3 months	DTaP IPV Hib MenC	Diphtheria, tetanus, pertussis Polio *H. influenzae* type b (Hib) Meningitis C
4 months	DTaP IPV Hib MenC PCV	Diphtheria, tetanus, pertussis Polio *H. influenzae* type b (Hib) Meningitis C Pneumococcal infection
12 months	Hib MenC	*Haemophilus influenza* type b (Hib) Meningitis C
13 months	MMR PCV	Measles, mumps and rubella Pneumococcal infection
3 years and 4 months	DTaP/IPV or dTaP/IPV MMR	Diphtheria, tetanus, pertussis Polio Measles, mumps and rubella
Girls aged 12-13 years	HPV	Cervical cancer caused by human papillomavirus types 16 and 18.
13 to 18 years old	Td IPV	Diphtheria, tetanus, Polio

Non-routine immunisations

Some children who may be at increased risk of certain diseases (e.g. tuberculosis and hepatitis B) may be given additional vaccines:

Age of child	Vaccine	Diseases protected against
At birth, for babies who are more likely to be exposed to TB	BCG	Tuberculosis
At birth, for babies whose mothers are hepatitis B positive	Hep B	Hepatitis B

Further information

- Further information on immunisations is available from:
- The NHS Immunisation information website http://www.immunisation.nhs.uk
- Immunisation against Infectious Diseases - 'The Green Book' http://www.dh.gov.uk/en/Publichealth/Healthprotection/Immunisation/Greenbook/dh_4097254

Notifiable diseases (ND)

- The statutory requirement for the notification of certain infectious diseases (e.g. cholera, diphtheria, smallpox and typhoid) started towards the end of the 19th century.
- Since then the list of diseases has expanded considerably (Table 5.20).
- Originally the head of the family or landlord had the responsibility of reporting the disease to the local 'Proper Officer'; this is now done by the attending medical practitioner.
- The prime purpose of the notifications system is to detect possible outbreaks or epidemics. Accuracy of diagnosis is secondary, and a clinical suspicion of a notifiable infection is all that is required. If a diagnosis later proves incorrect it can always be changed or cancelled
- In 1997 the responsibility for administering the notification of infectious diseases system was transferred to the Communicable Disease Surveillance Centre (CDSC), now the Health Protection Agency (HPA) Centre for Infections (CfI).
- The diseases listed below should be notified to the local Consultant in Communicable Disease Control (CCDC) under the Public Health (Infectious Diseases) Regulations 1988.

Table 5.20 Notifiable diseases

Acute encephalitis	Paratyphoid fever
Acute poliomyelitis	Plague
Anthrax	Rabies
Cholera	Relapsing fever
Diphtheria	Rubella
Dysentery	Scarlet fever
Food poisoning	Smallpox
Leprosy*	Tetanus
Leptospirosis	Tuberculosis
Malaria	Typhoid fever
Measles	Typhus fever
Meningitis	Viral haemorrhagic fever
Meningococcal septicaemia	Viral hepatitis
Mumps	Whooping cough
Ophthalmia neonatorum	Yellow fever

* Leprosy should be notified directly to the HPA CfI.

Bioterrorism

Biological warfare has a long and unpleasant history. Around 400BC the Scythians were attempting to poison their arrows with blood and manure and in the 14th century the Tarter catapulted the corpses of plague victims into the city of Kaffa with the intention of initiating an outbreak. The years after the Second World War saw a race to develop more effective biological agents before various treaties later led to the limitation and even destruction of biological weapons stockpiles by many nations (UK 1957, USA 1973). Today around 17 countries are suspected of having biological weapons programmes. The threat of biological warfare is seen as issuing not primarily from states but from independent organisations and terrorists. The term 'deliberate release' refers to any intentional spread of a biological or chemical agent. Such a release may be overt (e.g., a prior warning, or the release may be apparent, either due to the use of an explosive device, or because a suspicious substance is obviously visible) or covert (the release not becoming apparent until the first cases of disease arise). Only two proven deliberate releases have recently affected a large number of people: contamination of restaurant salads with *S. typhimurium* in Oregon in 1984 and dissemination of *Bacillus anthracis* via the US mail in 2001.

Organisms with the potential to be used as weapons agents

The ideal biological weapon agent has low visibility, high potency, is accessible with a long shelf-life, is relatively easy to deliver, and shows limited epidemic spread. A small amount of agent may be capable of killing a large number of people (particularly in a metropolitan environment) and creating a disproportionate level of fear and disruption – a key part of their attractiveness to terrorist organizations.

- Category A agents – those organisms easily disseminated or transmitted from person-to-person, with high mortality rates and potential for major public health impact and requiring special action for public health readiness:
  - anthrax (📖 p.298) – pulmonary anthrax presents with a severe febrile illness or sepsis with respiratory failure (massive mediastinal lymphadenopathy). The organism may be identified in blood cultures or sputum
  - smallpox (📖 p.490) – the previously vaccinated lose protection after 10–20 years. Vaccination provides moderate protection if given within 2–4 days of exposure. Disease may develop 1–3 weeks after exposure
  - botulism (📖 p.308) – toxin may be inhaled or food-borne. Antitoxin is available
  - plague (📖 p.361) – inhaled as aerosol, causing pneumonic plague
  - tularaemia (📖 p.366)
  - viral haemorrhagic fevers (📖 p.501).
- Category B agents – moderately easy to disseminate, moderate morbidity rates and low mortality rates, require enhancement of both diagnostic capacity and disease surveillance. They include: glanders (📖 p.349), melioidosis (📖 p.348), brucellosis (📖 p.359), psittacosis (📖 p.406), and Q fever (📖 p.398).

Recognizing an attack

In the absence of issued warnings or a very obvious release (e.g. explosive device), the first indicator of an outbreak may be a cluster of symptomatic cases. Such clusters may present acutely or over a period of days or weeks. Isolated fatalities due to undiagnosed febrile illness are not uncommon. Prompt epidemiological inquiry is essential. Features indicative of deliberate release include:

- an unusually large number of patients over a short time period
- cases that are linked by epidemiological or geographical features
- signs/symptoms that are unusual or very severe
- unknown cause or an identified cause unresponsive to normal therapy or unusual in the UK or where acquired.

Remember that symptoms may also be due to radiological or chemical contamination.

Responding to an attack

Consider the risk of transmission to or contamination of staff and other patients – it may be appropriate to isolate affected patients and use personal protective equipment. Decontamination of potentially exposed individuals is vital for suspected releases of *Bacillus anthracis*. Expert advice must be sought locally and the health protection unit informed. Empirical antibacterial prophylaxis is indicated for possible exposure to certain bacterial agents such as anthrax, plague, and tularaemia (ciprofloxacin) or brucella, burkholderia, and Q fever (e.g. doxycycline). National agencies have stockpiles of suitable antibiotics for such emergencies. Early cases should be managed according to the best available advice until more-detailed epidemiological information and laboratory tests are available. In the UK, management of all incidents is led by the police with involvement of other emergency and health services as appropriate. All microbiological testing of suspect material must be done in *specialist* laboratories.

More information

The HPA website has extensive information on the management of deliberate release incidents, including clinical and diagnostic algorithms, antibiotic protocols and guidelines for dealing with 'suspect packages'. See www.hpa.org.uk/infections/topics_az/deliberate_release/menu.htm.

Index

D

E

H

J

K

L

N

O

P

Q

R

U

V

W

X

Y

Z